Stochastic Processes for Insurance and Finance

Stochastic Processes for Insurance and Finance

Tomasz Rolski
Mathematical Institute, University of Wrocław, Poland

Hanspeter Schmidli
Department of Theoretical Statistics, Aarhus University, Denmark

Volker Schmidt
Faculty of Mathematics and Economics, University of Ulm, Germany

and

Jozef Teugels
Department of Mathematics, Catholic University of Leuven, Belgium

John Wiley & Sons
Chichester • New York • Weinheim • Brisbane • Singapore • Toronto

This paperback edition first published 2008

Registered office
John Wiley & Sons Ltd, The Atrium, Southern Gate, Chichester, West Sussex,
PO19 8SQ, United Kingdom

For details of our global editorial offices, for customer services and for information about how to apply for permission to reuse the copyright material in this book please see our website at www.wiley.com.

A catalogue record for this book is available from the British Library.

ISBN: 978-0-470-74363-8 (P/B)

Contents

Preface

This book is designed for a beginning or an intermediate graduate course in stochastic modelling. It is intended for a serious student in probability theory, statistics, actuarial sciences or financial mathematics. The overall objective is to make the basic concepts of stochastic modelling and insurance accessible to students and research workers in a comprehensive manner. Renewal theory, random walks, discrete and continuous-time Markov processes, martingale theory and point processes are among the major subjects treated. The selection of the topics has been largely made on the basis of their relevance within an actuarial or financial context. In this sense, the book is rather special. On the other hand, one could have written a similar textbook but with queueing theory or stochastic networks as the scrutinizing subject.

A few words are in order about the selection of topics. Space limitations have forced us to make a choice from stochastic processes, actuarial mathematics and the mathematics of finance. Also, each one of the authors, coming from four European countries, had a list of favourite topics when the writing project began.

One advantage of using insurance questions as guidelines in the selection of the topics is that the treated subjects gain in coherence. Another facet is that any important actuarial problem is highlighted from a variety of different stochastic angles. A possible disadvantage might be that important subjects are not duly treated. We consider the topics that are covered as the basic intersection of stochastic modelling, insurance mathematics and financial mathematics. As a result, we only give a few elements of branching processes or of jump-diffusion processes. In the same fashion, we do not cover credibility theory, IBNR claims (IBNR = incurred but not reported) or topics from advanced finance. On the other hand, we do treat some uncommon subjects like subexponentiality, phase-type distributions, piecewise deterministic Markov processes, stationary and marked point processes, etc. We very much hope that what has been covered will be sufficiently stimulating to encourage the reader to continue their efforts.

Some special features of this book are the following:

1. The first chapter gives the reader a bird's-eye view of the main themes treated in the book. We have made an attempt to introduce the principal concepts from insurance and finance in an intuitive fashion, avoiding rigour. This chapter should however convince the reader about the need for the mathematically sound treatment in the rest of the book.
2. The book is not covering the statistical aspects of stochastic models in insurance and finance. However, to emphasize the relevance of the stochastic models, a fair number of practical illustrations with real life data have been included.
3. In a similar fashion, some numerical and algorithmic procedures have been included since they make up a vital portion of current day practical approaches in insurance.
4. An attempt has been made to make the book self-contained. Only well-known results from analysis, probability and measure theory have been mentioned without proofs. A couple of times we allude to results that are only treated in a subsequent chapter. This happens most often when the proof of a result depends on more sophisticated material treated in the later part of the book.
5. Notes and comments at the end of each section include references to additional reading material. An extensive list of references is included. This bibliographical material should serve two purposes: helping the probabilistically trained reader to find their way in the actuarial and financial literature, while at the same time informing the practitioner on the sources from which to find the mathematical treatment of one or the other useful methodology.

Thanks to the unusual and unifying approach, many of the topics are put in a novel framework. While we do not claim originality, a sizable set of results and proofs appear here for the first time.

The numbering of chapters, sections, subsections, definitions, formulae, lemma's, and theorems is traditional. Chapters are subdivided into sections, while sections are further subdivided into subsections. References of the form $i.j.k$ refer to chapter i, section j, serial number k of subsections, definitions, lemma's, and theorems. If we refer to a formula, we write $(i.j.k)$. References to the literature are of the form: name(s) of the author(s) plus year; for example Smith (1723) or Kovalyov and Smith (1794).

We emphasize that the book has been conceived as a course text. In an attempt to keep the size of the book at a reasonable level, we decided not to include sections with exercises. In many places, the reader is, however, asked to provide additional arguments and parts of proofs. This, of course, is not sufficient. A subject like stochastic modelling not only requires routine-like skills; it also demands training and sharpening of accurate probabilistic intuition which can only be achieved by tackling nontrivial exercises. We are convinced that the best way to help students and teachers is by supplementing this textbook with a forthcoming *Teacher's Manual*.

This book would never have been finished without the help of many people. Individually, we take pleasure in thanking our home institutions for their indulgence and logistic support, needed during the many working sessions of the quartet. Each one of us extends his thanks to his students and colleagues who patiently read first, second or n-th drafts of the manuscript and who helped us whenever there were problems with the styling. Jointly, our appreciation goes to a large number of colleagues in academia and in insurance companies who helped us during the selection process of what and what not to include.

In particular we are grateful to Sabine Schlegel for her invaluable help during the whole process of the preparation of the book. Apart from reading the manuscript and helping to eliminate a number of errors, she provided the computer illustrations included in the text. Also, our contacts with J. Wiley & Sons through Helen Ramsey and Sharon Clutton have always been pleasant and constructive.

Further support and help is greatly acknowledged. TR thanks the Department of Stochastics at the University of Ulm for their hospitality and for creating optimal working conditions during his extended stays in Ulm. The latter were made possible by the financial support of the Deutsche Forschungsgemeinschaft and of the Institut für Finanz- und Aktuarwissenschaften Ulm. TR and VS want to express their gratitude to the Mathematisches Forschungsinstitut Oberwolfach and to the Volkswagen Stiftung. The programme *Research in Pairs* granted them a very useful five-week stay in Oberwolfach. TR and VS also acknowledge the financial support from the Exchange Programme between the Deutsche Forschungsgemeinschaft and the Polish Academy of Sciences.

Our deepest gratitude is reserved for four loving families, who have endured our preoccupation with this project for so long.

TOMASZ ROLSKI
HANSPETER SCHMIDLI
VOLKER SCHMIDT
JOZEF TEUGELS

August 1998

List of Principal Notation

A	event
A^{c}	complement of set A
$\boldsymbol{A}^{\top}$	transposition of matrix $\boldsymbol{A}$
$\mathbb{I}(A)$	indicator of event A
B	Borel set
$\mathcal{B}(\mathbb{R})$, $\mathcal{B}(E)$	Borel σ-algebra on $\mathbb{R}$, E
$\mathbb{C}$	complex plane
Cov	covariance
E	state space
$\boldsymbol{E}$	squared matrix with all entries equal to 1
$\boldsymbol{e}$	(row) vector with all components equal to 1
$\boldsymbol{e}_i = (0, \ldots, 0, 1, 0, \ldots, 0)$	
$\mathbb{E}$	expectation
$\mathbb{E}(X \mid \mathcal{G})$	conditional expectation
F, G	distributions; distribution functions
$F * G$	convolution
F^{*k}	k-th convolution of F; $F^{*1} = F$, $F^{*0} = \delta_0$
F^{s}	integrated tail distribution
$\overline{F}(x)$	tail function
$F^{-1}(y)$	generalized inverse function
$f(t)$	density function
$\mathcal{F}, \mathcal{F}_n, \mathcal{G}$	σ-algebras
$\hat{g}(s)$	generating function
$g_1(x) \sim g_2(x)$	asymptotical equivalence as $x \to \infty$
$H(t)$	renewal function
I	index of dispersion
$\boldsymbol{I}$	identity matrix
$\Im z$	imaginary part of z
i	imaginary unit
$L(x)$	slowly varying function as $x \to \infty$
$\hat{L}(s)$	Laplace transform
$\hat{l}(s)$	Laplace–Stieltjes transform
$\hat{m}(s)$	moment generating function

N	$\mathbb{N}$-valued random variable
$\{N(t)\}$	counting process
$\mathbb{N}$	set of nonnegative integers
$\emptyset$	empty set
$\mathbb{P}$	probability measure
$Q(y)$	quantile function
$\mathbb{Q}$	set of rational numbers
$\mathbb{R}$	real line
$\mathbb{R}_+$	nonnegative reals
$\Re z$	real part of z
S, T, U, V, W, X, Y, Z	random variables
$\mathcal{S}$	family of subexponential distributions
T_n	inter-occurrence times
$U_{(i)}, X_{(i)}$	i-th order statistic
u	initial risk reserve
Var	variance
$\{X(t)\}$	stochastic process
$\lfloor x \rfloor$	largest integer $\leq x$
$\lceil x \rceil$	smallest integer $\geq x$
$x \wedge y = \min\{x, y\}$	
$x \vee y = \max\{x, y\}$	
$\mathbb{Z}$	set of integers
β	premium rate
γ	adjustment coefficient
δ_a	Dirac measure
δ, $\delta(t)$	force of interest
η	relative safety loading
λ	claim occurrence rate
$\mu^{(n)}$	n-th moment
ν^-, ν^+	ladder epochs
$\Pi(X)$	premium for risk X
ρ	expected aggregate claim amount
τ	stopping time
$\tau_{\mathrm{d}}(u), \tau(u)$	ruin times
$\hat{\varphi}(t)$	characteristic function
$\psi(u)$, $\psi(u; x)$	ruin probabilities
σ_n	arrival epoch
Ω	basic space
$(\Omega, \mathcal{F}, \mathbb{P})$	probability space
$\leq_{\mathrm{sl}}$	stop-loss order
$\leq_{\mathrm{st}}$	stochastic order
$\sum_{i=1}^{j} a_i = 0$ if $j = 0$	
$\prod_{i=1}^{j} a_i = 1$ if $j = 0$	
$\inf \emptyset = \infty$	

CHAPTER 1

Concepts from Insurance and Finance

1.1 INTRODUCTION

Let us start out with a concrete example. Consider all policies covering fire of apartments in a suburb of a major city that are underwritten with a specific insurance company. The insured goods have a comparable value and the chances of fire are probably not very different from one building to another. Such a set of policies makes up a homogeneous portfolio.

Most client-related insurance businesses use such portfolios as basic building blocks. Properly compiled they make up branches within the insurance firm, like fire, automobile, theft, property, life, health, pension, etc. The branch *fire* contains many portfolios dealing with different types of risk. Detached houses, terraced houses, apartments, apartment buildings, shops, supermarkets, business premises and industrial sites constitute an incomplete but already varied set of *insurance risks* for which different premiums have to be designed. Indeed, for each of the mentioned portfolios, the probabilities of a fire might depend on the portfolio. Moreover, the resulting claim sizes may very well be incomparable.

In the chapters dealing with insurance aspects, we will restrict our attention to one specific portfolio. Such a portfolio is characterized by a number of ingredients of both a deterministic and a stochastic nature.

Among the first we mention the *starting position* and a *time period*. Usually, data referring to an insurance portfolio refer to a time span of one year in accordance with the bookkeeping of the company. Far more important is the *initial reserve* or *initial capital*. One interpretation of the latter is the amount of capital set aside to cover costs occurring during the initial period of the portfolio when the company has not yet received the yearly premiums. In the sequel the initial reserve will be denoted by u.

Among the elements that usually have a stochastic nature are the following:

- The *epochs of the claims*; denote them by $\sigma_1, \sigma_2, \ldots$. In some cases we consider an additional claim arrival epoch at time zero denoted by $\sigma_0 = 0$. Apart from the fact that the epochs form a nondecreasing sequence we do not in general assume anything specific about their interdependence. The random variables defined by $T_n = \sigma_n - \sigma_{n-1}$, $n \geq 1$, are called the *inter-occurrence times* in between successive claims.

- The *number of claims* up to time t is denoted by $N(t)$ where $N(t) = \sup\{n : \sigma_n \leq t\}$. The intrinsic relation between the sequence of claim arrivals $\{\sigma_0, \sigma_1, \sigma_2, \ldots\}$ and the counting process $\{N(t), t \geq 0\}$ is given by $\{N(t) = n\} = \{\sigma_n \leq t < \sigma_{n+1}\}$.
- The claim occurring at time σ_n has size U_n. The sequence $\{U_n, n = 1, 2, \ldots\}$ of consecutive *claim sizes* is often assumed to consist of independent and identically distributed random variables. However, other possibilities will show up in the text as well.
- The *aggregate claim amount* up to time t is given by $X(t) = \sum_{i=1}^{N(t)} U_i$ while $X(t) = 0$ if $N(t) = 0$. By its very definition, the aggregate claim amount is in general a *random sum* of random variables.
- The *premium income*. In the course of time 0 to t we assume that a total of $\Pi(t)$ has been received through premiums.
- The *risk reserve* at time t is then $R(t) = u + \Pi(t) - X(t)$.

The above setup allows flexibility in that an individual claim may mean a claim from an individual customer (e.g. third-liability insurance) or a claim caused by a single event (e.g. windstorm insurance).

In the following sections we will give more details on the concepts introduced above and on actuarial quantities linked to them. We also show by practical illustrations how the stochastic character of these elements can be formalized. This approach will then automatically serve as an invitation for a thorough probabilistic treatment by concrete stochastic processes later on in the book.

1.2 THE CLAIM NUMBER PROCESS

Let us start out by considering in more detail the claim number process $\{N(t), t \geq 0\}$, built on the claim epochs. We will always assume that the claim number process is a *counting process*. This means that we require the process $\{N(t), t \geq 0\}$ to satisfy the following three conditions. For all $t, h \geq 0$

- $N(0) = 0$,
- $N(t) \in \mathbb{N}$,
- $N(t) \leq N(t + h)$,

$N(t + h) - N(t)$ models the number of claims occurring in the time interval $(t, t+h]$. We note that realizations of a counting process (also called the *sample paths* or *trajectories*) are monotonically nondecreasing and right-continuous functions. Usually one assumes that the jumps are of size one so that multiple claim arrivals are excluded. But there are cases where this assumption is not valid. For example there is a vast proportion of road accidents where more than one person is injured. It is then useful to associate a *mark* to each jump epoch. The resulting generalization naturally leads to the notion of a *marked counting process* treated in Chapter 12.

Let us write $\{p_k(t), k = 0, 1, \ldots\}$ with

$$p_k(t) = \mathbb{P}(N(t) = k) = \mathbb{P}\left(\sum_{i=1}^{k} T_i \le t < \sum_{i=1}^{k+1} T_i\right)$$

for the probability function of the counting variable $N(t)$. In some cases, time t does not play a significant role; if so, the t-dependence will be dropped from the notation. This happens, for example, when the insurer is interested in the number of claims received in successive one-year periods.

A general computational expression for the probabilities $p_k(t)$ is impossible because we have not specified the interdependence between the epochs of the claims. We indicate some possible choices.

1.2.1 Renewal Processes

Depending on the type of portfolio, the insurer can make a variety of different assumptions on the sequence of inter-occurrence times $\{T_n, n \ge 1\}$. In some particular cases it might be useful to assume that this sequence is generated by a *renewal process* $\{\sigma_n, n \ge 1\}$ of claim arrival epochs, i.e. the random variables T_n are nonnegative, independent and have the same distribution as a *generic random variable* T. The distribution function of the inter-occurrence time is then denoted by $F_T(x) = \mathbb{P}(T \le x)$.

Because this model appears as a natural candidate for the underlying stochastic mechanism in a wealth of processes, we will spend substantial space on a thorough discussion of renewal processes in Chapter 6. One might think that a renewal process is a rather simple type of process. However, the reader may be surprised to see some of the highly nontrivial results that form the backbone of all applications of renewal theory. In Chapter 6 we will deal in more detail with the *Sparre Andersen model*, in which the claim number process is a (general) renewal process.

Mathematically the simplest renewal process is the *Poisson process* introduced in Section 5.2 where the generic random variable T is exponentially distributed. Poisson processes have particular properties that distinguish them from other renewal processes. The main reason for this extra structure is provided by the lack of memory property of the exponential distribution. The latter distribution plays a similar crucial role in actuarial applications of stochastic processes as the normal distribution does in statistics. In particular, for the Poisson process we have

$$p_k(t) = \mathrm{e}^{-\lambda t}\frac{(\lambda t)^k}{k!}, \qquad k = 0, 1, \ldots$$

and $\mathbb{E}\,N(t) = \lambda t$, $\operatorname{Var} N(t) = \lambda t$ for all $t \geq 0$, where $\lambda^{-1} = \mathbb{E}\,T$ is the mean inter-occurrence time. As a side result, the *index of dispersion* $\mathrm{I}(t) = \operatorname{Var} N(t)/\mathbb{E}\,N(t)$ is constant and equal to 1.

It should be mentioned that time is not to be considered as real time, but as *operational time*. Indeed, some seasonal effects might affect the claim reporting process and the portfolio does not have the same size over time. It is therefore advantageous to measure time via the expected mean number of claims. In Sections 1.2.2 and 1.2.3 we consider some possible deviations from the constant expected number of claims per unit of operational time.

1.2.2 Mixed Poisson Processes

As early as 1948 actuaries noticed that the variability in a portfolio, expressed for example by $\mathrm{I}(t)$, was often greater than 1, the value corresponding to the Poisson case. One reason is this. The risk is exposed to some environment, for example weather conditions in motor insurance. This environment is different each year and influences the number of claims from the portfolio. If one could know the environment completely, one would also know the mean number of claims Λ to be expected in a particular year. Because one hardly has any information on how the environment influences the mean number of claims, one can estimate the distribution F_Λ of Λ and model the environment via F_Λ. This observation naturally leads to the following representation for the claim number distribution:

$$p_k(t) = \mathbb{P}(N(t) = k) = \int_0^\infty \frac{\mathrm{e}^{-\lambda t}(\lambda t)^k}{k!}\,\mathrm{d}F_\Lambda(\lambda), \tag{1.2.1}$$

where $F_\Lambda(\lambda) = \mathbb{P}(\Lambda \leq \lambda)$ is the distribution function of the *mixing random variable* Λ.

Counting processes of this type are called *mixed Poisson processes*. They have appeared in this general form in the actuarial literature since 1959; see Thyrion (1959). A sound definition can be given using the theory of continuous-time Markov processes, the subject of Section 8.5. This treatment is inspired by Lundberg (1964). Alternatively, mixed Poisson processes can be introduced via the general theory of doubly stochastic Poisson processes studied in Section 12.2.

The special case where F_Λ is a gamma distribution has already been introduced by Ammeter (1948) and is known as the *Pólya process* or the *Pascal process*. Other choices are, however, possible. For example, the case where F_Λ is an inverse normal distribution has great use in insurance but also in geophysical modelling.

It is easy to prove that for the mixed Poisson model, $\mathrm{I}(t) = 1+t(\mathrm{Var}\,\Lambda/\mathbb{E}\,\Lambda)$, which equals 1 if and only if Λ is degenerate, the classical case of a homogeneous Poisson process. The above explanation means that it is not surprising that the mixed Poisson process has always been very popular among insurance modellers.

1.2.3 Some Other Models

Of course, neither of the above models exhausts the possibilities for the claim counting process. Let us just mention a few alternatives that will be discussed in more detail in the forthcoming chapters.

- *Recursively defined claim number distributions.* Panjer (1980) used a recursion relation for the probability function $\{p_k(t), k = 0, 1, \ldots\}$ of the number of claims $N(t)$ to derive a recursive relation for the distribution of the aggregate claim amount. In the actuarial literature we find an increasing number of papers dealing with variations on the following recursion formula

$$p_k(t) = \Big(a + \frac{b}{k}\Big) p_{k-1}(t), \qquad k = 2, 3, \ldots, \tag{1.2.2}$$

  where the quantities a and b may depend on the time variable t. The above class has been introduced by Sundt and Jewell (1981) in an attempt to gather a variety of classical claim number distributions under the same umbrella. Later on Willmot (1988) reconsidered equation (1.2.2) and added a number of overlooked solutions. We will deal with (1.2.2) in Section 4.3. Most often, time dependence in (1.2.2) is suppressed. Also, some authors include the value $k = 1$ in (1.2.2), hence lowering the number of free parameters in the model. Special cases of the above model are (shifted) versions of the Poisson, binomial, negative binomial (or Pascal) and logarithmic distribution. On the other hand, the recursion relation (1.2.2) can readily be generalized, and we find a variety of such extensions in the actuarial literature.
- *Processes with stationary or independent increments.* Another line of thinking is based on the following observations. The distribution of the *increment* $N(t+h) - N(t)$ of the counting process $\{N(t)\}$ may be the same for all $t \geq 0$. This means that, as a function of time, the counting process has *stationary increments*. This model is studied in Chapter 12, where $\{N(t)\}$ is said to be the counting process corresponding to a *stationary point process*. For example, the number of industrial fires in a small region might easily satisfy this type of stationarity condition, while the number of car accidents on a road, experiencing different types of weather conditions, may fail.

The counting process $\{N(t)\}$ may also satisfy the *independent increments* condition. This means that increments over disjoint time intervals are independent random variables. For example, car portfolios usually satisfy this condition. However, the number of severe accidents on oil platforms probably does not, since imposing stricter safety regulations will change the distribution of the number of similar accidents in later periods.

- *Compound Poisson processes.* In general it is a complicated matter to derive a closed expression for $p_k(t)$. However, in some cases it may be mathematically much more convenient to work with the *probability generating function* given by $\hat{g}_{N(t)}(s) = \mathbf{E}\, s^{N(t)}$. This is the case if claims are arriving with stationary *and* independent increments. It follows from arguments covered in Section 5.2.2 that in this case we have the relation

$$\hat{g}_{N(t)}(s) = \mathrm{e}^{-\lambda t(1-\hat{g}_V(s))}\,, \tag{1.2.3}$$

where $\hat{g}_V(s)$ is itself the probability generating function of a discrete random variable V, concentrated on the strictly positive integers. As such the counting process $\{N(t)\}$ could also be called a *discrete compound Poisson process.* The probabilities $p_k(t)$ are then given in the form

$$p_k(t) = \sum_{j=0}^{\infty} \mathrm{e}^{-\lambda t}\frac{(\lambda t)^j}{j!}p_{V,k}^{*j}\,,$$

where $\{p_{V,k}^{*j}, k = 1,2,\ldots\}$ is the j-fold convolution of the probability function $\{p_{V,k}, k = 1,2,\ldots\}$ with probability generating function $\hat{g}_V(s) = \mathbf{E}\, s^V$. Needless to say, the explicit evaluation of the above probabilities is mostly impossible because of the complicated nature of the convolutions. However, approximations and bounds are available. Some of them are discussed in Chapter 4. If the probability generating function $\hat{g}_V(s)$ is known, a numerical method which is based on the *inverse fast Fourier transform* can be used to compute the $p_k(t)$. This is discussed in Section 4.7. Kupper (1963) gives a nice application of discrete compound Poisson processes to claim counts. Assume that the number of accidents that have happened in a factory up to time t is denoted by $N'(t)$. Assume that $\{N'(t)\}$ follows a Poisson process with parameter λ as defined before. In the n-th accident the number of casualties is equal to V_n. The sequence $\{V_n, n \geq 1\}$ is assumed to be a sequence of independent and identically distributed discrete random variables with probability generating function $\hat{g}_V(s) = \sum_{k=1}^{\infty} p_{V,k}s^k$. Note that the inclusion of $p_{V,0}$ would hardly make sense. If the sequence $\{V_n, n \geq 1\}$ is independent of the process $\{N'(t), t \geq 0\}$, then the probability generating function of $N(t) = \sum_{i=1}^{N'(t)} V_i$, the total number of claims up to time t, is of the form (1.2.3). A proper choice of $\hat{g}_V(s)$ gives a variety of possible processes. Note in particular that this claim number process allows multiple jumps.

The compound Poisson process will constantly appear as a backbone model for the claims arrival process. We will recognize it in its own right in Chapters 4 and 5. Later on it will reappear as a special type of continuous-time Markov processes in Section 8.3.1, as a piecewise deterministic Markov process in Section 11.3 and again as a particular point process in Chapter 12.

- *Claim number distributions related to Markov chains.* Another model for the claim number process which generalizes the classical Poisson process can be provided by a (nonhomogeneous) *pure birth process* as treated in Section 8.5.4. Here the probabilities $p_k(t)$ satisfy the *Kolmogorov differential equations*, well known from the theory of continuous-time Markov processes. Explicitly,

$$\frac{\mathrm{d}p_k(t)}{\mathrm{d}t} = -q_k(t)p_k(t) + q_{k-1}(t)p_{k-1}(t) \tag{1.2.4}$$

  for some nonnegative functions $q_k(t)$, where we take $p_{-1}(t) = q_{-1}(t) = 0$. A further example of a claim number process related to a continuous-time Markov process is the *Markov-modulated Poisson process* studied in Chapter 12 in the general framework of marked point processes. Here, instead of considering a single mixing random variable as in (1.2.1), we truly investigate more general claim number processes where, for example, the claim arrival rate can fluctuate in time according to the realizations of a Markov chain.

1.3 THE CLAIM SIZE PROCESS

In most chapters of this book, we will assume that the sequence $U_1, U_2, \ldots$ of successive claim sizes consists of independent and identically distributed random variables generated by the distribution F_U of a generic random variable U. The n-th *moment* of the claim size distribution will be denoted by $\mu_U^{(n)} = \mathbb{E}(U^n) = \int_0^\infty x^n \,\mathrm{d}F_U(x)$. For $n = 1$ we simply write $\mu_U = \mu_U^{(1)}$. For the variance of the claim sizes we write $\operatorname{Var} U = \mu_U^{(2)} - (\mu_U)^2$.

1.3.1 Dangerous Risks

A nonnegative random variable or its distribution is frequently called a *risk*. In principle any distribution, concentrated on the nonnegative halfline, can be used as a claim size distribution. However, we will often make a mental distinction between "well-behaved" distributions and dangerous distributions with a *heavy tail*. Concepts like well-behaved or heavy-tailed distributions belong to the common vocabulary of actuaries. We will make a serious attempt to formalize them in a mathematically sound definition.

Roughly speaking, the class of well-behaved distributions consists of those distributions F with an exponentially bounded tail, i.e. $1 - F(x) \le c\,\mathrm{e}^{-ax}$ for some positive a and c and all $x \ge 0$. The condition means that large claims are not impossible, but the probability of their occurrence decreases exponentially fast to zero as the threshold x becomes larger and larger. As we will see in later chapters, this condition enhances the exponential-type behaviour of most important actuarial diagnostics like aggregate claim amount and ruin probabilities.

In Chapter 2 we will give a somewhat streamlined approach to heavy-tailed distributions. For such distributions there is no proper exponential bound and huge claims are getting more likely. A natural nonparametric class of heavy-tailed claim size distributions is the class $\mathcal{S}$ of *subexponential distributions* introduced and studied in Section 2.5. The class $\mathcal{S}$ has some extremely neat probabilistic properties that will be highlighted whenever possible. For example, the aggregate claim amount is mainly determined by the largest claim in the portfolio. From the practical point of view, however, the class $\mathcal{S}$ is too wide since it cannot be characterized by parameters having a useful interpretation. For this reason practitioners usually fall back on "weakly parametrized" subexponential distributions. For example, the *lognormal distribution* belongs to $\mathcal{S}$ and is extremely popular in modelling motor insurance claim data. However, for the case of fire or catastrophic event insurance, the *Pareto distribution* F with $1 - F(x) = (c/x)^\alpha$, for $x \ge c$, seems to be sufficiently flexible to cope with most practical examples. It is fortunate that, over the past few decades, the asymptotical and statistical properties of subexponential distributions have received considerable attention. In this textbook we will, however, mainly deal with asymptotic properties and only touch upon some of the resulting statistical issues.

1.3.2 The Aggregate Claim Amount

The *aggregate claim amount* at time t given by $X(t) = \sum_{i=1}^{N(t)} U_i$ has the distribution function

$$F_{X(t)}(x) = \mathbb{P}(X(t) \le x) = \mathbb{P}\Big(\sum_{i=1}^{N(t)} U_i \le x\Big)\,.$$

Sometimes the reference to $X(t)$ is omitted and then we write $F(x)$ for $F_{X(t)}(x)$. The derivation of an explicit formula for $F_{X(t)}(x)$ is almost always impossible. In any case, more specific knowledge on the interdependencies between and within the two processes $\{N(t)\}$ and $\{U_n\}$ is needed.

It is most often assumed that the processes $\{N(t)\}$ and $\{U_n\}$ are stochastically independent, and we usually will follow this practice. However, it is easy to imagine situations where the processes $\{N(t)\}$ and $\{U_n\}$ will not be independent. Suppose that we are considering a portfolio of road accidents.

In winter time there might be a large number of claims due to poor road conditions. However, most of these claims will be small since the weather conditions prevent high speeds. A similar kind of thinking applies to the summer period where the number of accidents will usually be rather small but some of them will result in severely large claims due to reckless driving at high speeds.

As another example, take the portfolio of accidents on oil platforms of a certain type. As soon as a severe accident happens on one of these platforms, improved safety measures will try to prevent similar accidents in the future. This action hopefully results in a subsequent drop in the number of accidents.

In Chapter 4 we will cover a variety of different approaches to determine the aggregate claim amount $X(t)$. They all have in common that they are geared to investigate the behaviour of $F_{X(t)}(x)$ for a fixed t, which henceforth will be mostly omitted.

As already stated, the most studied case, however, is that where the two processes $\{N(t)\}$ and $\{U_n\}$ are independent. Even then, calculation of $F_{X(t)}(x)$ remains a formidable undertaking. By an application of the law of total probability it is easy to see that we have the following fundamental relation

$$F_{X(t)}(x) = \mathbb{P}(X(t) \le x) = \sum_{k=0}^{\infty} p_k(t) F_U^{*k}(x)\,, \tag{1.3.1}$$

where F_U^{*k} refers to the k-th convolution of F_U with itself, i.e. $F_U^{*k}(x) = \mathbb{P}(U_1 + U_2 + \ldots + U_k \le x)$, the distribution function of the sum on k independent random variables with the same distribution as U. One of the unfortunate aspects in the applications of formula (1.3.1) is that k-fold convolutions are seldom calculable in closed form. One often needs to rely on approximations, expansions and/or numerical algorithms. In Section 4.4 we deal with recursive and numerical algorithms, while in Section 4.6 we look for approximations by easier compounds.

An alternative way of writing the above formula is given in terms of the *Laplace–Stieltjes transform* $\hat{l}_{X(t)}(s) = \mathbb{E}\,\mathrm{e}^{-sX(t)}$. Namely,

$$\hat{l}_{X(t)}(s) = \hat{g}_{N(t)}(\hat{l}_U(s))\,, \tag{1.3.2}$$

where $\hat{g}_{N(t)}(s)$ is the probability generating function of the number of claims $N(t)$ up to time t and $\hat{l}_U(s)$ is the Laplace–Stieltjes transform of U. But then inversion techniques as in Section 4.7 or even more specifically as in Section 5.5 may be useful.

If the number of claims is very large, then one can forecast that a central limit effect will be dominant. A *large-scale approximation* like

$$F_{X(t)}(x) \approx \Phi\left(\frac{x - \mathbb{E}\,X(t)}{\sqrt{\operatorname{Var} X(t)}}\right),$$

where $\Phi(x)$ denotes the standard normal distribution function, might be possible. In more concrete situations one can even replace $\mathbb{E}\,X(t)$ and $\operatorname{Var} X(t)$ by subsequent approximations. For example, if $\{N(t)\}$ is a renewal counting process, these approximations follow from the basic renewal theorems, as will be shown in Section 6.2.3.

On the other hand, the above large-scale approximation has often been shown to be rather unreliable. Practitioners have tried to remedy this shortcoming by using refined versions of the central limit theorem. For example, *Edgeworth expansions* and *Gram–Charlier series* resulted in significant but still not always satisfying improvements. Other approaches like the *normal-power approximations* or approximations using gamma-distributions have been tried out.

The main reason why results inspired by the central limit theorem often give poor results is that the number of those claims that *de facto* determine the entire portfolio is actually (and fortunately) very small. This is particularly true in the case of large claims. The outcome is that a genuine centralization of the claims is totally absent and any central limit approximation is meaningless.

1.3.3 Comparison of Risks

We have already alluded to the difference between light-tailed and heavy-tailed claim size distributions and we have tried to fit this intuitive idea into a mathematically sound definition. On the more methological level, one might like to compare risks and even order them according to some ranking criterion. A first and general attempt to *ordering of risks* is made in Section 3.2. On further locations, we investigate how this ordering translates into a subsequent ordering of compounds in Section 4.2.4 and of ruin functions in Section 5.4.5.

Running along sample paths, we can even compare full processes and decide which one is most apt to incorporate essential features of a risk situation. In Section 7.4.2 we compare discrete-time Markov chains, while Section 8.1.4 treats a similar comparison in the continuous-time setup. Another illustration of a comparison of entire processes is given in Section 12.2.4, where Markov modulated Poisson processes, Poisson cluster processes, mixed Poisson processes and merging renewal processes are put in proper perspective to homogeneous Poisson processes with the same arrival intensity.

1.4 SOLVABILITY OF THE PORTFOLIO

1.4.1 Premiums

We have used the abbreviation $\Pi(t)$ to denote the totality of premiums collected from the policy-holders within the portfolio up to time t. Usually premiums are individually collected once a year, but the insurer can safely assume that premium income is evenly spread over the year.

The determination of a functional expression for $\Pi(t)$ is one of the few things where the insurer can intervene in the overall process. The function $\Pi(t)$ should be determined in such a way that the solvability of the portfolio can be guaranteed. This requires $\Pi(t)$ to increase fast enough to cope with incoming claims. On the other hand, a very high value of $\Pi(t)$ may be rather undesirable since then rival insurance companies might attract clients by offering lower premiums while covering the same risk.

The whole area of determining the specific shape of the function $\Pi(t)$ is called *premium calculation* and constitutes an essential part of the actuarial know-how. For an exhaustive treatment, see Goovaerts, De Vylder and Haezendonck (1984). In Chapter 3 we cover a few and isolated *premium principles* underlying the general thinking behind premium calculations. From these general principles a wide set of possible candidates emerges. Note that most often these premiums are *nonrandom* even if their calculation involves information on the stochastic elements within the portfolio.

The most popular form of the premium function $\Pi(t)$ is

$$\Pi(t) = (1+\eta)\mathbb{E}\,N(t)\mathbb{E}\,U\,, \tag{1.4.1}$$

where $\mathbb{E}\,N(t)$ is the expected number of claims up to time t, while $\mathbb{E}\,U$ is the mean claim size. The constant η is the *safety loading* which has to take care, not only of the administrative costs from handling the portfolio, but also of the necessary gain that the company wants to make ultimately. Moreover, it is clear that η will be portfolio-dependent, as, for example, the higher the risk, the higher η has to be. Note that in the case of a Poisson process $\{N(t)\}$ the premium function $\Pi(t)$ given in (1.4.1) is of the form $\Pi(t) = \beta t$ for some constant $\beta > 0$.

The premium principle of the form (1.4.1) with η independent of the portfolio is called the *expected value principle.* Of course, the expected value principle does not keep track of the variability in the portfolio and so alternative principles include, for example, the variances of $N(t)$ and U. A few other possibilities are also covered in Chapter 3.

Another aspect of premium calculation is that not all individual policies in a portfolio must be charged with the same premium. For example, the by now classical *bonus-malus* premium calculation principle is widely spread in car insurance. Depending on the past history of a policy-holder, the insurer ranks the client in a certain *state* and charges an amount typical for that state. In the course of time the customer will move from one state to another, depending on his claim record. If we denote by X_n the state of a client at the beginning of year n, then the process $\{X_n, n \geq 0\}$ describes a jump process adequately modelled by a *discrete-time Markov chain.* No course on stochastic processes can be considered complete if it would lack a serious treatment of this important building block of stochastic modelling. We will deal with discrete-time Markov chains in general in Chapter 7 and with bonus-malus in Section 7.1.4 in particular. In Chapter 8 we will then continue this discussion by changing from discrete time to continuous time.

1.4.2 The Risk Reserve

Recall from the introduction that the risk reserve at time t is given by $R(t) = u + \Pi(t) - X(t)$, where u is the initial reserve. The quantity $R(t)$ is, of course, random. The ultimate hope would be to get some information on the distribution of $R(t)$ or, if this is impossible, on the first few moments of $R(t)$. As has already been pointed out, the distribution of $X(t)$ is very complicated. If we superimpose a possibly random character of the premium income, then the overall situation will become even worse.

There is one particular situation, however, where we can give a full treatment of the stochastic nature of the risk reserve. Assume that time is measured in integers. We then add all premiums collected in period n in one single number $\Pi_n = \Pi(n) - \Pi(n-1)$, which can even be assumed to be random. Similarly we add all claims arriving in period n and call this amount X_n, where $X_n = X(n) - X(n-1)$. The risk reserve after period n is then equal to

$$R_n = R(n) = u + \Pi(n) - X(n) = u - \sum_{i=1}^{n}(X_i - \Pi_i) .$$

The resulting stochastic process $\{S_n, n \geq 0\}$ of partial sums $S_n = \sum_{i=1}^{n} Y_i$ is called the (discrete-time) *claim surplus process*. In particular, $\{S_n\}$ is a *random walk* if we assume that the sequence $\{Y_n, n \geq 1\}$ with $Y_n = X_n - \Pi_n$ consists of independent and identically distributed random variables.

The definition of a random walk is very similar to that of a renewal process, the only difference being that now the generic random variable is no longer concentrated on the nonnegative halfline. The theory of random walks makes up a substantial chapter in a traditional course on stochastic modelling. We will cover the most important aspects of random walk theory in Chapter 6.

If premiums are nonrandom we can write

$$\mathbb{P}(R(t) \geq x) = \mathbb{P}(X(t) \leq u + \Pi(t) - x) = F_{X(t)}(u + \Pi(t) - x)$$

which shows how important it is to have workable expressions for the distribution of the aggregate claim amount.

1.4.3 Economic Environment

A number of problems in insurance and finance contain an economic element like interest, discounting or indexing. As such, a variety of subjects will be treated within an *economic environment.*

Consider data from car insurance. When calculating the premium, the insurer needs certain characteristics of the claim sizes, such as their mean value and variance. Just using the common estimators for mean and variance would lead to problems. Over the years, cars have become more expensive. Therefore, the claim data from some years ago cannot be compared directly with claim data of today and so need to be adjusted.

Suppose that we consider a time horizon of n years. Then, for each $i = 1, \ldots, n$, an index I_i for year i is defined which is related to the costs of that year, including, for instance, prices of new cars and repair costs. This index is then used to measure all claims in units of year 0. Denote by X_i the aggregate claim amount of year i. Then adequate estimators for mean and variance of the aggregate claim amount measured in units of year 0 are given by

$$\bar{X} = \frac{1}{n}\sum_{i=1}^{n} X_i/I_i\,, \qquad \bar{\sigma}^2 = \frac{1}{n-1}\sum_{i=1}^{n}(X_i/I_i - \bar{X})^2\,.$$

The corresponding estimates for year $n+1$ are then

$$\hat{X}_{n+1} = \hat{I}_{n+1}\bar{X}\,, \qquad \hat{\sigma}_{n+1} = \hat{I}^2_{n+1}\bar{\sigma}^2\,,$$

where $\hat{I}_{n+1}$ is an estimate of the index for year $n+1$.

The indices I_i are often expressed via the *interest rates* $r_i = I_i/I_{i-1} - 1$. This is particularly advantageous in *life insurance*. For example, in the case of a single premium payment, suppose that the dependant of the policy-holder will get a predefined lump sum x in the case of the death of the policy-holder within n years. The insurance company will invest the premium. Thus the value V of the payment will be different depending on the year of death. Let p_i denote the probability that the policy-holder dies in year i after the issue of the policy. Then, for a given sequence $r_1, \ldots, r_{n-1}$ of interest rates, the expected value $\mathbb{E}\,V$ of the contract is

$$\mathbb{E}\,V = p_1 x + p_2 x(1+r_1) + \ldots + p_n x(1+r_1)\ldots(1+r_{n-1})\,.$$

The problem with the above expression is that, at policy issue, the interest rates r_i are not known in advance. Therefore insurance companies use a *technical interest rate* r instead of the true interest rates r_i. This leads to a technical value V' with expectation

$$\mathbb{E}\,V' = p_1 x + p_2 x(1+r) + \ldots + p_n x(1+r)^{n-1}\,.$$

We will use this kind of economic setting in Section 7.3, where we deal with Markov chains with rewards. In Section 11.4 we deal with risk processes in an economic environment within the framework of piecewise deterministic Markov processes. Questions of insurance and finance connected with stochastic interest and discounting are discussed in Chapter 13.

1.5 REINSURANCE

Reinsurance is a prime activity of almost all insurers. We first explain why reinsurance may be necessary. Then we give a number of specific forms of reinsurance and point out how each one of them provides further support for the topics that will be treated in later chapters.

1.5.1 Need for Reinsurance

As phrased by Borch (1960), a company makes arrangements for reinsurance when it seeks to reduce the probability of suffering losses which may jeopardize the position of the company. Among the reasons to ponder reinsurance the insurer can think about the following.

- The appearance of *excessively large claims;* here we think about claims resulting from severe accidents as with nuclear power stations or cases of serious medical maltreatment. In other instances an insurer might be confronted with large claims coming from a policy involving very valuable items such as air carriers, oil tankers, dams and large building complexes.
- An unusually *large number of claims* or *clustering of claims*, whether large or not. Extensive forest fires may temporarily lead to a very large number of more or less large claims. Hurricanes, earthquakes and floods can cause similar explosions of the number of claims.
- Unexpected *changes in premium collection* as in the case of a sudden inflation or unforeseen increase in handling costs. Under these circumstances the company actually does not quite receive the premium income it had expected at the beginning of the book year.
- There are *legal restrictions* forcing the company to have reserves to cover a certain part of future claims. For a smaller company these restrictions would cause a noncompetitive premium. Taking reinsurance is a comfortable way of solving that problem.
- If a company can take reinsurance, it can also offer more services to its clients. Reinsurance can therefore be considered to *increase the capacity of the company.*

There exists a variety of reinsurance forms. What they all have in common is the desire to diminish the impact of the large claims. In what follows we provide the mathematical formulation of most of the currently employed *reinsurance treaties.*

1.5.2 Types of Reinsurance

Recall that we use $\{U_1, U_2, \ldots, U_{N(t)}\}$ as the sequence of successive claims up to time t when the underlying claim number process is $\{N(t), t \geq 0\}$. Further the aggregate claim amount over that period is $X(t) = \sum_{i=1}^{N(t)} U_i$.

On the basis of the past history of the portfolio or stimulated by one or more arguments under Section 1.5.1, the insurer will redesign the portfolio in such a way that he himself keeps a certain amount of the aggregate claim amount $X(t)$ while he looks for reinsurance for the remaining part. The amount that he keeps is called the *deductible* and will be denoted by $h(X(t))$ or $h(U_i)$, depending on whether the reinsurance form acts on the whole portfolio or on single claims. The remaining part, i.e. $X(t) - h(X(t))$ or $U_i - h(U_i)$ is the *reinsured part*. Acting like that, the insurer himself is taking an insurance with a reinsurance company and hence himself becomes a client. The *first line insurer* is still drafting the premiums to be asked from his own customers. Part of that premium now has to be transferred to the *second line insurer*, who has agreed to cover the risk at a negotiated premium. The second line insurer can then again redivide the risk and go to a third line company, hence building up a *reinsurance chain*, where at each step deductibles and corresponding premiums have to be negotiated and transmitted down the chain to the first underwriter.

Here are the most commonly used types of reinsurance. The first form of reinsurance is *proportional* or *quota-share reinsurance* where a certain proportion, say a, of the total portfolio is reinsured. This means that $h(x)$ takes the special form $h(x) = ax$. But then

$$h(X(t)) = aX(t) = a\sum_{i=1}^{N(t)} U_i = \sum_{i=1}^{N(t)} h(U_i)$$

where $0 < a < 1$ is the *proportionality factor*.

This form of reinsurance is very popular in almost all insurance branches, presumably because of its conceptual and administrative simplicity. Moreover this kind of reinsurance is often used at the start of smaller companies to broaden their chances in underwriting policies. In general the first line insurance cedes to the reinsurer a similarly determined proportion of the premiums.

From the distributional point of view the basic properties of $h(X(t))$ can be derived from the analogous properties for the case $a = 1$. Indeed, by its definition

$$\mathbb{P}(h(X(t)) \leq x) = \mathbb{P}(X(t) \leq x/a) = F_{X(t)}(x/a).$$

This formula shows again how important it is to get reliable and accurate formulae for the distribution of the aggregate claim amount.

Another important form of reinsurance is *excess-loss*, which is determined by a positive number b, called the *retention level*. The reinsured amount is then equal to $\sum_{i=1}^{N(t)} (U_i - b)_+$, where $x_+ = \max\{x, 0\}$. This reinsurance form covers the overshoot over the retention level b for all individual claims whether or not they are considered to be large. It is clear that excess-loss reinsurance limits the liability of the first line insurer. It appears as if the underwriter of the policy decides that he himself will cover all claims below the retention b.

Among the insurance branches where this type of reinsurance is used we mention in particular general liability, and to a lesser extent motor-liability and windstorm reinsurance. Excess-loss reinsurance is particularly interesting if a relatively small number of risks is insured, and the individual claim size distributions are heavy-tailed. Because of its very form, all claims have to be checked individually and as such this reinsurance contract leads to an expensive administration.

From the distributional angle, the excess-loss amount has a distribution which is completely similar to that of the aggregate claim amount but with a claim size distribution truncated at the retention. Again the importance of good approximations to the distribution $F_{X(t)}(x)$ is apparent.

A reinsurance policy that considers each claim as an integral part of the entire portfolio is determined by the quantity

$$\left(\sum_{i=1}^{N(t)} U_i - b\right)_+ = (X(t) - b)_+$$

where the *retention* b determines the *stop-loss reinsurance.* In this type of reinsurance, the small claims also show their influence on the total amount reinsured. In particular, when the number of claims is very large, the aggregate claim amount is highly dependent upon the small claims as well. On the other hand, stop-loss reinsurance seems a natural adaptation of the excess-loss treaty, but then to the portfolio as a whole.

Stop-loss reinsurance is used in windstorm and hail reinsurance and occasionally in fire insurance. Due to its form, stop-loss reinsurance is very simple to apply and does not require expensive individualized administration. In general one would not use stop-loss reinsurance unless the number of policies were large. Moreover we will show in Chapter 3 that stop-loss reinsurance has some desirable optimality properties.

From the distributional point of view one notices the necessity to derive compact expressions for the probabilities of overshooting a certain barrier. If one indeed considers the portfolio as a single policy, then again the approximations and bounds for the aggregate claim amount are of prime importance.

There are still other types of reinsurance contracts.

- Thinking especially about coverage against large claims, it seems desirable to look for treaties based on the largest claims in the portfolio. If we denote by $(U_{(1)}, U_{(2)}, \ldots, U_{(N(t))})$ the order statistics of the random vector

$(U_1, U_2, \ldots, U_{N(t)})$ of claim sizes, then

$$Z(t) = \sum_{i=1}^{r} U_{(N(t)-i+1)} - rU_{(N(t)-r)} = \sum_{i=1}^{N(t)} (U_i - U_{(N(t)-r)})_+$$

would make up a rather neat reinsurance treaty. It has been introduced by Thépaut (1950) and is called *excédent du coût moyen relatif* (ECOMOR). The amount $Z(t)$ covers only that part of the r largest claims that overshoots the random retention $U_{(N(t)-r)}$, where $U_{(N(t)-r)} = 0$ if $N(t) \leq r$. In some sense the ECOMOR treaty rephrases the excess-loss treaty but with a random retention at a large claim.
Not much is known about ECOMOR treaties and as such this reinsurance form has been largely neglected by most reinsurers. The main reason is the rather complicated form of $Z(t)$, which defies a simple treatment.

- A reinsurance form that is somewhat akin to proportional reinsurance is *surplus reinsurance*. Here the reinsured amount is determined individually and proportionally by the value of the insured object. The insurer is forced to introduce the value of the insured object as an extra unknown and basically random quantity. The overall value of the insured amount is the key factor when choosing this type of reinsurance.
- Of course there are possibilities for *combined reinsurance contracts*. For example, the insurer can first apply an excess-loss retention on the individual claims; at the next step he applies an additional stop-loss cover to the remaining excess over a certain retention.

We will not develop any systematic study of these last three reinsurance treaties and restrict ourselves to the basic theory.

1.6 RUIN PROBLEMS

Ruin theory has always been a vital part of actuarial mathematics. At first glance, some of the theoretically derived results seem to have limited scope in practical situations. Nevertheless, calculation of and approximation to ruin probabilities have been a constant source of inspiration and technique development in actuarial mathematics.

Assume an insurance company is willing to risk a certain amount u in a certain branch of insurance, i.e. if the claim surplus exceeds the level u some drastic action will have to be taken for that branch. Because in some sense this part of the business starts with the capital u we can safely call u the *initial capital*. The actuary now has to make some decisions, for instance which premium should be charged and which type of reinsurance to take, see Section 1.5. Often, the premium is determined by company policies and by tariffs of rivals. A possible criterion for optimizing the reinsurance treaty would be to minimize the probability that the claim surplus ever exceeds the level u. To be more specific, consider the risk reserve $R(t) = u + \Pi(t) - X(t)$ and define the random variable $\tau = \inf\{t \geq 0 : R(t) < 0\}$. The instant τ gives us the *ruin time* of the portfolio, where we interpret ruin in a technical sense. Of course, we should allow the possibility that no ruin ever occurs, which means that $\tau = \infty$. We should realize that τ is dependent on all the stochastic elements in the risk reserve process $\{R(t)\}$ as well as on the deterministic value u. For this reason one often singles out the latter quantity in the notation for the ruin time by writing $\tau(u)$ for τ. More specifically, the *survival* or *nonruin probability in finite time* will be defined and denoted by

$$\overline{\psi}(u;x) = \mathbb{P}\big(\inf_{0 \leq t \leq x} R(t) \geq 0\big) = \mathbb{P}(\tau(u) > x)$$

when we consider a *finite horizon* $x > 0$. The *survival probability* over an infinite time horizon is defined by the quantity

$$\overline{\psi}(u) = \mathbb{P}\big(\inf_{t \geq 0} R(t) \geq 0\big) = \mathbb{P}(\tau(u) = \infty) .$$

Alternative notations that are in constant use refer to the *ruin probabilities* which are defined by the equalities

$$\psi(u;x) = 1 - \overline{\psi}(u;x) \,, \qquad \psi(u) = 1 - \overline{\psi}(u) .$$

Each year the insurer of a portfolio has to negotiate a reinsurance contract. While an optimal strategy will depend on $R(t)$, the insurer has to apply the contract at the beginning of the year. Simultaneously, the future policies of the reinsurance companies have to be taken into account. The resulting problem is hard to solve. In one approximation procedure, the premium and the reinsurance treaty is fixed for the whole future and then the optimal reinsurance is chosen; see, for instance, Dickson and Waters (1997). Then this new reinsurance treaty is chosen as input and the procedure is repeated. To get rid of the dependence on the initial capital, an alternative approach is to consider the *adjustment coefficient* introduced in Chapter 5. The adjustment coefficient is some sort of *measure of risk*. Maximizing the adjustment coefficient is in some sense minimizing the risk for large initial capitals. This optimization procedure was, for instance, considered in Waters (1983).

Ruin theory is often restricted to the classical compound Poisson risk model. The latter model will therefore appear over and over again as a prime example. Unfortunately, because of its intrinsic simplicity, the compound Poisson model does not always give a good description of reality. There are more realistic but still tractable models in the literature, that hopefully will find their way into actuarial applications. We will review a number of these models. In Chapter 11 we introduce piecewise deterministic Markov processes to broaden the Markovian treatment of risk analysis. Interest and discounting can then be easily introduced as elements of economic environments. Another prospective area of actuarial and financial applications can be found in the theory of point processes that will be treated in Chapter 12.

As one can expect, ruin probabilities will depend heavily on the claim size distribution. If the latter is well-behaved the ruin probabilities will turn out to be typically exponentially bounded as the initial capital becomes large. However, when the claim size distribution has a heavy tail, then one single large claim may be responsible for the ultimate ruin of the portfolio. The reader will definitely appreciate how the class $\mathcal{S}$ of subexponential distributions provides a beautiful characterization for the heavy-tailed case as shown in Sections 5.4.3, 6.5.5 and in even more depth in Sections 12.6 and 13.2.4.

Phase-type distributions form another and versatile alternative class of distributions for which more or less explicit calculations are possible. They are treated in some detail in Section 8.2 and applied to ruin calculations later on.

The results for ruin probabilities on which we will focus in this book can be characterized by the following features.

- Only in the easiest cases will we succeed in getting explicit formulae for the ruin probabilities as in Section 5.6.
- As soon as we allow more complex models, one way out is to use approximations of *Cramér-Lundberg type* for large initial capital as in Sections 5.4, 6.5 or more generally in Chapters 11, 12 and 13. In all these models, a surprisingly different asymptotic behaviour of ruin probabilities is observed depending on whether light-tailed or heavy-tailed claim size distributions are considered.
- An alternative is to employ numerical procedures as in Section 5.5.
- A further and even more important theme of the text is the derivation of bounds for the ruin probabilities. A vast number of *Lundberg bounds* have been derived in the text, ranging from the simplest in Section 5.4.1 in the compound Poisson model, over Section 6.5.2 in the Sparre Andersen model, to the more general in Chapters 11, 12 and 13. Another type of bound is obtained in Section 12.2.4, where we are dealing with the comparison of ruin probabilities for risk reserve processes with identical safety loadings but with different volatilities.
- Finally, simulation methods are discussed in Section 9.2.5.

Quite a few of the above results have been derived using *martingale techniques*. Martingale theory makes up a vast portion of current day probabilistic modelling and cannot be left out of any serious treatment of risk analysis and financial mathematics. In Chapter 9 we treat the fairly easy discrete-time situation, making links to random walk theory and life insurance. The more demanding continuous-time martingale theory is treated in Chapter 10. This chapter is also vitally needed in the treatment of financial models; it nevertheless provides some unexpected links to finite-horizon ruin probabilities in Section 10.3.

1.7 RELATED FINANCIAL TOPICS

1.7.1 Investment of Surplus

In Section 1.4 we have already mentioned that the calculation of premiums constitutes an essential part of the actuarial know-how. Although these premiums are nonrandom for a certain time horizon, their calculation involves information on the stochastic elements within the portfolio and within the economic environment. However, even once the premiums are fixed, the future premium income of an insurer is not deterministic. For one reason, the number of customers may increase or decrease in time outside the control of the insurer. Since the insurer invests the surplus in financial markets, there still is another source of uncertainty which is caused by the random fluctuations of these markets. Notice that interest and inflation rates usually change in much smaller steps than the risk reserve of an insurance portfolio changes with the arriving claims. This situation is modelled in Section 13.2 by means of *perturbed risk processes*, which are defined as the sum of a usual risk reserve process $\{R(t)\}$ and a stochastic perturbation process. Typically, perturbed risk processes belong to the class of *jump-diffusion processes.* Their sample paths change discontinuously from time to time since jump epochs and jump sizes are random. In the intervals between the jumps they behave like the sample paths of a diffusion process; see Section 1.7.2. This class of stochastic processes is one of the main subjects of financial mathematics.

Over recent years, a number of textbooks have appeared that provide introductions to the mathematical theory of finance. We refer to the bibliographical notes in Chapter 13. An inclusion of all the technical material needed to do full justice to this contemporary subject would increase the size of our monograph substantially. On the other hand, we felt a need to at least include a streamlined introduction to the subject. We made a special attempt to show the clear links between the material already mentioned above and a subset of topics from the realm of financial stochastics.

1.7.2 Diffusion Processes

The description of the stochastic properties of stock prices, interest rates, etc. in terms of *diffusion processes* has been one of the great break throughs of stochastic thinking in a real-life context. It is by now generally accepted that an appropriate formulation for the time-evolution of a diffusion process $\{X(t)\}$ should be given in terms of a *stochastic differential equation* of the form

$$\mathrm{d}X(t) = a(t, X(t))\,\mathrm{d}t + \sigma(t, X(t))\,\mathrm{d}W(t)\,.$$

Here, $\{W(t)\}$ is the usual *Brownian motion*, which is at the core of stochastic analysis. Further, $a(t,x)$ takes care of the drift of the process, while $\sigma(t,x)$ describes the strength of the extraneous fluctuations caused by the Brownian motion. In Section 13.1 we give an abridged treatment of stochastic differential equations and we show existing links with martingales and Markov processes.

For specific choices of $a(t,x)$ and $\sigma(t,x)$ we arrive at a wealth of possible models. Here is a more concrete example. A major step forward in the use of stochastic calculus in finance came from an attempt to price options on a security. This led to the popular Black–Scholes formula, for which the Nobel price 1997 was won. In fact, the use of the Black–Scholes model for option pricing, induced a change in the economy so that stock prices in liquid markets became very close to a Black–Scholes model. Up to 1972, the Black–Scholes model for option prices was a bad approximation to real prices. In the above terminology, the price process takes the form $X(t) = \mathrm{e}^{\delta t}X^*(t)$, where δ refers to the *force of interest*, while $\{X^*(t)\}$ satisfies the stochastic differential equation with the choice $a(t,x) = (\mu - \delta)\,x$ and $\sigma(t,x) = \sigma\,x$. The drift is regulated by the difference between the expected *rate of return* μ and the force of interest δ, while the fluctuations are modelled by the *volatility* σ.

In Section 13.3.1 we will explain how trading stategies in a market with two financial goods can be developed once the option is chosen. In particular, we consider the case of a *European call option* where the option-holder has the right (but not the obligation) to buy an asset for a fixed price at a fixed point in time.

1.7.3 Equity Linked Life Insurance

Section 13.3.2 provides an inspiring link between the Black–Scholes model and stochastic modelling of life insurance.

From the Middle Ages, states and towns in Europe have been selling annuities. Life insurance mathematics arose in response to the need for evaluation of the price of these annuities. These calculations stimulated developments in different fields, like demography and probability theory. In London, early life tables were proposed by John Graunt in 1662. However, it is considered that the first life table in the modern sense was constructed by E. Haley in 1693. He used data from Breslau in Silesia, collected by the pastor and scientist Casper Neumann during the period 1687–1691. For further historical details, the reader is referred to the survey by Hald (1987).

The subsequent mathematical modelling of life insurance is based on probability theory, since in our modern terminology lifetimes are random variables. A classical life table contains the expected proportions of that part of a population of age a that reaches a later age c. If we use T_a as the stochastic notation for the remaining lifetime of a policy-holder of age a, then the life tables give us the necessary means to estimate (sometimes using interpolations) the distribution $\mathbb{P}(T_a \le y) = \mathbb{P}(T \le y \mid T > a)$, where T is the typical lifetime of any member of the population.

Under a classical life insurance contract, the insured benefit typically consists of one single payment – the sum insured. Other useful types of contracts in life insurance are life annuities, consisting of a series of payments which are made during the lifetime of the beneficiary. Furthermore, there are combinations of a life insurance contract with an annuity. The subsequent problem is then to follow such contracts from the instant of policy issue up to the death of the customer.

Take, for instance, the case where a customer of age a underwrites a classical *term insurance policy* where he gets the value $\max\{X(T_a), b\}$ at time T_a of his death. Note that this value is similar to the value of a European call option, but the payoff is now paid at a random time. Here, $b > 0$ is a guaranteed lower bound and $\{X(t)\}$ is a stochastic *price process* as described in Section 1.7.2. With the additional notions of discounting and interest, one is now equipped to develop a coherent theory of *equity linked life insurance* contracts. A detailed treatment of the above life insurance situation is given in Section 13.3.2 in the case of a constant force of interest $\delta \ge 0$. As there is no obvious reason why the force of interest should be kept constant, the final Sections 13.3.3 and 13.4 cover a few possible forms of a stochastic force of interest $\{\delta(t)\}$ via appropriate stochastic differential equations.

CHAPTER 2

Probability Distributions

2.1 RANDOM VARIABLES AND THEIR CHARACTERISTICS

2.1.1 Distributions of Random Variables

Random variables are basic concepts in probability theory. They are mathematical formalizations of random outcomes given by numerical values. An example of a random variable is the amount of a claim associated with the occurrence of an automobile accident. The numerical value of the claim is usually a function of many factors: the time of year, the type of car, the weather conditions, etc. One introduces a random variable X as a function defined on the set Ω of all possible outcomes. In many cases, neither the set Ω nor the function X need to be given explicitly. What is important is to know the probability law governing the random variable X or, in other words, its distribution; this is a prescription of how to evaluate probabilities $F(B)$ that X takes values in an appropriate subset B of the real line $\mathbb{R}$. Usually these subsets belong to the class $\mathcal{B}(\mathbb{R})$ of Borel subsets, which consists of all subsets of $\mathbb{R}$ resulting from countable unions and intersections of intervals.

For reasons of mathematical convenience, it is useful to consider a certain family $\mathcal{F}$ of subsets of Ω, called *events*. Furthermore, a probability $\mathbb{P}$ on $\mathcal{F}$ assigns to each event A from $\mathcal{F}$ its probability $\mathbb{P}(A)$. The crucial assumptions on $\mathcal{F}$ and $\mathbb{P}$ are the closeness of $\mathcal{F}$ with respect to countable unions and intersections of sets from $\mathcal{F}$. In the terminology of probability theory, $\mathcal{F}$ is a *σ-algebra* and the additivity property of $\mathbb{P}$ with respect to countable unions of disjoint sets from $\mathcal{F}$ make $\mathbb{P}$ a *probability measure.*

More formally, a *random variable* is a measurable mapping $X : \Omega \to \mathbb{R}$, i.e. the set $\{X \in B\} \stackrel{\text{def}}{=} \{\omega \in \Omega : X(\omega) \in B\}$ belongs to $\mathcal{F}$ for each $B \in \mathcal{B}(\mathbb{R})$. The *distribution* F of X is the mapping $F : \mathcal{B}(\mathbb{R}) \to [0, 1]$ defined by $F(B) = \mathbb{P}(X \in B)$. Furthermore, $F : \mathbb{R} \to [0, 1]$ with $F(x) = \mathbb{P}(X \le x)$ is called the *distribution function* of X. We use the same symbol F because there is a one-to-one correspondence between distributions and distribution functions. By $\overline{F}(x) = 1 - F(x)$ we denote the *tail* of F. We say that a distribution F is concentrated on the Borel set $B \in \mathcal{B}(\mathbb{R})$ if $F(B) = 1$. In actuarial applications, a nonnegative random variable is frequently called a *risk*.

There are two important but particular types of random variables – discrete and continuous ones. We say that X is *discrete* if there exists a denumerable subset $E = \{x_0, x_1, \ldots\}$ of $\mathbb{R}$ such that $\mathbb{P}(X \in E) = 1$. In this case, we define the *probability function* $p : E \to [0,1]$ by $p(x_k) = \mathbb{P}(X = x_k)$; the pair (E, p) gives a full probabilistic description of X. The most important subclass of nonnegative discrete random variables is the lattice case, in which $E \subset h\mathbb{N}$, i.e. $x_k = hk$ for some $h > 0$, where $\mathbb{N} = \{0, 1, \ldots\}$. We then simply write $p(x_k) = p_k$ and say that X is a *lattice* random variable.

On the other hand, we say that X is *absolutely continuous* if there exists a measurable function $f : \mathbb{R} \to \mathbb{R}_+$ such that $\int f(x)\,\mathrm{d}x = 1$ and $\mathbb{P}(X \in B) = \int_B f(x)\,\mathrm{d}x$ for each $B \in \mathcal{B}(\mathbb{R})$. We call f the *density function* of X.

The distribution of a discrete random variable is called *discrete.* If X is a lattice random variable, then we call its distribution *lattice*, otherwise we say that it is *nonlattice.* Analogously, the distribution of an absolutely continuous random variable is called *absolutely continuous.* Sometimes the distribution F of the random variable X is neither purely discrete nor absolutely continuous, but a *mixture* $F = \theta F_1 + (1-\theta)F_2$, where F_1 is a discrete distribution with probability function p, and F_2 is an absolutely continuous distribution with density function f; $0 \le \theta \le 1$.

In order to emphasize that we consider the distribution, distribution function, probability function, density function, etc. of a given random variable X, we shall use the notation $F_X, p_X, f_X, \ldots$. The index X is omitted if it is clear which random variable is meant.

For two random variables X and Y with the same distribution we write $X \stackrel{\mathrm{d}}{=} Y$. Furthermore, we write $(X_1, \ldots, X_n) \stackrel{\mathrm{d}}{=} (Y_1, \ldots, Y_n)$ if two vectors of random variables $(X_1, \ldots, X_n)$ and $(Y_1, \ldots, Y_n)$ have the same distribution, i.e.

$$\mathbb{P}(X_1 \in B_1, \ldots, X_n \in B_n) = \mathbb{P}(Y_1 \in B_1, \ldots, Y_n \in B_n)$$

for all $B_1, \ldots, B_n \in \mathcal{B}(\mathbb{R})$ or, equivalently,

$$\mathbb{P}(X_1 \le x_1, \ldots, X_n \le x_n) = \mathbb{P}(Y_1 \le x_1, \ldots, Y_n \le x_n)$$

for all $x_1, \ldots, x_n \in \mathbb{R}$. We say that $(X_1, \ldots, X_n)$ is *absolutely continuous* if there exists a nonnegative and integrable function $f : \mathbb{R}^n \to \mathbb{R}_+$ with

$$\int_{\mathbb{R}} \cdots \int_{\mathbb{R}} f(x_1, \ldots, x_n)\,\mathrm{d}x_n \ldots \mathrm{d}x_1 = 1$$

such that $\mathbb{P}(X_1 \in B_1, \ldots, X_n \in B_n) = \int_{B_1} \cdots \int_{B_n} f(x_1, \ldots, x_n)\,\mathrm{d}x_n \ldots \mathrm{d}x_1$ for all Borel sets $B_1, \ldots, B_n \in \mathcal{B}(\mathbb{R})$. The function f is called the *density* of $(X_1, \ldots, X_n)$. If the random vectors $(X_1, \ldots, X_{n-1})$ and $(X_1, \ldots, X_n)$ are absolutely continuous with densities $f_{X_1,\ldots,X_{n-1}}$ and $f_{X_1,\ldots,X_n}$ respectively and if $f_{X_1,\ldots,X_{n-1}}(x_1, \ldots, x_{n-1}) > 0$, then

$$f_{X_n|X_1,\ldots,X_{n-1}}(x_n \mid x_1, \ldots, x_{n-1}) = \frac{f_{X_1,\ldots,X_n}(x_1, \ldots, x_n)}{f_{X_1,\ldots,X_{n-1}}(x_1, \ldots, x_{n-1})}$$

is called the *conditional density* of X_n under the condition that $X_1 = x_1, \ldots, X_{n-1} = x_{n-1}$. For other basic notions related to the concepts of independence and conditioning, see also Section 2.1.3.

For two random variables X and Y with $\mathbb{P}(X = Y) = 1$ we simply write $X = Y$. Furthermore, for a sequence of random variables $X, X_1, X_2, \ldots$ with $\mathbb{P}(X = \lim_{n\to\infty} X_n) = 1$ we write $X = \lim_{n\to\infty} X_n$.

For n fixed and for all $k = 1, 2, \ldots, n$ and $\omega \in \Omega$, let $X_{(k)}(\omega)$ denote the k-th smallest value of $X_1(\omega), \ldots, X_n(\omega)$. The components of the random vector $(X_{(1)}, \ldots, X_{(n)})$ are called the *order statistics* of $(X_1, \ldots, X_n)$.

2.1.2 Basic Characteristics

Let X be a random variable and $g : \mathbb{R} \to \mathbb{R}$ a measurable mapping. We can then consider the random variable $g(X)$. For example, if X is an insurance risk, $g(X)$ can be that part of the risk taken by the (first) insurer, while $X - g(X)$ is the residual risk passed on to the reinsurer. In Chapter 3 more specific examples of reinsurance agreements are studied, such as

$$g(X) = \begin{cases} X & \text{if } X \le a \\ a & \text{if } X > a \end{cases}$$

(stop-loss reinsurance or, alternatively, excess-of-loss reinsurance, with retention level $a > 0$) and $g(X) = aX$ (proportional reinsurance; $0 < a < 1$).

Define the value $\mathbb{E}\, g(X)$ by

$$\mathbb{E}\, g(X) = \begin{cases} \sum_k g(x_k)p(x_k) & \text{if } X \text{ is discrete} \\ \\ \int_{-\infty}^{\infty} g(x)f(x)\,\mathrm{d}x & \text{if } X \text{ is absolutely continuous} \end{cases}$$

provided that $\sum_k |g(x_k)|p(x_k) < \infty$ and $\int_{-\infty}^{\infty} |g(x)|f(x)\,\mathrm{d}x < \infty$, respectively. If g is nonnegative, we use the symbol $\mathbb{E}\, g(X)$ whether finite or not. If the distribution of X is a mixture, then we define $\mathbb{E}\, g(X)$ by

$$\mathbb{E}\, g(X) = \theta \sum_k g(x_k)p(x_k) + (1-\theta) \int_{-\infty}^{\infty} g(x)f(x)\,\mathrm{d}x\,. \tag{2.1.1}$$

It is obviously convenient to consider the expectation $\mathbb{E}\, g(X)$ given by (2.1.1) in the more general framework of the *Stieltjes integral*

$$\mathbb{E}\, g(X) = \int_{-\infty}^{\infty} g(x)\,\mathrm{d}F(x)\,,$$

taken with respect to the distribution function $F : \mathbb{R} \to [0,1]$ of X. Alternatively, we could also use the *Lebesgue integral* $\mathbb{E}\, g(X) = \int_{\mathbb{R}} g(x) F(\mathrm{d}x)$, with respect to the distribution $F : \mathcal{B}(\mathbb{R}) \to [0,1]$ of X. Moreover, both expressions define the expectation for cases not of the form (2.1.1). In most of our applications, however, integrals will be of the form (2.1.1). For $g(x) = x$ the value $\mu = \mathbb{E}\, X$ is called the *mean*, the *expectation* or the *first moment* of X. For $g(x) = x^n$, $\mu^{(n)} = \mathbb{E}\,(X^n)$ is called the n-th *moment*. The *variance* of X is $\sigma^2 = \mathbb{E}\,(X-\mu)^2$ and $\sigma = \sqrt{\sigma^2}$ is the *standard deviation*. Sometimes the symbol $\operatorname{Var} X$ will be used instead of σ^2. The *coefficient of variation* is given by $\mathrm{cv}_X = \sigma/\mu$, the *index of dispersion* by $\mathrm{I}_X = \sigma^2/\mu$, and the *coefficient of skewness* by $\mathbb{E}\,(X-\mu)^3\sigma^{-3}$. For two random variables X, Y we define the *covariance* $\mathrm{Cov}(X,Y)$ by $\mathrm{Cov}(X,Y) = \mathbb{E}\,((X - \mathbb{E}\, X)(Y - \mathbb{E}\, Y))$ provided that $\mathbb{E}\, X^2, \mathbb{E}\, Y^2 < \infty$. Note that, equivalently, $\mathrm{Cov}(X,Y) = \mathbb{E}\, XY - \mathbb{E}\, X\, \mathbb{E}\, Y$. If $\mathrm{Cov}(X,Y) > 0$, then we say that X, Y are *positively correlated*. Similarly, X, Y are *negatively correlated* if $\mathrm{Cov}(X,Y) < 0$, while they are *uncorrelated* if $\mathrm{Cov}(X,Y) = 0$.

A *median* of the random variable X is any number $\zeta_{1/2}$ such that

$$\mathbb{P}(X \le \zeta_{1/2}) \ge \tfrac{1}{2}, \qquad \mathbb{P}(X \ge \zeta_{1/2}) \ge \tfrac{1}{2}.$$

We call $\kappa_X = \mathbb{E}\,|X - \zeta_{1/2}|$ the *absolute deviation* of X (from median $\zeta_{1/2}$).

Note that we can relate the expectation $\mathbb{E}$ with the probability $\mathbb{P}$ using the notion of the *indicator function* $\mathbb{I}(A) : \Omega \to \mathbb{R}$, which is given by

$$\mathbb{I}(A,\omega) = \begin{cases} 1 & \text{if } \omega \in A, \\ 0 & \text{otherwise.} \end{cases}$$

Thus, if $A \in \mathcal{F}$, then $\mathbb{I}(A)$ is a random variable and $\mathbb{P}(A) = \mathbb{E}\, \mathbb{I}(A)$. Furthermore, we use the notation $\mathbb{E}\,[X; A] = \mathbb{E}\,(X\,\mathbb{I}(A))$ for any random variable X and any event $A \in \mathcal{F}$.

If we want to emphasize the random variable X or its distribution F when using the mean, n-th moment, variance, etc. we write $\mu_X, \mu_X^{(n)}, \sigma_X^2, \ldots$ or $\mu_F, \mu_F^{(n)}, \sigma_F^2, \ldots$.

2.1.3 Independence and Conditioning

An important idea in probability theory is the concept of independence. The random variables $X_1, \ldots, X_n : \Omega \to \mathbb{R}$ are *independent* if

$$\mathbb{P}(X_1 \in B_1, \ldots, X_n \in B_n) = \prod_{k=1}^{n} \mathbb{P}(X_k \in B_k)$$

for all $B_1, \ldots, B_n \in \mathcal{B}(\mathbb{R})$ or, equivalently,

$$\mathbb{P}(X_1 \le x_1, \ldots, X_n \le x_n) = \prod_{k=1}^{n} \mathbb{P}(X_k \le x_k)$$

for all $x_1, \ldots, x_n \in \mathbb{R}$. An infinite sequence $X_1, X_2, \ldots$ is said to consist of independent random variables if each finite subsequence has this property. We say that two sequences $X_1, X_2, \ldots$ and $Y_1, Y_2, \ldots$ of random variables are independent if

$$\mathbb{P}\Big(\bigcap_{i=1}^{n}\{X_i \le x_i\} \cap \bigcap_{i=1}^{m}\{Y_i \le y_i\}\Big) = \mathbb{P}\Big(\bigcap_{i=1}^{n}\{X_i \le x_i\}\Big)\mathbb{P}\Big(\bigcap_{i=1}^{m}\{Y_i \le y_i\}\Big)$$

for all $n, m \in \mathbb{N}$ and $x_1, \ldots, x_n, y_1, \ldots, y_m \in \mathbb{R}$. For a sequence $X_1, X_2, \ldots$ of independent and identically distributed random variables, it is convenient to use the notion of a generic random variable X with the same distribution. Sometimes we say that a random variable or a sequence of random variables is independent of an event $A \in \mathcal{F}$, by which we mean that the random variables are independent of the indicator random variable $\mathbb{I}(A)$.

Let $A \in \mathcal{F}$. The *conditional probability* of an event A' given A is

$$\mathbb{P}(A' \mid A) = \begin{cases} \mathbb{P}(A' \cap A)/\mathbb{P}(A) & \text{if } \mathbb{P}(A) > 0, \\ 0 & \text{otherwise.} \end{cases}$$

In connection with this concept of conditional probability, an elementary and useful formula is given by the so-called *law of total probability.* Let $A, A_1, A_2, \ldots \in \mathcal{F}$ be a sequence of events such that $A_i \cap A_j = \emptyset$ for $i \ne j$ and $\sum_{i=1}^{\infty} \mathbb{P}(A_i) = 1$. Then,

$$\mathbb{P}(A) = \sum_{i=1}^{\infty} \mathbb{P}(A \mid A_i)\,\mathbb{P}(A_i)\,. \tag{2.1.2}$$

We will find use of the notion of the *conditional expectation* $\mathbb{E}\,(X \mid A)$ of a random variable X given that the event A has occurred. By this we mean the expectation taken with respect to the *conditional distribution* $F_{X|A}$, where $F_{X \mid A}(B) = \mathbb{P}(\{X \in B\} \cap A)/\mathbb{P}(A)$ for $B \in \mathcal{B}(\mathbb{R})$. In later chapters of the book we will use more general versions of the above conditional concepts. A typical example is the conditional expectation $\mathbb{E}\,(X \mid \mathcal{G})$ with respect to a sub-σ-algebra $\mathcal{G}$ of $\mathcal{F}$. Under the assumption that $\mathbb{E}\,|X| < \infty$, the conditional expectation is a mapping $\mathbb{E}\,(X \mid \mathcal{G}) : \Omega \to \mathbb{R}$ which is measurable with respect to the sub-σ-algebra $\mathcal{G}$ and for which $\mathbb{E}\,[\mathbb{E}\,(X \mid \mathcal{G}); A] = \mathbb{E}\,[X; A]$ for all $A \in \mathcal{G}$. The conditional expectation is unique in the sense that if Y is another $\mathcal{G}$-measurable random variable with $\mathbb{E}\,[Y; A] = \mathbb{E}\,[X; A]$ for all $A \in \mathcal{G}$ then $\mathbb{P}(Y = \mathbb{E}\,(X \mid \mathcal{G})) = 1$.

2.1.4 Convolution

The convolution operation for distributions allows us to compute the distribution of the sum $X+Y$ of two independent random variables X and Y from their respective distributions F and G. The *convolution* $F*G$ of two distribution functions F,G is defined by

$$F*G(x)=\int_{-\infty}^{\infty}F(x-u)\,\mathrm{d}G(u)\,,\qquad x\in\mathbb{R}\,. \tag{2.1.3}$$

Note that $F*G$ is absolutely continuous provided that at least one of the distributions F,G is absolutely continuous. If both X and Y have densities f and g, respectively, then the density of $X+Y$ is given by the (density) convolution $f*g(x)=\int_{-\infty}^{\infty}f(x-u)g(u)\,\mathrm{d}u$ for $x\in\mathbb{R}$. The (discrete) convolution of two probability functions $\{p_k;k\in\mathbb{N}\}$ and $\{p'_k;k\in\mathbb{N}\}$ is given by

$$(p*p')_k=\sum_{i,j\in\mathbb{N}:i+j=k}p_ip'_j\,,\qquad k\in\mathbb{N}\,.$$

The operation of convolution can still be defined for other types of functions like unbounded functions; what is important is that the integration in (2.1.3) can be performed. The n-fold convolution of F, denoted by F^{*n} is defined iteratively: for $n=0$, $F^{*0}(x)=\delta_0(x)$ with $\delta_0(x)=1$ if $x\ge 0$ and $\delta_0(x)=0$ if $x<0$ while for $n\ge 1$, $F^{*n}=F^{*(n-1)}*F=F*\ldots*F$ (n times). The n-fold convolution of other functions is similarly defined and denoted. For the tail of F^{*n} we write $\overline{F^{*n}}(x)=1-F^{*n}(x)$.

2.1.5 Transforms

Let $I=\{s\in\mathbb{R}:\mathbb{E}\,\mathrm{e}^{sX}<\infty\}$. Note that I is an interval which can be the whole real line $\mathbb{R}$, a halfline or even the singleton $\{0\}$. The *moment generating function* $\hat{m}:I\to\mathbb{R}$ of X is defined by $\hat{m}(s)=\mathbb{E}\,\mathrm{e}^{sX}$. Note the difference between the moment generating function and the *Laplace–Stieltjes transform* $\hat{l}(s)=\mathbb{E}\,\mathrm{e}^{-sX}=\int_{-\infty}^{\infty}\mathrm{e}^{-sx}\,\mathrm{d}F(x)$ of X or of the distribution function F of X. Besides the Laplace–Stieltjes transform, we sometimes consider the *Laplace transform* $\hat{L}(s)=\int_{-\infty}^{\infty}\mathrm{e}^{-sx}c(x)\,\mathrm{d}x$ of a function $c:\mathbb{R}\to\mathbb{R}_+$. Clearly, if the distribution function F is continuous with density f, then its Laplace–Stieltjes transform is equal to the Laplace transform of f. It is rather easy to prove that the Laplace–Stieltjes transform $\hat{l}(s)$ of a distribution on $\mathbb{R}_+$ is *completely monotone*, i.e. for all integer values of n, $(-1)^n\hat{l}^{(n)}(s)\ge 0$. A classical result from real analysis, known as *Bernstein's theorem*, states that, conversely, every completely monotone function $l(s)$ satisfying $l(0)=1$ is the Laplace–Stieltjes transform of a distribution on $\mathbb{R}_+$.

For lattice random variables on $\mathbb{N}$ with probability function $\{p_k,\ k\in\mathbb{N}\}$ we additionally use the notion of the *probability generating function* $\hat{g}:[-1,1]\to$

$\mathbb{R}$ defined by $\hat{g}(s) = \sum_{k=0}^{\infty} p_k s^k$. If $\{a_k,\ k \in \mathbb{N}\}$ is an arbitrary sequence of real numbers, not necessarily a probability function, we also define the *generating function* $\hat{g}(z)$ of $\{a_k\}$ by $\hat{g}(z) = \sum_{k=0}^{\infty} a_k z^k$ provided the sum is convergent for $z \in D$, where D is a subset of the complex plane $\mathbb{C}$. For example, a probability generating function can always be considered on the unit sphere $\{z \in \mathbb{C} : |z| \le 1\}$. If the generating function $\hat{g}(z)$ is well-defined for all $z \in \mathbb{C}$ with $|z| \le s_0$, then the supremum of all $s_0 \ge 1$ with this property is called the *radius of convergence* of $\hat{g}(z)$.

Let X be an arbitrary real-valued random variable with distribution F. The *characteristic function* $\hat{\varphi} : \mathbb{R} \to \mathbb{C}$ of X is given by

$$\hat{\varphi}(s) = \mathbb{E}\, \mathrm{e}^{\mathrm{i}sX}. \tag{2.1.4}$$

In particular, if F is absolutely continuous with density f, then

$$\hat{\varphi}(s) = \int_{-\infty}^{\infty} \mathrm{e}^{\mathrm{i}sx} f(x)\,\mathrm{d}x = \int_{-\infty}^{\infty} \cos(sx) f(x)\,\mathrm{d}x + \mathrm{i} \int_{-\infty}^{\infty} \sin(sx) f(x)\,\mathrm{d}x\,.$$

In this case, $\hat{\varphi}$ is the *Fourier transform* of f. If $X : \Omega \to \mathbb{N}$ is a discrete random variable with probability function $\{p_k\}$, then

$$\hat{\varphi}(s) = \sum_{k=0}^{\infty} \mathrm{e}^{\mathrm{i}sk} p_k\,. \tag{2.1.5}$$

If we want to emphasize the random variable X or its distribution F when using the moment generating function, Laplace–Stieltjes transform, Laplace transform, probability generating function or characteristic function, then we write $\hat{m}_X, \ldots, \hat{\varphi}_X$ or, alternatively, $\hat{m}_F, \ldots, \hat{\varphi}_F$.

Note that the formal relationships $\hat{g}(\mathrm{e}^s) = \hat{l}(-s) = \hat{m}(s)$ hold. Similarly $\hat{\varphi}(s) = \hat{g}(\mathrm{e}^{\mathrm{i}s})$. However, it is somewhat more delicate to decide for what arguments s the transforms $\hat{m}(s)$ and $\hat{l}(s)$ are well-defined. For example, if X is nonnegative, then $\hat{m}(s)$ is well-defined for all $s \le 0$ while $\hat{l}(s)$ is well-defined for all $s \ge 0$. But examples show that $\hat{m}(s)$ may be ∞ for all $s > 0$, while others show that $\hat{m}(s)$ is finite on $(-\infty, a)$ for some $a > 0$.

There is a one-to-one correspondence between distributions of random variables and their characteristic functions. For the moment generating function, Laplace–Stieltjes transform, Laplace transform or probability generating function additional assumptions have to be imposed. If the n-th moment of the random variable $|X|$ is finite, then the n-th derivatives $\hat{m}^{(n)}(0), \hat{l}^{(n)}(0)$ exist provided that the functions $\hat{m}(s), \hat{l}(s)$ are well-defined in a certain (complex) neighbourhood of $s = 0$. In this case, the following equation holds:

$$\mathbb{E}\, X^n = \hat{m}^{(n)}(0) = (-1)^n \hat{l}^{(n)}(0) = (-\mathrm{i})^n \hat{\varphi}^{(n)}(0)\,. \tag{2.1.6}$$

If $\hat{m}(s)$ and $\hat{l}(s)$ are well-defined only on $(-\infty, 0]$ and $[0, \infty)$, respectively, then the derivatives in (2.1.6) have to be replaced by one-sided derivatives, i.e.

$$\mathbb{E}\, X^n = \hat{m}^{(n)}(0-) = (-1)^n \hat{l}^{(n)}(0+)\,, \tag{2.1.7}$$

or by the corresponding derivative of the characteristic function $\hat{\varphi}$. If the range of X is a subset of $\mathbb{N}$ and if $\mathbb{E}\, X^n < \infty$, then

$$\mathbb{E}\, \left(X(X-1)\ldots(X-n+1)\right) = \hat{g}^{(n)}(1-)\,. \tag{2.1.8}$$

Another important property of $\hat{m}(s)$ (exactly the same holds for the transforms $\hat{l}(s)$ and $\hat{\varphi}(s)$) is the following. Let $X_1, \ldots, X_n$ be independent; then

$$\hat{m}_{X_1+\ldots+X_n}(s) = \prod_{k=1}^{n} \hat{m}_{X_k}(s)\,. \tag{2.1.9}$$

Similarly, if $X_1, \ldots, X_n$ are independent and take their values in $\mathbb{N}$, then

$$\hat{g}_{X_1+\ldots+X_n}(s) = \prod_{k=1}^{n} \hat{g}_{X_k}(s)\,. \tag{2.1.10}$$

More generally, for arbitrary measurable functions $g_1, \ldots, g_n : \mathbb{R} \to \mathbb{R}$ we have

$$\mathbb{E}\left(\prod_{k=1}^{n} g_k(X_k)\right) = \prod_{k=1}^{n} \mathbb{E}\, g_k(X_k)\,, \tag{2.1.11}$$

provided that $X_1, \ldots, X_n$ are independent.

Let $X, X_1, X_2, \ldots$ be real-valued random variables. We say that $\{X_n\}$ *converges weakly* (or *in distribution*) to X if $F_{X_n}(x) \to F_X(x)$ at all points of continuity of F_X. We write $X_n \stackrel{\rm d}{\to} X$. This definition is equivalent to the following one, which is called the *Helly–Bray theorem* , saying that $X_n \stackrel{\rm d}{\to} X$ if and only if

$$\lim_{n\to\infty} \mathbb{E}\, g(X_n) = \mathbb{E}\, g(X) \tag{2.1.12}$$

for each bounded continuous function $g : \mathbb{R} \to \mathbb{R}$. A sufficient condition for weak convergence is that $X_n \stackrel{\rm d}{\to} X$ if

$$\lim_{n\to\infty} \hat{m}_{X_n}(s) = \hat{m}_X(s) \tag{2.1.13}$$

for all inner points s of the interval where $\hat{m}_X(s)$ exists, provided this interval is different from the singleton $\{0\}$. The following two results are related to convergence in distribution and are known as *Slutsky's arguments.* If $X_n \stackrel{\rm d}{\to} X$ and $Y_n \stackrel{\rm d}{\to} c \in \mathbb{R}$, then

$$X_n Y_n \stackrel{\rm d}{\to} cX\,, \qquad X_n + Y_n \stackrel{\rm d}{\to} X + c\,. \tag{2.1.14}$$

Bibliographical Notes. Examples of textbooks with an elementary introduction to probability theory are Brémaud (1988), Krengel (1991) and Ross (1997a). More advanced background material on probability and measure theory can be found, for example, in Billingsley (1995), Breiman (1992), Feller (1968), Gänssler and Stute (1977), Karr (1993) and Pitman (1993). For Bernstein's theorem on completely monotone functions, see, for instance, Feller (1968). Slutsky's arguments can be found, for example, in Serfling (1980).

2.2 PARAMETRIZED FAMILIES OF DISTRIBUTIONS

In this section we introduce several classes of discrete or absolutely continuous distributions appearing in insurance and finance. They all are characterized by a finite set of parameters. Many of their properties are discussed in later sections. See also the tables at the end of the book where formulae with basic characteristics of these distributions are listed. Some special functions repeatedly appear later and are stated in Section 2.2.5.

2.2.1 Discrete Distributions

Among the discrete distributions we have:

- *Degenerate distribution* δ_a concentrated on $a \in \mathbb{R}$ with $p(x) = 1$ if $x = a$ and $p(x) = 0$ otherwise.
- *Bernoulli distribution* $\mathrm{Ber}(p)$ with $p_k = p^k(1-p)^{1-k}$ for $k = 0, 1$; $0 < p < 1$: the distribution of a random variable assuming the values 0 and 1 only.
- *Binomial distribution* $\mathrm{Bin}(n,p)$ with $p_k = \binom{n}{k}p^k(1-p)^{n-k}$ for $k = 0, 1 \ldots, n$; $n \in \mathbb{N}, 0 < p < 1$: the distribution of the sum of n independent and identically $\mathrm{Ber}(p)$-distributed random variables. Thus, $\mathrm{Bin}(n_1,p) * \mathrm{Bin}(n_2,p) = \mathrm{Bin}(n_1+n_2,p)$.
- *Poisson distribution* $\mathrm{Poi}(\lambda)$ with $p_k = \mathrm{e}^{-\lambda}\lambda^k/k!$ for $k = 0, 1 \ldots$; $0 < \lambda < \infty$: one of the building blocks of probability theory. Historically it appeared as a weak limit of binomial distributions $\mathrm{Bin}(n,p)$ for which $np \to \lambda$ as $n \to \infty$. The closure property of the class of Poisson distributions under convolution is important in that the sum of two independent Poisson-distributed random variables is again Poisson-distributed, that is $\mathrm{Poi}(\lambda_1) * \mathrm{Poi}(\lambda_2) = \mathrm{Poi}(\lambda_1 + \lambda_2)$.
- *Geometric distribution* $\mathrm{Geo}(p)$ with $p_k = (1-p)p^k$ for $k = 0, 1 \ldots$; $0 < p < 1$: the distribution with the discrete lack-of-memory property – that is, $\mathbb{P}(X \geq i + j \mid X \geq j) = \mathbb{P}(X \geq i)$ for all $i, j \in \mathbb{N}$ iff X is geometrically distributed.

- *Negative binomial distribution* or *Pascal distribution* $\mathrm{NB}(\alpha, p)$ with $p_k = \Gamma(\alpha+k)/(\Gamma(\alpha)\Gamma(k+1))(1-p)^\alpha p^k$ for $k = 0, 1, \ldots$; $\alpha > 0, 0 < p < 1$. Using

$$\binom{x}{0} = 1, \qquad \binom{x}{k} = \frac{x(x-1)\ldots(x-k+1)}{k!},$$

for $x \in \mathbb{R}$, $k = 1, 2, \ldots$ as a general notation, we have

$$p_k = \binom{\alpha+k-1}{k}(1-p)^\alpha p^k = \binom{-\alpha}{k}(1-p)^\alpha(-p)^k.$$

Moreover, for $\alpha = 1, 2, \ldots$, $\mathrm{NB}(\alpha, p)$ is the distribution of the sum of α independent and identically $\mathrm{Geo}(p)$-distributed random variables. This means in particular that the subclass of negative binomial distributions $\{\mathrm{NB}(\alpha, p), \alpha = 1, 2, \ldots\}$ is closed with respect to convolution, i.e. $\mathrm{NB}(\alpha_1, p) * \mathrm{NB}(\alpha_2, p) = \mathrm{NB}(\alpha_1 + \alpha_2, p)$ if $\alpha_1, \alpha_2 = 1, 2, \ldots$. Moreover, the latter formula holds for any $\alpha_1, \alpha_2 > 0$ and $0 < p < 1$.
- *Delaporte distribution* $\mathrm{Del}(\lambda, \alpha, p)$ with $\mathrm{Del}(\lambda, \alpha, p) = \mathrm{Poi}(\lambda) * \mathrm{NB}(\alpha, p)$ for $\lambda > 0$, $\alpha > 0$, $0 < p < 1$.
- *Logarithmic distribution* $\mathrm{Log}(p)$ with $p_k = p^k/(-k\log(1-p))$ for $k = 1, 2, \ldots$; $0 < p < 1$: limit of truncated negative binomial distributions; alternatively limit of the Engen distribution when $\theta \to 0$.
- *(Discrete) uniform distribution* $\mathrm{UD}(n)$ with $p_k = n^{-1}$ for $k = 1, \ldots, n$; $n = 1, 2, \ldots$.
- *Sichel distribution* $\mathrm{Si}(\theta, \lambda, a)$ with

$$p_k = \frac{\lambda^k}{k!}(1+2a)^{-\frac{1}{2}(\theta+k)}\frac{K_{\theta+k}\left(\lambda/a\sqrt{1+2a}\right)}{K_\theta(\lambda/a)}$$

for $k = 0, 1, \ldots$, where $K_\theta(x)$ denotes the Bessel function of the third kind (see Section 2.2.5 below); λ, $a > 0$, $\theta \in \mathbb{R}$.
- *Engen distribution* $\mathrm{Eng}(\theta, a)$ with

$$p_k = \frac{\theta}{1-(1-a)^\theta}\frac{a^k\Gamma(k-\theta)}{k!\Gamma(1-\theta)}, \qquad 0 < \theta,\ a < 1,\ k \geq 1.$$

2.2.2 Absolutely Continuous Distributions

Among the absolutely continuous distributions we have:

- *Normal distribution* $\mathrm{N}(\mu, \sigma^2)$ with $f(x) = (2\pi\sigma^2)^{-1/2}\mathrm{e}^{-(x-\mu)^2/(2\sigma^2)}$ for all $x \in \mathbb{R}$; $\mu \in \mathbb{R}, \sigma^2 > 0$: another building block of probability theory. It appears as the limit distribution of sums of an unboundedly increasing number of independent random variables each of which is asymptotically negligible. The class of normal distributions is closed with respect to convolution, that is $\mathrm{N}(\mu_1, \sigma_1^2) * \mathrm{N}(\mu_2, \sigma_2^2) = \mathrm{N}(\mu_1 + \mu_2, \sigma_1^2 + \sigma_2^2)$.

- *Exponential distribution* $\mathrm{Exp}(\lambda)$ with $f(x) = \lambda \mathrm{e}^{-\lambda x}$ for $x > 0$; $\lambda > 0$: basic distribution in the theory of Markov processes because of its continuous lack-of-memory property, i.e. $\mathbb{P}(X > t + s \mid X > s) = \mathbb{P}(X > t)$ for all $s, t \geq 0$ if X is exponentially distributed.
- *Erlang distribution* $\mathrm{Erl}(n, \lambda)$ with $f(x) = \lambda^n x^{n-1} \mathrm{e}^{-\lambda x}/(n-1)!$; $n = 1, 2, \ldots$: the distribution of the sum of n independent and identically $\mathrm{Exp}(\lambda)$-distributed random variables. Thus, $\mathrm{Erl}(n_1, \lambda) * \mathrm{Erl}(n_2, \lambda) = \mathrm{Erl}(n_1 + n_2, \lambda)$.
- *χ^2-distribution* $\chi^2(n)$ with $f(x) = (2^{n/2}\Gamma(n/2))^{-1} x^{(n/2)-1} \mathrm{e}^{-x/2}$ if $x > 0$ and $f(x) = 0$ if $x \leq 0$; $n = 1, 2, \ldots$: the distribution of the sum of n independent and identically distributed random variables, which are the squares of $\mathrm{N}(0, 1)$-distributed random variables. This means that $\chi^2(n_1) * \chi^2(n_2) = \chi^2(n_1 + n_2)$.
- *Gamma distribution* $\Gamma(a, \lambda)$ with $f(x) = \lambda^a x^{a-1} \mathrm{e}^{-\lambda x}/\Gamma(a)$ for $x \geq 0$; with shape parameter $a > 0$ and scale parameter $\lambda > 0$: if $a = n \in \mathbb{N}$, then $\Gamma(n, \lambda) = \mathrm{Erl}(n, \lambda)$. If $\lambda = 1/2$ and $a = n/2$, then $\Gamma(n/2, 1/2) = \chi^2(n)$.
- *Uniform distribution* $\mathrm{U}(a, b)$ with $f(x) = (b - a)^{-1}$ for $a < x < b$; $-\infty < a < b < \infty$.
- *Beta distribution* $\mathrm{Beta}(a, b, \eta)$ with

$$f(x) = \frac{x^{a-1}(\eta - x)^{b-1}}{B(a, b)\eta^{a+b-1}}$$

  for all $0 < x < \eta$, where $B(a, b)$ denotes the beta function (see Section 2.2.5); $a, b, \eta > 0$; if $a = b = 1$, then $\mathrm{Beta}(1, 1, \eta) = \mathrm{U}(0, \eta)$.
- *Inverse Gaussian distribution* $\mathrm{IG}(\mu, \lambda)$ with

$$f(x) = \left(\lambda/(2\pi x^3)\right)^{1/2} \exp\left(-\lambda(x - \mu)^2/(2\mu^2 x)\right)$$

  for all $x > 0$; $\mu \in \mathbb{R}, \lambda > 0$.
- *Extreme value distribution* $\mathrm{EV}(\gamma)$ with $F(x) = \exp(-(1 + \gamma x)_+^{-1/\gamma})$ for all $\gamma \in \mathbb{R}$ where the quantity $(1 + \gamma x)_+^{-1/\gamma}$ is defined as e^{-x} when $\gamma = 0$. The latter distribution is known as the *Gumbel distribution.* For $\gamma > 0$ we obtain the *Fréchet distribution* ; this distribution is concentrated on the half-line $(-1/\gamma, +\infty)$. Finally, for $\gamma < 0$ we obtain the *extremal-Weibull distribution*, a distribution concentrated on the half-line $(-\infty, -1/\gamma)$.

2.2.3 Parametrized Distributions with Heavy Tail

Note that, except the Fréchet distribution, each of the absolutely continuous distributions considered above has an *exponentially bounded tail*, i.e. for some $a, b > 0$ we have $\overline{F}(x) \leq a\mathrm{e}^{-bx}$ for all $x > 0$. Such distributions are sometimes said to have a light tail. In non-life insurance one is also interested in distributions with heavy tails. Formally we say that F has a *heavy tail* if its moment generating function $\hat{m}_F(s)$ fulfils $\hat{m}_F(s) = \infty$ for all $s > 0$. Typical absolutely continuous distributions with heavy tail are:

- *Logarithmic normal (or lognormal) distribution* $\mathrm{LN}(a,b)$ with $f(x) = (xb\sqrt{2\pi})^{-1}\exp\left(-(\log x - a)^2/(2b^2)\right)$ for $x > 0$; $a \in \mathbb{R}, b > 0$; if X is $\mathrm{N}(a,b)$-distributed, then e^X is $\mathrm{LN}(a,b)$-distributed.
- *Weibull distribution* $\mathrm{W}(r,c)$ with $f(x) = rcx^{r-1}\exp(-cx^r)$ for $x > 0$ where the shape parameter $r > 0$ and the scale parameter $c > 0$; $\overline{F}(x) = \exp(-cx^r)$; if $r \geq 1$, then $\mathrm{W}(r,c)$ has a light tail, otherwise $\mathrm{W}(r,c)$ is heavy-tailed.
- *Pareto distribution* $\mathrm{Par}(\alpha,c)$ with $f(x) = (\alpha/c)(c/x)^{\alpha+1}$ for $x > c$; with exponent or shape parameter $\alpha > 0$ and scale parameter $c > 0$, $\overline{F}(x) = (c/x)^\alpha$.
- *Pareto mixtures of exponentials* $\mathrm{PME}(\alpha)$ with

$$f(x) = \int_{(\alpha-1)/\alpha}^{\infty} \alpha^{-\alpha+1}(\alpha-1)^\alpha y^{-(\alpha+1)} y^{-1}\mathrm{e}^{-x/y}\,\mathrm{d}y$$

  for $x > 0$; $\alpha > 1$, a class of distributions with heavy tails having an explicit Laplace–Stieltjes transform for $\alpha = 2, 3, \ldots$:

$$\begin{aligned} \hat{l}(s) &= \sum_{i=1}^{\alpha}(-1)^{\alpha-i}\frac{\alpha}{i}\left(\frac{\alpha-1}{\alpha}\right)^{\alpha-i} s^{\alpha-i} \\ &\quad +(-1)^\alpha\alpha\left(\frac{\alpha-1}{\alpha}\right)^\alpha s^\alpha \log\left(1+\frac{\alpha}{(\alpha-1)s}\right). \end{aligned}$$

  Moreover, for arbitrary $\alpha > 1$,

$$\hat{l}(s) = \int_0^\infty \mathrm{e}^{-sx} f_\alpha(x)\,\mathrm{d}x = \alpha\left(\frac{\alpha-1}{\alpha}\right)^\alpha \int_0^{\alpha/(\alpha-1)} \frac{x^\alpha}{s+x}\,\mathrm{d}x$$

  and $\overline{F}(x) \sim \Gamma(\alpha+1)(\alpha - 1/\alpha)^\alpha x^{-\alpha}$ as $x \to \infty$, where the symbol $g_1(x) \sim g_2(x)$ means that $\lim_{x\to\infty} g_1(x)/g_2(x) = 1$.

Further parameterized families of heavy-tailed distributions being popular in insurance mathematics are:

- *Loggamma distribution* $\mathrm{L}\Gamma(a,\lambda)$ with $f(x) = \lambda^a/\Gamma(a)(\log x)^{a-1}x^{-\lambda-1}$ for $x > 1$; $\lambda, a > 0$; if X is $\Gamma(a,\lambda)$-distributed, then e^X is $\mathrm{L}\Gamma(a,\lambda)$-distributed.
- *Benktander type I distribution* $\mathrm{BenI}(a,b,c)$ with $\overline{F}(x) = cx^{-a-1}\mathrm{e}^{-b(\log x)^2}(a + 2b\log x)$ for $x > 1$; $a, b, c > 0$, chosen in such a way that F is a distribution on $\mathbb{R}_+$. The latter requires that $a(a+1) \geq 2b$ and $ac \leq 1$.
- *Benktander type II distribution* $\mathrm{BenII}(a,b,c)$ with tail function $\overline{F}(x) = cax^{-(1-b)}\exp(-(a/b)x^b)$ for $x > 1$; $a > 0, 0 < b < 1$ and $0 < c \leq a^{-1}\mathrm{e}^{a/b}$.

2.2.4 Operations on Distributions

Starting from these basic parametrized families of distributions, one can generate more distributions by means of the following operations.

Mixture Consider a sequence $F_1, F_2, \ldots$ of distributions on $\mathcal{B}(\mathbb{R})$ and a probability function $\{p_k,\ n = 1, 2, \ldots\}$. Then, the distribution $F = \sum_{k=1}^{\infty} p_k F_k$ is called a *mixture* of $F_1, F_2, \ldots$ with weights $p_1, p_2, \ldots$ If X is a random variable with distribution F, then F_k can be interpreted as the conditional distribution of X, and p_k as the probability that the conditional distribution F_k is selected. We can also have an uncountable family of distributions F_θ parametrized by θ, where θ is chosen from a certain subset Θ of $\mathbb{R}$ according to a distribution G concentrated on Θ. Formally, the mixture $F(x)$ of the family $\{F_\theta,\ \theta \in \Theta\}$ with *mixing distribution* G is given by $F(x) = \int_\Theta F_\theta(x)\,\mathrm{d}G(\theta)$, $x \in \mathbb{R}$.

Truncation Let X be a random variable with distribution F and let $C \in \mathcal{B}(\mathbb{R})$ be a certain subset of $\mathbb{R}$. The *truncated distribution* F_C is the conditional distribution of X given that the values of X are restricted to the set $C^{\mathrm{c}} = \mathbb{R} \setminus C$, i.e. $F_C(B) = \mathbb{P}(X \in B \mid X \notin C)$, $B \in \mathcal{B}(\mathbb{R})$. For example, if X is discrete with probability function $\{p_k, k = 0, 1, 2, \ldots\}$, then the *zero truncation* $F_{\{0\}}$ is given by the probability function $\{\mathbb{P}(X = k \mid X \geq 1), k = 1, 2, \ldots\}$. In particular, if X is Geo(p)-distributed, then the zero truncation is given by $\mathbb{P}(X = k \mid X \geq 1) = (1 - p)p^{k-1}$ for $k = 1, 2, \ldots$. In the present book we refer to this distribution as the *truncated geometric distribution*; we use the abbreviation TG(p).

Modification If the distribution F is discrete with probability function $\{p_k,\ k = 0, 1, \ldots\}$ and $0 < \theta < 1$, by the *θ-modification* G we mean the distribution $(1 - \theta)\delta_0 + \theta F$. That is, a random variable X with the modified distribution G is again discrete with probability function

$$\mathbb{P}(X = k) = \begin{cases} 1 - \theta + \theta p_0 & \text{if } k = 0, \\ \theta p_k & \text{if } k \geq 1. \end{cases}$$

In some cases, modification is the inverse operation of zero truncation. For example, the p-modification of TG(p) is Geo(p).

Shifting By *shifting* the argument of a distribution function $F(x)$ by some $a \in \mathbb{R}$, we get the new distribution function $G(x) = F(x + a)$. This means that if X has distribution F, then $X - a$ has distribution G.

Scaling By *scaling* the argument of a distribution function $F(x)$ by some $a > 0$, we get a new distribution function $G(x) = F(x/a)$. This means that if X has distribution F, then aX has distribution G.

Integrated Tail The notion of the *integrated tail distribution* F^{s} of a distribution F on the nonnegative halfline is defined by

$$F^{s}(x) = \frac{1}{\mu_F} \int_0^x \overline{F}(y)\,dy\,, \qquad x \geq 0\,,$$

provided $0 < \mu_F < \infty$. Sometimes F^s is called the *stationary excess distribution* or the *equilibrium distribution* of F.

2.2.5 Some Special Functions

We now collect some basic properties of a few special functions which will be used in later sections of this book.

- *Gamma function*

$$\Gamma(x) = \int_0^\infty t^{x-1} e^{-t}\,dt\,, \qquad x > 0\,.$$

Note that $\Gamma(x)$ is a continuous function. Furthermore, $\Gamma(1) = 1$ and $\Gamma(1/2) = \sqrt{\pi}$. Integration by parts shows that $\Gamma(x) = (x-1)\Gamma(x-1)$ and hence $\Gamma(n) = (n-1)!$.

- *Beta function*

$$B(a,b) = \int_0^1 t^{a-1}(1-t)^{b-1}\,dt = \int_0^\infty \frac{t^{a-1}}{(1+t)^{a+b}}\,dt\,, \qquad a,b > 0.$$

Note that $B(a,b) = \Gamma(a)\Gamma(b)/\Gamma(a+b) = B(b,a)$.

- *Modified Bessel function*

$$I_\nu(x) = \sum_{k=0}^\infty \frac{(x/2)^{2k+\nu}}{k!\Gamma(\nu+k+1)}\,, \qquad x \in \mathbb{R},\ \nu \in \mathbb{R}\,. \tag{2.2.1}$$

- *Bessel function of the third kind*

$$K_\theta(x) = \frac{1}{2}\int_0^\infty \exp\left(-\tfrac{1}{2}x\left(y+y^{-1}\right)\right) y^{\theta-1}\,dy\,, \qquad x > 0,\ \theta > 0\,. \tag{2.2.2}$$

- *Confluent hypergeometric functions*

$$M(a,b;x) = \sum_{n=0}^\infty \frac{a(a+1)\ldots(a+n-1)\,x^n}{b(b+1)\ldots(b+n-1)n!}\,, \qquad x \in \mathbb{R}\,, \tag{2.2.3}$$

where b cannot be a negative integer; if a is a negative integer, i.e. $a = -m$ for some positive $m = 1,2,\ldots$, then $M(a,b;x)$ is a polynomial of degree

m. Useful relationships are $\mathrm{d}/\mathrm{d}x M(a,b;x) = (a/b)M(a+1,b+1;x)$ and $M(a,b;x) = \mathrm{e}^x M(b-a,b;-x)$. An integral representation is

$$M(a,b;x) = \frac{\Gamma(b)}{\Gamma(b-a)\Gamma(a)} \int_0^1 \mathrm{e}^{xt} t^{a-1} (1-t)^{b-a-1} \, \mathrm{d}t \tag{2.2.4}$$

for $a, b > 0$. Another confluent hypergeometric function is

$$U(a,b;x) = \frac{\Gamma(1-b)}{\Gamma(1+a-b)} M(a,b;x) + \frac{\Gamma(b-1)}{\Gamma(a)} x^{1-b} M(1+a-b, 2-b; x) \,.$$

It admits the integral representation

$$U(a,b;x) = \frac{1}{\Gamma(a)} \int_0^\infty \mathrm{e}^{-xt} t^{a-1} (1+t)^{b-a-1} \, \mathrm{d}t \,, \qquad a, b > 0. \tag{2.2.5}$$

We end this section by recalling a family of orthogonal polynomials.

- The *generalized Laguerre polynomials* are given by

$$L_n^a(x) = \frac{\mathrm{e}^x x^{-a}}{n!} \frac{\mathrm{d}^n}{\mathrm{d}x^n} (\mathrm{e}^{-x} x^{n+a}) = \sum_{m=0}^{n} (-1)^m \binom{n+a}{n-m} \frac{x^m}{m!} \tag{2.2.6}$$

and for $a = 0$ we obtain the *Laguerre polynomials*, that is $L_n^0(x) = L_n(x)$. A useful identity is

$$\frac{\mathrm{d}}{\mathrm{d}x} L_n^a(x) = -L_{n-1}^{a+1}(x) \,. \tag{2.2.7}$$

The generalized Laguerre polynomials are related to the confluent hypergeometric functions by

$$M(-n, a+1, x) = \frac{n!}{(a+1)(a+2)\ldots(a+n)} L_n^a(x) \,.$$

The generating function $\sum_{n=0}^\infty L_n^a(x) z^n$ is given by

$$\sum_{n=0}^\infty L_n^a(x) z^n = (1-z)^{-a-1} \exp\left(\frac{xz}{z-1}\right) \,. \tag{2.2.8}$$

Bibliographical Notes. An exhaustive survey of distributions is given in the following volumes of Johnson and Kotz (1972), Johnson, Kotz and Balakrishnan (1994, 1995), and Johnson, Kotz and Kemp (1992, 1996). Most popular distributions in insurance mathematics are reviewed in books like Beard, Pentikäinen and Pesonen (1984), Conti (1992), and Hogg and Klugman (1984); see also Benktander (1963). Pareto mixtures of exponentials were introduced by Abate, Choudhury and Whitt (1994). For special functions and orthogonal functions we refer to Abramowitz and Stegun (1965).

2.3 ASSOCIATED DISTRIBUTIONS

In many applications of probability theory to actuarial problems, we need a concept that has received a lot of attention in risk theory under different names like associated distribution, Esscher transform, exponential tilting, etc. In later chapters of this book several variants of this concept will be used; see, for example, Sections 6.5.3, 9.2 and 10.2.6. Consider a real-valued (not necessarily nonnegative) random variable X with distribution F and moment generating function $\hat{m}_F(s) = \mathbb{E}\,\mathrm{e}^{sX} = \int_{-\infty}^{\infty} \mathrm{e}^{sx}\,\mathrm{d}F(x)$ for all $s \in \mathbb{R}$ for which this integral is finite. Let $s_F^- = \inf\{s \le 0 : \hat{m}_F(s) < \infty\}$ and $s_F^+ = \sup\{s \ge 0 : \hat{m}_F(s) < \infty\}$ be the lower and upper *abscissa of convergence*, respectively, of the moment generating function $\hat{m}_F(s)$. Clearly $s_F^- \le 0 \le s_F^+$. Assume now that $\hat{m}_F(s)$ is finite for a value $s \neq 0$. From the definition of the moment generating function we see that $\hat{m}_F(s)$ is a well-defined, continuous and strictly increasing function of $s \in (s_F^-, s_F^+)$ with value 1 at the origin. Furthermore,

$$\begin{aligned} \hat{m}_F(s) - 1 &= \int_{-\infty}^{\infty} (\mathrm{e}^{sx} - 1)\,\mathrm{d}F(x) \\ &= -s\int_{-\infty}^{0}\int_{x}^{0} \mathrm{e}^{sy}\,\mathrm{d}y\,\mathrm{d}F(x) + s\int_{0}^{\infty}\int_{0}^{x} \mathrm{e}^{sy}\mathrm{d}y\,\mathrm{d}F(x) \\ &= -s\int_{-\infty}^{0} F(y)\mathrm{e}^{sy}\,\mathrm{d}y + s\int_{0}^{\infty} \overline{F}(y)\mathrm{e}^{sy}\,\mathrm{d}y. \end{aligned} \tag{2.3.1}$$

This relation is useful in order to derive a necessary and sufficient condition that $\hat{m}_F(s) < \infty$ for some $s \neq 0$.

Lemma 2.3.1 *Assume that $\hat{m}_F(s_0) < \infty$ for some $s_0 > 0$. Then there exists $b > 0$ such that for all $x \ge 0$*

$$1 - F(x) \le b\,\mathrm{e}^{-s_0 x}\,. \tag{2.3.2}$$

Conversely, if (2.3.2) *is fulfilled, then $\hat{m}_F(s) < \infty$ for all $0 \le s < s_0$. Analogously, if $\hat{m}_F(s_0) < \infty$ for some $s_0 < 0$, then there exists $b > 0$ such that for all $x \le 0$*

$$F(x) \le b\,\mathrm{e}^{s_0 x}\,. \tag{2.3.3}$$

Conversely, if (2.3.3) *is fulfilled, then $\hat{m}_F(s) < \infty$ for all $s_0 < s \le 0$.*

Proof Assume that condition (2.3.2) is fulfilled. Then (2.3.1) leads to

$$\frac{\hat{m}_F(s) - 1}{s} \le \int_0^{\infty} (1 - F(y))\,\mathrm{e}^{sy}\,\mathrm{d}y \le \frac{b}{s_0 - s}$$

for $0 < s < s_0$, which shows that $\hat{m}_F(s)$ is finite at least for all $0 \le s < s_0$. Conversely, if $\hat{m}_F(s)$ is finite for a positive value $s = s_0$, then for any $x \ge 0$

$$\begin{aligned}\infty > \frac{\hat{m}_F(s_0) - 1}{s_0} &\ge -\int_{-\infty}^{0} F(y)\mathrm{e}^{s_0 y}\,\mathrm{d}y + \int_{0}^{x} \mathrm{e}^{s_0 y}(1 - F(y))\,\mathrm{d}y \\ &\ge -\frac{1}{s_0} + (1 - F(x))\frac{\mathrm{e}^{s_0 x} - 1}{s_0}.\end{aligned}$$

This means that for all $x \ge 0$, $1-F(x)$ is bounded by $b\,\mathrm{e}^{-s_0 x}$ for some constant b. The second part of the lemma can be proved similarly. □

Theorem 2.3.1 *If* $a^+ = \liminf_{x\to\infty} -x^{-1}\log \overline{F}(x) > 0$, *then*

$$a^+ = s_F^+ . \tag{2.3.4}$$

If $a^- = \limsup_{x\to-\infty} -x^{-1}\log F(x) < 0$, *then*

$$a^- = s_F^- . \tag{2.3.5}$$

Proof We show only (2.3.4). Let $\varepsilon > 0$ be such that $a^+ - \varepsilon > 0$. Then, there exists $x_0 > 0$ such that $-x^{-1}\log \overline{F}(x) \ge a^+ - \varepsilon$ for $x > x_0$, which is equivalent to $\overline{F}(x) \le \mathrm{e}^{-(a^+-\varepsilon)x}$ for $x > x_0$. Because $\varepsilon > 0$ was arbitrary we conclude from Lemma 2.3.1 that $\hat{m}_F(s) < \infty$ for all $s < a^+$. Conversely, suppose that $\hat{m}_F(s_0) < \infty$ for some $s_0 > a^+$. By Lemma 2.3.1, $\overline{F}(x) \le b\mathrm{e}^{-s_0 x}$ for some $b > 0$. Hence $-x^{-1}\log \overline{F}(x) \ge -x^{-1}\log b + s_0$ for $x > 0$, which yields $a^+ \ge s_0$. This contradicts $s_0 > a^+$. Therefore $s_F^+ = a^+$. □

From the above considerations we get $s_F^+ = \liminf_{x\to\infty} -x^{-1}\log \overline{F}(x)$ and $s_F^- = \limsup_{x\to-\infty} x^{-1}\log F(x)$ provided that the limits are nonzero. Note that for nonnegative random variables $s_F^- = -\infty$, and in this case we write $s_F^+ = s_F$.

Whenever $s_F^- < s_F^+$, an infinite family of related distributions can be associated with F. For each $t \in (s_F^-, s_F^+)$

$$\tilde{F}_t(x) = \frac{1}{\hat{m}_F(t)}\int_{-\infty}^{x} \mathrm{e}^{ty}\,\mathrm{d}F(y)\,, \qquad x \in \mathbb{R} \tag{2.3.6}$$

defines a proper distribution on $\mathbb{R}$ called an *associated distribution* to F. The distribution $\tilde{F}_t$ is also called an *Esscher transform* of F. The whole family $\{\tilde{F}_t; s_F^- < t < s_F^+\}$ is called the *class of distributions associated to* F.

Lemma 2.3.2 *Let* $s_F^- < t < s_F^+$. *Then the moment generating function of* $\tilde{F}_t$ *is*

$$\hat{m}_{\tilde{F}_t}(s) = \frac{\hat{m}_F(s+t)}{\hat{m}_F(t)} \tag{2.3.7}$$

for $s_F^- - t < s < s_F^+ - t$. Moreover, $\tilde{F}_t$ has all moments; in particular, the expectation $\mu_{\tilde{F}_t}$ of the associated distribution $\tilde{F}_t$ is given by

$$\mu_{\tilde{F}_t} = \frac{\hat{m}_F^{(1)}(t)}{\hat{m}_F(t)}, \tag{2.3.8}$$

while the variance $\sigma^2_{\tilde{F}_t}$ of $\tilde{F}_t$ is given by

$$\sigma^2_{\tilde{F}_t} = \frac{\hat{m}_F^{(2)}(t)\hat{m}_F(t) - (\hat{m}_F^{(1)}(t))^2}{(\hat{m}_F(t))^2}. \tag{2.3.9}$$

The *proof* is left to the reader to be shown as an exercise. Note also that the associated distribution $\tilde{F}_t$ has the following useful property.

Lemma 2.3.3 *For all $s_F^- < t < s_F^+$ and $n = 1, 2, \ldots$ the fundamental equality*

$$(\hat{m}_F(t))^n \, \mathrm{d}\tilde{F}_t^{*n}(x) = \mathrm{e}^{tx} \, \mathrm{d}F^{*n}(x) \tag{2.3.10}$$

holds and, consequently for all $x \in \mathbb{R}$,

$$1 - F^{*n}(x) = (\hat{m}_F(t))^n \int_x^\infty \mathrm{e}^{-ty} \, \mathrm{d}\tilde{F}_t^{*n}(y). \tag{2.3.11}$$

Proof Raise equation (2.3.7) to the n-th power for any fixed $n \in \mathbb{N}$. But clearly $(\hat{m}_{\tilde{F}_t}(s))^n$ is the moment generating function of $\tilde{F}_t^{*n}$ and so by unicity of the moment generating function the relation

$$(\hat{m}_{\tilde{F}}(t))^n \int_{-\infty}^\infty \mathrm{e}^{sx} \, \mathrm{d}\tilde{F}_t^{*n}(x) = (\hat{m}_F(s+t))^n = \int_{-\infty}^\infty \mathrm{e}^{sx} \, \mathrm{d}\Big(\int_{-\infty}^x \mathrm{e}^{ty} \, \mathrm{d}F^{*n}(y)\Big)$$

yields the fundamental equality (2.3.10). □

One of the important features of (2.3.11) is that on the right-hand side one has a *free* parameter t which is only restricted by the inequalities $s_F^- < t < s_F^+$. In practical applications a judicious choice of this parameter will rewrite intractable formulae into simpler ones. Let us illustrate the above procedure on a concrete example that will prove to be useful in a forthcoming application of renewal theory to an actuarial problem; see Section 6.2.3. One possible implication of (2.3.11) is that an exponential bound on the tail of a distribution results in an exponential rate of convergence in the weak law of large numbers.

Theorem 2.3.2 *Assume that μ_F exists and $s_F^+ > 0$. Then for all $0 < t < s_F^+$*

$$1 - F^{*n}(nx) =$$

$$(\mathrm{e}^{-tx}\hat{m}_F(t))^n \theta t\sqrt{n} \int_0^\infty \mathrm{e}^{-\theta t\sqrt{n}v} \big(\tilde{F}_t^{*n}(nx + \theta v\sqrt{n}) - \tilde{F}_t^{*n}(nx)\big) \, \mathrm{d}v, \tag{2.3.12}$$

where $\theta > 0$. Moreover, for each $x > \mu_F$ there exists $0 < c = c(x) < 1$ such that $1 - F^{*n}(nx) \le c^n$ for $n = 1, 2, \ldots$.

Proof Let $t \in (0, s_F^+)$. Then, by an integration by parts we can rewrite (2.3.11) in the form

$$1 - F^{*n}(x) = (\hat{m}_F(t))^n \Big(\mathrm{e}^{-tx}(1 - \tilde{F}_t^{*n}(x)) - t \int_x^\infty \mathrm{e}^{-ty}(1 - \tilde{F}_t^{*n}(y))\,\mathrm{d}y\Big).$$

Note, however, that $t \int_x^\infty \mathrm{e}^{-ty}\,\mathrm{d}y = \mathrm{e}^{-tx}$ for $0 < t < s_F^+$. Hence the first term on the right can be incorporated within the integration to give

$$1 - F^{*n}(x) = t(\hat{m}_F(t))^n \int_x^\infty \mathrm{e}^{-ty}\big(\tilde{F}_t^{*n}(y) - \tilde{F}_t^{*n}(x)\big)\,\mathrm{d}y\,. \tag{2.3.13}$$

We now replace x by nx and the variable of integration y by $nx + \theta\sqrt{n}v$ for some constant $\theta > 0$ to get (2.3.12). Note that the quantity $\tilde{F}_t^{*n}(nx + cv\sqrt{n}) - \tilde{F}_t^{*n}(nx)$ can be interpreted as a probability $\mathbf{P}(nx < S_n \le nx + cv\sqrt{n})$ where S_n is the sum of independent random variables $X_1, \ldots, X_n$ all with the same associated distribution $\tilde{F}_t$. Replacing this probability by 1 we get the upper bound $1 - F^{*n}(nx) \le (\mathrm{e}^{-tx}\hat{m}_F(t))^n$, where we still have the free parameter $t \in [0, s_F^+)$. Keeping $x > \mu_F$ fixed, we show that there has to be a positive value of t in this interval where the quantity $g(t, x) = \mathrm{e}^{-tx}\hat{m}_F(t)$ is strictly less than 1. To show that, note first that $g(0, x) = 1$. Further

$$\frac{\partial}{\partial t} g(t, x) = g(t, x)\Big(\frac{\hat{m}_F^{(1)}(t)}{\hat{m}_F(t)} - x\Big), \tag{2.3.14}$$

which equals the value $\mu_F - x < 0$ at the origin. So, $\frac{\partial}{\partial t} g(t, x)_{|t=0} < 0$. Since $\frac{\partial}{\partial t} g(t, x)$ is continuous in t, $\frac{\partial}{\partial t} g(t, x)$ is negative for some positive values of t, and hence $g(t, x)$ is strictly decreasing in t to the right of $t = 0$. □

Assume now additionally that F is nondegenerate. Then there exists a uniquely determined $t \in [0, s_F^+]$ which is optimal in the sense that the value of $g(t, x) = \mathrm{e}^{-tx}\hat{m}_F(t)$ becomes minimal. Namely, (2.3.14) together with a little algebra shows that

$$\frac{\partial^2}{\partial t^2} g(t, x) = g(t, x)\left(\frac{\hat{m}_F^{(2)}(t)\hat{m}_F(t) - (\hat{m}_F^{(1)}(t))^2}{(\hat{m}_F(t))^2} + \Big(x - \frac{\hat{m}_F^{(1)}(t)}{\hat{m}_F(t)}\Big)^2\right),$$

which is then positive for all $t \in [0, s_F^+)$ since the first summand on the right can be seen as the variance of a nondegenerate distribution. Indeed, from (2.3.9) the variance of $\tilde{F}_t$ is known and this variance is positive except for the case where $\tilde{F}_t$ is degenerate, a possibility that we have excluded by the assumption that F is nondegenerate. Hence in Theorem 2.3.2 we can put $c = \inf_{0<t\le s_F^+} g(t, x)$, where (2.3.14) implies that the value of t which realizes this infimum can be uniquely determined by $x = \hat{m}_F^{(1)}(t)/\hat{m}_F(t)$ provided that this equation has a solution in $(0, s_F^+]$.

Note that besides the upper bound for $1-F^{*n}(nx)$ shown in Theorem 2.3.2, one can derive an asymptotic expression for $1-F^{*n}(nx)$ as $n \to \infty$. We need the following result, which is known as a *local limit theorem*.

Lemma 2.3.4 *Let $X_1, X_2, \ldots$ be nonlattice real-valued random variables which are independent and identically distributed such that $\mathbb{E} X_1 = 0$ and $\sigma^2 = \mathbb{E}(X_1)^2 < \infty$. Then $\lim_{n\to\infty} \sigma\sqrt{2\pi n}\,\mathbb{P}\Big(\sum_{j=1}^n X_j \in I\Big) = |I|$ for any finite interval I with length $|I|$.*

The *proof* is omitted. It can be found, for example, in Breiman (1992) and Petrov (1975).

We now prove an explicit asymptotic estimate for the tail of the associated distribution.

Theorem 2.3.3 *Assume that F is nonlattice. If μ_F exists and $s_F^+ > 0$, then for each $x > \mu_F$ such that a solution $t_0 = t_0(x)$ to $x = \hat{m}_F^{(1)}(t)/\hat{m}_F(t)$ exists,*

$$\lim_{n\to\infty} (1 - F^{*n}(nx))\sqrt{n}\left(\mathrm{e}^{-xt_0}\hat{m}_F(t_0)\right)^{-n} = \frac{1}{\theta t_0}\,; \tag{2.3.15}$$

here $\theta = \hat{m}_F(t_0)^{-1}\sqrt{2\pi\big(\hat{m}_F^{(2)}(t_0)\hat{m}_F(t_0) - (\hat{m}_F^{(1)}(t_0))^2\big)}$.

Proof We start from representation (2.3.13) to get

$$1 - F^{*n}(x) = t(\hat{m}_F(t))^n \int_x^\infty \mathrm{e}^{-ty}\big(\tilde{F}_t^{*n}(y) - \tilde{F}_t^{*n}(x)\big)\,\mathrm{d}y$$

where we replace x by nx and put $y = nx + w/t$. Shifting a factor from the right-hand side to the left we obtain the expression

$$(\mathrm{e}^{-tx}\hat{m}_F(t))^{-n}(1 - F^{*n}(nx)) = \int_0^\infty \mathrm{e}^{-w}\left(\tilde{F}_t^{*n}(nx + \frac{w}{t}) - \tilde{F}_t^{*n}(nx)\right)\mathrm{d}w\,.$$

Note again that the quantity $\tilde{F}_t^{*n}(nx + \frac{w}{t}) - \tilde{F}_t^{*n}(nx)$ can be interpreted as a probability $\mathbb{P}(nx < S_n^{(t)} \le nx + \frac{w}{t})$, where $S_n^{(t)}$ is the sum of n independent random variables all with the same distribution as $S_1^{(t)}$. To apply Lemma 2.3.4, choose $t = t_0$ in such a way that $x = \mathbb{E}\, S_1^{(t_0)}$ and define $X_1 = S_1^{(t_0)} - x$. We then obtain

$$\theta\sqrt{n}(\mathrm{e}^{-tx}\hat{m}_F(t))^{-n}(1 - F^{*n}(nx)) = \int_0^\infty \mathrm{e}^{-w}(\theta\sqrt{n})\,\mathbb{P}\Big(\sum_{j=1}^n X_j \in (0\,,\,\frac{w}{t})\Big)\,\mathrm{d}w\,, \tag{2.3.16}$$

where $\theta^2 = 2\pi \mathrm{Var}\, X_1$ or, more explicitly,

$$\theta = \hat{m}_F(t_0)^{-1}\sqrt{2\pi\big(\hat{m}_F^{(2)}(t_0)\hat{m}_F(t_0) - (\hat{m}_F^{(1)}(t_0))^2\big)}\,.$$

If we are entitled to apply the bounded convergence theorem, then on the right-hand side of (2.3.16) we obtain

$$\lim_{n\to\infty} \int_0^\infty \mathrm{e}^{-w}(\theta\sqrt{2\pi n})\, \mathbb{P}\Big(\sum_{j=1}^n X_j \in (0\,,\,\frac{w}{t})\Big)\,\mathrm{d}w = \int_0^\infty \mathrm{e}^{-w}\frac{w}{t_0}\,\mathrm{d}w = \frac{1}{t_0}\,,$$

which proves (2.3.15). The proof that the bounded convergence theorem can be applied is left to the reader as an exercise. □

Bibliographical Notes. The Esscher transforms method was developed to approximate the aggregate claim amount distribution (see, for example, Esscher (1932)), where the concept of associated distributions is attributed to Lundberg (1930). For a concise treatment see Jensen (1995). Recently, in Gerber and Shiu (1996), an extension of the method of Esscher transforms was studied in changing probability measures for a certain class of stochastic processes that model security prices. The exposition on the abscissa of convergence follows Section 5.5 of Widder (1971).

2.4 DISTRIBUTIONS WITH MONOTONE HAZARD RATES

Instead of considering parametrized families of distributions it is sometimes more appropriate to deal with classes of distributions of nonnegative random variables which can be described by qualitative properties of some characteristics. For example, hazard rates and, equivalently, mortality rates reflect the conditional probability of dying at age x, given that age x is reached. Another related characteristic is the distribution of the remaining lifetime after age x. We study these and other characteristics and the classes of distributions defined through them. Clearly, hazard rates are important in life insurance mathematics. In a mathematically equivalent form this notion is also considered in other areas of insurance, e.g. in fire insurance, but also in survival analysis and reliability theory. Hazard rates of most distributions sampled from real data do not possess global monotonicity properties, but local monotonicity is often observed and has a natural explanation. On the other hand, many parametrized families of (theoretical) distributions, like gamma, uniform and Weibull distributions, have global monotonicity properties.

2.4.1 Discrete Distributions

We first consider the case of discrete random variables taking their values on a lattice, $\mathbb{N}$ say. Let $\{p_k\}$ be the probability function of an $\mathbb{N}$-valued random variable X. A possible interpretation of X is that it measures the year of death of an individual belonging to a certain population. Define $r_n = \mathbb{P}(X \geq n) = p_n + p_{n+1} + \ldots$ and, for all $n \in \mathbb{N}$ such that $r_n > 0$,

$$m_n = \mathbb{P}(X = n \mid X \geq n) = \frac{p_n}{r_n}. \tag{2.4.1}$$

The quotient m_n in (2.4.1) is called the *hazard rate* of $\{p_k\}$ in the n-th period. It can be interpreted as the conditional probability that an individual, who survived $n-1$ years, dies in the n-th year. Clearly, the graph of $\{m_n\}$ can have different shapes. For example, in life insurance one usually observes data giving hazard rates whose graph has the form shown in Figure 2.4.1. Usually, hazard rates are locally but not globally monotone.

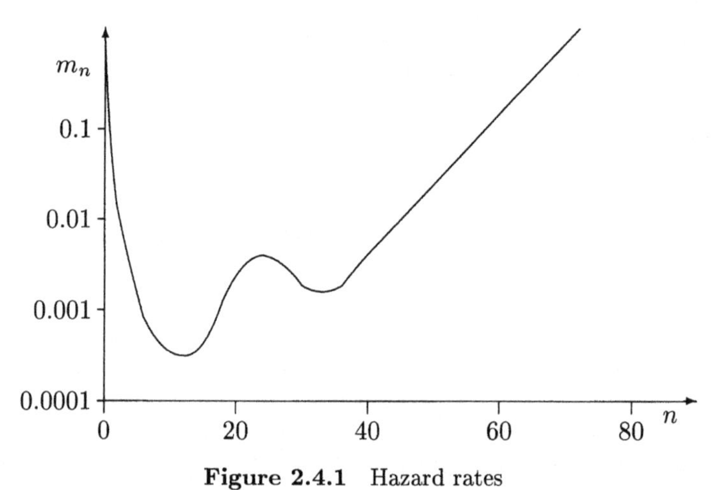

Figure 2.4.1 Hazard rates

In fire insurance one speaks of "extinction rates" instead of hazard rates. Most fires are stopped at the very beginning, i.e. at this stage the extinction rate is relatively large. However, if the early extinction fails, then the extinction rate, i.e. the chance of stopping the fire, soon decreases. Another interpretation of X, considered in reliability theory, is that X measures the lifetime of a technical system until breakdown; then one speaks of "failure rates" instead of hazard rates.

Hazard rates for discrete distributions often turn out to be monotonically increasing or decreasing. Throughout this book we call a function $g : B \to \mathbb{R}$ defined on $B \subset \mathbb{R}$ *increasing* if $g(x) \leq g(y)$ for all $x, y \in B$ such that $x \leq y$ and *decreasing* if $g(x) \geq g(y)$ for all $x, y \in B$ such that $x \leq y$. Obviously, in this terminology, a constant function is both increasing and decreasing. This is the case with the hazard rates of the geometric distribution, where it is easily shown that $m_n = 1 - p$ for all $n \in \mathbb{N}$. This property illustrates the *lack-of-memory* property, which says that a geometrically distributed X satisfies

$$\mathbb{P}(X \geq i + j \mid X \geq j) = \mathbb{P}(X \geq i) \tag{2.4.2}$$

for all $i, j \in \mathbb{N}$. For theoretical purposes it is convenient to distinguish classes of distributions having special monotonicity properties.

We say that the probability function $\{p_k\}$ or, alternatively, the corresponding distribution is:

- $\mathrm{IHR_d}$ if the sequence $\{m_n\}$ is increasing (where IHR comes from *Increasing Hazard Rate* and "d" means *discrete*),
- $\mathrm{DHR_d}$ if the sequence $\{m_n\}$ is decreasing (where DHR comes from *Decreasing Hazard Rate*).

We leave it to the reader to show that a mixture of distributions whose probability functions are $\mathrm{DHR_d}$ is again $\mathrm{DHR_d}$. Hence all mixtures of geometric distributions are $\mathrm{DHR_d}$. The proof that the property $\mathrm{IHR_d}$ is closed with respect to convolution needs additional concepts (see the bibliographical notes below). It is often rather difficult to decide straightforwardly whether a distribution is $\mathrm{IHR_d}$ or $\mathrm{DHR_d}$. However, there are sufficient conditions which in many cases can be checked more easily. We say that a probability function $\{p_k\}$ is *logconvex* if $p_{k+1}^2 \leq p_{k+2}p_k$ for all $k \in \mathbb{N}$ and *logconcave* if $p_{k+1}^2 \geq p_{k+2}p_k$ for all $k \in \mathbb{N}$.

Theorem 2.4.1 (a) *If* $\{p_k\}$ *is logconcave, then it is* $\mathrm{IHR_d}$.
(b) *If* $\{p_k\}$ *is logconvex, then it is* $\mathrm{DHR_d}$.

Proof Let

$$b_n = \begin{cases} p_{n+1}/p_n & \text{if } p_n > 0, \\ 0 & \text{otherwise.} \end{cases} \tag{2.4.3}$$

Assume that the probability function $\{p_k\}$ is logconcave. Then, it is easy to see from the definition of concavity that the sequence $\{b_n\}$ is decreasing. Thus, because of

$$m_n^{-1} = 1 + \frac{p_{n+1}}{p_n} + \frac{p_{n+2}}{p_n} + \ldots = 1 + b_n + b_n b_{n+1} + b_n b_{n+1} b_{n+2} + \ldots , \tag{2.4.4}$$

the sequence $\{m_n\}$ is increasing. The proof of (b) is similar. □

Corollary 2.4.1 *Each Poisson distribution is* $\mathrm{IHR_d}$.

Proof In view of Theorem 2.4.1 it suffices to observe that the probability function of any Poisson distribution is logconcave. □

Corollary 2.4.2 *The negative binomial distribution is* $\mathrm{IHR_d}$ *if* $\alpha > 1$, *and* $\mathrm{DHR_d}$ *if* $0 \leq \alpha < 1$. *For* $\alpha = 1$, $m_n = 1 - p$ *for all* $n \in \mathbb{N}$.

Proof The criterion given in Theorem 2.4.1 requires to check the monotonicity properties of the sequence $\{b_n\}$ given in (2.4.3). For the negative binomial distribution we have

$$b_n = \frac{\frac{\Gamma(\alpha+n+1)}{\Gamma(\alpha)(n+1)!}(1-p)^\alpha p^{n+1}}{\frac{\Gamma(\alpha+n)}{\Gamma(\alpha)n!}(1-p)^\alpha p^n} = \frac{\Gamma(\alpha+n+1)}{\Gamma(\alpha+n)(n+1)}\, p = \frac{\alpha+n}{n+1}\, p\,,$$

which shows that $\{b_n\}$ is increasing for $0 < \alpha < 1$ (and also $\{p_k\}$ logconvex) and decreasing for $\alpha > 1$ (and also $\{p_k\}$ logconcave). □

2.4.2 Absolutely Continuous Distributions

Now we consider a nonnegative random variable X with absolutely continuous distribution F and density function f. For example, X can be interpreted as a random lifetime sampled from very short periods. Drawing an analogy between the discrete and continuous case, we define the *hazard rate function* $m(t)$ by

$$m(t) = \frac{f(t)}{1-F(t)} \qquad \text{if } F(t) < 1\,. \tag{2.4.5}$$

The formal infinitesimal interpretation

$$m(t)\,\mathrm{d}t = \frac{f(t)\,\mathrm{d}t}{1-F(t)} = \mathbb{P}(X - t \in \mathrm{d}t \mid X > t)$$

explains this terminology and shows the similarity to the definition (2.4.1) of $\{m_n\}$. In life insurance mathematics, $m(t)$ given in (2.4.5) is called the *mortality rate function* of F.

We say that the distribution F is IHR if $m(t)$ is increasing, and DHR if $m(t)$ is decreasing. Analogous to the geometric distribution considered above, the only absolutely continuous distribution F which is both IHR and DHR is the exponential distribution. Furthermore, the exponential distribution can be characterized by the continuous version of the *lack-of-memory* property (2.4.2). Namely, X is exponentially distributed if and only if for all $s, t \geq 0$

$$\mathbb{P}(X > t + s \mid X > s) = \mathbb{P}(X > t)\,. \tag{2.4.6}$$

In later chapters of this book we make use of the following lemma, which is of independent interest. It establishes a "randomized" version of the lack-of-memory property (2.4.6) of exponential distributions.

Lemma 2.4.1 *Let X be an exponentially distributed random variable. If W is a random variable and $A \in \mathcal{F}$ an event such that X is independent of W and A, then for all $x \in \mathbb{R}$*

$$\mathbb{P}(X > x + W \mid \{X > W > 0\} \cap A) = \mathbb{P}(X > x)\,. \tag{2.4.7}$$

Proof Let X be exponentially distributed with parameter λ. Using the law of total probability and the lack-of-memory property (2.4.6) of exponential distributions we have for all $x \geq 0$

$$\begin{aligned}\mathbb{P}(X > x + W \mid \{X > W > 0\} \cap A) &= \frac{\mathbb{P}(\{X > x + W, W > 0\} \cap A)}{\mathbb{P}(\{X > W > 0\} \cap A)} \\ = \frac{\int_0^\infty \mathbb{P}(X > x + w)\mathbb{P}(W \in \mathrm{d}w \mid A)}{\mathbb{P}(X > W > 0 \mid A)} &= \frac{\mathrm{e}^{-\lambda x} \int_0^\infty \mathrm{e}^{-\lambda w}\mathbb{P}(W \in \mathrm{d}w \mid A)}{\mathbb{P}(X > W > 0 \mid A)} \\ = \frac{\mathrm{e}^{-\lambda x} \int_0^\infty \mathbb{P}(X > w)\mathbb{P}(W \in \mathrm{d}w \mid A)}{\mathbb{P}(X > W > 0 \mid A)} &= \mathrm{e}^{-\lambda x}.\end{aligned}$$

For $x < 0$, (2.4.7) is obvious. □

Referring to (2.4.6), the notions of IHR and DHR can be introduced in a slightly different way (without using the assumption that F is absolutely continuous). Define the *residual hazard distribution* F_t at t by

$$F_t(x) = \mathbb{P}(X - t \leq x \mid X > t) \tag{2.4.8}$$

if $F(t) < 1$. Note that $F_t(x) = (F(t + x) - F(t))(1 - F(t))^{-1}$. Consequently, for the expectation μ_{F_t} of F_t we get

$$\mu_{F_t} = (1 - F(t))^{-1} \int_t^\infty \overline{F}(x)\,\mathrm{d}x \qquad \text{if } F(t) < 1\,. \tag{2.4.9}$$

Since $\mu_{F_t} = \mathbb{E}(X - t \mid X > t)$, the function $\mu_F(t) = \mu_{F_t}$, $t \geq 0$ is called *the mean residual hazard function.* It turns out that $m(t)$ is increasing or decreasing if and only if the family $\{F_t\}$ of residual hazard distributions is stochastically decreasing or increasing, respectively. Here a distribution F is called *stochastically smaller (larger)* than a distribution G if

$$\overline{F}(x) \leq (\geq)\, \overline{G}(x) \tag{2.4.10}$$

for all $x \in \mathbb{R}$. In this case, we write $F \leq_{\mathrm{st}} G$ and $F \geq_{\mathrm{st}} G$, respectively. Furthermore, we write $X \leq_{\mathrm{st}} Y$ if X, Y have distributions F, G respectively such that $F \leq_{\mathrm{st}} G$.

Theorem 2.4.2 *The distribution* F *is* IHR (DHR) *if and only if, for all* $t_1 \leq t_2$,

$$F_{t_1} \geq_{\mathrm{st}} (\leq_{\mathrm{st}}) F_{t_2}\,. \tag{2.4.11}$$

Proof Note that $\overline{F}(x) = \exp(-\int_0^x m(s)\,\mathrm{d}s)$ and hence we get $\overline{F_t}(x) = \exp(-\int_t^{t+x} m(s)\,\mathrm{d}s)$. Thus, for each x, the function $\overline{F_t}(x)$ is decreasing in t if and only if $m(t)$ is increasing. The proof for DHR is analogous. □

Similarly to the discrete case, a mixture of DHR distributions is again DHR. We leave it to the reader to show this as an exercise. For example, a mixture of exponential distributions is DHR. For IHR, see the bibliographical notes.

Note that there is another interpretation of the distribution F_t if X is considered to be the accumulated loss in one year and the reinsurance treaty is stop loss. That is, if X exceeds a fixed level t (first risk), the reinsurer pays the exceess amount $X - t$. Therefore F_t is the conditional distribution of the compensation and μ_{F_t} is the conditional stop-loss premium. We discuss these matters in Chapter 3.

Next we show how to weaken the monotonicity conditions IHR and DHR or, equivalently, how to enlarge the corresponding classes of distributions. Such weaker conditions are not only useful in fitting models to real data but also give rise to more theoretical results.

In reliability theory, the following classes of distributions larger than those of IHR and DHR distributions are introduced. Following (2.4.10), a distribution F is called NBU (*New Better than Used*) if

$$F_t \leq_{\rm st} F \tag{2.4.12}$$

for all $t \geq 0$. Analogously, F is called NWU (*New Worse than Used*) if

$$F_t \geq_{\rm st} F \tag{2.4.13}$$

for all $t \geq 0$. Since $F_0 = F$, Theorem 2.4.2 yields that a distribution is NBU (NWU) provided that it is IHR (DHR).

A distribution F is called NBUE (*New Better than Used in Expectation*) if $\mu_{F_t} \leq \mu_F$ for all $t \geq 0$. Analogously, one says that F is NWUE (*New Worse than Used in Expectation*) if $\mu_{F_t} \geq \mu_F$ for all $t \geq 0$.

Denoting the sets of distributions with the property IHR, NBU, ... by the same symbol IHR, NBU, ..., respectively, we have the inclusions IHR $\subset$ NBU $\subset$ NBUE and DHR $\subset$ NWU $\subset$ NWUE.

Another important class of distributions consists of the distributions with heavy tails. As they are of special interest in reinsurance mathematics, we will consider them in Section 2.5.

Bibliographical Notes. The notion of hazard rate is one of the measures of mortality used in life insurance. For a detailed account, see Benjamin and Pollard (1993) and also Gerber (1995). Classes of distributions with some monotonicity property were developed in reliability theory, and the general theory is summarized in Barlow and Proschan (1965,1975). In insurance mathematics such classes were studied in Heilmann (1988), and Heilmann and Schröter (1991), for example. A proof of the fact that the classes $\mathrm{IHR_d}$ and IHR are closed with respect to convolution can be found in Barlow and Proschan (1965,1975).

2.5 HEAVY-TAILED DISTRIBUTIONS

2.5.1 Definition and Basic Properties

In this section we study classes of distributions of nonnegative random variables such that $\hat{m}(s) = \infty$ for all $s > 0$. We call them *heavy-tailed distributions*. Prominent examples of heavy-tailed distributions are the lognormal, Pareto and Weibull distributions with shape parameter smaller than 1. For example, actuaries believe that lognormal distributions are plausible models for motor insurance, while they feel that Pareto distributions are apt to model fire claim data.

Let $\alpha_F = \limsup_{x\to\infty} M(x)/x$, where $M(x) = -\log \overline{F}(x)$ is the *hazard function* of F. This terminology is motivated by the following fact. If F has a continuous density, then $M(x)$ is differentiable and $\mathrm{d}M(x)/\mathrm{d}x = m(x)$, where $m(x)$ is the hazard rate function considered in Section 2.4. The proof is left to the reader as an exercise. In this section we consider distributions on $\mathbb{R}_+$ fulfilling $F(0-) = 0$.

Theorem 2.5.1 *If $\alpha_F = 0$, then F is heavy-tailed.*

Proof Suppose that $\alpha_F = 0$. Then $\lim_{x\to\infty} M(x)/x = 0$. Thus, for each $\varepsilon > 0$ there exists an $x' > 0$ such that $M(x) \le \varepsilon x$ for all $x \ge x'$. Therefore for some $c > 0$ we have $\overline{F}(x) \ge c\mathrm{e}^{-\varepsilon x}$ for all $x \ge 0$ and hence

$$\int_0^\infty \mathrm{e}^{sx}\overline{F}(x)\,\mathrm{d}x = \infty \tag{2.5.1}$$

for all $s \ge \varepsilon$. Since $\varepsilon > 0$ is arbitrary, (2.5.1) holds for all $s > 0$, which means that F is heavy-tailed. □

Remark For a heavy-tailed distribution F we have

$$\lim_{x\to\infty} \mathrm{e}^{sx}\overline{F}(x) = \infty \tag{2.5.2}$$

for all $s > 0$. We leave it to the reader to show this as an exercise.

2.5.2 Subexponential Distributions

Note that the term "subexponentiality" is motivated by (2.5.2), which is, however, used to single out distributions from a smaller class of heavy-tailed distributions. A distribution F on $\mathbb{R}_+$ is said to be *subexponential* if

$$\lim_{x\to\infty} \frac{1 - F^{*2}(x)}{1 - F(x)} = 2\,. \tag{2.5.3}$$

Let $\mathcal{S}$ denote the class of all subexponential distributions. We show later that the following important (parametrized) families of distributions are in $\mathcal{S}$: the lognormal distributions, Pareto distributions and Weibull distributions with shape parameter smaller than 1.

A direct consequence of (2.5.3) is that $\overline{F}(x) > 0$ for all $x \geq 0$. However, not all distributions with this property are subexponential. Note, for example, that trivially the exponential distribution is not subexponential because in this case $(1 - F^{*2}(x))/(1 - F(x)) = e^{-\lambda x}(1 + \lambda x)/e^{-\lambda x} \to \infty$ as $x \to \infty$. On the other hand, it is easy to see that if F is subexponential and X_1, X_2 are independent and identically distributed random variables with distribution F, then we have for $x \to \infty$ that

$$\mathbb{P}(X_1 + X_2 > x) \sim \mathbb{P}(\max\{X_1, X_2\} > x), \tag{2.5.4}$$

since $\mathbb{P}(\max\{X_1, X_2\} > x) = 1 - F^2(x) = (1 - F(x))(1 + F(x))$ and hence

$$1 = \lim_{x\to\infty} \frac{1 - F^{*2}(x)}{2(1 - F(x))} = \lim_{x\to\infty} \frac{1 - F^{*2}(x)}{(1 + F(x))(1 - F(x))} = \lim_{x\to\infty} \frac{1 - F^{*2}(x)}{1 - F^2(x)}.$$

The following identity is obvious:

$$\frac{\overline{F^{*2}}(x)}{\overline{F}(x)} = 1 + \int_0^x \frac{\overline{F}(x-y)}{\overline{F}(x)}\, \mathrm{d}F(y), \tag{2.5.5}$$

from which we obtain that always

$$\liminf_{x\to\infty} \frac{\overline{F^{*2}}(x)}{\overline{F}(x)} \geq 2. \tag{2.5.6}$$

The proof of (2.5.6) is left to the reader. Note that (2.5.6) implies that the limit value 2 in (2.5.3) is minimal. Furthermore, (2.5.5) yields two useful properties of subexponential distributions.

Lemma 2.5.1 *If $F \in \mathcal{S}$, then for all $x' > 0$,*

$$\lim_{x\to\infty} \frac{\overline{F}(x - x')}{\overline{F}(x)} = 1 \tag{2.5.7}$$

and

$$\lim_{x\to\infty} \int_0^x \frac{\overline{F}(x-y)}{\overline{F}(x)}\, \mathrm{d}F(y) = 1. \tag{2.5.8}$$

Proof For $x' \leq x$, identity (2.5.5) yields

$$\begin{aligned}
\frac{\overline{F^{*2}}(x)}{\overline{F}(x)} &= 1 + \int_0^{x'} \frac{\overline{F}(x-y)}{\overline{F}(x)}\, \mathrm{d}F(y) + \int_{x'}^x \frac{\overline{F}(x-y)}{\overline{F}(x)}\, \mathrm{d}F(y) \\
&\geq 1 + F(x') + \frac{\overline{F}(x-x')}{\overline{F}(x)}(F(x) - F(x')),
\end{aligned}$$

which gives

$$1 \leq \frac{\overline{F}(x-x')}{\overline{F}(x)} \leq \Big(\frac{\overline{F^{*2}}(x)}{\overline{F}(x)} - 1 - F(x')\Big)(F(x) - F(x'))^{-1}.$$

This completes the proof of (2.5.7) because, for $F \in \mathcal{S}$, the right-hand side of the last inequality tends to 1 as $x \to \infty$. The limit in (2.5.8) is an immediate consequence of (2.5.5). □

Lemma 2.5.2 *Let $F \in \mathcal{S}$ and F' be a distribution with $F'(0) = 0$ such that $\lim_{x\to\infty} \overline{F'}(x)/\overline{F}(x) = c$ for some $c \in [0, \infty)$. Then*

$$\lim_{x\to\infty} \frac{\overline{F * F'}(x)}{\overline{F}(x)} = 1 + c\,. \tag{2.5.9}$$

Proof We have to show that

$$\lim_{x\to\infty} \frac{\int_0^x \overline{F'}(x-y)\,\mathrm{d}F(y)}{\overline{F}(x)} = c\,. \tag{2.5.10}$$

Choose $\varepsilon > 0$. There exists x_0 such that $\overline{F'}(x) \leq (c+\varepsilon)\overline{F}(x)$ for $x \geq x_0$. Then

$$\begin{aligned}
\frac{\int_0^x \overline{F'}(x-y)\,\mathrm{d}F(y)}{\overline{F}(x)} &\leq (c+\varepsilon)\frac{\int_0^{x-x_0} \overline{F}(x-y)\,\mathrm{d}F(y)}{\overline{F}(x)} + \frac{\overline{F}(x-x_0) - \overline{F}(x)}{\overline{F}(x)} \\
&\leq (c+\varepsilon)\frac{\int_0^{x} \overline{F}(x-y)\,\mathrm{d}F(y)}{\overline{F}(x)} + \frac{\overline{F}(x-x_0) - \overline{F}(x)}{\overline{F}(x)}\,.
\end{aligned}$$

The latter expression tends by Lemma 2.5.1 to $c+\varepsilon$ as $x \to \infty$. Thus

$$\lim_{x\to\infty} \frac{\int_0^x \overline{F'}(x-y)\,\mathrm{d}F(y)}{\overline{F}(x)} \leq c\,.$$

Similarly it follows that

$$\lim_{x\to\infty} \frac{\int_0^x \overline{F'}(x-y)\,\mathrm{d}F(y)}{\overline{F}(x)} \geq c\,.$$

This proves (2.5.10) and also the lemma. □

The next result shows that the class of subexponential distributions is incorporated within the class of heavy-tailed distributions.

Theorem 2.5.2 *Each $F \in \mathcal{S}$ is heavy-tailed.*

Proof Let $F \in \mathcal{S}$. By Theorem 2.5.1 it suffices to demonstrate that $\alpha_F = 0$. Taking the logarithmic version of (2.5.7) we have

$$\lim_{x\to\infty}(\log \overline{F}(x-y) - \log \overline{F}(x)) = \lim_{x\to\infty}(M(x) - M(x-y)) = 0$$

for all $y \ge 0$. Hence, for all $\varepsilon > 0$, there exists $x_0 > 0$ such that for all $x \ge x_0$ we have $M(x) - M(x-1) < \varepsilon$. By iteration we get

$$M(x) \le M(x-1) + \varepsilon \le M(x-2) + 2\varepsilon \le \ldots \le M(x-n) + n\varepsilon\,,$$

where n is such that $x_0 \le x - n < x_0 + 1$. Thus

$$M(x) \le \sup_{x_0 - 1 \le x' \le x_0} M(x') + (x - x_0)\varepsilon\,, \qquad x \ge x_0\,.$$

Since ε is arbitrary we have $\lim_{x\to\infty} M(x)/x = 0$. □

Using Lemma 2.5.2 we get the following characterization of subexponential distributions.

Theorem 2.5.3 *Let F be a distribution on $\mathbb{R}_+$. Then, $F \in \mathcal{S}$ if and only if for each $n = 2, 3, \ldots$*

$$\lim_{x\to\infty} \frac{\overline{F^{*n}}(x)}{\overline{F}(x)} = n\,. \tag{2.5.11}$$

Proof The proof is by induction on $n = 2, 3 \ldots$. Assume that F is subexponential. Then (2.5.11) holds for $n = 2$ by definition. Suppose that (2.5.11) holds for $n - 1$. Then, Lemma 2.5.2 with $F' = F^{*(n-1)}$ yields the assertion. □

We recommend the reader to show the following natural extension of (2.5.4) to an arbitrary (finite) number of random variables with subexponential distribution: if $X_1, \ldots, X_n$ are independent and identically distributed with distribution $F \in \mathcal{S}$, then $\mathbb{P}(\sum_{i=1}^n X_i > x) \sim \mathbb{P}(\max_{1\le i\le n} X_i > x)$ as $x \to \infty$. Furthermore, Theorem 2.5.3 immediately yields that for distributions of the form $F(x) = \sum_{k=0}^n p_k G^{*k}(x)$, where $\{p_0, p_1, \ldots, p_n\}$ is a probability function and G a subexponential distribution, we have

$$\lim_{x\to\infty} \frac{\overline{F}(x)}{\overline{G}(x)} = \sum_{k=0}^n k p_k\,. \tag{2.5.12}$$

Such *compound distributions* F are important in insurance mathematics and will be studied later, for instance in Chapter 4. For example, ruin functions of some risk processes can be expressed by compound distributions. To study the asymptotic behaviour of ruin functions in the case of subexponential claim size distributions (in Sections 5.4.3, 6.5.5 and 12.6) we need an extended version of (2.5.12) for compound distributions of type $F(x) = \sum_{k=0}^{\infty} p_k G^{*k}(x)$ where $\{p_0, p_1, \ldots\}$ is a probability function. In connection with this the following lemma is useful.

Lemma 2.5.3 *If $F \in \mathcal{S}$, then for each $\varepsilon > 0$ there exists a constant $c < \infty$ such that for all $n \geq 2$*

$$\frac{\overline{F^{*n}}(x)}{\overline{F}(x)} \leq c(1+\varepsilon)^n, \qquad x \geq 0. \tag{2.5.13}$$

Proof Let $\alpha_n = \sup_{x\geq 0}(\overline{F^{*n}}(x)/\overline{F}(x))$. Note that (2.5.11) implies $\alpha_n < \infty$. Furthermore, $\overline{F^{*(n+1)}}(x) = \overline{F}(x) + F * \overline{F^{*n}}(x)$. Thus, for all $a < \infty$,

$$\begin{aligned}
\alpha_{n+1} &\leq 1 + \sup_{0\leq x\leq a} \int_0^x \frac{\overline{F^{*n}}(x-y)}{\overline{F}(x)} \, \mathrm{d}F(y) \\
&\quad + \sup_{x\geq a} \int_0^x \frac{\overline{F^{*n}}(x-y)}{\overline{F}(x-y)} \frac{\overline{F}(x-y)}{\overline{F}(x)} \, \mathrm{d}F(y) \\
&\leq 1 + c_a + \alpha_n \sup_{x\geq a} \frac{\overline{F^{*2}}(x) - \overline{F}(x)}{\overline{F}(x)},
\end{aligned}$$

where $c_a = 1/\overline{F}(a) < \infty$. Since $F \in \mathcal{S}$, for each $\varepsilon > 0$ we can choose a such that $\alpha_{n+1} \leq 1 + c_a + \alpha_n(1+\varepsilon)$. Hence

$$\begin{aligned}
\alpha_n &\leq (1+c_a) + (1 + c_a + \alpha_{n-2}(1+\varepsilon))(1+\varepsilon) \leq \ldots \\
&\leq (1+c_a)(1 + (1+\varepsilon) + \ldots + (1+\varepsilon)^{n-2}) + (1+\varepsilon)^{n-1} \\
&\leq (1+c_a)\varepsilon^{-1}(1+\varepsilon)^n,
\end{aligned}$$

which implies (2.5.13). □

Theorem 2.5.4 *Let $F(x) = \sum_{k=0}^{\infty} p_k G^{*k}(x)$, where $\{p_0, p_1, \ldots\}$ is a probability function and $G \in \mathcal{S}$. If $\sum_{n=1}^{\infty} p_n(1+\varepsilon)^n < \infty$ for some $\varepsilon > 0$, then*

$$\lim_{x\to\infty} \frac{\overline{F}(x)}{\overline{G}(x)} = \sum_{k=0}^{\infty} k p_k. \tag{2.5.14}$$

Proof The assertion immediately follows from Lemma 2.5.3 and the dominated convergence theorem. □

We close this section showing subexponentiality for an important class of distributions, containing Pareto distributions and other parametrized families of distributions like loggamma distributions. We first need a definition. We say that a positive function $L : \mathbb{R}_+ \to (0, \infty)$ is a *slowly varying function of x*

at ∞ if for all $y > 0$, $L(xy)/L(x) \to 1$ as $x \to \infty$. Examples of such functions are $|\log^p x|$, and functions converging to a positive limit as $x \to \infty$. Note that (2.5.7) gives in particular that, if $F \in \mathcal{S}$, then $\overline{F}(\log x)$ is a slowly varying function of x at ∞. We now say that the distribution F is *Pareto-type* with exponent $\alpha > 0$ if $\overline{F}(x) \sim L(x)x^{-\alpha}$ as $x \to \infty$ for a slowly varying function $L(x)$. In the literature, Pareto-type distributions are also called *distributions with regular varying tails.*

Theorem 2.5.5 *If F is Pareto-type, then $F \in \mathcal{S}$.*

Proof Let X, X_1 and X_2 be independent and identically distributed risks with Pareto-type distribution F. Note that $\{X_1 + X_2 > x\}$ implies that for $\varepsilon \in (0,1)$

$$\{X_1 > (1-\varepsilon)x\} \quad \text{or} \quad \{X_2 > (1-\varepsilon)x\} \quad \text{or} \quad \{X_1 > \varepsilon x \quad \text{and} \quad X_2 > \varepsilon x\},$$

which yields $\mathbb{P}(X_1 + X_2 > x) \le 2\mathbb{P}(X > (1-\varepsilon)x) + (\mathbb{P}(X > \varepsilon x))^2$. Hence

$$\limsup_{x\to\infty} \frac{\mathbb{P}(X_1 + X_2 > x)}{L(x)x^{-\alpha}} \le 2(1-\varepsilon)^{-\alpha}.$$

Since $\varepsilon > 0$ is arbitrary, $\limsup_{x\to\infty} \overline{F^{*2}}(x)/\overline{F}(x) \le 2$. However, in view of (2.5.6) this gives $\lim_{x\to\infty} \overline{F^{*2}}(x)/\overline{F}(x) = 2$ and the proof is completed. □

2.5.3 Criteria for Subexponentiality and the Class $\mathcal{S}^*$

In most cases it is not an easy task to prove directly that a given distribution is subexponential. In Theorem 2.5.5 we were able to verify subexponentiality for Pareto-type distributions. However, for future applications in risk theory, we need the integrated tail of the distribution F to be subexponential rather than the distribution itself. Recall that for a distribution F of a nonnegative random variable with finite expectation $\mu > 0$, the *integrated tail distribution* F^{s} is given by

$$F^{\mathrm{s}}(x) = \begin{cases} 0 & \text{if } x \le 0, \\ \mu^{-1} \displaystyle\int_0^x \overline{F}(y)\,\mathrm{d}y & \text{if } x > 0. \end{cases} \tag{2.5.15}$$

It seems to be not yet known whether $F \in \mathcal{S}$ and $0 < \mu < \infty$ imply $F^{\mathrm{s}} \in \mathcal{S}$ in general. Thus, it is useful to have conditions for a distribution with finite expectation to be subexponential jointly with its integrated tail distribution. On the other hand, there exist examples of distributions F on $\mathbb{R}_+$ such that $F^{\mathrm{s}} \in \mathcal{S}$, but $F \notin \mathcal{S}$.

We now show that, for a certain subset $\mathcal{S}^*$ of $\mathcal{S}$ which is defined below, $F \in \mathcal{S}^*$ implies $F^{\mathrm{s}} \in \mathcal{S}$. Throughout this section we only consider distributions F on $\mathbb{R}_+$ such that $F(0) = 0, F(x) < 1$ for all $x \in \mathbb{R}_+$.

Definition 2.5.1 (a) *We say that F belongs to the class $\mathcal{S}^*$ if F has finite expectation μ and*

$$\lim_{x\to\infty}\int_0^x \frac{\overline{F}(x-y)}{\overline{F}(x)}\overline{F}(y)\,\mathrm{d}y = 2\mu\,. \tag{2.5.16}$$

(b) *We say that F belongs to $\mathcal{L}$ if for all $y \in \mathbb{R}$*

$$\lim_{x\to\infty}\frac{\overline{F}(x-y)}{\overline{F}(x)} = 1\,. \tag{2.5.17}$$

Note that Lemma 2.5.1 implies $\mathcal{S} \subset \mathcal{L}$. Class $\mathcal{L}$ will serve to show that class $\mathcal{S}^*$ of distributions on $\mathbb{R}_+$ has some desired properties. We leave it to the reader to show as an exercise that all distribution functions with hazard rate functions tending to 0 are in $\mathcal{L}$. We also have the identity

$$\int_0^x \frac{\overline{F}(x-y)}{\overline{F}(x)}\overline{F}(y)\,\mathrm{d}y = \int_0^{x/2}\ldots + \int_{x/2}^{x}\ldots = 2\int_0^{x/2}\frac{\overline{F}(x-y)}{\overline{F}(x)}\overline{F}(y)\,\mathrm{d}y\,, \tag{2.5.18}$$

from which we get that (2.5.16) is equivalent to

$$\lim_{x\to\infty}\int_0^{x/2}\frac{\overline{F}(x-y)}{\overline{F}(x)}\overline{F}(y)\,\mathrm{d}y = \mu\,. \tag{2.5.19}$$

We now study the relationship between $\mathcal{S}^*$ and $\{F : F \in \mathcal{S} \text{ and } F^{\mathrm{s}} \in \mathcal{S}\}$. For this we need three lemmas. In the first we give an equivalence relation for subexponential distributions.

Lemma 2.5.4 *Let F, G be two distributions on $\mathbb{R}_+$ and assume that there exists a constant $c \in (0,\infty)$ such that*

$$\lim_{x\to\infty}\frac{\overline{G}(x)}{\overline{F}(x)} = c\,. \tag{2.5.20}$$

Then, $F \in \mathcal{S}$ if and only if $G \in \mathcal{S}$.

Proof Suppose $F \in \mathcal{S}$ and consider a distribution G for which (2.5.20) holds. Remember that from (2.5.6) we always have $\liminf_{x\to\infty}\overline{G^{*2}}(x)/\overline{G}(x) \ge 2$. Thus, recalling the identity

$$\frac{\overline{G^{*2}}(x)}{\overline{G}(x)} = 1 + \int_0^x \frac{\overline{G}(x-y)}{\overline{G}(x)}\,\mathrm{d}G(y)\,,$$

it suffices to show that

$$\limsup_{x\to\infty}\int_0^x \frac{\overline{G}(x-y)}{\overline{G}(x)}\,\mathrm{d}\,G(y) \le 1\,. \tag{2.5.21}$$

Let $a > 0$ be a fixed number. Note that for $x > a$, the function $I_1(x) = \int_{x-a}^{x} \overline{G}(x-y)/\overline{G}(x)\,\mathrm{d}G(y)$ can be bounded by

$$0 \le I_1(x) \le \frac{G(x) - G(x-a)}{\overline{G}(x)} = \frac{\overline{G}(x-a)}{\overline{G}(x)} - 1\,.$$

Thus, using Lemma 2.5.1 and (2.5.20), we have

$$\lim_{x\to\infty} I_1(x) = \lim_{x\to\infty} \frac{\overline{F}(x)}{\overline{F}(x-a)} \frac{\overline{G}(x-a)}{\overline{G}(x)} - 1 = 0\,.$$

Choose now $\varepsilon > 0$ and $a \ge 0$ such that $c - \varepsilon \le \overline{G}(x)/\overline{F}(x) \le c + \varepsilon$ for all $x \ge a$. Then for the function $I_2(x) = \int_0^{x-a} \overline{G}(x-y)/\overline{G}(x)\,\mathrm{d}G(y)$ we have

$$\begin{aligned}
I_2(x) &\le \frac{c+\varepsilon}{c-\varepsilon} \int_0^{x-a} \frac{\overline{F}(x-y)}{\overline{F}(x)}\,\mathrm{d}G(y) \\
&\le \frac{c+\varepsilon}{c-\varepsilon} \frac{G(x) - \int_0^x G(x-y)\,\mathrm{d}F(y)}{\overline{F}(x)} \\
&= \frac{c+\varepsilon}{c-\varepsilon} \frac{\overline{F}(x) - \overline{G}(x) + \int_0^x \overline{G}(x-y)\,\mathrm{d}F(y)}{\overline{F}(x)} \\
&= \frac{c+\varepsilon}{c-\varepsilon} \frac{\overline{F}(x) - \overline{G}(x) + \int_0^{x-a} \overline{G}(x-y)\,\mathrm{d}F(y) + \int_{x-a}^{x} \overline{G}(x-y)\,\mathrm{d}F(y)}{\overline{F}(x)} \\
&\le \frac{c+\varepsilon}{c-\varepsilon} \Big\{ \frac{\overline{F}(x) - (c-\varepsilon)\overline{F}(x) + (c+\varepsilon)\int_0^{x-a} \overline{F}(x-y)\mathrm{d}\,F(y)}{\overline{F}(x)} \\
&\qquad\qquad + \frac{\overline{F}(x-a) - \overline{F}(x)}{\overline{F}(x)} \Big\}\,.
\end{aligned}$$

Again using Lemma 2.5.1, this gives $\limsup_{x\to\infty} I_2(x) \le (1+2\varepsilon)(c+\varepsilon)/(c-\varepsilon)$, i.e. (2.5.21) follows because $\varepsilon > 0$ is arbitrary. □

The above lemma justifies the following definition. Two distribution functions F and G on $\mathbb{R}_+$ are said to be *tail-equivalent* if $\lim_{x\to\infty} \overline{G}(x)/\overline{F}(x) = c$ for some $0 < c < \infty$. This will be denoted by $G \sim^{\mathrm{t}} F$. It turns out that for distributions from $\mathcal{S}^*$, condition (2.5.20) can be weakened.

Lemma 2.5.5 *Let $F, G \in \mathcal{L}$. Suppose there exist $c_-, c_+ \in (0,\infty)$ such that*

$$c_- \le \frac{\overline{G}(x)}{\overline{F}(x)} \le c_+ \tag{2.5.22}$$

for all $x \ge 0$. Then, $F \in \mathcal{S}^*$ if and only if $G \in \mathcal{S}^*$.

Proof Suppose $F \in \mathcal{S}^*$. Then (2.5.22) implies that G has finite expectation. Furthermore, for fixed $v > 0$ and $x > 2v$, we have

$$\int_0^{x/2} \frac{\overline{G}(x-y)}{\overline{G}(x)} \overline{G}(y)\,\mathrm{d}y = \int_0^{v} \frac{\overline{G}(x-y)}{\overline{G}(x)} \overline{G}(y)\,\mathrm{d}y + \int_v^{x/2} \frac{\overline{G}(x-y)}{\overline{G}(x)} \overline{G}(y)\,\mathrm{d}y\,.$$

Recalling that $G \in \mathcal{L}$, for $v \geq y \geq 0$,

$$1 \leq \frac{\overline{G}(x-y)}{\overline{G}(x)} \leq \frac{\overline{G}(x-v)}{\overline{G}(x)} \longrightarrow 1$$

as $x \to \infty$ and hence $\sup_{x \geq 0} \overline{G}(x-v)/\overline{G}(x) < \infty$. Thus, by the dominated convergence theorem,

$$\lim_{x\to\infty} \int_0^{v} \frac{\overline{G}(x-y)}{\overline{G}(x)} \overline{G}(y)\,\mathrm{d}y = \int_0^{v} \overline{G}(y)\,\mathrm{d}y\,,$$

and so it suffices to show that

$$\lim_{v\to\infty} \limsup_{x\to\infty} \int_v^{x/2} \frac{\overline{G}(x-y)}{\overline{G}(x)} \overline{G}(y)\,\mathrm{d}y = 0\,. \tag{2.5.23}$$

Using (2.5.22) we have

$$\int_v^{x/2} \frac{\overline{G}(x-y)}{\overline{G}(x)} \overline{G}(y)\,\mathrm{d}y \leq \frac{c_+^2}{c_-} \int_v^{x/2} \frac{\overline{F}(x-y)}{\overline{F}(x)} \overline{F}(y)\,\mathrm{d}y\,.$$

This gives (2.5.23) because, by (2.5.16) and the dominated convergence theorem,

$$\begin{aligned}
&\limsup_{x\to\infty} \int_v^{x/2} \frac{\overline{F}(x-y)}{\overline{F}(x)} \overline{F}(y)\,\mathrm{d}y \\
&= \limsup_{x\to\infty} \Big(\int_0^{x/2} \frac{\overline{F}(x-y)}{\overline{F}(x)} \overline{F}(y)\,\mathrm{d}y - \int_0^{v} \frac{\overline{F}(x-y)}{\overline{F}(x)} \overline{F}(y)\,\mathrm{d}y \Big) \\
&= \mu - \int_0^{v} \overline{F}(y)\,\mathrm{d}y\,.
\end{aligned}$$

□

It can be proved that for a distribution function F with hazard rate function $m_F(x)$, we have $F \in \mathcal{L}$ if $\lim_{x\to\infty} m_F(x) = 0$. A certain conversion of this statement is given in the following lemma.

Lemma 2.5.6 *For each $F \in \mathcal{L}$ there exists a distribution $G \in \mathcal{L}$ with $\overline{F} \sim^{\mathrm{t}} \overline{G}$ such that its hazard function $M_G(x) = -\log \overline{G}(x)$ and its hazard rate function $m_G(x) = \mathrm{d}M_G(x)/\mathrm{d}x$ have the following properties: $M_G(x)$ is continuous and almost everywhere differentiable with the exception of points in $\mathbb{N}$, and $\lim_{x\to\infty} m_G(x) = 0$.*

Proof We define G by the following hazard function M_G via the formula $\overline{G}(x) = \mathrm{e}^{-M_G(x)}$. Let the function M_G be continuous such that $M_G(n) = M_F(n)$ for all $n \in \mathbb{N}$, and M_G piecewise linear in $[n, n+1]$, for all $n \in \mathbb{N}$. Then M_G is differentiable with the exception of points from $\mathbb{N}$, where we put $m_G(n) = 0$. Otherwise $m_G(x) = M_F(n+1) - M_F(n)$ for $x \in (n, n+1)$. To see that $\lim_{x\to\infty} m_G(x) = 0$ it suffices to observe that

$$\lim_{n\to\infty} (M_F(n+1) - M_F(n)) = \lim_{n\to\infty} \log\Big(\frac{\overline{F}(n)}{\overline{F}(n+1)}\Big) = 0\,,$$

because $F \in \mathcal{L}$. Moreover, $\overline{F} \sim^{\mathrm{t}} \overline{G}$ since

$$|M_F(x) - M_G(x)| \le M_F(\lfloor x \rfloor + 1) - M_F(\lfloor x \rfloor) \to 0$$

as $x \to \infty$, which immediately yields

$$\lim_{x\to\infty} \frac{\overline{F}(x)}{\overline{G}(x)} = \lim_{x\to\infty} \exp(-M_F(x) + M_G(x)) = 1\,. \qquad \square$$

Remark A consequence of Lemmas 2.5.4, 2.5.5 and 2.5.6 is that to check subexponentiality for $F \in \mathcal{L}$ it suffices to verify this for G, which is tail-equivalent to F and for which $\lim_{x\to\infty} m_G(x) = 0$. Moreover, if G^s belongs to $\mathcal{S}$, then F^{s} belongs to $\mathcal{S}$, too. The proof is left to the reader.

We use the idea from the above remark in the proof of the following theorem.

Theorem 2.5.6 *If $F \in \mathcal{S}^*$, then $F \in \mathcal{S}$ and $F^{\mathrm{s}} \in \mathcal{S}$.*

Proof We show first that $F \in \mathcal{L}$. For fixed $v > 0$ and $x > 2v$, we have

$$\begin{aligned}\int_0^{x/2} \frac{\overline{F}(x-y)}{\overline{F}(x)} \overline{F}(y)\,\mathrm{d}y &= \int_0^{v} \frac{\overline{F}(x-y)}{\overline{F}(x)} \overline{F}(y)\,\mathrm{d}y + \int_v^{x/2} \frac{\overline{F}(x-y)}{\overline{F}(x)} \overline{F}(y)\,\mathrm{d}y \\ &\ge \mu\Big\{F^{\mathrm{s}}(v) + \frac{\overline{F}(x-v)}{\overline{F}(x)}(F^{\mathrm{s}}(x/2) - F^{\mathrm{s}}(v))\Big\}\end{aligned}$$

because

$$\begin{aligned}\int_0^{v} \frac{\overline{F}(x-y)}{\overline{F}(x)} \overline{F}(y)\,\mathrm{d}y &\ge \int_0^{v} \overline{F}(y)\,\mathrm{d}y, \\ \int_0^{x/2} \frac{\overline{F}(x-y)}{\overline{F}(x)} \overline{F}(y)\,\mathrm{d}y &\ge \frac{\overline{F}(x-v)}{\overline{F}(x)} \int_v^{x/2} \overline{F}(y)\,\mathrm{d}y.\end{aligned}$$

Hence

$$1 \le \frac{\overline{F}(x-v)}{\overline{F}(x)} \le \Big\{\frac{1}{\mu}\int_0^{x/2} \frac{\overline{F}(x-y)}{\overline{F}(x)} \overline{F}(y)\,\mathrm{d}y - F^{\mathrm{s}}(v)\Big\}(F^{\mathrm{s}}(x/2) - F^{\mathrm{s}}(v))^{-1}\,.$$

Since (2.5.16) holds, the right-hand side tends to 1 as $x \to \infty$. This proves that $F \in \mathcal{L}$. Therefore, in view of Lemmas 2.5.4, 2.5.5 and 2.5.6, we can assume without loss of generality that F has density f and hazard rate function $m_F(x)$ with $\lim_{x\to\infty} m_F(x) = \lim_{x\to\infty} f(x)/\overline{F}(x) = 0$. Consequently, for some $x_0 \geq 0$, we have $m_F(x) \leq 1$ for all $x \geq x_0$, that is,

$$f(x) \leq \overline{F}(x) \tag{2.5.24}$$

for $x \geq x_0$. Note that

$$\int_0^x \frac{\overline{F}(x-y)}{\overline{F}(x)}\overline{F}(y)\,\mathrm{d}y = 2\int_0^v \frac{\overline{F}(x-y)}{\overline{F}(x)}\overline{F}(y)\,\mathrm{d}y + \int_v^{x-v} \frac{\overline{F}(x-y)}{\overline{F}(x)}\overline{F}(y)\,\mathrm{d}y$$

for fixed $v > 0$ and all $x > 2v$. Thus, by the same argument as used in the proof of Lemma 2.5.5, we have

$$\lim_{v\to\infty}\limsup_{x\to\infty}\int_v^{x-v} \frac{\overline{F}(x-y)}{\overline{F}(x)}\overline{F}(y)\,\mathrm{d}y = 0\,.$$

This and (2.5.24) give

$$\lim_{v\to\infty}\limsup_{x\to\infty}\int_v^{x-v} \frac{\overline{F}(x-y)}{\overline{F}(x)}f(y)\,\mathrm{d}y = 0\,. \tag{2.5.25}$$

Using (2.5.5) and integration by parts, we have

$$\begin{aligned}
\frac{\overline{F^{*2}}(x)}{\overline{F}(x)} &= 1 + \int_0^x \frac{\overline{F}(x-y)}{\overline{F}(x)}\,\mathrm{d}F(y) \\
&= 1 + 2\int_0^v \frac{\overline{F}(x-y)}{\overline{F}(x)}\,\mathrm{d}F(y) + \frac{\overline{F}(x-v)\overline{F}(v) - \overline{F}(x)}{\overline{F}(x)} \\
&\quad + \int_v^{x-v} \frac{\overline{F}(x-y)}{\overline{F}(x)}f(y)\,\mathrm{d}y
\end{aligned}$$

for $x > 2v$. Hence (2.5.17) and (2.5.25) give $\limsup_{x\to\infty} \overline{F^{*2}}(x)/\overline{F}(x) \leq 2$, from which we conclude that $F \in \mathcal{S}$ because the reverse inequality always holds. It remains to show that $F^{\mathrm{s}} \in \mathcal{S}$. Clearly,

$$\overline{(F^{\mathrm{s}})^{*2}}(x) = \frac{1}{\mu^2}\int_x^\infty \int_0^t \overline{F}(t-y)\overline{F}(y)\,\mathrm{d}y\,\mathrm{d}t\,.$$

On the other hand, by (2.5.16), for each $\varepsilon > 0$ there exists $x_0 \geq 0$ such that

$$2\mu(1-\varepsilon)\overline{F}(t) \leq \int_0^t \overline{F}(t-y)\overline{F}(y)\,\mathrm{d}y \leq 2\mu(1+\varepsilon)\overline{F}(t)\,, \qquad t > x_0\,.$$

Integrating these inequalities over (x, ∞) and dividing by $\overline{F^s}(x)$ we find that $2(1-\varepsilon) \le \overline{(F^s)^{*2}}(x)/\overline{F^s}(x) \le 2(1+\varepsilon)$ for $x > x_0$. Thus, $F^s \in \mathcal{S}$ since ε is arbitrary. □

Corollary 2.5.1 *Assume that the hazard rate function $m_F(x)$ of F exists and $\mu < \infty$. If $\limsup_{x\to\infty} x m_F(x) < \infty$, then $F \in \mathcal{S}$ and $F^s \in \mathcal{S}$.*

Proof Clearly $\lim_{x\to\infty} m_F(x) = 0$, which implies $\lim_{x\to\infty}(M_F(x) - M_F(x - y)) = 0$ for all $y \in \mathbb{R}$. Hence $F \in \mathcal{L}$. Using (2.5.18), for $x > 2v$ we have

$$\begin{aligned}\int_0^x \frac{\overline{F}(x-y)}{\overline{F}(x)}\overline{F}(y)\,dy &= 2\int_0^{x/2} \frac{\overline{F}(x-y)}{\overline{F}(x)}\overline{F}(y)\,dy \\ &\le 2\int_0^v \frac{\overline{F}(x-y)}{\overline{F}(x)}\overline{F}(y)\,dy + 2\frac{\overline{F}(x/2)}{\overline{F}(x)}\int_v^\infty \overline{F}(y)\,dy\,. \qquad (2.5.26)\end{aligned}$$

We next show that

$$\limsup_{x\to\infty} \frac{\overline{F}(x/2)}{\overline{F}(x)} < \infty \qquad (2.5.27)$$

because then, by (2.5.26) and $F \in \mathcal{L}$,

$$\lim_{x\to\infty} \int_0^x \frac{\overline{F}(x-y)}{\overline{F}(x)}\overline{F}(y)\,dy \le 2\mu\,.$$

Since the reverse inequality is always satisfied, we have $F \in \mathcal{S}^*$. This yields $F \in \mathcal{S}$ and $F^s \in \mathcal{S}$ by Theorem 2.5.6. To prove (2.5.27) note that by the assumption of the corollary, there exist c and x_0 such that $x m_F(x) \le c$ for $x \ge x_0$ and hence

$$\begin{aligned}\limsup_{x\to\infty}(M_F(x) - M_F(x/2)) &= \limsup_{x\to\infty} \int_{x/2}^x m(y)\,dy \\ &\le c\limsup_{x\to\infty} \int_{x/2}^x \frac{dy}{y} = c\log 2 < \infty\,.\end{aligned}$$ □

In the case that $\limsup_{x\to\infty} x m_F(x) = \infty$, one can use the following criterion for $F \in \mathcal{S}^*$.

Theorem 2.5.7 *Assume that the hazard rate function $m_F(x)$ of F exists and is ultimately decreasing to 0. If $\int_0^\infty \exp(x m_F(x))\overline{F}(x)\,dx < \infty$ then $F \in \mathcal{S}^*$.*

Proof Since $\exp(x m_F(x)) \ge 1$, the integrability condition implies that $\mu < \infty$. Suppose $m_F(x)$ is decreasing on $[v, \infty)$ for some $v \ge 0$ and define

$$m'(x) = \begin{cases} m_F(v) & \text{if } x \in [0, v), \\ m_F(x) & \text{if } x \in [v, \infty). \end{cases}$$

Let $\overline{F}'(x) = \exp(-\int_0^x m'(t)\,dt)$. It is straightforward to check that $c_- \le \overline{F}(x)/\overline{F'}(x) \le c_+$ for all $x \ge 0$ and for some $c_-, c_+ > 0$. Furthermore, it is not difficult to show that a distribution function with hazard rate function tending to 0 belongs to $\mathcal{L}$. Consequently, Lemma 2.5.5 implies that $F \in \mathcal{S}^*$ if and only if $F' \in \mathcal{S}^*$. Moreover, the function $\exp(xm_F(x))\overline{F}(x)$ is integrable if and only if the function $\exp(xm'(x))\overline{F'}(x)$ is integrable as well. Thus we can assume without loss of generality that $m_F(x)$ is decreasing on $[0,\infty)$. Since

$$\int_0^x \frac{\overline{F}(x-y)}{\overline{F}(x)}\overline{F}(y)\,dy = 2\int_0^{x/2} \exp(M_F(x) - M_F(x-y) - M_F(y))\,dy$$

for all $x \ge 0$, it suffices to show that

$$\lim_{x\to\infty}\int_0^{x/2} \exp(M_F(x) - M_F(x-y) - M_F(y))\,dy = \mu\,. \tag{2.5.28}$$

The monotonicity of $m_F(x)$ implies that

$$1 \le \exp(ym_F(x)) \le \exp(M_F(x) - M_F(x-y)) \le \exp(ym_F(x/2))$$

for $0 \le y \le x/2$. This gives

$$\begin{aligned}\int_0^{x/2}\overline{F}(y)\,dy &\le \int_0^{x/2} \exp(M_F(x) - M_F(x-y) - M_F(y))\,dy \\ &\le \int_0^{x/2} \exp(ym_F(x/2) - M_F(y))\,dy\,,\end{aligned}$$

where the lower bound and the upper bound tend to μ as $x \to \infty$. For the upper bound, note that $\exp(ym_F(x/2) - M_F(y)) \le \exp(ym_F(y) - M_F(y))$ for $0 \le y \le x/2$, and that $m_F(x/2) \to 0$ as $x \to \infty$. Now apply the dominated convergence theorem to prove (2.5.28). □

Examples 1. For the Weibull distribution $F = \mathrm{W}(r,c)$ with $0 < r < 1, c > 0$ we have $\overline{F}(x) = \exp(-cx^r)$ and $m_F(x) = crx^{r-1}$. Hence $\lim_{x\to\infty} xm_F(x) = \infty$ and Corollary 2.5.1 cannot be applied. But, the function $\exp(xm_F(x))\overline{F}(x) = \exp(c(r-1)x^r)$ is integrable and so $F = \mathrm{W}(r,c) \in \mathcal{S}^*$ by Theorem 2.5.7.

2. Consider the standard lognormal distribution $F = \mathrm{LN}(0,1)$. Let $\Phi(x)$ be the standard normal distribution function with density denoted by $\phi(x)$. Then, F has the tail and hazard rate functions

$$\overline{F}(x) = 1 - \Phi(\log x)\,, \qquad m_F(x) = \frac{\phi(\log x)}{x(1-\Phi(\log x))}\,.$$

Furthermore, $\phi(x) \sim x(1-\Phi(x))$ as $x \to \infty$. This follows from the fact that

$$\frac{e^{-x^2/2}}{(2\pi)^{1/2}}\frac{1}{x} = \frac{1}{(2\pi)^{1/2}}\int_x^\infty e^{-y^2/2}\Big(1 + \frac{1}{y^2}\Big)\,dy$$

$$
\begin{aligned}
> \; & 1-\Phi(x) > \frac{1}{(2\pi)^{1/2}}\int_x^\infty \mathrm{e}^{-y^2/2}\Big(1-\frac{3}{y^4}\Big)\,\mathrm{d}y \\
= \; & \frac{\mathrm{e}^{-x^2/2}}{(2\pi)^{1/2}}\Big(\frac{1}{x}-\frac{1}{x^3}\Big).
\end{aligned}
$$

Thus, we have $\mathrm{e}^{xm_F(x)}\overline{F}(x) \sim x(1-\Phi(\log x))$ as $x \to \infty$. For $x \to \infty$, the function

$$
x(1-\Phi(\log x)) \sim \frac{x\phi(\log x)}{\log x}
$$

is integrable, because $\phi(\log x) = 2\pi^{-1/2}x^{-(\log x)/2}$ and $\int_1^\infty x^{1-(\log x)/2}\,\mathrm{d}x < \infty$. Hence the standard lognormal distribution LN(0, 1) belongs to $\mathcal{S}^*$ and therefore F and F^{s} are subexponential. The case of a general lognormal distribution can be proved analogously.

To show that the integrated tail distribution of Pareto-type distributions is subexponential, we need the following result, known as *Karamata's theorem.* We state this theorem without proof, for which we refer to Feller (1971).

Theorem 2.5.8 *If $L_1(x)$ is a slowly varying function and locally bounded in $[x_0, \infty)$ for some $x_0 > 0$, then for $\alpha > 1$*

$$
\int_x^\infty y^{-\alpha}L_1(y)\,\mathrm{d}y = x^{-\alpha+1}L_2(x)\,, \tag{2.5.29}
$$

where $L_2(x)$ is also a slowly varying function of x at ∞ and moreover $\lim_{x\to\infty} L_1(x)/L_2(x) = \alpha - 1$. If $L_1(y)/y$ is integrable, then the result also holds for $\alpha = 1$.

As proved in Section 2.5.2, every Pareto-type distribution F with exponent $\alpha > 1$ is subexponential. We now get that the corresponding integrated tail distribution F^{s} is also subexponential, because Theorem 2.5.8 implies that $\overline{F}^{\mathrm{s}}(x) = x^{-\alpha+1}L_2(x)$ is Pareto-type too. This yields that many distributions, like Pareto and loggamma distributions as well as Pareto mixtures of exponentials studied in the next section, have the desired property that $F \in \mathcal{S}$ and $F^{\mathrm{s}} \in \mathcal{S}$.

2.5.4 Pareto Mixtures of Exponentials

Heavy-tailed distributions like the lognormal, Pareto or Weibull distributions lack tractable formulae for their Laplace–Stieltjes transforms. We now discuss a class of subexponential distributions F_α with tail behaviour $\overline{F}_\alpha(x) \sim cx^{-\alpha}$ as $x \to \infty$, mean 1 and an explicitly given Laplace–Stieltjes transform for $\alpha = 2, 3, \ldots$. Such distributions can be useful for numerical experiments. Clearly, to have the mean equal to 1, we must assume that $\alpha > 1$. For each $\alpha > 1$, let F_α be the mixture of the family of exponential distributions

$\{\mathrm{Exp}(\theta^{-1}), \theta > 0\}$ with respect to the mixing distribution $\mathrm{Par}(\alpha, (\alpha-1)/\alpha)$. Explicitly, F_α has the density function

$$f_\alpha(x) = \int_{(\alpha-1)/\alpha}^{\infty} \theta^{-1} \mathrm{e}^{-\theta^{-1}x} \alpha \left(\frac{\alpha-1}{\alpha}\right)^{\alpha} \theta^{-(\alpha+1)} \, \mathrm{d}\theta \,. \tag{2.5.30}$$

The distribution with density $f_\alpha(x)$ is called a *Pareto mixture of exponentials* and is denoted by $\mathrm{PME}(\alpha)$. Basic properties of the distribution $\mathrm{PME}(\alpha)$ are studied in the following theorem.

Theorem 2.5.9 *Let F_α be the Pareto mixture* $\mathrm{PME}(\alpha)$. *Then*
(a) *if $\alpha > n$ the n-th moment $\mu_\alpha^{(n)}$ of F_α is*

$$\mu_\alpha^{(n)} = \frac{n!}{\alpha - n} \alpha \left(\frac{\alpha-1}{\alpha}\right)^n .$$

(b) *the Laplace–Stieltjes transform $\hat{l}_\alpha(s)$ of F_α is*

$$\hat{l}_\alpha(s) = \int_0^\infty \mathrm{e}^{-sx} f_\alpha(x) \, \mathrm{d}x = \alpha \left(\frac{\alpha-1}{\alpha}\right)^{\alpha} \int_0^{\alpha/(\alpha-1)} \frac{x^\alpha}{s+x} \, \mathrm{d}x \,.$$

Proof (a) By inserting (2.5.30) into (2.1.1) we have

$$\begin{aligned}
\mu_\alpha^{(n)} &= \int_0^\infty x^n f_\alpha(x) \, \mathrm{d}x \\
&= \int_0^\infty x^n \Big(\int_{(\alpha-1)/\alpha}^\infty \theta^{-1} \mathrm{e}^{-\theta^{-1}x} \alpha \left(\frac{\alpha-1}{\alpha}\right)^{\alpha} \theta^{-(\alpha+1)} \, \mathrm{d}\theta \Big) \, \mathrm{d}x \\
&= \int_{(\alpha-1)/\alpha}^\infty \Big(\int_0^\infty x^n \theta^{-1} \mathrm{e}^{-\theta^{-1}x} \, \mathrm{d}x \Big) \alpha \left(\frac{\alpha-1}{\alpha}\right)^{\alpha} \theta^{-(\alpha+1)} \, \mathrm{d}\theta \\
&= \frac{n!}{\alpha-n} \alpha \left(\frac{\alpha-1}{\alpha}\right)^n .
\end{aligned}$$

(b) Analogously, we have

$$\begin{aligned}
\hat{l}_\alpha(s) &= \int_0^\infty \mathrm{e}^{-sx} \Big(\int_{(\alpha-1)/\alpha}^\infty \theta^{-1} \mathrm{e}^{-\theta^{-1}x} \alpha \left(\frac{\alpha-1}{\alpha}\right)^{\alpha} \theta^{-(\alpha+1)} \, \mathrm{d}\theta \Big) \, \mathrm{d}x \\
&= \alpha \left(\frac{\alpha-1}{\alpha}\right)^{\alpha} \int_{(\alpha-1)/\alpha}^\infty \frac{\theta^{-2}}{\theta^{-1}+s} \frac{\mathrm{d}\theta}{\theta^\alpha} \\
&= \alpha \left(\frac{\alpha-1}{\alpha}\right)^{\alpha} \int_0^{\alpha/(\alpha-1)} \frac{y^\alpha}{s+y} \, \mathrm{d}y \,,
\end{aligned}$$

where in the last equation we used the substitution $\theta^{-1} = y$. □

Corollary 2.5.2 *For* $n = 2, 3, \ldots$

$$\begin{aligned}\hat{l}_n(s) &= \sum_{i=1}^{n}(-1)^{n-i}\frac{n}{i}\left(\frac{n-1}{n}\right)^{n-i} s^{n-i} \\ &\quad +(-1)^n n\left(\frac{n-1}{n}\right)^n s^n \log\Big(1+\frac{n}{(n-1)s}\Big).\end{aligned}$$

Proof By inspection we can verify that

$$\frac{x^n}{s+x} = \sum_{i=1}^{n}(-1)^{n-i}x^{i-1}s^{n-i} + (-1)^n\frac{s^n}{s+x}$$

for all $n = 2, 3, \ldots$. Furthermore,

$$\int_0^{n/(n-1)} x^{i-1}\,\mathrm{d}x = \frac{1}{i}\left(\frac{n}{n-1}\right)^i$$

and

$$\int_0^{n/(n-1)} \frac{1}{s+x}\,\mathrm{d}x = \log\Big(1+\frac{n}{(n-1)s}\Big).$$ □

We now prove that for each $\alpha > 1$ the Pareto mixture F_α is subexponential. It turns out that F_α is Pareto-type and, consequently, we get that $F_\alpha \in \mathcal{S}$ by Theorem 2.5.5. The following auxiliary result is useful.

Lemma 2.5.7 *For each* $\alpha > 1$, *the tail function of* F_α *has the form*

$$\overline{F}_\alpha(x) = \alpha\left(\frac{\alpha-1}{\alpha}\right)^\alpha x^{-\alpha}\int_0^{\alpha x/(\alpha-1)} v^{\alpha-1}\mathrm{e}^{-v}\,\mathrm{d}v\,.$$

Proof By (2.5.30) we have

$$\begin{aligned}\overline{F}_\alpha(x) &= \int_x^\infty\Big(\int_{(\alpha-1)/\alpha}^\infty \theta^{-1}\mathrm{e}^{-\theta^{-1}y}\alpha\left(\frac{\alpha-1}{\alpha}\right)^\alpha\theta^{-(\alpha+1)}\,\mathrm{d}\theta\Big)\,\mathrm{d}y \\ &= \int_{(\alpha-1)/\alpha}^\infty\Big(\int_x^\infty \theta^{-1}\mathrm{e}^{-\theta^{-1}y}\,\mathrm{d}y\Big)\alpha\left(\frac{\alpha-1}{\alpha}\right)^\alpha\theta^{-(\alpha+1)}\,\mathrm{d}\theta \\ &= \int_{(\alpha-1)/\alpha}^\infty \alpha\left(\frac{\alpha-1}{\alpha}\right)^\alpha\theta^{-(\alpha+1)}\mathrm{e}^{-\theta^{-1}x}\,\mathrm{d}\theta \\ &= \alpha\left(\frac{\alpha-1}{\alpha}\right)^\alpha x^{-\alpha}\int_0^{\alpha x/(\alpha-1)} v^{\alpha-1}\mathrm{e}^{-v}\,\mathrm{d}v\,,\end{aligned}$$

where we used the substitution $v = x\theta^{-1}$. □

Theorem 2.5.10 *Let $\alpha > 1$. Then F_α and the integrated tail distribution $F^{\rm s}_\alpha$ are Pareto-type distributions and, consequently, F_α and $F^{\rm s}_\alpha$ are subexponential.*

Proof Note that

$$L(x) = \alpha \left(\frac{\alpha-1}{\alpha}\right)^{\alpha} \int_0^{\alpha x/(\alpha-1)} v^{\alpha-1} {\rm e}^{-v}\, {\rm d}v$$

is a slowly varying function of x at ∞, because it is bounded and non-decreasing. Hence by Lemma 2.5.7, F_α is Pareto-type and, by Theorem 2.5.5, F_α is subexponential. $F^{\rm s} \in \mathcal{S}$ is obtained in the same way. □

Remark Note that $\lim_{x\to\infty} \overline{F}_\alpha(x)/(cx^{-\alpha}) = 1$ where $c = \Gamma(\alpha+1)\left(\frac{\alpha-1}{\alpha}\right)^{\alpha}$. Indeed, from Lemma 2.5.7 we get

$$\lim_{x\to\infty} \frac{\overline{F}_\alpha(x)}{x^{-\alpha}} = \alpha\left(\frac{\alpha-1}{\alpha}\right)^{\alpha} \int_0^\infty v^{\alpha-1}{\rm e}^{-v}\,{\rm d}v = \Gamma(\alpha+1)\left(\frac{\alpha-1}{\alpha}\right)^{\alpha},$$

because $\Gamma(\alpha) = \int_0^\infty v^{\alpha-1}{\rm e}^{-v}\,{\rm d}v$ and $\Gamma(\alpha+1) = \alpha\Gamma(\alpha)$.

Bibliographical Notes. The class of subexponential distributions on $\mathbb{R}_+$ has been introduced by Chistyakov (1964) and independently by Chover, Ney and Wainer (1973). Theorem 2.5.5 is from Feller (1971), Section VIII.8. References concerning the evidence of heavy-tailed distributions in practical insurance are, for example: Andersson (1971), Benckert and Jung (1974), Benckert and Sternberg (1958), Keller and Klüppelberg (1991), Mandelbrot (1964), Mikosch (1997), Resnick (1997), Shpilberg (1977). Basic properties of the class $\mathcal{S}$ are reviewed in Athreya and Ney (1972); see also Cline (1987), Cline and Resnick (1988) and the survey paper by Beirlant and Teugels (1992). Further criteria for subexponentiality can be found in Teugels (1975) and Pitman (1980). The properties of $\mathcal{S}^*$ stated in Section 2.5.3 are due to Klüppelberg (1988). The class of Pareto mixtures of exponentials was introduced and studied in Abate, Choudhury and Whitt (1994). Subexponential distributions on the whole real line have been considered in Grübel (1984); see also Grübel (1983).

2.6 DETECTION OF HEAVY-TAILED DISTRIBUTIONS

2.6.1 Large Claims

It goes without saying that the detection of dangerous claim size distributions is one of the main worries of the practicing actuary. Most practitioners have some personal concept of what they would call a *large claim.* However, a mathematically sound formulation is not always obvious. We need to introduce a bit of notation. Denote by $\{U_i, 1 \leq i \leq n\}$ the successive claims in a portfolio. The total claim amount is then $X_n = U_1 + U_2 + \ldots + U_n$. Recall that by $U_{(1)}, \ldots, U_{(n)}$ we denote the sequence of ordered claims with

$$\min_{1\leq i\leq n} U_i = U_{(1)} \leq U_{(2)} \leq \ldots \leq U_{(n)} = \max_{1\leq i\leq n} U_i \,.$$

Often, a claim is called large when the total claim amount is predominantly determined by it. This rather vague formulation can be interpreted in a variety of ways. Let us give a number of possible examples.

- One sometimes hears that a claim within a portfolio is large if a value of that size is only experienced every so many years. It needs no explanation that this kind of description can hardly be forged into a workable definition.
- Another interpretation could be that the ratio of $U_{(n)}$ and X_n is too large. This could be phrased as the condition that $U_{(n)}/X_n \stackrel{\mathrm{d}}{=} Z$, where the distribution of Z has most mass near one. If there are no excessive claims then we expect $U_{(n)}$ to play an increasingly lesser role in the total X_n.
- More generally, a claim is called large if it consumes more than a fair portion p of the total claim amount. This means that we call $U_{(m)}$ large if $m \geq \min\left\{k : U_{(k)} > pX_n\right\}$.
- When the practitioner tries to estimate the mean and/or variance of the claim size distribution, he will use resampling techniques to obtain a reliable estimate. However it happens that the successive sample values are not averaging out to a limiting value. One possible and theoretically understandable reason is that the mean and/or the variance of the claim size distribution do not exist because there is too much mass in the tail. A possible parametrized distribution causing this type of phenomenon is any Pareto-type distribution with small exponent α.

Let us now turn to a number of definitions of large claims that are mathematically sound. However, due to a variety of reasons, such definitions are hard to verify statistically.

- The total claim amount is large because the largest claim is so. Mathematically this can be interpreted as

$$\mathbb{P}\left(X_n > x\right) \sim \mathbb{P}\left(U_{(n)} > x\right), \qquad x \to \infty \,. \tag{2.6.1}$$

As has been explained in Section 2.5.2, this concept leads naturally to the notion of subexponentiality. The class $\mathcal{S}$ is known to contain a wide set of possible candidates. However, the statistical verification of the statement $F = F_U \in \mathcal{S}$ is far from trivial. Let us try to explain why. We learned from Theorem 2.5.3 that $F \in \mathcal{S}$ if and only if (2.6.1) holds for $n = 2$. The practitioner can use the sample values $\{U_i, 1 \leq i \leq n\}$ to check whether or not the limit of the expression $\left(1 - F^{*2}(x)\right) / \left(1 - (F(x))^2\right)$ tends to the numerical value 1 as $x \to \infty$. So, we need to replace F by its empirical analogue, the *empirical distribution* F_n, defined by

$$F_n(x) = n^{-1} \max\left\{i : U_{(i)} \leq x\right\} \tag{2.6.2}$$

for all $x \in \mathbb{R}$. So, for each x, the value $nF_n(x)$ equals the number of sample values on or to the left of the point x. Put in a different fashion we have

$$\{F_n(x) = kn^{-1}\} = \{U_{(k)} \leq x < U_{(k+1)}\}\,. \tag{2.6.3}$$

If we replace F by F_n in the definition of subexponentiality then we still need to take x very large. The only way is to replace the variable x by a large order statistic, like the maximum. Hence we need to verify whether or not $\left(1 - F_n^{*2}(U_{(n)})\right) / \left(1 - (F_n(U_{(n)}))^2\right)$ is in any way close to 1. Without any further information on F, this is a hard problem for which no satisfactory solution exists. As a consequence, most actuaries that want to model claim sizes in a specific portfolio will select their favourite and duly parametrized member from $\mathcal{S}$.

- In view of Theorem 2.5.1, our definition of heavy-tailed distributions can be coined in the requirement that $\alpha_F = 0$, i.e.

$$\limsup_{x\to\infty} \frac{-\log(1 - F(x))}{x} = 0\,.$$

If we want to verify this hypothesis, we consider its empirical analogue. Assume that we take an order statistic with a large index, $n-k$, say, where n is large and k is such that $k/n \to 0$. Then we need to verify whether or not

$$\limsup_{n\to\infty} \frac{-\log\left(1 - F_n(U_{(n-k)})\right)}{U_{(n-k)}} = \limsup_{n\to\infty} \frac{\log \frac{n}{k}}{U_{(n-k)}} = 0\,. \tag{2.6.4}$$

However, this condition is statistically unverifiable because of the limes superior in (2.6.4).

- A sufficient condition for a heavy-tailed distribution can be given in terms of the mean residual hazard function $\mu_F(x) = \mu_{F_x}$ defined in (2.4.9), i.e.

$$\mu_F(x) = \mathbb{E}\left(U - x \mid U > x\right) = \int_x^\infty \frac{1 - F(y)}{1 - F(x)}\,\mathrm{d}y\,. \tag{2.6.5}$$

Indeed, it is not difficult to show that if $\mu_F(x) \to \infty$ as $x \to \infty$, then $\alpha_F = 0$. To verify statistically whether the distribution F is heavy-tailed or not we suggest looking at the empirical analogue to the mean residual hazard function, i.e. for a similar choice of k and n as above,

$$\mu_n\left(U_{(n-k)}\right) = \int_{U_{(n-k)}}^\infty \frac{1 - F_n(y)}{1 - F_n(U_{(n-k)})}\,dy\,.$$

Use equation (2.6.3) to rewrite this in the form

$$\begin{aligned}\mu_n\left(U_{(n-k)}\right) &= \frac{n}{k}\sum_{i=n-k}^{n-1}\int_{U_{(i)}}^{U_{(i+1)}}(1-F_n(y))\,dy \\ &= \frac{1}{k}\sum_{i=n-k}^{n-1}(n-i)\left(U_{(i+1)}-U_{(i)}\right).\end{aligned}$$

Rewriting the last sum we arrive at the *empirical mean residual hazard function*

$$\mu_n(U_{(n-k)}) = \frac{1}{k}\Big(\sum_{j=n-k+1}^{n} U_{(j)} - kU_{(n-k)}\Big) = \frac{1}{k}\sum_{j=n-k+1}^{n}\left(U_{(j)} - U_{(n-k)}\right).$$

A possible interpretation of the latter quantity is the average overshoot of the $k-1$ largest overshoots over the level $U_{(n-k)}$. Again it is statistically not obvious how to verify a condition that essentially says that the quantity $\mu_n(U_{(n-k)})$ has to go to ∞.

We deduce from the above explanations that it seems an ill-posed problem to verify statistically whether or not a distribution is heavy-tailed. Assume, however, that we are willing to sacrifice the rigidity of a formal definition. What we want to check is whether or not a claim size distribution has to have a heavier tail than some standard reference distribution. In statistics one usually compares distributions to a normal reference. Looking at our definition of heavy-tailed distribution it seems far more realistic to compare a distribution with an exponential reference. A distribution F with a lighter tail than an exponential, i.e. an F that satisfies the inequality $1 - F(x) \le ce^{-ax}$ for all $x \ge 0$ and for some constants $a, c > 0$, will automatically have a strictly positive value of α_F. A distribution that satisfies the opposite inequality for all $a > 0$ will have $\alpha_F = 0$.

We now give two methods that allow us to compare a sample with that from a standard distribution, in particular the exponential distribution. One method is the Q-Q plot or the *Quantile-Quantile plot*; the other refers to the mean residual hazard.

2.6.2 Quantile Plots

The general philosophy of quantile plots relies on the observation that linearity in a graph cannot only be easily checked by eye but can be quantified by a correlation coefficient. To explain the essentials of the method, we start by considering the (standard) exponential distribution G with the tail of the latter given by $\overline{G}(x) = \exp(-x)$ for $x \ge 0$. We want to know whether a sampled claim size distribution F is of the same basic form as G, save perhaps for a scale factor. More specifically, we want to know whether $\overline{F}(x) = \exp(-\lambda x)$ for some $\lambda > 0$ is an acceptable model for F. The answer has to rely on the data $U_1 = u_1, U_2 = u_2, \ldots, U_n = u_n$ which we have at our disposal. The parameter λ just adds some flexibility to our procedure.

For an increasing and right-continuous function $F(x)$, we define the *generalized inverse function* $F^{-1}(y)$ by

$$F^{-1}(y) = \inf\{x : \; F(x) \geq y\} . \tag{2.6.6}$$

If F is a distribution function, then function Q_F defined by $Q_F(y) = F^{-1}(y)$ is called the *quantile function* of F. Simultaneously, we construct the empirical version of the quantile function by considering the generalized inverse of the empirical distribution as defined in (2.6.2). More specifically, $Q_n(y) = Q_{F_n}(y)$, so that for the ordered sample $U_{(1)} \leq U_{(2)} \leq \ldots \leq U_{(n)}$ we have

$$\{Q_n(y) = U_{(k)}\} = \{(k-1)n^{-1} < y \leq kn^{-1}\} .$$

For the standard exponential distribution G the quantile function has a simple form $Q_G(y) = -\log(1-y)$ if $0 < y < 1$. If we want to compare the sample values with those of a standard exponential then it suffices to compare the two quantile functions. To do exactly that, we plot the two functions Q_G and Q_n in an orthogonal coordinate system. If our data are coming from a (not necessarily standard) exponential distribution, then we expect the resulting graph to show a straight line pattern since the quantile function of the exponential distribution with parameter λ is given by $Q_F(y) = -\lambda^{-1}\log(1-y)$. The slope of the line would be given by λ^{-1}, offering us a possible estimate for this unknown parameter. If the data are coming from a distribution with a heavier tail than the exponential, then we expect the graph to increase faster than a straight line; if the tail is less heavy, then the increase will be slower. The resulting *quantile plot* immediately tells us whether or not our data are coming from a distribution which is close to an exponential.

From the definition of the empirical quantile function $Q_n(y)$ it follows that the quantile plot will be a nondecreasing left-continuous step function; its only points of increase are situated at the values $\{k/n, 1 \leq k \leq n\}$ which form a lattice on the positive horizontal axis. It therefore would suffice to just plot the graph at the points $y \in \{k/n, 1 \leq k \leq n\}$. However, there is a slight problem at the right extreme of the picture since for $k = n$ we have $Q_G(k/n) = \infty$. For this reason one often applies a *continuity correction* by graphing the scatter plot at the points $\{k/(n+1), 1 \leq k \leq n\}$. We will stick to this practice in our examples at the end of this section.

Apart from the visual conclusion obtained from the Q-Q plot we can also derive quantitative information from the graph. If the exponential distribution seems acceptable then we can fit a straight line through the scatter plot by using a traditional least-squares algorithm. The slope λ^{-1} of the straight line should be chosen to minimize the sum of squares $\sum_{k=1}^{n}\left(U_{(k)}+\lambda^{-1}\log\left(1-k/(n+1)\right)\right)^2$. This yields the classical formula for the least-squares statistic $\hat{\lambda}^{-1}$

$$\hat{\lambda}^{-1}=\sum_{k=1}^{n}U_{(k)}Q_G(k/(n+1))\Big/\sum_{k=1}^{n}\Big(Q_G(k/(n+1))\Big)^2 .$$

The fit itself can then be quantified by looking at the practical value of the empirical correlation coefficient $r(u_1,\ldots,u_n)$ based on the experimental data $u_1,u_2,\ldots,u_n$. Note that $r(u_1,\ldots,u_n)$ is given by the formula

$$r(u_1,\ldots,u_n)=\frac{\sum_{k=1}^{n}(u_{(k)}-\bar{u})(Q_G(\frac{k}{n+1})-\bar{Q}_G)}{\sqrt{\sum_{k=1}^{n}(Q_G(\frac{k}{n+1})-\bar{Q}_G)^2\sum_{k=1}^{n}(u_{(k)}-\bar{u})^2}},$$

where $\bar{u}=n^{-1}\sum_{k=1}^{n}u_k=n^{-1}\sum_{k=1}^{n}u_{(k)}$ and

$$\bar{Q}_G=\frac{1}{n}\sum_{k=1}^{n}Q_G\Big(\frac{k}{n+1}\Big)=-\frac{1}{n}\sum_{k=1}^{n}\log\Big(1-\frac{k}{n+1}\Big).$$

As is known, $|r(u_1,\ldots,u_n)|\le 1$, while $r(u_1,\ldots,u_n)=\pm 1$ if and only if the points $(1,u_1),\ldots,(n,u_n)$ lie on a straight line.

In actuarial practice it often happens that data are truncated on the left, on the right or even on both sides. For example a reinsurance company will often not know the values of the claims that have been covered by the first line insurance under a retention. Suppose that the claim U is exponentially distributed with parameter λ. Then, for $a>0$, the *truncated exponential distribution* $F_{[0,a]}$ is of the form

$$\overline{F}_{[0,a]}(x)=\mathbb{P}(U>x\mid U>a)=\frac{\mathbb{P}(U>x)}{\mathbb{P}(U>a)}=\mathrm{e}^{-\lambda(x-a)},\qquad x>a.$$

The corresponding quantile function is given by

$$Q_{[0,a]}(y)=a-\frac{1}{\lambda}\log(1-y),\qquad 0<y<1.$$

If data come from a truncated exponential distribution, then the intercept of the Q-Q plot at the origin $y=0$ will give an estimate of the parameter a. If data are not well represented by an exponential distribution, then of course we can suggest other candidates. Among the most popular candidates in an actuarial context are the normal and lognormal distributions, the Pareto distribution and, to a lesser extent, the Weibull distribution. We shortly deal with all four cases separately.

- The *normal quantile plot.* Recall that we denote the standard normal distribution function by

$$\Phi(x) = \frac{1}{\sqrt{2\pi}} \int_{-\infty}^{x} e^{-y^2/2}\, dy\,.$$

Let $\Phi^{-1}(y)$ be the corresponding quantile function. A standard normal quantile plot will graph the points

$$\left\{ \left(\Phi^{-1}(k/(n+1)), U_{(k)}\right),\ 1 \le k \le n \right\}. \tag{2.6.7}$$

Note that the general normal distribution $N(\mu, \sigma^2)$ has the quantile function $Q(y) = \mu + \sigma\Phi^{-1}(y)$. We leave it to the reader to show this as an exercise. Thus, if there results a straight line pattern in the plot (2.6.7), then the slope of the line will provide an estimate for the parameter σ while the intercept at 0 will estimate the parameter μ.
- The *lognormal quantile plot.* This is easily defined since U will be lognormal distributed if and only if $\log U$ is normal distributed. Hence the scatter plot will be given by $\{(\Phi^{-1}(k/(n+1)), \log U_{(k)}),\ 1 \le k \le n\}$. The lognormal distribution frequently shows up when dealing with car claim data, as will be illustrated later.
- The *Pareto quantile plot* is an important actuarial tool. Recall that for a claim U with Pareto distribution $\text{Par}(\alpha, c)$ we have $\mathbb{P}(U \le x) = 1 - (x/c)^{-\alpha}$ for $x \ge c$, so that $\log Q(y) = \log c - \alpha^{-1} \log(1 - y)$, which resembles the truncated exponential distribution. The Pareto quantile plot is obtained by plotting the graph of the points $\{(-\log(1 - k/(n+1)), \log U_{(k)}),\ 1 \le k \le n\}$. If the data come from a Pareto distribution, then the above graph will have a linear shape with intersect $\log c$ and slope α^{-1}. Note that the Pareto quantile plot is also useful when we have a Pareto-type distribution, i.e. a distribution with tail of the form $\mathbb{P}(U > x) \sim x^{-\alpha} L(x)$, where L is slowly varying at ∞. The plot will then show a linear trend for the data points to the right of the plot. The Pareto distribution is popular among actuaries when modelling fire claim data or other data with very heavy tails.
- The *Weibull quantile plot.* Recall that in this case $\overline{F}(x) = \exp(-cx^r)$. The quantile function is obviously $Q(y) = \left(-c^{-1} \log(1 - y)\right)^{1/r}$ for $0 < y < 1$. If we take the logarithm of this expression once more, then we find that $\log Q(y) = -r^{-1} \log c + r^{-1} \log(-\log(1 - y))$, which automatically leads to the Weibull quantile plot $\{(\log(-\log(1 - k/(n+1))), \log U_{(k)}),\ 1 \le k \le n\}$. Under the Weibull model we expect a straight line behaviour where the slope estimates the parameter r^{-1}. Further the intercept estimates the quantity $-r^{-1} \log c$.

Figure 2.6.1 gives these four Q–Q plots for 227 industrial accident data collected in Belgium over the year 1992. The claim sizes have been given

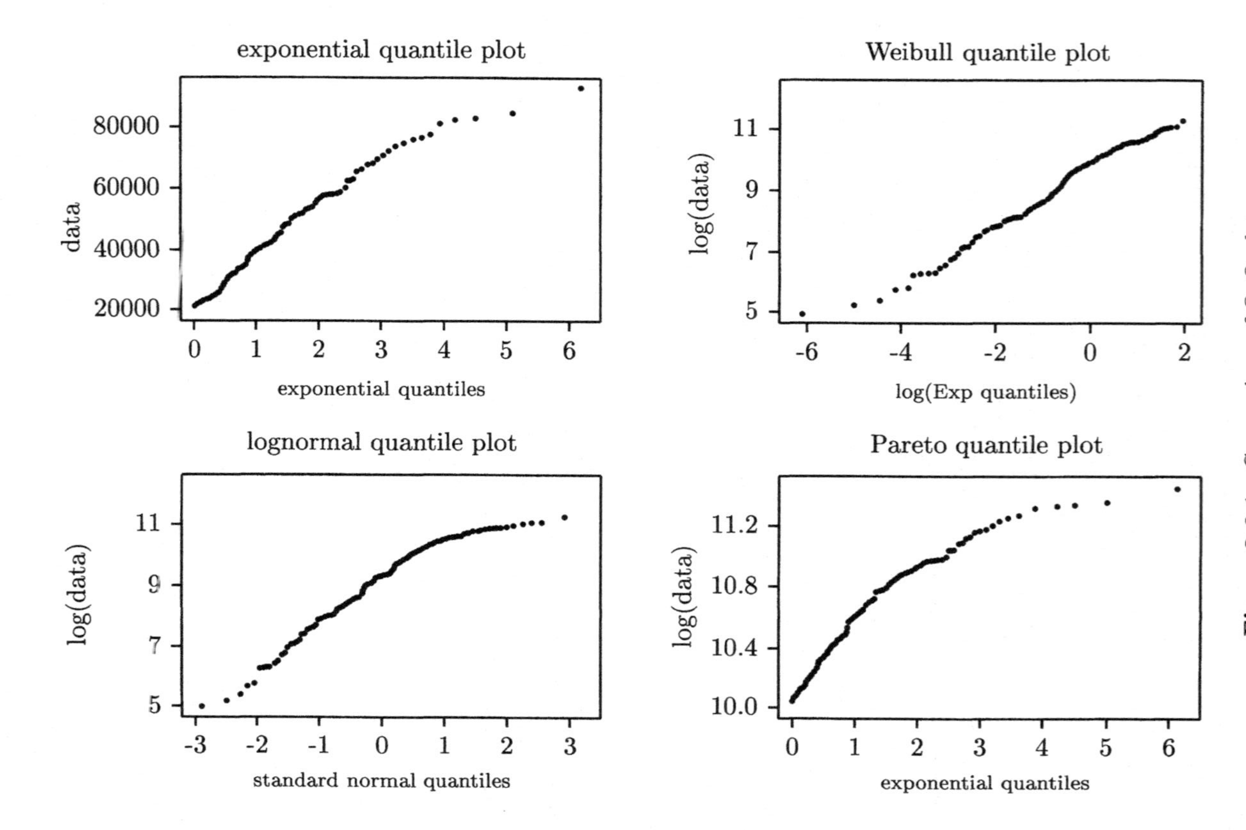

Figure 2.6.1 Comparison of Q–Q plots

in multiples of 1000 BEF with a prior deductible of 23.000 BEF. Because we use real data, there is no reason to believe that one of our four distributions gives a perfect fit over the entire range of the data. By comparing a slate of possible candidates, we will, however, get a better feeling for the overall structure of the portfolio. For example, the exponential distribution seems to do well for the smaller claims but not for the largest values in the portfolio. For the not very small claims, the Weibull distribution fits much better. Both the lognormal and the Pareto distributions seem to overestimate the importance of the larger claims in the portfolio since both show a concave bending away from a straight line fit.

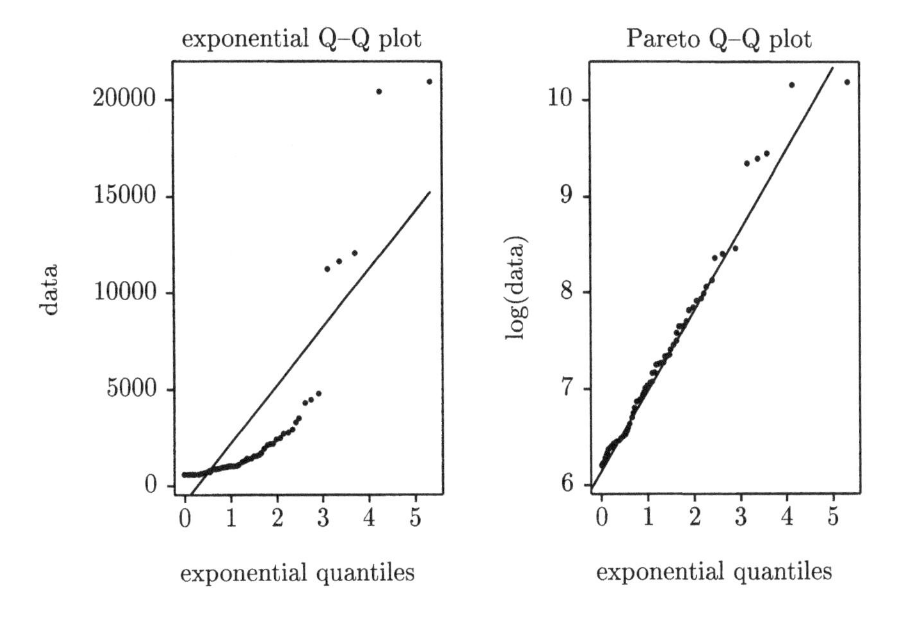

Figure 2.6.2 Q–Q plots

Figure 2.6.2 shows two plots of 105 Norwegian industrial fire claim data (in 1000 Kroner). They have been collected in a combined portfolio over the year 1971 but only the values above 500 Kronen have been reported. The straight line fits are obtained by a classical linear regression procedure. We clearly see that the Pareto distribution fits very well with a correlation coefficient of 0.9917, whereas the exponential distribution provides a very poor fit even when the correlation coefficient equals 0.8445. From the slope of the straight line in the Pareto-plot we can infer that the value of α should be about 1.25. It has often been reported that industrial fire claim data are well modelled by a Pareto distribution with an α-value close to 1. There is no need to stress again that taking insurance or reinsurance on industrial fire portfolios is a very delicate undertaking.

Figure 2.6.3 shows a lognormal Q–Q plot for motor insurance claim data. The data result from a random selection of 1250 out of a large portfolio that resulted from a combination of car claim portfolios of six leading car insurance companies for 1989. Out of the 1250, 14 values had to be discarded because of incompleteness of the data. The resulting plot generally shows that a lognormal distribution seems like a good choice. A straight line fit can be made by an obvious linear regression. Note that in the middle of the log(claim) values we see a rather long flat stretch. This value corresponds to an upper limit that is often applied when an older car is classified as *total loss*. One of the six companies applied a slightly higher value than the others; as a result there is a second smaller flat stretch to the right of the first. The straight line fit deteriorates at both ends of the picture. On the left there is an obvious truncation for the claim sizes. On the right, however, the values seem to increase faster than we expect under a lognormal assumption. This indicates that the upper part of the claim values might be better modelled by a distribution that has an even heavier tail than the lognormal distribution. For example, a Pareto distribution could be more appropriate. A closer look at the data shows that the upper right tail of the picture is crucially determined not only by the usual amounts for material damage but by legal and administrative settlements. The latter costs are constant for the major and central part of the claim portfolio. Leaving out the settlements costs reveals an even better fit by a lognormal distribution.

2.6.3 Mean Residual Hazard Function

Another global method to discover heavy-tailed distributions relies on the mean residual hazard function $\mu_F(x)$ considered in (2.6.5). It is often an instructive exercise to evaluate explicitly or asymptotically the form of the function $\mu_F(x)$. If we start again with the exponential distribution, then obviously $\mu_F(x) = \lambda^{-1}$. Conversely, if $\mu_F(x)$ is constant, then F is exponential. We leave it to the reader to show this as an exercise.

In order to see whether the claim distribution F is comparable to an exponential, we use the empirical analogue $\mu_n(U_{(n-k)})$ to $\mu_F(x)$ which has been introduced in Section 2.6.1. When the distribution F of U has a heavier tail than the exponential distribution, then the empirical mean residual hazard function $\{\mu_n(U_{(n-k)}),\ 0 < k \le n\}$ will consistently stay above the analogous function for the exponential. In particular, when the empirical mean residual hazard function tends to ∞ as $n \to \infty$, then this is the case of a heavy-tailed distribution as indicated in Section 2.6.1. For example, the mean residual hazard function of the Pareto distribution is typical. Then $\mu_{\mathsf{Par}(\alpha,c)}(x) = \int_x^\infty (c/y)^{\alpha+1}(c/x)^{-(\alpha+1)}\,dy = x/\alpha$, which increases linearly.

If the tail of the distribution F is lighter than that of any exponential,

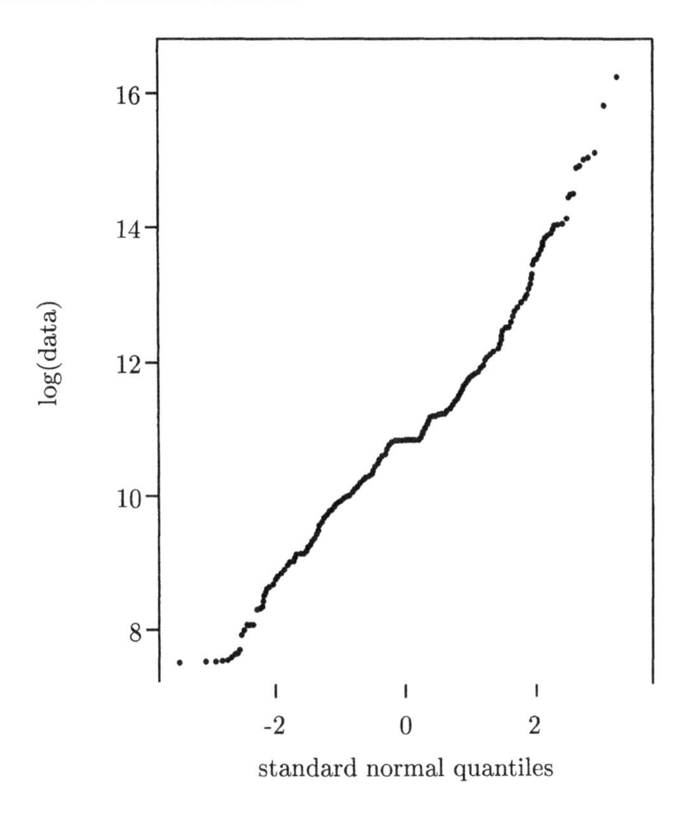

Figure 2.6.3 Lognormal Q–Q plot

then the mean residual hazard function $\mu_F(x)$ will typically decrease. For a concrete example let us take U to be uniformly distributed on (a, b). Then, for $a < x < b$,

$$\mu_{U(a,b)}(x) = \int_x^b \frac{(b-y)/(b-a)}{(b-x)/(b-a)}\,dy = \frac{(b-x)}{2}$$

is linearly decreasing over the interval $(0, b)$. When plotting the empirical mean residual hazard function we can hope to recognize the shape of one of the standard pictures. In Figure 2.6.4 a number of mean residual hazard functions are depicted in the same coordinates. The heavy-tailed distributions like the Pareto distribution, the lognormal distribution and the Weibull distribution with $r < 1$ show a clear upward trend. Light-tailed distributions such as the uniform distribution show a decreasing profile, while the exponential distribution and the Weibull distribution with $0 < r < 1$ are intermediate examples.

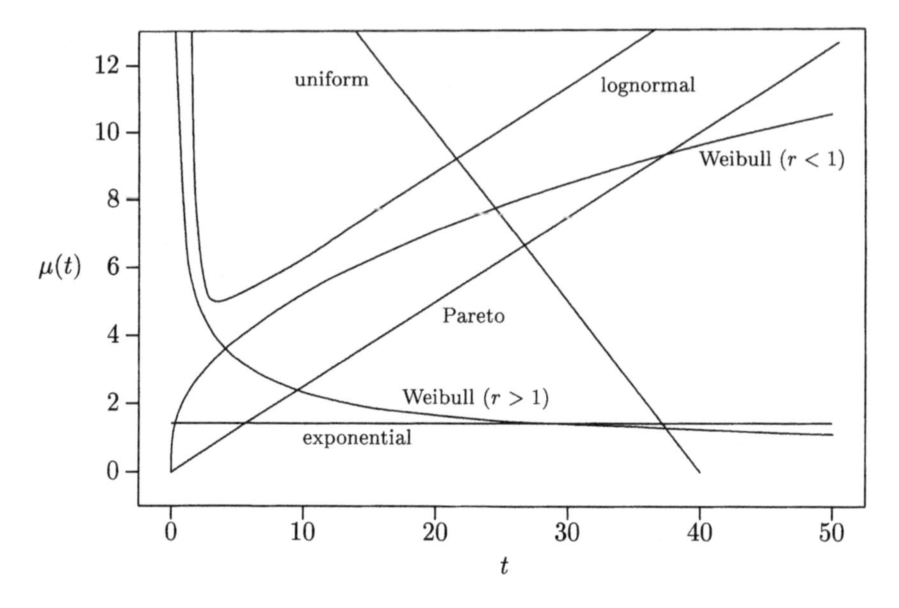

Figure 2.6.4 Mean residual hazard functions

2.6.4 Extreme Value Statistics

In this section we collect a few results concerning extreme value statistics. Extreme value theory is a vast subject in probability and statistics. As we only want to highlight the role of extremal claims in insurance mathematics, we will not prove a number of the following results. However, we provide the necessary intuitive background to make the results as natural as possible. Assume that $\{U_1, U_2, \ldots, U_n\}$ is a set of independent and identically distributed claims all with claim size distribution F. If we order the claim sizes, then the maximum $U_{(n)} = \max\{U_1, U_2, \ldots, U_n\}$ gives the largest claim in the sample. If the underlying distribution F is not concentrated on a bounded set, then the maximum $U_{(n)}$ will ultimately tend to ∞ as $n \to \infty$. Hence, we can hope to find norming and centring constants that provide an asymptotic approximation to the distribution of $U_{(n)}$. To be more precise we look for constants $a_n > 0$ and $b_n \in \mathbb{R}$ for which $a_n^{-1}(U_{(n)} - b_n) \stackrel{\mathrm{d}}{\to} Y$ where Y is assumed to be nondegenerate. The answer to this limiting problem is that the distribution of Y, up to some scaling and shifting parameter, belongs to a one-parameter family of distributions $G_\gamma = \mathrm{EV}(\gamma)$, the extreme value distributions introduced in Section 2.2.2. Recall that G_γ is given by $G_\gamma(x) = \mathbb{P}(Y \le x) = \exp(-(1+\gamma x)_+^{-1/\gamma})$ for all $x \in \mathbb{R}$. For $\gamma = 0$ we interpret the exponent as e^{-x}, which gives the Gumbel distribution $G_0(x) = \exp(-\mathrm{e}^{-x})$.

A necessary and sufficient condition on F to end up with a specific extremal distribution G_γ is that $F \in \mathcal{C}_\gamma$, a condition that is expressed in terms of the *inverse quantile function* $I_F : [1, \infty) \to \mathbb{R}_+$ of F. This function is defined by $I_F(x) = F^{-1}(1 - x^{-1})$, where F^{-1} denotes the generalized inverse function introduced in (2.6.6). Then $F \in \mathcal{C}_\gamma$ if and only if

$$\lim_{x\to\infty} \frac{I_F(xt) - I_F(x)}{c_F(x)} = \int_1^t w^{\gamma-1} \mathrm{d}w\,, \tag{2.6.8}$$

for some (ultimately) positive and measurable function $c_F(x)$ and all $t > 1$. The function $c_F(x)$ is called the *auxiliary function* of F. In the next theorem we show how the above condition comes into the picture. To simplify the proof we additionally assume that F is absolutely continuous.

Theorem 2.6.1 *Assume that $U_1, U_2, \ldots$ is a sequence of independent and identically distributed random variables with absolutely continuous distribution F satisfying* (2.6.8)*. Then $(c_F(n))^{-1}(U_{(n)} - I_F(n)) \stackrel{\mathrm{d}}{\to} Y$ as $n \to \infty$, where Y has distribution $G_\gamma = \mathrm{EV}(\gamma)$.*

Proof Let $h(x)$ be any real-valued, continuous and bounded function on $\mathbb{R}$. Since $\{U_{(n)} \le x\}$ is equivalent to $\bigcap_{1\le i\le n}\{U_i \le x\}$ we find that $\mathbf{P}(U_{(n)} \le x) = F^n(x)$ by independence. Introduce the suggested centring and the norming as well as the substitution $F(x) = 1 - (w/n)$ to find the expression

$$\begin{aligned} \mathbf{E}\, h(a_n^{-1}(U_{(n)} - b_n)) &= \int_{-\infty}^{\infty} h\Big(\frac{x - b_n}{a_n}\Big)\, \mathrm{d}F^n(x) \\ &= n \int_{-\infty}^{\infty} h\Big(\frac{x - b_n}{a_n}\Big) F^{n-1}(x)\, \mathrm{d}F(x) \\ &= \int_0^n h\Big(\frac{I_F(n/w) - b_n}{a_n}\Big)\Big(1 - \frac{w}{n}\Big)^{n-1} \mathrm{d}w\,. \end{aligned}$$

From condition (2.6.8) we see that we should take $b_n = I_F(n)$ and $a_n = c_F(n)$. The bounded convergence theorem implies that the limit of the last expression, as $n \to \infty$, exists and equals

$$\lim_{n\to\infty} \mathbf{E}\, h\Big(\frac{U_{(n)} - U_F(n)}{c_F(n)}\Big) = \int_0^\infty h\Big(\int_1^{w^{-1}} v^{\gamma-1}\, \mathrm{d}v\Big) \mathrm{e}^{-w}\, \mathrm{d}w.$$

By the Helly–Bray theorem (see (2.1.12)) the latter is equivalent to

$$\begin{aligned} &\lim_{n\to\infty} \mathbf{P}\Big(\frac{U_{(n)} - U_F(n)}{c_F(n)} \le x\Big) \\ &\quad = \int_0^\infty \mathbb{I}\Big(\int_1^{w^{-1}} v^{\gamma-1}\, \mathrm{d}v \le x\Big)\, \mathrm{e}^{-w}\, \mathrm{d}w = \mathbf{P}(Y \le x) \end{aligned}$$

as can be easily recovered. □

We have to stress the fact that the converse of Theorem 2.6.1 is also valid but its proof is outside the scope of this book.

For actuarial applications the most interesting case is that where $\gamma > 0$. Then, condition (2.6.8) takes the form

$$\lim_{x \to \infty} \frac{I_F(xt) - I_F(x)}{c_F(x)} = \frac{t^\gamma - 1}{\gamma} \tag{2.6.9}$$

which can be transformed into the equivalent condition $I_F(x) \sim x^\gamma L(x)$, where $L(x)$ is slowly varying at ∞. Hence, for $\gamma > 0$ we fall back on the class of Pareto-type distributions, introduced in Section 2.5.2. The fact that this condition is sufficient is easily proved by taking $c_F(x) = \gamma x^\gamma L(x)$.

Bibliographical Notes. For the limit theory of extremes and other order statistics references abound. We refer to Bingham, Goldie and Teugels (1987), De Haan (1970) and Resnick (1987). Most of the properties about regularly varying functions that have been referred to can be found in these books and their references. The approach in Section 2.6.1 is from Beirlant and Teugels (1996), streamlining previous fundamental results in extreme value theory by Fisher–Tippett and Gnedenko. Further statistical techniques for analysing heavy-tailed distributions can be found, for instance, in Adler, Feldman and Taqqu (1997), Embrechts, Klüppelberg and Mikosch (1997), and Reiss and Thomas (1997). Empirical processes and their use in goodness-of-fit statistics are dealt with in d'Agostino and Stephens (1986) and in Shorack and Wellner (1986). Benktander and Segerdahl (1960) seems to be the first paper which applies mean excess plots to actuarial data. The first paper which uses quantile plots in this context is Benckert and Jung (1974).

CHAPTER 3

Premiums and Ordering of Risks

3.1 PREMIUM CALCULATION PRINCIPLES

In this section we study on rules how to fix an adequate price, called a *premium*, for a family of risks X to be insured. The investigation of such rules is an essential element of actuarial science. Clearly, premiums cannot be too low because this would result in unacceptably large losses for the insurer. On the other hand, premiums cannot be too high either because of competition between insurers. Consider a certain family of risks X. A *premium calculation principle* is a rule that determines the premium as a functional, assigning a value $\Pi(F_X) \in \mathbb{R} \cup \{\pm\infty\}$ to the risk distribution F_X. Following our notational convention we usually write $\Pi(X)$ instead of $\Pi(F_X)$. Typically, the premium $\Pi(X)$ depends on certain characteristics of F_X like the expectation $\mathbb{E}\,X$ or the variance $\operatorname{Var} X$. For easy application, a premium calculation principle should require as little as possible information on the distribution of the risk X. For example the simplest premium principle is the (pure) *net premium principle* $\Pi(X) = \mathbb{E}\,X$. The difference $\Pi(X) - \mathbb{E}\,X$ is called the *safety loading*. The safety loading should be positive unless the distribution of X is concentrated at a single point. Otherwise, in the long run, ruin occurs with probability 1 even in the case of very large (though finite) initial reserves.

Recall that, throughout this book, a risk is modelled as a nonnegative random variable. However, sometimes it is convenient to define the value $\Pi(X)$ also for real-valued (not necessarily nonnegative) random variables X.

3.1.1 Desired Properties of "Good" Premiums

Before we survey some of the most common principles of premium calculations, we discuss the general properties which one associates with the idea of a "good" premium principle. Usually, the premium $\Pi(X)$ is finite; a risk X (or its distribution F_X) is then called *insurable*. Let X, Y, Z be arbitrary risks for which the premiums below are well-defined and finite. We have the following list of desirable properties:

- *no unjustified safety loading* if, for all constants $a \geq 0$, $\Pi(a) = a$,
- *proportionality* if, for all constants $a \geq 0$, $\Pi(aX) = a\Pi(X)$,
- *subadditivity* if $\Pi(X + Y) \leq \Pi(X) + \Pi(Y)$,
- *additivity* if $\Pi(X + Y) = \Pi(X) + \Pi(Y)$,
- *consistency* if, for all $a \geq 0$, $\Pi(X + a) = \Pi(X) + a$,
- *preservation of stochastic order* if $X \leq_{\rm st} Y$ implies $\Pi(X) \leq \Pi(Y)$,
- *compatibility under mixing* if, for all $p \in [0,1]$ and for all Z, $\Pi(X) = \Pi(Y)$ implies $\Pi(pF_X + (1-p)F_Z) = \Pi(pF_Y + (1-p)F_Z)$.

Note that an additive premium calculation principle with no unjustified safety loading is also consistent. Typically, additivity is required for independent risks. The subadditivity of a premium principle implies that policyholders cannot gain advantage from splitting a risk into pieces. We also remark that in general $\Pi(X + Y)$ depends on the joint distribution of X and Y.

3.1.2 Basic Premium Principles

One of the simplest premium calculation principles is the

- *expected value principle.* For some $a \geq 0$

$$\Pi(X) = (1 + a)\mathbb{E} X \,, \tag{3.1.1}$$

provided that $\mathbb{E} X < \infty$. For $a = 0$ we get the *net premium principle.*

The expected value principle looks fair but it does not take into account the variability of the underlying risk X, and this may be dangerous to the insurer. In an attempt to overcome this disadvantage, one introduces principles where the safety loading $\Pi(X) - \mathbb{E} X$ depends on the variability of X. For some constant $a > 0$ one has the

- *variance principle*

$$\Pi(X) = \mathbb{E} X + a{\rm Var}\, X \,, \tag{3.1.2}$$

- *standard deviation principle*

$$\Pi(X) = \mathbb{E} X + a\sqrt{{\rm Var}\, X} \,, \tag{3.1.3}$$

- *modified variance principle*

$$\Pi(X) = \begin{cases} \mathbb{E} X + a{\rm Var}\, X / \mathbb{E} X & \text{if } \mathbb{E} X > 0, \\ 0 & \text{if } \mathbb{E} X = 0, \end{cases} \tag{3.1.4}$$

- *exponential principle*

$$\Pi(X) = a^{-1} \log \mathbb{E}\, {\rm e}^{aX} \,. \tag{3.1.5}$$

Note that the variance principle is additive for uncorrelated risks, whereas it is subadditive if X and Y are negatively correlated. The standard deviation principle and the modified variance principle are proportional. They are also subadditive provided that $\mathrm{Cov}(X, Y) \le 0$. The proof of these properties is left to the reader. Unfortunately, the disadvantage of the premium calculation principles given by (3.1.2)–(3.1.4) is that they are not monotone with respect to stochastic ordering.

We will study the exponential principle in Section 3.2.4. We will show that it can be characterized by rather natural conditions. Notice, however, that the exponential principle is not suitable for heavy-tailed risks. In the present section we first prove the monotonicity of the exponential principle with respect to the parameter $a > 0$. In connection with this, we need the following classical *Lyapunov inequality.*

Lemma 3.1.1 *For all* $0 < v < w$ *and a nonnegative random variable* X,

$$(\mathbb{E}\, X^v)^{1/v} \le (\mathbb{E}\, X^w)^{1/w}. \tag{3.1.6}$$

The inequality (3.1.6) *is strict unless* X *is concentrated at a single point.*

Proof Take the convex function $h(x) = x^{w/v}$ $(x \ge 0)$ and apply Jensen's inequality to $Y = X^v$. This gives $\mathbb{E}\, X^w = \mathbb{E}\, h(Y) \ge h(\mathbb{E}\, Y) = (\mathbb{E}\, X^v)^{w/v}$. □

Theorem 3.1.1 *Consider a risk* X *with* $\mathbb{E}\, \mathrm{e}^{a_0 X} < \infty$ *for some* $a_0 > 0$. *Then* (a) $\Pi_a(X) = a^{-1} \log \mathbb{E}\, \mathrm{e}^{aX}$ *is a strictly increasing function of* $a \in (0, a_0]$ *provided that* F_X *is not concentrated on a single point,*
(b)

$$\lim_{a \to 0+} a^{-1} \log \mathbb{E}\, \mathrm{e}^{aX} = \mathbb{E}\, X\,, \tag{3.1.7}$$

(c) *if* $\mathbb{E}\, \mathrm{e}^{aX} < \infty$ *for all* $a > 0$,

$$\lim_{a \to \infty} a^{-1} \log \mathbb{E}\, \mathrm{e}^{aX} = r_F = \sup\{x : \mathbb{P}(X \le x) < 1\}\,. \tag{3.1.8}$$

Proof Assume that F_X is not concentrated at a single point. From Lyapunov's inequality (3.1.6) we have $\left(\mathbb{E}\, \mathrm{e}^{vX}\right)^{1/v} < \left(\mathbb{E}\, \mathrm{e}^{wX}\right)^{1/w}$ for $0 < v < w \le a_0$, i.e. $\Pi_a(X)$ is strictly increasing on $(0, a_0]$. Using $\log(1+x) = x + o(x)$, $x \to 0$, we get $a^{-1} \log \mathbb{E}\, \mathrm{e}^{aX} = a^{-1}(\mathbb{E}\, \mathrm{e}^{aX} - 1 + o(\mathbb{E}\, \mathrm{e}^{aX} - 1))$, which proves (3.1.7) since $\lim_{a \to 0+} a^{-1}\left(\mathbb{E}\, \mathrm{e}^{aX} - 1\right) = \mathbb{E}\, X$. To prove (3.1.8) notice first that since $X \le r_F$ we have $a^{-1} \log \mathbb{E}\, \mathrm{e}^{aX} \le a^{-1} \log \exp(a r_F) = r_F$. To prove the reverse inequality, let $0 < \delta < r_F$. From $X' \le X$, where $X' = 0$ if $X \le \delta$ and $X' = \delta$ if $X > \delta$, we have $\mathbb{E}\, \mathrm{e}^{aX} \ge F(\delta) + (1 - F(\delta))\mathrm{e}^{a\delta} \ge (1 - F(\delta))\mathrm{e}^{a\delta}$. Hence $\Pi_a(X) \ge a^{-1}(\log(1 - F(\delta)) + a\delta)$ and consequently $\lim_{a \to \infty} \Pi_a(X) \ge \delta$, which completes the proof because $0 < \delta < r_F$ is arbitrary. □

A further modification of the net premium principle is the

- *risk-adjusted principle*

$$\Pi(X) = \int_0^\infty (1 - F_X(x))^{1/p}\,\mathrm{d}x \tag{3.1.9}$$

for some $p \geq 1$. Assume that F_X has a density. Then, the premium given by (3.1.9) can be interpreted as the net premium of another risk Y with tail function $\overline{F}_Y(x) = (1 - F_X(x))^{1/p}$ and with the proportionally lower hazard rate function $m_Y(t) = \mathrm{d}/\mathrm{d}t \log \overline{F}_Y(t) = p^{-1} m_X(t)$. Thus, Y can be seen as the risk corresponding to X after deflating the hazard rate function of X by the constant factor p^{-1}. This is consistent with the practice of adding a safety margin to the mortality rates in life insurance. It is easily seen that the risk-adjusted principle has a nonnegative safety loading, but no unjustified safety loading. Moreover, this premium principle is proportional, consistent, subadditive and monotone with respect to stochastic ordering of risks. The formal proof of these properties is left to the reader.

3.1.3 Quantile Function: Two More Premium Principles

Before introducing another premium principle, we first study the concept of the quantile function defined in Section 2.6.2.

Lemma 3.1.2 *Let $F(x)$ be an increasing and right-continuous function. The generalized inverse function $F^{-1}(y)$ has the following properties:*
(a) *$F^{-1}(y)$ is increasing,*
(b) *$y \leq F(x)$ if and only if $F^{-1}(y) \leq x$.*

Proof Part (a) is an immediate consequence of the definition. To show part (b) observe that, by (2.6.6), $y \leq F(x)$ yields $F^{-1}(y) \leq x$. Now assume that $F^{-1}(y) \leq x$. Then, because of the monotonicity of F, there exists a sequence $\{x_n\}$ such that $x_n \downarrow x$ and $F(x_n) \geq y$ for all n. This gives $F(x) \geq y$ because F is right-continuous. □

Theorem 3.1.2 *Let F be a distribution function. If the random variable $Z : \Omega \to [0,1]$ is uniformly distributed on $[0,1]$, then the random variable $F^{-1}(Z)$ has distribution function F.*

Proof Observe that the mapping $F^{-1} : \mathbb{R} \to \mathbb{R}$ is measurable since it is increasing. Thus $F^{-1}(Z)$ is a random variable and, because of Lemma 3.1.2,

$$\mathbb{P}(F^{-1}(Z) \leq x) = \mathbb{P}(Z \leq F(x)) = F(x)\,, \qquad x \in \mathbb{R}\,.$$ □

Let $0 < \varepsilon < 1$. Using the notion of the quantile function F_X^{-1} of a risk X we define the

- *ε-quantile principle*

$$\Pi(X) = F_X^{-1}(1-\varepsilon)\,, \tag{3.1.10}$$

i.e. the smallest premium such that the probability of a loss is at most ε.

Note that $F_X^{-1}(1/2)$ is a median of X. Consider the expected absolute deviation $\kappa_X = \mathbb{E}\,|X - F_X^{-1}(1/2)|$. Then, for arbitrary risks X, Y we have

$$\kappa_{X+Y} \le \kappa_X + \kappa_Y \tag{3.1.11}$$

A modification of the standard deviation principle is the

- *absolute deviation principle*

$$\Pi(X) = \mathbb{E}\,X + a\kappa_X\,, \tag{3.1.12}$$

where $a \ge 0$ is some constant. From (3.1.11) and (3.1.12) we see that the absolute deviation principle is subadditive. Moreover, this premium principle is proportional and consistent and if $a \le 1$ it is monotone with respect to stochastic ordering. Note that Theorem 3.1.2 gives $\mathbb{E}\,X = \int_0^1 F_X^{-1}(z)\,\mathrm{d}z$ and

$$\kappa_X = \int_0^{1/2} \left(F_X^{-1}(1/2) - F_X^{-1}(z)\right)\,\mathrm{d}z + \int_{1/2}^1 \left(F_X^{-1}(z) - F_X^{-1}(1/2)\right)\,\mathrm{d}z\,.$$

Hence, for the absolute deviation principle,

$$\Pi(X) = \int_0^{1/2} F_X^{-1}(z)(1-a)\,\mathrm{d}z + \int_{1/2}^1 F_X^{-1}(z)(1+a)\,\mathrm{d}z\,. \tag{3.1.13}$$

Bibliographical Notes. More details on premium calculation principles, including a detailed discussion of their properties, can be found, for example, in Bühlmann (1980), Denneberg (1990), Gerber (1979), Goovaerts, De Vylder and Haezendonck (1982, 1984), Kaas, van Heerwaarden and Goovaerts (1994), Ramsay (1994), Reich (1986) and Wang (1995, 1996).

3.2 ORDERING OF DISTRIBUTIONS

3.2.1 Concepts of Utility Theory

We begin with some concepts from utility theory. Assume that a utility $v(x)$ is related to some wealth of x currency units. It is plausible that utility is growing with wealth and so we suppose that the function $v(x)$ is increasing. Next, the increments of $v(x)$ for small values of x should exceed those for large values of x (because giving a bank note to a poor person makes more sense than giving it to a millionaire). Therefore we impose the condition that $v(x)$ is increasing and concave.

Consider two $\mathbb{N}$-valued random variables X and Y, which can be interpreted as the risky outcomes of currency units (gain, profit, reward, etc.) under two different types of decisions. Assume that we are able to compare the expected utilities $\mathbb{E}\, v(X)$, $\mathbb{E}\, v(Y)$ and that

$$\mathbb{E}\, v(X) \le \mathbb{E}\, v(Y)\,. \tag{3.2.1}$$

Then, with respect to expected utility, the decision corresponding to Y is better than that corresponding to X. A function $v : (b_1, b_2) \to \mathbb{R}$ which is increasing and concave on a certain interval $(b_1, b_2) \subset \mathbb{R}$ is called a *utility function.* Possible examples of utility functions are

- $v(x) = x\,,$
- $v(x) = (1 - \mathrm{e}^{-ax})/a, \qquad a > 0,$
- $v(x) = -(a - x)_+, \qquad a \in \mathbb{R},$

where $x_+ = \max\{0, x\}$. If the random variables X, Y describe risks or losses under two types of decisions, then the reasoning is different. In this case, the utility of X and Y is given by $v(-X)$ and $v(-Y)$, respectively, and the risk X is preferred to Y if $\mathbb{E}\, v(-X) \ge \mathbb{E}\, v(-Y)$. With the notation $w(x) = -v(-x)$, this inequality is equivalent to

$$\mathbb{E}\, w(X) \le \mathbb{E}\, w(Y)\,, \tag{3.2.2}$$

where the function $w(x)$ is increasing and convex. A function $w : (b_1, b_2) \to \mathbb{R}$ which is increasing and convex on (b_1, b_2) is called a *loss function.* Examples of useful loss functions are

- $w(x) = x\,,$
- $w(x) = \mathrm{e}^{ax}, \qquad a > 0,$
- $w(x) = (x - a)_+, \qquad a \in \mathbb{R}.$

In real insurance problems, we often do not know the explicit form of the underlying utility or loss functions. However, in some cases one can show that the inequalities (3.2.1) and (3.2.2) hold for all possible utility or loss functions, respectively. This motivates the following three orderings of distributions. We precede the formal definitions by some general remarks on orderings.

By an ordering we mean a *partial ordering* $\prec$ of a set $\mathcal{X}$ which is a binary relation on $\mathcal{X}$ fulfilling

- $x \prec x$ (*reflexivity*),
- $\{x \prec y, y \prec z\}$ implies $x \prec z$ (*transitivity*),
- $\{x \prec y, y \prec x\}$ implies $x = y$ (*antisymmetry*).

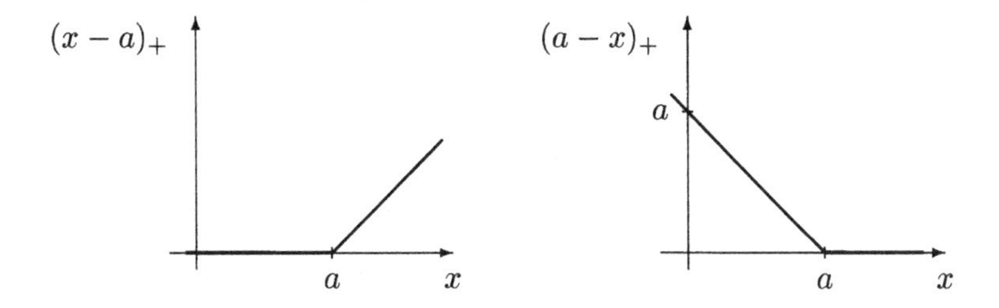

Figure 3.2.1 The "angles" $(x-a)_+$ and $(a-x)_+$

In the following we will mainly concentrate on the case where $\mathcal{X}$ is the set of distributions of real-valued random variables. In later chapters we will also compare distributions of random vectors and of stochastic processes. In connection with orderings of distributions of real-valued random variables, the functions $x \to (x-a)_+$ and $x \to (a-x)_+$ depicted in Figure 3.2.1 play a central role. In Hardy, Littlewood and Pólya (1929) these functions have been called "angles". They satisfy the identity

$$(x-a)_+ - (a-x)_+ = x - a\,. \tag{3.2.3}$$

Let F be a distribution on $\mathbb{R}$. Then, for all $a \in \mathbb{R}$, integration by parts gives the following useful identities:

$$\int_{-\infty}^{\infty} (x-a)_+ \,\mathrm{d}F(x) = \int_a^{\infty} \overline{F}(x)\,\mathrm{d}x, \tag{3.2.4}$$

$$\int_{-\infty}^{\infty} (a-x)_+ \,\mathrm{d}F(x) = \int_{-\infty}^{a} F(x)\,\mathrm{d}x\,, \tag{3.2.5}$$

provided that $\int_{-\infty}^{\infty} |x|\,\mathrm{d}F(x) < \infty$. For any partial ordering $\prec$ of distributions considered in this book, we will write $X \prec Y$ if $F_X \prec F_Y$ holds for the corresponding distributions.

Definition 3.2.1 *Let X, Y be two real-valued random variables.*
(a) *We say that X is stochastically dominated by (or stochastically smaller than) Y and we write $X \le_{\mathrm{st}} Y$ if for all increasing functions $g : \mathbb{R} \to \mathbb{R}$*

$$\mathbb{E}\, g(X) \le \mathbb{E}\, g(Y)\,, \tag{3.2.6}$$

provided the expectations $\mathbb{E}\, g(X), \mathbb{E}\, g(Y)$ exist and are finite.
(b) *Assume that $\mathbb{E}\, X_+, \mathbb{E}\, Y_+ < \infty$. We say that X is smaller than Y in stop-loss order and we write $X \le_{\mathrm{sl}} Y$ if (3.2.6) holds for all increasing convex functions $g : \mathbb{R} \to \mathbb{R}$ provided the expectations $\mathbb{E}\, g(X), \mathbb{E}\, g(Y)$ exist and are finite.*

(c) *Assume that* $\mathbb{E}(-X)_+, \mathbb{E}(-Y)_+ < \infty$. *We say that* X *is smaller than* Y *in increasing-concave order and we write* $X \leq_{\rm icv} Y$ *if* (3.2.6) *holds for all increasing concave functions* $g : \mathbb{R} \to \mathbb{R}$ *provided the expectations* $\mathbb{E}\, g(X), \mathbb{E}\, g(Y)$ *exist and are finite.*

Literally speaking, relation (3.2.6) as well as the symbols $\leq_{\rm st}, \leq_{\rm sl}$ and $\leq_{\rm icv}$ mean "not larger than"; for simplicity we will say "smaller than", not excluding equality. This is in agreement with the notions of an increasing, decreasing, convex and concave function, introduced in Section 2.4 in the same weak sense. In other areas of applied probability like reliability or queueing theory, one usually says *increasing-convex order* instead of *stop-loss order*, writing $\leq_{\rm icx}$ instead of $\leq_{\rm sl}$. In insurance mathematics, however, the notion of stop-loss order is quite common due to its connection with reinsurance; see Section 3.3 and the remark following Theorem 3.2.2.

3.2.2 Stochastic Order

Next we show that the notion of *stochastic order* given in Definition 3.2.1 is equivalent to that introduced in Section 2.4. Recall that the symbol $X \leq Y$ means that the random variables X and Y are defined on a common probability space $(\Omega, \mathcal{F}, \mathbb{P})$ and that $X(\omega) \leq Y(\omega)$ for almost all $\omega \in \Omega$, i.e. $\mathbb{P}(X \leq Y) = 1$. For the generalized inverse function $F^{-1}(x)$ see (2.6.6). Part (b) of the following characterization of stochastic ordering is sometimes called the *coupling theorem* for $\leq_{\rm st}$. It formalizes the useful fact that stochastic dominance can always be expressed by comparing realizations of certain auxiliary random variables with the same distributions, where these random variables are defined on a common probability space.

Theorem 3.2.1 *The following statements are equivalent:*
(a) $X \leq_{\rm st} Y$.
(b) *There exist a probability space* $(\Omega', \mathcal{F}', \mathbb{P}')$ *and two random variables* X', Y' *defined on it such that* $X' \leq Y'$, $X \stackrel{\rm d}{=} X'$ *and* $Y \stackrel{\rm d}{=} Y'$.
(c) *For all* $x \in \mathbb{R}$,

$$\overline{F}_X(x) \leq \overline{F}_Y(x) . \tag{3.2.7}$$

Proof To show that (c) follows from (a) it suffices to insert the increasing function $g(t) = \delta_x(t)$ in (3.2.6), i.e. $g(t) = 0$ for $t < x$ and $g(t) = 1$ for $t \geq x$. Assume now that (c) holds. Consider a random variable Z which is uniformly distributed on $[0, 1]$ and defined on some probability space $(\Omega', \mathcal{F}', \mathbb{P}')$. Put $X' = F_X^{-1}(Z)$ and $Y' = F_Y^{-1}(Z)$. From Theorem 3.1.2, we get that $X' \stackrel{\rm d}{=} X$ and $Y' \stackrel{\rm d}{=} Y$. Moreover, from (2.6.6) and (3.2.7) we have

$$F_X^{-1}(Z) = \min\{t : F_X(t) \geq Z\} \leq \min\{t : F_Y(t) \geq Z\} = F_Y^{-1}(Z) ,$$

i.e. $X' \le Y'$ and statement (b) follows. Finally, from (b) we get that $\mathbb{E}\,g(X) = \mathbb{E}\,g(X') \le \mathbb{E}\,g(Y') = \mathbb{E}\,g(Y)$ for any increasing function $g : \mathbb{R} \to \mathbb{R}$, i.e. $X \le_{\mathrm{st}} Y$ holds. □

The notion of stochastic dominance can be naturally extended to (multidimensional) distributions of random vectors. Property (c) of stochastic ordering given in Theorem 3.2.1 has to be reformulated but the coupling property (b) essentially remains unchanged. However, the proof becomes much more complicated; see Section 7.4.1. In the special case that two random vectors $(X_1, \ldots, X_n)$ and $(Y_1, \ldots, Y_n)$ have independent components for which $X_i \le_{\mathrm{st}} Y_i$ for all $i = 1, \ldots, n$, it is relatively easy to derive a multidimensional analogue to the coupling property. Indeed, it suffices to consider n independent random variables $Z_1, \ldots, Z_n$ on some common probability space $(\Omega', \mathcal{F}', \mathbb{P}')$ which are uniformly distributed on $[0, 1]$. Then, using Theorem 3.1.2, we get that the random vectors $(X_1, \ldots, X_n)$ and $(X'_1, \ldots, X'_n)$ have the same distribution, where $X'_i = F^{-1}_{X_i}(Z_i)$ for $i = 1, \ldots, n$. Analogously, the random vectors $(Y_1, \ldots, Y_n)$ and $(Y'_1, \ldots, Y'_n)$ have the same distribution, where $Y'_i = F^{-1}_{Y_i}(Z_i)$ for $i = 1, \ldots, n$. Moreover, $X'_i \le Y'_i$ for all $i = 1, \ldots, n$.

3.2.3 Stop-Loss Order

Similar to part (c) of Theorem 3.2.1, the next characterization of the stop-loss order holds.

Theorem 3.2.2 *The following statements are equivalent:*
(a) $X \le_{\mathrm{sl}} Y$.
(b) *For all* $x \in \mathbb{R}$,

$$\mathbb{E}\,(X - x)_+ \le \mathbb{E}\,(Y - x)_+ \,. \tag{3.2.8}$$

Proof To show that (b) follows from (a) it suffices to insert the increasing and convex function $g(t) = (t - x)_+$ in (3.2.6). Assume now that $g : \mathbb{R} \to \mathbb{R}$ is increasing and convex. First we consider the case $g(-\infty) > -\infty$. Then

$$g(x) = g(-\infty) + \int_{-\infty}^{x} g^+(t)\,\mathrm{d}t = g(-\infty) + \int_{-\infty}^{x} (x - t)\,\mathrm{d}g^+(t)$$

for all $x \in \mathbb{R}$, where $g^+(t)$ is the right derivative of $g(t)$. Thus,

$$\begin{aligned}\int_{-\infty}^{\infty} g(x)\,\mathrm{d}F_X(x) &= g(-\infty) + \int_{-\infty}^{\infty} \Big(\int_{-\infty}^{\infty} (x - y)_+\,\mathrm{d}g^+(y) \Big)\,\mathrm{d}F_X(x) \\ &= g(-\infty) + \int_{-\infty}^{\infty} \int_{-\infty}^{\infty} (x - y)_+\,\mathrm{d}F_X(x)\,\mathrm{d}g^+(y)\end{aligned}$$

by Fubini's theorem. Hence, (3.2.8) implies (3.2.6) under the additional assumption that $g(-\infty) > -\infty$. In the general case, take $g(t)$ increasing and convex, but $g(-\infty) = -\infty$ and put $g_n(t) = \max\{-n, g(t)\}$ for $n \in \mathbb{N}$. The functions $g_n(t)$ are increasing and convex with $g_n(-\infty) > -\infty$. Assuming (3.2.8), the first part of the proof tells us that

$$\int_{-\infty}^{\infty} g_n(x)\,\mathrm{d}F_X(x) \le \int_{-\infty}^{\infty} g_n(x)\,\mathrm{d}F_Y(x)\,, \qquad n \in \mathbb{N}\,.$$

The proof is completed using the monotone convergence theorem. □

Remark The characterization given in (3.2.8) explains the name of stop-loss ordering for $\le_{\mathrm{sl}}$ as used in insurance mathematics. For if X is a risk to be insured with a retention level x, then the net premium $\mathbb{E}\,(X-x)_+$ is called the *stop-loss premium.*

Similarly to part (b) of Theorem 3.2.1, there is also a *coupling theorem* for the stop-loss order; see Baccelli and Brémaud (1994). Here, the notion of the conditional expectation $\mathbb{E}\,(Y' \mid X')$ is required. This is a random variable that can be interpreted as the expectation of Y' taken with respect to a certain "random" conditional distribution depending on the actual value of X'.

Theorem 3.2.3 *The following statements are equivalent:*
(a) $X \le_{\mathrm{sl}} Y$.
(b) *There exist a probability space* $(\Omega', \mathcal{F}', \mathbb{P}')$ *and two random variables* X', Y' *defined on it such that* $X' \le \mathbb{E}\,(Y' \mid X')$ *and* $X \stackrel{\mathrm{d}}{=} X'$, $Y \stackrel{\mathrm{d}}{=} Y$.

The *proof* that statement (b) follows from (a) is rather difficult and goes beyond the scope of our book. We therefore omit it and refer the reader to the bibliographical notes. On the other hand, assuming (b), for all $x \in \mathbb{R}$ it follows from Jensen's inequality for conditional expectations that

$$\begin{aligned}\mathbb{E}\,(X-x)_+ &\le \mathbb{E}\,(\mathbb{E}\,(Y' \mid X') - x)_+ \le \mathbb{E}\,(\mathbb{E}\,((Y'-x)_+ \mid X'))\\ &= \mathbb{E}\,(Y'-x)_+ = \mathbb{E}\,(Y-x)_+\,.\end{aligned}$$

Because of Theorem 3.2.2 this gives (a). □

We next derive a sufficient condition for the stop-loss ordering to hold between two random variables. This so-called *cut criterion* appears in the literature under various names and is attributed to numerous authors (e.g. as Ohlin's lemma or the Karlin–Novikoff cut criterion).

We begin with a simple auxiliary result.

Lemma 3.2.1 *Suppose* $h : \mathbb{R} \to \mathbb{R}$ *is a measurable function such that* $\int_{-\infty}^{\infty} |h(t)|\,\mathrm{d}t < \infty$ *and* $\int_{-\infty}^{\infty} h(t)\,\mathrm{d}t \ge 0$. *If* $h(t) \le 0$ *for all* $t < t_0$ *and* $h(t) \ge 0$ *for all* $t > t_0$ *for some* $t_0 \in \mathbb{R}$, *then* $\int_x^{\infty} h(t)\,\mathrm{d}t \ge 0$ *for all* $x \in \mathbb{R}$.

Proof Note that the function $x \mapsto \int_x^{\infty} h(t)\,\mathrm{d}t$ is continuous and increasing on $(-\infty, t_0)$, decreasing on (t_0, ∞) and nonnegative at $-\infty$. □

The *one-cut criterion* for the stop-loss ordering of two random variables X and Y with distributions F_X and F_Y is given next.

Theorem 3.2.4 *Suppose* $\mathbb{E}|X| < \infty$, $\mathbb{E}|Y| < \infty$ *and* $\mathbb{E}X \le \mathbb{E}Y$. *If, for some* $t_0 \in \mathbb{R}$, $F_X(t) \le F_Y(t)$ *for* $t < t_0$ *and* $F_X(t) \ge F_Y(t)$ *for* $t > t_0$ *then* $X \le_{\rm sl} Y$.

Proof Put $h(t) = F_X(t) - F_Y(t) = 1 - F_Y(t) - (1 - F_X(t))$. Then, $h(t) \le 0$ for all $t < t_0$ and $h(t) \ge 0$ for all $t > t_0$. Moreover, integration by parts gives

$$\begin{aligned}
&\int_{-\infty}^{\infty} h(t)\,\mathrm{d}t \\
&= -\int_{-\infty}^{0} F_Y(t)\,\mathrm{d}t + \int_0^{\infty} \overline{F_Y}(t)\,\mathrm{d}t - \Big(-\int_{-\infty}^{0} F_X(t)\,\mathrm{d}t + \int_0^{\infty} \overline{F_X}(t)\,\mathrm{d}t\Big) \\
&= \mathbb{E}Y - \mathbb{E}X \ge 0\,.
\end{aligned}$$

Thus, by Lemma 3.2.1, the proof is finished. □

We continue by studying some basic *extremality* properties of the stop-loss order.

Theorem 3.2.5 *Let* X *be an arbitrary random variable with distribution* F *and expectation* μ. *Then*

$$\delta_\mu \le_{\rm sl} F\,. \tag{3.2.9}$$

Proof Note that $\delta_\mu(t) \le F(t)$ for all $t < \mu$ and $\delta_\mu(t) \ge F(t)$ for all $t > \mu$. Thus, the assertion follows from Theorem 3.2.4. □

Inequality (3.2.9) means that within the class of all distributions with a fixed expectation the degenerate distribution is minimal with respect to stop-loss order. Further, we apply Jensen's inequality to see that

$$\int_{-\infty}^{\infty} v(x)\,\mathrm{d}F(x) \le \int_{-\infty}^{\infty} v(x)\,\mathrm{d}\delta_\mu(x) \tag{3.2.10}$$

holds for all concave functions $v : \mathbb{R} \to \mathbb{R}$. In terms of utility theory, this means that deterministic wealth maximizes the expected utility in the class of all random wealths with the same expectation. Next, we look for the minimal and maximal expected utility $\mathbb{E}\,v(X)$ within the class of all (nonnegative) wealths X with fixed first and second moments $\mu = \mu^{(1)}$ and $\mu^{(2)}$, where the utility function $v : \mathbb{R}_+ \to \mathbb{R}$ is assumed to be a 2-*concave function.* By this we mean that $v(x)$ is bounded from above and can be represented either as

$$v(x) = v(+\infty) - \int_{-\infty}^{\infty} (t - x)_+^2\,\mathrm{d}\eta(t) \tag{3.2.11}$$

for all $x \in \mathbb{R}_+ = [0,\infty)$, or as a monotone limit of such functions. Here $(t-x)_+^2 = ((t-x)_+)^2$ and $\eta(t)$ is an increasing and right-continuous function with $\int_{-\infty}^{\infty} t^2 \, \mathrm{d}\eta(t) < \infty$. Clearly, 2-concave functions form a subclass of the class of all concave functions. An example of a 2-concave function is given by $v(x) = 1 - e^{-ax}; a > 0$. According to (3.2.10), the degenerate distribution δ_μ maximizes the expected utility $\mathbb{E}\, v(X)$. Moreover, for any given 2-concave utility function $v(x)$, the distribution F_{m} given by

$$F_{\mathrm{m}} = \left(1 - \frac{\mu^2}{\mu^{(2)}}\right) \delta_0 + \frac{\mu^2}{\mu^{(2)}} \, \delta_{\mu^{(2)}/\mu} \, ,$$

minimizes $\mathbb{E}\, v(X)$ within the class of all distributions of nonnegative random variables with fixed first and second moments.

Theorem 3.2.6 *If $v(x)$ is a 2-concave function and X is a nonnegative random variable with distribution F and expectation μ, then*

$$\int_0^\infty v(x) \, \mathrm{d}F_{\mathrm{m}}(x) \le \int_0^\infty v(x) \, \mathrm{d}F(x) \le \int_0^\infty v(x) \, \mathrm{d}\delta_\mu(x) \, . \tag{3.2.12}$$

Proof The upper bound is nothing but (3.2.10). To get the lower bound, it suffices to prove that $\int_0^t (t-x)^2 \mathrm{d}F_{\mathrm{m}}(x) \ge \int_0^t (t-x)^2 \mathrm{d}F(x)$ for all $t \ge 0$ because

$$\int_0^\infty v(x) \, \mathrm{d}F_{\mathrm{m}}(x) = v(+\infty) - \int_{-\infty}^\infty \Big(\int_0^t (t-x)^2 \mathrm{d}F_{\mathrm{m}}(x)\Big) \mathrm{d}\eta(t)$$

and

$$\int_0^\infty v(x) \, \mathrm{d}F(x) = v(+\infty) - \int_{-\infty}^\infty \Big(\int_0^t (t-x)^2 \mathrm{d}F(x)\Big) \mathrm{d}\eta(t)$$

for functions of the form (3.2.11). Observe that

$$\int_0^t (t-x)^2 \mathrm{d}F_{\mathrm{m}}(x) = \begin{cases} t^2(1-\mu^2/\mu^{(2)}) & \text{if } t < \mu^{(2)}/\mu \, , \\ t^2 - 2t\mu + \mu^{(2)} & \text{if } t \ge \mu^{(2)}/\mu \, . \end{cases}$$

Thus, it suffices to show that

$$\int_0^t (t-x)^2 \mathrm{d}F(x) \le \begin{cases} t^2(1-\mu^2/\mu^{(2)}) & \text{if } t < \mu^{(2)}/\mu, \\ t^2 - 2t\mu + \mu^{(2)} & \text{if } t \ge \mu^{(2)}/\mu. \end{cases}$$

Assume first that $t \ge \mu^{(2)}/\mu$. Then

$$\int_0^t (t-x)^2 \mathrm{d}F(x) \le \int_0^\infty (t-x)^2 \mathrm{d}F(x) = t^2 - 2t\mu + \mu^{(2)}.$$

On the other hand, using the substitution $z = t^{-1}$, it is easily seen that

$$\int_0^t (t-x)^2 \mathrm{d}F(x) = \mathbb{E}\,(t-X)_+^2 \le t^2 \left(1 - \mu^2/\mu^{(2)}\right)$$

for all $t < \mu^{(2)}/\mu$ if and only if $\mathbb{E}(1-zX)_+^2 \le 1-\mu^2/\mu^{(2)}$ for all $z > \mu/\mu^{(2)}$. Since

$$\mathbb{E}(1-zX)_+^2 \le \mathbb{E}(1-z_0X)_+^2 \le \mathbb{E}(1-z_0X)^2 = 1-\frac{\mu^2}{\mu^{(2)}}$$

for $z > z_0 = \mu/\mu^{(2)}$, the proof is complete. □

3.2.4 The Zero Utility Principle

In this section we take $v : \mathbb{R} \to \mathbb{R}$ to be a strictly increasing utility function. Given this utility function, the *zero utility principle* defines the premium $\Pi(X)$ of a risk X as the solution to the equation

$$\mathbb{E}\,(v(\Pi(X)-X)) = v(0)\,. \tag{3.2.13}$$

This means that the insurer fixes the premium at such a level that, with respect to the utility function $v(x)$, the utility $v(0)$ of the initial surplus $x=0$ is equal to the expected utility of the surplus $\Pi(X)-X$ resulting from insuring the risk X at the premium $\Pi(X)$. Note that the premium $\Pi(X)$ is the same for all utility functions $av(x)+b$, where $a > 0$ and $b \in \mathbb{R}$. The formal proof of this property is left to the reader.

A premium principle Π is said to be *monotone* with respect to an order $\prec$ in a certain family of distribution functions if $F_X \prec F_Y$ implies $\Pi(X) \le \Pi(Y)$.

Theorem 3.2.7 *The zero utility principle given by* (3.2.13) *has nonnegative safety loading and no unjustified safety loading. Moreover, it is monotone with respect to stop-loss order* $\le_{\rm sl}$.

Proof From (3.2.13) and from the strict monotonicity of v, we immediately get $\Pi(x) = x$, i.e. Π has no unjustified safety loading. Moreover, Jensen's inequality gives $v(0) = \mathbb{E}\,(v(\Pi(X)-X)) \le v(\Pi(X)-\mathbb{E}\,X)$. Thus, $v(0) \le v(\Pi(X)-\mathbb{E}\,X)$, which gives $\Pi(X)-\mathbb{E}\,X \ge 0$. To prove that Π is monotone with respect to stop-loss order, consider two risks X, X' such that $X \le_{\rm sl} X'$. Then, by Definition 3.2.1 of the stop-loss order, $\mathbb{E}\,v(c-X) \ge \mathbb{E}\,v(c-X')$ for each increasing concave function $v(x)$ and for all $c \in \mathbb{R}$ because $-v(c-x)$ is an increasing and convex function of x. Hence $\Pi(X) \le \Pi(X')$. □

By way of an exercise, the reader should show that the exponential principle $\Pi_a(X) = a^{-1}\log \mathbb{E}\,{\rm e}^{aX}$ solves the equation (3.2.13) for the utility function $v(x) = (1-{\rm e}^{-ax})/a$ provided that $a > 0$ is such that $\mathbb{E}\,{\rm e}^{aX} < \infty$. Furthermore, the exponential principle is additive for independent risks.

Under some regularity conditions on the utility function, the exponential principle is characterized as the only zero utility principle which is additive for independent risks.

Theorem 3.2.8 *Assume that the utility function $v : \mathbb{R} \to \mathbb{R}$ is twice continuously differentiable and that*

$$v(0) = 0\,, \qquad v^{(1)}(0) = 1\,, \qquad v^{(2)}(0) = -a \tag{3.2.14}$$

for some $a \geq 0$. Let Π be the zero utility principle defined for nonnegative random variables by (3.2.13) *and assume that*

$$\Pi(X + Y) = \Pi(X) + \Pi(Y) \tag{3.2.15}$$

for all independent and insurable risks X, Y. Then

$$\Pi(X) = \begin{cases} a^{-1} \log \mathbb{E}\, \mathrm{e}^{aX} & \text{if } a > 0, \\ \mathbb{E}\, X & \text{if } a = 0. \end{cases}$$

Before proving Theorem 3.2.8 we show the following auxiliary result. For arbitrary fixed $b > 0$ and $p \in [0, 1]$, we consider a two-point risk X_b with distribution $p\delta_b + (1 - p)\delta_0$. Next we consider the premium $\varphi_b(p) = \Pi(p\delta_b + (1 - p)\delta_0)$ for this risk defined by the zero utility principle (3.2.13), i.e. $\varphi_b(p)$ is the solution to

$$pv(\varphi_b(p) - b) + (1 - p)v(\varphi_b(p)) = 0\,. \tag{3.2.16}$$

Lemma 3.2.2 *Under the assumptions of Theorem* 3.2.8 *the function $\varphi_b : [0, 1] \to \mathbb{R}_+$ is twice continuously differentiable.*

Proof We apply the implicit functions theorem to the bivariate function $F(x, y) = xv(y - b) + (1 - x)v(y)$. This function is continuous, has continuous second-order partial derivatives and $F_y(x, y) \neq 0$ for all $(x, y) \in [0, 1] \times \mathbb{R}$. Thus, the theorem on implicit functions (see Theorems 17.1.1 and 17.4.1 in Hille (1966)) yields that $\varphi_b(p)$ is twice continuously differentiable on $[0, 1]$. □

We also need a result from differential equations.

Lemma 3.2.3 *Under the assumptions of Theorem* 3.2.8, *the only solution to*

$$v(t + h) - v^{(1)}(h)v(t) - v(h)v^{(1)}(t) - av(h)v(t) = 0 \tag{3.2.17}$$

is $v(t) = (1 - \mathrm{e}^{-at})/a$ when $a > 0$, and $v(t) = t$ when $a = 0$.

Proof By inspection, one can see that $v(t) = (1 - \mathrm{e}^{-at})/a$ and $v(t) = t$ are solutions to (3.2.17) in the cases that $a > 0$ and $a = 0$, respectively. It remains to show that there are no other solutions to (3.2.17) which satisfy (3.2.14). Assume first that $a > 0$. Since $v(t)$ is concave and $v(0) = 0$, we have $v(2t) \leq 2v(t)$ for all $t \geq 0$. Thus, for $h = t > 0$, (3.2.17) gives

$$v^{(1)}(t) = \frac{v(2t) - a(v(t))^2}{2v(t)} \leq \frac{2v(t) - a(v(t))^2}{2v(t)} = 1 - \frac{a}{2}v(t)\,.$$

This means that $v(t)$ must be bounded because otherwise we would have $v^{(1)}(t) < 0$ for sufficiently large t, which is impossible since $v(t)$ is increasing. Hence $\lim_{t\to\infty} v^{(1)}(t) = 0$ and $\lim_{t\to\infty} v(t+h)/v(t) = 1$ for all $h \in \mathbb{R}$. Now, dividing both sides of (3.2.17) by $v(t)$ and letting $t \to \infty$, we arrive at the linear differential equation $1 - v^{(1)}(h) - av(h) = 0$. From the theory of ordinary linear differential equations (see, for example, Hille (1966)) we know that this equation has exactly one solution satisfying $v(0) = 0$. If $a = 0$, then (3.2.17) takes the form

$$v(t+h) - v^{(1)}(h)v(t) - v(h)v^{(1)}(t) = 0\,. \tag{3.2.18}$$

Assume for the moment that $v(t)$ is bounded. Then, in the same way as before, we would get that $v^{(1)}(h) = 1$ for all $h \in \mathbb{R}$, which leads to a contradiction. Thus, $\lim_{t\to\infty} v(t) = \infty$ and, consequently, $\lim_{t\to\infty} v(t+h)/v(t) = 1$ since $v(h) + v(t) \ge v(t+h) \ge v(t)$. Furthermore, $\lim_{t\to\infty} v^{(1)}(t)/v(t) = 0$, since $v^{(1)}(t)$ is decreasing. Now, dividing both sides of (3.2.18) by $v(t)$ and letting $t \to \infty$, we see that $v^{(1)}(h) = 1$ for all $h \in \mathbb{R}$, i.e. $v(h) = h$. □

Proof of Theorem 3.2.8. Taking the derivative with respect to p in (3.2.16), and letting $p = 1$, we get

$$\varphi_b^{(1)}(1) = v(b)\,. \tag{3.2.19}$$

From the assumed additivity property we obtain

$$\begin{aligned}&\Pi((p\delta_h + (1-p)\delta_0) * (q\delta_t + (1-q)\delta_0))\\ &\quad = \Pi(p\delta_h + (1-p)\delta_0) + \Pi(q\delta_t + (1-q)\delta_0) = \varphi_h(p) + \varphi_t(q)\end{aligned}$$

for $h, t > 0$ and $0 \le p, q \le 1$. Therefore, by (3.2.13),

$$\begin{aligned}&pqv(\varphi_h(p) + \varphi_t(q) - h - t) + p(1-q)v(\varphi_h(p) + \varphi_t(q) - h)\\ &\quad + (1-p)qv(\varphi_h(p) + \varphi_t(q) - t) + (1-p)(1-q)v(\varphi_h(p) + \varphi_t(q)) = 0\end{aligned} \tag{3.2.20}$$

is obtained. Differentiating this equation twice, first with respect to p and then with respect to q, and setting $p = q = 1$, we get

$$\begin{aligned}&(v(0) + v^{(1)}(0)\varphi_t^{(1)}(1)) + (v^{(1)}(0)\varphi_h^{(1)}(1) + v^{(2)}(0)\varphi_h^{(1)}(1)\varphi_t^{(1)}(1))\\ &\quad + (-v(t)) + (-v^{(1)}(t)\varphi_h^{(1)}(1)) + (-v(h) - v^{(1)}(h)\varphi_t^{(1)}(1)) + (v(t+h)) = 0\,.\end{aligned}$$

Together with (3.2.14) and (3.2.19) this yields

$$v(t+h) - v^{(1)}(h)v(t) - v(h)v^{(1)}(t) - av(h)v(t) = 0 \tag{3.2.21}$$

for all $h, t > 0$. To prove that (3.2.21) holds for all $h, t < 0$ we can use similar arguments. Namely, taking in (3.2.16) the derivative with respect to p and setting $p = 0$, we obtain

$$\varphi_b^{(1)}(0) = -v(-b)\,. \tag{3.2.22}$$

Differentiating (3.2.20) twice, first with respect to p and then with respect to q, setting $p = q = 0$, and using (3.2.14) and (3.2.22), we get (3.2.21) for all $h, t < 0$. By Lemma 3.2.3 this finishes the proof. □

Bibliographical Notes. Surveys on various orderings of distributions are given, for example, in Marshall and Olkin (1979), Mosler and Scarsini (1993), Shaked and Shanthikumar (1993), Stoyan (1983) and Szekli (1995). For the proof of the necessity part of Theorem 3.2.3, see, for example, Baccelli and Brémaud (1994) or Lindvall (1992). Sometimes the implication (a) $\Rightarrow$ (b) of Theorem 3.2.3 is called *Strassen's theorem.* Results like Theorem 3.2.4 can be traced back to Hardy, Littlewood and Pólya (1929) or Karamata (1932); see also Ohlin (1969). Some generalizations are given in Karlin and Novikoff (1963). An application of Theorem 3.2.4 to optimal reinsurance structures is considered in Hesselager (1993); see also Section 3.3. Results like those given in Section 3.2.4 can be found, for example, in Gerber (1979) for the zero utility principle, and in Kaas, van Heerwaarden and Goovaerts (1994) for the so-called mean value principle, where further characterizations of the exponential principle are discussed.

3.3 SOME ASPECTS OF REINSURANCE

If a risk X is too dangerous (for instance if X has large variance), the insurer may want to transfer part of the risk X to another insurer. This risk transfer from a first insurer to another insurance company is called *reinsurance.* The first insurer that transfers (part of) his risk is called a *cedant.* Often the reinsurance company does the same, i.e. it passes part of its own risk to a third company, and so on. By passing on parts of risks, large risks are split into a number of smaller portions taken up by different risk carriers. This procedure of risk exchange makes large claims less dangerous to the individual insurers, while the total risk remains the same.

A reinsurance contract specifies the part $X - h(X)$ of the claim amount X which has to be compensated by the reinsurer, after taking off the retained amount $h(X)$. Here $h : \mathbb{R}_+ \to \mathbb{R}_+$, the *retention function,* is assumed to have the following properties:

- $h(x)$ and $x - h(x)$ are increasing,
- $0 \le h(x) \le x$ and in particular $h(0) = 0$.

It is reasonable to suppose that both the retention function $h(x)$ and the *compensation function* $k(x) = x - h(x)$ are increasing, i.e. with the growing claim size, both parts contribute more. In practice, retention functions are often continuous or even locally smooth, but we do not require such properties in this section. Possible choices of retention functions $h(x)$ are

- $h(x) = ax$ for the *proportional contract*, where $0 < a \le 1$,
- $h(x) = \min\{a, x\}$ for the *stop-loss contract*, where $a > 0$.

Consider risks of the form $X = \sum_{i=1}^{N} U_i$, where N is an $\mathbb{N}$-valued random variable and where the nonnegative random variables U_i are interpreted as local risks. We can model *local reinsurance* with local retention functions $h_i(x)$ as follows: for the i-th claim of size U_i the part $U_i - h_i(U_i)$ is carried by the reinsurer. The local retention functions $h_i(x)$ are assumed to have the same properties as their global alternatives $h(x)$. Let $v(x)$ be a utility function. If $\Pi(X)$ is the premium paid to the reinsurer, the utility of the reinsurer is $v(\Pi(X) - \sum_{i=1}^{N} k_i(U_i))$, where $k_i(x) = x - h_i(x)$ are the corresponding local compensation functions. If a global contract is used with retention function $h(x)$ the utility of the reinsurer is $v(\Pi(X) - k(X))$, where $k(x) = x - h(x)$ is the corresponding compensation function. The following result suggests that in some cases a global contract is better for the reinsurer.

Theorem 3.3.1 *Let X be an insurable risk of the form $X = \sum_{i=1}^{N} U_i$ and let $v : \mathbb{R} \to \mathbb{R}$ be an increasing and concave function. For a local reinsurance with compensation functions k_i ($i = 1, 2, \ldots$), there exists a function $k(x)$ such that*

$$\mathbb{E}\, k(X) = \mathbb{E}\left(\sum_{i=1}^{N} k_i(U_i)\right)$$

and $\mathbb{E}\, v(\Pi(X) - \sum_{i=1}^{N} k_i(U_i)) \le \mathbb{E}\, v(\Pi(X) - k(X))$.

Proof For each $x \ge 0$, define

$$k(x) = \mathbb{E}\left(\sum_{i=1}^{N} k_i(U_i) \mid X = x\right). \tag{3.3.1}$$

By the definition of conditional expectation $\mathbb{E}\, k(X) = \mathbb{E}\,(\sum_{i=1}^{N} k_i(U_i))$. Since the function $v(\Pi(X) - x)$ of x is concave, we can apply Jensen's inequality for conditional expectation to obtain

$$\begin{aligned} \mathbb{E}\, u\Big(\Pi(X) - \sum_{i=1}^{N} k_i(U_i)\Big) &= \mathbb{E}\left(\mathbb{E}\left(u\Big(\Pi(X) - \sum_{i=1}^{N} k_i(U_i)\Big) \mid X\right)\right) \\ &\le \mathbb{E}\, u\left(\Pi(X) - \mathbb{E}\left(\sum_{i=1}^{N} k_i(U_i) \mid X\right)\right) = \mathbb{E}\, v(\Pi(X) - k(X)). \end{aligned}$$

□

Remark The following example shows that (3.3.1) does not always give a compensation function. Take $N = 2$ and suppose that $\mathbb{P}(U_1 = 4) + \mathbb{P}(U_1 = 5) = 1$ and $\mathbb{P}(U_2 = 5) + \mathbb{P}(U_2 = 7) = 1$, where all four probabilities are assumed to be strictly positive. If $k_1(x) = x/2$ and $k_2(x) = x/8$, then it is easily seen that the function $k(x)$ given in (3.3.1) is not increasing.

Under appropriate conditions, however, $k(x)$ defined in (3.3.1) is a compensation function. To show this, we first extend the notion of *stochastic order* as given in Definition 3.2.1. Let $\boldsymbol{X} = (X_1, \ldots, X_n)$ and $\boldsymbol{Y} = (Y_1, \ldots, Y_n)$ be random vectors taking values in $\mathbb{R}^n$. We say that $\boldsymbol{X}$ is *stochastically dominated* by (or *stochastically smaller* than) $\boldsymbol{Y}$ and we write $\boldsymbol{X} \leq_{\mathrm{st}} \boldsymbol{Y}$ if for each measurable function $g : \mathbb{R}^n \to \mathbb{R}$ which is increasing in each of its arguments,

$$\mathbb{E}\, g(\boldsymbol{X}) \leq \mathbb{E}\, g(\boldsymbol{Y}), \tag{3.3.2}$$

provided that the expectations $\mathbb{E}\, g(\boldsymbol{X})$, $\mathbb{E}\, g(\boldsymbol{Y})$ exist and are finite. By $\boldsymbol{X}_t$ we mean a random vector with the same distribution as the conditional distribution of $\boldsymbol{X} = (X_1, \ldots, X_n)$ given $\sum_{i=1}^n X_i = t$.

In the next lemma, sometimes called *Efron's theorem*, a special class of functions is considered. A function $f : \mathbb{R} \to \mathbb{R}_+$ is said to be a *Pólya frequency function of order* 2, or PF_2 for short, if

$$\det \begin{pmatrix} f(x_1 - y_1) & f(x_1 - y_2) \\ f(x_2 - y_1) & f(x_2 - y_2) \end{pmatrix} \geq 0 \tag{3.3.3}$$

whenever $x_1 \leq x_2$ and $y_1 \leq y_2$. We leave it to the reader to show that the gamma distribution $\Gamma(a, \lambda)$ with $a \geq 1$, the uniform distribution $\mathrm{U}(a, b)$, and the Weibull distribution $\mathrm{W}(r, c)$ with $r \geq 1$ are PF_2.

Lemma 3.3.1 *Let $X_1, \ldots, X_n$ be independent and nonnegative random variables with densities $f_1, \ldots, f_n$, respectively. If f_i is* PF_2 *for all $i = 1, \ldots, n$, then $\boldsymbol{X}_{t_1} \leq_{\mathrm{st}} \boldsymbol{X}_{t_2}$ for all $0 \leq t_1 \leq t_2$.*

The *proof* of Lemma 3.3.1 goes beyond the scope of the book; we refer to the bibliographical notes. Sufficient conditions for $k(x)$ in (3.3.1) to be a compensation function are contained in the following consequence of Lemma 3.3.1.

Theorem 3.3.2 *Let $n \geq 1$ be a fixed natural number and $k_1, \ldots, k_n$ arbitrary compensation functions. Consider the risk $X = \sum_{i=1}^N U_i$, where $N = n$ is deterministic, $U_1, \ldots, U_n$ are independent and each U_i is continuous with* PF_2 *density function f_{U_i}. Then the function k defined in* (3.3.1) *is a compensation function.*

Proof By Lemma 3.3.1 we immediately get that $k(x)$ is increasing. However,

$$x - k(x) = \mathbb{E}\Big(x - \sum_{i=1}^n k_i(U_i) \mid X = x\Big) = \mathbb{E}\Big(\sum_{i=1}^n (U_i - k_i(U_i)) \mid X = x\Big)$$

and also $g(\boldsymbol{u}) = \sum_{i=1}^n (u_i - k_i(u_i))$ is increasing in each argument. Hence, Lemma 3.3.1 also implies that $x - k(x)$ is increasing. Furthermore, using the monotonicity of conditional expectation and the fact that k_i is a compensation function, we have $0 \le \mathbb{E}(\sum_{i=1}^n k_i(U_i) \mid X = x) \le \mathbb{E}(\sum_{i=1}^n U_i \mid X = x) = x$ for each $x \ge 0$. Thus, $0 \le k(x) \le x$ and in particular $k(0) = 0$. □

For a given risk X, a reinsurance contract with retention function $h(x)$ is said to be *compatible* with respect to a premium calculation principle Π if

$$\Pi(X) = \Pi(h(X)) + \Pi(X - h(X)). \tag{3.3.4}$$

For example, the proportional contract is compatible with respect to the expected-value and standard-deviation principles, but not with respect to the variance and modified variance principles, unless the variance $\mathrm{Var}\, X$ of X vanishes. The stop-loss contract has the same compatibility properties, with the exception that in general it is not compatible with respect to the standard-deviation principle. Further, both the proportional contract and the stop-loss contract are compatible with respect to the absolute deviation principle given in (3.1.12). The compatibility of the stop-loss contract (and of other reinsurance contracts) with respect to the absolute deviation principle follows from the following property of quantile functions.

Lemma 3.3.2 *Let $v, w : \mathbb{R} \to \mathbb{R}$ be two increasing functions. Then, for each real-valued random variable Z,*

$$F^{-1}_{v(Z)+w(Z)}(z) = F^{-1}_{v(Z)}(z) + F^{-1}_{w(Z)}(z), \qquad 0 \le z \le 1. \tag{3.3.5}$$

Proof For any two functions $x(t), x'(t)$, by $x \circ x'(t)$ denotes the superposition $x(x'(t))$. Note that

$$F^{-1}_{v(Z)} = v \circ F^{-1}_Z . \tag{3.3.6}$$

Indeed, by Lemma 3.1.2b, for all $z \in (0, 1)$ and $t \in \mathbb{R}$

$$\begin{aligned}
F^{-1}_{v(Z)}(z) \le t &\iff F_{v(Z)}(t) \ge z \iff \mathbb{P}(v(Z) \le t) \ge z \\
&\iff \mathbb{P}(Z \le v^{-1}(t)) \ge z \iff F_Z(v^{-1}(t)) \ge z \\
&\iff F^{-1}_Z(z) \le v^{-1}(t) \iff v(F^{-1}_Z(z)) \le t
\end{aligned}$$

which gives (3.3.6). In the same vain we obtain $F^{-1}_{v(Z)} = v \circ F^{-1}_Z$ and $F^{-1}_{(v+w)(Z)} = (v + w) \circ F^{-1}_Z$, yielding

$$F^{-1}_{v(Z)+w(Z)} = (v + w) \circ F^{-1}_Z = v \circ F^{-1}_Z + w \circ F^{-1}_Z = F^{-1}_{v(Z)} + F^{-1}_{w(Z)} .$$ □

We say that two risks X, Y are *comonotonic* if there exist two increasing functions $v, w : \mathbb{R} \to \mathbb{R}$, and a probability space with a random variable Z defined on it such that $(X, Y) \overset{\mathrm{d}}{=} (v(Z), w(Z))$.

Lemma 3.3.3 *Let* Π *be the absolute deviation principle and let* X, Y *be arbitrary comonotonic risks with* $\mathbb{E}\,X, \mathbb{E}\,Y < \infty$. *Then* $\Pi(X+Y) = \Pi(X) + \Pi(Y)$.

Proof Using (3.1.13) and (3.3.5) we have

$$
\begin{aligned}
\Pi(X+Y) &= \int_0^{1/2} F_{X+Y}^{-1}(z)(1-a)\,\mathrm{d}z + \int_{1/2}^1 F_{X+Y}^{-1}(z)(1+a)\,\mathrm{d}z \\
&= \int_0^{1/2} F_{v(Z)+w(Z)}^{-1}(z)(1-a)\,\mathrm{d}z + \int_{1/2}^1 F_{v(Z)+w(Z)}^{-1}(z)(1+a)\,\mathrm{d}z \\
&= \int_0^{1/2} F_{v(Z)}^{-1}(z)(1-a)\,\mathrm{d}z + \int_{1/2}^1 F_{v(Z)}^{-1}(z)(1+a)\,\mathrm{d}z \\
&\quad + \int_0^{1/2} F_{w(Z)}^{-1}(z)(1-a)\,\mathrm{d}z + \int_{1/2}^1 F_{w(Z)}^{-1}(z)(1+a)\,\mathrm{d}z \\
&= \Pi(X) + \Pi(Y)\,.
\end{aligned}
$$

□

Theorem 3.3.3 *For each retention function* $h(x)$, *the corresponding reinsurance contract is compatible with respect to the absolute deviation principle.*

Proof Since the functions $h(x)$ and $k(x)$ are increasing, the random variables $h(X)$ and $k(X) = X - h(X)$ are comonotonic. Hence by Lemma 3.3.3 we get (3.3.4) for the absolute deviation principle. □

Bibliographical Notes. Lemma 3.3.1 has been derived in Efron (1965). More recent references to Efron's theorem are Shanthikumar (1987) and Daduna and Szekli (1996). Further effects on premium calculation of splitting a risk into two or more components have been studied for instance in Hürlimann (1994a,b), Mack (1997) and Michaud (1994); see also Hürlimann (1995). The compatibility of reinsurance contracts with respect to the absolute deviation principle has been investigated in Denneberg (1990). For a discussion of reinsurance premium calculation without arbitrage, see Albrecht (1992) and Venter (1991).

CHAPTER 4

Distributions of Aggregate Claim Amount

4.1 INDIVIDUAL AND COLLECTIVE MODEL

In this chapter we study different concepts related to aggregate claim amounts. As in this chapter we assume that the time horizon is fixed we do not include the time parameter. Traditionally, computing or approximating (graduating) the distribution function of the aggregate claim amount has been one of the central points in insurance mathematics. More recently, in the era of computers, approximation methods often lost their practical value. On the other hand, numerical methods like recursions or numerical inversion of Fourier transforms are becoming more important and produce excellent results for the case of a finite range of values. Nevertheless, bounds and asymptotic techniques, like the study of the tail behaviour of the distribution of the aggregate claim amount are still of interest. In order to investigate the distribution of the aggregate claim amount it is customary to consider one of the following two models.

Individual Model Consider a portfolio consisting of n policies with individual risks $U_1, \ldots, U_n$ over a given time period (one year, say). We assume that the nonnegative random variables $U_1, \ldots, U_n$ are independent, but not necessarily identically distributed. Let the distribution F_{U_i} of U_i be the mixture $F_{U_i} = (1 - \theta_i)\delta_0 + \theta_i F_{V_i}$, where $0 < \theta_i \leq 1$ and where F_{V_i} is the distribution of a (strictly) positive random variable V_i, $i = 1, \ldots, n$. In actuarial applications, the probabilities θ_i are small and can be interpreted as the probability that the i-th policy produces a positive claim V_i. The *aggregate claim amount* in this model, which we call the *individual model*, is $X^{\text{ind}} = \sum_{i=1}^{n} U_i$ with distribution $F_{U_1} * \ldots * F_{U_n}$. A portfolio is called *homogeneous* if $F_{V_1} = \ldots = F_{V_n}$.

Collective Model We suppose that a portfolio consists of a number of anonymous policies which we do not observe separately. The total number N of claims occurring in a given period is random. Further, the claim sizes

U_i are (strictly) positive and are assumed to form a sequence $U_1, U_2, \ldots$ of independent and identically distributed random variables. We also assume that the sequence $U_1, U_2, \ldots$ of individual claim sizes is independent of the claim number N. Typically, N has a Poisson, binomial or negative binomial distribution, but other choices are possible. This model is called the *collective model*, and the aggregate claim amount is the random variable $X^{\rm col} = \sum_{i=1}^N U_i$. Here and throughout the whole book we use the convention that $\sum_{i=1}^0 U_i = 0$.

The idea is to approximate the individual model by a suitably chosen collective model if the size n of the portfolio is large. This is done because the collective model is often mathematically easier to handle. In this connection, a crucial problem is how to specify the parameters of the collective model to have a good approximation. In Section 4.6 we study such approximations by collective models, in particular by Poisson compounds.

Bibliographical Notes. For a more extensive discussion of the practical background of the individual and the collective model investigated in insurance mathematics we refer, for instance, to Albrecht (1981), Beard, Pentikäinen and Pesonen (1984), Bohman and Esscher (1963), Bowers, Gerber, Hickman, Jones and Nesbitt (1986), Bühlmann (1970), Daykin, Pentikäinen and Pesonen (1994), Gerber (1979), Goovaerts, Kaas, van Heerwaarden and Bauwelinckx (1990), Heilmann (1988), Mack (1997), Panjer and Willmot (1992), Straub (1988) and Sundt (1993).

4.2 COMPOUND DISTRIBUTIONS

4.2.1 Definition and Elementary Properties

Suppose that we want to evaluate the total payment over a period (one year, month, week etc.) from a portfolio, either using the individual or the collective model. Let N be a nonnegative integer-valued random variable and $U_1, U_2, \ldots$ a sequence of nonnegative random variables. Then the random variable

$$X = \begin{cases} \sum_{i=1}^N U_i & \text{if } N \geq 1, \\ 0 & \text{if } N = 0, \end{cases} \tag{4.2.1}$$

is called *compound* and describes the aggregate claim amount in the individual model as well as in the collective model. We assume throughout this chapter that the random variables $N, U_1, U_2, \ldots$ are independent. If not stated otherwise, we also assume that $U_1, U_2, \ldots$ are identically distributed. The latter assumption is sometimes omitted when considering the individual model. We say that X has a *compound distribution* determined by the (compounding) probability function $\{p_k,\ k \in \mathbb{N}\}$ of N and by the distribution

F_U of U_i if the distribution of X is given by

$$F_X = \sum_{k=0}^{\infty} p_k F_U^{*k}, \tag{4.2.2}$$

where F_U^{*k} denotes the k-fold convolution of F_U. Note that compound distributions form a special class of mixtures of distributions as considered in Section 2.2.4.

We first show how to express the Laplace–Stieltjes transform $\hat{l}_X(s)$ in terms of the generating function $\hat{g}_N(s)$ and the Laplace–Stieltjes transform $\hat{l}_U(s)$.

Theorem 4.2.1 *For each $s \ge 0$,*

$$\hat{l}_X(s) = \hat{g}_N(\hat{l}_U(s)). \tag{4.2.3}$$

Proof We apply the law of total probability, using the fact that the random variables $\exp(-s\sum_{i=1}^{k} U_i)$ and N are independent and that $e^{-sU_1}, e^{-sU_2}, \ldots$ are independent and identically distributed. We get for each $s \ge 0$

$$\begin{aligned}
\mathbb{E}\exp\Bigl(-s\sum_{i=1}^{N} U_i\Bigr) &= \sum_{k=0}^{\infty} \mathbb{E}\Bigl(\exp\Bigl(-s\sum_{i=1}^{N} U_i\Bigr) \Bigm| N = k\Bigr)\mathbb{P}(N = k)\\
&= \sum_{k=0}^{\infty} \mathbb{E}\Bigl(\exp\Bigl(-s\sum_{i=1}^{k} U_i\Bigr)\Bigr)\mathbb{P}(N = k) = \sum_{k=0}^{\infty} \mathbb{E}\Bigl(\prod_{i=1}^{k} e^{-sU_i}\Bigr)\mathbb{P}(N = k)\\
&= \sum_{k=0}^{\infty} (\mathbb{E}\, e^{-sU})^k \mathbb{P}(N = k) = \hat{g}_N(\mathbb{E}\, e^{-sU}).
\end{aligned}$$

See (2.1.11) for the last but one equality. □

Remark Similarly, the moment generating function $\hat{m}_X(s)$ of X is well-defined at least for $s \le 0$ and can be expressed as $\hat{m}_X(s) = \hat{g}_N(\hat{m}_U(s))$. If the claim sizes are lattice (e.g. take values in $\mathbb{N}$), then X is also lattice and we can determine the generating function of X by $\hat{g}_X(s) = \hat{g}_N(\hat{g}_U(s))$; $-1 < s < 1$.

From Theorem 4.2.1 we easily get formulae for the first two moments of X.

Corollary 4.2.1 *Assume that the relevant moments exist. Then*

$$\mathbb{E}\, X = \mathbb{E}\, N \mathbb{E}\, U, \quad \operatorname{Var} X = \operatorname{Var} N(\mathbb{E}\, U)^2 + \mathbb{E}\, N \operatorname{Var} U. \tag{4.2.4}$$

Proof To use formulae (2.1.7) and (2.1.8) compute the first derivative of $\hat{l}_X(s)$ at $s = 0+$ to obtain

$$\mathbb{E}\, X = -\mathrm{d}/\mathrm{d}s\, \hat{l}_X(s)|_{s=0+} = -\hat{g}_N^{(1)}(1)\hat{l}_U^{(1)}(0+) = \mathbb{E}\, N \mathbb{E}\, U.$$

Further, the second derivative at $s = 0+$ yields

$$\begin{aligned}\mathbb{E}(X^2) = \mathrm{d}^2/\mathrm{d}s^2\, \hat{l}_X(s)|_{s=0+} &= \hat{g}_N^{(1)}(1)\hat{l}_U^{(2)}(0+) + \hat{g}_N^{(2)}(1)(\hat{l}_U^{(1)}(0+))^2 \\ &= \mathbb{E}(N^2)(\mathbb{E}\,U)^2 + \mathbb{E}\,N \mathrm{Var}\,U\,.\end{aligned}$$

Thus, $\mathrm{Var}\,X = \mathbb{E}(X^2) - (\mathbb{E}\,X)^2 = \mathrm{Var}\,N(\mathbb{E}\,U)^2 + \mathbb{E}\,N\mathrm{Var}\,U$. $\square$

Remark Note that the first identity in (4.2.4) is a special case of Wald's identity (9.1.33) given in Section 9.1.6. Furthermore, the second identity shows that the variance of the compound X consists of two components: one is induced by the variance of the compounding random variable N, the other by the variance of the summands $U_1, U_2, \ldots$.

Note also that for each distribution F_U of a nonnegative random variable U we have $F_U^{*k}(x) \le F_U^k(x)$ for all $k \in \mathbb{N}$, $x \in \mathbb{R}$. This entails for the compound distribution F_X in (4.2.2) that $F_X(x) \le \sum_{k=0}^{\infty} p_k F_U^k(x) = \hat{m}_N(F_U(x))$. Thus,

$$\frac{1 - F_X(x)}{1 - F_U(x)} \ge \frac{1 - \hat{m}_N(F_U(x))}{1 - F_U(x)}\,. \tag{4.2.5}$$

If the claims $U_1, U_2, \ldots$ are unbounded but finite, that is $F_U(x) < 1$ for all $x > 0$ and $\lim_{x\to\infty} F_U(x) = 1$, then (4.2.5) implies the inequality

$$\liminf_{x\to\infty} \frac{1 - F_X(x)}{1 - F_U(x)} \ge \mathbb{E}\,N\,, \tag{4.2.6}$$

since $\hat{m}_N^{(1)}(1-) = \mathbb{E}\,N$.

Equation (4.2.6) clearly shows that if the tail of the claim size distribution F_U converges dangerously slowly to zero, then also the aggregate claim amount will suffer from a similar drawback. If we require some further specific properties of the underlying ingredients $\{p_k\}$ and F_U, then the above inequality will actually turn into an asymptotic equality. To be more specific, recall Theorem 2.5.4, where it has been shown that if the generating function $\hat{g}_N(s)$ of the claim number N is analytic at the point $s = 1$ then $\lim_{x\to\infty} \overline{F_X}(x)/\overline{F_U}(x) = \mathbb{E}\,N$ if the claim size distribution is subexponential. However, in what follows in this chapter we will mainly be interested in cases where the claim size distribution is not heavy-tailed. This will then lead to exponential-type bounds.

In many cases it is rather difficult to determine a compound distribution analytically, i.e. in terms of a closed formula. Although by Theorem 4.2.1 we are able to determine transforms of compound distributions, we are unable to invert them analytically except for a few rather specific cases. However, as mentioned in Section 4.2.3 below, it is of great importance for actuarial purposes to know the probability $\mathbb{P}(X > x)$ of the compound X exceeding a given level x. For these reasons, several numerical methods have been developed to compute probabilistic characteristics of a compound distribution, using, for example, recursive algorithms, asymptotic techniques, bounds and further numerical approximation methods. They will be discussed in the following sections.

4.2.2 Three Special Cases

In actuarial applications three cases are of special interest:

- *Poisson compounds* where N has a Poisson distribution; in this case the distribution of the compound, determined by $\lambda = \mathbb{E}N$ and by the distribution F_U, is called a *compound Poisson distribution* with characteristics (λ, F_U).
- *Pascal* (or *negative binomial*) *compounds* with compounding distribution $\text{NB}(\alpha, p)$.
- *Geometric compounds* where N has a geometric distribution; in this case the distribution of the compound is determined by $p = 1 - \mathbb{P}(N = 0)$ and by the distribution F_U and is called a *compound geometric distribution.*

Recall that the binomial, Poisson, negative binomial, normal, Erlang and χ^2-distributions are closed under convolutions, as stated in Section 2.2. We now show that Poisson compounds share this useful property.

Theorem 4.2.2 *For some $n \in \mathbb{N}$, let $X = X_1 + \ldots + X_n$ be the sum of the independent Poisson compounds $X_1, \ldots, X_n$ with characteristics (λ_j, F_j) $(j = 1, \ldots, n)$. Then X has a compound Poisson distribution with characteristics (λ, F) given by*

$$\lambda = \sum_{j=1}^{n} \lambda_j \qquad \textit{and} \qquad F = \sum_{j=1}^{n} \frac{\lambda_j}{\lambda} F_j \,. \tag{4.2.7}$$

Proof Consider the representation $X_j = \sum_{i=1}^{N_j} U_{ij}$ of X_j where N_j has a Poisson distribution with parameter λ_j, and where $U_{1j}, U_{2j}, \ldots$ are independent and identically distributed with distribution F_j. Let U_j be a generic random variable with distribution F_j. Since N_j has generating function $\hat{g}_{N_j}(s) = \exp(-\lambda_j(1-s))$, we get from Theorem 4.2.1 that the Laplace–Stieltjes transform $\hat{l}_X(s)$ of X is given by

$$\hat{l}_X(s) = \prod_{j=1}^{n} \exp(-\lambda_j(1 - \hat{l}_{U_j}(s))) = \exp\Big(-\lambda\Big(1 - \sum_{j=1}^{n} \frac{\lambda_j}{\lambda} \hat{l}_{U_j}(s)\Big)\Big) \,.$$

To complete the proof note that the Laplace–Stieltjes transform of F is given by $\sum_{j=1}^{n} \frac{\lambda_j}{\lambda} \hat{l}_{U_j}(s)$ and that, consequently, $\exp(-\lambda(1 - \sum_{j=1}^{n} \frac{\lambda_j}{\lambda} \hat{l}_{U_j}(s)))$ is the Laplace–Stieltjes transform of a Poisson compound with characteristics

(λ, F). Now use the one-to-one correspondence between distributions and their Laplace–Stieltjes transforms. □

Examples 1. Consider the following special case of a Poisson compound. If the compounding probability function $\{p_k\}$ is Poisson with parameter λ and if the claim size distribution F_U is exponential, i.e. $F_U = \text{Exp}(\delta)$ for some $\delta > 0$, then the distribution function F_X of the aggregate claim amount has the form $F_X(x) = \sum_{n=0}^{\infty} \mathrm{e}^{-\lambda}(\lambda^n/n!)F_U^{*n}(x)$, where

$$F_U^{*n}(x) = \frac{\delta^n}{(n-1)!}\int_0^x \mathrm{e}^{-v\delta}v^{n-1}\,\mathrm{d}v$$

for $n = 1, 2, \ldots$. Note that the distribution function $F_X(x)$ has a jump of size $\mathrm{e}^{-\lambda}$ at the origin and is differentiable in the interval $(0, \infty)$ with

$$F_X^{(1)}(x) = \mathrm{e}^{-(\lambda+x\delta)}\sum_{n=1}^{\infty}\frac{1}{n!(n-1)!}\frac{1}{x}(x\lambda\delta)^n, \qquad x > 0.$$

Using definition (2.2.1) of the modified Bessel function $I_1(z)$ we see that

$$F_X^{(1)}(x) = \mathrm{e}^{-(\lambda+x\delta)}2\sqrt{\lambda\delta/x}\,I_1(2\sqrt{\lambda x\delta}), \qquad x > 0. \tag{4.2.8}$$

2. Another example with an explicit expression for the density part $F_X^{(1)}(x)$ is obtained for Pascal compounds with compounding distribution $\text{NB}(\alpha, p)$ and exponential claim size distribution $F_U = \text{Exp}(\delta)$. Here the aggregate claim amount X has distribution function

$$F_X(x) = \sum_{n=0}^{\infty}\binom{\alpha+n-1}{n}(1-p)^{\alpha}p^nF_U^{*n}(x).$$

As a special case assume that the parameter $\alpha = m$, a strictly positive integer, and that the claim size distribution is the same as above, i.e. $\overline{F}_U(x) = \exp(-\delta x)$. Then we find that $F_X(x)$ has a jump of size $(1-p)^m$ at the origin. For the density part we need a bit more work. Introducing the densities of F_U^{*n} in the above formula we find that

$$F_X^{(1)}(x) = (1-p)^m p\delta\exp(-\delta x)\sum_{n=1}^{\infty}\binom{m+n-1}{n}\frac{1}{(n-1)!}(p\delta x)^{n-1} \tag{4.2.9}$$

for all $x > 0$. Put $y = p\delta x$. Then the series in (4.2.9), call it $R(y)$, can be tackled by the use of the confluent hypergeometric function defined in (2.2.3):

$$\begin{aligned} R(y) &= \frac{\mathrm{d}}{\mathrm{d}y}\sum_{n=0}^{\infty}\binom{m+n-1}{n}\frac{y^n}{n!} = \frac{\mathrm{d}}{\mathrm{d}y}\sum_{n=0}^{\infty}\frac{m(m+1)\ldots(m+n-1)}{1\cdot 2\cdot\ldots\cdot n}\frac{y^n}{n!} \\ &= \frac{\mathrm{d}}{\mathrm{d}y}M(m, 1; y). \end{aligned}$$

We now use that

$$\frac{\mathrm{d}}{\mathrm{d}y} M(m,1;y) = mM(m+1,2;y)\,, \quad M(m+1,2;y) = \mathrm{e}^y M(-m+1,2;-y)$$

to write $R(y) = mM(m+1,2;y) = m\mathrm{e}^y M(-m+1,2;-y)$. The latter factor $M(-m+1,2;-y)$ is a polynomial. It can be transformed into a generalized Laguerre polynomial defined in (2.2.6) by

$$\binom{m}{m-1} M(-m+1,2;y) = L^1_{m-1}(-y)\,.$$

Hence we ultimately find that, for $x > 0$,

$$F^{(1)}_X(x) = -p\delta(1-p)^m \exp(-(1-p)\delta x) L^{(1)}_m(-p\delta x)\,, \tag{4.2.10}$$

where $L^{(1)}_m(x) = \mathrm{d}/\mathrm{d}x L^0_m(x)$ denotes the derivative of the Laguerre polynomial $L_m(x) = L^0_m(x)$. Note that in the last equality we applied a classical identity (2.2.7) for Laguerre polynomials, in that $\mathrm{d}/\mathrm{d}x\; L^0_m(x) = -L^1_{m-1}(x)$.

4.2.3 Some Actuarial Applications

There are a variety of reasons why information on the distribution of the aggregate claim amount is of prime importance in actuarial practice. Let us illustrate its use by three examples.

1. In our first example we consider the risk reserve $R(t)$ of a portfolio at time t where we assume that the random variable $R(t)$ is given by $R(t) = u + \Pi(t) - X(t)$. Here $u \geq 0$ is the initial reserve, $\Pi(t)$ is the totality of premiums collected up to time t and $X(t)$ is the aggregate claim amount up to time t, i.e. $X(t) = \sum_{i=1}^{N(t)} U_i$, where $N(t)$ is the number of claims up to time t. The random fluctuation in time of the risk reserve process $\{R(t), t \geq 0\}$ is one of the main topics investigated in later chapters of this book. Assume now that we can derive a transparent upper bound for the tail function $\overline{F}_{X(t)}(x) = \mathbb{P}(X(t) > x)$ of the aggregate claim amount $X(t)$ at time t. So, $\overline{F}_{X(t)}(x) \leq g_+(t,x)$. The actuary wants to safeguard himself against a deficit at the end of the year, say at time point t_0. To do this, he allows a very small probability of at most ε that at t_0 the risk reserve is negative. More precisely, he puts $g_+(t_0, x) = \varepsilon$. Solve this equation for $x = x(t_0, \varepsilon)$. If the actuary chooses $u + \Pi(t_0) \geq x(t_0, \varepsilon)$, then

$$\begin{aligned}
\mathbb{P}(R(t_0) < 0) &= \mathbb{P}(X(t_0) > u + \Pi(t_0)) = \overline{F}_{X(t_0)}(u + \Pi(t_0)) \\
&\leq \overline{F}_{X(t_0)}(x(t_0,\varepsilon)) \leq g_+(t_0, x(t_0,\varepsilon)) = \varepsilon\,.
\end{aligned}$$

If we can solve the equation $g_+(t,x) = \varepsilon$ for all $x = x(t,\varepsilon)$ and arbitrary $t \geq 0$, then u can be chosen by the equality $u = x(0,\varepsilon)$. The premium $\Pi(t)$ that has to be collected by time t should then satisfy the inequality $\Pi(t) \geq x(t,\varepsilon) - x(0,\varepsilon)$.

2. In the next example we consider a risk X with distribution F_X and we assume that X is the aggregate claim amount over a certain period of time. One of the popular premium principles is the stop-loss principle, already considered in Chapter 3. We look at a slightly more general situation if we look at the *generalized stop-loss premium* $\Pi_{a,m}(X)$. This is defined by

$$\Pi_{a,m}(X) = \int_a^\infty (x-a)^m \, \mathrm{d}F_X(x)\,, \qquad a \geq 0,\ m \in \mathbb{N}\,. \tag{4.2.11}$$

In terms of expectations, this reads $\Pi_{a,m}(X) = \mathbb{E}\left((X-y)_+\right)^m$. Often one restricts attention to the special case where $m = 1$, the usual stop-loss premium. Note that also the case $m = 0$ is of particular interest since $\Pi_{a,0}(X) = \mathbb{P}(X > a) = \overline{F}_X(a)$ is the tail function of the aggregate claim amount X. Instead of (4.2.11) we can use an alternative definition of the generalized stop-loss premium derived from integration by parts. If $\mathbb{E}\,X^m < \infty$, then it is easily seen that for $m = 1, 2, \ldots$

$$\Pi_{a,m}(X) = m\int_0^\infty v^{m-1}\overline{F}_X(a+v)\,\mathrm{d}v = m\int_0^\infty v^{m-1}\Pi_{a+v,0}(X)\,\mathrm{d}v\,. \tag{4.2.12}$$

Thus, if we know a way to handle the tail function of the aggregate claim amount X, then a simple additional integration gives us full insight into the generalized stop-loss premium.

3. Let us turn to reinsurance. One of the most frequently applied reinsurance treaties is proportional reinsurance; see Section 3.3. On the basis of past experience, the first insurer wants to buy reinsurance from a reinsurance company. If the insurer has experienced an aggregate claim amount X in a certain year, then he might decide to reinsure a proportional part of next year's total claim amount, say $Z = aX$ for $0 < a < 1$. It is obvious that

$$\mathbb{P}(Z \leq x) = \mathbb{P}(X \leq xa^{-1}) = F_X(xa^{-1})\,.$$

In particular the (pure) net premium for this reinsurance contract is $\mathbb{E}\,Z = a\mathbb{E}\,X$. Now assume that we are able to derive a simple monotone lower bound for the tail function of X, say $\mathbb{P}(X \geq x) = \overline{F}_X(x) \geq g_-(x)$. The first insurer wants to avoid that the price for this reinsurance - his own premium - will be excessively large if he chooses a too large. One way of estimating this proportion goes as follows. First determine a value x_0 such that $\mathbb{P}(Z \geq x_0) < \varepsilon$ for a given ε. If the insurer takes a to satisfy the inequality $g_-(x_0/a) < \mathbb{P}(Z > x_0) < \varepsilon$, then by inversion $a < x_0/g_-^{-1}(\varepsilon)$, where $g_-^{-1}(x)$ is the generalized inverse function of $g_-(x)$.

4.2.4 Ordering of Compounds

In Section 4.2.3 we saw that, for actuarial purposes, it is useful to know lower and upper bounds for the tail function of a compound. This is closely related to the idea of ordering of risks. We now compare compounds with respect to stochastic and stop-loss orderings. Further bounds for compound distributions will be studied in Section 4.5.

Lemma 4.2.1 *Let $U_1, U_2, \ldots, U_n$ and $U'_1, U'_2, \ldots, U'_n$ be two sequences of independent and identically distributed random variables.*
(a) *If $U \le_{\mathrm{st}} U'$ then $\sum_{j=1}^n U_j \le_{\mathrm{st}} \sum_{j=1}^n U'_j$.*
(b) *If $U \le_{\mathrm{sl}} U'$ then $\sum_{j=1}^n U_j \le_{\mathrm{sl}} \sum_{j=1}^n U'_j$.*

The *proof* is left to the reader.

Theorem 4.2.3 *Consider two compounds $X = \sum_{i=1}^N U_i$ and $X' = \sum_{i=1}^{N'} U'_i$.*
(a) *If $N \le_{\mathrm{st}} N'$ and $U \le_{\mathrm{st}} U'$ then $X \le_{\mathrm{st}} X'$.*
(b) *If $N \le_{\mathrm{sl}} N'$ and $U \le_{\mathrm{sl}} U'$ then $X \le_{\mathrm{sl}} X'$.*

Proof Define $a_n = \mathbb{E}\, h(\sum_{j=1}^n U_j)$ for $n = 0, 1, \ldots$. If $h : \mathbb{R}_+ \to \mathbb{R}_+$ is increasing then the sequence $\{a_n\}$ is increasing. To prove (a) we take $h(x) = \mathbb{I}(x \ge b)$ for some fixed but arbitrary b. Then by Lemma 4.2.1a we have $a_n = \mathbb{E}\, h(\sum_{j=1}^n U_j) \le \mathbb{E}\, h(\sum_{j=1}^n U'_j) = a'_n$. Thus,

$$\mathbb{E}\, h\Big(\sum_{j=1}^N U_j\Big) = \sum_{n=1}^\infty p_n a_n \le \sum_{n=1}^\infty p'_n a_n \le \sum_{n=1}^\infty p'_n a'_n = \mathbb{E}\, h\Big(\sum_{j=1}^{N'} U'_j\Big).$$

To show (b), take now $h(x) = (x - b)_+$ for some fixed but arbitrary b. This function is increasing and convex. With a_n as before and Lemma 4.2.1b

$$a_n = \mathbb{E}\, h\Big(\sum_{j=1}^n U_j\Big) \le \mathbb{E}\, h\Big(\sum_{j=1}^n U'_j\Big) = a'_n.$$

We show that $\{a_n\}$ is a convex sequence that is $a_n + a_{n+2} \ge 2a_{n+1}$ for all $n = 0, 1, \ldots$. For n fixed, define the function $k(x) = \mathbb{E}\, h(\sum_{j=1}^n U_j + x)$. Since k is convex we have $k(x + y) + k(0) \ge k(x) + k(y)$ for $x, y \ge 0$. This gives $\mathbb{E}\, k(U_{n+1} + U_{n+2}) + k(0) \ge \mathbb{E}\, k(U_{n+1}) + \mathbb{E}\, k(U_{n+2})$, from which the convexity of $\{a_n\}$ follows. Now

$$\mathbb{E}\, h\Big(\sum_{j=1}^N U_j\Big) = \sum_{n=1}^\infty p_n a_n \le \sum_{n=1}^\infty p'_n a_n \le \sum_{n=1}^\infty p'_n a'_n = \mathbb{E}\, h\Big(\sum_{j=1}^{N'} U_j\Big).$$ □

4.2.5 The Larger Claims in the Portfolio

In this section we collect a few results in connection with the larger claims that are part of the aggregate claim amount. Recall that for the claims $U_1, \ldots, U_n$

the largest claim is denoted by $U_{(n)} = \max\{U_1, \ldots, U_n\}$. The second largest claim is $U_{(n-1)}$ and in general the k-th largest claim is $U_{(n-k+1)}$. Consider now a random number N of claims $U_1, \ldots, U_N$, where $U_1, U_2, \ldots$ are independent and identically distributed and independent of N. Fix the integer k and ask for the distribution of the k-th largest claim. We put $U_{(N-k+1)} = 0$ if $N < k$.

Theorem 4.2.4 *The distribution of the k-th largest claim is given by*

$$\mathbb{P}(U_{(N-k+1)} \geq x) = \frac{1}{(k-1)!} \int_x^\infty \hat{g}_N^{(k)}(F_U(y))(1 - F_U(y))^{k-1} \, \mathrm{d}F_U(y). \tag{4.2.13}$$

Proof Fix the integer $k = 1, 2 \ldots$. One possibility is that not even k claims have occurred yet and then $N < k$. This event happens with probability $\mathbb{P}(N \leq k-1) = 1 - r_k$, where $r_k = p_k + p_{k+1} + \ldots$ and $p_k = \mathbb{P}(N = k)$. Hence $\mathbb{P}(U_{(N-k+1)} \leq x) = 1 - r_k + \sum_{n=k}^\infty \mathbb{P}(U_{(N-k+1)} \leq x \mid N = n) p_n$ by the law of total probability. The conditional probability equals

$$\mathbb{P}(U_{(n-k+1)} \leq x) = \frac{n!}{(n-k)!(k-1)!} \int_0^x (1 - F_U(y))^{k-1} F_U^{n-k}(y) \, \mathrm{d}F_U(y).$$

Indeed, any of the n claims can play the role of the requested order statistic $U_{(n-k+1)}$. Assume the latter takes a value in an interval $[y, y + \mathrm{d}y)$ which happens with probability $\mathrm{d}F_U(y)$. The remaining $n - 1$ claims can be binomially distributed into $n - k$ smaller than U_{n-k+1} and $k - 1$ larger than U_{n-k+1}. The probability to have a claim smaller than the one in $[y, y + \mathrm{d}y)$ equals $F_U(y)$, while the claim will be larger with probability $1 - F_U(y)$. The remaining sum can be written in a simplified version thanks to the expression

$$\hat{g}_N^{(k)}(s) = \sum_{n=k}^\infty \frac{n!}{(n-k)!} p_n s^{n-k}$$

which can be derived from the definition of the generating function in Section 2.1.5 whenever $|s| < 1$. We therefore obtain the expression

$$\mathbb{P}(U_{(N-k+1)} \leq x) = 1 - r_k + \frac{1}{(k-1)!} \int_0^x \hat{g}_N^{(k)}(F_U(y))(1 - F_U(y))^{k-1} \, \mathrm{d}F_U(y).$$

Changing to complementary events gives the statement of the theorem. □

Note in particular that the distribution of the largest claim in a portfolio has the rather transparent tail function $\mathbb{P}(U_{(N)} > x) = 1 - \hat{g}_N(F_U(x))$, a formula that caused excitement in actuarial circles when it was discovered.

Another consequence of Theorem 4.2.4 is that we can obtain formulae for the moments of the k-th largest order statistics. Namely, using (4.2.13) we find that

$$\begin{aligned}\mathbb{E}\left(U_{(N-k+1)}\right)^n &= -\int_0^\infty x^n \mathrm{d}\mathbb{P}(U_{(N-k+1)} > x) \\ &= \frac{1}{(k-1)!}\int_0^\infty y^n \hat{g}_N^{(k)}(F_U(y))(1-F_U(y))^{k-1}\,\mathrm{d}F_U(y)\,.\end{aligned}$$

It is obvious that the explicit evaluation of the above integrals in closed form is often impossible. Even in the simplest cases the resulting integrals can seldom be expressed by simple functions.

Example Let the claim number N be Poisson distributed with mean λ while the claim size distribution F_U is exponential with mean δ^{-1}. Then $\hat{g}_N(z) = \mathrm{e}^{-\lambda(1-z)}$ and $\hat{g}_N^{(k)}(z) = \lambda^k \mathrm{e}^{-\lambda(1-z)}$. Further, $F_U(y) = 1-\mathrm{e}^{-\delta y}$. Changing $\mathrm{e}^{-\delta y} = x$ we derive that

$$\mathbb{E}\left(U_{(N-k+1)}\right)^n = \frac{\lambda^k}{\delta^n (k-1)!}\int_0^1 \mathrm{e}^{-\lambda x}(-\log x)^n x^{k-1}\,\mathrm{d}x$$

which cannot be further simplified.

Similarly, the case of a Pareto distribution F_U does not simplify. In the latter case we assume that $F_U(x) = 1 - x^{-\alpha}$ for $x > 1$. As before we have now

$$\mathbb{E}\left(U_{(N-k+1)}\right)^n = \frac{\lambda^k}{(k-1)!}\int_0^1 \mathrm{e}^{-\lambda x} x^{k-1-(n/\alpha)}\,\mathrm{d}x$$

which only exists when $n < \alpha k$. Note that the existence not only depends on n but also on the relative position k of the order statistic $U_{(N-k+1)}$.

To show the difference between the two situations above and their influence on the aggregate claim amount we consider the mean $\mathbb{E}\,U_{(N)}$ of the largest claim on the total portfolio. For a proper and simple comparison we take $\lambda = 100$ and $\delta = \alpha - 1 = 1$, i.e. the means of the claim size distributions coincide. In Table 4.2.1 we compare the mean aggregate claim amount $\mathbb{E}\,X = \mathbb{E}\,N\mathbb{E}\,U = 100$ with the percentage that the largest claims are expected to contribute to this mean. We use the abbreviations

$$a_k = \frac{100^k}{(k-1)!}\int_0^1 \mathrm{e}^{-100x}(-\log x)^n x^{k-1}\,\mathrm{d}x$$

and

$$b_k = \frac{100^k}{(k-1)!}\int_0^1 \mathrm{e}^{-100x} x^{k-3/2}\,\mathrm{d}x$$

corresponding to the exponential, respectively the Pareto, case. To have an even better comparison we also consider the cumulative values $a'_k = \sum_{j\le k} a_j$

and $b'_k = \sum_{j\le k} b_j$ of the percentages that the largest claims contribute within the aggregate claim amount. For example, the largest claim in the Pareto case can be expected to be more than three times larger than in the exponential case. Furthermore, the eleven highest claims in the Pareto case take on the average 2/3 of the total, while this is only 1/3 in the exponential case.

k	a_k	a'_k	b_k	b'_k
1	5.18	5.18	17.72	17.72
2	4.18	9.36	8.86	26.58
3	3.68	13.06	6.65	33.23
4	3.35	16.40	5.54	38.77
5	3.10	19.50	4.85	43.62
6	2.90	22.39	4.36	47.98
7	2.73	25.13	4.00	51.98
8	2.58	27.72	3.71	55.69
9	2.46	30.18	3.48	58.17
10	2.35	32.53	3.29	62.46
11	2.25	34.78	3.12	65.58
12	2.16	36.94	2.98	68.56

Table 4.2.1 Comparison of the largest claims

Bibliographical Notes. The material presented in Section 4.2 is standard. For another version of (4.2.10) see Panjer and Willmot (1981). Ordering of compounds as in Theorem 4.2.3 is considered, for instance, in Borovkov (1976), Goovaerts, De Vylder and Haezendonck (1982,1984), Kaas, van Heerwaarden and Goovaerts (1994), Jean-Marie and Liu (1992) and Rolski (1976). Stop-loss ordering for portfolios of dependent risks are studied in Dhaene and Goovaerts (1996,1997) and Müller (1997).

4.3 CLAIM NUMBER DISTRIBUTIONS

As before, let N denote the number of claims incurred by the insurance company in a given time period. In Section 4.2 we saw that the probability function $\{p_k\}$ of N is an important element of compound distributions. In this section we collect some popular examples of claim number distributions and their properties. The reason to use these particular examples has often been based on actuarial intuition. More sound arguments will come from our study in later chapters in connection with models for the time-dependent behaviour of portfolios.

4.3.1 Classical Examples; Panjer's Recurrence Relation

There are at least three particular cases that are immediately applicable in insurance. The *Poisson distribution* $\mathrm{Poi}(\lambda)$ is by far the most famous example of a claim number distribution. Recall that then $p_k = e^{-\lambda}\lambda^k/k!$ for $k \in \mathbb{N}$, where $\lambda = \mathbb{E}\,N$ is the mean number of claims. Moreover, $\mathrm{Var}\,N = \lambda$ so that the index of dispersion $\mathrm{I}_N = \mathrm{Var}\,N/\mathbb{E}\,N$ equals 1.

In the next two examples we show an *overdispersed distribution* (index of dispersion greater than 1) and an *underdispersed distribution* (index of dispersion less than 1).

The *negative binomial* or *Pascal distribution* $\mathrm{NB}(\alpha, p)$ is another favoured claim number distribution. Recall that then

$$p_k = \binom{\alpha+k-1}{k}(1-p)^\alpha p^k\,, \qquad k \in \mathbb{N}\,,$$

where $\alpha > 0$ and $p \in (0,1)$. Now, $\mathbb{E}\,N = \alpha p/(1-p)$ and $\mathrm{Var}\,N = \alpha p/(1-p)^2$. Notice that $\mathrm{I}_N = p^{-1} > 1$. The overdispersion of the Pascal distribution is one reason for its popularity as an actuarial model.

The *binomial distribution* $\mathrm{Bin}(n, p)$ is an underdispersed distribution. Assume that an insurance policy covers the total breakdown of a vital component in a computer system. At the beginning of the year the portfolio covers n identically manufactured and newly installed components in a computer park. The probability that an arbitrary but fixed component will fail before time t is given by a distribution function $G(t)$. The probability p_k that the number N of breakdowns up to time t is equal to k is given by

$$p_k = \binom{n}{k}p^k(1-p)^{n-k}\,, \qquad 0 \le k \le n\,,$$

where $p = G(t)$. For such a *binomial model* the mean number of claims equals $\mathbb{E}\,N = np$, while $\mathrm{Var}\,N = np(1-p)$. The index of dispersion is now $\mathrm{I}_N = 1-p$, which is less than 1. The latter result is intuitive since the larger the value of t, the smaller is $1-p = 1-G(t)$, and hence the smaller the dispersion.

We can check that the probability functions $\{p_k\}$ of the Poisson, negative binomial and binomial distribution all fulfil *Panjer's recurrence relation*

$$p_k = \left(a + \frac{b}{k}\right)p_{k-1}\,, \qquad k = 1, 2, \ldots\,, \tag{4.3.1}$$

where $a < 1$ and $b \in \mathbb{R}$ are some constants. The following result shows that there are no other distributions which satisfy (4.3.1).

Theorem 4.3.1 *Suppose that the probability function $\{p_k\}$ of the $\mathbb{N}$-valued random variable N satisfies* (4.3.1). *Then $\{p_k\}$ is the probability function of*

a binomial, Poisson or negative binomial distribution. More specifically:
(a) *If* $a = 0$, *then* $b = \lambda > 0$, *and* N *has the Poisson distribution* $\text{Poi}(\lambda)$.
(b) *If* $0 < a < 1$, *then* $a + b > 0$, *and* N *has the negative binomial distribution* $\text{NB}(\alpha, p)$ *with* $p = a$ *and* $\alpha = 1 + bp^{-1}$.
(c) *If* $a < 0$, *then* $b = -a(n+1)$ *for some* $n \in \mathbb{N}$, *and* N *has the binomial distribution* $\text{Bin}(n, p)$ *with* $p = a(a-1)^{-1}$ *and* $n = -1 - ba^{-1}$.

Proof First take the case $a = 0$. Then $b > 0$ is needed. Inserting the left-hand side of (4.3.1) repeatedly into its right-hand side, we see that $p_k = p_0(b^k/k!)$ and hence, since $\sum_{k=0}^{\infty} p_k = 1$, we deduce that $p_0 = \mathrm{e}^{-b}$. Thus, $\{p_k\}$ is the probability function of the Poisson distribution $\text{Poi}(b)$. Assume now that $a \neq 0$. Then by repeated application of the recursion (4.3.1) we find that

$$p_k = p_0(\Delta + k - 1)(\Delta + k - 2)\ldots(\Delta + 1)\Delta\frac{a^k}{k!}, \qquad k \in \mathbb{N},$$

where $\Delta = ba^{-1} + 1$. Since $\sum_{k=0}^{\infty} c(c+1)\ldots(c+k-1)z^k/k! = (1-z)^{-c}$ for $|z| < 1$ and all $c \in \mathbb{R}$, we see that formally

$$p_k = \binom{\Delta + k - 1}{k} a^k (1-a)^{\Delta}, \qquad k \in \mathbb{N}.$$

Note that for $0 < a < 1$ the expression on the right is positive for all $k \in \mathbb{N}$ if and only if $\Delta > 0$ or equivalently $b > -a$. This is the case if and only if we are in the negative binomial situation. For $a < 0$ the probabilities in (4.3.1) are nonnegative only if $b = -a(n+1)$ for some $n \in \mathbb{N}$. This means that $-\Delta \in \mathbb{N}$, which leads to the binomial case. □

It is rather obvious that (4.3.1) can be weakened so that many other specific distributions appear as candidates. For example, if we start recursion (4.3.1) only at $k = 2$ rather than at $k = 1$ then p_0 appears as a free parameter while the other terms are given by a shifted version of the distributions appearing in Theorem 4.3.1. However, there are a few new solutions as well.

4.3.2 Discrete Compound Poisson Distributions

Assume that claims occur in bulk, where the number of bulks N' occurring in a given period of time follows a Poisson distribution with parameter λ. Each bulk consists of a random number of claims so that the total number of claims is of the form $N = \sum_{i=1}^{N'} Z_i$, where Z_i denotes the number of claims in the i-th bulk. Assume that $\{Z_i,\ i = 1, 2, \ldots\}$ is a sequence of independent random variables with values in $\{1, 2, \ldots\}$ and with common probability function $\{p_k^Z,\ k \geq 1\}$. If the claim numbers $Z_1, Z_2, \ldots$ are independent of the bulk number N' then $\hat{g}_N(s) = (-\lambda(1 - \hat{l}_Z(s)))$. From Corollary 4.2.1 we have $\mathbb{E}\,N = \lambda\,\mathbb{E}\,Z$ and $\text{Var}\,N = \lambda\,\mathbb{E}\,(Z^2)$. The total number of claims N henceforth follows a *(discrete) compound Poisson distribution.* The probability function $\{p_k\}$ of N is given by

$$p_k = \sum_{n=0}^{\infty} \mathrm{e}^{-\lambda} \frac{\lambda^n}{n!} \left(p^Z\right)_k^{*n}, \qquad k \in \mathbb{N}, \tag{4.3.2}$$

where $\{(p^Z)_k^{*n},\ k = 1,2,\ldots\}$ is the n-fold convolution of $\{p_k^Z\}$. Needless to say that the explicit evaluation of the probabilities p_k in (4.3.2) is mostly impossible because of the complicated nature of the convolutions. However there exists a simple recursive procedure for these probabilities shown in Section 4.4.2.

In a few special cases it is possible to determine the probabilities p_k directly. Consider the following example. Assume that Z is governed by the truncated geometric distribution TG(p), i.e. $\hat{l}_Z(s) = (1-p)\mathrm{e}^{-s}(1-p\,\mathrm{e}^{-s})^{-1}$ with $0 < p < 1$. The Laplace transform of N is then

$$\hat{l}_N(s) = \mathrm{e}^{-\lambda} \exp\Big(\frac{\lambda(1-p)\mathrm{e}^{-s}}{1-p\mathrm{e}^{-s}}\Big).$$

The probabilities p_k can be evaluated in terms of the generalized Laguerre polynomials $L_{k-1}^1(x)$ using formula (2.2.8). Note that

$$\frac{\mathrm{d}}{\mathrm{d}z} \mathrm{e}^{(xz)/(z-1)} = -x(1-z)^{-2}\mathrm{e}^{(xz)/(z-1)}$$

and $\hat{g}_N(s) = \mathrm{e}^{-\lambda}\mathrm{e}^{(xz)/(z-1)}$, where $x = -\lambda(1-p)/p$ and $z = ps$. This yields

$$p_k = \mathbb{P}(N = k) = \begin{cases} \mathrm{e}^{-\lambda} & \text{if } k = 0, \\ \mathrm{e}^{-\lambda} \frac{\lambda(1-p)p^{k-1}}{k} L_{k-1}^1\left(-\frac{\lambda(1-p)}{p}\right) & \text{if } k \geq 1. \end{cases} \tag{4.3.3}$$

The distribution given in (4.3.3) is called the *Pólya–Aeppli distribution.*

4.3.3 Mixed Poisson Distributions

Imagine a situation where the counting variable N consists of two or more different subvariables that individually follow a Poisson distribution but with a different parameter value. In motor insurance, for example, one might like to make a difference between male and female car owners; or the insurer may use layers in the age structure of his insured drivers. In general one assumes that the claims come from a heterogeneous group of policyholders; each one of them produces claims according to a Poisson distribution Poi(λ), where the intensity of claims λ varies from one policy to another according to an intensity distribution.

In mathematical terms this means that the parameter λ for a subvariable should be considered as the outcome of a random variable Λ in the sense that

$\mathbb{P}(N = k \mid \Lambda = \lambda) = \mathrm{e}^{-\lambda}\lambda^k/k!$ for each $k \in \mathbb{N}$. The random variable Λ itself is called the *mixing* or *structure variable*. Its distribution is called the *mixing* or *structure distribution* and denoted by F; $F(\lambda) = \mathbb{P}(\Lambda \le \lambda)$. Furthermore, we say that N has a *mixed Poisson distribution* with mixing distribution F.

We get for the (unconditional) probability function $\{p_k\}$ of N

$$p_k = \int_0^\infty \mathrm{e}^{-\lambda}\frac{\lambda^k}{k!}\,\mathrm{d}F(\lambda)\,, \qquad k \in \mathbb{N}\,,$$

which gives $\hat{l}_N(s) = \int_0^\infty \exp(-\lambda(1-\mathrm{e}^{-s}))\,\mathrm{d}F(\lambda)$. The latter relation immediately yields $\mathbb{E}\,N = \int_0^\infty \lambda\,\mathrm{d}F(\lambda) = \mathbb{E}\,\Lambda$ and similarly $\mathrm{Var}\,N = \mathbb{E}\,\Lambda + \mathrm{Var}\,\Lambda$. This expression for the variance $\mathrm{Var}\,N$ shows that among all mixed Poisson distributions with fixed expectation, the Poisson distribution has the smallest variance. Also note that the index of dispersion is easily found to be equal to $\mathrm{I}_N = 1 + (\mathrm{Var}\,\Lambda/\mathbb{E}\,\Lambda) = 1 + \mathrm{I}_\Lambda$ which is minimal for the Poisson distribution. In other words, a mixed Poisson distribution is overdispersed provided that the mixing distribution F is not degenerate. The latter property has been at the origin of the success of mixed Poisson distributions in actuarial data fitting.

Let us give an example which shows the flexibility of the mixed Poisson model. Further examples are discussed in the bibliographical notes at the end of this section. Assume that Λ has the gamma distribution $\Gamma(\alpha, c)$. By a simple calculation we obtain

$$p_k = \binom{\alpha+k-1}{k}\left(\frac{c}{1+c}\right)^\alpha \left(\frac{1}{1+c}\right)^k .$$

We immediately recognize the negative binomial distribution $\mathrm{NB}(\alpha, 1/(1+c))$ which appeared in Section 4.3.1.

Bibliographical Notes. Since Panjer (1981), a variety of generalizations of the recursion (4.3.1) have appeared in the actuarial literature. Occasionally, one considers a somewhat larger class of claim number distributions assuming that (4.3.1) merely holds for $k = r, r+1, \ldots$, where $r \ge 1$ is an arbitrary fixed natural number; see, for example, Sundt and Jewell (1981) and Willmot (1988). Note that the logarithmic distribution belongs to that class with $r = 2$. For more general recursions and related results, see Sundt (1992). Because of their importance in forthcoming evaluations of compound distributions we have treated some of these claim number distributions here, leaving further examples to the exercises. We also remark that Sichel (1971) introduced a distribution that can be obtained as a mixed Poisson distribution with mixing distribution F being a *generalized inverse Gaussian distribution*. This means that the density $f(x)$ of F has the form

$$f(\lambda) = \frac{\mathrm{d}F(\lambda)}{\mathrm{d}\lambda} = \frac{c^{-b}\lambda^{b-1}}{2K_b(c/a)}\exp\left(-\frac{\lambda^2+c^2}{2a\lambda}\right),$$

where the three parameters a, b and c are nonnegative. The function K_b is the modified Bessel function of the third kind. The case where $b = -1/2$ is particularly interesting since then the general inverse Gaussian distribution simplifies to the classical inverse Gaussian distribution. The resulting mixed distribution is called the *Sichel distribution* or *Poisson-inverse Gauss distribution*. Willmot (1987) illustrates the usefulness of the Poisson-inverse Gauss distribution as an alternative to the negative binomial distribution. In particular the distribution has been fitted to automobile frequency data. Ruohonen (1988) advocates the *Delaporte distribution* which had been introduced in Delaporte (1960) for claim number data involving car insurance. This distribution is obtained as a mixed Poisson distribution with a shifted gamma mixing distribution whose density $f(x)$ has the form

$$f(\lambda) = \frac{b^a}{\Gamma(a)}(\lambda - c)^{a-1} \exp(-(\lambda - c)b), \qquad \lambda > c.$$

More recently the Delaporte distribution has been considered by Willmot and Sundt (1989) and Schröter (1990). The probability function $\{p_k\}$ of this distribution can be given in terms of degenerate hypergeometric functions. A comprehensive survey on mixed Poisson distributions is given in Grandell (1997).

4.4 RECURSIVE COMPUTATION METHODS

We now discuss recursive methods to compute the distribution of the aggregate claim amount for individual and collective risk models. We first assume that the individual claim amounts are discrete random variables, say random integer multiples of some monetary unit. Note that continuous individual claim amounts can also be analysed by these methods provided that the claim amounts are previously given under discretization. An alternative approach to computing the distribution function of a continuous compound uses the integral equations stated in Section 4.4.3.

4.4.1 The Individual Model: De Pril's Algorithm

Consider the following individual model which describes a portfolio of n independent insurance policies. Suppose that each policy has an individual claim amount which is a random integer multiple of some monetary unit. Furthermore, suppose that the portfolio can be divided into a number of classes by gathering all policies with the same claim probability and the same conditional claim amount distribution. Let n_{ij} be the number of policies with claim probability $\theta_j < 1$ and with conditional claim amount probabilities $p_1^{(i)}, \ldots, p_{m_i}^{(i)}$, i.e. the individual claim amount distribution F_{ij} for each policy of this class is given by the mixture $F_{ij} = (1-\theta_j)\delta_0 + \theta_j \sum_{k=1}^{m_i} p_k^{(i)} \delta_k$. The generating function $\hat{g}(s)$ of the aggregate claim amount X^{ind} in this model is

$$\hat{g}(s) = \prod_{i=1}^{a} \prod_{j=1}^{b} \Big(1 - \theta_j + \theta_j \sum_{k=1}^{m_i} p_k^{(i)} s^k\Big)^{n_{ij}} \tag{4.4.1}$$

where a is the number of possible conditional claim amount distributions and b is the number of different claim probabilities. Using (4.4.1) we can calculate the probabilities $p_k = \mathbb{P}(X^{\text{ind}} = k)$ recursively for $k = 0, 1, \ldots, m$, where $m = \sum_{i=1}^{a} \sum_{j=1}^{b} n_{ij} m_i$ is the maximal aggregate claim amount.

Theorem 4.4.1 *The probability function $\{p_k\}$ of X^{ind} can be computed by*

$$p_0 = \prod_{i=1}^{a} \prod_{j=1}^{b} (1-\theta_j)^{n_{ij}}\,, \qquad p_k = \frac{1}{k} \sum_{i=1}^{a} \sum_{j=1}^{b} n_{ij} v_{ij}(k) \tag{4.4.2}$$

for $k = 1, \ldots, m$, where

$$v_{ij}(k) = \frac{\theta_j}{1-\theta_j} \sum_{l=1}^{m_i} p_l^{(i)} (l p_{k-l} - v_{ij}(k-l)) \tag{4.4.3}$$

for $k = 1, \ldots, m$ and $v_{ij}(k) = 0$ otherwise.

Proof Letting $s = 0$ in (4.4.1) leads to the first formula. In order to prove the second, take the derivative on both sides of (4.4.1). This gives

$$\begin{aligned} \hat{g}^{(1)}(s) &= \hat{g}(s) \frac{\mathrm{d} \log \hat{g}(s)}{\mathrm{d}s} = \hat{g}(s) \sum_{i=1}^{a} \sum_{j=1}^{b} n_{ij} \frac{\theta_j \sum_{k=1}^{m_i} p_k^{(i)} k s^{k-1}}{1 - \theta_j + \theta_j \sum_{k=1}^{m_i} p_k^{(i)} s^k} \\ &= \sum_{i=1}^{a} \sum_{j=1}^{b} n_{ij} \frac{\theta_j \hat{g}_i^{(1)}(s) \hat{g}(s)}{1 - \theta_j + \theta_j \hat{g}_i(s)} = \sum_{i=1}^{a} \sum_{j=1}^{b} n_{ij} V_{ij}(s)\,, \end{aligned}$$

where

$$V_{ij}(s) = \frac{\theta_j \hat{g}_i^{(1)}(s) \hat{g}(s)}{1 - \theta_j + \theta_j \hat{g}_i(s)} \tag{4.4.4}$$

and $\hat{g}_i(s) = \sum_{k=1}^{m_i} p_k^{(i)} s^k$. Taking the derivative of order $k-1$ and inserting $s = 0$ yields

$$p_k = \frac{1}{k!} \frac{\mathrm{d}^{k-1}}{\mathrm{d}s^{k-1}} \hat{g}^{(1)}(s)|_{s=0} = \frac{1}{k!} \sum_{i=1}^{a} \sum_{j=1}^{b} n_{ij} V_{ij}^{(k-1)}(s)|_{s=0} = \frac{1}{k} \sum_{i=1}^{a} \sum_{j=1}^{b} n_{ij} v_{ij}(k)\,,$$

where $v_{ij}(k) = ((k-1)!)^{-1} V_{ij}^{(k-1)}(s)|_{s=0}$. Furthermore, (4.4.3) is obtained by differentiating the following expression $k-1$ times, which is equivalent to

(4.4.4): $V_{ij}(s) = \theta_j(1-\theta_j)^{-1}\big(\hat{g}_i^{(1)}(s)\hat{g}(s) - \hat{g}_i(s)V_{ij}(s)\big)$ and letting $s=0$. By differentiation,

$$\begin{aligned}
V_{ij}^{(k-1)}(s)|_{s=0} &= \frac{\theta_j}{1-\theta_j}\sum_{l=0}^{k-1}\binom{k-1}{l}\big(\hat{g}_i^{(l+1)}(s)\hat{g}^{(k-1-l)}(s) \\
&\qquad - \hat{g}_i^{(l)}(s)V_{ij}^{(k-1-l)}(s)|_{s=0}\big) \\
&= \frac{\theta_j}{1-\theta_j}\Big(\sum_{l=1}^{k}\binom{k-1}{l}l!\,p_l^{(i)}(k-l)!\,p_{k-l} \\
&\qquad -\sum_{l=1}^{k-1}\binom{k-1}{l}l!\,p_l^{(i)}V_{ij}^{(k-1-l)}(s)|_{s=0}\Big) \\
&= (k-1)!\,\frac{\theta_j}{1-\theta_j}\Big(kp_k^{(i)}p_0 + \sum_{l=1}^{k-1}p_l^{(i)}(lp_{k-l} - v_{ij}(k-l))\Big) \\
&= (k-1)!\,\frac{\theta_j}{1-\theta_j}\sum_{l=1}^{m_i}p_l^{(i)}(lp_{k-l} - v_{ij}(k-l))\,.
\end{aligned}$$

This completes the proof. □

In the special case of an individual model which describes a portfolio of independent life insurance policies, i.e. $p_i^{(i)} = 1$, we get the following result.

Corollary 4.4.1 *If $p_i^{(i)} = 1$ for all $i = 1,\ldots,a$, then the probability function $\{p_k\}$ of $X^{\rm ind}$ can be computed by* (4.4.2) *where*

$$v_{ij}(k) = \frac{\theta_j}{1-\theta_j}(ip_{k-i} - v_{ij}(k-i))\,. \tag{4.4.5}$$

Notice that the result of Corollary 4.4.1 is an efficient reformulation of another recursive scheme which is usually called *De Pril's algorithm.*

Corollary 4.4.2 *If $p_i^{(i)} = 1$ for all $i = 1,\ldots,a$, then the probability function $\{p_k\}$ of $X^{\rm ind}$ satisfies the recursion formula*

$$p_k = \frac{1}{k}\sum_{i=1}^{\min\{a,k\}}\sum_{l=1}^{\lfloor k/i\rfloor} c_{il}p_{k-li}\,, \tag{4.4.6}$$

where $\lfloor x\rfloor = \max\{n \in \mathbb{N} : n \le x\}$ *and*

$$c_{il} = (-1)^{l+1}i\sum_{j=1}^{b} n_{ij}\left(\frac{\theta_j}{1-\theta_j}\right)^l. \tag{4.4.7}$$

Proof Define $\tilde{p}_k$ for $k = 0, 1 \ldots, m$ by $\tilde{p}_0 = p_0$ and

$$\tilde{p}_k = \frac{1}{k} \sum_{i=1}^{\min\{a,k\}} \sum_{l=1}^{\lfloor k/i \rfloor} c_{il} \tilde{p}_{k-li} \tag{4.4.8}$$

for $k \geq 1$. Inserting (4.4.7) into (4.4.8) gives

$$\tilde{p}_k = \frac{1}{k} \sum_{i=1}^{\min\{a,k\}} \sum_{j=1}^{b} n_{ij} \tilde{v}_{ij}(k)\,, \tag{4.4.9}$$

where $\tilde{v}_{ij}(k) = i \sum_{l=1}^{\lfloor k/i \rfloor} (-1)^{l+1} \left(\theta_j/(1-\theta_j)\right)^l \tilde{p}_{k-li}$ for $k = 1, 2, \ldots, m$ and $\tilde{v}_{ij}(k) = 0$ otherwise. Furthermore, utilizing

$$\begin{aligned}
\tilde{v}_{ij}(k) &= \frac{\theta_j}{1-\theta_j} \left(i\tilde{p}_{k-i} - i \sum_{l=2}^{\lfloor k/i \rfloor} (-1)^{(l-1)+1} \left(\frac{\theta_j}{1-\theta_j} \right)^{l-1} \tilde{p}_{k-i-(l-1)i} \right) \\
&= \frac{\theta_j}{1-\theta_j} \left(i\tilde{p}_{k-i} - i \sum_{l=1}^{\lfloor (k-i)/i \rfloor} (-1)^{l+1} \left(\frac{\theta_j}{1-\theta_j} \right)^{l} \tilde{p}_{k-i-li} \right) \\
&= \frac{\theta_j}{1-\theta_j} \left(i\tilde{p}_{k-i} - \tilde{v}_{ij}(k-i) \right), \qquad (4.4.10)
\end{aligned}$$

and comparing (4.4.2)–(4.4.3) with (4.4.9)–(4.4.10), we see that $\tilde{p}_k = p_k$. □

In practical applications, the coefficients c_{il} defined in (4.4.7) will be close to zero for all l large enough, since the claim probabilities θ_j will be small. This means that the recursion formula (4.4.6) can be used in an approximate way by truncating the inner summation. If the coefficients c_{il} are neglected for $l > r$ the following *r-th order approximation* $p_{k,r}$ to p_k is obtained:

$$p_{0,r} = \prod_{i=1}^{a} \prod_{j=1}^{b} (1-\theta_j)^{n_{ij}}, \qquad p_{k,r} = \frac{1}{k} \sum_{i=1}^{\min\{a,k\}} \sum_{l=1}^{\min\{r,\lfloor k/i \rfloor\}} c_{il} p_{k-li,r}\,. \tag{4.4.11}$$

4.4.2 The Collective Model: Panjer's Algorithm

In this section we consider a compound $X = \sum_{i=1}^{N} U_i$ satisfying the following assumptions. Let the claim sizes $U_1, U_2, \ldots$ be (discrete) independent and identically distributed random variables taking their values on a lattice. We also assume that the sequence $U_1, U_2, \ldots$ of claim sizes is independent of the claim number N. Without further loss of generality we can and will assume that $\mathbb{P}(U_i \in \mathbb{N}) = 1$. We denote the probability function of $U_1, U_2, \ldots$ by $\{q_k, k \in \mathbb{N}\}$. Besides this we suppose that the probability function

$\{p_k, k \in \mathbb{N}\}$ of the number of claims N satisfies Panjer's recurrence relation (4.3.1). Theorem 4.3.1 shows that (4.3.1) is exactly fulfilled for the three most important parametric families of distributions $\mathrm{B}(n,p)$, $\mathrm{Poi}(\lambda)$, $\mathrm{NB}(r,p)$. Furthermore, it is not difficult to show that for the logarithmic distribution $\mathrm{Log}(p)$, the recursion formula (4.3.1) holds for $k = 2, 3, \ldots$.

We need the following auxiliary result.

Lemma 4.4.1 *For any $j, k \in \mathbb{N}$ and $n = 1, 2, \ldots$,*

$$\mathbb{E}\Big(U_1 \,\Big|\, \sum_{i=1}^{n} U_i = j\Big) = \frac{j}{n} \tag{4.4.12}$$

and

$$\mathbb{P}\Big(U_1 = k \,\Big|\, \sum_{i=1}^{n} U_i = j\Big) = \frac{q_k q_{j-k}^{*(n-1)}}{q_j^{*n}}, \tag{4.4.13}$$

where $\{q_k^{*n}\}$ denotes the n-fold convolution of $\{q_k\}$.

Proof Since $U_1, U_2, \ldots$ are identically distributed, we have for $X = \sum_{i=1}^{n} U_i$

$$n\mathbb{E}(U_1 \mid X = j) = \sum_{k=1}^{n} \mathbb{E}(U_k \mid X = j) = \mathbb{E}(X \mid X = j) = j\,.$$

This yields (4.4.12). Moreover, because $U_1, U_2, \ldots$ are independent, we have

$$\begin{aligned}\mathbb{P}(U_1 = k \mid U_1 + \ldots + U_n = j) &= \frac{\mathbb{P}(U_1 = k, U_2 + \ldots + U_n = j - k)}{\mathbb{P}(U_1 + \ldots + U_n = j)} \\ &= \frac{\mathbb{P}(U_1 = k)\,\mathbb{P}(U_2 + \ldots + U_n = j - k)}{\mathbb{P}(U_1 + \ldots + U_n = j)} = \frac{q_k q_{j-k}^{*(n-1)}}{q_j^{*n}}\,. \qquad \square\end{aligned}$$

In the following theorem we state a recursive method which is called *Panjer's algorithm* and which can be used to calculate the probabilities $p_k^X = \mathbb{P}(X = k)$ of the compound $X = \sum_{i=1}^{N} U_i$ provided that (4.3.1) holds.

Theorem 4.4.2 *Assume that* (4.3.1) *is fulfilled. Then*

$$p_j^X = \begin{cases} \hat{g}_N(q_0), & \text{for } j = 0, \\ (1 - aq_0)^{-1} \sum_{k=1}^{j} \left(a + bkj^{-1}\right) q_k p_{j-k}^X, & \text{for } j = 1, 2, \ldots. \end{cases} \tag{4.4.14}$$

Proof For $j = 0$ we have $p_0^X = p_0 + p_1 q_0 + p_2 (q_0)^2 + \ldots = \hat{g}_N(q_0)$. Let $j \geq 1$ and note that then $q_j^{*0} = 0$. Thus, using (4.3.1) and (4.4.12) we get

$$p_j^X \;=\; p_0 q_j^{*0} + \sum_{n=1}^{\infty} p_n q_j^{*n} = \sum_{n=1}^{\infty} \left(a + \frac{b}{n}\right) p_{n-1} q_j^{*n}$$

$$
\begin{aligned}
&= \sum_{n=1}^{\infty}\Big(a + b\mathbb{E}\Big(\frac{U_1}{j} \,\Big|\, \sum_{i=1}^{n} U_i = j\Big)\Big)p_{n-1}q_j^{*n} \\
&= \sum_{n=1}^{\infty}\Big(a + b\sum_{k=0}^{j}\frac{k}{j}\mathbb{P}\Big(U_1 = k \,\Big|\, \sum_{i=1}^{n} U_i = j\Big)\Big)p_{n-1}q_j^{*n}.
\end{aligned}
$$

Because of (4.4.13), this yields

$$
\begin{aligned}
p_j^X &= \sum_{n=1}^{\infty}\sum_{k=0}^{j}\left(a + b\frac{k}{j}\right) q_k q_{j-k}^{*(n-1)} p_{n-1} \\
&= \sum_{k=0}^{j}\left(a + b\frac{k}{j}\right) q_k \sum_{n=1}^{\infty} q_{j-k}^{*(n-1)} p_{n-1} \\
&= \sum_{k=0}^{j}\left(a + b\frac{k}{j}\right) q_k p_{j-k}^X = a q_0 p_j^X + \sum_{k=1}^{j}\left(a + b\frac{k}{j}\right) q_k p_{j-k}^X. \qquad \square
\end{aligned}
$$

Example We show how to compute the *stop-loss transform* $\Pi_n = \mathbb{E}\,(X-n)_+$ for $n = 0, 1, \ldots$ by means of Theorem 4.4.2. Note that

$$
\mathbb{E}\,(X - n)_+ = \sum_{m=n+1}^{\infty} (m - n)_+ \, p_n^X = \sum_{m=n}^{\infty} \overline{F}(m)\,,
$$

which gives $\mathbb{E}\,(X - n)_+ = \sum_{m=n-1}^{\infty} \overline{F}(m) - \overline{F}(n-1)$, where $\overline{F}(m) = \sum_{j=m+1}^{\infty} p_j^X$. Hence, $\Pi_0 = \mathbb{E}\,X$ and $\Pi_n = \Pi_{n-1} + F(n-1) - 1$ for $n = 1, 2, \ldots$, where $F(n-1) = \sum_{j=0}^{n-1} p_j^X$ and the p_j^X can be calculated recursively by (4.4.14).

4.4.3 A Continuous Version of Panjer's Algorithm

Panjer's algorithm stated in Theorem 4.4.2 has the disadvantage that the individual claim sizes need to take their values on a lattice. In order to overcome this drawback we derive an integral equation for the distribution F_X of the aggregate claim amount. It holds for an arbitrary (not necessarily discrete) distribution F_U of individual claim sizes provided that the compounding probability function $\{p_k\}$ satisfies Panjer's recursion (4.3.1).

Theorem 4.4.3 *If the compounding probability function $\{p_k\}$ is governed by the recursion (4.3.1) with parameters a and b and $F_U(0) = 0$, then the compound distribution $F_X = \sum_{k=0}^{\infty} p_k F_U^{*k}$ satisfies the integral equation*

$$
F_X(x) = p_0 + aF_U * F_X(x) + b\int_0^x v \int_0^{x-v} \frac{\mathrm{d}F_X(y)}{v+y}\,\mathrm{d}F_U(v)\,, \qquad x > 0\,. \tag{4.4.15}
$$

If $F_U(0) = \alpha > 0$, *then* $F_X(0) = ((1-a)/(1-a\alpha))^{(a+b)a^{-1}}$, *where this expression is interpreted as* $e^{(\alpha-1)b}$ *if* $a = 0$.

Proof Assume first that $F_U(0) = 0$. Since $F_U^{*0}(x) = \delta_0(x)$, using (4.3.1) we can write successively

$$F_X(x) = p_0 + \sum_{n=0}^{\infty} p_{n+1} F_U^{*(n+1)}(x) = p_0 + \sum_{n=0}^{\infty}\Big(a + \frac{b}{n+1}\Big) p_n F_U^{*(n+1)}(x)$$

and equivalently $F_X(x) = p_0 + aF_U * F_X(x) + bG(x)$, where

$$G(x) = \sum_{n=0}^{\infty} \frac{p_n}{n+1} F_U^{*(n+1)}(x).$$

In order to prove (4.4.15) we still have to derive an alternative expression for the function $G(x)$. Note that for independent and identically distributed random variables $U_1, U_2, \ldots$ with distribution F_U, the following identities hold:

$$\begin{aligned}
&(n+1)\int_0^x v \int_0^{x-v} \frac{\mathrm{d}F_U^{*n}(y)}{v+y}\,\mathrm{d}F_U(v)\\
&= (n+1)\mathbb{E}\left[\frac{U_1}{U_1+\ldots U_{n+1}};\ U_1+\ldots U_{n+1} \le x\right]\\
&= \sum_{i=1}^{n+1}\mathbb{E}\left[\frac{U_i}{U_1+\ldots U_{n+1}};\ U_1+\ldots U_{n+1} \le x\right]\\
&= \mathbb{E}\left[\frac{U_1+\ldots U_{n+1}}{U_1+\ldots U_{n+1}};\ U_1+\ldots U_{n+1} \le x\right] = F_U^{*(n+1)}(x).
\end{aligned}$$

Thus, $\int_0^x v \int_0^{x-v} (v+y)^{-1}\,\mathrm{d}F_U^{*n}(y)\,\mathrm{d}F_U(v) = F_U^{*(n+1)}(x)/(n+1)$ and consequently

$$G(x) = \sum_{n=0}^{\infty} p_n \int_0^x v \int_0^{x-v} \frac{\mathrm{d}F_U^{*n}(y)}{v+y}\,\mathrm{d}F_U(v) = \int_0^x v \int_0^{x-v} \frac{\mathrm{d}F_X(y)}{v+y}\,\mathrm{d}F_U(v).$$

This proves equation (4.4.15). Now it remains to show the result on the atom at the origin. If the distribution F_U has an atom $\alpha \in (0,1)$ at the origin, then also F_X has an atom at the origin with size $g(\alpha) = \sum_{n=0}^{\infty} p_n \alpha^n$. Following the same kind of argument as before we see that $g(\alpha) = p_0 + a\alpha g(\alpha) + b\int_0^{\alpha} g(u)\,du$. Rewriting this equation in the form of a differential equation for the auxiliary function $z(\alpha) = \int_0^{\alpha} g(v)\,dv$ gives $z^{(1)}(\alpha)(1-a\alpha) = bz(\alpha) + p_0$. This yields $z(\alpha) = c(1-a\alpha)^{-ba^{-1}} - b^{-1}p_0$ for some constant c. Note, however, that $g(\alpha) = z^{(1)}(\alpha)$. If we put $\alpha = 0$ then we have $g(0) = p_0$. Hence $g(\alpha) =$

$((1-a)/(1-a\alpha))^{(a+b)a^{-1}}$ when $a \neq 0$ and $g(\alpha) = \mathrm{e}^{(\alpha-1)b}$ when $a = 0$. This finishes the proof of the theorem. □

We now additionally assume that the distribution F_U of the individual claim sizes is absolutely continuous and that the density $f_U(x)$ of F_U is bounded. Then the aggregate claim distribution F_X can be decomposed in a discrete part which is the atom at the origin, and in an absolutely continuous part, that is $F_X(B) = p_0\delta_0(B) + \int_B \tilde{f}_X(x)\,\mathrm{d}x$ for all Borel sets $B \in \mathcal{B}(\mathbb{R})$, where $\tilde{f}_X(x) = \sum_{k=1}^{\infty} p_k f_U^{*k}(x)$ and $f_U^{*k}(x)$ is the density of F_U^{*k}. In order to derive an integral equation for $\tilde{f}_X(x)$, we need the following representation formula for the function:

$$g_k(x) = \int_0^x y f_U(y) f_U^{*k}(x-y)\,\mathrm{d}y\,. \tag{4.4.16}$$

Lemma 4.4.2 *For $x > 0$, $k = 1, 2, \ldots$,*

$$g_k(x) = \frac{x}{k+1} f_U^{*(k+1)}(x)\,. \tag{4.4.17}$$

Proof We show (4.4.17) by induction with respect to k. Note that

$$x f_U^{*2}(x) - g_1(x) = \int_0^x (x-y) f_U(y) f_U(x-y)\,\mathrm{d}y = g_1(x)$$

where the substitution $z = x - y$ is used. Thus, (4.4.17) holds for $k = 1$. Suppose now that (4.4.17) has been proved for $n = 1, \ldots, k-1$. Then

$$\begin{aligned}
x f_U^{*(k+1)}(x) - g_k(x) &= \int_0^x (x-y) f_U(y) f_U^{*k}(x-y)\,\mathrm{d}y \\
&= k \int_0^x f_U(y) g_{k-1}(x-y)\,\mathrm{d}y \\
&= k \int_0^x f_U(y) \int_0^{x-y} z f_U(z) f_U^{*(k-1)}(x-y-z)\,\mathrm{d}z\,\mathrm{d}y \\
&= k \int_0^z z f_U(z) \int_0^{x-z} f_U(y) f_U^{*(k-1)}(x-z-y)\,\mathrm{d}y\,\mathrm{d}z \\
&= k \int_0^x z f_U(z) f_U^{*k}(x-z)\,\mathrm{d}z = k g_k(x)\,.
\end{aligned}$$

This shows that (4.4.17) holds for $n = k$, too. □

Theorem 4.4.4 *If $\{p_k\}$ is governed by recursion* (4.3.1) *with parameters a, b and p_0, and F_U is absolutely continuous with bounded density $f(x)$, then the density $\tilde{f}_X(x)$ of the absolutely continuous part of the compound distribution*

$F_X = \sum_{k=0}^{\infty} p_k F_U^{*k}$ *satisfies*

$$\tilde{f}_X(x) - \frac{1}{x}\int_0^x (ax+by) f_U(y) \tilde{f}_X(x-y)\,\mathrm{d}y = p_1 f_U(x)\,, \qquad x>0\,. \quad (4.4.18)$$

Moreover, the function $\tilde{f}_X(x)$ *is the only solution to* (4.4.18) *in the set of all integrable functions on* $(0,\infty)$.

Proof Using (4.3.1) and (4.4.17) we can write for $x>0$

$$\frac{1}{x}\int_0^x (ax+by) f_U(y)\tilde{f}_X(x-y)\,\mathrm{d}y = \frac{1}{x}\sum_{k=1}^{\infty} p_k \int_0^x (ax+by) f_U(y) f_U^{*k}(x-y)\,\mathrm{d}y$$

$$= \sum_{k=1}^{\infty} p_k\Big(a+\frac{b}{k+1}\Big) f_U^{*(k+1)}(x) = \sum_{k=1}^{\infty} p_{k+1} f_U^{*(k+1)}(x) = \tilde{f}_X(x) - p_1 f_U(x)\,.$$

This shows that $\tilde{f}_X(x)$ solves (4.4.18). Since $f_U^{*k}(x)\ge 0$ and $\int_0^\infty f_U^{*k}(x)\,\mathrm{d}x = 1$ for each $k\ge 1$, the function $\tilde{f}_X(x)$ is integrable. It remains to show that $\tilde{f}_X(x)$ is the only integrable function solving (4.4.18). For an arbitrary integrable function $g:(0,\infty)\to\mathbb{R}$ we define the mapping $g\to \boldsymbol{A}g$ by

$$(\boldsymbol{A}g)(x) = \frac{1}{x}\int_0^x (ax+by) f_U(y) g(x-y)\,\mathrm{d}y\,, \qquad x>0\,.$$

Then using (4.3.1) and (4.4.17) we have

$$p_k(\boldsymbol{A}f_U^{*k})(x) = p_{k+1} f_U^{*(k+1)}(x) \quad (4.4.19)$$

for all $x>0$ and $k=1,2,\ldots$. Note that $g(x)=\tilde{f}_X(x)$ fulfils

$$g(x) = (\boldsymbol{A}g)(x) + p_1 f_U(x)\,. \quad (4.4.20)$$

Now let $g(x)$ be any integrable function which fulfils (4.4.20). By induction, (4.4.19) implies that $(\boldsymbol{A}^n g)(x) = g(x) - \sum_{k=1}^{n} p_k f_U^{*k}(x)$ for all $n=1,2,\ldots$ and for each integrable solution $g(x)$ to (4.4.20). However, it is not difficult to show that $\lim_{n\to\infty}(\boldsymbol{A}^n g)(x) = 0$ for $x>0$ and for each integrable function $g:(0,\infty)\to\mathbb{R}$. Hence, $g(x)=\sum_{k=1}^{\infty} p_k f_U^{*k}(x) = \tilde{f}_X(x)$. □

Example Consider the compound $X=\sum_{i=1}^{N} U_i$, where N has the negative binomial distribution $\mathrm{NB}(2,p)$ with $\mathbb{E}\,N=100$ and U_i is $\mathrm{Exp}(\delta)$-distributed. Then, using (4.2.10) the distribution of X can be computed analytically:

$$\overline{F}_X(x) = \mathrm{e}^{-(1-p)\delta x}(1-(1-p)^2 + p^2(1-p)\delta x)\,, \qquad x\ge 0\,.$$

In Table 4.4.1 the tail function $\overline{F}(x)=\mathbb{P}(X>x)$ of this "exact" distribution of X is compared with Panjer's approximation from Theorem 4.4.2, where we put $\delta = 1$. Here the exponentially distributed claim amount U has been transformed into a discrete random variable according to the rule $U_h = h\lfloor U/h \rfloor$, where $\lfloor x \rfloor = \max\{n \in \mathbb{N} : n \le x\}$. In the fourth column of Table 4.4.1, an alternative approximation to $\overline{F}(x)$ is stated which uses the continuous version of Panjer's algorithm given in Theorem 4.4.4. We first computed $\tilde{f}_X(kh)$, $1 \le k \le 1000$, from (4.4.18) using the trapezoid method to perform the integration. Then we used the approximation $\int_{(k-1)h}^{kh} \tilde{f}_X(x)\,dx \approx h\tilde{f}_X(kh)$ and $\overline{F}_X(kh) \approx \overline{F}_X((k-1)h) - h\tilde{f}_X(kh)$.

x	exact	discrete Panjer $h = 0.01$	continuous Panjer $h = 0.01$
0.2	0.999 457 6849	0.999 467 0538	0.999 087 9445
0.4	0.999 285 7597	0.999 297 1743	0.998 915 6727
0.6	0.999 099 8702	0.999 113 4705	0.998 729 4395
0.8	0.998 900 1291	0.998 916 0536	0.998 529 3566
1.0	0.998 686 6458	0.998 705 0343	0.998 315 5350
2.0	0.997 416 9623	0.997 449 7336	0.997 044 1967
4.0	0.993 909 1415	0.993 980 7697	0.993 533 2344
6.0	0.989 196 0212	0.989 319 4873	0.988 817 1510
8.0	0.983 375 4927	0.983 563 2134	0.982 993 7890
10.0	0.976 539 7011	0.976 803 5345	0.976 155 2479

Table 4.4.1 Panjer's algorithm for a negative binomial compound

Bibliographical Notes. The recursive computation method considered in Theorem 4.4.1 has been given in Dhaene and Vandebroek (1995) extending the algorithm established in Waldmann (1994) for the individual life insurance model (see Corollary 4.4.1). Another recursive procedure for computing the probability function of the aggregate claim amount in the individual model with arbitrary positive claim amounts has been derived in De Pril (1989). The recursion formula (4.4.6) is due to De Pril (1986). The efficiency of these algorithms as well as recursive procedures for approximate computation of the p_k as in (4.4.11), are discussed, for example, in De Pril (1988), Dhaene and De Pril (1994), Dhaene and Vandebroek (1995), Kuon, Reich and Reimers (1987), Waldmann (1994,1995). The recursion formula (4.4.14) is due to Panjer (1980, 1981); see also Adelson (1966). Recursions for the evaluation of further related compound distributions are developed, for example, in Schröter (1990) and Willmot and Sundt (1989), where a kind of twofold Panjer-type algorithm is considered. The application of Panjer's algorithm for computing the stop-loss transform $\Pi_n = \mathbb{E}(X-n)_+$ is considered in Gerber (1982). For reviews see Dhaene, Willmot and Sundt (1996), Dickson (1995) and Panjer and Wang (1993).

4.5 LUNDBERG BOUNDS

In this section we investigate the asymptotic behaviour of the tail $\overline{F}_X(x) = \mathbb{P}(X > x)$ of the compound $X = \sum_{i=1}^N U_i$ when x becomes large. As usual, we assume that the random variables $N, U_1, U_2, \ldots$ are independent and that $U_1, U_2, \ldots$ are identically distributed with distribution F_U. In addition, we will assume in this section that F_U has a light tail; the subexponential case has already been mentioned at the end of Section 4.2.1.

4.5.1 Geometric Compounds

First we consider the case where N has a geometric distribution with parameter $p \in (0,1)$. Then the compound geometric distribution F_X is given by

$$F_X(x) = \sum_{k=0}^{\infty}(1-p)p^k F_U^{*k}(x). \tag{4.5.1}$$

Writing the first summand in (4.5.1) separately, we get

$$F_X = (1-p)\delta_0 + p\, F_U * F_X \tag{4.5.2}$$

which is called a *defective renewal equation* or a *transient renewal equation.* Such equations are analysed in Section 6.1.4, where in Lemma 6.1.2 it is shown that the (bounded) solution F_X to (4.5.2) is uniquely determined. Moreover, replacing the distribution F_X on the right-hand side of (4.5.2) by the term $(1-p)\delta_0 + pF_U * F_X$ and iterating this procedure, we obtain

$$F_X(x) = \lim_{n\to\infty} F_n(x)\,, \qquad x \ge 0\,, \tag{4.5.3}$$

where F_n is defined by the recursion

$$F_n = (1-p)\delta_0 + pF_U * F_{n-1} \tag{4.5.4}$$

for all $n \ge 1$ and F_0 is an arbitrary (initial) distribution on $\mathbb{R}_+$.

Assume additionally that the distribution F_U is such that

$$\hat{m}_U(\gamma) = p^{-1} \tag{4.5.5}$$

has a solution $\gamma > 0$, which is usually called the *adjustment coefficient* If no explicit solution to (4.5.5) is available, then it is not difficult to solve (4.5.5) numerically. A statistical method for estimating γ from empirical data will be discussed in Section 4.5.3 below.

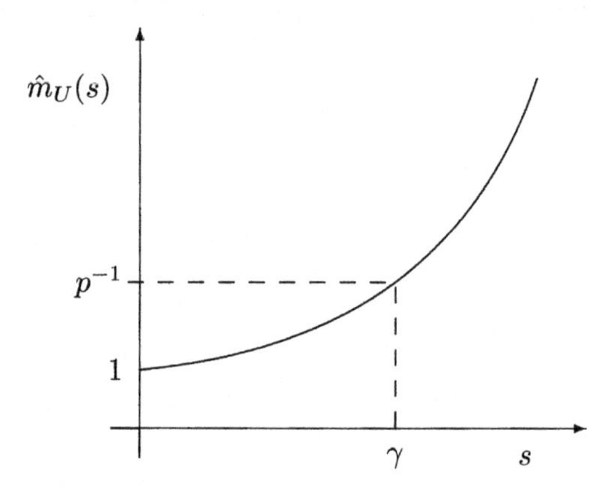

Figure 4.5.1 Moment generating function

If we say that (4.5.5) has a solution $\gamma > 0$, then we tacitly assume that $\hat{m}_U(s) < \infty$ for all $s \leq \gamma$. The existence of a positive solution γ to (4.5.5) is guaranteed if the abscissa of convergence of F_U is positive, that is $\alpha_U = \limsup_{x\to\infty} -x^{-1}\log \overline{F}_U(x) > 0$, and if $\hat{m}_U(\alpha_U) = \infty$, as is usually the case. The uniqueness of such a solution $\gamma > 0$ to (4.5.5) can be seen from Figure 4.5.1.

We get the following lower and upper *Lundberg-type bounds* for the tail of the compound distribution F_X. Let $x_0 = \sup\{x : F_U(x) < 1\}$.

Theorem 4.5.1 *If X is a geometric compound with characteristics (p, F_U) such that* (4.5.5) *admits a positive solution γ, then*

$$a_- e^{-\gamma x} \leq \overline{F}_X(x) \leq a_+ e^{-\gamma x}, \qquad x \geq 0, \tag{4.5.6}$$

where

$$a_- = \inf_{x\in[0,x_0)} \frac{e^{\gamma x}\overline{F}_U(x)}{\int_x^\infty e^{\gamma y}\, dF_U(y)}, \qquad a_+ = \sup_{x\in[0,x_0)} \frac{e^{\gamma x}\overline{F}_U(x)}{\int_x^\infty e^{\gamma y}\, dF_U(y)}. \tag{4.5.7}$$

Proof To get the upper bound in (4.5.6), we aim to find an initial distribution F_0 such that the corresponding distribution F_1 defined in (4.5.4) for $n = 1$ satisfies

$$F_1(x) \geq F_0(x), \qquad x \geq 0. \tag{4.5.8}$$

Then $F_U * F_1(x) \ge F_U * F_0(x)$ for $x \ge 0$ and, by induction, $F_{n+1}(x) \ge F_n(x)$ for all $x \ge 0$ and $n \in \mathbb{N}$. In view of (4.5.3), this means that

$$\overline{F}_X(x) \le \overline{F}_0(x), \qquad x \ge 0. \tag{4.5.9}$$

Let $F_0(x) = 1 - a\mathrm{e}^{-\gamma x} = (1-a)\delta_0(x) + aG(x)$, where $a \in (0,1]$ is some constant and $G(x) = 1 - \exp(-\gamma x)$. Inserting this into (4.5.4) we get

$$\begin{aligned} F_1(x) &= 1 - p + p\Big((1-a)F_U(x) + a\int_0^x G(x-y)\,\mathrm{d}F_U(y)\Big) \\ &= 1 - p + p\Big(F_U(x) - a\int_0^x \mathrm{e}^{-\gamma(x-y)}\,\mathrm{d}F_U(y)\Big), \end{aligned}$$

for all $x \ge 0$. Since we want to arrive at (4.5.8) we look for a such that

$$1 - p + p\Big(F_U(x) - a\int_0^x \mathrm{e}^{-\gamma(x-y)}\,\mathrm{d}F_U(y)\Big) \ge 1 - a\mathrm{e}^{-\gamma x}, \qquad x \ge 0. \tag{4.5.10}$$

This inequality can be simplified to $a(1 - p\int_0^x \mathrm{e}^{\gamma y}\,\mathrm{d}F_U(y)) \ge p\mathrm{e}^{\gamma x}\overline{F}_U(x)$, which is trivial for $x \ge x_0$. Using

$$1 = p\int_0^\infty \mathrm{e}^{\gamma y}\,\mathrm{d}F_U(y) = p\int_0^x \mathrm{e}^{\gamma y}\,\mathrm{d}F_U(y) + p\int_x^\infty \mathrm{e}^{\gamma y}\,\mathrm{d}F_U(y)$$

we notice that (4.5.10) is equivalent to $ap\int_x^\infty \mathrm{e}^{\gamma y}\,\mathrm{d}F_U(y) \ge p\mathrm{e}^{\gamma x}\overline{F}_U(x)$. Hence, setting $a_+ = \sup_{x\in[0,x_0)} \mathrm{e}^{\gamma x}\overline{F}_U(x)(\int_x^\infty \mathrm{e}^{\gamma y}\,\mathrm{d}F_U(y))^{-1}$ we get (4.5.8) and consequently (4.5.9). From this the upper bound in (4.5.6) follows. The lower bound in (4.5.6) can be similarly derived. □

Corollary 4.5.1 *Suppose that F_U is exponential with parameter δ. Then $\gamma = \delta(1-p)$, $a_- = a_+ = p$ and, consequently,*

$$\overline{F}_X(x) = p\mathrm{e}^{-(1-p)\delta x}, \qquad x \ge 0. \tag{4.5.11}$$

Proof Note that $F_U(x) = 1 - \exp(-\delta x)$ implies that the solution $\gamma > 0$ to (4.5.5) satisfies the equation $p\delta(\delta - \gamma)^{-1} = 1$. Thus, $\gamma = \delta(1-p)$. Moreover, by inspection we get $\mathrm{e}^{\gamma x}\overline{F}_U(x)(\int_x^\infty \mathrm{e}^{\gamma y}\,\mathrm{d}F_U(y))^{-1} = p$ for $x \ge 0$. □

A generalization of formula (4.5.11) is given in Lemma 8.3.2 below, for the case that F_U is a phase-type distribution.

Remark For the constants a_- and a_+ appearing in Theorem 4.5.1, we obviously have $a_- \le a_+$. Moreover, since $\int_x^\infty \mathrm{e}^{\gamma y}\,\mathrm{d}F_U(y) \ge \mathrm{e}^{\gamma x}\overline{F}_U(x)$, we conclude that $a_+ \le 1$. Furthermore, note that

$$\frac{\int_x^\infty \mathrm{e}^{\gamma y}\,\mathrm{d}F_U(y)}{\mathrm{e}^{\gamma x}\overline{F}_U(x)} = \mathbb{E}\,(\mathrm{e}^{\gamma(U-x)} \mid U > x) = \int_0^\infty \mathrm{e}^{\gamma y}\,\mathrm{d}F_{U,x}(y)\,,$$

where the residual hazard distribution $F_{U,x}$ at x was defined in (2.4.8). If F_U is DHR, then by Theorem 2.4.2, the value of $\mathbb{E}\left(e^{\gamma(U-x)} \mid U > x\right)$ is increasing in x. Analogously, for F_U being IHR the above is decreasing in x. Hence $a_+ = 1/\hat{m}_U(\gamma) = p$ for F_U being DHR (by our definition it cannot have an atom at zero) and, if F_U is IHR, $a_+ = \lim_{x\to\infty} e^{\gamma x}\overline{F}_U(x)(\int_x^\infty e^{\gamma y}\, dF_U(y))^{-1}$.

4.5.2 More General Compound Distributions

Generalizing the previous results, we now only assume the existence of a constant $\theta \in (0,1)$ such that the distribution of N is stochastically smaller than the $(1-p_0)$-modification of the truncated geometric distribution $\mathrm{TG}(\theta)$. More precisely,

$$\{p_k\} \le_{\rm st} \{p'_k\} = p_0\delta_0 + (1-p_0)\mathrm{TG}(\theta)\,, \tag{4.5.12}$$

which is equivalent to $r_{j+1} \le (1-p_0)\theta^j$ for all $j \in \mathbb{N}$, where $r_j = \sum_{k=j}^\infty p_k$. Note that the inequality (4.5.12) holds provided that

$$r_{j+1} \le \theta r_j\,, \qquad j \ge 1\,, \tag{4.5.13}$$

because $r_1 = 1 - p_0 = \sum_{k=1}^\infty p'_k$ and because (4.5.13) yields

$$r_{j+1} \le \theta^j r_1 = \sum_{k=j}^\infty (1-\theta)\theta^k r_1 = \sum_{k=j+1}^\infty (1-p_0)(1-\theta)\theta^{k-1} \tag{4.5.14}$$

for all $j \ge 1$. However, in general, (4.5.13) does not follow from (4.5.12).

Theorem 4.5.2 *If* (4.5.12) *holds for some* $\theta < 1$, *then*

$$\overline{F}_X(x) \le \frac{1-p_0}{\theta} a_+ e^{-\gamma x}\,, \qquad x \ge 0\,, \tag{4.5.15}$$

where $\gamma > 0$ *fulfils* $1 = \theta\hat{m}_U(\gamma)$ *and* a_+ *is defined by* (4.5.7).

Proof Since for $x \ge 0$, the tail $\overline{F}_X(x)$ can be represented by

$$\begin{aligned}\overline{F}_X(x) &= \sum_{k=0}^\infty p_k \sum_{j=0}^{k-1}\left(F_U^{*j}(x) - F_U^{*(j+1)}(x)\right)\\ &= \sum_{j=0}^\infty\sum_{k=j+1}^\infty p_k\left(F_U^{*j}(x) - F_U^{*(j+1)}(x)\right),\end{aligned}$$

we get from (4.5.14)

$$\begin{aligned}\overline{F}_X(x) &\le \frac{1-p_0}{\theta}\sum_{j=0}^\infty\sum_{k=j+1}^\infty (1-\theta)\theta^k\left(F_U^{*j}(x) - F_U^{*(j+1)}(x)\right)\\ &= \frac{1-p_0}{\theta}\left(1 - \sum_{k=0}^\infty (1-\theta)\theta^k F_U^{*k}(x)\right).\end{aligned}$$

Applying Theorem 4.5.1 to the right-hand side, we get (4.5.15). □

The logconcave probability functions $\{p_k\}$ form a subclass of $\mathrm{IHR_d}$ and lead to further improvements.

Corollary 4.5.2 *If $\{p_k\}$ is logconcave with $p_0 + p_1 < 1$, then*

$$\overline{F}_X(x) \le \frac{(1-p_0)^2}{1-p_0-p_1}\, a_+ \mathrm{e}^{-\gamma x}, \qquad x \ge 0, \tag{4.5.16}$$

where a_+ is defined as in (4.5.7) *and $\gamma > 0$ is the solution to the equation*

$$\hat{m}_U(\gamma) = \frac{1-p_0}{1-p_0-p_1}. \tag{4.5.17}$$

Proof In view of Theorem 4.5.2 it suffices to show that (4.5.13) holds with $\theta = (1-p_0-p_1)(1-p_0)^{-1}$. Clearly we have $r_2 r_1^{-1} = (1-p_0-p_1)(1-p_0)^{-1}$. Note that the ratio r_{j+1}/r_j is decreasing because, by assumption, the ratio p_{j+1}/p_j is decreasing and that

$$\frac{r_{j+1}}{r_j} = 1 - \frac{1}{1 + \frac{p_{j+1}}{p_j} + \frac{p_{j+2}}{p_{j+1}}\frac{p_{j+1}}{p_j} + \frac{p_{j+3}}{p_{j+2}}\frac{p_{j+2}}{p_{j+1}}\frac{p_{j+1}}{p_j} + \ldots}.$$

Hence (4.5.13) holds and consequently (4.5.12) holds as well. The statement then follows from Theorem 4.5.2. □

Note that, unless the counting variable N is geometrically distributed, the exponent γ in the bounds (4.5.15) and (4.5.16) is not optimal. This can be seen from the following example. Consider the tail $\overline{F}(x)$ of the compound $X = \sum_{i=1}^N U_i$, where N has the negative binomial distribution $\mathrm{NB}(2,p)$ and U is $\mathrm{Exp}(\delta)$-distributed. If $\gamma > 0$ is computed from (4.5.17), then

$$\gamma = \frac{2(1-p)^2\delta}{2-p} < (1-p)\delta.$$

However, the advantage of bounds like (4.5.15) and (4.5.16) is their simplicity and wide applicability. Moreover, in many cases these bounds seem to be satisfactory.

4.5.3 Estimation of the Adjustment Coefficient

Let us have a closer look at a slight generalization of equation (4.5.5),

$$\hat{m}_U(s) = c > 1, \tag{4.5.18}$$

which appears over and over again in risk theory. Assume that we have a sample $U_1, U_2, \ldots, U_n$ of n independent claim sizes with common distribution

F_U. How do we get an estimate for the unknown solution $\gamma > 0$ to (4.5.18) based on this sample? A rather natural procedure is to replace the moment generating function $\hat{m}_U(s)$ by its empirical analogue, the *empirical moment generating function* $\hat{m}_{F_n}(s)$, which is obtained as the moment generating function of the empirical distribution F_n. The latter has been defined in Section 2.6.1 by $F_n(x) = n^{-1} \max\{i : U_{(i)} \le x\}$. Hence $\hat{m}_{F_n}(s) = \int_0^\infty e^{sx}\, \mathrm{d}F_n(x) = n^{-1}\sum_{i=1}^n \mathrm{e}^{sU_i}$. It seems natural to define an estimator $\hat{\gamma}_n$ for γ by the equation

$$\hat{m}_{F_n}(\hat{\gamma}_n) = \frac{1}{n}\sum_{i=1}^n \mathrm{e}^{\hat{\gamma}_n U_i} = c\,. \tag{4.5.19}$$

We show that this procedure can indeed be followed. The proof is based on the following consistency property.

Lemma 4.5.1 *Assume that $\alpha_U > 0$ and $\hat{m}_U(\alpha_U) = \infty$. Then for any closed interval $I \subset (-\infty, \alpha_U)$ and any $k \in \mathbb{N}$*

$$\lim_{n\to\infty} \sup_{s\in I} |\hat{m}_{F_n}^{(k)}(s) - \hat{m}_U^{(k)}(s)| = 0\,. \tag{4.5.20}$$

The *proof* of Lemma 4.5.1 is left to the reader. Note that (4.5.20) implies that $\hat{\gamma}_n \to \gamma$ with probability 1.

Theorem 4.5.3 *If 2γ lies inside the region of convergence of the moment generating function $\hat{m}_U(s)$, then*

$$\sqrt{n}(\hat{\gamma}_n - \gamma) \xrightarrow{\mathrm{d}} \mathrm{N}\Big(0, \frac{\hat{m}_U(2\gamma) - (\hat{m}_U(\gamma))^2}{(\hat{m}_U^{(1)}(\gamma))^2}\Big) = \mathrm{N}\Big(0, \frac{\mathrm{Var}\,\mathrm{e}^{\gamma U}}{(\mathbb{E}\,(U\mathrm{e}^{\gamma U}))^2}\Big)\,.$$

Proof We start from the two equations (4.5.18) and (4.5.19) and write

$$0 = c - c = \frac{1}{n}\sum_{i=1}^n \left((\mathrm{e}^{\hat{\gamma}_n U_i} - \mathrm{e}^{\gamma U_i}) - (\mathrm{e}^{\gamma U_i} - \mathbb{E}\,\mathrm{e}^{\gamma U})\right)\,.$$

We now represent the difference $\mathrm{e}^{\hat{\gamma}_n U_i} - \mathrm{e}^{\gamma U_i}$ as an integral of the form $\int_a^b \mathrm{e}^x\,\mathrm{d}x$ and apply the mean value theorem to get

$$\mathrm{e}^{\hat{\gamma}_n U_i} - \mathrm{e}^{\gamma U_i} = \mathrm{e}^{\Theta_{n,i} U_i}(\hat{\gamma}_n - \gamma)U_i\,,$$

where $\Theta_{n,i}$ lies in the interval determined by $\hat{\gamma}_n$ and γ. Thus, we get equivalently

$$\sum_{i=1}^n \mathrm{e}^{\Theta_{n,i}U_i}(\hat{\gamma}_n - \gamma)U_i + \sum_{i=1}^n \left(\mathrm{e}^{\gamma U_i} - \mathbb{E}\,\mathrm{e}^{\gamma U}\right) = 0\,.$$

Solving for the difference $\hat{\gamma}_n - \gamma$ we get the fundamental equation

$$\sqrt{n}(\hat{\gamma}_n - \gamma) = -\frac{n^{-1/2}\sum_{i=1}^{n}\left(e^{\gamma U_i} - \mathbb{E}\, e^{\gamma U}\right)}{n^{-1}\sum_{i=1}^{n} e^{\Theta_{n,i}U_i}U_i} \,. \tag{4.5.21}$$

In the numerator we can immediately apply the central limit theorem since that quantity converges in distribution to a normal distribution with variance equal to that of $e^{\gamma U}$, i.e. $\mathrm{Var}\, e^{\gamma U} = \hat{m}_U(2\gamma) - (\hat{m}_U(\gamma))^2$. The application of the central limit theorem is, however, only possible if the value 2γ lies in the region of convergence of $\hat{m}_U(s)$. By Lemma 4.5.1 we have $\hat{\gamma}_n \to \gamma$ and consequently $\Theta_{n,i} \to \gamma$. Thus, the denominator in (4.5.21) converges by the law of large numbers and the bounds on $\Theta_{n,i}$ to the quantity $\mathbb{E}\,(e^{\gamma U}U) = \hat{m}_U^{(1)}(\gamma)$, which always exists, since within the region of convergence the moment generating function $\hat{m}_U(s)$ is infinitely often differentiable. The above, together with a Slutsky argument in (2.1.14), gives the desired result. □

The variance of the limiting normal distribution in Theorem 4.5.3 depends both on the unknown value of γ and on the moment generating function $\hat{m}_U(s)$ that is usually unknown as well. In practice one will, of course, use the result in an empirical form where the unknown quantity γ is replaced by the sample variable $\hat{\gamma}_n$, while $\hat{m}_U(s)$ is similarly replaced by $\hat{m}_{F_n}(s)$.

Bibliographical Notes. The results of this section are essentially due to Willmot (1994,1997a,1997b) and Willmot and Lin (1994,1997a); see also Willmot and Lin (1997b). The idea of the actual proof of Theorem 4.5.1 is from Bergmann and Stoyan (1976). An algorithm for the numerical solution of (4.5.5) can be found in Hipp and Michel (1990). Other bounds for compounds, asymptotically better than those given in this section, can be found in Runnenburg and Goovaerts (1985); see also Kalashnikov (1997). For the case of Pascal compounds, we refer to an application of renewal theory. Theorem 4.5.3 is due to Csörgő and Teugels (1990).

4.6 APPROXIMATION BY COMPOUND DISTRIBUTIONS

Consider the aggregate claim amount $X^{\mathrm{ind}} = \sum_{i=1}^{n} U_i$ in the individual model, where we assume that the random variables $U_1, \ldots, U_n$ are independent, but not necessarily identically distributed. As in Section 4.1, let the distribution F_{U_i} of U_i be the mixture $F_{U_i} = (1-\theta_i)\delta_0 + \theta_i F_{V_i}$, where $0 < \theta_i \le 1$ and F_{V_i} is the distribution of some (strictly) positive random variable V_i. Remember that in actuarial applications the weight θ_i is small for each i and can be interpreted as the probability that the i-th risk produces a positive claim.

If we want to compute the percentile premiums for the risk X^{ind}, one has to know the distribution of X^{ind}. Without the help of computers, this is a difficult task, especially for large portfolios. However, in some concrete situations the computation of the distribution of the aggregate claim amount in collective models is much easier. This is particularly true when N follows a Poisson, binomial or negative binomial distribution. Hence, a lot of effort has gone into finding the best possible fit of the individual model by a collective model. In Section 4.4.2, we explained the reason why the compounds appearing in such collective models are computationally tractable. Let us start from a collective model with the aggregate claim amount $X^{\text{col}} = \sum_{i=1}^{N} U_i'$ but where $U_1', U_2', \ldots$ are independent and identically distributed. We have to evaluate the quality of the approximation between X^{ind} and X^{col}. Possible choices to define a distance between these two quantities are, for example, the *supremum distance*

$$d_{\text{SD}}(X^{\text{ind}}, X^{\text{col}}) = \sup_{x \geq 0} \left| \mathbb{P}(X^{\text{ind}} \leq x) - \mathbb{P}(X^{\text{col}} \leq x) \right| , \tag{4.6.1}$$

the *stop-loss distance*

$$d_{\text{SL}}(X^{\text{ind}}, X^{\text{col}}) = \sup_{x \geq 0} \left| \mathbb{E}\,(X^{\text{ind}} - x)_+ - \mathbb{E}\,(X^{\text{col}} - x)_+ \right| \tag{4.6.2}$$

or the *total variation distance*

$$d_{\text{TV}}(X^{\text{ind}}, X^{\text{col}}) = \sup_{B \in \mathcal{B}(\mathbb{R})} \left| \mathbb{P}(X^{\text{ind}} \in B) - \mathbb{P}(X^{\text{col}} \in B) \right| . \tag{4.6.3}$$

Clearly, $d_{\text{SD}}(X^{\text{ind}}, X^{\text{col}}) \leq d_{\text{TV}}(X^{\text{ind}}, X^{\text{col}})$.

The Compound Poisson Approximation The idea is to approximate each random variable U_i by a Poisson compound Y_i with characteristics (θ_i, F_{V_i}). Note that $\mathbb{E}\, U_i = \mathbb{E}\, Y_i$. Taking $Y_1, \ldots, Y_n$ independent, Theorem 4.2.2 implies that $Y = Y_1 + \ldots + Y_n$ is a Poisson compound with characteristics (λ, F) given by

$$\lambda = \sum_{i=1}^{n} \theta_i \, , \qquad F = \sum_{i=1}^{n} \frac{\theta_i}{\lambda} F_{V_i} \, . \tag{4.6.4}$$

The Compound Binomial Approximation Here, the compound is binomial with N $\text{Bin}(n,p)$-distributed, where $p = \lambda/n$, U_i' has the distribution F, and λ, F are as in the compound Poisson approximation.

The Compound Negative Binomial Approximation Now N is taken to be $\text{NB}(n, p/(1+p))$-distributed, where $p = \lambda/n$, U_i' has the distribution F, and λ, F are as in the compound Poisson approximation.

We leave it to the reader to check that in all the cases considered above $\mathbb{E}\, X^{\text{ind}} = \mathbb{E}\, X^{\text{col}}$ and $\text{Var}\, X^{\text{ind}} < \text{Var}\, X^{\text{col}}$. Furthermore, the variances of the aggregate claim amounts in the approximating collective models are in the following ascending order: binomial model, Poisson model, negative binomial model.

4.6.1 The Total Variation Distance

Let F and G be two arbitrary distributions on $\mathbb{R}$ and let

$$d_{\text{TV}}(F,G) = \sup_{B\in\mathcal{B}(\mathbb{R})} |F(B) - G(B)|\,. \tag{4.6.5}$$

Note that the mapping $(F,G) \to d_{\text{TV}}(F,G)$ is a *metric*, i.e. $d_{\text{TV}}(F,G) = 0$ if and only if $F = G$, $d_{\text{TV}}(F,G) = d_{\text{TV}}(G,F)$ and $d_{\text{TV}}(F_1,F_2) \le d_{\text{TV}}(F_1,F_3) + d_{\text{TV}}(F_3,F_2)$ for arbitrary distributions F_1, F_2, F_3 on $\mathbb{R}$.

Lemma 4.6.1 *An equivalent form of* (4.6.5) *is*

$$d_{\text{TV}}(F,G) = \sup_{B\in\mathcal{B}(\mathbb{R})} (F(B) - G(B))\,.$$

Proof The assertion follows directly from

$$F(B) - G(B) \le |F(B) - G(B)| = \max\{F(B) - G(B), F(B^c) - G(B^c)\}\,,$$

where B^c denotes the complement of the set B. □

Lemma 4.6.2 *Let* $F = \sum_{i=0}^{\infty} p_i\delta_i$ *and* $G = \sum_{i=0}^{\infty} q_i\delta_i$ *be two discrete distributions. Then* $d_{\text{TV}}(F,G) = 2^{-1}\sum_{i=0}^{\infty} |p_i - q_i|$.

Proof Let $C = \{i : p_i \ge q_i\}$. Then $F(C) - G(C) \ge F(B) - G(B)$ for all $B \in \mathcal{B}(\mathbb{R})$. Therefore we have

$$\begin{aligned}
\sum_{i=0}^{\infty} |p_i - q_i| &= \sum_{i\in C}(p_i - q_i) + \sum_{n\in C^c}(q_n - p_n) \\
&= F(C) - G(C) + G(C^c) - F(C^c) \\
&= 2(F(C) - G(C)) \ge 2(F(B) - G(B))
\end{aligned}$$

for all $B \in \mathcal{B}(\mathbb{R})$. This completes the proof in view of Lemma 4.6.1. □

Let X and Y denote random variables with distributions F and G, respectively; then we also write $d_{\text{TV}}(X,Y)$ instead of $d_{\text{TV}}(F,G)$.

Lemma 4.6.3 *Let* $X_1,\ldots,X_n$ *and* $Y_1,\ldots,Y_n$ *be two sequences of independent random variables. Then*

$$d_{\text{TV}}\Big(\sum_{i=1}^{n} X_i, \sum_{j=1}^{n} Y_j\Big) \le \sum_{i=1}^{n} d_{\text{TV}}(X_i,Y_i)\,. \tag{4.6.6}$$

Proof By F_i and G_i we denote the distributions of X_i and Y_i, respectively; $1 \le i \le n$. First we show that (4.6.6) holds for $n = 2$. Namely,

$$
\begin{aligned}
& d_{\mathrm{TV}}(X_1 + X_2, Y_1 + Y_2) \\
= \; & \sup_B \left| \int_{-\infty}^{\infty} F_1(B - x) F_2(\mathrm{d}x) - \int_{-\infty}^{\infty} G_1(B - x) G_2(\mathrm{d}x) \right| \\
= \; & \sup_B \left| \int_{-\infty}^{\infty} F_1(B - x) F_2(\mathrm{d}x) - \int_{-\infty}^{\infty} G_1(B - x) F_2(\mathrm{d}x) \right. \\
& \left. + \int_{-\infty}^{\infty} G_1(B - x) F_2(\mathrm{d}x) - \int_{-\infty}^{\infty} G_1(B - x) G_2(\mathrm{d}x) \right| \\
= \; & \sup_B \left| \int_{-\infty}^{\infty} F_1(B - x) F_2(\mathrm{d}x) - \int_{-\infty}^{\infty} G_1(B - x) F_2(\mathrm{d}x) \right. \\
& \left. + \int_{-\infty}^{\infty} F_2(B - x) G_1(\mathrm{d}x) - \int_{-\infty}^{\infty} G_2(B - x) G_1(\mathrm{d}x) \right| \\
= \; & \sup_B \left| \int_{-\infty}^{\infty} (F_1(B - x) - G_1(B - x)) F_2(\mathrm{d}x) \right. \\
& \left. + \int_{-\infty}^{\infty} (F_2(B - x) - G_2(B - x)) G_1(\mathrm{d}x) \right| \\
\le \; & \sup_B |F_1(B) - G_1(B)| + \sup_B |F_2(B) - G_2(B)| \,.
\end{aligned}
$$

The general case follows by induction, replacing F_2 by $F_2 * \ldots * F_n$ and G_2 by $G_2 * \ldots * G_n$ in the above. □

4.6.2 The Compound Poisson Approximation

We next investigate the distance between the distribution of the risk $X = \sum_{i=1}^n U_i$ in the individual model and the distribution of the Poisson compound $Y = \sum_{i=1}^n Y_i$ with the characteristics (λ, F) given in (4.6.4), where the $Y_1, \ldots, Y_n$ are independent Poisson compounds with characteristics $(\theta_1, F_{V_1}), \ldots, (\theta_n, F_{V_n})$ respectively.

Theorem 4.6.1 *The following upper bound holds:*

$$
d_{\mathrm{TV}}(X, Y) \le \sum_{i=1}^n \theta_i^2 \,. \tag{4.6.7}
$$

Proof We first compare the distributions of U_i and Y_i. For each $B \in \mathcal{B}(\mathbb{R})$,

$$
\begin{aligned}
& \mathbb{P}(U_i \in B) - \mathbb{P}(Y_i \in B) \\
= \; & (1 - \theta_i)\delta_0(B) + \theta_i F_{V_i}(B)
\end{aligned}
$$

$$- \left(\mathrm{e}^{-\theta_i}\delta_0(B) + \theta_i \mathrm{e}^{-\theta_i} F_{V_i}(B) + \sum_{k=2}^{\infty} \frac{\theta_i^k}{k!} \mathrm{e}^{-\theta_i} (F_{V_i})^{k*}(B) \right)$$
$$\leq (1-\theta_i)\delta_0(B) + \theta_i F_{V_i}(B) - \mathrm{e}^{-\theta_i}\delta_0(B) - \theta_i \mathrm{e}^{-\theta_i} F_{V_i}(B) .$$

Thus, using $1 - \theta_i \leq \mathrm{e}^{-\theta_i}$ we obtain

$$\mathbb{P}(U_i \in B) - \mathbb{P}(Y_i \in B) \leq \theta_i F_{V_i}(B)(1 - \mathrm{e}^{-\theta_i}) \leq \theta_i^2 F_{V_i}(B) \leq \theta_i^2 .$$

Lemmas 4.6.1 and 4.6.3 now imply $d_{\mathrm{TV}}(X,Y) \leq \sum_{i=1}^n d_{\mathrm{TV}}(U_i, Y_i) \leq \sum_{i=1}^n \theta_i^2$. This proves the theorem. □

4.6.3 Homogeneous Portfolio

Theorem 4.6.1 tells us that the compound Poisson approximation is good if the probability θ_i that the i-th policy produces a positive claim is small for all $i = 1, \ldots, n$, in comparison to the number n of policies. If it is possible to group the risks $U_1, \ldots, U_n$ into almost identically distributed compound risks $\sum_{i=1}^{n_1} U_i, \sum_{i=n_1+1}^{n_2} U_i, \ldots, \sum_{i=n_k+1}^{n} U_i$, then still another bound for the approximation error $d_{\mathrm{TV}}(X,Y)$ can be given. This grouping procedure is sometimes called a *homogenization* of the portfolio. For homogeneous portfolios we get a better bound than in (4.6.7) if $\sum_{i=1}^n \theta_i > 1$. Since in practice we deal with large portfolios, this condition seems to be realistic.

Theorem 4.6.2 *For a homogeneous portfolio, i.e.* $F_{V_1} = \ldots = F_{V_n} = F$,

$$d_{\mathrm{TV}}(X,Y) \leq \frac{\sum_{i=1}^n \theta_i^2}{\sum_{i=1}^n \theta_i} . \tag{4.6.8}$$

In the *proof* of Theorem 4.6.2 we use the following auxiliary results. Let $I_1, \ldots, I_n$ be independent Bernoulli 0-1-variables with $\mathbb{P}(I_i = 1) = \theta_i$. Furthermore, take N Poisson distributed with parameter $\lambda = \sum_{i=1}^n \theta_i$ and independent of $I_1, \ldots, I_n$. Define $N' = \sum_{i=1}^n I_i$ with $p'_k = \mathbb{P}(N' = k)$ and probability generating function $\hat{g}_{N'}(s) = \prod_{i=1}^n (1-\theta_i+\theta_i s)$. Then the following is true.

Lemma 4.6.4 *If* $F_{V_1} = \ldots = F_{V_n} = F$, *then*

$$F_{U_1} * \ldots * F_{U_n} = \sum_{i=1}^n p'_i F^{*i} . \tag{4.6.9}$$

Proof We show that the Laplace–Stieltjes transforms of the distributions on both sides of (4.6.9) coincide. Indeed, for the transform $\hat{l}_{U_1+\ldots+U_n}(s)$ we have

$$\hat{l}_{U_1+\ldots+U_n}(s) = \prod_{i=1}^n \hat{l}_{U_i}(s) = \prod_{i=1}^n \left((1-\theta_i) + \theta_i \int_0^\infty \mathrm{e}^{-sx} \, \mathrm{d}F(x) \right)$$

$$= \sum_{j=0}^{n} p'_j \Big(\int_0^\infty e^{-sx} \, dF(x) \Big)^j ,$$

where the last expression is the Laplace–Stieltjes transform of $\sum_{j=1}^n p'_j F^{*j}$. This proves the lemma. □

Lemma 4.6.5 *If* $F_{V_1} = \ldots = F_{V_n} = F$, *then*

$$d_{\mathrm{TV}}(X, Y) \le d_{\mathrm{TV}}(N', N) . \tag{4.6.10}$$

Proof Let $C = \{i \in \{1, \ldots, n\} : p'_i \ge p_i\}$, where $p_i = \mathbb{P}(N = i)$. Then from Lemma 4.6.4 we get that, for each $B \in \mathcal{B}(\mathbb{R})$,

$$\mathbb{P}(X \in B) - \mathbb{P}(Y \in B) = \sum_{i=0}^{n} p'_i F^{*i}(B) - \sum_{i=0}^{\infty} p_i F^{*i}(B) .$$

Thus,

$$\begin{aligned} \mathbb{P}(X \in B) - \mathbb{P}(Y \in B) &\le \sum_{i=1}^{n} (p'_i - p_i) F^{*i}(B) \le \sum_{i \in C} (p'_i - p_i) \\ &= \mathbb{P}(N' \in C) - \mathbb{P}(N \in C) \le d_{\mathrm{TV}}(N', N) , \end{aligned}$$

where the last inequality follows from Lemma 4.6.1. Taking the supremum over all $B \in \mathcal{B}(\mathbb{R})$ and using Lemma 4.6.1 again, we get (4.6.10). □

Proof of Theorem 4.6.2. In view of Lemma 4.6.5 it suffices to show that

$$d_{\mathrm{TV}}(N', N) \le \frac{\sum_{i=1}^n \theta_i^2}{\sum_{i=1}^n \theta_i} . \tag{4.6.11}$$

Let $B \subset \mathbb{N}$ be an arbitrary set of natural numbers. Furthermore, let $C_k = \{0, 1, \ldots, k-1\}$ and define

$$g(k) = \frac{1}{k p_k} \big(\mathbb{P}(N \in B \cap C_k) - \mathbb{P}(N \in B) \mathbb{P}(N \in C_k) \big) \tag{4.6.12}$$

for $k = 1, 2, \ldots$; we set $g(0) = 0$. Since $(k+1) p_{k+1} = \lambda p_k$, we have

$$\begin{aligned} \lambda g(k+1) - k g(k) &= \frac{1}{p_k} \big(\mathbb{P}(N \in B \cap C_{k+1}) - \mathbb{P}(N \in B) \mathbb{P}(N \in C_{k+1}) \big) \\ &\quad - \frac{1}{p_k} \big(\mathbb{P}(N \in B \cap C_k) - \mathbb{P}(N \in B) \mathbb{P}(N \in C_k) \big) \\ &= \frac{1}{p_k} \big(\mathbb{P}(N \in B \cap \{k\}) - \mathbb{P}(N \in B) \mathbb{P}(N = k) \big) \\ &= \delta_k(B) - \mathbb{P}(N \in B) . \end{aligned} \tag{4.6.13}$$

Consequently,

$$\begin{aligned}\mathbb{P}(N' \in B) - \mathbb{P}(N \in B) &= \mathbb{E}\,(\delta_{N'}(B) - \mathbb{P}(N \in B)) \\ &= \mathbb{E}\,(\lambda g(N'+1) - N'g(N'))\,.\end{aligned}$$

With the notation $N^{(i)} = I_1 + \ldots + I_{i-1} + I_{i+1} + \ldots + I_n$ this yields

$$\mathbb{P}(N' \in B) - \mathbb{P}(N \in B) = \sum_{i=1}^{n} \mathbb{E}\,(\theta_i g(N'+1) - I_i g(N'))$$

$$\begin{aligned}&= \sum_{i=1}^{n} \mathbb{E}\left((1-\theta_i)\theta_i g(N^{(i)}+1) + \theta_i^2 g(N^{(i)}+2) - \theta_i g(N^{(i)}+1)\right) \\ &= \sum_{i=1}^{n} \theta_i^2 \mathbb{E}\left(g(N^{(i)}+2) - g(N^{(i)}+1)\right).\end{aligned}$$

From the last equation, we see that for (4.6.11) it suffices to show that $g(k+1) - g(k) \le \lambda^{-1}$ holds for all $k = 1, 2, \ldots$. Since from (4.6.13) we get

$$\lambda(g(k+1) - g(k)) = \delta_k(B) - \mathbb{P}(N \in B) + (k - \lambda)g(k)\,,$$

it suffices to show that

$$(k - \lambda)g(k) \le \mathbb{P}(N \in B)\,, \qquad k = 1, 2, \ldots. \tag{4.6.14}$$

First consider the case $k > \lambda$. Then

$$\begin{aligned}&\mathbb{P}(N \in B \cap C_k) - \mathbb{P}(N \in B)\mathbb{P}(N \in C_k) \\ &\quad\le \mathbb{P}(N \in B)\mathbb{P}(N \notin C_k) = \mathbb{P}(N \in B)\mathbb{P}(N \ge k) \\ &\quad= \mathbb{P}(N \in B)\; p_k \sum_{i=0}^{\infty} \frac{\lambda^i}{(i+k)!} k! \le k p_k \frac{\mathbb{P}(N \in B)}{k - \lambda}\,.\end{aligned}$$

Together with the definition (4.6.12) of $g(k)$, this results in (4.6.14) for $k > \lambda$. On the other hand, if $k \le \lambda$, then (4.6.12) yields

$$\begin{aligned}(k - \lambda)g(k) &= \frac{\lambda - k}{k p_k}\left(\mathbb{P}(N \in B)\mathbb{P}(N \in C_k) - \mathbb{P}(N \in B \cap C_k)\right) \\ &\le \frac{\lambda - k}{k p_k}\mathbb{P}(N \in B)\mathbb{P}(N \in C_k)\end{aligned}$$

and thus (4.6.14) for $k \le \lambda$ because $(\lambda - k)\mathbb{P}(N \le k-1) \le k p_k$. □

4.6.4 Higher-Order Approximations

We show how the compound Poisson approximation discussed in Sections 4.6.2 and 4.6.3 can still be refined further. A numerical example will be given in Section 4.7.

For notational case, we first consider the special case of a portfolio consisting of a single policy ($n = 1$). Then the distribution F of the claim amount X can be given in the form $F = (1-\theta)\delta_0 + \theta F_V$, where $0 < \theta \le 1$ and F_V is the distribution of some positive random variable V. For the characteristic function $\hat{\varphi}(s)$ of X we have

$$\hat{\varphi}(s) = 1 + \theta(\hat{\varphi}_V(s) - 1)\,. \tag{4.6.15}$$

If $\theta|\hat{\varphi}_V(s) - 1| < 1$, then we can rewrite (4.6.15) in the following way:

$$\hat{\varphi}(s) = \exp(\log(1 + \theta(\hat{\varphi}_V(s) - 1))) = \exp\Big(\sum_{k=1}^{\infty} \frac{(-1)^{k+1}}{k}\theta^k(\hat{\varphi}_V(s) - 1)^k\Big). \tag{4.6.16}$$

Thus, for $\theta|\hat{\varphi}_V(s) - 1| < 1$, we can approximate $\hat{\varphi}(s)$ by

$$\hat{\varphi}_r(s) = \exp\Big(\sum_{k=1}^{r} \frac{(-1)^{k+1}}{k}\theta^k(\hat{\varphi}_V(s) - 1)^k\Big), \qquad r \ge 1\,. \tag{4.6.17}$$

We prove that $\hat{\varphi}_r(s)$ is the characteristic function of a certain signed measure H_r on $\mathbb{R}$, i.e. $\hat{\varphi}_r(s) = \int e^{ist}\,dH_r(t)$. It turns out that H_r is a good approximation to F if r is large enough. Here and in the following, under the notion of a (finite) *signed measure* M we understand a σ-additive set function $M : \mathcal{B}(\mathbb{R}) \to \mathbb{R}$ for which $|M(B)| < \infty$ for all Borel sets $B \in \mathcal{B}(\mathbb{R})$.

An application of the one-to-one correspondence between distributions and their characteristic functions to (4.6.17) shows that H_1 is the compound Poisson distribution with characteristics (θ, F_V). This fact provides a first-order approximation to F. Analogously, H_r is called the r-th order *Kornya approximation* to F. We show that H_r is of the compound-Poisson type with certain characteristics (λ_r, G_r), where $\lambda_r > 0$ and G_r is a signed measure.

Theorem 4.6.3 *For each $r \ge 1$, $\hat{\varphi}_r(s)$ is the characteristic function of the signed measure*

$$H_r = e^{-\lambda_r} \sum_{j=0}^{\infty} \frac{\lambda_r^j}{j!} G_r^{*j}\,; \tag{4.6.18}$$

here, $\lambda_r = \sum_{k=1}^{r} \theta^k/k$, while G_r is a signed measure on $\mathbb{R}$ with $G_r(B) = 0$ for all $B \in \mathcal{B}((-\infty, 0])$ which solves the equation

$$\sum_{k=1}^{r} \frac{(-1)^{k+1}}{k}\theta^k(F_V - \delta_0)^{*k} = \lambda_r(G_r - \delta_0)\,. \tag{4.6.19}$$

Before proving Theorem 4.6.3 we state some elementary but useful properties of signed measures. Let $M_1, M_2 : \mathcal{B}(\mathbb{R}) \to \mathbb{R}$ be two signed measures. The *convolution* $M_1 * M_2$ of M_1 and M_2 is the signed measure given by

$$M_1 * M_2(B) = \iint \mathbb{I}_B(x+y)\mathrm{d}M_1(x)\mathrm{d}M_2(y)\,, \qquad B \in \mathcal{B}(\mathbb{R})\,.$$

The n-fold convolution M^{*n} of a signed measure M is defined recursively by $M^{*0} = \delta_0$, and $M^{*n} = M^{*(n-1)} * M, n \ge 1$. It is easily seen that the *exponential set function* $\exp(M)$ of a signed measure M, i.e. $\exp(M) = \sum_{k=0}^{\infty}(k!)^{-1}M^{*k}$, is a well-defined (finite) signed measure. We leave it to the reader to prove this. It is also well known that each signed measure M can be represented as the difference $M = M_+ - M_-$ of two (nonnegative) measures M_+, M_-, called the *Hahn–Jordan decomposition* of M. The *total variation* of M is given by $||M|| = M_+(\mathbb{R}) + M_-(\mathbb{R})$. We will need the following auxiliary result.

Lemma 4.6.6 *Let M, M_1, M_2 be arbitrary signed measures on $\mathbb{R}$. Then*
(a) $\exp(\hat{\varphi}_M(s))$ *is the characteristic function of* $\exp(M)$,
(b) $\exp(M_1 + M_2) = \exp(M_1) * \exp(M_2)$,
(c) $||M_1 * M_2|| \le ||M_1||\,||M_2||$,
(d) $||\exp(M) - \delta_0|| \le \mathrm{e}^{||M||} - 1$.

The *proof* of Lemma 4.6.6 is left to the reader.

Proof of Theorem 4.6.3. Consider the signed measure

$$H_r = \exp\Big(\sum_{k=1}^{r} \frac{(-1)^{k+1}}{k}\theta^k (F_V - \delta_0)^{*k}\Big). \tag{4.6.20}$$

Then, by Lemma 4.6.6a, $\hat{\varphi}_r(s)$ is the characteristic function of H_r. Note that

$$\sum_{k=1}^{r} \frac{(-1)^{k+1}}{k}\theta^k (F_V - \delta_0)^{*k}(\{0\}) = -\sum_{k=1}^{r}\frac{\theta^k}{k}$$

and $\sum_{k=1}^{r}(-1)^{k+1}(\theta^k/k)(F_V - \delta_0)^{*k}(B) = 0$ for all $B \in \mathcal{B}((-\infty,0))$. Thus, there exists a signed measure G_r on $\mathbb{R}$ satisfying (4.6.19) and such that $G_r(B) = 0$ for all $B \in \mathcal{B}((-\infty,0])$. Now, by Lemma 4.6.6b, (4.6.19) and (4.6.20) give $H_r = \exp(\lambda_r G_r - \lambda_r \delta_0) = \mathrm{e}^{-\lambda_r}\sum_{j=0}^{\infty}(\lambda_r^j/j!)G_r^{*j}$. □

For the more general case of the individual model describing a portfolio of n policies, we can approximate the distribution F of the aggregate claim amount $X^{\mathrm{ind}} = \sum_{i=1}^{n} U_i$ in a completely analogous manner. Defining

$$H_r = \exp\Big(\sum_{i=1}^{n}\sum_{k=1}^{r} \frac{(-1)^{k+1}}{k}\theta_i^k (F_{V_i} - \delta_0)^{*k}\Big) \tag{4.6.21}$$

we can derive the following error bound.

Theorem 4.6.4 *For $i = 1, \ldots, n$, let $\theta_i < \frac{1}{2}, \xi_i = \frac{1}{r+1}(2\theta_i)^{r+1}(1-2\theta_i)^{-1}$, and $\xi = \sum_{i=1}^n \xi_i$. Then $d_{\rm TV}(F, H_r) \le {\rm e}^{\xi} - 1$.*

Proof Note that, by (4.6.16), we have the following representation of F:

$$F = \exp\Big(\sum_{i=1}^n \sum_{k=1}^\infty \frac{(-1)^{k+1}}{k} \theta_i^k (F_{V_i} - \delta_0)^{*k}\Big).$$

Thus, using Lemma 4.6.6,

$$\begin{aligned}
d_{\rm TV}(F, H_r) &= \|F - H_r\| \\
&= \Big\|\exp\Big(\sum_{i=1}^n \sum_{k=1}^\infty \frac{(-1)^{k+1}}{k} \theta_i^k (F_{V_i} - \delta_0)^{*k}\Big) \\
&\qquad - \exp\Big(\sum_{i=1}^n \sum_{k=1}^r \frac{(-1)^{k+1}}{k} \theta_i^k (F_{V_i} - \delta_0)^{*k}\Big)\Big\| \\
&= \Big\|F * \Big(\delta_0 - \exp\Big(\sum_{k=r+1}^\infty \sum_{i=1}^n \frac{(-1)^{k+1}}{k} \theta_i^k (F_{V_i} - \delta_0)^{*k}\Big)\Big)\Big\| \\
&\le \|F\| \Big\|\delta_0 - \exp\Big(\sum_{k=r+1}^\infty \sum_{i=1}^n \frac{(-1)^{k+1}}{k} \theta_i^k (F_{V_i} - \delta_0)^{*k}\Big)\Big\| \\
&= \Big\|\exp\Big(\sum_{k=r+1}^\infty \sum_{i=1}^n \frac{(-1)^{k+1}}{k} \theta_i^k (F_{V_i} - \delta_0)^{*k}\Big) - \delta_0\Big\| \\
&\le \exp\Big(\Big\|\sum_{k=r+1}^\infty \sum_{i=1}^n \frac{(-1)^{k+1}}{k} \theta_i^k (F_{V_i} - \delta_0)^{*k}\Big\|\Big) - 1.
\end{aligned}$$

Since $\|\sum_{k=r+1}^\infty \sum_{i=1}^n (-1)^{k+1} k^{-1} \theta_i^k (F_{V_i} - \delta_0)^{*k}\| \le \xi$, the proof of the statement is complete. □

Similarly to the case $n = 1$ considered in Theorem 4.6.3, the Kornya approximation H_r defined in (4.6.21) can be represented as a signed measure of the compound-Poisson type with some characteristics (λ_r, G_r). For example,

$$\lambda_2 = \sum_{i=1}^n \Big(\theta_i + \frac{\theta_i^2}{2}\Big), \quad \lambda_3 = \sum_{i=1}^n \Big(\theta_i + \frac{\theta_i^2}{2} + \frac{\theta_i^3}{3}\Big)$$

and

$$G_2 = \frac{1}{\lambda_2} \sum_{i=1}^n \Big((\theta_i + \theta_i^2) F_{V_i} - \frac{1}{2} \theta_i^2 F_{V_i}^{*2}\Big),$$

$$G_3 = \frac{1}{\lambda_3} \sum_{i=1}^n \Big((\theta_i + \theta_i^2 + \theta_i^3) F_{V_i} - \Big(\frac{\theta_i^2}{2} + \theta_i^3\Big) F_{V_i}^{*2} + \frac{\theta_i^3}{3} F_{V_i}^{*3}\Big).$$

Inserting these formulae into (4.6.18) and applying the Fourier transform method (see Section 4.7) to the signed measure H_r of the compound-Poisson type, reliable approximations can be obtained. For a practical illustration, see Figure 4.7.2.

Bibliographical Notes. The compound Poisson approximations given in Theorems 4.6.1 and 4.6.2 have been derived in Barbour and Hall (1984). More error bounds can be found in Gerber (1984). For example, it is shown that for the compound Poisson approximation to the individual model $d_{\rm SL}(X^{\rm ind}, X^{\rm col}) \le \sum_{i=1}^n \theta_i^2 \mu_{V_i}$. Further, in the individual model with deterministic claim sizes V_i, $d_{\rm SL}(X^{\rm ind}, X^{\rm col}) \le 2^{-1} \sum_{i=1}^n \theta_i^2 \mu_{V_i}$. The total variation distance between $\mathrm{Bin}(n,p)$ and the distribution of $\sum_{i=1}^n I_i$, where $I_1, \ldots, I_n$ are independent Bernoulli random variables with $\mathbb{P}(I_i = 1) = \theta_i$ and $p = \sum_{i=1}^n \theta_i$ is given in Theorem 9.E in Barbour, Holst and Janson (1992). Bounds with respect to the metric $d_{\rm SD}$ are studied in Hipp (1985) and Kuon, Radke and Reich (1991). Further bounds are given in de Pril and Dhaene (1992), where the individual risk model is approximated by compound Poisson models with more general characteristics than those given in (4.6.4); see also Barbour, Chen and Loh (1992) and Dhaene and Sundt (1997). Higher-order approximations to the compound-Poisson type have been introduced in Kornya (1983) and Hipp (1986); see also Dufresne (1996) and Hipp and Michel (1990).

4.7 INVERTING THE FOURIER TRANSFORM

Besides the algorithms presented in Section 4.4, we can use a formula for inverting the Fourier transform when calculating the probability function of a discrete random variable. In particular, this inversion formula can be used to calculate the probability function of the compound $\sum_{k=1}^N U_k$ when the claim sizes $U_1, U_2, \ldots$ are lattice. In comparison to Panjer's algorithm, this method has the advantage that no special assumptions on the distribution of N are needed. Recall that the characteristic function of an $\mathbb{N}$-valued random variable with probability function $\{p_k\}$ is given by the Fourier transform

$$\hat{\varphi}(s) = \sum_{k=0}^{\infty} \mathrm{e}^{\mathrm{i}sk} p_k\,, \qquad s \in \mathbb{R}. \tag{4.7.1}$$

It is useful to introduce the Fourier transform in a more general situation. For some fixed $n \in \mathbb{N}$, consider a sequence $p_0, p_1, \ldots, p_{n-1}$ of arbitrary real numbers. The Fourier transform $\hat{\varphi}_{(n)}(s)$ of $\{p_0, \ldots, p_{n-1}\}$ is defined by

$$\hat{\varphi}_{(n)}(s) = \sum_{k=0}^{n-1} \mathrm{e}^{\mathrm{i}sk} p_k\,, \qquad s \in \mathbb{R}. \tag{4.7.2}$$

Clearly, if $\{p_0, \ldots, p_{n-1}\}$ is a probability function, then $\hat{\varphi}_{(n)}(s)$ is the characteristic function of $\{p_0, p_1, \ldots, p_{n-1}\}$. We now show that the row vector $\boldsymbol{p} = (p_0, p_1, \ldots, p_{n-1})$ can be calculated from limited information on the Fourier transform $\hat{\varphi}_{(n)}(s)$. Assume that the values $\hat{\varphi}_{(n)}(s_j)$ of $\hat{\varphi}_{(n)}$ are given at the points $s_j = 2\pi j/n$ for $j = 0, 1, \ldots, n-1$. Write these values of the Fourier transform $\hat{\varphi}_{(n)}(s)$ as a row vector $\boldsymbol{\varphi} = (\hat{\varphi}_{(n)}(s_0), \ldots, \hat{\varphi}_{(n)}(s_{n-1}))$. Then

$$\boldsymbol{\varphi}^\top = \boldsymbol{F}\boldsymbol{p}^\top, \tag{4.7.3}$$

where $\boldsymbol{F}$ is the $n \times n$ matrix given by

$$\boldsymbol{F} = \left(\mathrm{e}^{\mathrm{i}s_j k}\right)_{j,k=0,\ldots,n-1} = \begin{pmatrix} 1 & 1 & \cdots & 1 \\ 1 & \mathrm{e}^{\mathrm{i}2\pi/n} & \cdots & \mathrm{e}^{\mathrm{i}2\pi(n-1)/n} \\ \vdots & \vdots & \vdots & \vdots \\ \mathrm{e}^{\mathrm{i}s_j 0} & \mathrm{e}^{\mathrm{i}s_j 1} & \cdots & \mathrm{e}^{\mathrm{i}s_j(n-1)} \\ \vdots & \vdots & \vdots & \vdots \\ 1 & \mathrm{e}^{\mathrm{i}2\pi(n-1)/n} & \cdots & \mathrm{e}^{\mathrm{i}2\pi(n-1)^2/n} \end{pmatrix}$$

and $\boldsymbol{x}^\top$ is the transpose of $\boldsymbol{x}$. Note that the rows of $\boldsymbol{F}$ form orthogonal vectors because

$$\sum_{k=0}^{n-1} \mathrm{e}^{\mathrm{i}s_j k}\overline{\mathrm{e}^{\mathrm{i}s_\ell k}} = \sum_{k=0}^{n-1} \mathrm{e}^{\mathrm{i}(s_j - s_\ell)k} = \begin{cases} \dfrac{\mathrm{e}^{\mathrm{i}(s_j-s_\ell)n} - 1}{\mathrm{e}^{\mathrm{i}(s_j-s_\ell)} - 1} = 0 & \text{for } j \neq \ell, \\ n & \text{for } j = \ell. \end{cases}$$

Therefore the inverse matrix $\boldsymbol{F}^{-1}$ of $\boldsymbol{F}$ is given by

$$\boldsymbol{F}^{-1} = \left(\frac{1}{n}\mathrm{e}^{-\mathrm{i}s_k j}\right)_{j,k=0,\ldots,n-1}.$$

Thus, from (4.7.3) we get $\boldsymbol{p}^\top = \boldsymbol{F}^{-1}\boldsymbol{\varphi}^\top$, i.e. for all $k = 0, \ldots, n-1$

$$p_k = \frac{1}{n}\sum_{j=0}^{n-1} \hat{\varphi}_{(n)}(s_j)\mathrm{e}^{-\mathrm{i}s_j k}. \tag{4.7.4}$$

Note that computing $\boldsymbol{p}$ directly from $\boldsymbol{\varphi}$ requires n^2 operations. However, if $n = 2^m$ for some nonnegative integer m, then there exist algorithms with complexity of order $n \log_2 n$. For this reason, the corresponding procedure for computing $p_0, p_1, \ldots, p_{n-1}$ on the basis of (4.7.4) is called the *fast Fourier transform* (FFT) or the inverse fast Fourier transform (IFFT). These procedures are available in many software packets like MAPLE, MATHEMATICA or MATLAB.

	Amount at risk					
θ_i	1	2	3	4	5	6
0.02	3	1	-	1	2	-
0.03	-	-	2	3	-	2
0.04	-	4	-	-	1	-
0.05	-	3	1	1	-	1
0.06	-	-	3	-	-	1

Table 4.7.1 A sample portfolio of 29 policies

k	r_k	k	r_k	k	r_k
0	1.00000	7	0.19154	14	0.01866
1	0.67176	8	0.15469	15	0.01246
2	0.65166	9	0.11344	16	0.00807
3	0.53802	10	0.08019	17	0.00529
4	0.43065	11	0.05778	18	0.00336
5	0.35297	12	0.04053	19	0.00211
6	0.28675	13	0.02674	20	0.00133

Table 4.7.2 The distribution of aggregate claims

Suppose now that we know the Fourier transform $\hat{\varphi}(s) = \sum_{k=0}^{\infty} \mathrm{e}^{\mathrm{i}sk} p_k$ of an infinite, summable sequence $p_0, p_1, \ldots$ (that is $\sum_{k=0}^{\infty} |p_k| < \infty$). Then it is possible to obtain an approximation to the first n terms $p_0, \ldots, p_{n-1}$ of the sequence $p_0, p_1, \ldots$ by sampling the Fourier transform $\hat{\varphi}(s)$ at the points $s_j = 2\pi j/n$, $j = 0, 1, \ldots, n-1$. Since for each $k = 1, 2, \ldots$ the function $\{\mathrm{e}^{\mathrm{i}sk},\ s \geq 0\}$ has the period 2π, we obtain

$$\hat{\varphi}(s_j) = \sum_{k=0}^{\infty} p_k \mathrm{e}^{\mathrm{i}s_j k} = \sum_{\ell=0}^{\infty} \sum_{k=0}^{n-1} p_{k+n\ell} \mathrm{e}^{\mathrm{i}s_j(k+n\ell)} = \sum_{k=0}^{n-1} \tilde{p}_k \mathrm{e}^{\mathrm{i}s_j k},$$

where $\tilde{p}_k = \sum_{\ell=0}^{\infty} p_{k+n\ell}$ for $k = 0, 1, \ldots, n-1$. The values $\tilde{p}_0, \tilde{p}_1, \ldots, \tilde{p}_{n-1}$ can be calculated from $\{\hat{\varphi}(s_j), j = 0, 1, \ldots, n-1\}$ by the same argument which led to (4.7.4), i.e.

$$\tilde{p}_k = \frac{1}{n} \sum_{j=0}^{n-1} \hat{\varphi}(s_j) \mathrm{e}^{-\mathrm{i}s_j k}. \tag{4.7.5}$$

Note that $\tilde{p}_k$ approximates p_k for each $k = 0, \ldots, n-1$, since by the assumed summability of $p_0, p_1, \ldots$, the error $\tilde{p}_k - p_k = \sum_{\ell=1}^{\infty} p_{k+n\ell}$ becomes arbitrarily

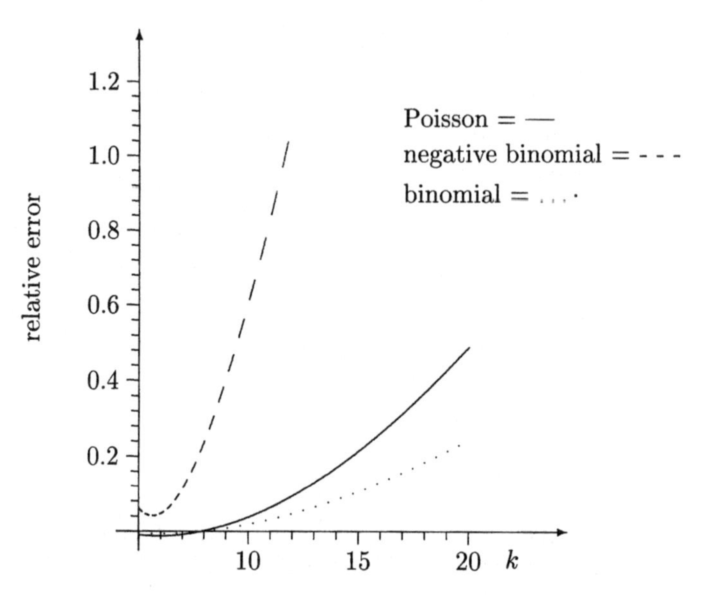

Figure 4.7.1 Approximation by compound distributions

small if n tends to infinity. When bounds on the tail of $\{p_k\}$ are available, we can even estimate the error $\tilde{p}_k - p_k$.

We turn to the compound $X = \sum_{k=1}^{N} U_k$, where $U_1, U_2, \ldots$ are independent and identically distributed $\mathbb{N}$-valued random variables. Assume that the generating function $\hat{g}_N$ of N and the values $\{\hat{\varphi}_U(s_j), j = 0, 1, \ldots, n-1\}$ of the characteristic function $\hat{\varphi}_U(t)$ are known. Then the values $\{\hat{\varphi}_X(s_j),\ j = 0, 1, \ldots, n-1\}$ of the characteristic function $\hat{\varphi}_X(s)$ can be computed from the formula $\hat{\varphi}_X(s) = \hat{g}_N(\hat{\varphi}_U(s))$, $(s \in \mathbb{R})$, which can be derived as in the proof of Theorem 4.2.1. Thus, the probability function $\{p_k\}$ of X can be calculated or, at least, approximated in the way given above.

Example Consider the following portfolio of 29 life insurance policies which is defined in Table 4.7.1 following a proposal in Gerber (1979). For this reason we will call it *Gerber's portfolio.* The characteristic function of the individual risk of each policy has the form $\hat{\varphi}_k(s) = (1 - \theta_k) + \theta_k e^{i\mu_k s}$, $k = 1, 2, \ldots, 29$, where θ_k is the probability of death and μ_k is the amount of risk for the k-th policy. From Table 4.7.1 we see that the aggregate claim amount $X^{\mathrm{ind}} = \sum_{k=1}^{29} U_k$ takes values in the set $\{0, \ldots, 96\}$. Let $\{p_k, k = 0, 1, \ldots, 96\}$ denote its probability function. The corresponding characteristic function is given by $\hat{\varphi}(s) = \prod_{k=1}^{29} \hat{\varphi}_k(s)$. The computation of the $r_k = p_k + p_{k+1} + \ldots$, $k = 0, 1, \ldots$, uses the Fourier transform method and is shown in Table 4.7.2.

Other possibilities to compute the r_k are provided by approximations

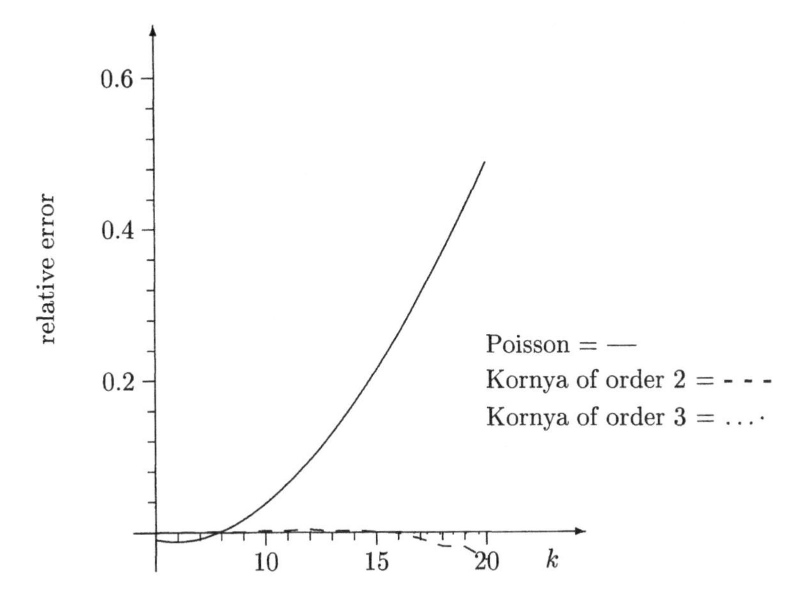

Figure 4.7.2 Higher-order Kornya approximation, $\lambda_1 = 1.09$, $\lambda_2 = 1.11325$, $\lambda_3 = 1.113976$

by Poisson, negative binomial and binomial compounds or by higher-order (Kornya) approximations, as proposed in Section 4.6. In Figures 4.7.1 and 4.7.2 we present the relative errors of these approximations with respect to the r_k as computed by the Fourier transform method. For Gerber's portfolio, the relative errors of second-order and third-order Kornya approximations are practically negligible.

Bibliographical Notes. The computation of the probability function of a discrete compound by inverting the Fourier transform is reviewed in Buchwalder, Chevallier and Klüppelberg (1993), Embrechts and Klüppelberg (1994) and Kaas (1987). There are other numerical methods to compute compound distributions where the claim sizes are of lattice type; see, for example, the algorithms considered in Sections 4.4.1 and 4.4.2. If one wants to use these methods in the case of general claim sizes, one first has to approximate the claim sizes by lattice alternatives. Hipp and Michel (1990) discuss such discretization techniques as well as the error caused by them. Using a linear upper bound for the concentration function of a distribution function, they derive a linear bound for the discretization error when a compound with characteristics $(\{p_k\}, F)$ is replaced by a compound with characteristics $(\{p_k\}, F_h)$ and where F_h is a discrete distribution defined by $F_h(x) = F(h(i+1))$ for $hi \le x < h(i+1)$; $h > 0$. Second-order approximations are derived in Embrechts, Grübel and Pitts (1994). In den Iseger, Smith and Dekker (1997) an approximation method is proposed which is based on cubic splines. Panjer and Lutek (1983) examine various discretization methods of the claim size in order to calculate stop-loss premiums recursively; see also Kaas (1987). Bühlmann (1984) compares the efficiency of Panjer's algorithm versus the Fourier transform method.

CHAPTER 5

Risk Processes

5.1 TIME-DEPENDENT RISK MODELS

In the preceding chapter we considered a risk X, which was an aggregate claim over a single period. For simplicity we thought of the period being one year, but other lengths of time would have been possible. Now we consider the reserve at the end of a number of such periods of equal length. Hence, we suppose that a sequence $X_1, X_2, \ldots$ of risks is given which are assumed to be independent and identically distributed, where X_n is the aggregate claim over the n-th period $(n-1, n]$. Moreover, we assume that X_n takes values in $\mathbb{N}$ and that the common probability function is $\{p_k\} = \{\mathbb{P}(X_n = k)\}$. Apart from the sequence $X_1, X_2, \ldots$ we take into account that constant premiums are collected during each period. For simplicity we take the premium equal to 1. Finally, suppose that the initial reserve is equal to $u \in \mathbb{N}$. Then, the reserve R_n after the n-th period is

$$R_n = u + n - \sum_{i=1}^{n} X_i \,. \tag{5.1.1}$$

The sequence $\{R_n, \, n \in \mathbb{N}\}$ describes the evolution of the reserve checked at the end of each period. We will call this process the *discrete-time risk reserve process*. Stochastic processes of this type in discrete and continuous time are one of the main subjects of interest in this book. In particular, we search for the probability that the risk reserve process $\{R_n\}$ ever falls below a certain critical level.

5.1.1 The Ruin Problem

Obviously, if $p_0 + p_1 < 1$, then the risk reserve R_n can be negative for some $n \in \mathbb{N}$. This event $\{R_1 < 0\} \cup \{R_2 < 0\} \cup \ldots$ is called the (technical) *ruin* of the portfolio. Some knowledge of the probability that ruin occurs can be helpful for determining the solvency of a portfolio.

Formally, the probability of ruin is defined in the following way. Consider the epoch $\tau_{\rm d}(u) = \min\{n \geq 1 : \ R_n < 0\}$ when the risk reserve process becomes negative for the first time and where we put $\tau_{\rm d}(u) = \infty$ if $R_n \geq 0$ for all $n \in \mathbb{N}$. Thus, $\tau_{\rm d}(u)$ is an extended random variable and it is called the *time of ruin* (or *ruin time*). Furthermore,

$$\psi(u) = \mathbb{P}(\{R_1 < 0\} \cup \{R_2 < 0\} \cup \ldots) = \mathbb{P}(\tau_{\rm d}(u) < \infty)$$

is called the (infinite-horizon) ***ruin probability*** for the initial reserve $u \in \mathbb{N}$. If $\psi(u)$ is seen as a function of the variable u, then $\psi(u)$ is called the *ruin function*. Another characteristic is the so-called *finite-horizon ruin probability* $\psi(u; n) = \mathbb{P}(\tau_{\rm d}(u) \leq n)$ which is the probability that ruin occurs not later than after the n-th period, $n \in \mathbb{N}$. Unfortunately, in many cases it is difficult to express the finite-horizon ruin probabilities $\psi(u; n)$ in a closed form. The infinite-horizon ruin probabilities $\psi(u)$ are mathematically simpler. As will be seen in later sections of the book, bounds and approximations to $\psi(u)$ are often available, even for more general risk models.

Instead of the risk reserve process, it is sometimes preferable to consider the *claim surplus process* $\{S_n\}$ defined by

$$S_n = \sum_{i=1}^{n} X_i - n\,, \qquad n \in \mathbb{N}\,. \tag{5.1.2}$$

Then,

$$\psi(u) = \mathbb{P}(\{S_1 > u\} \cup \{S_2 > u\} \cup \ldots) = \mathbb{P}(\max\{S_1, S_2, \ldots\} > u)\,. \tag{5.1.3}$$

With the notation $Y_i = X_i - 1$, we have

$$S_n = \sum_{i=1}^{n} Y_i\,, \tag{5.1.4}$$

where the $Y_1, Y_2, \ldots$ are independent and identically distributed. Note that the random variables $Y_1, Y_2, \ldots$ do not need to have the special form $Y_i = X_i - 1$. We can consider a more general sequence $Y_1, Y_2, \ldots$ of independent and identically distributed random variables. In particular, the aggregate claims $X_1, X_2, \ldots$ can have a continuous distribution and the premiums collected during each period can be random, modelled as a sequence $X_1', X_2', \ldots$ of nonnegative independent and identically distributed random variables. Then, $Y_n = X_n - X_n'$. A discrete-time stochastic process $\{S_n, n \in \mathbb{N}\}$ defined by the sums $S_n = \sum_{i=1}^n Y_i$ of arbitrary independent and identically distributed (not necessarily integer-valued) random variables $Y_1, Y_2, \ldots$ is called a *random walk*. Random walks will be considered in more detail in Section 6.3. Furthermore, as before, for each $u \geq 0$, we can consider the ruin time $\tau_{\rm d}(u) = \min\{n \geq 1 : \ S_n > u\}$ and the ruin function $\psi(u)$ defined by $\psi(u) = \mathbb{P}(\tau_{\rm d}(u) < \infty)$.

In a random walk setting, the ruin probability $\psi(u)$ is the probability that the random walk $\{S_n\}$ (strictly) exceeds the level u. The finite-horizon ruin probability $\psi(u;n) = \mathbb{P}(\tau_{\rm d}(u) \le n)$ is defined similarly.

From the strong law of large numbers we have that $S_n/n \to \mathbb{E}\,Y$. This tells us that $S_n \to \infty$ provided that $\mathbb{E}\,Y > 0$. Thus, in this case (5.1.3) implies that $\psi(u) \equiv 1$ for each $u \in \mathbb{N}$. It can be shown that this result extends to the case where $\mathbb{E}\,Y = 0$, but the proof given in Section 6.1 requires more advanced tools. However, if $\mathbb{E}\,Y < 0$, then $S_n \to -\infty$, i.e. the maximum $\max\{S_1, S_2, \ldots\}$ is finite with probability 1 and therefore $\psi(u) < 1$ for all $u \in \mathbb{N}$.

5.1.2 Computation of the Ruin Function

In this section we consider the random walk $\{S_n\}$ with generic increment $Y = X - 1$. We assume $\mathbb{E}\,Y < 0$. Note that for the maximum $M = \max\{0, S_1, S_2, \ldots\}$ of $\{S_n\}$ we have $\psi(u) = \mathbb{P}(M > u)$. Furthermore, $M = \max\{0, Y_1, Y_1 + \tilde{S}_1, Y_1 + \tilde{S}_2, \ldots\}$ where $\tilde{S}_1 = Y_2$, $\tilde{S}_2 = Y_2 + Y_3, \ldots$, i.e.

$$M \stackrel{\rm d}{=} (M + Y)_+ \,. \tag{5.1.5}$$

Thus, with the notations $\hat{g}_M(s) = \mathbb{E}\,s^M, \hat{g}_X(s) = \mathbb{E}\,s^X$, and $\rho = \mathbb{E}\,X$ we get the following result.

Theorem 5.1.1 (a) *The generating function of M is*

$$\hat{g}_M(s) = \frac{(1-\rho)(1-s)}{\hat{g}_X(s) - s}\,, \qquad s \in (-1,1)\,. \tag{5.1.6}$$

(b) *The maximum M of the random walk $\{S_n\}$ has the same distribution as the geometric compound $\sum_{i=1}^N U_i$ specified by $\rho = \mathbb{P}(N \ge 1)$ and $\mathbb{P}(U = k) = \mathbb{P}(X > k)/\mathbb{E}\,X$, $k \in \mathbb{N}$.*

Proof (a) From (5.1.5) we have

$$\begin{aligned}
\hat{g}_M(s) &= \mathbb{E}\,s^M = \mathbb{E}\,s^{(M+X-1)_+} \\
&= \mathbb{E}\left[s^{(M+X-1)_+}; M + X - 1 \ge 0\right] + \mathbb{E}\left[s^{(M+X-1)_+}; M + X - 1 = -1\right] \\
&= \frac{1}{s}\sum_{k=1}^{\infty} s^k \mathbb{P}(M + X = k) + \mathbb{P}(M + X = 0) \\
&= s^{-1}\hat{g}_M(s)\hat{g}_X(s) + (1 - s^{-1})\mathbb{P}(M + X = 0)\,,
\end{aligned}$$

i.e.

$$\hat{g}_M(s) = \frac{(s-1)\mathbb{P}(M + X = 0)}{s - \hat{g}_X(s)}\,. \tag{5.1.7}$$

Since $\lim_{s\uparrow 1}\hat{g}_M(s) = 1$ and $\lim_{s\uparrow 1}\hat{g}_X^{(1)}(s) = \mathbb{E}X$, L'Hospital's rule gives $1 = \mathbb{P}(M+X=0)/(1-\rho)$. Thus, (5.1.6) follows from (5.1.7). To prove part (b), observe that the generating function $\hat{g}_U(s)$ of U is given by

$$\hat{g}_U(s) = \frac{1}{\mathbb{E}X}\frac{1-\hat{g}_X(s)}{1-s}. \tag{5.1.8}$$

Hence, from (5.1.6) we get

$$\hat{g}_M(s) = \frac{(1-\rho)(1-s)}{\hat{g}_X(s)-s} = \frac{(1-\rho)(1-s)}{(s-1)\rho\hat{g}_U(s)+(1-s)} = \frac{1-\rho}{1-\rho\hat{g}_U(s)}.$$

Let $\tilde{M} = \sum_{i=1}^N U_i$. Note that $\mathbb{E}(s^{\tilde{M}} \mid N=n) = (\hat{g}_U(s))^n$. Then

$$\begin{aligned}\hat{g}_{\tilde{M}}(s) &= \mathbb{E}\,s^{\tilde{M}} = \sum_{n=0}^{\infty}\mathbb{E}(s^{\tilde{M}} \mid N=n)\,\mathbb{P}(N=n)\\ &= \sum_{n=0}^{\infty}(\hat{g}_U(s))^n(1-\rho)\rho^n = \hat{g}_M(s).\end{aligned}$$

This completes the proof because of the one-to-one correspondence between distributions on $\mathbb{N}$ and their generating functions. □

5.1.3 A Dual Queueing Model

The intrinsic goal of mathematical modelling is to capture the most important facts of a real problem into a stochastic framework. For this reason, mathematical models are rather universal. It may happen that one and the same model can describe different situations which seemingly do not have too much in common. A nice example where this is clearly visible stems from queueing theory.

It turns out that the simple risk model considered above can be used to analyse a problem of data transmission in a computer network. Suppose that at each time $n \in \mathbb{N}$ a random number of data packets arrives at a node of the network; in each time-slot $[n, n+1)$ one packet can be transmitted provided that at the beginning of this slot at least one packet is waiting. Let X_n be the number of packets arriving at $n-1$. As before, we assume that the $\mathbb{N}$-valued random variables $X_1, X_2, \ldots$ are independent and identically distributed with common probability function $\{p_k\}$. Suppose that just before time $n=0$ there are L_0 packets waiting for transmission. We may assume that L_0 is random, but independent of the $X_1, X_2, \ldots$. After one unit of time the number of packets decreases by 1. But at the same time X_1 new packets arrive and so $(L_0 + X_1 - 1)_+$ packets are waiting just before the beginning of the next slot. In general, the number L_n of packets just before the beginning of the $(n+1)$-th slot fulfils the recurrence relation

$$L_n = (L_{n-1} + X_n - 1)_+ \,, \tag{5.1.9}$$

where L_n is called the *queue length* at time n. We leave it to the reader to show that the solution to (5.1.9) is

$$L_n = \max\{0, Y_n, Y_{n-1} + Y_n, \ldots, L_0 + Y_1 + \ldots + Y_n\} \,, \tag{5.1.10}$$

where $Y_i = X_i - 1$.

An important auxiliary result is the following *duality property* expressing a certain invariance property under *inversion of time.*

Lemma 5.1.1 *Let $Y_1, Y_2, \ldots$ be independent and identically distributed. Then, $(Y_n, Y_{n-1} + Y_n, \ldots, Y_1 + \ldots + Y_n) \stackrel{\mathrm{d}}{=} (Y_1, Y_1 + Y_2, \ldots, Y_1 + \ldots + Y_n)$ for all $n = 1, 2, \ldots$.*

Proof Since the random variables $Y_1, Y_2, \ldots$ are independent and identically distributed, the random vectors $(Y_1, Y_2, \ldots, Y_n)$ and $(Y_n, Y_{n-1}, \ldots, Y_1)$ have the same distribution for every $n = 1, 2, \ldots$. Thus, the random vectors $(Y_1, Y_1 + Y_2, \ldots, Y_1 + \ldots + Y_n)$ and $(Y_n, Y_n + Y_{n-1}, \ldots, Y_n + \ldots + Y_1)$ are identically distributed. This completes the proof. □

Now we are ready to state a relationship between the queueing model of data transmission and the risk model considered above. It says that the limit distribution of the queue lengths L_n can be expressed by the ruin function $\psi(u)$. For simplicity, we assume that $L_0 = 0$.

Theorem 5.1.2 *Let $L_0 = 0$, $\rho = \mathbb{E}\, X < 1$ and $\psi(u) = \mathbb{P}(M > u)$. Then,*

$$\lim_{n\to\infty} \mathbb{P}(L_n \le u) = 1 - \psi(u) \,, \qquad u \in \mathbb{N} \,. \tag{5.1.11}$$

Proof Using Lemma 5.1.1, we get from (5.1.10) that

$$\begin{aligned} \mathbb{P}(L_n > u) &= \mathbb{P}(\max\{0, Y_n, Y_{n-1} + Y_n, \ldots, Y_1 + \ldots + Y_n\} > u) \\ &= \mathbb{P}(\max\{0, Y_1, Y_1 + Y_2, \ldots, Y_1 + \ldots + Y_n\} > u) \,. \end{aligned}$$

Since $\max\{0, Y_1, Y_1 + Y_2, \ldots, Y_1 + \ldots + Y_n\} \le \max\{0, Y_1, Y_1 + Y_2, \ldots, Y_1 + \ldots + Y_{n+1}\}$, the limit $\lim_{n\to\infty} \mathbb{P}(L_n > u)$ exists. Thus, monotone convergence yields (5.1.11). □

5.1.4 A Risk Model in Continuous Time

The risks $X_1, X_2, \ldots$ considered in the preceding sections of this chapter can take the value 0 with a positive probability p_0. This means that in the corresponding period no claim occurs. An equivalent model would be obtained by recording the (random) indices of the periods where claims occur, together with the corresponding (strictly positive) aggregate claim amounts. We make this statement more precise. Consider the 0-1 sequence $\{I_n\} = \{\mathbb{I}(X_n > 0),\ n = 1,2,\ldots\}$ and define the positions of ones and the distances between them in the sequence $\{I_n\}$: $\sigma_0 = 0$, $\sigma_n = \min\{k > \sigma_{n-1} : I_k = 1\}$ and $T_n = \sigma_n - \sigma_{n-1}$; $n \geq 1$. The number of ones in the sequence $I_1,\ldots,I_n$ is $N(n) = \sum_{k=1}^{\infty} \mathbb{I}(\sigma_k \leq n)$. Let $U_1, U_2, \ldots$ be a sequence of independent and identically distributed random variables with distribution being equal to the conditional distribution of X provided that $X > 0$. Moreover, let the sequences $\{N(n)\}$ and $\{U_n\}$ be independent. Then, for the risk reserve process $\{R_n\}$ defined in (5.1.1), we have $R_n \stackrel{\mathrm{d}}{=} R'_n$, where $R'_n = u + n - \sum_{i=1}^{N(n)} U_i$. Note that an even stronger property holds. Namely, for each $n = 1,2,\ldots$ we have $(R_1,\ldots,R_n) \stackrel{\mathrm{d}}{=} (R'_1,\ldots,R'_n)$.

We can weaken the assumption that the random variables T and U are $\mathbb{N}$-valued assuming only that they are nonnegative. Then, a more general risk reserve model is defined as follows. We are given

- random epochs $\sigma_1, \sigma_2, \ldots$ with $0 < \sigma_1 < \sigma_2 < \ldots$ at which the claims occur, where the random variables σ_n can be discrete or continuous,
- the corresponding positive (individual or aggregate) claim sizes $U_1, U_2, \ldots$,
- the initial risk reserve $u \geq 0$, and
- the premiums which are collected at a constant rate $\beta > 0$, so that the premium income is a linear function of time.

Besides the sequence $\{(T_n, U_n)\}$ of inter-occurrence times and claim sizes, there are other but equivalent ways to describe the process of arriving claims. One such possibility is to consider the sequence $\{(\sigma_n, U_n)\}$ of arrival epochs σ_n and corresponding claim sizes U_n, where $\sigma_n = \sum_{i=1}^{n} T_i$. Sometimes, the random sequence $\{\sigma_n\}$ is called a *point process* and $\{(\sigma_n, U_n)\}$ a *marked point process*; see Chapter 12. Still another approach is based on the *cumulative arrival process* $\{X(t), t \geq 0\}$

$$X(t) = \sum_{k=1}^{\infty} U_k \mathbb{I}(\sigma_k \leq t) = \sum_{k=1}^{N(t)} U_k\,, \tag{5.1.12}$$

where $X(t)$ is the aggregate amount of all claims arriving in the interval $(0, t]$ and the *counting process* $\{N(t), t \geq 0\}$ is given by

$$N(t) = \sum_{k=1}^{\infty} \mathbb{I}(\sigma_k \leq t)\,. \tag{5.1.13}$$

The *risk reserve process* $\{R(t),\ t \geq 0\}$ is then given by

$$R(t) = u + \beta t - \sum_{i=1}^{N(t)} U_i\,, \tag{5.1.14}$$

while the *claim surplus process* $\{S(t),\ t \geq 0\}$ is

$$S(t) = \sum_{i=1}^{N(t)} U_i - \beta t\,. \tag{5.1.15}$$

The *time of ruin* $\tau(u) = \min\{t : R(t) < 0\} = \min\{t : S(t) > u\}$ is the first epoch when the risk reserve process becomes negative or, equivalently, when the claim surplus process crosses the level u. We will mainly be interested in the *ruin probabilities* $\psi(u;x) = \mathbb{P}(\tau(u) \leq x)$ and $\psi(u) = \lim_{x\to\infty} \psi(u;x) = \mathbb{P}(\tau(u) < \infty)$. Here $\psi(u;x)$ is called the *finite-horizon ruin probability* and $\psi(u)$ the *infinite-horizon ruin probability.* Alternatively, $\psi(u)$ can be called the *probability of ultimate ruin.* We will further need the notion of the *survival probability* $\overline{\psi}(u) = 1 - \psi(u)$.

There is a relationship between infinite-horizon ruin probabilities of risk models in discrete time and in continuous time. To get $\tau(u)$ it is sufficient to check the claim surplus process $\{S(t)\}$ at the embedded epochs σ_k ($k = 1, 2, \ldots$); see Figure 5.1.1. Indeed, the largest value $M = \max_{t\geq 0} S(t)$ of the

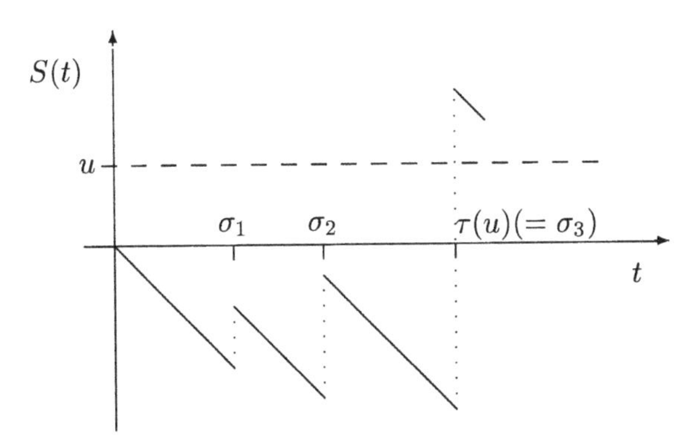

Figure 5.1.1 Claim surplus process

claim surplus process can be given by $M = \max_{n\geq 0} \sum_{k=1}^{n} (U_k - \beta T_k)$ and consequently

$$\psi(u) = \mathbb{P}(M > u)\,. \tag{5.1.16}$$

The representation formula (5.1.16) gives us the possibility to interpret the ruin function $\psi(u)$ as the tail function of the stationary waiting time in a *single-server system* of queueing theory.

Note, however, that one has to be careful when comparing finite-horizon ruin probabilities in discrete time with those in continuous time because in general

$$\mathbb{P}\Big(\max_{0\le t\le x} S(t) > u\Big) \ne \mathbb{P}\Big(\max_{0\le n\le x} \sum_{k=1}^{n}(U_k - \beta T_k) > u\Big).$$

Anyhow, in order to keep the notation simple we will use the same symbol for the finite-horizon ruin function in the continuous-time risk model as in the discrete-time risk model, i.e.

$$\psi(u;x) = \mathbb{P}(\tau(u) \le x) = \mathbb{P}\Big(\max_{0\le t\le x} S(t) > u\Big).$$

Apart from the time of ruin $\tau(u)$, there are other characteristics related to the concept of technical ruin. The *overshoot* above the level u of the random walk $\{S_n\}$ crossing this level for the first time is defined by

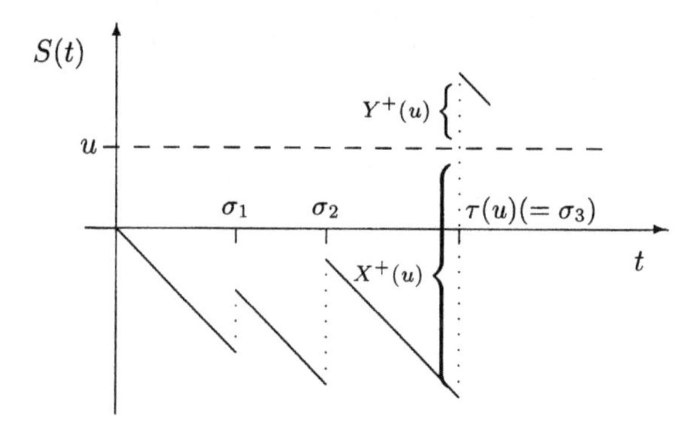

Figure 5.1.2 Severity of ruin and surplus prior to ruin

$$Y^+(u) = \begin{cases} S(\tau(u)) - u & \text{if } \tau(u) < \infty, \\ \infty & \text{if } \tau(u) = \infty. \end{cases}$$

Note that it is possible to express $Y^+(u)$ in terms of the risk reserve process:

$$Y^+(u) = \begin{cases} -R(\tau(u)) & \text{if } \tau(u) < \infty, \\ \infty & \text{if } \tau(u) = \infty. \end{cases}$$

In other words, $Y^+(u)$ can be interpreted as the *severity of ruin* at time $\tau(u)$; see Figure 5.1.2.

Another quantity of interest is the *surplus prior to ruin* given by

$$X^+(u) = \begin{cases} u - S(\tau(u)-) & \text{if } \tau(u) < \infty, \\ \infty & \text{if } \tau(u) = \infty. \end{cases}$$

Clearly, $X^+(u) + Y^+(u)$ is the *size of the claim causing ruin* at time $\tau(u)$. In order to determine the joint distribution of $X^+(u), Y^+(u)$, we will consider the *multivariate ruin function* $\psi(u, x, y)$ given by

$$\psi(u,x,y) = \mathbb{P}(\tau(u) < \infty, X^+(u) \le x, Y^+(u) > y), \tag{5.1.17}$$

where $u, x, y \ge 0$ or its dual

$$\varphi(u,x,y) = \mathbb{P}(\tau(u) < \infty, X^+(u) > x, Y^+(u) > y), \tag{5.1.18}$$

when the latter is more convenient. Another characteristic related to the severity of ruin is the time $\tau'(u) = \inf\{t : t > \tau(u), R(t) > 0\}$ at which the risk reserve process $\{R(t)\}$ crosses the level zero from below for the first time after the ruin epoch $\tau(u)$. Then

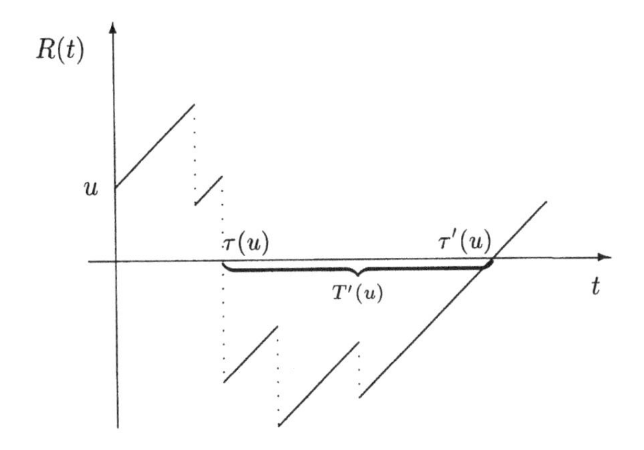

Figure 5.1.3 Time in the red

$$T'(u) = \begin{cases} \tau'(u) - \tau(u) & \text{if } \tau(u) < \infty, \\ 0 & \text{if } \tau(u) = \infty, \end{cases}$$

is the *time in the red* (see Figure 5.1.3), the amount of time the risk reserve process $\{R(t)\}$ stays below zero after the ruin time $\tau(u)$. It is clear that $T'(u)$ does not fully describe the severity of ruin, because it does not carry any information about the behaviour of the risk reserve between $\tau(u)$ and $\tau'(u)$. However, for the insurer it makes a difference whether $\{R(t)\}$ remains slightly below zero for a long time, or whether the *total maximal deficit*

$$Z^+(u) = \max\{-R(t) : \tau(u) \le t\}$$

after $\tau(u)$ is large. In the latter case, all successive times in the red are taken into account. We can finally consider the *maximal deficit*

$$Z_1^+(u) = \max\{-R(t) : \tau(u) \le t \le \tau'(u)\}$$

during the first period in the red, that is between $\tau(u)$ and $\tau'(u)$.

Bibliographical Notes. Results of the type as in Theorem 5.1.1 can be found in many places, such as Daley and Rolski (1984), Feller (1968), Gerber (1988) and Shiu (1989). A recursive method to calculate the finite-horizon ruin probabilities $\psi(u;n)$ in the discrete-time risk model has been proposed in De Vylder and Goovaerts (1988).

5.2 POISSON ARRIVAL PROCESSES

The risk reserve process $\{R(t)\}$ in continuous time has been defined by (5.1.14) in Section 5.1.4. We now consider the special case of the risk model where the claim sizes $\{U_n\}$ are independent and identically distributed and independent of the sequence $\{\sigma_n\}$ of claim occurrence epochs. Furthermore, we assume that the sequence $\{\sigma_n\}$ forms a *Poisson point process.* By this we mean that the inter-occurrence times $T_n = \sigma_n - \sigma_{n-1}$ are independent and (identically) exponentially distributed. These assumptions lead to the classical *compound Poisson model* of risk theory. While this model does not really suit the needs of the actuary, it is a skeleton for more adequate generalizations discussed in later chapters of the book. Thanks to its nice properties, the classical compound Poisson model is the most studied model in the literature.

5.2.1 Homogeneous Poisson Processes

Let $\{T_n\}$ be a sequence of independent random variables with exponential distribution Exp(λ); $\lambda > 0$. Then, the counting process $\{N(t)\}$ is called a *homogeneous Poisson process* with *intensity* λ. As such, the process is a special kind of renewal process, a topic discussed in Chapter 6. In the present chapter we simply omit the adjective "homogeneous", since we do not yet deal with other types of claim occurrence processes.

A basic property of Poisson processes is that they have *independent and stationary increments.* These notions are defined first.

Definition 5.2.1 *A real-valued stochastic process $\{X(t), t \geq 0\}$ is said to have*
(a) *independent increments if for all $n = 1, 2, \ldots$ and $0 \leq t_0 < t_1 < \ldots < t_n$, the random variables $X(0), X(t_1) - X(t_0), X(t_2) - X(t_1), \ldots, X(t_n) - X(t_{n-1})$ are independent,*
(b) *stationary increments if for all $n = 1, 2, \ldots$, $0 \leq t_0 < t_1 < \ldots < t_n$ and $h \geq 0$, the distribution of $(X(t_1+h) - X(t_0+h), \ldots, X(t_n+h) - X(t_{n-1}+h))$ does not depend on h.*

We leave it to the reader to prove that, if a process $\{X(t)\}$ has independent increments, then $\{X(t)\}$ has stationary increments whenever the distribution of the univariate random variable $X(t+h) - X(h)$ does not depend on h.

We now give some equivalent definitions of a Poisson process.

Theorem 5.2.1 *Let $\{N(t), t \geq 0\}$ be a counting process. Then the following statements are equivalent:*
(a) *$\{N(t)\}$ is a Poisson process with intensity λ.*
(b) *For all $t \geq 0$, $n = 1, 2, \ldots$ the random variable $N(t)$ has distribution* Poi(λt) *and, given $\{N(t) = n\}$, the random vector $(\sigma_1, \ldots, \sigma_n)$ has the same distribution as the order statistics of n independent points uniformly distributed on $[0, t]$.*
(c) *$\{N(t)\}$ has independent increments such that $\mathbb{E}\, N(1) = \lambda$ and for all $t \geq 0$, $n = 1, 2, \ldots$, given $\{N(t) = n\}$, the random vector $(\sigma_1, \ldots, \sigma_n)$ has the same distribution as the order statistics of n independent points uniformly distributed on $[0, t]$.*
(d) *$\{N(t)\}$ has stationary and independent increments and satisfies as $h \downarrow 0$,*

$$\mathbb{P}(N(h) = 0) = 1 - \lambda h + o(h)\,, \qquad \mathbb{P}(N(h) = 1) = \lambda h + o(h)\,. \tag{5.2.1}$$

(e) *$\{N(t)\}$ has stationary and independent increments and, for each fixed $t \geq 0$, the random variable $N(t)$ is* Poi(λt) *distributed.*

Proof (a)$\Rightarrow$(b) Note that (a) implies that $\sigma_n = \sum_{i=1}^n T_i$ is the sum of n independent random variables with exponential distribution Exp(λ), i.e. σ_n has distribution Erl(n, λ). Thus $\mathbb{P}(N(t) = 0) = \mathbb{P}(\sigma_1 > t) = \mathrm{e}^{-\lambda t}$ and

$$\begin{aligned}
\mathbb{P}(N(t) = n) &= \mathbb{P}(N(t) \geq n) - \mathbb{P}(N(t) \geq n+1) \\
&= \mathbb{P}(\sigma_n \leq t) - \mathbb{P}(\sigma_{n+1} \leq t) \\
&= \int_0^t \frac{\lambda^n v^{n-1}}{(n-1)!} \mathrm{e}^{-\lambda v}\, \mathrm{d}v - \int_0^t \frac{\lambda^{n+1} v^n}{n!} \mathrm{e}^{-\lambda v}\, \mathrm{d}v \\
&= \int_0^t \frac{\mathrm{d}}{\mathrm{d}v}\Big(\frac{(\lambda v)^n}{n!} \mathrm{e}^{-\lambda v}\Big)\, \mathrm{d}v = \frac{(\lambda t)^n}{n!} \mathrm{e}^{-\lambda t}
\end{aligned}$$

for $n \geq 1$, which shows that $N(t)$ has distribution Poi(λt). Furthermore, since for $t_0 = 0 \leq t_1 \leq \ldots \leq t_n \leq t \leq t_{n+1}$ the joint density $f_{\sigma_1, \ldots, \sigma_{n+1}}(t_1, \ldots, t_{n+1})$ of $\sigma_1, \ldots, \sigma_{n+1}$ is given by

$$f_{\sigma_1, \ldots, \sigma_{n+1}}(t_1, \ldots, t_{n+1}) = \prod_{k=1}^{n+1} \lambda \mathrm{e}^{-\lambda(t_k - t_{k-1})} = \lambda^{n+1} \mathrm{e}^{-\lambda t_{n+1}}\,,$$

the joint conditional density $f_{\sigma_1, \ldots, \sigma_n}(t_1, \ldots, t_n \mid N(t) = n)$ of $\sigma_1, \ldots, \sigma_n$ given $N(t) = n$ is then

$$\begin{aligned}
&f_{\sigma_1, \ldots, \sigma_n}(t_1, \ldots, t_n \mid N(t) = n) \\
&= \frac{\int_t^\infty \lambda^{n+1} \mathrm{e}^{-\lambda t_{n+1}}\, \mathrm{d}t_{n+1}}{\int_0^t \int_{t_1}^t \cdots \int_{t_{n-1}}^t \int_t^\infty \lambda^{n+1} \mathrm{e}^{-\lambda t_{n+1}}\, \mathrm{d}t_{n+1} \cdots \mathrm{d}_{t_1}} = \frac{n!}{t^n}\,.
\end{aligned}$$

This is the density of the order statistics of n independent random variables uniformly distributed on $[0,t]$.

(b)⇒(c) From (b) it clearly follows that $\mathbb{E}\,N(1) = \lambda$. Furthermore, let $x_k \in \mathbb{N}$ and $t_0 = 0 < t_1 < \ldots < t_n$. Then (b) implies that for $x = x_1 + \ldots + x_n$,

$$\begin{aligned}
&\mathbb{P}\Big(\bigcap_{k=1}^{n} \{N(t_k) - N(t_{k-1}) = x_k\}\Big) \\
&= \mathbb{P}\Big(\bigcap_{k=1}^{n} \{N(t_k) - N(t_{k-1}) = x_k\} \;\Big|\; N(t_n) = x\Big)\mathbb{P}(N(t_n) = x) \\
&= \frac{(\lambda t_n)^x}{x!}\mathrm{e}^{-\lambda t_n} \frac{x!}{x_1! \ldots x_n!} \prod_{k=1}^{n} \Big(\frac{t_k - t_{k-1}}{t_n}\Big)^{x_k} \\
&= \prod_{k=1}^{n} \frac{(\lambda(t_k - t_{k-1}))^{x_k}}{x_k!}\mathrm{e}^{-\lambda(t_k - t_{k-1})}
\end{aligned}$$

and therefore $\{N(t)\}$ has independent increments.

(c)⇒(d) We now assume that, under $\{N(t_n + h) = m\}$, the random vector $(\sigma_1, \ldots, \sigma_m)$ has the same distribution as the order statistics of m independent points uniformly distributed on $[0, t_n + h]$. Hence, for $x_k \in \mathbb{N}$, $t_0 = 0 < t_1 < \ldots < t_n$ and $h > 0$

$$\begin{aligned}
&\mathbb{P}\Big(\bigcap_{k=1}^{n} \{N(t_k + h) - N(t_{k-1} + h) = x_k\} \;\Big|\; N(t_n + h) = m\Big) \\
&= \mathbb{P}\Big(\bigcap_{k=1}^{n} \{N(t_k) - N(t_{k-1}) = x_k\} \;\Big|\; N(t_n + h) = m\Big).
\end{aligned}$$

Thus the law of total probability yields that $\{N(t)\}$ has stationary increments. Furthermore, the conditional uniformity property of statement (c) implies that, for $0 < h < 1$,

$$\begin{aligned}
\mathbb{P}(N(h) = 0) &= \sum_{k=0}^{\infty} \mathbb{P}(N(h) = 0, N(1) - N(h) = k) \\
&= \sum_{k=0}^{\infty} \mathbb{P}(N(1) = k)\; \mathbb{P}(N(1) - N(h) = k \mid N(1) = k) \\
&= \sum_{k=0}^{\infty} \mathbb{P}(N(1) = k)(1 - h)^k.
\end{aligned}$$

Thus,

$$\frac{1}{h}(1 - \mathbb{P}(N(h) = 0)) = \frac{1}{h}\Big(1 - \sum_{k=0}^{\infty} \mathbb{P}(N(1) = k)(1 - h)^k\Big)$$

$$= \sum_{k=1}^{\infty} \mathbb{P}(N(1)=k)\frac{1-(1-h)^k}{h}\,.$$

Recall that $(1-h)^k \geq 1-kh$ for all $0<h<1$, whenever $k=1,2,\ldots$. This means that the functions $g_h(k)=h^{-1}(1-(1-h)^k)$ in the last sum have the uniform bound $g(k)=k$. Moreover, this bound is integrable since

$$\sum_{k=1}^{\infty} k\mathbb{P}(N(1)=k) = \mathbb{E}\,N(1) = \lambda < \infty\,.$$

By interchanging sum and limit we obtain $\lim_{h\to 0} h^{-1}\mathbb{P}(N(h)>0)=\lambda$ from which the first part of (5.2.1) follows. In the same way we get that

$$\lim_{h\to 0}\frac{1}{h}\mathbb{P}(N(h)=1) = \lim_{h\to 0}\sum_{k=1}^{\infty}\mathbb{P}(N(1)=k)\,k(1-h)^{k-1} = \lambda\,,$$

which is equivalent to the second part of (5.2.1).
(d)$\Rightarrow$(e) Let $n\in\mathbb{N}$, $t\geq 0$ and $p_n(t)=\mathbb{P}(N(t)=n)$. Then for $h>0$,

$$p_0(t+h) = \mathbb{P}(N(t)=0, N(t+h)-N(t)=0) = p_0(t)(1-\lambda h+o(h)) \quad (5.2.2)$$

and for $t\geq h>0$,

$$p_0(t) = p_0(t-h)(1-\lambda h+o(h))\,. \quad (5.2.3)$$

This implies that $p_0(t)$ is continuous on $(0,\infty)$ and right-continuous at $t=0$. Rearranging terms in (5.2.2) and (5.2.3) we see that

$$\frac{p_0(t+h)-p_0(t)}{h} = -\lambda p_0(t)+o(1)$$

for $h\geq -t$ because $p_0(t-h)=p_0(t)+o(1)$. Thus $p_0(t)$ is differentiable and fulfils the differential equation

$$p_0^{(1)}(t) = -\lambda p_0(t) \quad (5.2.4)$$

for $t>0$. Since $p_0(0)=\mathbb{P}(N(0)=0)=1$, the only solution to (5.2.4) is

$$p_0(t) = \mathrm{e}^{-\lambda t}, \qquad t\geq 0\,. \quad (5.2.5)$$

In order to show that

$$p_n(t) = \frac{(\lambda t)^n}{n!}\mathrm{e}^{-\lambda t}, \qquad t\geq 0 \quad (5.2.6)$$

holds for all $n\in\mathbb{N}$, we can proceed as in the proof of (5.2.5). Namely, observe that (5.2.1) implies that $\mathbb{P}(N(h)>1)=o(h)$ as $h\downarrow 0$. Now use (5.2.5) and

induction on n to arrive at (5.2.6). The details of this part of the proof are left to the reader.
(e)⇒(a) Let $b_0 = 0 \le a_1 < b_1 \le \ldots \le a_n < b_n$. Then, (e) implies that

$$
\begin{aligned}
&\mathbb{P}\Big(\bigcap_{k=1}^{n} \{a_k < \sigma_k \le b_k\}\Big) \\
&= \mathbb{P}\Big(\bigcap_{k=1}^{n-1} \{N(a_k) - N(b_{k-1}) = 0, N(b_k) - N(a_k) = 1\} \\
&\qquad\qquad \cap \{N(a_n) - N(b_{n-1}) = 0, N(b_n) - N(a_n) \ge 1\}\Big) \\
&= e^{-\lambda(a_n - b_{n-1})}(1 - e^{-\lambda(b_n - a_n)}) \prod_{k=1}^{n-1} e^{-\lambda(a_k - b_{k-1})}\lambda(b_k - a_k)e^{-\lambda(b_k - a_k)} \\
&= (e^{-\lambda a_n} - e^{-\lambda b_n})\lambda^{n-1} \prod_{k=1}^{n-1} (b_k - a_k) \\
&= \int_{a_1}^{b_1} \ldots \int_{a_n}^{b_n} \lambda^n e^{-\lambda y_n} \, dy_n \ldots dy_1 \\
&= \int_{a_1}^{b_1} \int_{a_2 - x_1}^{b_2 - x_1} \ldots \int_{a_n - x_1 - \ldots - x_{n-1}}^{b_n - x_1 - \ldots - x_{n-1}} \lambda^n e^{-\lambda(x_1 + \ldots + x_n)} \, dx_n \ldots dx_1 .
\end{aligned}
$$

Thus, the joint density of $\sigma_1, \sigma_2 - \sigma_1, \ldots, \sigma_n - \sigma_{n-1}$ is given by

$$
\mathbb{P}\Big(\bigcap_{k=1}^{n} \{\sigma_k - \sigma_{k-1} \in dx_k\}\Big) = \lambda^n e^{-\lambda(x_1 + \ldots + x_n)} \, dx_1 \ldots dx_n ,
$$

and therefore these random variables are independent and have a distribution $\text{Exp}(\lambda)$, i.e. $\{N(t)\}$ is a Poisson process with intensity λ. □

5.2.2 Compound Poisson Processes

We continue to assume that the inter-occurrence times $\{T_n\}$ are exponentially distributed with parameter $\lambda > 0$ or that the counting process $\{N(t)\}$ is a Poisson process with intensity λ. Let the claim sizes $\{U_n\}$ be independent and identically distributed with distribution F_U and let $\{U_n\}$ be independent of $\{N(t)\}$. Then the cumulative arrival process $\{X(t),\ t \ge 0\}$ defined in (5.1.12) is called a *compound Poisson process* with characteristics (λ, F_U), i.e. with intensity λ and jump size distribution F_U. This terminology is motivated by the property that $X(t)$ has a compound Poisson distribution with characteristics $(\lambda t, F_U)$. Since $X(t) = \sum_{i=1}^{N(t)} U_i$, it suffices to observe that, by the result of Theorem 5.2.1, $N(t)$ is Poisson distributed with parameter λt.

The next result follows from Theorem 5.2.1.

Corollary 5.2.1 *Let $\{X(t)\}$ be a compound Poisson process with characteristics (λ, F_U). Then,*
(a) *the process $\{X(t)\}$ has stationary and independent increments,*
(b) *the moment generating function of $X(t)$ is given by*

$$\hat{m}_{X(t)}(s) = \mathrm{e}^{\lambda t(\hat{m}_U(s)-1)}, \tag{5.2.7}$$

and the mean and variance by

$$\mathbb{E}\, X(t) = \lambda t \mu_U\,, \qquad \operatorname{Var} X(t) = \lambda t \mu_U^{(2)}\,. \tag{5.2.8}$$

Proof (a) We have to show that, for all $n = 1, 2, \ldots, h \geq 0$ and $0 \leq t_0 < t_1 < \ldots < t_n$, the random variables

$$\sum_{i_1=N(t_0+h)+1}^{N(t_1+h)} U_{i_1}\,, \ldots, \sum_{i_n=N(t_{n-1}+h)+1}^{N(t_n+h)} U_{i_n}$$

are independent and that their distribution does not depend on h. Since the sequence $\{U_n\}$ consists of independent and identically distributed random variables which are independent of $\{T_n\}$, we have

$$\begin{aligned}
&\mathbb{P}\Big(\sum_{i_1=N(t_0+h)+1}^{N(t_1+h)} U_{i_1} \leq x_1\,, \ldots, \sum_{i_n=N(t_{n-1}+h)+1}^{N(t_n+h)} U_{i_n} \leq x_n \Big) \\
&\quad = \sum_{k_1,\ldots,k_n \in \mathbb{N}} \prod_{j=1}^{n} F_U^{*k_j}(x_j) \mathbb{P}(N(t_1+h) - N(t_0+h) = k_1\,, \\
&\qquad\qquad \ldots, N(t_n+h) - N(t_{n-1}+h) = k_n)
\end{aligned}$$

for all $x_1, \ldots, x_n \geq 0$. Therefore, it suffices to recall that by Theorem 5.2.1, the Poisson (counting) process $\{N(t)\}$ has independent and stationary increments.
(b) The proof of (5.2.7) and (5.2.8) is left to the reader as an exercise. □

Bibliographical Notes. The material covered in Section 5.2 can be found in a large number of textbooks, such as Billingsley (1995). For a discussion of Poisson processes in the context of risk theory, see also Schmidt (1996).

5.3 RUIN PROBABILITIES: THE COMPOUND POISSON MODEL

In the sequel of this chapter we consider the compound Poisson model. The risk reserve process $\{R(t), t \geq 0\}$ is defined in (5.1.14) and claims occur according to a compound Poisson process with characteristics (λ, F_U). The most frequently used property of the process $\{R(t)\}$ is the independence and stationarity of its increments. Considering $\{R(t)\}$ from time t onwards is like restarting a risk reserve process with an identically distributed claim arrival process but with initial reserve $R(t)$. In particular, if $R(t) = y$ and ruin has not yet occurred by time t, then the (conditional) ruin probability is $\psi(y)$. Furthermore, considering $\{R(t)\}$ from the first claim occurrence epoch σ_1 on is like starting a risk reserve process with initial reserve $R(t + \sigma_1 - U_1)$.

Let $\{S_n, n \geq 0\}$ be the random walk given by

$$S_n = \sum_{i=1}^{n} Y_i\,, \qquad Y_i = U_i - \beta T_i\,. \tag{5.3.1}$$

In Theorem 6.3.1 we will show that $\limsup_{n\to\infty} S_n = \infty$ if $\mathbb{E} Y \geq 0$. Thus, (5.1.16) implies that $\psi(u) \equiv 1$ in this case. Let us therefore assume that $\mathbb{E} Y < 0$, i.e. $\beta > \lambda\mu$, where $\mu = \mu_U$ denotes the expected claim size. Recall that β is the premium income in the unit time interval and that $\lambda\mu$ is the expected aggregate claim over the unit time interval (see (5.2.8)). The condition

$$\beta > \lambda\mu \tag{5.3.2}$$

is therefore called the *net profit condition.* Throughout the rest of this chapter we will assume (5.3.2). Note that in this case $\lim_{n\to\infty} S_n = -\infty$, since from the strong law of large numbers we have $S_n/n \to \mathbb{E} Y < 0$. Thus, the maximum of $\{S_n\}$ is finite. Using (5.1.16) we get $\lim_{u\to\infty} \psi(u) = 0$. Moreover, we will see later in Theorem 5.3.4 that this implies $\psi(u) < 1$ for all $u \geq 0$.

5.3.1 An Integro-Differential Equation

In this section we study the survival probability $\overline{\psi}(u) = 1 - \psi(u)$. We show that $\overline{\psi}(u)$ is differentiable everywhere on $\mathbb{R}_+$ with the exception of an at most countably infinite set of points. Furthermore, we prove that $\overline{\psi}(u)$ fulfils an integro-differential equation.

Theorem 5.3.1 *The survival function $\overline{\psi}(u)$ is continuous on $\mathbb{R}_+$ with right and left derivatives $\overline{\psi}_+^{(1)}(u)$ and $\overline{\psi}_-^{(1)}(u)$, respectively. Moreover*

$$\beta\overline{\psi}_+^{(1)}(u) = \lambda\Big(\overline{\psi}(u) - \int_0^u \overline{\psi}(u-y)\,\mathrm{d}F_U(y)\Big) \tag{5.3.3}$$

and

$$\beta\overline{\psi}_-^{(1)}(u) = \lambda\Big(\overline{\psi}(u) - \int_0^{u-} \overline{\psi}(u-y)\,\mathrm{d}F_U(y)\Big)\,. \tag{5.3.4}$$

Proof As mentioned before, considering $\{R(t)\}$ from the first claim occurrence epoch σ_1 is like considering a risk reserve process with initial reserve $R(t + \sigma_1 - U_1)$. Thus, conditioning on the first claim occurrence epoch σ_1, we obtain

$$\overline{\psi}(u) = \mathrm{e}^{-\lambda h}\overline{\psi}(u+\beta h) + \int_0^h \int_0^{u+\beta t} \overline{\psi}(u+\beta t - y)\,\mathrm{d}F_U(y)\,\lambda \mathrm{e}^{-\lambda t}\,\mathrm{d}t \tag{5.3.5}$$

for all $h, u \geq 0$. Letting $h \downarrow 0$, (5.3.5) implies that $\overline{\psi}(u)$ is a right-continuous function. Moreover, rearrange the terms in (5.3.5) to write

$$\begin{aligned}\beta\frac{\overline{\psi}(u+\beta h) - \overline{\psi}(u)}{\beta h}\\ = \frac{1-\mathrm{e}^{-\lambda h}}{h}\overline{\psi}(u+\beta h) - \frac{1}{h}\int_0^h \int_0^{u+\beta t} \overline{\psi}(u+\beta t - y)\,\mathrm{d}F_U(y)\,\lambda \mathrm{e}^{-\lambda t}\,\mathrm{d}t\,.\end{aligned}$$

This shows that $\overline{\psi}(u)$ is differentiable from the right and (5.3.3) follows as $h \downarrow 0$. For $h \leq \beta^{-1}u$, (5.3.5) can be rewritten in the form

$$\overline{\psi}(u-\beta h) = \mathrm{e}^{-\lambda h}\overline{\psi}(u) + \int_0^h \int_0^{u-\beta(h-t)} \overline{\psi}(u-\beta(h-t)-y)\,\mathrm{d}F_U(y)\,\lambda \mathrm{e}^{-\lambda t}\,\mathrm{d}t\,,$$

which implies that $\overline{\psi}(u)$ is also left-continuous and that the left derivative of $\overline{\psi}(u)$ exists and fulfils (5.3.4). □

An immediate consequence of Theorem 5.3.1 is that the continuous function $\overline{\psi}(u)$ is differentiable everywhere except for the countable set, where $F_U(y)$ is not continuous. The importance of this fact is that it implies

$$\int_u^\infty \overline{\psi}^{(1)}(v)\,\mathrm{d}v = \overline{\psi}(u)\,, \qquad u \geq 0\,.$$

In the terminology of measure theory, this means that $\overline{\psi}(u)$ is absolutely continuous with respect to the Lebesgue measure.

Note that in general (5.3.3) cannot be solved analytically. However, one can compute the survival probability $\overline{\psi}(u)$ in (5.3.3) numerically.

Example Assume that the claim sizes are exponentially distributed with parameter δ. Then the net profit condition (5.3.2) takes the form $\delta\beta > \lambda$. Furthermore, (5.3.3) can be solved analytically. The survival function $\overline{\psi}(u)$ is differentiable everywhere and satisfies the integral equation

$$\beta\overline{\psi}^{(1)}(u) = \lambda\Big(\overline{\psi}(u) - \mathrm{e}^{-\delta u}\int_0^u \overline{\psi}(y)\delta \mathrm{e}^{\delta y}\,\mathrm{d}y\Big)\,. \tag{5.3.6}$$

This equation implies that $\overline{\psi}^{(1)}(u)$ is differentiable and that

$$\beta\overline{\psi}^{(2)}(u) = \lambda\Big(\overline{\psi}^{(1)}(u) + \delta e^{-\delta u}\int_0^u \overline{\psi}(y)\delta e^{\delta y}\,dy - \delta\overline{\psi}(u)\Big) = (\lambda - \delta\beta)\overline{\psi}^{(1)}(u)\,.$$

The general solution to this differential equation is

$$\overline{\psi}(u) = c_1 - c_2 e^{-(\delta-\lambda/\beta)u}, \tag{5.3.7}$$

where $c_1, c_2 \in \mathbb{R}$. Since $\lim_{u\to\infty}\overline{\psi}(u) = 1$ it follows that $c_1 = 1$. Plugging (5.3.7) into (5.3.6) yields

$$\begin{aligned} & c_2(\delta\beta - \lambda)e^{-(\delta-\lambda/\beta)u} \\ = \; & \lambda\Big(1 - c_2 e^{-(\delta-\lambda/\beta)u} - (1 - e^{-\delta u}) + c_2 e^{-\delta u}\frac{\beta\delta}{\lambda}(e^{\lambda u/\beta} - 1)\Big) \end{aligned}$$

from which $c_2 = \lambda(\beta\delta)^{-1}$ is obtained. Thus,

$$\psi(u) = \frac{\lambda}{\beta\delta}e^{-(\delta-\lambda/\beta)u}. \tag{5.3.8}$$

5.3.2 An Integral Equation

Equation (5.3.3) is not easily solved because it involves both the derivative and an integral of $\overline{\psi}(u)$. It would be more convenient to get rid of the derivative. Indeed, integrating (5.3.3) we arrive at the following result.

Theorem 5.3.2 *The ruin function $\psi(u)$ satisfies the integral equation*

$$\beta\psi(u) = \lambda\Big(\int_u^\infty \overline{F}_U(x)\,dx + \int_0^u \psi(u-x)\overline{F}_U(x)\,dx\Big). \tag{5.3.9}$$

Proof We integrate (5.3.3) over the interval $(0, u]$. This gives

$$\begin{aligned} \frac{\beta}{\lambda}(\overline{\psi}(u) - \overline{\psi}(0)) &= \frac{1}{\lambda}\int_0^u \beta\overline{\psi}_+^{(1)}(x)\,dx \\ &= \int_0^u \overline{\psi}(x)\,dx - \int_0^u\int_0^x \overline{\psi}(x-y)\,dF_U(y)\,dx \\ &= \int_0^u \overline{\psi}(x)\,dx - \int_0^u\int_y^u \overline{\psi}(x-y)\,dx\,dF_U(y) \\ &= \int_0^u \overline{\psi}(x)\,dx - \int_0^u\int_0^{u-y} \overline{\psi}(x)\,dx\,dF_U(y) \\ &= \int_0^u \overline{\psi}(x)\,dx - \int_0^u\int_0^{u-x} dF_U(y)\,\overline{\psi}(x)\,dx \end{aligned}$$

$$= \int_0^u \overline{\psi}(x)(1 - F_U(u-x))\,\mathrm{d}x$$

$$= \int_0^u \overline{\psi}(u-x)\overline{F}_U(x)\,\mathrm{d}x\,,$$

i.e.

$$\frac{\beta}{\lambda}(\overline{\psi}(u) - \overline{\psi}(0)) = \int_0^u \overline{\psi}(u-x)\overline{F}_U(x)\,\mathrm{d}x\,. \tag{5.3.10}$$

Now, letting $u \to \infty$, (5.3.10) implies $\beta(1 - \overline{\psi}(0)) = \lambda \int_0^\infty \overline{F}_U(x)\,\mathrm{d}x = \lambda\mu$. This follows from $\overline{\psi}(u) \to 1$ and an application of the dominated convergence theorem on the right-hand side of (5.3.10). Thus,

$$\overline{\psi}(0) = 1 - \frac{\lambda\mu}{\beta}\,, \qquad \psi(0) = \frac{\lambda\mu}{\beta}\,. \tag{5.3.11}$$

Now, replacing $\overline{\psi}(u)$ by $1 - \psi(u)$, from (5.3.10) and (5.3.11) we have $\beta\psi(u) = \mu - \lambda\int_0^u (1 - \psi(u-x))\overline{F}_U(x)\,\mathrm{d}x$, which is equivalent to (5.3.9). □

Note that (5.3.11) shows that the ruin probability $\psi(u)$ at $u = 0$ only depends on the expected claim size μ and not on the specific form of the claim size distribution F_U. We further remark that (5.3.9) is called a *defective renewal equation* with respect to the unknown ruin function $\psi(u)$. In Section 6.1.4 we will analyse such equations in a more general context.

5.3.3 Laplace Transforms, Pollaczek–Khinchin Formula

In this section we compute the Laplace transforms

$$\hat{L}_\psi(s) = \int_0^\infty \psi(u)\mathrm{e}^{-su}\,\mathrm{d}u\,, \qquad \hat{L}_{\overline{\psi}}(s) = \int_0^\infty \overline{\psi}(u)\mathrm{e}^{-su}\,\mathrm{d}u\,.$$

Note that both integrals make sense for all $s > 0$. Furthermore, we have

$$\hat{L}_\psi(s) = \int_0^\infty (1 - \overline{\psi}(u))\,\mathrm{e}^{-su}\,\mathrm{d}u = \frac{1}{s} - \hat{L}_{\overline{\psi}}(s)\,. \tag{5.3.12}$$

Theorem 5.3.3 *The Laplace transforms $\hat{L}_{\overline{\psi}}(s)$ and $\hat{L}_\psi(s)$ are given by*

$$\hat{L}_{\overline{\psi}}(s) = \frac{\beta - \lambda\mu}{\beta s - \lambda(1 - \hat{l}_U(s))}\,, \qquad s > 0 \tag{5.3.13}$$

and

$$\hat{L}_\psi(s) = \frac{1}{s} - \frac{\beta - \lambda\mu}{\beta s - \lambda(1 - \hat{l}_U(s))}\,, \qquad s > 0\,. \tag{5.3.14}$$

Proof Let $s > 0$. Integrating by parts we find that $\int_0^\infty \overline{\psi}_+^{(1)}(u)\mathrm{e}^{-su}\,\mathrm{d}u = -\overline{\psi}(0) + s\hat{L}_{\overline{\psi}}(s)$. Furthermore,

$$\int_0^\infty \int_0^u \overline{\psi}(u-y)\,\mathrm{d}F_U(y)\mathrm{e}^{-su}\,\mathrm{d}u = \int_0^\infty \int_y^\infty \overline{\psi}(u-y)\mathrm{e}^{-su}\,\mathrm{d}u\,\mathrm{d}F_U(y)$$

$$= \int_0^\infty \int_0^\infty \overline{\psi}(u)\mathrm{e}^{-s(u+y)}\,\mathrm{d}u\,\mathrm{d}F_U(y) = \hat{L}_{\overline{\psi}}(s)\hat{l}_U(s)\,.$$

In order to complete the proof we multiply (5.3.3) by e^{-su} and integrate over $(0,\infty)$. Then, we see that the Laplace transform $\hat{L}_{\overline{\psi}}(s)$ satisfies the equation $\beta(s\hat{L}_{\overline{\psi}}(s) - \overline{\psi}(0)) = \lambda\hat{L}_{\overline{\psi}}(s)(1-\hat{l}_U(s))$, which is equivalent to (5.3.13) in view of (5.3.11). Now, (5.3.14) immediately follows from (5.3.12). □

Example In the case of exponentially distributed claims (with $\mu = \delta^{-1}$), (5.3.13) gives

$$\hat{L}_{\overline{\psi}}(s) = \frac{\beta-\lambda/\delta}{\beta s - \lambda(1-\delta/(\delta+s))} = \frac{\beta-\lambda/\delta}{s(\beta-\lambda/(\delta+s))} = \frac{(\beta-\lambda/\delta)(\delta+s)}{s(\beta(\delta+s)-\lambda)}$$

and, by (5.3.12),

$$\begin{aligned}\hat{L}_{\psi}(s) &= \frac{1}{s} - \frac{(\beta-\lambda/\delta)(\delta+s)}{s(\beta(\delta+s)-\lambda)} = \frac{\beta(\delta+s)-\lambda-(\beta-\lambda/\delta)(\delta+s)}{s(\beta(\delta+s)-\lambda)} \\ &= \frac{\lambda}{\delta}\frac{1}{\beta(\delta+s)-\lambda} = \frac{\lambda}{\delta\beta}\frac{1}{\delta-\lambda/\beta+s}\,.\end{aligned}$$

Hence, by comparison with the Laplace–Stieltjes transform of the exponential distribution we realize that $\psi(u) = \lambda(\delta\beta)^{-1}\mathrm{e}^{-(\delta-\lambda/\beta)u}$, in accordance with (5.3.8).

Although equation (5.3.9) is simpler than (5.3.3), it is generally difficult to solve it in closed form. However, (5.3.9) leads to a formula for $\psi(u)$ in the form of an infinite series of convolutions. In this connection, we need the integrated tail distribution F_U^{s} of F_U. Remember that F_U^{s} is given by

$$F_U^{\mathrm{s}}(x) = \frac{1}{\mu}\int_0^x \overline{F}_U(y)\,\mathrm{d}y\,, \qquad x \ge 0\,. \tag{5.3.15}$$

The representation formula for $\psi(u)$ derived in the next theorem is called the *Pollaczek–Khinchin formula.*

Theorem 5.3.4 *For each $u \ge 0$,*

$$\psi(u) = \Big(1-\frac{\lambda\mu}{\beta}\Big)\sum_{n=1}^\infty \Big(\frac{\lambda\mu}{\beta}\Big)^n \overline{(F_U^{\mathrm{s}})^{*n}}(u)\,. \tag{5.3.16}$$

Proof Taking Laplace transforms on both the sides of (5.3.9) gives

$$\hat{L}_\psi(s) = \frac{\lambda\mu}{\beta}\hat{L}_{\overline{F^s_U}}(s) + \frac{\lambda\mu}{\beta}\hat{L}_\psi(s)\cdot\hat{L}_{\overline{F^s_U}}(s)\,.$$

Thus, $\hat{L}_\psi(s) = \lambda\mu\beta^{-1}\hat{L}_{\overline{F^s_U}}(s)(1-\lambda\mu\beta^{-1}\hat{L}_{\overline{F^s_U}}(s))^{-1}$, i.e. $\hat{L}_\psi(s)$ is the Laplace transform of the tail function of a geometric compound with characteristics $(\lambda\mu\beta^{-1}, F^s_U)$. Again, the one-to-one correspondence between functions and their Laplace transforms yields (5.3.16). □

Besides the case of exponentially distributed claim sizes, where (5.3.16) has been written in closed form (see (5.3.8)), there are other claim size distributions for which (5.3.16) simplifies. One important class of such claim size distributions is provided by the phase-type distributions, discussed in Chapter 8.

The infinite series representation given in (5.3.16) is particularly useful for theoretical considerations. However, it is also useful for numerical approximations to the ruin probability $\psi(u)$ since (5.3.16) shows that $1-\psi(u)$ is the distribution function of a geometric compound. After discretization of the distribution F^s_U, Panjer's algorithm described in Section 4.4.2 will yield a numerical approximation to $\psi(u)$. A completely different method for numerical computation of $\psi(u)$ is based on the numerical inversion of the Laplace transform. This method will be discussed in Section 5.5. Further, Sections 5.4.1 and 5.4.2 treat useful bounds and approximations to $\psi(u)$ derived from (5.3.16), provided that the claim size distribution admits an adjustment coefficient. Finally, in Section 5.4.3, formula (5.3.16) will be crucial in deriving interesting asymptotic expressions of $\psi(u)$ as $u\to\infty$ when the claim size distribution is heavy-tailed.

5.3.4 Severity of Ruin

We now want to analyse further what happens if ruin occurs. Consider the ruin probabilities $\varphi(u,x,y) = \mathbb{P}(\tau(u)<\infty, X_+(u)>x, Y_+(u)>y)$ where $X_+(u) = R(\tau(u)-)$ and $Y_+(u) = -R(\tau(u))$ is the surplus just before and at the ruin time $\tau(u)$ respectively. Remember that the random variable $Y_+(u)$ is also called *severity of ruin*.

Since in general we were not able to find an explicit formula for $\psi(u)$ there is no hope of achieving this goal for $\varphi(u,x,y)$. But it is possible to derive integro-differential and integral equations for $\varphi(u,x,y)$. Moreover, we will be able to find $\varphi(x,y)=\varphi(0,x,y)$ explicitly.

We will proceed as in Section 5.3.1. Condition on the first claim occurrence epoch and on the size of that claim to find that $\varphi(u,x,y)$ satisfies

$$\varphi(u,x,y) \quad = \quad \mathrm{e}^{-\lambda h}\varphi(u+\beta h,x,y)$$

$$+\int_0^h\Big(\int_0^{u+\beta t}\varphi(u+\beta t-v,x,y)\,\mathrm{d}F_U(v)$$
$$+\,\mathbb{I}(u+\beta t>x)(1-F_U(u+\beta t+y))\Big)\lambda\mathrm{e}^{-\lambda t}\,\mathrm{d}t\,,$$

for all $h,u,x,y>0$. Thus $\varphi(u,x,y)$ is right-continuous and differentiable from the right with respect to u. Furthermore,

$$\beta\frac{\partial^+}{\partial u}\varphi(u,x,y)$$
$$=\lambda\Big(\varphi(u,x,y)-\int_0^u\varphi(u-v,x,y)\,\mathrm{d}F_U(v)-\mathbb{I}(u\ge x)\overline{F}_U(u+y)\Big)\,.$$

Analogously, $\varphi(u,x,y)$ is left-continuous and differentiable from the left with respect to u, and satisfies

$$\beta\frac{\partial^-}{\partial u}\varphi(u,x,y)=\lambda\Big(\varphi(u,x,y)-\int_0^{u-}\varphi(u-v,x,y)\,\mathrm{d}F_U(v)$$
$$-\,\mathbb{I}(u>x)\overline{F}_U((u+y)-)\Big)\,.$$

Thus the set of points u where the partial derivative $(\partial/\partial u)\varphi(u,x,y)$ does not exist is countable and therefore $\varphi(u,x,y)$ is absolutely continuous in u.

Proceeding as in Section 5.3.2 we obtain the integral equation

$$\beta(\varphi(u,x,y)-\varphi(0,x,y))$$
$$=\lambda\Big(\int_0^u\varphi(u-v,x,y)\,\overline{F}_U(v)\,\mathrm{d}v-\mathbb{I}(u\ge x)\int_{x+y}^{u+y}\overline{F}_U(v)\,\mathrm{d}v\Big).\qquad(5.3.17)$$

We now let $u\to\infty$. Note that $\int_0^\infty(1-F_U(v))\,\mathrm{d}v=\mu$ allows us to interchange integration and limit on the right-hand side of (5.3.17). Since $0\le\varphi(u,x,y)\le\psi(u)$ we find that $\lim_{u\to\infty}\varphi(u,x,y)=0$ and therefore

$$\varphi(0,x,y)=\frac{\lambda}{\beta}\int_{x+y}^\infty\overline{F}_U(v)\,\mathrm{d}v\,.\qquad(5.3.18)$$

Note that (5.3.18) is a generalization of (5.3.11). Another proof of (5.3.18) which is valid for more general risk models will be given in Chapter 12.

In Section 6.3, we will show how (5.3.18) provides another interpretation to the integrated tail distribution F_U^{s} as the *ladder height distribution* of the random walk $\{S_n\}$ given in (5.3.1).

Bibliographical Notes. The classical compound Poisson risk model was introduced by Filip Lundberg (1903) and extensively studied by Harald Cramér (1930,1955). It is therefore often called the *Cramér–Lundberg model.*

In particular, Theorems 5.3.1, 5.3.2 and 5.3.3 go back to these two authors. From the mathematical point of view, the ruin function $\psi(u)$ of the compound Poisson model is equivalent to the tail function of the stationary distribution of virtual waiting time in an M/GI/1 queue. Thus, formula (5.3.16) is equivalent to the celebrated *Pollaczek–Khinchin formula* of queueing theory; see, for example, Asmussen (1987), Baccelli and Brémaud (1994), Franken, König, Arndt and Schmidt (1982) and Prabhu (1965). It also is a special case of a more general result on the distribution of the maximum of a random walk with negative drift, see Theorem 6.3.3. Further details on the equivalence between characteristics of queueing and risk processes can be found, for example, in the books by Asmussen (1987) and Prabhu (1965). In risk theory, (5.3.16) is often called *Beekman's formula.* The notion of severity of ruin was introduced in Gerber, Goovaerts and Kaas (1987). Recursive algorithms for the calculation of the joint and marginal distributions of the surplus just before ruin and the severity of ruin can be found in Dickson, dos Reis and Waters (1995). For further results on the distribution of the ruin time, the surplus just before ruin and the severity of ruin, see Dickson (1992), Dickson and Waters (1992), Dufresne and Gerber (1988), Frey and Schmidt (1996) and Gerber and Shiu (1997). The duration of negative surplus and the maximal deficit during this time have been investigated in Dickson and dos Reis (1996,1997), dos Reis (1993) and Picard (1994). The compound Poisson risk model has been extended in several directions. Some of them will be discussed in later chapters of this book. For some other extensions we will refer to the literature. Notice that a compound Poisson process has finitely many jumps in bounded time intervals. Examples of claim arrival processes with stationary and independent increments and with infinitely many jumps in bounded intervals have been studied, for instance, in Dickson and Waters (1993), Dufresne and Gerber (1993) and Dufresne, Gerber and Shiu (1991). These processes are called *gamma processes* and belong to the larger class of Lévy processes. We return to this later in Chapters 12 and 13.

5.4 BOUNDS, ASYMPTOTICS AND APPROXIMATIONS

We have seen that it is generally difficult to determine the function $\psi(u)$ explicitly from formula (5.3.16). Therefore, bounds and approximations to the ruin probability $\psi(u)$ are requested. Besides this, knowledge of the asymptotic behaviour of $\psi(u)$ as $u \to \infty$ can also be useful in order to get information about the nature of the underlying risks.

5.4.1 Lundberg Bounds

Since the claim surplus $S(t)$ at time $t \geq 0$ has a shifted compound Poisson distribution with characteristics $(\lambda t, F_U)$ and the shift is on $-\beta t$, the moment generating function of $S(t)$ is

$$\hat{m}_{S(t)}(s) = \mathbb{E}\, e^{sS(t)} = \exp\left(t(\lambda(\hat{m}_U(s) - 1) - \beta s)\right).$$

If $\hat{m}_U(s_0) < \infty$ for some $s_0 > 0$, then the function $\theta(s) = \lambda(\hat{m}_U(s) - 1) - \beta s$ is infinitely often differentiable in the interval $(-\infty, s_0)$. In particular

$$\theta^{(2)}(s) = \lambda \hat{m}_U^{(2)}(s) = \lambda \mathbb{E}\left(U^2 e^{sU}\right) > 0, \tag{5.4.1}$$

which shows that $\theta(s)$ is a convex function. For the first derivative $\theta^{(1)}(s)$ at $s = 0$ we have

$$\theta^{(1)}(0) = \lambda \hat{m}_U^{(1)}(0) - \beta = \lambda\mu - \beta < 0. \tag{5.4.2}$$

It is easily seen that $\theta(0) = 0$. Moreover, there may exist a second root of

$$\theta(s) = 0. \tag{5.4.3}$$

If such a root $s \neq 0$ exists, then it is unique and strictly positive. We call this solution, if it exists, the *adjustment coefficient* or the *Lundberg exponent* and denote it by γ.

Note that the adjustment coefficient exists in the following situation.

Lemma 5.4.1 *Assume that there exists $s_\infty \in \mathbb{R} \cup \{\infty\}$ such that $\hat{m}_U(s) < \infty$ if $s < s_\infty$ and $\lim_{s\uparrow s_\infty} \hat{m}_U(s) = \infty$. Then there exists a unique positive solution γ to the equation* (5.4.3).

Proof By the above considerations it is enough to show that $\theta(s)$ tends to infinity as $s \uparrow s_\infty$. The case $s_\infty < \infty$ is obvious. Thus assume that $s_\infty = \infty$. Choose $x' > 0$ such that $F_U(x') < 1$. Then

$$\hat{m}_U(s) = \int_0^\infty e^{sx}\, dF_U(x) \geq e^{sx'} \overline{F}_U(x'),$$

which tends faster to infinity than any linear function. □

The existence of the adjustment coefficient is important because it allows uniform upper and lower exponential bounds for the ruin function $\psi(u)$. Let $x_0 = \sup\{x : F_U(x) < 1\}$.

Theorem 5.4.1 *Assume that the adjustment coefficient $\gamma > 0$ exists. Then,*

$$a_- e^{-\gamma u} \leq \psi(u) \leq a_+ e^{-\gamma u} \tag{5.4.4}$$

for all $u \geq 0$, where

$$a_- = \inf_{x\in[0,x_0)} \frac{e^{\gamma x}\int_x^\infty \overline{F}_U(y)\, dy}{\int_x^\infty e^{\gamma y}\overline{F}_U(y)\, dy}, \qquad a_+ = \sup_{x\in[0,x_0)} \frac{e^{\gamma x}\int_x^\infty \overline{F}_U(y)\, dy}{\int_x^\infty e^{\gamma y}\overline{F}_U(y)\, dy}.$$

Proof In view of Theorem 5.3.4, we can apply Theorem 4.5.1 with F_U replaced by F_U^{s} and $p = \rho = \lambda\beta^{-1}\mu$. This gives (5.4.4) because

$$\int_0^\infty \mathrm{e}^{sy}\,\mathrm{d}F_U^{\mathrm{s}}(y) = \frac{1}{\mu_U}\int_0^\infty \mathrm{e}^{sy}\,\overline{F}_U(y)\,\mathrm{d}y = \frac{\hat{m}_U(s)-1}{s\mu_U}\,.$$

Hence the positive root of $\rho^{-1} = \hat{m}_{F_U^{\mathrm{s}}}(\gamma)$ is the positive root of (5.4.3). □

Results as in Theorem 5.4.1 are known in risk theory as *two-sided Lundberg bounds* for the ruin function $\psi(u)$. Alternatively, an easy application of integral equation (5.3.9) leads to (5.4.4). Moreover, for all $u \ge 0$,

$$\psi(u)\begin{cases} < & a_+\mathrm{e}^{-\gamma u} \quad \text{if } a_+ > \psi(0),\\ > & a_-\mathrm{e}^{-\gamma u} \quad \text{if } a_- < \psi(0).\end{cases} \tag{5.4.5}$$

This can be shown in the following way. Note that

$$a_+ \ge \frac{\int_0^\infty \overline{F}_U(y)\,\mathrm{d}y}{\int_0^\infty \mathrm{e}^{\gamma y}\overline{F}_U(y)\,\mathrm{d}y} = \frac{\mu}{\gamma^{-1}(\hat{m}_U(\gamma)-1)} = \frac{\lambda\mu}{\beta} = \psi(0)\,,$$

and analogously $a_- \le \psi(0)$. Let $b \ge a_+$ such that $b > \psi(0)$. We prove indirectly that $\psi(u) < b\mathrm{e}^{-\gamma u}$ for all $u \ge 0$. Assume the contrary. Denote $u_0 = \inf\{u \ge 0 : \psi(u) \ge b\mathrm{e}^{-\gamma u}\}$. Since $\psi(u)$ is continuous we have $\psi(u_0) = b\mathrm{e}^{-\gamma u_0}$. Furthermore, since $\psi(0) < b$ we can conclude that $u_0 > 0$ and $\psi(u_0 - x) < b\mathrm{e}^{-\gamma(u_0-x)}$ for $0 < x \le u_0$. Note that by the definition of a_+, we have

$$\int_{u_0}^\infty \overline{F}_U(y)\,\mathrm{d}y \le b\int_{u_0}^\infty \mathrm{e}^{-\gamma(u_0-y)}\overline{F}_U(y)\,\mathrm{d}y\,.$$

Considering equation (5.3.9) for $u = u_0$, this gives

$$\begin{aligned}
\beta\psi(u_0) &= \lambda\Big(\int_{u_0}^\infty \overline{F}_U(x)\,\mathrm{d}x + \int_0^{u_0}\psi(u_0-x)\overline{F}_U(x)\,\mathrm{d}x\Big)\\
&< \lambda\Big(\int_{u_0}^\infty \overline{F}_U(x)\,\mathrm{d}x + \int_0^{u_0} b\mathrm{e}^{-\gamma(u_0-x)}\overline{F}_U(x)\,\mathrm{d}x\Big)\\
&\le \lambda\int_0^\infty b\mathrm{e}^{-\gamma(u_0-x)}\overline{F}_U(x)\,\mathrm{d}x = b\lambda\mathrm{e}^{-\gamma u_0}\int_0^\infty\int_x^\infty \mathrm{e}^{\gamma x}\,\mathrm{d}F_U(y)\,\mathrm{d}x\\
&= b\lambda\mathrm{e}^{-\gamma u_0}\int_0^\infty\int_0^y \mathrm{e}^{\gamma x}\,\mathrm{d}x\,\mathrm{d}F_U(y) = b\lambda\mathrm{e}^{-\gamma u_0}\int_0^\infty \frac{1}{\gamma}(\mathrm{e}^{\gamma y}-1)\,\mathrm{d}F_U(y)\\
&= b\frac{\lambda}{\gamma}\mathrm{e}^{-\gamma u_0}(\hat{m}_U(\gamma)-1) = b\beta\mathrm{e}^{-\gamma u_0}\,,
\end{aligned}$$

which leads to a contradiction. Thus the strict upper bound in (5.4.5) is proved. The lower bound follows analogously.

5.4.2 The Cramér–Lundberg Approximation

In Section 5.4.1 we have found exponential upper and lower bounds for the ruin function $\psi(u)$. We are now interested in the asymptotic behaviour of $\psi(u)e^{\gamma u}$. The question is whether $\psi(u)e^{\gamma u}$ converges to a limit or fluctuates between two bounds as $u \to \infty$. We will see that the limit $\lim_{u\to\infty} \psi(u)e^{\gamma u}$ exists. However, to show this we need the following auxiliary result.

Lemma 5.4.2 *Assume that the function $z_1 : \mathbb{R}_+ \to (0,\infty)$ is increasing and let $z_2 : \mathbb{R}_+ \to \mathbb{R}_+$ be decreasing, such that*

$$\int_0^\infty z_1(x)z_2(x)\,dx < \infty \tag{5.4.6}$$

and

$$\lim_{h\to 0} \sup\{z_1(x+y)/z_1(x) : x \ge 0, 0 \le y \le h\} = 1\,. \tag{5.4.7}$$

Then, for $z(x) = z_1(x)z_2(x)$ and for each distribution F on $\mathbb{R}_+$, the equation

$$g(u) = z(u) + \int_0^u g(u-v)\,dF(v)\,, \qquad u \ge 0\,, \tag{5.4.8}$$

admits a unique locally bounded solution such that

$$\lim_{u\to\infty} g(u) = \begin{cases} \mu_F^{-1} \displaystyle\int_0^\infty z(u)\,du & \text{if } \mu_F < \infty, \\ 0 & \text{if } \mu_F = \infty. \end{cases} \tag{5.4.9}$$

Note that Lemma 5.4.2 is a version of the so-called *key renewal theorem.* Furthermore, (5.4.8) is called a *renewal equation.* A more detailed discussion of notions and results from renewal theory is given in Chapter 6.

Theorem 5.4.2 *Assume that the adjustment coefficient $\gamma > 0$ exists. If $\hat{m}_U^{(1)}(\gamma) < \infty$, then*

$$\lim_{u\to\infty} \psi(u)e^{\gamma u} = \frac{\beta - \lambda\mu}{\lambda \hat{m}_U^{(1)}(\gamma) - \beta}\,. \tag{5.4.10}$$

If $m_U^{(1)}(\gamma) = \infty$, then $\lim_{u\to\infty} \psi(u)e^{\gamma u} = 0$.

Proof Multiplying (5.3.9) by $e^{\gamma u}$ yields

$$\psi(u)e^{\gamma u} = \frac{\lambda}{\beta}\Big(e^{\gamma u}\int_u^\infty \overline{F}_U(x)\,dx + \int_0^u \psi(u-x)e^{\gamma(u-x)}\overline{F}_U(x)e^{\gamma x}\,dx\Big). \tag{5.4.11}$$

It follows from the definition of γ that

$$\int_0^\infty \lambda\beta^{-1}\overline{F}_U(x)e^{\gamma x}\,dx = 1\,, \tag{5.4.12}$$

and therefore (5.4.11) is a renewal equation. The mean value of the integrating distribution $\mathrm{d}F(x) = \lambda\beta^{-1}\overline{F}_U(x)\mathrm{e}^{\gamma x}\,\mathrm{d}x$ is

$$\int_0^\infty x\lambda\beta^{-1}\overline{F}_U(x)\mathrm{e}^{\gamma x}\,\mathrm{d}x = \lambda\hat{m}_U^{(1)}(\gamma) - \beta/(\beta\gamma)\,, \tag{5.4.13}$$

if $m_U^{(1)}(\gamma) < \infty$, and ∞ otherwise. It is easily seen that the function

$$z(u) = \frac{\lambda}{\beta}\mathrm{e}^{\gamma u}\int_u^\infty \overline{F}_U(x)\,\mathrm{d}x \tag{5.4.14}$$

can be factored in the way given in Lemma 5.4.2. We leave it to the reader to show this as an exercise. Since

$$\int_0^\infty \frac{\lambda}{\beta}\mathrm{e}^{\gamma u}\int_u^\infty \overline{F}_U(x)\,\mathrm{d}x\,\mathrm{d}u = \frac{\beta-\lambda\mu}{\beta\gamma} \tag{5.4.15}$$

the assertion now follows from Lemma 5.4.2. □

The asymptotic result obtained in Theorem 5.4.2 for the ruin probability $\psi(u)$ gives rise to the so-called *Cramér–Lundberg approximation*

$$\psi_{\mathrm{app}}(u) = \frac{\beta-\lambda\mu}{\lambda\hat{m}_U^{(1)}(\gamma)-\beta}\mathrm{e}^{-\gamma u}. \tag{5.4.16}$$

The following numerical investigation shows that the above approximation works quite well even for small values of u.

Example Let $\beta = \lambda = 1$ and $F_U(x) = 1 - \frac{1}{3}(\mathrm{e}^{-x} + \mathrm{e}^{-2x} + \mathrm{e}^{-3x})$. In this example we use the expected inter-occurrence time as the time unit and the premium per unit time as the monetary unit. The mean value of claim sizes is $\mu = 0.611111$, i.e. the net profit condition (5.3.2) is fulfilled. Furthermore, computing the Laplace transform $\hat{L}_\psi(s)$ and inverting it, we get

$$\psi(u) = 0.550790\mathrm{e}^{-0.485131u} + 0.0436979\mathrm{e}^{-1.72235u} + 0.0166231\mathrm{e}^{-2.79252u}\,. \tag{5.4.17}$$

On the other hand, (5.4.16) implies that in this case the Cramér–Lundberg approximation to $\psi(u)$ is $\psi_{\mathrm{app}}(u) = 0.550790\mathrm{e}^{-0.485131u}$. By comparison to the exact formula given in (5.4.17), the accuracy of this approximation can be analysed. Table 5.4.1 shows the ruin function $\psi(u)$, its Cramér–Lundberg approximation $\psi_{\mathrm{app}}(u)$ and the relative error $(\psi_{\mathrm{app}}(u)-\psi(u))/\psi(u)$ multiplied by 100. Note that the relative error is below 1% for $u \geq 1.71358 = 2.8\mu$.

Remark In the case of exponentially distributed claim sizes, the constant on the right-hand side of (5.4.10) is $(\beta\delta)^{-1}\lambda$. Thus the Cramér–Lundberg approximation (5.4.10) becomes exact in this case. Vice versa, assume that

u	0	0.25	0.5	0.75	1
$\psi(u)$	0.6111	0.5246	0.4547	0.3969	0.3479
$\psi_{\mathrm{app}}(u)$	0.5508	0.4879	0.4322	0.3828	0.3391
Er	-9.87	-6.99	-4.97	-3.54	-2.54

u	1.25	1.5	1.75	2	2.25
$\psi(u)$	0.3059	0.2696	0.2379	0.2102	0.1858
$\psi_{\mathrm{app}}(u)$	0.3003	0.2660	0.2357	0.2087	0.1849
Er	-1.82	-1.32	-0.95	-0.69	-0.50

Table 5.4.1 Cramér–Lundberg approximation to ruin probabilities

the Cramér–Lundberg approximation is exact, i.e. there exists a constant $c \ge 0$ such that $\psi(u) = ce^{-\gamma u}$ for all $u \ge 0$. Then from (5.3.13) we have

$$\frac{\beta - \lambda\mu}{\beta s - \lambda(1 - \hat{l}_U(s))} = \frac{1}{s} - \frac{c}{\gamma + s}.$$

A rearrangement of the terms in this equation yields

$$\begin{aligned}\hat{l}_U(s) &= 1 - \frac{\beta s(\gamma + s - cs) - s(\gamma + s)(\beta - \lambda\mu)}{\lambda(\gamma + s - cs)} \\ &= 1 + \frac{(\beta c - \lambda\mu)s^2 - \lambda\mu\gamma s}{\lambda\gamma + (\lambda - \lambda c)s}.\end{aligned}$$

Since $\lim_{s\to\infty} \hat{l}_U(s) = 0$, we find that $c = \lambda\mu(\beta)^{-1}$ and $\gamma = \mu^{-1} - \lambda\beta^{-1}$. Thus the claim sizes must be exponentially distributed.

5.4.3 Subexponential Claim Sizes

In Section 5.4.2 we found the asymptotic behaviour of the ruin function $\psi(u)$ when the initial risk reserve u tends to infinity. However our result was limited to claim sizes for which the tail of the distribution function decreases exponentially fast. For many applications such an assumption is unrealistic. For instance, data from motor third liability insurance, fire insurance or catastrophe insurance (earthquakes, flooding etc.) clearly show heavy tail behaviour. In particular, Pareto, lognormal and loggamma distributions are popular in actuarial mathematics.

In Section 2.5, we have shown that several families of heavy-tailed claim size distributions belong to the class of subexponential distributions. It turns out (see Section 2.5.3) that also their integrated tail distributions are subexponential. Note that in such a case the Pollaczek–Khinchin formula (5.3.16) implies that the ruin function $\psi(u)$ decreases more slowly than any exponential function. Indeed, by (2.5.2) and (5.3.16), we have for $u \to \infty$

$$\psi(u)\mathrm{e}^{su} \geq (1 - \lambda\mu\beta^{-1})\lambda\mu\beta^{-1}\overline{F^{\mathrm{s}}_U}(u)\mathrm{e}^{su} \longrightarrow \infty$$

for all $s > 0$. This simple result indicates that, in the case of heavy-tailed claim sizes, the asymptotic behaviour of $\psi(u)$ is very different from that in Theorem 5.4.2. If the integrated tail distribution F^{s}_U is subexponential, then we have the following result.

Theorem 5.4.3 *Let $\rho = \lambda\mu\beta^{-1}$ and assume that $F^{\mathrm{s}}_U \in \mathcal{S}$. Then*

$$\lim_{u\to\infty} \frac{\psi(u)}{1 - F^s_U(u)} = \frac{\rho}{1-\rho}\,. \tag{5.4.18}$$

Proof From (5.3.16) we know that $\psi(u)$ is the tail function of a geometric compound with characteristics (ρ, F^{s}_U). Note that for the probability function $\{p_0, p_1, \ldots\}$ with $p_n = (1-\rho)\rho^n$, there exists some $\varepsilon > 0$ such that $\sum_{n=1}^{\infty} p_n(1+\varepsilon)^n < \infty$. Thus, (5.4.18) follows from Theorem 2.5.4 since

$$\sum_{k=1}^{\infty} kp_k = \sum_{k=1}^{\infty} k(1-\rho)\rho^k = \frac{\rho}{1-\rho}\,. \qquad \square$$

The above theorem suggests the approximation

$$\psi_{\mathrm{app}}(u) = \frac{\rho}{1-\rho}(1 - F^s_U(u))\,. \tag{5.4.19}$$

Note that the quantity ρ captures all the information on the claim number process one needs to know.

Examples 1. Assume that the claim sizes are $\mathrm{Par}(\alpha, c)$ distributed. In order to have a finite mean (which is necessary by the net profit condition (5.3.2)) we must have $\alpha > 1$. The integrated tail distribution F^{s}_U is readily obtained as

$$F^{\mathrm{s}}_U(x) = \begin{cases} (\alpha-1)x/\alpha c & \text{if } x \leq c, \\ 1 - \alpha^{-1}\,(x/c)^{-(\alpha-1)} & \text{if } x > c. \end{cases}$$

By Theorem 2.5.5, F^{s}_U is subexponential. Thus, Theorem 5.4.3 leads to the following approximation to the ruin probability $\psi(u)$:

$$\psi_{\mathrm{app}}(u) = \frac{\rho}{\alpha(1-\rho)}\left(\frac{u}{c}\right)^{-(\alpha-1)}$$

for $u > c$. Details are left to the reader.

2. Let $\beta = 1$, $\lambda = 9$ and $F_U(x) = 1 - (1+x)^{-11}$, where we use the premium as the monetary unit. The integrated tail distribution F^{s}_U is readily obtained:

$F_U^s(x) = 1-(1+x)^{-10}$. From Theorem 2.5.5 we conclude that both $F_U \in \mathcal{S}$ and $F_U^s \in \mathcal{S}$. Approximation (5.4.19) then reads $\psi_{\text{app}}(u) = 9(1+u)^{-10}$. Table 5.4.2 gives some values of $\psi(u)$ and of the approximation $\psi_{\text{app}}(u) = 9(1+u)^{-10}$ as well as 100 times the relative error. The "exact values" of $\psi(u)$ were calculated using Panjer's algorithm described in Section 4.4.2. In order to get a discrete approximation to the claim size distribution, this distribution was discretized with bandwidth $h = 10^{-3}$, i.e. $q_k = \mathbb{P}(k/1000 \le U < (k+1)/1000)$. Consider

u	$\psi(u)$	$\psi_{\text{app}}(u)$	Er
1	0.364	8.79×10^{-3}	-97.588
2	0.150	1.52×10^{-4}	-99.898
3	6.18×10^{-2}	8.58×10^{-6}	-99.986
4	2.55×10^{-2}	9.22×10^{-7}	-99.996
5	1.05×10^{-2}	1.49×10^{-7}	-99.999
10	1.24×10^{-4}	3.47×10^{-10}	-100
20	1.75×10^{-8}	5.40×10^{-13}	-99.997
30	2.50×10^{-12}	1.10×10^{-14}	-99.56
40	1.60×10^{-15}	6.71×10^{-16}	-58.17
50	1.21×10^{-16}	7.56×10^{-17}	-37.69

Table 5.4.2 Approximation to ruin probabilities for subexponential claims

for instance the initial risk reserve $u = 20$. Then the ruin probability $\psi(u)$ is 1.75×10^{-8}, which is so small that it is not interesting for practical purposes. However, the approximation error is still almost 100%. Thus, in the case of heavy-tailed claim sizes, the approximation (5.4.19) can be poor, even for large values of u.

Note that (5.3.18) and (5.4.19) imply that for u (very) large the ruin probability $\psi(u)$ is $(\beta - \lambda\mu)^{-1}\lambda\mu$ times the probability that the first ladder height of the random walk $\{S_n\}$ considered in (5.3.1) exceeds u. But $(\beta - \lambda\mu)^{-1}\lambda\mu$ is the expected number of ladder epochs of $\{S_n\}$. Intuitively this means that, for u large, the ruin will occur if one of the ladder heights is larger than u.

5.4.4 Approximation by Moment Fitting

We now present two further methods for getting approximations to the ruin function $\psi(u)$. The first one is based on replacing the risk reserve process $\{R(t)\}$ by another risk process $\{R'(t)\}$ such that for some $n \ge 1$ the moments up to order n of certain characteristics of $\{R(t)\}$ and $\{R'(t)\}$ coincide. Furthermore, $\{R'(t)\}$ is chosen in such a way that the ruin function $\psi'(u)$ of $\{R'(t)\}$ is easier to determine than $\psi(u)$.

The second approximation method replaces the distribution function $1 - \beta\psi(u)/(\lambda\mu)$ by a simpler distribution function such that the moments up to order n coincide.

The De Vylder Approximation In the case of exponentially distributed claim sizes, the ruin function is available $\psi(u)$ explicitly by (5.3.8). The idea of the De Vylder approximation is to replace $\{R(t)\}$ by $\{R'(t)\}$, where $\{R'(t)\}$ has exponentially distributed claim sizes (with parameter δ') and

$$\mathbb{E}\left((R(t))^k\right) = \mathbb{E}\left((R'(t))^k\right) \tag{5.4.20}$$

for $k = 1, 2, 3$ and $t \geq 0$. Note that

$$\mathbb{E}\,(R(t) - u) = (\beta - \lambda\mu)t\,, \qquad \operatorname{Var} R(t) = \operatorname{Var}\,(u + \beta t - R(t)) = \lambda\mu_U^{(2)} t\,,$$

and

$$\mathbb{E}\,((R(t) - \mathbb{E}\,(R(t)))^3) = -\mathbb{E}\,((u + \beta t - R(t) - \mathbb{E}\,(u + \beta t R(t)))^3) = -\lambda\mu_U^{(3)} t\,.$$

This implies that (5.4.20) holds if

$$(\beta - \lambda\mu)t = \left(\beta' - \frac{\lambda'}{\delta'}\right)t\,, \qquad \lambda\mu_U^{(2)} t = \frac{2\lambda'}{(\delta')^2}t\,, \qquad \lambda\mu_U^{(3)} t = \frac{6\lambda'}{(\delta')^3}t\,.$$

Thus, the parameters $(\delta', \lambda', \beta')$ are given by

$$\delta' = \frac{3\mu_U^{(2)}}{\mu_U^{(3)}}\,, \qquad \lambda' = \frac{\lambda\mu_U^{(2)}(\delta')^2}{2} = \frac{9(\mu_U^{(2)})^3}{2(\mu_U^{(3)})^2}\lambda\,, \tag{5.4.21}$$

and

$$\beta' = \beta - \lambda\mu + \frac{\lambda'}{\delta'} = \beta - \lambda\mu + \frac{3(\mu_U^{(2)})^2}{2\mu_U^{(3)}}\lambda\,. \tag{5.4.22}$$

Consequently, using (5.3.8), we derive the *De Vylder approximation* to the ruin probability $\psi(u)$:

$$\psi_{\text{app}}(u) = \frac{\lambda'}{\delta'\beta'}\mathrm{e}^{-(\delta' - \lambda'/\beta')u}. \tag{5.4.23}$$

Of course, the approximation $\psi_{\text{app}}(u)$ given in (5.4.23) is equal to $\psi(u)$ in the case of exponentially distributed claims. However, the numerical example discussed at the end of this section shows that the approximation (5.4.23) is quite accurate for nonexponential claim size distributions as well.

The Beekman–Bowers Approximation Consider the distribution function $F(x) = 1 - \beta\psi(x)/(\lambda\mu)$. Then, by (5.3.11) we have that $F(0) = 0$. Thus

$F(x)$ is the distribution function of a positive random variable Z. Moreover, by Theorems 5.3.1 and 5.3.3, the distribution F of Z is absolutely continuous and has moment generating function

$$\begin{aligned}\hat{m}_Z(s) &= \int_0^\infty \mathrm{e}^{sx} \frac{\beta}{\lambda\mu} \overline{\psi}_+^{(1)}(x)\,\mathrm{d}x \\ &= \frac{\beta}{\lambda\mu}\Big(-s\frac{\beta-\lambda\mu}{-\beta s-\lambda(1-\hat{m}_U(s))} - (1-\lambda\mu\beta^{-1})\Big)\,.\end{aligned}$$

In the derivation of the last equation, we used integration by parts. Thus,

$$\hat{m}_Z(s) = 1 - \frac{\beta}{\lambda\mu} + \frac{\beta(\beta-\lambda\mu)}{\lambda\mu}\,\frac{s}{\beta s - \lambda(\hat{m}_U(s)-1)}\,. \tag{5.4.24}$$

The idea of the Beekman–Bowers approximation is to approximate the distribution function F by the distribution function $F'(u)$ of a $\Gamma(a',\delta')$-distributed random variable such that the first two moments coincide. This means that we have to determine the first two moments of F. Assume that $\mu_U^{(3)} < \infty$. Then, by (5.4.24) the moment generating function $\hat{m}_Z(s)$ is twice differentiable and the first derivative of $\hat{m}_Z(s)$ is

$$\hat{m}_Z^{(1)}(s) = \frac{\beta(\beta-\lambda\mu)}{\mu}\,\frac{s\hat{m}_U^{(1)}(s) - (\hat{m}_U(s)-1)}{(\beta s-\lambda(\hat{m}_U(s)-1))^2}\,. \tag{5.4.25}$$

Now, using $\lim_{s\to 0} s^{-1}(\beta s - \lambda(\hat{m}_U(s)-1)) = \beta - \lambda\mu$, we find

$$\lim_{s\to 0}\frac{s\hat{m}_U^{(1)}(s) - (\hat{m}_U(s)-1)}{s^2} = \lim_{s\to 0}\frac{\hat{m}_U^{(1)}(s) + s\hat{m}_U^{(2)}(s) - \hat{m}_U^{(1)}(s)}{2s} = \frac{\mu_U^{(2)}}{2}\,,$$

and thus

$$\mathbb{E}\,Z = \frac{\beta(\beta-\lambda\mu)}{\mu}\,\frac{\mu_U^{(2)}}{2(\beta-\lambda\mu)^2} = \frac{\beta\mu_U^{(2)}}{2\mu(\beta-\lambda\mu)}\,. \tag{5.4.26}$$

Differentiating both sides of (5.4.25), we get for the second derivative of $\hat{m}_Z(s)$,

$$\begin{aligned}\hat{m}_Z^{(2)}(s) &= \frac{\beta(\beta-\lambda\mu)}{\mu}\,\frac{1}{(\beta s-\lambda(\hat{m}_U(s)-1))^3} \\ &\qquad\times\Big(s\hat{m}_U^{(2)}(s)(\beta s-\lambda(\hat{m}_U(s)-1)) \\ &\qquad -2(s\hat{m}_U^{(1)}(s)-(\hat{m}_U(s)-1))(\beta-\lambda\hat{m}_U^{(1)}(s))\Big)\,.\end{aligned}$$

Since

$$\lim_{s\to 0} s^{-3}\left(s\hat{m}_U^{(2)}(s)(\beta s - \lambda(\hat{m}_U(s)-1)\right)$$

$$
\begin{aligned}
&\quad - 2(s\hat{m}_U^{(1)}(s) - (\hat{m}_U(s) - 1))(\beta\lambda\hat{m}_U^{(1)}(s))) \\
&= \lim_{s\to 0} \frac{1}{3s^2}\left((\hat{m}_U^{(2)}(s) + s\hat{m}_U^{(3)}(s))(\beta s - \lambda(\hat{m}_U(s) - 1))\right. \\
&\quad + s\hat{m}_U^{(2)}(s)(\beta - \lambda\hat{m}_U^{(1)}(s)) - 2s\hat{m}_U^{(2)}(s)(\beta - \lambda\hat{m}_U^{(1)}(s)) \\
&\quad \left.+ 2(s\hat{m}_U^{(1)}(s) - (\hat{m}_U(s) - 1))\lambda\hat{m}_U^{(2)}(s)\right) \\
&= \frac{\mu_U^{(3)}}{3}(\beta - \lambda\mu) + \lambda\mu_U^{(2)} \lim_{s\to 0} \frac{s\hat{m}_U^{(1)}(s) - (\hat{m}_U(s) - 1)}{s^2} \\
&= \frac{\mu_U^{(3)}}{3}(\beta - \lambda\mu) + \frac{\lambda}{2}(\mu_U^{(2)})^2,
\end{aligned}
$$

the second moment of Z becomes

$$
\begin{aligned}
\mathbb{E}(Z^2) &= \frac{\beta(\beta - \lambda\mu)}{\mu} \frac{\mu_U^{(3)}(\beta - \lambda\mu)/3 + \lambda(\mu_U^{(2)})^2/2}{(\beta - \lambda\mu)^3} \\
&= \frac{\beta}{\mu}\Big(\frac{\mu_U^{(3)}}{3(\beta - \lambda\mu)} + \frac{\lambda(\mu_U^{(2)})^2}{2(\beta - \lambda\mu)^2}\Big). \qquad (5.4.27)
\end{aligned}
$$

Using (5.4.26) and (5.4.27), we get that the parameters a' and δ' of the approximating Gamma distribution $\Gamma(a', \delta')$ are given by

$$
\frac{a'}{\delta'} = \frac{\beta\mu_U^{(2)}}{2\mu(\beta - \lambda\mu)}, \qquad \frac{a'(a'+1)}{(\delta')^2} = \frac{\beta}{\mu}\Big(\frac{\mu_U^{(3)}}{3(\beta - \lambda\mu)} + \frac{\lambda(\mu_U^{(2)})^2}{2(\beta - \lambda\mu)^2}\Big).
$$

Thus, for these parameters a' and δ', the *Beekman–Bowers approximation* to the ruin probability $\psi(u) = \lambda\mu\beta^{-1}(1 - F(u))$ is

$$
\psi_{\mathrm{app}}(u) = \frac{\lambda\mu}{\beta}(1 - F'(u)), \qquad (5.4.28)
$$

where $F'(u)$ is the distribution function of the Gamma distribution $\Gamma(a', \delta')$.

Remark Let Z' be $\Gamma(a', \delta')$ distributed. If $2a' \in \{1, 2, \ldots\}$, then $2\delta' Z'$ is $\chi^2_{2a'}$-distributed. Thus, approximating $2a'$ by a natural number allows the computation of $F'(u)$ by using standard statistical tables and software for χ^2-distributions.

Example Let us again consider the example from Section 5.4.2 with $\lambda = \beta = 1$ and $F_U(x) = 1 - \frac{1}{3}(\mathrm{e}^{-x} + \mathrm{e}^{-2x} + \mathrm{e}^{-3x})$. The moments of the claim size U are $\mu = \mu_U^{(1)} = 0.611111$, $\mu_U^{(2)} = 0.907407$ and $\mu_U^{(3)} = 2.32407$. The parameters of the De Vylder approximation are

$$
\delta' = 1.17131, \qquad \lambda' = 0.622472, \qquad \beta' = 0.920319.
$$

For the Beekman–Bowers approximation we have to solve the equations

$$\frac{a'}{\delta'} = 1.90909\,, \qquad \frac{a'(a'+1)}{\delta'^2} = 7.71429\,,$$

which yields the parameters $a' = 0.895561$ and $\delta' = 0.469104$. Thus, using the gamma distribution $\Gamma(a', \delta')$ we can calculate the Beekman–Bowers approximation (5.4.28) to the ruin probability $\psi(u)$. Alternatively, it is also possible to approximate $\psi(u)$ by χ^2-distributions although $2a' = 1.79112$ is not near to any natural number. Following the remark above, we can

u	0	0.25	0.5	0.75	1
$\psi(u)$	0.6111	0.5246	0.4547	0.3969	0.3479
DV	0.5774	0.5102	0.4509	0.3984	0.3520
Er (in %)	-5.51	-2.73	-0.86	0.38	1.18
BB1	0.6111	0.5227	0.4553	0.3985	0.3498
Er (in %)	0.00	-0.35	0.12	0.42	0.54
BB2	0.6111	0.5105	0..4456	0.3914	0.3450
Er (in %)	0.00	-2.68	-2.02	-1.38	-0.83
u	1.25	1.5	1.75	2	2.25
$\psi(u)$	0.3059	0.2696	0.2379	0.2102	0.1858
DV	0.3110	0.2748	0.2429	0.2146	0.1896
Er (in %)	1.67	1.95	2.07	2.09	2.03
BB1	0.3076	0.2709	0.2387	0.2106	0.1859
Er (in %)	0.54	0.47	0.34	0.19	0.04
BB2	0.3046	0.2693	0.2383	0.2110	0.1869
Er (in %)	-0.42	-0.11	0.18	0.40	0.59

Table 5.4.3 Approximation by moment fitting

interpolate between the two distributions χ_1^2 and χ_2^2. This yields

$$F'_{\text{app}}(u) = 0.20888\chi_1^2(2\delta' u) + 0.79112\chi_2^2(2\delta' u)\,. \tag{5.4.29}$$

The weights are $2 - 2a'$ and $2a' - 1$, respectively, which is motivated by $2a' = (2-2a')\,1 + (2a'-1)\,2$. In Table 5.4.3 a variety of approximations to $\psi(u)$ are given for some realistic values u. We observe that these approximations work quite well: DV denotes the De Vylder approximation, BB1 the Beekman–Bowers approximation and BB2 gives the values obtained by interpolation (5.4.29). The relative error Er is given in percent.

Another approximation method, based on moment fitting, relies on a diffusion approximation to the ruin function $\psi(u)$ of the compound Poisson model, see also the bibliographical notes to Section 5.6.

5.4.5 Ordering of Ruin Functions

We compare the ruin functions $\psi(u)$ and $\psi'(u)$ of two compound Poisson models with arrival rates λ and λ', premium rates β and β', and claim size distributions F_U and $F_{U'}$, respectively. If we suppose that

$$\lambda \le \lambda', \quad \mu_U \le \mu_{U'}, \quad \beta \ge \beta' \tag{5.4.30}$$

and

$$F_U^{\mathrm{s}} \le_{\mathrm{st}} F_{U'}^{\mathrm{s}}, \tag{5.4.31}$$

then we immediately get $\psi(u) \le \psi'(u)$ for all $u \ge 0$. It suffices to recall that the right-hand side of (5.3.16) is the tail function of a geometric compound with characteristics (ρ, F_U^{s}) and to use Theorem 4.2.3a. It turns out that (5.4.31) can be replaced by a slightly weaker condition.

Theorem 5.4.4 *If $\lambda \le \lambda'$ and $\beta \ge \beta'$ and if $U \le_{\mathrm{sl}} U'$, then $\psi(u) \le \psi'(u)$ for all $u \ge 0$.*

Proof By Theorem 3.2.2 we have that $U \le_{\mathrm{sl}} U'$ is equivalent to

$$\int_x^\infty \overline{F}_U(y)\,\mathrm{d}y \le \int_x^\infty \overline{F}_{U'}(y)\,\mathrm{d}y$$

for all $x \ge 0$. We also have $\mu_U \le \mu_{U'}$. This gives

$$\overline{F_{U'}^{\mathrm{s}}}(x) = \frac{1}{\mu_{U'}}\int_x^\infty \overline{F}_{U'}(y)\,\mathrm{d}y \ge \frac{\theta}{\mu_U}\int_x^\infty \overline{F}_U(y)\,\mathrm{d}y = \theta\overline{F_U^{\mathrm{s}}}(x), \tag{5.4.32}$$

where $\theta = \mu_U(\mu_{U'})^{-1} \le 1$. Let $I_1, I_2, \ldots$ be a sequence of independent and identically distributed indicator random variables with $\mathbb{P}(I = 0) = 1 - \theta$ and $\mathbb{P}(I = 1) = \theta$. Furthermore, let $\tilde{U}_1, \tilde{U}_2, \ldots$ and $\tilde{U'}_1, \tilde{U'}_2, \ldots$ be sequences of independent and identically distributed random variables with distributions F_U^{s} and $F_{U'}^{\mathrm{s}}$ respectively, and independent of $\{I_n\}$. Finally, assume that N', N, N_θ are random variables which are geometrically distributed with parameters $(\lambda'\mu_{U'})/\beta'$,$(\lambda\mu_U)/\beta$, $\theta(\lambda'\mu_{U'})/\beta'$, respectively, and independent of $\{I_n\}, \{\tilde{U}_k\}, \{\tilde{U'}_k\}$. Recall that $\psi(u) = \mathbb{P}(\sum_{n=1}^N U_n > u)$ and $\psi'(u) = \mathbb{P}(\sum_{n=1}^{N'} U'_n > u)$ and notice that we can write (5.4.32) as

$$\tilde{U'} \ge_{\mathrm{st}} I\tilde{U}. \tag{5.4.33}$$

Then we get $\sum_{k=1}^{N'} \tilde{U}'_k \ge_{\mathrm{st}} \sum_{k=1}^{N'} I_k\tilde{U}_k$ by Theorem 4.2.3a and (5.4.33). It now remains to show that

$$\sum_{k=1}^{N'} I_k\tilde{U}_k \ge_{\mathrm{st}} \sum_{k=1}^{N} \tilde{U}_k. \tag{5.4.34}$$

Passing to Laplace–Stieltjes transforms we can verify that $\sum_{k=1}^{N'} I_k \tilde{U}_k \stackrel{\mathrm{d}}{=} \sum_{k=1}^{N_\theta} \tilde{U}_k$ and, by Theorem 4.2.3a, we get that $\sum_{k=1}^{N_\theta} \tilde{U}_k \geq_{\mathrm{st}} \sum_{k=1}^{N} \tilde{U}_k$ since $N_\theta \geq_{\mathrm{st}} N$. This completes the proof. □

Corollary 5.4.1 *If* F_U *is* NBUE, *then for all* $u \geq 0$

$$\psi(u) \leq \frac{\lambda \mu_U}{\beta} \mathrm{e}^{-(\mu_U^{-1} - \lambda\beta^{-1})u} . \tag{5.4.35}$$

Similarly, the reversed inequality in (5.4.35) *is true if* F_U *is* NWUE.

Proof Suppose that F_U is NBUE and let $F_{U'} = \mathrm{Exp}(\delta)$, where $\delta = \mu_U^{-1}$. This means that $\mathbb{E}\,(U - x \mid U > x) \leq \mu_U$ for all $x \geq 0$, which can be rewritten as

$$\frac{\mathrm{d}}{\mathrm{d}x} \log \int_x^\infty \overline{F}_U(y)\,\mathrm{d}y \leq -\mu_U^{-1}, \qquad x \geq 0 .$$

Integrating both sides yields

$$\log \int_x^\infty \overline{F}_U(y)\,\mathrm{d}y \leq -\mu_U^{-1} x + \log \mu_U , \qquad x \geq 0 .$$

Hence, $\int_x^\infty \overline{F}_U(y)\,\mathrm{d}y \leq \mu_U \mathrm{e}^{-\delta x}$, $x \geq 0$, or equivalently $U \leq_{\mathrm{sl}} U'$. Applying now Theorem 5.4.4 with $\lambda = \lambda'$ and $\beta = \beta'$ and recalling formula (5.3.8) we obtain that $\psi(u) \leq \psi'(u) = \lambda\mu_U\beta^{-1}\mathrm{e}^{-(\delta - \lambda/\beta)u}$, $u \geq 0$. Similar considerations are valid for F_U being NWUE with the reversed inequality in (5.4.35). □

Bibliographical Notes. One-sided bounds of the type $\psi(u) \leq \mathrm{e}^{-\gamma u}$ as well as asymptotic relations $\psi(u) \sim c\mathrm{e}^{-\gamma u}$ for large u-values have been studied by Filip Lundberg (1926,1932,1934). The modern approach to these estimations is due to Cramér (1955). By means of martingale techniques one-sided inequalities were also derived in Gerber (1973) and Kingman (1964) in the settings of risk and queueing theories, respectively. In Taylor (1976), two-sided bounds of the form (5.4.4) were obtained for the ruin function $\psi(u)$. The renewal approach to Theorem 5.4.2 is due to Feller (1971). Graphical and numerical techniques to estimate the adjustment coefficient abound. For an existence argument, see Mammitzsch (1986). Recently, statistical techniques have become available as well. A first attempt using stochastic approximation can be found in Herkenrath (1986). The approach used in Csörgő and Teugels (1990) employs the notion of the empirical moment generating function as defined in (4.5.3); see also Pitts, Grübel and Embrechts (1996). Other approaches identify the adjustment coefficient as the abscissa of convergence of a Laplace transform as done in Deheuvels and Steinebach (1990) or use intermediate order statistics as in Csörgő and Steinebach (1991). For a bootstrap version, see Embrechts and Mikosch (1991). A Hill-type estimate has been proposed in Richter, Steinebach and Taube (1993). Procedures that apply to more general risk models can be found in Christ and Steinebach (1995) and in Schmidli (1997b). Bounds for the adjustment coefficient are given in Gerber (1979). Theorem 5.4.3 goes back to Teugels and Veraverbeke (1973) and to Embrechts and Veraverbeke (1982); see also Asmussen, Schmidli and Schmidt (1999), Cohen (1973), Embrechts and Villaseñor (1988), Klüppelberg (1989) and Pakes (1975). In the special case of Pareto distributed claim sizes, the result of Theorem 5.4.3 was obtained by von Bahr (1975), while Thorin and Wikstad (1977) dealt with the lognormal distribution; see also Ramsay and Usabel (1997) and Seal (1980). Higher-order asymptotic expansions can be found, for example, in Willekens and Teugels (1992). Simulation of ruin probabilities for subexponential claim sizes was considered in Asmussen and Binswanger (1997). In Klüppelberg and Stadtmüller (1998), the asymptotic behaviour of the ruin function $\psi(u)$ has been investigated for the compound Poisson model with heavy-tailed claim size distribution and interest rates. It turns out that, in this case, $\psi(u)$ is asymptotically proportional to the tail function $\overline{F}_U(u)$ of claim sizes as $u \to \infty$; this is in contrast to the result in Theorem 5.4.3 for the model without interest. More details on risk models with interest are given in Section 11.4. Large deviation results for the claim surplus process in the compound Poisson model with heavy-tailed claim sizes have been derived in Asmussen and Klüppelberg (1996) and Klüppelberg and Mikosch (1997). The De Vylder approximation was introduced in De Vylder (1978). The Beekman–Bowers approximation can be found in Beekman (1969). Theorem 5.4.4 is from Daley and Rolski (1984). For orderings of risks and results like (5.4.31), see Pellerey (1995). Asymptotic ordering of risks and ruin probabilities has been studied, for example, in Asmussen, Frey, Rolski and Schmidt (1995) and Klüppelberg (1993).

5.5 NUMERICAL EVALUATION OF RUIN FUNCTIONS

In this section, we discuss an algorithm for the *numerical inversion* of Laplace transforms which makes use of Fourier transforms. Recall that the Fourier transform $\hat{\varphi}(s)$ of a function $g : \mathbb{R} \to \mathbb{R}$ is defined by $\hat{\varphi}(s) = \int_{-\infty}^{\infty} \mathrm{e}^{\mathrm{i}sx} g(x)\,\mathrm{d}x$. The following lemma shows how to invert the Fourier transform.

Lemma 5.5.1 *Let $g : \mathbb{R} \to \mathbb{R}$ be a measurable function such that $\int_{-\infty}^{\infty} |g(t)|\,\mathrm{d}t < \infty$ and let $\hat{\varphi}(s)$ be its Fourier transform.*

(a) *If* $\int_{-\infty}^{\infty} |\hat{\varphi}(s)|\,\mathrm{d}s < \infty$, *then* $g(t)$ *is a bounded continuous function and*

$$g(t) = \frac{1}{2\pi}\int_{-\infty}^{\infty} \mathrm{e}^{-ist}\hat{\varphi}(s)\,\mathrm{d}s\,, \qquad t \in \mathbb{R}\,. \tag{5.5.1}$$

(b) *If* $g(t)$ *is a continuous function with locally bounded total variation, then*

$$g(t) = \frac{1}{2\pi}\lim_{T\to\infty}\int_{-T}^{T} \mathrm{e}^{-ist}\hat{\varphi}(s)\,ds\,, \qquad t \in \mathbb{R}\,. \tag{5.5.2}$$

The *proof* of Lemma 5.5.1 is omitted. For part (a) a proof can be found, for example, in Feller (1971), Chapter XV.3, and for (b) in Doetsch (1950).

Another auxiliary result which we need is the *Poisson summation formula.* Recall that for a continuous periodic function $g_p(t)$, having locally bounded variation with period $2\pi/h$, the corresponding Fourier series $\sum_{k=-\infty}^{\infty} c_k \mathrm{e}^{ikht}$, where $c_k = (h/2\pi)\int_{-\pi/h}^{\pi/h} g_p(t)\mathrm{e}^{-ikht}\,\mathrm{d}t$, converges to $g_p(t)$ and this uniformly with respect to the variable t. Actually, for pointwise convergence of the Fourier series it suffices to assume that $g_p(t)$ is of locally bounded variation and that $g_p(t) = (1/2)(g_p(t+) + g_p(t-))$. In the following, let

$$g_p(t) = \sum_{k=-\infty}^{\infty} g\Big(t + \frac{2\pi k}{h}\Big)\,.$$

Then, assuming that $g_p(t)$ is well-defined for all t, the function $g_p(t)$ is periodic with period $2\pi/h$.

Lemma 5.5.2 *Let* $\hat{\varphi}(s)$ *be the Fourier transform* $\hat{\varphi}(s) = \int_{-\infty}^{\infty} \mathrm{e}^{ist}g(t)\,\mathrm{d}t$ *of an absolutely integrable continuous function* $g : \mathbb{R} \to \mathbb{R}$. *If the Fourier series corresponding to* $g_p(t)$ *converges pointwise to* $g_p(t)$, *then*

$$\sum_{k=-\infty}^{\infty} g\big(t + \frac{2\pi k}{h}\big) = \frac{h}{2\pi}\sum_{k=-\infty}^{\infty} \hat{\varphi}(kh)\mathrm{e}^{-ikht}\,. \tag{5.5.3}$$

Proof By the assumption we can represent the function $g_p(t)$ by the Fourier series $g_p(t) = \sum_{k=-\infty}^{\infty} c_k \mathrm{e}^{ikht}$, where

$$\begin{aligned} c_k &= \frac{h}{2\pi}\int_{-\pi/h}^{\pi/h} g_p(t)\mathrm{e}^{-ikht}\,\mathrm{d}t = \frac{h}{2\pi}\int_{-\pi/h}^{\pi/h}\Big(\sum_{n=-\infty}^{\infty} g\big(t + \frac{2\pi n}{h}\big)\Big)\mathrm{e}^{-ikht}\,\mathrm{d}t \\ &= \frac{h}{2\pi}\int_{-\infty}^{\infty} g(t)\mathrm{e}^{-ikht}\,\mathrm{d}t = \frac{h}{2\pi}\hat{\varphi}(-kh). \end{aligned}$$

Hence $\sum_{k=-\infty}^{\infty} g(t + 2\pi k h^{-1}) = (h/2\pi)\sum_{k=-\infty}^{\infty} \hat{\varphi}(kh)\mathrm{e}^{-ikht}$. □

We now discuss an algorithm for numerical evaluation of the Laplace transform

$$\hat{L}(z) = \int_0^\infty \mathrm{e}^{-zu} c(u)\,\mathrm{d}u \tag{5.5.4}$$

for some class of functions $c : \mathbb{R}_+ \to \mathbb{R}_+$. Note that if $c(u)$ is a bounded and measurable function, then its Laplace transform $\hat{L}(z)$ is well-defined for all complex numbers z with $\Re z > 0$. In this section we will assume that $c(u)$ is a bounded continuous function on $\mathbb{R}_+$ with locally bounded variation.

The numerical method discussed below can be used to compute the ruin function $\psi(u)$ in the case when its Laplace transform is known but does not allow analytical inversion. Sometimes the analytical inversion of the Laplace transform of $\psi(u)$ is possible, as in the exponential case as shown in Section 5.3.3. Inversion is also possible for the hyperexponential claim size distribution from Section 5.4.2, and for some further examples discussed at the end of the present section.

We use Lemma 5.5.1 to derive a formula for $c(u)$ in terms of its Laplace transform.

Theorem 5.5.1 *For all $u \ge 0$ and $x > 0$*

$$c(u) = \frac{2\mathrm{e}^{ux}}{\pi} \lim_{T\to\infty} \int_0^T \cos(uy)\Re\hat{L}(x+\mathrm{i}y)\,\mathrm{d}y\,. \tag{5.5.5}$$

Proof Let $x > 0$ be fixed. Consider the function $g : \mathbb{R} \to \mathbb{R}$ defined by

$$g(t) = \begin{cases} \mathrm{e}^{-xt}c(t) & \text{if } t \ge 0, \\ g(-t) & \text{if } t < 0, \end{cases} \tag{5.5.6}$$

with $x \ge 0$ and $b = \int_0^\infty \mathrm{e}^{-xt}c(t)\,\mathrm{d}t < \infty$. The corresponding Fourier transform is then given by $\hat{\varphi}(s) = \hat{L}(x-\mathrm{i}s) + \hat{L}(x+\mathrm{i}s) = 2\Re\hat{L}(x+\mathrm{i}s)$. Since $g(t)$ given in (5.5.6) is an even function, we have $\hat{\varphi}(s) = \Re\hat{\varphi}(s)$. Lemma 5.5.1b yields

$$\begin{aligned} g(t) &= \frac{1}{2\pi}\lim_{T\to\infty}\int_{-T}^T \mathrm{e}^{-\mathrm{i}ty}\hat{\varphi}(y)\,\mathrm{d}y = \frac{1}{2\pi}\lim_{T\to\infty}\int_{-T}^T \Re\left[\mathrm{e}^{-\mathrm{i}ty}\hat{\varphi}(y)\right]\mathrm{d}y \\ &= \frac{1}{2\pi}\lim_{T\to\infty}\int_{-T}^T \cos(ty)\hat{\varphi}(y)\,\mathrm{d}y = \frac{1}{\pi}\lim_{T\to\infty}\int_0^T \cos(ty)\hat{\varphi}(y)\,\mathrm{d}y \\ &= \frac{2}{\pi}\lim_{T\to\infty}\int_0^T \cos(ty)\Re\hat{L}(x+\mathrm{i}y)\,\mathrm{d}y\,. \end{aligned}$$

That is, $c(u) = 2\mathrm{e}^{ux}\pi^{-1}\lim_{T\to\infty}\int_0^T \cos(uy)\Re\hat{L}(x+\mathrm{i}y)\,\mathrm{d}y$ for $u, x \ge 0$. $\square$

For practical applications of Theorem 5.5.1, it remains to numerically compute the integral in (5.5.5). This integral has to be approximated by a suitably chosen sum. For instance, we can use the infinite trapezoidal rule with a (possibly small) discretization width $h > 0$ to derive the following approximation to $c(u)$:

$$c_{\rm app}(u) = \frac{2e^{ux}}{\pi} h\Big(\tfrac{1}{2}\Re\hat{L}(x) + \sum_{k=1}^{\infty}\cos(ukh)\Re\hat{L}(x+ikh)\Big). \tag{5.5.7}$$

For reasons that will be explained later, we put $h = \pi/(2u)$, $x = a/(2u)$ for some $a > 0$. Then

$$\begin{aligned}
&\frac{2e^{ux}}{\pi} h\Big(\tfrac{1}{2}\Re\hat{L}(x) + \sum_{k=1}^{\infty}\cos(ukh)\Re\hat{L}(x+ikh)\Big)\\
&= \frac{e^{a/2}}{2u}\Big(\Re\hat{L}\Big(\frac{a}{2u}\Big) + 2\sum_{k=1}^{\infty}\cos\Big(k\frac{\pi}{2}\Big)\Re\hat{L}\Big(\frac{a+i\pi k}{2u}\Big)\Big).
\end{aligned}$$

Now note that

$$\cos\Big(k\frac{\pi}{2}\Big) = \begin{cases} 0 & \text{if } k \text{ is odd,}\\ (-1)^{k/2} & \text{if } k \text{ is even,}\end{cases}$$

and thus we only have to consider even ks. We arrive at the discretization error

$$\begin{aligned}
d(u) &= |c(u) - c_{\rm app}(u)|\\
&= \Big|c(u) - \frac{e^{a/2}}{2u}\Big(\Re\hat{L}\Big(\frac{a}{2u}\Big) - 2\sum_{k=1}^{\infty}(-1)^{k+1}\Re\hat{L}\Big(\frac{a+i2\pi k}{2u}\Big)\Big)\Big|.
\end{aligned} \tag{5.5.8}$$

It turns out that the discretization error can be controlled.

Theorem 5.5.2 *For all $u \ge 0$,*

$$\begin{aligned}
c(u) &= \frac{e^{a/2}}{2u}\Big(\Re\hat{L}\Big(\frac{a}{2u}\Big) - 2\sum_{k=1}^{\infty}(-1)^{k+1}\Re\hat{L}\Big(\frac{a+i2k\pi}{2u}\Big)\Big)\\
&\quad - \sum_{k=1}^{\infty} e^{-ak}c((2k+1)u).
\end{aligned} \tag{5.5.9}$$

Proof Consider the function $g(t)$ defined in (5.5.6), i.e.

$$g(t) = \begin{cases} e^{-xt}c(t) & \text{for } t \ge 0,\\ g(-t) & \text{for } t < 0.\end{cases}$$

In this case the sum $g_p(t) = \sum_{k=-\infty}^{\infty} g(t + 2\pi k/h)$ converges uniformly in t. Thus, $g_p(t)$ is well-defined, continuous and of locally bounded total variation. The assumptions of Lemma 5.5.2 are therefore satisfied. We denote the Fourier transform of $g(t)$ by $\hat{\varphi}(s)$. Let $x = a/(2u)$ and $h = \pi/u$. Then, recalling that $\hat{L}$ is the Laplace transform of $c(x)$, we have $\hat{\varphi}\left(k\frac{\pi}{u}\right) = \hat{L}(x - \mathrm{i}kh)$, since

$$\begin{aligned}\hat{\varphi}\Big(\frac{k\pi}{u}\Big) &= \int_0^\infty \mathrm{e}^{\mathrm{i}(k\pi/u+\mathrm{i}a/(2u))t}c(t)\,\mathrm{d}t \\ &= \int_0^\infty \mathrm{e}^{-(a/(2u)-\mathrm{i}k\pi/u)t}c(t)\,\mathrm{d}t = \hat{L}((a-\mathrm{i}2k\pi)/(2u)).\end{aligned}$$

Thus (5.5.3) takes the form

$$\begin{aligned}\sum_{k=0}^\infty \mathrm{e}^{-a(2k+1)/2}c((2k+1)u) &= \frac{1}{2u}\sum_{k=-\infty}^\infty \hat{\varphi}\Big(\frac{k\pi}{u}\Big)\mathrm{e}^{-\mathrm{i}k\pi} \\ &= \frac{1}{2u}\sum_{k=-\infty}^\infty (-1)^k\hat{L}\Big(\frac{a-\mathrm{i}2k\pi}{2u}\Big),\end{aligned}$$

recalling that $\mathrm{e}^{-\mathrm{i}k\pi} = (-1)^k$. Since

$$\hat{\varphi}(s) = \int_{-\infty}^\infty \mathrm{e}^{\mathrm{i}st}g(t)\,\mathrm{d}t = \hat{L}(x-\mathrm{i}s) + \hat{L}(x+\mathrm{i}s) = 2\Re\hat{L}(x+\mathrm{i}s),$$

this gives

$$\begin{aligned}c(u) &= \frac{\mathrm{e}^{a/2}}{2u}\sum_{k=-\infty}^\infty (-1)^k\hat{L}\Big(\frac{a+\mathrm{i}2k\pi}{2u}\Big) \\ &\quad - \mathrm{e}^{a/2}\sum_{k=1}^\infty \mathrm{e}^{-a(2k+1)/2}c((2k+1)u) \\ &= \frac{\mathrm{e}^{a/2}}{2u}\Re\hat{L}\Big(\frac{a}{2u}\Big) + \frac{\mathrm{e}^{a/2}}{u}\sum_{k=1}^\infty (-1)^k\Re\hat{L}\Big(\frac{a+\mathrm{i}2k\pi}{2u}\Big) \\ &\quad - \sum_{k=1}^\infty \mathrm{e}^{-ak}c((2k+1)u),\end{aligned}$$

which completes the proof. □

Comparing (5.5.9) with (5.5.8), we get for the *discretization error* $d(u)$ in (5.5.8): $d(u) = \sum_{k=1}^\infty \mathrm{e}^{-ak}c((2k+1)u)$. Since $0 \le c(u) \le 1$,

$$d(u) \le \frac{\mathrm{e}^{-a}}{1-\mathrm{e}^{-a}}, \tag{5.5.10}$$

for $a > 0$. This suggests that a should be chosen as large as possible. However, if a is too large, other numerical problems can occur, e.g. rounding errors resulting from multiplication by the factor $\mathrm{e}^{a/2}$.

Note that the discretization width $h = \pi/(2u)$ has been chosen to end up with an alternating series (5.5.9), provided that the terms $\Re\hat{L}((a+\mathrm{i}2\pi k)/(2u))$ have the same sign for all $k \geq 1$. The *Euler summation* method, applied to alternating series, can now be used to accelerate convergence.

Let $\{a_k\}$ be a decreasing sequence of positive numbers such that $a_k \to 0$ as $k \to \infty$. With the notation

$$\Delta a_k = a_k - a_{k+1}\,, \qquad \Delta^{n+1} a_k = \Delta(\Delta^n a_k)\,, \qquad \Delta^0 a_k = a_k\,,$$

we can represent the n-th partial sum $s_n = \sum_{k=1}^{n}(-1)^{k+1}a_k$ of the infinite series $s = \sum_{k=1}^{\infty}(-1)^{k+1}a_k$ by

$$s_n = \Big(\frac{a_1}{2} + \frac{1}{2}\sum_{k=1}^{n-1}(-1)^{k+1}\Delta a_k\Big) + (-1)^{n+1}\frac{a_n}{2}\,. \tag{5.5.11}$$

Thus, $s'_n = a_1 2^{-1} + 2^{-1}\sum_{k=1}^{n-1}(-1)^{k+1}\Delta a_k \to s$ as $n \to \infty$. Moreover, if the sequence $\{\Delta a_k\}$ is decreasing, the sequence $\{s'_n\}$ is also alternating and (5.5.11) implies that the remainder $c_n = s - s_n$ can be bounded by

$$\begin{aligned} |c_n| &= (-1)^n c_n = (-1)^{n+1}(s_n - s) \\ &= \Big[(-1)^{n+1}\Big(\frac{a_1}{2} + \frac{1}{2}\sum_{k=1}^{n-1}(-1)^{k+1}\Delta a_k - s\Big)\Big] + \frac{a_n}{2} \\ &= [(-1)^{n+1}(s'_n - s)] + \frac{a_n}{2} \leq \frac{a_n}{2}\,, \end{aligned}$$

because then the term in the brackets is negative. Thus, using the same argument as above, for the remainder $c'_n = s - s'_n$ we have $|c'_n| < \Delta a_n/4$, provided that the sequences $\{\Delta a_k\}$ and $\{\Delta^2 a_k\}$ are decreasing. By iteration we derive a representation for the alternating series $s = \sum_{k=1}^{\infty}(-1)^{k+1}a_k$. More precisely, for each $n \in \mathbb{N}$, we have

$$s = \sum_{k=0}^{n-1}\frac{\Delta^k a_1}{2^{k+1}} + \frac{1}{2^n}\sum_{k=1}^{\infty}(-1)^{k+1}\Delta^n a_k\,, \tag{5.5.12}$$

which suggests that the first sum in (5.5.12) is a good approximation to s even for moderate values of n.

The *Euler transformation* $\{a_1, a_2, a_3, \ldots\} \mapsto \{\Delta^0 a_1, \Delta^1 a_1, \Delta^2 a_1, \ldots\}$ can also be used when computing the series s under much weaker conditions. It is for example not necessary to assume that the a_k are positive or monotone.

Lemma 5.5.3 *Let $\{a_k\}$ be an arbitrary sequence of real numbers. If the series $s = \sum_{k=1}^{\infty}(-1)^{k+1}a_k$ converges, then the Euler-transformed series*

$\sum_{k=0}^{\infty} \Delta^k a_1/2^{k+1}$ *converges to the same limit. Under the additional assumption that, for each* $n \in \mathbb{N}$, *the sequence* $\{\Delta^n a_k\}$ *is decreasing in* k,

$$\left| s - \sum_{k=0}^{n-1} \Delta^k a_1/2^{k+1} \right| \leq \frac{\Delta^n a_1}{2^n}.$$

The *proof* of Lemma 5.5.3 can be found, for example, in Johnsonbaugh (1979).

Usually, a good approximation to $s = \sum_{k=0}^{\infty}(-1)^k a_k$ is obtained when the alternating summation $a_0 - a_1 + a_2 - \ldots$ is combined with Euler summation. So, $s = s_n + (-1)^{n+1} \sum_{k=0}^{\infty} 2^{-(k+1)} \Delta^k a_{n+1}$ is approximated by

$$C(m,n) \stackrel{\text{def}}{=} s_n + (-1)^{n+1} \sum_{k=0}^{m-1} 2^{-(k+1)} \Delta^k a_{n+1}\,.$$

An induction argument shows that

$$C(m,n) = \sum_{k=0}^{m} \binom{m}{k} 2^{-m} s_{n+k}\,. \tag{5.5.13}$$

Hence, an application of the approximation $C(m,n)$ given in (5.5.13) to the series in (5.5.9) results in the approximation to $c(u)$

$$C(u,m,n) \stackrel{\text{def}}{=} \sum_{k=0}^{m} \binom{m}{k} 2^{-m} s_{n+k}(u)\,, \tag{5.5.14}$$

where

$$s_n(u) = \frac{\mathrm{e}^{a/2}}{2u} \Re \hat{L}\Big(\frac{a}{2u}\Big) + \frac{\mathrm{e}^{a/2}}{u} \sum_{k=1}^{n} (-1)^k \Re \hat{L}\Big(\frac{a + \mathrm{i}2\pi k}{2u}\Big).$$

Examples Consider the compound Poisson model specified by the characteristics (λ, F_U) and let $\psi(u)$ be its ruin function. Further, $\rho = (\lambda \mathbb{E}\, U)/\beta$. The Laplace transform $\hat{L}(s) = \int_0^\infty \exp(-su)\psi(u)\,du$ is known by (5.3.14). The proposed method can lead to two types of errors, resulting from discretization and from truncation. Choosing $a = 18.5$, the discretization error is less than 10^{-8} (computed from (5.5.10)). In Table 5.5.1 we present the results of the numerical computations for three cases of claim size distributions. We always take $\rho = 0.75$ and $\beta = 1$:

(a) $F_U = p\text{Exp}(a_1) + (1-p)\text{Exp}(a_2)$ with $p = 2/3$, $a_1 = 2$ and $a_2 = 1/2$, i.e. U is hyperexponentially distributed,

(b) F_U is the gamma distribution $\Gamma(1/2, 1/2)$ with the density function $f_U(x) = (2\pi x^3)^{1/2} \exp(-x/2)$ (then $\mu = 1$ and $\sigma^2 = 2$),

(c) F_U is the Pareto mixture of exponentials PME(2).

Hyper-exponentially distributed claim sizes

u	Exact	Numerical inversion	u	Exact	Numerical inversion
0.1	0.73192117	0.73192119	5.0	0.32004923	0.32004975
0.3	0.69927803	0.69927809	6.0	0.27355552	0.27355607
0.5	0.67037954	0.67037965	7.0	0.23382315	0.23382369
1.0	0.60940893	0.60940913	8.0	0.19986313	0.19986366
2.0	0.51446345	0.51446378	9.0	0.17083569	0.17083620
3.0	0.43843607	0.43843650	10.0	0.14602416	0.14602464

$\Gamma(1/2, 1/2)$-distributed claim sizes

u	Exact	Numerical inversion	u	Exact	Numerical inversion
0.1	0.733833531	0.733833534	5.0	0.322675414	0.322675411
0.3	0.705660848	0.705660851	6.0	0.274442541	0.274442538
0.5	0.680115585	0.680115587	7.0	0.233464461	0.233464459
1.0	0.622928580	0.622928581	8.0	0.198626710	0.198626707
2.0	0.526512711	0.526512711	9.0	0.168998278	0.168998276
3.0	0.446685586	0.446685585	10.0	0.143794910	0.143794907

Pareto mixture of exponentials PME(2)

u	Numerical inversion	Asymptotic approximation	u	Numerical inversion	Asymptotic approximation
1.0	0.60382220	1.5	20.0	0.11036	0.075
2.0	0.50796380	0.75	30.0	0.07060	0.05
3.0	0.43828568	0.5	40.0	0.05062	0.0375
5.0	0.34156802	0.3	50.0	0.03899	0.03
6.0	0.30629948	0.25	60.0	0.03151	0.025
7.0	0.27682399	0.21429	70.0	0.02635	0.02143
8.0	0.25183704	0.1875	80.0	0.02260	0.01875
9.0	0.23040797	0.16667	90.0	0.01976	0.01667
10.0	0.21185227	0.15	100.0	0.01754	0.015

Table 5.5.1 Ruin probability $\psi(u)$

For the numerical computations of $\psi(u)$ presented in Table 5.5.1 we used (5.5.14) with $m = 11$, $n = 15$ and $a = 18.5$. We then have that $|C(m, n+1) - C(m, n)|$ is of order 10^{-9}. Note however that for cases (a) and (b) there exist explicit solutions. For example, for the case (a) the ruin function $\psi(u)$ is given by

$$\psi(u) = 0.75\left(0.935194\mathrm{e}^{-0.15693u} + 0.0648059\mathrm{e}^{-1.59307u}\right).$$

We leave it to the reader to show this as an exercise. Furthermore, it is shown in Abate and Whitt (1992) that, for case (b),

$$\begin{aligned}\psi(u) &= \rho(\alpha_1\mathrm{e}^{-\delta u} + \rho^{-1}\overline{\Phi}(\sqrt{u}) - \alpha_1\mathrm{e}^{-\delta u}\overline{\Phi}(((1+8\rho)^{1/2}-1)\sqrt{u}/2) \\ &\quad -\alpha_2\mathrm{e}^{\kappa u}\overline{\Phi}(((1+8\rho)^{1/2}+1)\sqrt{u}/2)),\end{aligned}$$

where $\Phi(x)$ is the distribution function of the normal distribution, and

$$\begin{aligned}\alpha_1 &= 1 - \frac{2(1-\rho)}{\sqrt{1+8\rho}(1+2\rho+\sqrt{1+8\rho})}, & \alpha_2 &= \alpha_1 + \frac{1-2\rho}{\rho}, \\ \delta &= \frac{\sqrt{1+8\rho}-(4\rho-1)}{4}, & \kappa &= 2\rho+\delta-1/2.\end{aligned}$$

Bibliographical Notes. The method to compute the ruin function $\psi(u)$ by numerical inversion of its Laplace transform was presented in Abate and Whitt (1992); see also Abate and Whitt (1995), Choudhury, Lucantoni and Whitt (1994), O'Cinneide (1997). Another inversion technique for Laplace transforms has been studied in Jagerman (1978, 1982). For numerical inversions of characteristic functions, see, for example, Bohman (1975).

5.6 FINITE-HORIZON RUIN PROBABILITIES

We now show that for two special claim size distributions, the finite-horizon ruin probability $\psi(u;x)$ can be given in a relatively simple form. Furthermore, we will express the survival probability $\overline{\psi}(u;x) = 1 - \psi(u;x)$ in terms of the aggregate claim amount distribution and derive Seal's formulae.

5.6.1 Deterministic Claim Sizes

In this section we give a recursive method to calculate the finite-horizon ruin probability $\psi(u;x)$ for the case of deterministic claim sizes. Recall that $\tau(u)$ denotes the time of ruin, and that $\lfloor x \rfloor$ is the integer part of x.

Theorem 5.6.1 *Assume that $\mathbb{P}(U = \mu) = 1$ for some $\mu > 0$, i.e. the claim sizes are deterministic. Let $\phi(u;x,y) = \mathbb{P}(\tau(u) > x, R(x) = y)$.*

(a) *Then for* $y > 0$,

$$\phi(u;0,y) = \delta_u(y)\,, \tag{5.6.1}$$

and if $u/\mu, x\beta/\mu \in \{1,2,\ldots\}$, *then for* y *such that* $y/\mu \in \{1,2,\ldots\}$,

$$\phi(u;x,y) = \sum_{j=0}^{u/\mu} \mathrm{e}^{-\lambda\mu/\beta} \frac{(\lambda\mu/\beta)^j}{j!} \phi(u-(j-1)\mu; x-\mu/\beta, y)\,. \tag{5.6.2}$$

(b) *Let* $z = \lfloor u/\mu \rfloor + 1 - u/\mu$ *and assume that* $(u+\beta x)/\mu \in \{1,2,\ldots\}$. *Then for* y *such that* $y/\mu \in \{1,2,\ldots\}$,

$$\phi(u;x,y) = \sum_{j=0}^{\lfloor u/\mu \rfloor} \mathrm{e}^{-\lambda\mu z/\beta} \frac{(\lambda\mu z/\beta)^j}{j!} \phi(u-(j-z)\mu; x-\mu z/\beta, y)\,. \tag{5.6.3}$$

(c) *Let* $u, x \ge 0$ *be arbitrary and let* $z = (u+x\beta)/\mu - \lfloor (u+x\beta)/\mu \rfloor$. *Then, the survival probability* $\overline{\psi}(u;x)$ *can be obtained from*

$$\overline{\psi}(u;x) = \begin{cases} \displaystyle\sum_{k=1}^{\lfloor (u+x\beta)/\mu \rfloor} \phi(u; x - z\mu/\beta, k\mu) \mathbb{P}(N(z\mu) \le \beta k) & \textit{if } \beta x \ge z\mu, \\ \mathbb{P}(N(t)\mu > u) & \textit{otherwise.} \end{cases} \tag{5.6.4}$$

Proof Consider the counting process $\{\tilde{N}(t)\}$ where $\tilde{N}(t) = N(\mu\beta^{-1}t)$, i.e. $\{\tilde{N}(t)\}$ is a Poisson process with rate $\lambda\mu/\beta$. Using the transformation

$$\sum_{i=1}^{N(t)} U_i - \beta t = \mu \Bigg(\sum_{i=1}^{\tilde{N}(\beta\mu^{-1}t)} \mu^{-1} U_i - \beta\mu^{-1} t \Bigg)\,, \tag{5.6.5}$$

it is enough to prove the theorem in the case $\beta = \mu = 1$ which is assumed in the following. Furthermore, it is easy to see that then the multivariate survival function $\phi(u;x,y)$ can be analysed by considering the risk reserve process $\{R(t)\}$ at those times t only where $R(t)$ is an integer. If $R(t) = 0$, then ruin has occurred before t because $\mathbb{P}(N(t) - N(t-) > 0) = 0$. We subdivide the remaining part of the proof into three steps.
(a) Formula (5.6.1) is obvious. Let $x > 0$. There are j claims in the time interval $(0,1]$ with probability $\exp(-\lambda)\,\lambda^j/(j!)$. The risk reserve at time 1 is equal to $u + 1 - N(1)$. Ruin occurs in $(0,1]$ if and only if $u + 1 - N(1) \le 0$. Thus, (5.6.2) follows using the law of total probability and the independence properties of the compound Poisson model.
(b) Note that $R(x) = u + x - N(x) \in \mathbb{Z}$. The first epoch $t > 0$ where $R(t) \in \mathbb{Z}$ is z. In the time interval $(0,z]$, there are j claims with probability $\exp\{-\lambda z\}\,(\lambda z)^j/(j!)$. The risk reserve at time z is equal to $u + z - N(z)$. Ruin

occurs in $(0,z]$ if and only if $u+z-N(z)\le 0$. Now (5.6.3) follows by the same argument as used in case (a).
(c) Assume first that $x\ge z$. If ruin has not yet occurred, then the risk reserve at time $x-z$ is an integer between 1 and $u+(x-z)=\lfloor u+x\rfloor$. If $R(x-z)=k$ and ruin has not occurred up to time $x-z$, then ruin does not occur in the interval $(x-z,x]$ if and only if $N(x)-N(x-z)\le k$. This yields (5.6.4). If $x<z$ then $R(t)\notin \mathbb{N}$ for all $0\le t\le x$. Thus, ruin occurs if and only if $N(x)>u$. □

Remark Note that in (5.6.2) and (5.6.3) it is possible to sum over $y\in\{1,2,\ldots\}$. Thus, formulae (5.6.2) and (5.6.3) remain true if $\phi(u;x,y)$ is replaced by $\overline{\psi}(u;x)$. However, the use of the multivariate survival function $\phi(u;x,y)$ is necessary in order to be able to compute the finite-horizon ruin probability $\psi(u;x)=1-\overline{\psi}(u;x)$ in the case where $R(x)/\mu\notin\mathbb{Z}$. Indeed, if ruin has occurred before the next time $\tilde{x}$ after x where $R(\tilde{x})/\mu\in\mathbb{Z}$ but not before $\tilde{x}-\mu/\beta$, then we cannot decide whether ruin occurred before or after x.

5.6.2 Seal's Formulae

We now express the survival probability $\overline{\psi}(u;x)$ in terms of the distribution $F_{X(t)}$ of the aggregate claim amount $X(t)=\sum_{i=1}^{N(t)}U_i$. If U has density f_U then the distribution function of $X(t)$ can be expressed as

$$F_{X(t)}(x)=\mathrm{e}^{-\lambda x}+\int_0^x \tilde{f}_{X(t)}(y)\,\mathrm{d}y\,,\qquad x\ge 0\,,$$

where $\tilde{f}_{X(t)}(y)=\sum_{n=1}^{\infty}((\lambda t)^n/n!)\mathrm{e}^{-\lambda t}f_U^{*n}(y)$. Formulae (5.6.6) and (5.6.7) below are known in actuarial mathematics as *Seal's formulae*; see also Theorem 10.3.5.

Theorem 5.6.2 *Assume that* $\mathbb{P}(U>0)=1$.
(a) *Then, for initial risk reserve* $u=0$,

$$\overline{\psi}(0;x)=\frac{1}{\beta x}\mathbb{E}\,(R(x)_+)=\frac{1}{\beta x}\int_0^{\beta x}F_{X(x)}(y)\,\mathrm{d}y\,. \tag{5.6.6}$$

(b) *If* $u>0$ *and* U *has density* $f_U(y)$ *then*

$$\overline{\psi}(u;x)=F_{X(x)}(u+\beta x)-\beta\int_0^u\overline{\psi}(0,x-y)\tilde{f}_{X(y)}(u+\beta y)\,\mathrm{d}y\,. \tag{5.6.7}$$

Proof Assume first that $u=0$. Then

$$\begin{aligned}\overline{\psi}(0;x)&=\mathbb{P}\Big(\bigcap_{t\le x}\{R(t)\ge 0\}\Big)=\mathbb{E}\Big(\mathbb{P}\Big(\bigcap_{t\le x}\{X(t)\le\beta t\}\mid X(x)\Big)\Big)\\&=\mathbb{E}\Big(\Big(1-\frac{X(x)}{\beta x}\Big)_+\Big)=\frac{1}{\beta x}\mathbb{E}\,R(x)_+\end{aligned}$$

where we used the following property of the compound Poisson process $\{X(t)\}$. For all $x > 0$ and $y \in \mathbb{N}$, we have

$$\mathbb{P}\Big(\bigcap_{t\le x}\{X(t) \le \beta t\} \mid X(x) = y\Big) = \Big(1 - \frac{y}{\beta t}\Big)_+ ,$$

see also Theorem 10.3.5. Thus, the first part of (5.6.6) is proved. The second equality in (5.6.6) follows readily using integration by parts. Let now $u > 0$. We leave it to the reader to show that $\overline{\psi}(u;x)$ satisfies the integro-differential equation

$$\beta\frac{\partial}{\partial u}\overline{\psi}(u;x) - \frac{\partial}{\partial x}\overline{\psi}(u;x) + \lambda\int_0^u \overline{\psi}(u-y;x)\,\mathrm{d}F_U(y) - \lambda\overline{\psi}(u;x) = 0\,. \quad (5.6.8)$$

Let $\hat{L}_{\overline{\psi}}(s;x) = \int_0^\infty \mathrm{e}^{-su}\overline{\psi}(u;x)\,\mathrm{d}u$ denote the Laplace transform of $\overline{\psi}(u;x)$. Multiplying (5.6.8) by e^{-su} and integrating over $(0,\infty)$, we obtain

$$\beta(s\hat{L}_{\overline{\psi}}(s;x) - \overline{\psi}(0;x)) - \frac{\partial}{\partial x}\hat{L}_{\overline{\psi}}(s;x) + \lambda\hat{L}_{\overline{\psi}}(s;x)(\hat{l}_U(s) - 1) = 0\,,$$

as in the proof of Theorem 5.3.3. From the theory of ordinary differential equations (see, for example, Hille (1966)) we know that the general solution to this differential equation is

$$\hat{L}_{\overline{\psi}}(s;x) = \Big(c - \int_0^x \overline{\psi}(0;y)\beta\mathrm{e}^{-(\beta s+\lambda(\hat{l}_U(s)-1))y}\,\mathrm{d}y\Big)\mathrm{e}^{(\beta s+\lambda(\hat{l}_U(s)-1))x} \quad (5.6.9)$$

for some constant c. Putting $x = 0$ yields

$$c = \hat{L}_{\overline{\psi}}(s;0) = \int_0^\infty \mathrm{e}^{-su}\overline{\psi}(u;0)\,\mathrm{d}u = s^{-1}\,.$$

Note that $s^{-1}\exp\{\lambda t(\hat{l}_U(s) - 1)\}$ is the Laplace transform of $F_{X(t)}(v)$. Thus $\mathrm{e}^{\beta st}s^{-1}\exp\{\lambda t(\hat{l}_U(s) - 1)\}$ is the Laplace transform of $F_{X(t)}(v + \beta t)$ and $\mathrm{e}^{\beta st}\exp\{\lambda t(\hat{l}_U(s) - 1)\}$ is the Laplace transform of $\tilde{f}_{X(t)}(v + \beta t)$. For the second term in (5.6.9), this gives

$$\begin{aligned}
&\int_0^x \overline{\psi}(0;y)\beta\mathrm{e}^{(\beta s+\lambda(\hat{l}_U(s)-1))(x-y)}\,\mathrm{d}y \\
&\quad= \int_0^x\int_0^\infty \overline{\psi}(0;y)\beta\mathrm{e}^{-sv}\,\tilde{f}_{X(x-y)}(v + \beta(x-y))\,\mathrm{d}v\,\mathrm{d}y \\
&\quad= \int_0^\infty\int_0^x \overline{\psi}(0,x-y)\beta\tilde{f}_{X(y)}(v + \beta y)\,\mathrm{d}y\,\mathrm{e}^{-sv}\,\mathrm{d}v
\end{aligned}$$

which is the Laplace transform of the second term in (5.6.7). □

Note that $F_{X(x)}(u+\beta x)$ is the probability that $R(x) \geq 0$. This is, of course, necessary for survival. The second term in (5.6.7) must therefore be the probability that $R(x) \geq 0$ and ruin has occurred up to time x. If ruin occurs up to time x but $R(x) \geq 0$, then there must be a last (random) point Y before x for which $R(Y) = 0$. Then, formally we have

$$\begin{aligned}\mathbb{P}(Y \in \mathrm{d}y) &= \mathbb{P}\Big(R(\mathrm{d}y) = 0, \bigcap_{y<s\leq x} \{R(s) \geq 0\}\Big) \\ &= \mathbb{P}(X(y) \in u + \mathrm{d}(\beta y))\,\overline{\psi}(0; x-y) \\ &= \beta \tilde{f}_{X(y)}(u+\beta y)\overline{\psi}(0; x-y)\,\mathrm{d}y\,,\end{aligned}$$

because of the independence properties of the compound Poisson model. Thus the second term in (5.6.7) can be interpreted as conditioning on the last time y before x where $R(y) = 0$.

5.6.3 Exponential Claim Sizes

In this subsection we deal with the case of exponentially distributed claim sizes, that is $F_U = \mathrm{Exp}(\delta)$ for some $\delta > 0$. Before embarking on the detailed calculations, we introduce an auxiliary function that is intimately connected to the modified Bessel function $I_0(x)$ introduced in Section 2.2.1:

$$J(x) = \sum_{n=0}^{\infty} \frac{z^n}{n!n!} = I_0(2\sqrt{x})\,. \tag{5.6.10}$$

The next lemma collects some useful formulae involving the function $J(z)$.

Lemma 5.6.1 *The following relations are valid:*
(a) $xJ^{(2)}(x) + J^{(1)}(x) - J(x) = 0\,.$
(b) *For* $a, b, c \in \mathbb{R}$ *with* $a \neq 0$,

$$\int_0^c \mathrm{e}^{-bw} J^{(1)}(aw)\,\mathrm{d}w = \frac{1}{a}\,\mathrm{e}^{-bc} J(ac) - \frac{1}{a} + \frac{b}{a}\int_0^c \mathrm{e}^{-bw} J(aw)\,\mathrm{d}w\,.$$

(c) *For* $a, b, c \in \mathbb{R}$ *with* $b \neq 0$,

$$\int_0^c w\mathrm{e}^{-bw} J^{(1)}(aw)\,\mathrm{d}w = -\frac{c}{b}\mathrm{e}^{-bc}\, J^{(1)}(ac) + \frac{1}{b}\int_0^c \mathrm{e}^{-bw} J(aw)\,\mathrm{d}w\,.$$

(d) *For* $s > 0$ *and* $z \in \mathbb{R}$,

$$\int_0^{\infty} \mathrm{e}^{-sw} J(zw)\,\mathrm{d}w = s^{-1}\mathrm{e}^{z/s}\,.$$

Proof All statements of this lemma are elementary. The equation in (a) is the classical differential equation for the (modified) Bessel function. The formula in (b) is proved by integration by parts; (c) also follows from differentiation and an application of (a). Finally, (d) gives the Laplace transform of $J(x)$. Through Fubini's theorem we can write

$$\begin{aligned}\int_0^\infty \mathrm{e}^{-sw} J(zw)\,\mathrm{d}w &= \sum_{n=0}^\infty \frac{z^n}{n!n!}\int_0^\infty \mathrm{e}^{-sw} w^n\,\mathrm{d}w \\ &= \sum_{n=0}^\infty \frac{z^n}{n!n!}\frac{n!}{s^{n+1}} = s^{-1}\mathrm{e}^{z/s}\,.\end{aligned}$$ □

The main result of this section is the following formula for the finite-horizon ruin probability $\psi(u;x)$ in the case of exponentially distributed claim sizes.

Theorem 5.6.3 *Assume that $F_U(x) = 1 - \mathrm{e}^{-\delta x}$ for all $x \geq 0$. Then*

$$\psi(u;x) = 1 - \mathrm{e}^{-\delta u - (1+c)\lambda x} g(\delta u + c\lambda x, \lambda x)\,, \tag{5.6.11}$$

where $c = \delta\beta/\lambda$ and

$$g(z,\theta) = J(\theta z) + \theta J^{(1)}(\theta z) + \int_0^z \mathrm{e}^{z-v} J(\theta v)\,\mathrm{d}v - \frac{1}{c}\int_0^{c\theta} \mathrm{e}^{c\theta - v} J(zc^{-1}v)\,\mathrm{d}v\,. \tag{5.6.12}$$

The *proof* of Theorem 5.6.3 is subdivided into three steps. We first reformulate expression (4.2.8) for the distribution of the aggregate claim amount in terms of the function $J(x)$ introduced in (5.6.10). Then we derive formula (5.6.11) for the special case when the initial risk reserve $u = 0$. By using Seal's second formula (5.6.7) and a rather intricate argument we finally treat the general case.

In Section 4.2.2 we derived a compact formula for the distribution of the aggregate claim amount $X(x)$ when the claim number process is Poisson distributed with parameter λ and the claim size distribution is given by $F_U(y) = 1 - \mathrm{e}^{-\delta y}$. Then by (4.2.8)

$$F_{X(x)}(y) = \mathrm{e}^{-\lambda x}\left(1 + \sqrt{\lambda\delta x}\int_0^y \mathrm{e}^{-v\delta} I_1\left(2\sqrt{\lambda\delta v x}\right)\right)\frac{\mathrm{d}v}{\sqrt{v}}\,.$$

However it is easy to show that $I_1(z) = I_0^{(1)}(z)$ and hence we can rewrite the function I_1 in terms of $J^{(1)}$. This leads to the expression

$$F_{X(x)}(y) = \mathrm{e}^{-\lambda x}\left(1 + \lambda\delta x\int_0^y \mathrm{e}^{-v\delta} J^{(1)}(\lambda\delta v x)\,\mathrm{d}v\right). \tag{5.6.13}$$

Using Lemma 5.6.1b we can still give an other expression for $F_{X(x)}(y)$:

$$F_{X(x)}(y) = \mathrm{e}^{-\lambda x - \delta y} J(\lambda\delta xy) + \mathrm{e}^{-\lambda x} \int_0^{\delta y} \mathrm{e}^{-w} J(\lambda xw)\, \mathrm{d}w.$$

Note in particular that this equation can be written in the form

$$F_{X(x)}(y) = \nu(\lambda x, \delta y) \tag{5.6.14}$$

where

$$\nu(\theta, \eta) = \mathrm{e}^{-\theta-\eta} J(\theta\eta) + \mathrm{e}^{-\theta} \int_0^{\eta} \mathrm{e}^{-w} J(\theta w)\, \mathrm{d}w\,. \tag{5.6.15}$$

From (5.6.13) we also have an expression for $\tilde{f}_{X(x)}(y) = \partial/\partial y\; F_{X(x)}(y), y > 0$:

$$\tilde{f}_{X(x)}(y) = \lambda\delta x\, \mathrm{e}^{-(\lambda x + \delta y)} J^{(1)}(\lambda\delta xy)\,. \tag{5.6.16}$$

We now treat the case $u = 0$. To do that we apply Seal's formula (5.6.6) when calculating the survival probability $\overline{\psi}(0; x) = (\beta x)^{-1} \int_0^{\beta x} F_{X(x)}(y)\, \mathrm{d}y$, where the integrand has been obtained before. The calculations go as follows. Start from (5.6.13) to write

$$\begin{aligned} \overline{\psi}(0; x) &= \frac{1}{\beta x} \int_0^{\beta x} \mathrm{e}^{-\lambda x}\, \mathrm{d}y + \frac{\lambda\delta}{\beta} \mathrm{e}^{-\lambda x} \int_0^{\beta x} \int_0^{y} \mathrm{e}^{-v\delta} J^{(1)}\,(\lambda\delta v x)\, \mathrm{d}v\, \mathrm{d}y \\ &= \mathrm{e}^{-\lambda x} + \frac{\lambda\delta}{\beta} \mathrm{e}^{-\lambda x} \int_0^{\beta x} \mathrm{e}^{-\delta v} (\beta x - v) J^{(1)}(\lambda\delta x v)\, \mathrm{d}v\,. \end{aligned}$$

Now use statements (b) and (c) of Lemma 5.6.1 to arrive at the expression

$$\begin{aligned} &\mathrm{e}^{\lambda x(1+c)}\, \overline{\psi}(0; x) \\ &= J(\lambda\beta\delta x^2) + \lambda x J^{(1)}(\lambda\beta\delta x^2) + \frac{c-1}{c} \int_0^{c\lambda x} \mathrm{e}^{c\lambda x - w} J(\lambda x w)\, \mathrm{d}w \\ &= g(c\lambda x, \lambda x)\,. \end{aligned}$$

Thus,

$$\overline{\psi}(0; x) = \mathrm{e}^{-\lambda x(1+c)} g(c\lambda x, \lambda x) \tag{5.6.17}$$

where $c = \delta\beta/\lambda$ and $g(z, \theta)$ is defined by (5.6.12). This proves the statement of Theorem 5.6.3 for $u = 0$.

Before turning to the general case we derive an alternative expression for the quantity $g(z, \theta)$ introduced in (5.6.12). First note that for $c > 0$ and $a \in \mathbb{R}$

$$\int_0^{c} \mathrm{e}^{c-w} J(aw)\, \mathrm{d}w = c \int_0^{1} \mathrm{e}^{c(1-v)} J(acv)\, \mathrm{d}v$$

$$
\begin{aligned}
&= \sum_{s=0}^{\infty} \frac{a^s}{s!} \sum_{m=0}^{\infty} \frac{c^m}{m!} \int_0^1 (1-v)^m v^s \, dv = \sum_{s=0}^{\infty} \frac{a^s}{s!} \sum_{m=0}^{\infty} \frac{c^{m+s+1}}{(m+s+1)!} \\
&= \sum_{s=0}^{\infty} \frac{a^s}{s!} \sum_{n=s+1}^{\infty} \frac{c^n}{n!} = \sum_{n=1}^{\infty} \frac{c^n}{n!} \sum_{s=0}^{n-1} \frac{a^s}{s!} = \sum_{n=1}^{\infty} \frac{c^n}{n!} e_{n-1}(a) \,,
\end{aligned}
$$

where we used the abbreviation $e_m(z) = \sum_{s=0}^{m} z^s/(s!)$. We use this representation formula twice to rewrite $g(z,\theta)$ from (5.6.12) in a series expansion. This yields

$$
g(z,\theta) = \sum_{n=0}^{\infty} \frac{z^n \theta^n}{n!n!} + \theta \sum_{n=0}^{\infty} \frac{z^n \theta^n}{n!(n+1)!} + \sum_{n=1}^{\infty} \frac{z^n}{n!} e_{n-1}(\theta) - \frac{1}{c} \sum_{n=1}^{\infty} \frac{c^n \theta^n}{n!} e_{n-1}(z/c) \,,
$$

or equivalently

$$
g(z,\theta) = \sum_{n=0}^{\infty} \frac{z^n}{n!} e_{n+1}(\theta) - \frac{1}{c} \sum_{n=1}^{\infty} \frac{c^n \theta^n}{n!} e_{n-1}(z/c) \,. \tag{5.6.18}
$$

To simplify the notation a little, we introduce the function

$$
g_c(z) = c\, g(\xi, z/c) \,, \tag{5.6.19}
$$

which will be rather useful in the sequel. Using (5.6.18) we can write $g_c(z)$ in a power series with respect to z. As the calculations are tiresome but elementary we leave them to the reader as an exercise. We arrive at

$$
g_c(z) = \sum_{r=0}^{\infty} C_r(c)\, z^r \,, \tag{5.6.20}
$$

where

$$
C_r(c) = \begin{cases}
c & \text{if } r = 0, \\[1ex]
\dfrac{c^{1-m}}{m!m!} + (c-1) \displaystyle\sum_{n=0}^{m-1} \dfrac{c^{-n}}{n!(2m-n)!} & \text{if } r = 2m > 0, \\[1ex]
\dfrac{c^{-m}}{m!(m+1)!} + (c-1) \displaystyle\sum_{n=0}^{m} \dfrac{c^{-n}}{n!(2m-n+1)!} & \text{if } r = 2m+1.
\end{cases}
$$

We now turn to the case of an arbitrary initial reserve $u \geq 0$. Recall Seal's second formula (5.6.7) but written in the form

$$
F_{X(x)}(u+\beta x) - \overline{\psi}(u;x) = \beta \int_0^x \overline{\psi}(0; x-y) \tilde{f}_{X(y)}(u+\beta y) \, dy \,.
$$

On the right hand side we have expressions for both $\overline{\psi}(0; x-y)$ and $\tilde{f}_{X(y)}(u+\beta y)$. Now, using (5.6.16) and (5.6.17), after a little algebra we obtain the intermediate expression

$$\begin{aligned} & F_{X(x)}(u+\beta x) - \overline{\psi}(u;x) \\ = & \frac{\beta\delta\lambda}{c} \mathrm{e}^{\lambda x(1+c)-\delta u} \int_0^x g_c(c\lambda(x-y))\, y\, J^{(1)}(\lambda\delta y(u+\beta y))\, \mathrm{d}y \\ = & \mathrm{e}^{\lambda x(1+c)-\delta u} \int_0^{\lambda x} g_c(c(\lambda x - y))\, v\, J^{(1)}(v(\delta u + cv))\, \mathrm{d}v\,. \end{aligned}$$

Hence by (5.6.14), we can write that

$$\overline{\psi}(u;x) = \nu(\lambda x, \delta u + c\lambda x) - q(\lambda x, \delta u)\,, \tag{5.6.21}$$

where $\nu(\theta,\eta)$ is given by (5.6.15) and

$$q(\theta,\eta) = \mathrm{e}^{(-\eta+\theta(1+c))} \int_0^\theta v g_c(c(\theta - v)) J^{(1)}(v(\eta + cv))\, \mathrm{d}v\,. \tag{5.6.22}$$

It remains to rewrite $q(\theta,\eta)$ in such a way that we arrive at formula (5.6.11). We prepare this in the next lemma.

Lemma 5.6.2 *For the function $g_c(\xi)$ given in* (5.6.19) *and* (5.6.20), *respectively, the following identities hold:*
(a1) *For $\ell \in \mathbb{N}$ and $\lambda \in \mathbb{R}$*

$$\sum_{r=0}^{\ell} C_{2r}(\lambda) \frac{(-1)^{\ell-r}(r+\ell)!}{(\ell-r)!} \lambda^r = \lambda^{\ell+1}\,. \tag{5.6.23}$$

(a2) *For $\ell \in \mathbb{N}$ and $\lambda \in \mathbb{R}$*

$$\sum_{r=0}^{\ell} C_{2r+1}(\lambda) \frac{(-1)^{\ell-r}(r+\ell+1)!}{(\ell-r)!} \lambda^r = \lambda^{\ell+1}\,. \tag{5.6.24}$$

(b) *For $\lambda, w > 0$*

$$\int_0^w g_c(\lambda v) J(\lambda v(v-w))\, \mathrm{d}v = \mathrm{e}^{\lambda w} - 1\,. \tag{5.6.25}$$

(c) *For $\lambda, x, y > 0$*

$$\int_0^x g_\lambda(\lambda v) J\left((x-v)(y-\lambda v)\right)\, \mathrm{d}v = \mathrm{e}^{\lambda x} \int_0^{\lambda x} \mathrm{e}^{-v} J(\frac{y}{\lambda} v)\, \mathrm{d}v\,. \tag{5.6.26}$$

Proof (a1) For $\ell = 0$, the proof is trivial. For $\ell > 0$, we introduce the explicit expression of the quantities $C_{2r}(\lambda)$. The relation to be proved is then

$$\lambda \sum_{r=0}^{\ell} \frac{(-1)^{\ell-r}(r+\ell)!}{r!r!(\ell-r)!} + (\lambda-1) \sum_{r=1}^{\ell} \frac{(-1)^{\ell-r}(r+\ell)!}{(\ell-r)!} \sum_{s=0}^{r-1} \frac{\lambda^{r-s}}{s!(2r-s)!} = \lambda^{\ell+1} .$$

The first sum can be rewritten in the form

$$\begin{aligned} \sum_{r=0}^{\ell} \frac{(-1)^{\ell-r}(r+\ell)!}{r!r!(\ell-r)!} &= \sum_{r=0}^{\ell} (-1)^{\ell-r} \binom{\ell+r}{r} \binom{\ell}{r} \\ &= \sum_{r=0}^{\ell} (-1)^{\ell} \binom{-\ell-1}{r} \binom{\ell}{\ell-r} = (-1)^{\ell} \binom{-1}{\ell} = 1 . \end{aligned}$$

Using this outcome above, we still need to prove the identity

$$\sum_{r=1}^{\ell} \sum_{n=0}^{r-1} \frac{(-1)^{\ell-r}(r+\ell)!}{(\ell-r)!n!(2r-n)!} \lambda^{r-n} = \frac{\lambda^{\ell+1}-\lambda}{\lambda-1} .$$

Rearrange the left-hand side into a power series in λ to obtain

$$\sum_{m=1}^{\ell} \lambda^m \sum_{r=m}^{\ell} \frac{(-1)^{\ell-r}(r+\ell)!}{(\ell-r)!(r-m)!(r+m)!} = \frac{\lambda^{\ell+1}-\lambda}{\lambda-1} = \sum_{m=1}^{\ell} \lambda^m .$$

Put $r = m + j$. Now, it only remains to prove the identity

$$\sum_{r=m}^{\ell} \frac{(-1)^{\ell-r}(r+\ell)!}{(\ell-r)!(r-m)!(r+m)!} = \sum_{j=0}^{\ell-m} \frac{(-1)^{\ell-j-m}(m+\ell+j)!}{(\ell-m-j)!j!(2m+j)!} = 1 .$$

Replace $\ell = m + k$. Then the expression turns out to be

$$\sum_{j=0}^{k} \frac{(-1)^{k-j}(k+2m+j)!}{j!(k-j)!(2m+j)!} = \sum_{j=0}^{k} (-1)^{k-j} \binom{k+2m+j}{k} \binom{k}{j} = 1$$

which follows from a classical identity for the binomial coefficients, i.e. when $0 \le k \le n$ and $t \in \mathbb{R}$

$$\sum_{j=0}^{k} (-1)^j \binom{k}{j} \binom{t+j}{n} = (-1)^k \binom{t}{n-k} .$$

The proof of (a2) is left to the reader as an exercise. To show (b), we apply expression (5.6.19) and the series expansion of the function $J(x)$ given in

(5.6.10). Then we reduce the remaining integral to a Beta function. More explicitly,

$$
\begin{aligned}
&\int_0^w g_c(\lambda v) J(\lambda v(v-w))\,\mathrm{d}v \\
&\quad = \sum_{r=0}^{\infty} C_r(\lambda) \sum_{n=0}^{\infty} \frac{\lambda^{r+n}}{n!n!} \int_0^w v^{r+n}(v-w)^n\,\mathrm{d}v \\
&\quad = \sum_{r=0}^{\infty} C_r(\lambda) \sum_{n=0}^{\infty} \frac{(-1)^n \lambda^{r+n}}{n!n!} w^{r+2n+1} \int_0^1 t^{r+n}(1-t)^n\,\mathrm{d}t \\
&\quad = \sum_{r=0}^{\infty} C_r(\lambda) \sum_{n=0}^{\infty} \frac{(-1)^n \lambda^{r+n}(r+n)!}{n!(r+2n+1)!} w^{r+2n+1}.
\end{aligned}
$$

To continue, we are forced to split the summation over r into even and odd values. We then rewrite both terms by inverting the summations. For the even terms we obtain

$$
\begin{aligned}
&\sum_{r=0}^{\infty} C_{2r}(\lambda) \sum_{n=0}^{\infty} \frac{(-1)^n \lambda^{2r+n}(2r+n)!}{n!(2r+2n+1)!} w^{2r+2n+1} \\
&\quad = \sum_{r=0}^{\infty} C_{2r}(\lambda) \sum_{\ell=r}^{\infty} \frac{(-1)^{\ell-r} \lambda^{r+\ell}(\ell+r)!}{(\ell-r)!(2\ell+1)!} w^{2\ell+1} \\
&\quad = \sum_{\ell=0}^{\infty} \frac{w^{2\ell+1}}{(2\ell+1)!} \lambda^{\ell} \sum_{r=0}^{\ell} C_{2r}(\lambda) \frac{(-1)^{\ell-r}(r+\ell)!}{(\ell-r)!} \lambda^r \\
&\quad = \sum_{\ell=0}^{\infty} \frac{(w\lambda)^{2\ell+1}}{(2\ell+1)!},
\end{aligned}
$$

by an appeal to (5.6.23). We leave it to the reader to similarly prove that

$$
\sum_{r=0}^{\infty} C_{2r+1}(\lambda) \sum_{n=0}^{\infty} \frac{(-1)^n \lambda^{2r+n+1}(2r+n+1)!}{n!(2r+2n+2)!} w^{2r+2n+2} = \sum_{\ell=0}^{\infty} \frac{(w\lambda)^{2\ell+2}}{(2\ell+2)!},
$$

with a reference to formula (5.6.24). This then proves the required expression in (b). Next we show statement (c). As both sides of expression (5.6.26) are convolutions, the identity will be proved if the Laplace transforms of both sides coincide. Let $s > \lambda$. On the left-hand side we have

$$
\begin{aligned}
&\int_0^{\infty} \mathrm{e}^{-sx} \int_0^x g_\lambda(\lambda z) J\left((x-z)(y-\lambda z)\right)\,\mathrm{d}z\,\mathrm{d}x \\
&\quad = \int_0^{\infty} g_\lambda(\lambda z) \int_z^{\infty} \mathrm{e}^{-sx} J\left((x-z)(y-\lambda z)\right)\,\mathrm{d}x\,\mathrm{d}z
\end{aligned}
$$

$$
\begin{aligned}
&= \int_0^\infty \mathrm{e}^{-sz} g_\lambda(\lambda z) \int_0^\infty \mathrm{e}^{-sv} J\left(v(y-\lambda z)\right)\,\mathrm{d}v\,\mathrm{d}z \\
&= \int_0^\infty \mathrm{e}^{-sz}\mathrm{e}^{(y-\lambda z)/s} g_\lambda(\lambda z)\,\mathrm{d}z
\end{aligned}
$$

where we used statement (d) of Lemma 5.6.1. The calculation of the Laplace transform of the right hand side is even easier and leads to $s^{-1}(s-\lambda)^{-1}\lambda \mathrm{e}^{y/s}$. Hence we need to prove the equality

$$
\int_0^\infty \mathrm{e}^{-(s+\lambda s^{-1})v}\mathrm{e}^{(y-\lambda v)/s} g_\lambda(\lambda v)\,\mathrm{d}v \;=\; \frac{\lambda}{s-\lambda}\,. \tag{5.6.27}
$$

However this equation follows from (5.6.25) by taking Laplace transforms. We leave it to the reader to verify this. □

To finish the proof of Theorem 5.6.3, take partial derivatives with respect to y in (5.6.26) to obtain

$$
\int_0^x g_\lambda(\lambda v)(x-v)J^{(1)}\left((x-v)(y-\lambda v)\right)\,\mathrm{d}v = \frac{1}{\lambda}\mathrm{e}^{\lambda x}\int_0^{\lambda x} \mathrm{e}^{-v} v J^{(1)}\big(\frac{y}{\lambda}v\big)\,\mathrm{d}v\,.
$$

Identify this expression with the right-hand side of (5.6.22) to find

$$
\begin{aligned}
\mathrm{e}^{\eta+(1+c)\theta} q(\eta,\theta) &= \frac{1}{c}\mathrm{e}^{c\theta}\int_0^{c\theta} \mathrm{e}^{-v} v J^{(1)}\Big(v\big(\theta+\frac{\eta}{c}\big)\Big)\,\mathrm{d}v \\
&= -\theta J^{(1)}(\theta(\eta+c\theta)) + \frac{1}{c}\mathrm{e}^{c\theta}\int_0^{c\theta} \mathrm{e}^{-v} J\big(\frac{\eta+c\theta}{c}v\big)\,\mathrm{d}v
\end{aligned}
$$

where we used again statement (b) of Lemma 5.6.1. Combining the latter expression with formulae (5.6.15) and (5.6.21) the result of Theorem 5.6.3 follows.

An alternative representation formula for the finite-horizon ruin probability $\psi(u;x)$ is given by the following result.

Theorem 5.6.4 *Assume that $F_U(x) = 1-\mathrm{e}^{-\delta x}$ for all $x \ge 0$. Then*

$$
\psi(u;x) = c^{-1}\mathrm{e}^{-(c-1)c^{-1}\delta u} - \mathrm{e}^{-\delta u-(1+c)x}\pi^{-1}\int_0^\pi g(\delta u, \lambda x, y)\,\mathrm{d}y\,, \tag{5.6.28}
$$

where $c = \delta\beta/\lambda$ and

$$
g(w,\theta,y) = 2\sqrt{c}\frac{\mathrm{e}^{(2\sqrt{c}\theta+w/\sqrt{c})\cos y}}{1+c-2\sqrt{c}\cos y}\big(\sin y\,\sin(y+\frac{w}{\sqrt{c}}\sin y)\big)\,. \tag{5.6.29}
$$

Proof We only give a sketch of the proof. The crucial link between the two expressions for the function $\psi(u;x)$ is provided by the following integral

$$\begin{aligned} J(b,v,t) &= \frac{1}{\pi}\int_0^\pi \frac{\mathrm{e}^{b(2t+v)\cos y}}{1+b^2-2b\cos y}\sin y\,\sin(y+vb\sin y)\,\mathrm{d}y \\ &= \frac{b}{2}\mathrm{e}^{t+b^2(t+v)} - \frac{1}{2b}\mathrm{e}^{v+(b^2+1)t} \qquad (5.6.30) \\ &- \frac{b}{2}\sum_{n=1}^{\infty}\frac{t^n}{n!}e_{n-1}(b^2(t+v)) + \frac{1}{2b}\sum_{n=0}^{\infty}\frac{(t+v)^n}{n!}e_{n+1}(b^2t)\,. \end{aligned}$$

When proving this formula, we use the following expansion:

$$\mathrm{e}^{\eta\sigma\cos\theta}\sin(\alpha\theta+\varepsilon\eta(\rho-\sigma)\sin\theta) = \sum_{k=0}^{\infty}\frac{(\eta\sigma)^k}{k!}\sin((\alpha-\varepsilon k)\theta+\varepsilon\eta\rho\sin\theta)\,, \qquad (5.6.31)$$

where α,η,θ,ρ and ξ are arbitrary real numbers and $\varepsilon=\pm1$. Formula (5.6.31) is most easily proved by relying on the complex form of the sinus function. For it can easily be checked that

$$\begin{aligned} \mathrm{e}^{\eta\xi\cos\theta}\sin(\alpha\theta+\varepsilon\eta(\rho-\xi)\sin\theta) &= \frac{1}{2\mathrm{i}}(\exp(\mathrm{i}(\alpha\theta+\varepsilon\eta\rho\sin\theta)+\eta\xi\mathrm{e}^{-\mathrm{i}\varepsilon\theta}) \\ &\quad -\exp(-\mathrm{i}(\alpha\theta+\varepsilon\eta\rho\sin\theta)+\eta\xi\mathrm{e}^{\mathrm{i}\varepsilon\theta}))\,. \end{aligned}$$

Write twice that

$$\exp\left(\eta\xi\mathrm{e}^{\mathrm{i}\varepsilon\theta}\right) = \sum_{k=0}^{\infty}\frac{(\eta\xi)^k}{k!}\mathrm{e}^{\mathrm{i}\varepsilon k\theta}$$

and collect the coefficients of $(\eta\xi)^k$ to arrive at the requested formula (5.6.31). In proving (5.6.30) apply (5.6.31) with the choice $\alpha=1,\varepsilon=-1,\eta=b,\rho=t$ and $\xi=t+v$ to get

$$J(b,v,t) = \sum_{n=0}^{\infty}\frac{(b(t+v))^n}{n!}\frac{1}{\pi}\int_0^\pi\frac{\sin y\,\mathrm{e}^{bt\cos y}\sin((n+1)y-bt\sin y)}{1+b^2-2b\cos y}\,\mathrm{d}y\,.$$

Apply (5.6.31) again but now with the choice $\alpha=n+1,\varepsilon=1,\eta=b,\rho=0$ and $\xi=t$. We get now

$$J(b,v,t) = \sum_{n=0}^{\infty}\sum_{k=0}^{\infty}\frac{(b(t+v))^n}{n!}\frac{(bt)^k}{k!}\frac{1}{\pi}\int_0^\pi\frac{\sin y\,\sin((n+1-k)y)}{1+b^2-2b\cos y}\,\mathrm{d}y\,.$$

The remaining integral can be found, for example, via residual calculus:

$$\frac{1}{\pi}\int_0^\pi\frac{\sin y\,\sin(my)}{1+b^2-2b\cos y}\,\mathrm{d}y = \begin{cases} 2^{-1}b^m & \text{if } m\ge 1,\\ 0 & \text{if } m=0,\\ -2^{-1}b^{-m} & \text{if } m\le -1.\end{cases}$$

After a number of steps we arrive at the expression

$$J(b,v,t) = \frac{b}{2}\sum_{n=0}^{\infty}\frac{(b^2(t+v))^n}{n!}e_n(t) - \frac{1}{2b}\sum_{n=0}^{\infty}\frac{(t+v)^n}{n!}\sum_{k=n+2}^{\infty}\frac{(b^2t)^k}{k!}\,.$$

In the first sum we use the obvious identity

$$\sum_{n=0}^{\infty}\frac{x^n}{n!}e_n(y) + \sum_{n=1}^{\infty}\frac{y^n}{n!}e_{n-1}(x) = \mathrm{e}^{x+y}$$

with $x = b^2(t+v)$ and $y = t$. In the second sum we can write

$$\sum_{k=n+2}^{\infty}\frac{(b^2t)^k}{k!} = \mathrm{e}^{b^2t} - \mathrm{e}_{n+1}(b^2t)\,.$$

This proves (5.6.30). The link with the function $g(z,\theta)$ from (5.6.12) is very easy and reads as follows:

$$g(z,\theta) = \frac{2}{\sqrt{c}}J(c^{-1/2}, z - c\theta, c\theta) + \mathrm{e}^{z+\theta} - c^{-1}\mathrm{e}^{(z/c)+c\theta}\,.$$

Substitution of $z = \delta u + c\lambda x$ and $\theta = \lambda x$ leads to the desired formula. □

By way of conclusion we like to point out that (5.6.18) provides a third expression for the ruin probability $\psi(u;x)$. Furthermore, using formula (5.6.28) for the finite-horizon ruin probability $\psi(u;x)$ in the compound Poisson model with exponentially distributed claim sizes, we also get a De Vylder approximation to $\psi(u;x)$:

$$\psi_{\mathrm{app}}(u;x) = \frac{\lambda'}{\delta'\beta'}\mathrm{e}^{-(\delta'-\lambda'/\beta')u} - \frac{1}{\pi}\int_0^{\pi} g(y)\,\mathrm{d}y \qquad (5.6.32)$$

where β', δ', λ' are given in (5.4.21) and (5.4.22), and $g(y)$ is given in (5.6.29).

Bibliographical Notes. Formulae (5.6.1) and (5.6.2) as well as Theorem 5.6.2 can be found in Seal (1974). Note however that in the context of queueing theory, Theorem 5.6.2 goes back to Takács (1962). Other recursive methods to calculate the finite-horizon ruin probabilities $\psi(u;x)$ can be found in Dickson and Waters (1991) and Stanford and Stroinski (1994). Under the assumption that the claim sizes are integer-valued, Picard and Lefevre (1997) show that $\psi(u;x)$ can be expressed in terms of generalized Appell polynomials. The proof of Theorem 5.6.4 is inspired by Arfwedson (1950), whose proof relies on the solution to a partial differential equation; see also Arfwedson (1955). Our method is elementary. A proof of Theorem 5.6.4 which uses the queueing-theoretic approach is given in Asmussen (1984). For normal-type approximations to $\psi(u;x)$, see Asmussen (1984), Malinovskii (1994) and von Bahr (1974).

CHAPTER 6

Renewal Processes and Random Walks

In this chapter, we consider the risk reserve process $\{R(t), t \geq 0\}$ in continuous time as it has been introduced in Section 5.1.4. Unless otherwise stated, we assume that the claim counting process $\{N(t)\}$ is a renewal counting process, i.e. $\{N(t)\}$ is governed by a sequence of independent and identically distributed inter-occurrence times $\{T_n\}$ with a common distribution F_T. Furthermore, the sequence $\{U_n\}$ of claim sizes consists of independent and identically distributed random variables with distribution F_U and is independent of $\{T_n\}$. This model is called the *Sparre Andersen model.* A case of particular interest is the classical compound Poisson model which was studied in Chapter 5; there, $\{N(t)\}$ was a Poisson process with intensity λ. We recall that then $F_T(x) = 1 - \mathrm{e}^{-\lambda x}$, $x \geq 0$.

6.1 RENEWAL PROCESSES

6.1.1 Definition and Elementary Properties

Let $T_1, T_2, \ldots$ be a sequence of nonnegative, independent and identically distributed random variables. The sequence $\{\sigma_n, n \in \mathbb{N}\}$ with $\sigma_0 = 0$ and $\sigma_n = T_1 + \ldots + T_n$ for $n = 1, 2, \ldots$ is called a *renewal point process* and σ_n is the n-th *renewal epoch.*

To avoid trivialities we assume that the inter-renewal distances $T_1, T_2, \ldots$ are not concentrated at zero, that is $\mathbb{P}(T = 0) < 1$. As in Chapter 5 we may think about the $T_1, T_2, \ldots$ as inter-occurrence times of claims. Another and mathematically equivalent description of the renewal process $\{\sigma_n\}$ is given in terms of the *renewal counting process* $\{N(t), t \geq 0\}$, where

$$N(t) = \sum_{n=1}^{\infty} \mathbb{I}(\sigma_n \leq t) \tag{6.1.1}$$

is the number of renewal epochs in the interval $(0,t]$. The equivalence of the two processes $\{\sigma_n\}$ and $\{N(t)\}$ follows from the fact that

$$N(t) = n \qquad \text{if and only if} \qquad \{\sigma_n \le t < \sigma_{n+1}\}. \tag{6.1.2}$$

Note that $\{N(t), t \ge 0\}$ is a continuous-time process with piecewise constant right-continuous trajectories.

Suppose that the generic inter-renewal distance T has distribution F and expectation μ. In some cases, F will be a *defective distribution*, which for the distribution function of the nonnegative random variable T means that $\lim_{x\to\infty} F(x) < 1$. That is, T can be infinite with the positive probability $1 - F(\infty)$, where the symbol $F(\infty)$ is defined by $F(\infty) = \lim_{x\to\infty} F(x)$. The resulting renewal process is called *terminating*. We will indeed see below that $N(\infty) = \lim_{t\to\infty} N(t)$ is finite with probability 1 and $\sigma_{N(\infty)}$ is a geometric compound.

Theorem 6.1.1 (a) *If $F(\infty) = 1$, then the trajectories of $\{N(t), t \ge 0\}$ are increasing to infinity with probability 1. Moreover, with probability 1*

$$\lim_{t\to\infty} \frac{N(t)}{t} = \begin{cases} \mu^{-1} & \text{if } \mu < \infty, \\ 0 & \text{if } \mu = \infty. \end{cases} \tag{6.1.3}$$

(b) *For a terminating renewal process, $N(\infty)$ is finite with probability 1. Moreover, $\sigma_{N(\infty)}$ is a geometric compound with characteristics $(F(\infty), \tilde{F})$, where $\tilde{F}(x) = F(x)/F(\infty)$.*

Proof We first show part (a). The limit $\lim_{t\to\infty} N(t)$ exists in $(0,\infty]$ because the trajectories of the process $\{N(t)\}$ are increasing. Since $\mathbb{P}(\sigma_n < \infty) = 1$ for every $n \in \mathbb{N}$, we get from (6.1.2) that $\lim_{t\to\infty} N(t) = \infty$ with probability 1. From the strong law of large numbers we get that $\lim_{n\to\infty} n^{-1}\sigma_n = \mu > 0$ with probability 1. Consequently, $\lim_{t\to\infty} N(t)^{-1}\sigma_{N(t)} = \mu$ with probability 1. From (6.1.2) we have $\sigma_{N(t)} \le t < \sigma_{N(t)+1}$. Thus, dividing by $N(t)$ we have

$$\frac{\sigma_{N(t)}}{N(t)} \le \frac{t}{N(t)} \le \frac{\sigma_{N(t)+1}}{N(t)+1}\,\frac{N(t)+1}{N(t)}$$

and (6.1.3) follows. To prove part (b), note that $N(\infty) = \lim_{t\to\infty} N(t) = N(\infty) = \min\{n : T_n = \infty\} - 1$, i.e. $N(\infty)$ has a geometric distribution with $\mathbb{P}(N(\infty) = 0) = 1 - F(\infty)$. By the law of total probability we now easily get that the random variable $\sigma_{N(\infty)}$ is a geometric compound with characteristics $(F(\infty), \tilde{F})$. $\square$

In Chapter 4 we discussed methods to determine the distribution of the aggregate claim amount $\sum_{i=1}^{N(t)} U_i$. In connection with this it is important to know the distribution of the claim number $N(t)$. Unfortunately, in many cases it is impossible to determine this distribution explicitly. The following *central limit theorem* gives a possible approximation to the claim number distribution, provided that t is large enough.

Theorem 6.1.2 *If* $0 < \operatorname{Var} T < \infty$, *then for each* $x \in \mathbb{R}$

$$\lim_{t\to\infty} \mathbb{P}\Big(\frac{N(t) - t\mu^{-1}}{\sqrt{ct}} \le x\Big) = \Phi(x) \tag{6.1.4}$$

where $c = \mu^{-3}\operatorname{Var} T$ *and* $\Phi(x)$ *is the standard normal distribution function.*

Proof By the usual central limit theorem for sums of independent and identically distributed random variables we have

$$\lim_{n\to\infty} \mathbb{P}\Big(\frac{\sigma_n - n\mu}{\sqrt{n\operatorname{Var} T}} \le x\Big) = \Phi(x)\,, \tag{6.1.5}$$

uniformly in $x \in \mathbb{R}$. Furthermore, using (6.1.2) we can write

$$\begin{aligned}\mathbb{P}\Big(\frac{N(t) - t\mu^{-1}}{\sqrt{ct}} \le x\Big) &= \mathbb{P}(N(t) \le x\sqrt{ct} + t\mu^{-1})\\ &= \mathbb{P}(\sigma_{m(t)+1} \ge t) = \mathbb{P}\Big(\frac{\sigma_{m(t)+1} - \mu(m(t)+1)}{\sqrt{(m(t)+1)\operatorname{Var} T}} \ge \frac{t - \mu(m(t)+1)}{\sqrt{(m(t)+1)\operatorname{Var} T}}\Big)\,,\end{aligned}$$

where $m(t) = \lfloor x\sqrt{ct} + t\mu^{-1}\rfloor$. Since $\lim_{t\to\infty} m(t) = \infty$, it suffices to show that

$$\lim_{t\to\infty} \frac{t - \mu m(t)}{\sqrt{m(t)\operatorname{Var} T}} = -x\,,$$

bearing in mind that $1 - \Phi(-x) = \Phi(x)$ and that the convergence in (6.1.5) is uniform in $x \in \mathbb{R}$. Note that $m(t) = x\sqrt{ct} + t\mu^{-1} + \varepsilon(t)$, where $0 \le |\varepsilon(t)| < 1$. Thus

$$\frac{t - \mu m(t)}{\sqrt{m(t)\operatorname{Var} T}} = \frac{t - \mu x\sqrt{ct} - t - \mu\varepsilon(t)}{\sqrt{m(t)\operatorname{Var} T}} \longrightarrow_{t\to\infty} -x\,. \qquad \square$$

The following *law of small numbers* approximates the distribution of the number of claims reported to a reinsurer in the interval $(0,t]$ under the stop-loss contract with retention level a. As in Theorem 6.1.2 the result gives an asymptotic approximation which works well if a and t are large enough.

Theorem 6.1.3 *Assume that* $\mu < \infty$. *For each* $a > 0$, *consider the claim counting process* $\{N_a(t), t \ge 0\}$ *with* $N_a(t) = \sum_{i=1}^{N(t)} \mathbb{I}(U_i > b)$. *Let* $a(t)$ *be a function such that* $\mu^{-1}\overline{F}_U(a(t))t \to \lambda$ *as* $t \to \infty$, *for some* $\lambda > 0$. *Then the random variable* $N_{a(t)}(t)$ *is asymptotically* $\operatorname{Poi}(\lambda)$*-distributed, i.e. for each*

$$\lim_{t\to\infty} \mathbb{P}\left(N_{a(t)}(t) = k\right) = \frac{\lambda^k}{k!}\,\mathrm{e}^{-\lambda}\,, \qquad k \in \mathbb{N}. \tag{6.1.6}$$

Proof We use Theorem 6.1.1a and the compound Poisson approximation considered in Section 4.6. Note that

$$\begin{aligned}
\mathbb{P}(N_{a(t)}(t)=k) &= \sum_{n=0}^{\infty}\mathbb{P}\Big(\sum_{i=1}^{n}\mathbb{I}(U_i>a(t))=k\Big)\mathbb{P}(N(t)=n) \\
&\le \sum_{n=0}^{\infty}\Big|\mathbb{P}\Big(\sum_{i=1}^{n}\mathbb{I}(U_i>a(t))=k\Big) \\
&\qquad -\exp\left(-n\overline{F}_U(a(t))\right)\frac{\left(-n\overline{F}_U(a(t))\right)^k}{k!}\Big|\,\mathbb{P}(N(t)=n) \\
&\qquad +\sum_{n=0}^{\infty}\exp\left(-n\overline{F}_U(a(t))\right)\frac{\left(-n\overline{F}_U(a(t))\right)^k}{k!}\mathbb{P}(N(t)=n)\,.
\end{aligned}$$

Thus, putting $\theta_i=\overline{F}_U(a(t))$ and $G=\delta_1$, Theorem 4.6.2 gives

$$\mathbb{P}(N_{a(t)}(t)=k)\le \overline{F}_U(a(t))+\mathbb{E}\Big(\frac{\left(N(t)\overline{F}_U(a(t))\right)^k}{k!\exp\left(N(t)\overline{F}_U(a(t))\right)}\Big).$$

Analogously, we get

$$-\overline{F}_U(a(t))+\mathbb{E}\Big(\frac{\left(N(t)\overline{F}_U(a(t))\right)^k}{k!\exp\left(N(t)\overline{F}_U(a(t))\right)}\Big)\le \mathbb{P}(N_{a(t)}(t)=k)\,.$$

Now, using (6.1.3), the dominated convergence theorem gives (6.1.6). □

6.1.2 The Renewal Function; Delayed Renewal Processes

Equation (6.1.3) motivates the term *intensity* of the renewal process for the quantity μ^{-1}. The mean number of renewals $H(t)=\mathbb{E}N(t)$ as a function of t is called the (zero-deleted) *renewal function* of $\{N(t)\}$. Since $N(t)=\sum_{n=1}^{\infty}\mathbb{I}(\sigma_n\le t)$ we have

$$H(t)=\mathbb{E}\sum_{n=1}^{\infty}\mathbb{I}(\sigma_n\le t)=\sum_{n=1}^{\infty}\mathbb{E}\,\mathbb{I}(\sigma_n\le t)=\sum_{n=1}^{\infty}F^{*n}(t)\,. \tag{6.1.7}$$

Sometimes we need to include the renewal epoch at 0. Consider the random measure N given by $N(B)=\sum_{n=0}^{\infty}\mathbb{I}(\sigma_n\in B)$, $B\in\mathcal{B}(\mathbb{R}_+)$, where $N(B)$ counts the number of renewal epochs (including zero) in the set B. The *renewal measure* H is then defined by

$$H(B)=\mathbb{E}N(B)=\sum_{n=0}^{\infty}F^{*n}(B)\,,\qquad B\in\mathcal{B}(\mathbb{R}_+)\,. \tag{6.1.8}$$

We will write $N_0(t) = N([0,t])$ and $H_0(t) = H([0,t])$ instead of $N(t) = N((0,t])$ and $H(t) = H((0,t])$. From Theorem 6.1.4 we get that $H(B)$ is finite for bounded $B \in \mathcal{B}(\mathbb{R}_+)$. For some distributions, (6.1.7) can be used to derive the renewal function $H(t)$ explicitly.

Examples 1. If the inter-renewal times $T_1, T_2, \ldots$ are constant and equal to λ^{-1}, then (6.1.7) yields $H(t) = \lfloor \lambda t \rfloor \le \lambda t$.

2. If the inter-renewal times $T_1, T_2, \ldots$ are exponentially distributed with parameter λ, then the corresponding renewal counting process $\{N(t),\ t \ge 0\}$ is a Poisson process. Note that in this case the random variables $N(t)$ are Poisson distributed with parameter λt. Hence $H(t) = \lambda t$ for all $t \ge 0$ as shown in Section 5.2.

3. For another example where the renewal function can be determined explicitly, take $T_1, T_2, \ldots$ Bernoulli distributed with generic variable T such that $\mathbb{P}(T = 1) = p = 1 - \mathbb{P}(T = 0)$. In this case it is convenient to have zero counted. Indeed, at each time $t \in \mathbb{N}$ we can have multiple renewal epochs forming batches $Y_0, Y_1, \ldots$ that are independent and follow a zero truncated geometrical distribution with $\mathbb{P}(Y = k) = p(1-p)^{k-1}$ for $k = 1, 2, \ldots$. Thus $N([0,t]) = \sum_{i=0}^{\lfloor t \rfloor} Y_i$ and

$$H_0(t) = \lfloor t \rfloor \mathbb{E}\, Y \le \frac{tp}{1-p}\,. \tag{6.1.9}$$

Theorem 6.1.4 (a) *If $F(\infty) = 1$, then $H(t) = H_0(t) - 1 < \infty$ for all $t < \infty$. Moreover, $H(t)$ is increasing and $\lim_{t\to\infty} H(t) = \infty$.*
(b) *For a terminating renewal process, $H(\infty) = \lim_{t\to\infty} H(t) < \infty$.*

Proof Let $F(\infty) = 1$. Without essential loss of generality suppose that $F(1) > 0$ and define the Bernoulli distributed random variables T_i^{ber} by

$$T_i^{\mathrm{ber}} = \begin{cases} 0 & \text{if } T_i \le 1\,, \\ 1 & \text{if } T_i > 1\,. \end{cases}$$

Since $T_i^{\mathrm{ber}} \le T_i$, we have for the corresponding renewal counting process $\{N^{\mathrm{ber}}([0,t]),\ t \ge 0\}$ that $N([0,t]) \le N^{\mathrm{ber}}([0,t])$ for each $t \ge 0$. Hence, from (6.1.9) we get $H_0(t) \le H_0^{\mathrm{ber}}(t) < \infty$ for all $t < \infty$. Part (b) follows from Theorem 6.1.1 using the monotone convergence theorem. $\square$

From Theorem 6.1.4 we get that $H(B)$ is finite for bounded sets $B \in \mathcal{B}(\mathbb{R}_+)$. Since the probability measures F^{*n} are concentrated on $[0,\infty)$, the measure H is also concentrated on $[0,\infty)$.

With the exception of a few special cases, the renewal function $H(t)$ cannot be given in simple form. However the Laplace–Stieltjes transform $\hat{l}_H(s) = \int_0^\infty \mathrm{e}^{-sx}\, \mathrm{d}H(x)$ of H can always be expressed in terms of the Laplace–Stieltjes transform of F.

Theorem 6.1.5 *For all $s > 0$,*

$$\hat{l}_H(s) = \frac{\hat{l}_F(s)}{1 - \hat{l}_F(s)}\,. \tag{6.1.10}$$

Proof From (6.1.7) we get

$$\hat{l}_H(s) = \int_0^\infty \mathrm{e}^{-sx}\,\mathrm{d}\Big(\sum_{n=1}^\infty F^{*n}(x)\Big) = \sum_{n=1}^\infty (\hat{l}_F(s))^n = \frac{\hat{l}_F(s)}{1-\hat{l}_F(s)}\,,$$

where the geometric series converges because $\hat{l}_F(s) < 1$ for $s > 0$. □

For $\mathbb{N}$-valued random variables $T_1, T_2, \ldots$ we can also study the *renewal sequence* $\{h_n\}$, defined as the mean number of renewals at n, i.e.

$$h_n = \mathbb{E}\sum_{k=1}^\infty \mathbb{I}(\sigma_k = n) = \sum_{k=1}^\infty p_n^{*k}\,, \qquad n \in \mathbb{N}\,,$$

where $\{p_n^{*k}, n \in \mathbb{N}\}$ is the probability function of the k-fold convolution of $\{p_n\}$. If $\mathbb{P}(T = 0) = 0$, then $h_n \le 1$ since we can have at most one renewal epoch at n. If however $p = \mathbb{P}(T = 0) > 0$, then the conditional distribution of the number of renewal epochs at n is modified geometric under the condition that there is at least one renewal epoch at n. Hence, in both cases the renewal sequence is bounded.

In view of Theorem 6.1.1a we can conjecture that the renewal function $H(t) = \mathbb{E}\,N(t)$ will show an asymptotic linear behaviour, similar to (6.1.3). To show that this conjecture is correct, we use the following auxiliary result which is called *Wald's identity* for renewal processes. A more general version of this identity will be proved in Chapter 9 using martingale techniques; see Corollary 9.1.1.

Lemma 6.1.1 *Let $g : \mathbb{R}_+ \to \mathbb{R}_+$ be a measurable function. Then for all $t \ge 0$*

$$\mathbb{E}\Big(\sum_{i=1}^{N(t)+1} g(T_i)\Big) = \mathbb{E}\,g(T)(\mathbb{E}\,N(t) + 1)\,. \tag{6.1.11}$$

Proof In order to prove (6.1.11) we define the auxiliary random variables

$$Y_i = \begin{cases} 0 & \text{if } i > N(t)+1, \\ 1 & \text{if } i \le N(t)+1. \end{cases}$$

Then by the monotone convergence theorem

$$\mathbb{E}\Big(\sum_{i=1}^{N(t)+1} g(T_i)\Big) = \mathbb{E}\Big(\sum_{i=1}^\infty g(T_i)Y_i\Big) = \sum_{i=1}^\infty \mathbb{E}\,(g(T_i)Y_i)\,.$$

The remaining expectation can be further evaluated by the fact that $g(T_i)$ and Y_i are independent. This is true since the Bernoulli variable Y_i will be independent of T_i if and only if the event $\{Y_i = 1\}$ is independent of T_i. However, $\{Y_i = 1\} = \{N(t) + 1 \geq i\} = \{\sigma_{i-1} \leq t\}$ and this last event is independent of T_i. Hence we find

$$\begin{aligned}\mathbb{E}\Big(\sum_{i=1}^{N(t)+1} g(T_i)\Big) &= \sum_{i=1}^{\infty} \mathbb{E}\, g(T_i) \mathbb{E}\, Y_i = \mathbb{E}\, g(T) \sum_{i=1}^{\infty} \mathbb{P}(Y_i = 1) \\ &= \mathbb{E}\, g(T) \sum_{i=1}^{\infty} \mathbb{P}(N(t) + 1 \geq i) = \mathbb{E}\, g(T)(\mathbb{E}\, N(t) + 1)\,. \end{aligned}$$

□

We are now in a position to prove the *elementary renewal theorem.*

Theorem 6.1.6 *Assume that* $F(\infty) = 1$. *Then,*

$$\lim_{t\to\infty} \frac{H(t)}{t} = \begin{cases} \mu^{-1} & \text{if } \mu < \infty, \\ 0 & \text{if } \mu = \infty. \end{cases} \tag{6.1.12}$$

Proof Suppose first $\mu < \infty$. Using Theorem 6.1.1a, we get by Fatou's lemma

$$\mu^{-1} = \mathbb{E} \liminf_{t\to\infty} t^{-1} N(t) \leq \liminf_{t\to\infty} t^{-1} \mathbb{E}\, N(t) = \liminf_{t\to\infty} t^{-1} H(t)\,.$$

However, we also have

$$\mu^{-1} \geq \limsup_{t\to\infty} t^{-1} H(t)\,. \tag{6.1.13}$$

In order to show this we consider the truncated inter-occurrence time $T_n' = T_n \wedge b$ for some $b > 0$ such that $\mathbb{E}\, T' > 0$. Let $\{N'(t), t \geq 0\}$ denote the corresponding counting process with

$$N'(t) = \sum_{n=1}^{\infty} \mathbb{I}(\sigma_n' \leq t)\,, \qquad \sigma_n' = \sum_{i=1}^{n} T_i'\,, \tag{6.1.14}$$

and $H'(t) = \mathbb{E}\, N'(t)$. Then, $N'(t) \geq N(t)$ and consequently $H'(t) \geq H(t)$ for each $t \geq 0$. This and Lemma 6.1.1 give

$$\begin{aligned}\limsup_{t\to\infty} t^{-1} H(t) &\leq \limsup_{t\to\infty} t^{-1} H'(t) \leq \limsup_{t\to\infty} t^{-1} \frac{\mathbb{E}\, \sigma'_{N'(t)+1}}{\mathbb{E}\, T'} \\ &\leq \limsup_{t\to\infty} t^{-1} \mathbb{E}\left(\sigma'_{N'(t)} + T'_{N'(t)+1}\right) (\mathbb{E}\, T')^{-1} \\ &\leq \limsup_{t\to\infty} t^{-1} (t + b) (\mathbb{E}\, T')^{-1} = (\mathbb{E}\, T')^{-1}\,. \end{aligned}$$

Thus, (6.1.13) follows since $\lim_{b\to\infty} \mathbb{E}\, T' = \mathbb{E}\, T$. If $\mu = \infty$, then we get $\limsup_{t\to\infty} H(t)/t = 0$ in the same way as (6.1.13). □

We now consider the following slight generalization of the renewal model introduced in Section 6.1.1. Assume that $T_1, T_2, \ldots$ is a sequence of independent nonnegative random variables and that $T_2, T_3, \ldots$ are identically distributed with distribution F. Note that T_1 can have an arbitrary distribution F_1, which need not be equal to F. Then $\{\sigma_n, n \ge 1\}$ with $\sigma_n = T_1 + \ldots + T_n$ is called a *delayed renewal point process.* Defining $N(t)$ as in (6.1.1), $\{N(t), t \ge 0\}$ is called a *delayed renewal counting process.* The case when F_1 is equal to the integrated tail distribution F^{s} of F is of particular interest.

Theorem 6.1.7 *Assume that* $\mu < \infty$ *and that*

$$F_1(x) = F^{\mathrm{s}}(x) = \frac{1}{\mu}\int_0^x \overline{F}(y)\,\mathrm{d}y \tag{6.1.15}$$

for all $x \ge 0$*. Then, the renewal function* $H(t) = \mathbb{E}\, N(t)$ *is given by*

$$H(t) = \mu^{-1} t\,. \tag{6.1.16}$$

Proof Analogously to (6.1.7) we have

$$H(t) = \sum_{n=1}^{\infty} (F_1 * F^{*(n-1)})(t)\,. \tag{6.1.17}$$

Taking Laplace–Stieltjes transforms of this equation we get

$$\hat{l}_H(s) = \frac{\hat{l}_{F_1}(s)}{1-\hat{l}_F(s)} = \frac{\hat{l}_{F^{\mathrm{s}}}(s)}{1-\hat{l}_F(s)}$$

as in the proof of Theorem 6.1.5. Use the fact that $\hat{l}_{F^{\mathrm{s}}}(s) = (1-\hat{l}_F(s))(s\mu)^{-1}$ to see that $\hat{l}_H(s) = (s\mu)^{-1} = \mu^{-1}\int_0^\infty \mathrm{e}^{-sv}\,\mathrm{d}v$. This yields (6.1.16) because of the one-to-one correspondence between renewal functions and their Laplace–Stieltjes transforms. □

For a delayed renewal process which satisfies (6.1.15), an even stronger statement than the one in Theorem 6.1.7 is valid. To formulate it properly, we introduce the *excess* $T(t) = \sigma_{N(t)+1} - t$ at time $t \ge 0$.

Theorem 6.1.8 *Under the assumptions of Theorem* 6.1.7*, the delayed renewal counting process* $\{N(t), t \ge 0\}$ *has stationary increments.*

Proof Since the random variables $T_1, T_2, \ldots$ are independent, the joint distribution of the increments

$$(N(t_1+t) - N(t_0+t), \ldots, N(t_n+t) - N(t_{n-1}+t))$$

does not depend on t if the distribution of the excess $T(t)$ has this property. Thus, in view of Definition 5.2.1b, it amply suffices to show that $\mathbb{P}(T(t) \le x) = F^{\mathrm{s}}(x)$ for $x \ge 0$, independently of t. We have

$$\begin{aligned}
\mathbb{P}(T(t) \le x) &= \sum_{n=1}^{\infty} \mathbb{P}(\sigma_{n-1} \le t < \sigma_n \le t + x) \\
&= F^{\mathrm{s}}(t+x) - F^{\mathrm{s}}(t) + \sum_{n=1}^{\infty} \int_0^t (F(t+x-y) - F(t-y)) \, \mathrm{d}(F^{\mathrm{s}} * F^{*(n-1)})(y) \\
&= F^{\mathrm{s}}(t+x) - F^{\mathrm{s}}(t) + \mu^{-1} \int_0^t (F(t+x-y) - F(t-y)) \, \mathrm{d}y = F^{\mathrm{s}}(x) ,
\end{aligned}$$

where we used (6.1.16) in the third equality. □

6.1.3 Renewal Equations and Lorden's Inequality

We continue our discussion of (nondelayed) renewal processes with a number of results that turn out to be useful in connection with actuarial problems. In many applications we meet the *renewal equation*

$$g(x) = z(x) + \int_0^x g(x-v) \, \mathrm{d}F(v) , \tag{6.1.18}$$

where $z : \mathbb{R} \to \mathbb{R}_+$ is a locally bounded function vanishing on $(-\infty, 0)$ and F is a distribution on $\mathbb{R}_+$. If F is a defective distribution, then we call (6.1.18) a *defective renewal equation*. First we show that, whether F is defective or not, the (locally bounded) solution to the renewal equation is unique. Let $H_0(x) = \sum_{n=0}^{\infty} F^{*n}(x)$ be the renewal function corresponding to F.

Lemma 6.1.2 *The only solution $g(x)$ to* (6.1.18) *which is vanishing for $x < 0$ and bounded on finite intervals is given by*

$$\begin{aligned}
g(x) &= \sum_{k=0}^{\infty} \int_0^x z(x-v) \, \mathrm{d}F^{*k}(v) \\
&= \int_0^{\infty} z(x-v) \, \mathrm{d}H_0(v) , \qquad x \ge 0 .
\end{aligned} \tag{6.1.19}$$

Proof Note that the series $\sum_{k=0}^{\infty} \int_0^x |z(x-v)| \, \mathrm{d}F^{*k}(v)$ converges to a finite limit for each $x \ge 0$ because

$$\sum_{k=0}^{\infty} \int_0^x |z(x-v)| \, \mathrm{d}F^{*k}(v) \le H_0(x) \sup_{0 \le v \le x} |z(v)|$$

and $H_0(x) < \infty$ by Theorem 6.1.4. Thus the function $g(x)$ given in (6.1.19) is well-defined and locally bounded. It is easily seen that the function $g(x)$ solves the equation (6.1.18). Assume now that $g'(x)$ is another solution to (6.1.18) vanishing for $u < 0$ and bounded on finite intervals. Then using (6.1.18) we have

$$g(x) - g'(x) = \int_0^x (g(x-v) - g'(x-v))\,\mathrm{d}F(v)\,.$$

Inserting (6.1.18) repeatedly into the right-hand side of this expression, we can prove by induction on n that

$$\begin{aligned} |g(x) - g'(x)| &= \Big|\int_0^x (g(x-v) - g'(x-v))\,\mathrm{d}F^{*n}(v)\Big| \\ &\le \int_0^x |g(x-v) - g'(x-v)|\,\mathrm{d}F^{*n}(v) \\ &\le F^{*n}(x)\sup\{|g(v) - g'(v)| : 0 \le v \le x\}\,, \end{aligned}$$

for all $n = 1, 2, \ldots$ and $x \ge 0$. Thus $g(x) = g'(x)$ because Theorem 6.1.4 implies that $F^{*n}(x) \to 0$ as $n \to \infty$. □

Examples We illustrate the general solution (6.1.19) to the renewal equation (6.1.18) by a number of different choices of the function $z(x)$.

1. Take $z(x) = 1$ for all $x \ge 0$. Then the equation $g(x) = 1 + \int_0^x g(x-v)\,\mathrm{d}F(v)$ has the unique solution $g(x) = H_0(x)$.

2. Consider the expected number $H_0(x) - H_0(x-y)$ of renewal epochs in the interval $(x-y, x]$, where $y \ge 0$ is kept fixed. Then using the result of Lemma 6.1.2, we obtain a renewal equation for the function $g(x) = H_0(x) - H_0(x-y)$, i.e.

$$g(x) = (\delta_0(x) - \delta_0(x-y)) + \int_0^x g(x-v)\,\mathrm{d}F(v)\,. \tag{6.1.20}$$

3. Take $z(x) = F(x)$. Then the unique solution (6.1.19) to (6.1.18) has the form $g(x) = \sum_{n=1}^{\infty} F^{*n}(x) = H(x)$.

4. As another example assume that F is nondefective and has a finite mean μ. Now take $z(u)$ equal to the integrated tail distribution F^{s} of F, i.e. $z(x) = \mu^{-1}\int_0^x (1 - F(v))\,\mathrm{d}v$. Using (6.1.17) and (6.1.19), Theorem 6.1.7 then gives $g(x) = x/\mu$.

We next show that the renewal function H_0 is *subadditive.*

Lemma 6.1.3 *For all* $t, s \ge 0$

$$H_0(t+v) \le H_0(t) + H_0(v)\,. \tag{6.1.21}$$

Proof Consider the excess $T(t) = \sigma_{N(t)+1} - t$ at time t. Note that

$$N_0(t+v) = N_0(t) + N_0(t+v) - N_0(t) = N_0(t) + N'(v)\,,$$

where $\{N'(v), v \geq 0\}$ is a delayed renewal counting process with $F_1(x) = \mathbb{P}(T(t) \leq x)$ and with inter-renewal distance distribution F. Since $\delta_0 \leq_{\rm st} F_1$, we have $N'(v) \leq_{\rm st} N_0(v)$. Thus, (6.1.21) is proved. □

We close this section on elementary properties of the renewal functions $H(t)$ and $H_0(t)$ by proving *Lorden's inequality.* This inequality yields estimates for the speed of convergence in the elementary renewal theorem; see Theorem 6.1.6.

Theorem 6.1.9 *If the second moment $\mu^{(2)}$ of F is finite, then*

$$0 \leq H_0(t) - \frac{t}{\mu} \leq \frac{\mu^{(2)}}{\mu^2}\,. \tag{6.1.22}$$

Proof Using (6.1.11), the subadditivity property (6.1.21) of $H_0(t)$ can be easily employed to prove that also the expected excess $\mathbb{E}\,T(t) = \mu H_0(t) - t$ is subadditive, i.e. $\mathbb{E}\,T(t+v) \leq \mathbb{E}\,T(t) + \mathbb{E}\,T(v)$ for all $t, v \geq 0$. From a graphical representation of this excess over the interval $[0,t]$ (see Figure 6.1.1) we obtain

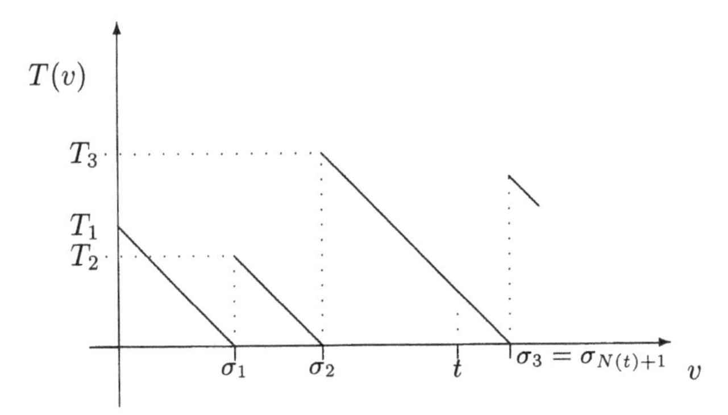

Figure 6.1.1 The excess $T(v)$ at time v

the following equality:

$$\int_0^t T(v)\,\mathrm{d}v = \tfrac{1}{2} \sum_{i=1}^{N(t)+1} T_i^2 - \tfrac{1}{2} T^2(t)\,.$$

Take expectations of these expressions, apply Fubini's theorem on the left and (6.1.11) on the right to obtain the intermediate formula

$$\int_0^t \mathbb{E}\,T(v)\,\mathrm{d}v = \tfrac{1}{2}\,\mu^{(2)} H_0(t) - \tfrac{1}{2}\mathbb{E}\,T^2(t)\,. \tag{6.1.23}$$

On the left hand side we apply the subadditivity of $\mathbb{E}\, T(t)$ to get

$$\int_0^t \mathbb{E}\, T(v)\, \mathrm{d}v = \int_0^{t/2} (\mathbb{E}\, T(v) + \mathbb{E}\, T(t-v))\, \mathrm{d}v \geq \frac{t}{2}\mathbb{E}\, T(t)\,.$$

Thus, using $\mathbb{E}\, T^2(t) \geq (\mathbb{E}\, T(t))^2$, (6.1.23) gives

$$t\, \mathbb{E}\, T(t) \leq \frac{\mu^{(2)}}{\mu}(\mathbb{E}\, T(t) + t) - (\mathbb{E}\, T(t))^2\,.$$

The resulting quadratic inequality for $\mathbb{E}\, T(t)$ can easily be put into the form $(\mathbb{E}\, T(t) + t)(\mathbb{E}\, T(t) - \mu^{(2)}/\mu) \leq 0$, from which (6.1.22) follows by using $\mathbb{E}\, T(t) = \mu H_0(t) - t$ once more. □

6.1.4 Key Renewal Theorem

Besides the asymptotic linearity property stated in Theorem 6.1.6, a much stronger result can be shown for the asymptotic behaviour of the renewal function $H(t)$ as $t \to \infty$. The following limit theorem, called *Blackwell's renewal theorem,* says roughly that the renewal measure defined in (6.1.8) asymptotically behaves like the Lebesgue measure.

Theorem 6.1.10 *Assume F is nonlattice with $\mu < \infty$. Then, for each $y \geq 0$*

$$\lim_{x\to\infty} (H_0(x) - H_0(x-y)) = \mu^{-1} y\,. \tag{6.1.24}$$

The *proof* of Theorem 6.1.10 goes beyond the scope of this book. We therefore omit it and refer, for example, to the books by Daley and Vere-Jones (1988) and Resnick (1992), where a probabilistic proof of this theorem is given. The proof there uses a coupling method by comparing the renewal function $H_0(t)$ with the renewal function of a delayed renewal process which satisfies (6.1.15).

In Section 6.1.3, (6.1.20), we mentioned that the difference $g(x) = H_0(x) - H_0(x-y)$ can be seen as solution to the renewal equation (6.1.18) with $z(x) = \delta_0(x) - \delta_0(x-y)$. Thus, (6.1.24) can be written in the form

$$\lim_{x\to\infty} g(x) = \mu^{-1} \int_0^\infty z(v)\, \mathrm{d}v\,. \tag{6.1.25}$$

We next study the asymptotic behaviour of the solution $g(x)$ to (6.1.18) as $x \to \infty$, when $z(x)$ is nonnegative and satisfies some integrability property. More specifically, for each $h > 0$, define the upper integral sum $\overline{z}(h) = h\sum_{n=1}^\infty \sup\{z(x) : (n-1)h \leq x \leq nh\}$ and the lower integral sum $\underline{z}(h) = h\sum_{n=1}^\infty \inf\{z(x) : (n-1)h \leq x \leq nh\}$. The function $z(x)$ is called *directly Riemann integrable* if $\overline{z}(h) < \infty$ for all $h > 0$ and if

$$\lim_{h\to 0} (\overline{z}(h) - \underline{z}(h)) = 0\,. \tag{6.1.26}$$

The following lemma gives a sufficient but useful condition for direct Riemann integrability.

Lemma 6.1.4 *Let* $z_1 : \mathbb{R}_+ \to (0, \infty)$ *be increasing while* $z_2 : \mathbb{R}_+ \to \mathbb{R}_+$ *is decreasing, such that*

$$\int_0^\infty z_1(x) z_2(x)\, \mathrm{d}x < \infty \tag{6.1.27}$$

and

$$\lim_{h \to 0} \sup\Big\{ \frac{z_1(x+y)}{z_1(x)} : x \geq 0, 0 \leq y \leq h \Big\} = 1\,. \tag{6.1.28}$$

Then, the product $z(x) = z_1(x) z_2(x)$ *is directly Riemann integrable.*

Proof With the notation $c(h) = \sup\{z_1(x+y)/z_1(x) : x \geq 0, 0 \leq y \leq h\}$ for $h > 0$, we have

$$\begin{aligned} \sup\{z(x) : (n-1)h \leq x \leq nh\} &\leq z_1(nh) z_2((n-1)h) \\ &\leq c(2h) z_1((n-2)h) z_2((n-1)h) \end{aligned}$$

for $n = 2, 3, \ldots$. Since $z_1((n-2)h) z_2((n-1)h) \leq z_1(x) z_2(x)$ for $n = 2, 3, \ldots$ and $(n-2)h \leq x \leq (n-1)h$, this gives

$$\overline{z}(h) \leq h \sup\{z(x) : 0 \leq x \leq h\} + c(2h) \int_0^\infty z_1(x) z_2(x)\, \mathrm{d}x\,.$$

Similarly, we obtain $\underline{z}(h) \geq (c(2h))^{-1} \int_h^\infty z_1(x) z_2(x)\, \mathrm{d}x$. Thus (6.1.27) and (6.1.28) imply (6.1.26). □

Remark Each directly Riemann integrable function is also Riemann integrable in the usual sense. However, the converse statement is not true, as can be seen from the following example. For each $n = 1, 2, \ldots$, we consider the function $z_n : \mathbb{R} \to \mathbb{R}_+$ with

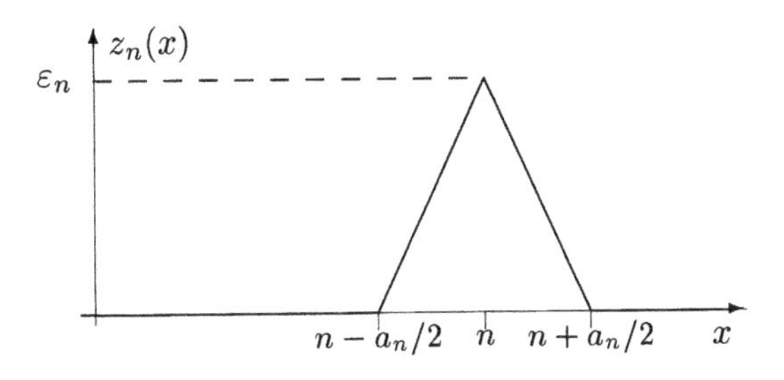

Figure 6.1.2 The function $z_n(x)$

$$z_n(x) = \begin{cases} \varepsilon_n - a_n{}^{-1} 2\varepsilon_n |n - x| & \text{if } x \in (n - a_n/2, n + a_n/2), \\ 0 & \text{otherwise,} \end{cases}$$

where $2^{-1} > a_1 > a_2 > \ldots > 0$ and $\lim_{n\to\infty} a_n = 0, \lim_{n\to\infty} \varepsilon_n = \infty$ and $\sum_{n=1}^{\infty} \varepsilon_n a_n < \infty$. Then, for $z(x) = \sum_{n=1}^{\infty} z_n(x)$, we have $\int_0^\infty z(x)\,\mathrm{d}x = 2^{-1}\sum_{n=1}^{\infty} \varepsilon_n a_n < \infty$, but $\overline{z}(h) \geq h\sum_{n=1}^{\infty} \varepsilon_n = \infty$ for all $h > 0$.

To formulate the next result, assume that F is a nonlattice distribution, that is $F(x)$ does not increase on a lattice only. We also suppose that F is proper (nondefective). The following theorem is known in the literature as the *key renewal theorem.*

Theorem 6.1.11 *Let $z(x)$ be directly Riemann integrable. Then for the solution $g(x) = \int_0^\infty z(x-v)\,\mathrm{d}H_0(v)$ to the renewal equation* (6.1.18)

$$\lim_{x\to\infty} g(x) = \begin{cases} {\mu_F}^{-1}\int_0^\infty z(v)\,\mathrm{d}v & \text{if } \mu < \infty, \\ 0 & \text{if } \mu = \infty. \end{cases} \tag{6.1.29}$$

Proof Take $h > 0$ arbitrary but fixed and approximate $z(x)$ from below and from above by the step functions

$$z_*(x) = \sum_{n=1}^{\infty} \underline{m}_n \delta_x([(n-1)h, nh))\,, \qquad z^*(x) = \sum_{n=1}^{\infty} \overline{m}_n \delta_x([(n-1)h, nh))\,,$$

where we used the abbreviations $\underline{m}_n = \inf\{z(x) : (n-1) \leq xh^{-1} \leq n\}$ and $\overline{m}_n = \sup\{z(x) : (n-1) \leq xh^{-1} \leq n\}$ for $n \geq 1$. Obviously for all $x \geq 0$, $z_*(x) \leq z(x) \leq z^*(x)$. By the monotonicity of the renewal function $H_0(x)$ and the positivity of $z(x)$ we also have

$$\int_0^x z_*(x-v)\,\mathrm{d}H_0(v) \leq \int_0^x z(x-v)\,\mathrm{d}H_0(v) \leq \int_0^x z^*(x-v)\,\mathrm{d}H_0(v)\,.$$

Take the quantity on the right of the above inequalities. Then by the additivity of the integral we find that

$$\begin{aligned} \int_0^x z^*(x-v)\,\mathrm{d}H_0(v) &= \sum_{n=1}^{\infty} \overline{m}_n \int_0^x \delta_{x-v}([(n-1)h, nh))\,\mathrm{d}H_0(v) \\ &= \sum_{n=1}^{\infty} \overline{m}_n \left(H_0(x-(n-1)h) - H_0(x-nh)\right)\,, \end{aligned}$$

where we put $H_0(t) = 0$ for $t < 0$. Note that by the result of Theorem 6.1.10, the differences $H_0(x-(n-1)h) - H_0(x-nh)$ are uniformly bounded for all $x \geq 0$ and $n \in \mathbb{N}$. Furthermore, $\sum_{n=1}^{\infty} \overline{m}_n < \infty$ by the direct Riemann integrability of $z(x)$. Hence by an application of the dominated convergence theorem, Theorem 6.1.10 gives

$$\limsup_{x\to\infty} \int_0^x z(x-v)\,\mathrm{d}H_0(v) \leq \limsup_{x\to\infty} \int_0^x z^*(x-v)\,\mathrm{d}H_0(v) = \frac{h}{\mu}\sum_{n=1}^{\infty} \overline{m}_n\,.$$

In a similar fashion we find that

$$\liminf_{x\to\infty} \int_0^x z(x-v)\,\mathrm{d}H_0(v) \ge \liminf_{x\to\infty} \int_0^x z_*(x-v)\,\mathrm{d}H_0(v) = \frac{h}{\mu}\sum_{n=1}^{\infty} \underline{m}_n\,.$$

If we now let $h \downarrow 0$, then (6.1.29) follows by condition (6.1.26) because

$$\lim_{h\downarrow 0} h \sum_{n=1}^{\infty} \underline{m}_n = \lim_{h\downarrow 0} h \sum_{n=1}^{\infty} \overline{m}_n = \int_0^{\infty} z(v)\,\mathrm{d}v\,.$$

□

6.1.5 Another Look at the Aggregate Claim Amount

We consider anew the aggregate claim amount $X(t) = \sum_{i=1}^{N(t)} U_i$ introduced in Section 5.1.4. Here, $\{U_n\}$ is the sequence of independent claim sizes with common distribution U. Further, $\{N(t), t \ge 0\}$ is the claim counting process given by $N(t) = \sum_{i=1}^{\infty} \mathbb{I}(\sigma_i \le t)$; $\sigma_n = \sum_{i=1}^{n} T_i$ and where the sequence $\{\sigma_n, n \in \mathbb{N}\}$ is an (undelayed) renewal point process with inter-occurrence times T_n following a distribution F. In this section we do not assume that U is nonnegative.

In order to keep a renewal structure, we have to assume that the sequence $\{(T_n, U_n)\}$ consists of independent and identically distributed random vectors. But, we do not exclude that U_n may eventually depend on T_n. For example, in health insurance the amount to be paid out can depend on the time since the last payment. We assume, however, that $\mathbb{E}\,[|U|; T = 0] = 0$. This excludes the possibility that one has to pay a bill even when no time has elapsed since the last payment.

Note that $\{X(t), t \ge 0\}$ is also called a *renewal reward process.* We cannot expect that we can get very precise asymptotics for $X(t)$ as $t \to \infty$, especially when we do not specify the dependence structure between U and T. However, we have the following general results.

Theorem 6.1.12 *Assume that* $F(0) = 0$, $\mu_F = \mathbb{E}\,T < \infty$ *and* $\mathbb{E}\,|U| < \infty$. *Then, with probability* 1

$$\lim_{t\to\infty} t^{-1} X(t) = \lim_{t\to\infty} t^{-1}\mathbb{E}\,X(t) = \mu_F^{-1}\mathbb{E}\,U\,; \tag{6.1.30}$$

if F is nonlattice, then

$$\lim_{t\to\infty} \mathbb{E}\,(X(t+h) - X(t)) = \mu_F^{-1} h \mathbb{E}\,U\,, \qquad h \ge 0\,. \tag{6.1.31}$$

Proof We start with the first equality in (6.1.30). Observe that, as in the proof of Theorem 6.1.1, the strong law of large numbers for sums of independent random variables yields $\lim_{t\to\infty}(N(t))^{-1}\sum_{i=1}^{N(t)} U_i = \mathbb{E}\,U$ since $N(t)\to\infty$ as $t\to\infty$. This and (6.1.3) imply that

$$\lim_{t\to\infty} t^{-1}X(t) = \mu_F^{-1}\mu_U\,. \tag{6.1.32}$$

To derive the second equality in (6.1.30), decompose $X(t)$ into two parts:

$$X(t) = \sum_{i=1}^{N(t)} U_i \mathbb{I}(U_i > 0) + \sum_{i=1}^{N(t)} U_i \mathbb{I}(U_i \le 0)\,.$$

Considering the sums on the right-hand side separately, we can proceed as in the proof of Theorem 6.1.6 to obtain the second equality in (6.1.30). The details are left to the reader. We are now going to prove (6.1.31). Using analogous arguments as in the proof of Wald's identity (6.1.11), we find that $\mathbb{E}\,(\sum_{i=1}^{N(t)+1} U_i) = \mathbb{E}\,U H_0(t)$. Hence $\mathbb{E}\,X(t) = \mathbb{E}\,U H_0(t) - \mathbb{E}\,U_{N(t)+1}$. To evaluate the remaining expectation, we condition on the number of renewal epochs σ_n that have occurred up to time t. More specifically,

$$\begin{aligned}\mathbb{E}\,U_{N(t)+1} &= \sum_{n=0}^{\infty} \mathbb{E}\,[U_{N(t)+1}; N(t) = n] = \sum_{n=0}^{\infty} \mathbb{E}\,[U_{n+1}; \sigma_n \le t < \sigma_{n+1}] \\ &= \sum_{n=0}^{\infty} \int_0^t \mathbb{E}\,[U_{n+1}; t - v < T_{n+1}]\,\mathrm{d}F^{*n}(v) = \int_0^t \mathbb{E}\,[U; T > t - v]\,\mathrm{d}H_0(v)\,,\end{aligned}$$

where $z(t) = \mathbb{E}\,[U; T > t]$. This gives $\mathbb{E}\,U_{N(t)+1} = \int_0^t z(t-v)\,\mathrm{d}H_0(v)$ for all $t > \infty$. Noticing that $z(0) = \mathbb{E}\,U$ we find that

$$\mathbb{E}\,X(t) = \int_0^t (z(0) - z(t-v))\,\mathrm{d}H_0(v) = -\int_0^t H_0(t-v)\,\mathrm{d}z(v)\,.$$

Now,

$$\begin{aligned}\mathbb{E}\,(X(t+h) - X(t)) &= -\int_0^t (H_0(t+h-v) - H_0(t-v))\,\mathrm{d}z(v) \\ &\quad - \int_t^{t+h} H_0(t+h-v)\,\mathrm{d}z(v)\,.\end{aligned}$$

Using Theorem 6.1.10 we can apply the dominated convergence theorem to the first integral to obtain the expression $h\mathbb{E}\,U\mu_F^{-1}$ for the limit of this integral as $t\to\infty$. Using the monotonicity of $H_0(t)$, the second term can be dominated by $H_0(h)\,|z(t) - z(t+h)|$, which tends to 0 as $t\to\infty$. □

Bibliographical Notes. The results presented in this section and further details on renewal processes can be found, for example, in Asmussen (1987),

Daley and Vere-Jones (1988), Feller (1971) and Resnick (1992). A comprehensive survey on the coupling method which is a useful tool for proving Theorem 6.1.10 is given in Lindvall (1992). For discrete-time renewal theory we refer to Feller (1968) and Kingman (1972). The notion of direct Riemann integrability was introduced in Feller (1971). An earlier version of the key renewal theorem is due to W. Smith and therefore sometimes we meet the notion of *Smith's theorem* for Theorem 6.1.11. Note also that many results of the present section hold true under weaker mathematical assumptions. For example, in the proof of Theorem 6.1.3 we did not explicitly use the assumption that the inter-occurrence times are independent. We only used the compound Poisson approximation considered in Section 4.6 and the statement of Theorem 6.1.1a, a special case of a more general ergodic theorem. See also Chapter 12, where we discuss results for more general point processes with not necessarily independent inter-occurrence times. Further details of the general theory of point processes on the real line can be found, for example, in Baccelli and Brémaud (1994), Daley and Vere-Jones (1988), König and Schmidt (1992) and Last and Brandt (1995).

6.2 EXTENSIONS AND ACTUARIAL APPLICATIONS

In this section we derive some extensions to the basic results from renewal theory presented in Section 6.1. They will prove to be useful in the investigation of actuarial problems.

6.2.1 Weighted Renewal Functions

We study a generalization of Theorem 6.1.6 for the *weighted renewal function*

$$A(x) = \sum_{n=0}^{\infty} a_n F^{*n}(x) , \tag{6.2.1}$$

where F is a nondefective distribution on $\mathbb{R}_+$ with mean μ and $\{a_n, n \in \mathbb{N}\}$ is a sequence of nonnegative numbers. By the weak law of large numbers

$$\lim_{n\to\infty} F^{*n}(ny) \to \begin{cases} 0 & \text{if } y < \mu, \\ 1 & \text{if } y > \mu. \end{cases} \tag{6.2.2}$$

Intuitively, the individual summands in (6.2.1) will contribute a value near to 1 if $x > n\mu$ or if $n < \lfloor x/\mu \rfloor$, and a value near to 0 otherwise. This suggests an estimate for (6.2.1) of the form $\sum_{n=0}^{\lfloor x/\mu \rfloor} a_n$. We will prove this conjecture under some appropriate requirements on the sequence $\{a_n, n \in \mathbb{N}\}$.

Theorem 6.2.1 *Assume that*

$$\lim_{\delta\downarrow 0}\limsup_{r\to\infty}\frac{\sum_{n=r+1}^{\lfloor (r+1)(1+\delta)\rfloor} a_n}{\sum_{n=0}^{r} a_n} = 0 \tag{6.2.3}$$

and

$$a(s) = \sum_{n=0}^{\infty} a_n s^n \text{ converges for } |s| < 1. \tag{6.2.4}$$

If $0 < \mu < \infty$, *then as* $x \to \infty$

$$A(x) \sim \sum_{n=0}^{\lfloor x/\mu \rfloor} a_n\,. \tag{6.2.5}$$

Proof It is pretty obvious that (6.2.5) holds if $\sum_{n=0}^{\infty} a_n < \infty$. Indeed, both sides of (6.2.5) converge to the same constant by the monotone convergence theorem. So, assume $\sum_{n=0}^{\infty} a_n = \infty$. Let $\varepsilon \in (0, \mu)$ be fixed and define $m_x = \lfloor x/\mu \rfloor$ for any $x > 0$. With this definition of m_x we can write

$$\begin{aligned} A(x) &= \sum_{n=0}^{\infty} a_n \mathbf{P}(\sigma_n \le x) \\ &= \sum_{n=0}^{m_x} a_n - \sum_{n=0}^{m_x} a_n \mathbf{P}(\sigma_n > x) + \sum_{n=m_x+1}^{\infty} a_n \mathbf{P}(\sigma_n \le x) \end{aligned}$$

and hence we see that we need to show that

$$\lim_{x\to\infty}\Big(\sum_{n=0}^{m_x} a_n\Big)^{-1}\Big(-\sum_{n=0}^{m_x} a_n \mathbf{P}(\sigma_n > x) + \sum_{n=m_x+1}^{\infty} a_n \mathbf{P}(\sigma_n \le x)\Big) = 0\,.$$

We further subdivide the two sums in two parts by considering the quantities

$$I_1(x) = \sum_{n=0}^{m_-(x)} a_n \mathbf{P}(\sigma_n > x)\,, \qquad I_2(x) = \sum_{n=m_-(x)+1}^{m_x} a_n \mathbf{P}(\sigma_n > x)\,,$$

$$J_1(x) = \sum_{n=m_x+1}^{m_+(x)+1} a_n \mathbf{P}(\sigma_n \le x)\,, \qquad J_2(x) = \sum_{n=m_+(x)+2}^{\infty} a_n \mathbf{P}(\sigma_n \le x)\,,$$

where

$$m_-(x) = \Big\lfloor \frac{m_x}{1 + (\varepsilon/\mu)} \Big\rfloor, \qquad m_+(x) = \Big\lfloor \frac{m_x + 1}{1 - (\varepsilon/\mu)} \Big\rfloor - 1\,.$$

We prove that each one of these four sums is $o(\sum_{n=0}^{m_x} a_n)$ as $x \to \infty$ or equivalently as $m_x \to \infty$. Note that by the definitions of $m_x, m_-(x)$ and $m_+(x)$ we have a number of inequalities that will be used a couple of times:

$$m_x \mu \le x < (m_x + 1)\mu\,, \tag{6.2.6}$$

$$m_-(x)(1+\mu^{-1}\varepsilon) \le m_x < (m_-(x)+1)(1+\mu^{-1}\varepsilon)\,, \tag{6.2.7}$$

$$(m_+(x)+1)(1-\mu^{-1}\varepsilon) \le m_x + 1 < (m_-(x)+2)(1-\mu^{-1}\varepsilon)\,. \tag{6.2.8}$$

We consider first $I_1(x)$. Since $n \le m_-(x)$ in this summation, we can write $x \ge m_x\mu > m_-(x)(1+\varepsilon/\mu)\mu \ge (\mu+\varepsilon)n$. Hence

$$I_1(x) \le \sum_{n=0}^{m_-(x)} a_n \mathbf{P}(n^{-1}\sigma_n - \mu > \varepsilon)\,,$$

which is of the form $\sum_{n=0}^{m_-(x)} a_n b_n$ with $b_n = \mathbf{P}(\sigma_n/n - \mu > \varepsilon)$. By the weak law of large numbers $b_k \to 0$ as $k \to \infty$; see (6.2.2). Hence for the given $\varepsilon > 0$, choose m' large enough to have $b_k < \varepsilon$ as soon as $k > m'$. Then

$$\sum_{n=0}^{m_-(x)} a_n b_n = \sum_{n=0}^{m'} a_n b_n + \sum_{n=m'+1}^{m_-(x)} a_n b_n \le \sum_{n=0}^{m'} a_n + \varepsilon \sum_{n=0}^{m_-(x)} a_n\,.$$

Dividing by $\sum_{n=0}^{m_x} a_n$, which tends to ∞ with x, we see that

$$\limsup_{x\to\infty} \frac{I_1(x)}{\sum_{n=0}^{m_x} a_n} < \varepsilon\,.$$

Next consider $I_2(x)$. The values of σ_n might be too close to x, and hence we do not expect help from the probabilities involved. Instead we need a condition on the sequence $\{a_n\}$. By inequality (6.2.7) on m_x we can write

$$\frac{I_2(x)}{\sum_{n=0}^{m_x} a_n} \le \frac{\sum_{n=m_-(x)+1}^{(m_-(x)+1)(1+(\varepsilon/\mu))} a_n}{\sum_{n=0}^{m_-(x)} a_n}\,.$$

Hence condition (6.2.3) on the sequence $\{a_n\}$ can be applied with $r = m_-(x)$ and $\delta = \varepsilon/\mu$. We therefore deduce that

$$\lim_{\varepsilon\downarrow 0} \limsup_{x\to\infty} \frac{I_2(x)}{\sum_{n=0}^{m_x} a_n} = 0\,.$$

We now turn to $J_1(x)$. From the definition of $m_+(x)$ we can write

$$J_1(x) \le \sum_{n=m_x+1}^{(m_x+1)(1+\delta')} a_n\,,$$

with $\delta' = (2\varepsilon)/(\mu-\varepsilon)$. Again apply (6.2.3) with $r = m_x$ and $\delta = \delta'$. Hence, $J_1(x) = o(\sum_{n=0}^{m_x} a_n)$. Finally we consider $J_2(x)$. In this summation we have

$n \geq m_+(x) + 2$, and $x < (m_x + 1)\mu \leq (m_+(x) + 2)(\mu - \varepsilon)$ by (6.2.6) and by the definition of $m_+(x)$. Hence

$$J_2(x) = \sum_{n=m_+(x)+2}^{\infty} a_n \mathbb{P}(\sigma_n \leq x) \leq \sum_{n=m_+(x)+2}^{\infty} a_n \mathbb{P}(\sigma_n/n - \mu < -\varepsilon).$$

Since the inter-occurrence time T is nonnegative, we can apply Theorem 2.3.2 to the random variable $n\mu - \sigma_n$ to see that there exists a constant $c < 1$ such that $\mathbb{P}(\sigma_n/n - \mu < -\varepsilon) < c^n$ for all sufficiently large $n \in \mathbb{N}$. Thus, the sum $\sum_{m_+(x)+2}^{\infty} a_n c^n$ is convergent by condition (6.2.4). Since $\sum_{n=0}^{m_x} a_n \to \infty$ also $J_2(x) = o(\sum_{n=0}^{m_x} a_n)$ as $x \to \infty$. □

Examples Let us illustrate Theorem 6.2.1 by a series of examples, where $A(x) = \sum_{n=0}^{\infty} a_n F^{*n}(x)$ and μ is the expectation of F.

1. In order to obtain the elementary renewal theorem derived in Section 6.1.2, we take $a_n = 1$ for all $n \geq 0$. Then $A(x) = H_0(x)$ and by Theorem 6.2.1 we have $H_0(x) \sim x\mu^{-1}$ provided that $\mu < \infty$. Recall that in Theorem 6.1.6 we proved this result by relying on Wald's identity. Furthermore, considering the truncated inter-occurrence times $T'_n = T_n \wedge b$ for some $b > 0$ which leads to the *shadow renewal process* $\{N'(t), t \geq 0\}$ defined in (6.1.14), we showed in Theorem 6.1.6 that the condition $\mu < \infty$ is not necessary, that is $\lim_{x\to\infty} H_0(x)/x = 0$ if $\mu = \infty$.

2. Another example of a weighted renewal function is obtained by the choice $a_0 = 0$, $a_n = n^{-1}$ for $n \geq 1$. In this case we check that assumptions (6.2.3) and (6.2.4) are fulfilled using $\sum_{k=1}^{n} 1/k \sim \log n$ as $n \to \infty$. Then we get the asymptotic behaviour of the *harmonic renewal function*,

$$A(x) \sim \sum_{n=1}^{\lfloor x/\mu \rfloor} \frac{1}{n} \sim \log\, x\,, \qquad x \to \infty\,.$$

3. We now study the weighted renewal function with a power-like sequence $\{a_n\}$, where $a_n \sim cn^{-d}$ for some $c > 0$ and $0 \leq d < 1$. In this case we also assume $\mu < \infty$. Using $\sum_{n=1}^{m} n^{-d} \sim m^{1-d}/(1-d)$, we can check that conditions (6.2.3) and (6.2.4) hold. Moreover as $x \to \infty$

$$A(x) \sim \sum_{n=1}^{\lfloor x/\mu \rfloor} a_n \sim c\mu^{d-1}(1-d)^{-1}x^{1-d}. \tag{6.2.9}$$

We leave it to the reader to show (6.2.9). Note that for the case $d = 0$ and $c = 1$ we find the elementary renewal theorem again. On the other end of the scale, $d = 1$ and $c = 1$ do not yield the harmonic renewal case unless we interpret $(x^y - 1)/y$ as $\log x$ when $y = 0$.

6.2.2 A Blackwell-Type Renewal Theorem

In this section we prove a theorem that could be called of *Blackwell type*. More precisely, we will show that we can derive a Blackwell-type renewal theorem for the weighted renewal function $A(x) = \sum_{n=0}^{\infty} a_n F^{*n}(x)$ if we already have an elementary renewal theorem for the allied function $B(x) = \sum_{n=0}^{\infty} b_n F^{*n}(x)$. Here the two sequences $\{a_n\}$ and $\{b_n\}$ are linked by the relation

$$b_n = (n+1)a_{n+1} - na_n \,, \qquad n \ge 0 \,, \tag{6.2.10}$$

or equivalently by the *Cesàro averages* $a_n = n^{-1} \sum_{j=0}^{n-1} b_j$, $n \ge 1$. In this section we assume that $\sum_{n=0}^{\infty} a_n s^n$ is convergent at least for $|s| < 1$.

Theorem 6.2.2 *Let F be a nonlattice distribution on $\mathbb{R}_+$ such that $0 < \mu_F < \infty$. Assume that the sequence $\{b_n\}$ is nonnegative. If for all $y \in \mathbb{R}$, $B(x+y) \sim B(x)$ as $x \to \infty$, then for all $y \in \mathbb{R}$*

$$A(x+y) - A(x) \sim yB(x)/x \,, \qquad x \to \infty \,. \tag{6.2.11}$$

In the *proof* of Theorem 6.2.2 we use an auxiliary result. To simplify the notation, for a function of bounded variation G on $\mathbb{R}_+$ we put

$$G'(x) = \int_0^x y \, \mathrm{d}G(y) \,. \tag{6.2.12}$$

Note that in terms of Laplace–Stieltjes transforms this means that

$$\hat{l}_{G'}(s) = -s^{-1} \hat{l}_G^{(1)}(s) \,. \tag{6.2.13}$$

We first derive a link between the function $A(x)$ and the zero-deleted renewal function $H_0(x)$ generated by the distribution F.

Lemma 6.2.1 *For the weighted renewal functions $A(x), B(x)$ with $\{a_n\}$ and $\{b_n\}$ linked by* (6.2.10),

$$A'(x) = B * F' * H_0(x) \,, \qquad x \ge 0 \,. \tag{6.2.14}$$

Proof It suffices to prove that the Laplace–Stieltjes transforms of both sides of (6.2.14) coincide. By (6.2.13) we have $\hat{l}_{A'}(s) = -s^{-1}\hat{l}_A^{(1)}(s)$, where in turn

$$\hat{l}_A(s) = \int_0^\infty \mathrm{e}^{-sx} \, \mathrm{d}\Big(\sum_{n=0}^{\infty} a_n F^{*n}(x)\Big) = \sum_{n=0}^{\infty} a_n (\hat{l}_F(s))^n = a(\hat{l}_F(s)) \,,$$

and $a(z) = \sum_{n=0}^{\infty} a_n z$. Hence $\hat{l}_{A'}(s) = -s^{-1}\hat{l}_F^{(1)}(s) a^{(1)}(\hat{l}_F(s))$. However, using (6.2.10) we find the relationship $(1-z)a^{(1)}(z) = b(z)$ for $|z| < 1$, where $b(z) = \sum_{n=0}^{\infty} b_n z^n$. But then $\hat{l}_{A'}(s) = b(\hat{l}_F(s))(-\hat{l}_F^{(1)}(s))\,(s(1-\hat{l}_F(s)))^{-1}$ and, using Theorem 6.1.5, we see that this is the Laplace–Stieltjes transform of the right-hand side of (6.2.14), since as before $\hat{l}_B(s) = \hat{b}(\hat{l}_F(s))$. □

Proof of Theorem 6.2.2 Let $A'(x)$ and $F'(x)$ be the functions which are induced by $A(x)$ and $F(x)$, respectively, according to the abbreviation (6.2.12). The monotonicity of $A(x)$ then implies for $0 < x < x + y$

$$\frac{1}{x+y}\left(A'(x+y) - A'(x)\right) \le A(x+y) - A(x) \le \frac{1}{x}\left(A'(x+y) - A'(x)\right).$$

Since $\lim_{x\to\infty} B(x+y)/B(x) = 1$ for all $y \ge 0$, the dominated convergence theorem shows that

$$B * F'(x) \sim \mu_F B(x) \tag{6.2.15}$$

as $x \to \infty$. Thus, it suffices to prove that for each $y > 0$

$$\lim_{x\to\infty} \frac{1}{B * F'(x)}\left(A'(x+y) - A'(x)\right) = \frac{y}{\mu_F}. \tag{6.2.16}$$

To show this we will use the result of Lemma 6.2.1. Put $R(x) = B * F'(x)$ for convenience, so that $A'(x) = R * H_0(x)$. Thus, for each $x_0 \in [0, x]$

$$\begin{aligned} A'(x+y) - A'(x) &= \int_x^{x+y} H_0(x+y-z)\,\mathrm{d}R(z) \\ &\quad + \int_{x-x_0}^{x} \left(H_0(x+y-z) - H_0(x-z)\right)\mathrm{d}R(z) \\ &\quad + \int_0^{x-x_0} \left(H_0(x+y-z) - H_0(x-z)\right)\mathrm{d}R(z). \end{aligned}$$

Denote the three integrals on the right by $I_1(x)$, $I_2(x)$ and $I_3(x)$, respectively. We first estimate $I_1(x)$. By the monotonicity of $R(x)$ we find

$$0 \le \frac{I_1(x)}{R(x)} \le H_0(y)\Big(\frac{R(x+y)}{R(x)} - 1\Big).$$

Further, for $I_3(x)$, take any $\varepsilon > 0$ and choose x_0 so large that by Theorem 6.1.10, $\mu_F^{-1}y - \varepsilon \le H_0(v+y) - H_0(v) \le \mu_F^{-1}y + \varepsilon$ for all $v > x_0$. Then

$$\Big(\frac{y}{\mu_F} - \varepsilon\Big)\frac{R(x-x_0)}{R(x)} \le \frac{I_3(x)}{R(x)} \le \Big(\frac{y}{\mu_F} + \varepsilon\Big)\frac{R(x-x_0)}{R(x)}.$$

Finally, for $I_2(x)$ and the same x_0 we apply the subadditivity of $H_0(x)$ proved in Lemma 6.1.3 to find that $|I_2(x)/R(x)| \le H_0(y)(1 - R(x-x_0)/R(x))$. Now let $x \to \infty$. Then the contributions $I_2(x)$ and $I_3(x)$ disappear since by (6.2.15) we have $R(x+y)/R(x) \sim B(x+y)/B(x) \to 1$ for each $y \ge 0$ fixed and $x \to \infty$. Thus, by letting $\varepsilon \downarrow 0$, we get (6.2.16). □

6.2.3 Approximation to the Aggregate Claim Amount

As an application of weighted renewal functions to actuarial problems we deal with estimates for the aggregate claim amount, that is we consider again the compound distribution $F_X = \sum_{k=0}^{\infty} p_k F_U^{*k}$, where F_X is the distribution of the aggregate claim amount accumulated by a claim size distribution F_U. Recall from the discussion of the Lundberg bounds in Section 4.5 how exponential estimates for the tail function $\overline{F}_X(x)$ heavily depended on the existence of a solution $\gamma > 0$ to an equation related to the moment generating function $\hat{m}_{F_U}(s)$. Recall the notation $s_{F_U}^+ = \sup\{s \geq 0 : \hat{m}_{F_U}(s) < \infty\}$ introduced in Section 2.3. If a compound geometric distribution is considered, i.e. $p_k = (1-p)p^k$ for some $p \in (0,1)$, and if $\hat{m}_{F_U}(s_{F_U}^+) = \infty$, as is usually the case, then the existence of a unique solution to (4.5.5) is guaranteed, whatever the value of p.

In this section we will show that a similar procedure can be used to derive asymptotic expressions of $\overline{F}_X(x)$ for large x. Starting out with a distribution F_U that has an exponentially bounded tail, we use $\{\tilde{F}_{U,t}, t \in (s_{F_U}^-, s_{F_U}^+)\}$, the family of distributions associated with F_U which has been defined in Section 2.3, to get an alternative expression for the tail of the compound distribution F_X.

Consider the weighted renewal function

$$M_s(y) = \sum_{k=1}^{\infty} p_k \left(\hat{m}_{F_U}(s)\right)^k \tilde{F}_{U,s}^{*k}(y) \tag{6.2.17}$$

for $s \in (s_{F_U}^-, s_{F_U}^+)$. Recall that $r_0 = (\limsup_{n\to\infty}(p_n)^{1/n})^{-1}$ is the radius of convergence of $\sum_{n=0}^{\infty} p_n s^n$. Then we have the following representation formula.

Lemma 6.2.2 *Assume that the generating function $\hat{g}(z) = \sum_{k=0}^{\infty} p_k z^k$ of the claim number distribution $\{p_k\}$ has radius of convergence r_0, which is larger than 1. Take $r \in (1, r_0)$ and assume that there exists a positive value γ for which $\hat{m}_{F_U}(\gamma) = r$. Then the compound distribution F_X is given by*

$$1 - F_X(x) = \gamma \mathrm{e}^{-\gamma x} \int_0^{\infty} \mathrm{e}^{-\gamma v} \left(M_\gamma(x+v) - M_\gamma(x)\right) \mathrm{d}v\,. \tag{6.2.18}$$

Proof By our assumptions we have $s_{F_U}^+ > 0$. Thus, for any $s \in (0, s_{F_U}^+)$ we can write

$$\begin{aligned} 1 - F_X(x) &= \sum_{k=1}^{\infty} p_k (1 - {F_U}^{*k}(x)) \\ &= \sum_{k=1}^{\infty} p_k \left(\hat{m}_{F_U}(s)\right)^k \int_x^{\infty} \mathrm{e}^{-sy}\, \mathrm{d}\tilde{F}_{U,s}^{*k}(y) = \int_x^{\infty} \mathrm{e}^{-sy}\, \mathrm{d}M_s(y) \end{aligned}$$

with $M_s(y)$ as defined in (6.2.17). Now, changing variables in the last integral yields the formula

$$\mathrm{e}^{sx}(1 - F_X(x)) = \int_0^\infty \mathrm{e}^{-sv} M_s(x + \mathrm{d}v) . \tag{6.2.19}$$

We rewrite this expression by performing an integration by parts. Let w be any large but finite positive number. Then

$$\begin{aligned}
&\int_0^w \mathrm{e}^{-\gamma v} M_\gamma(x + \mathrm{d}v) \\
&= \mathrm{e}^{-\gamma w} M_\gamma(x + w) - M_\gamma(x) + \gamma \int_0^w \mathrm{e}^{-\gamma v} M_\gamma(x + v)\, \mathrm{d}v \\
&= \mathrm{e}^{-\gamma w} \big(M_\gamma(x + w) - M_\gamma(x)\big) + \gamma \int_0^w \mathrm{e}^{-\gamma v} \big(M_\gamma(x + v) - M_\gamma(x)\big)\, \mathrm{d}v .
\end{aligned}$$

We next show that, due to the fact that $\gamma > 0$, the first summand disappears as $w \to \infty$. Let $\delta \in (0, \gamma)$ be arbitrary but fixed. Then

$$\hat{m}^k_{\tilde{F}_{U,\gamma}}(-\delta) \geq \int_0^{x+w} \mathrm{e}^{-\delta y}\, \mathrm{d}\tilde{F}^{*k}_{U,\gamma}(y) \geq \mathrm{e}^{-\delta(x+w)}\, \tilde{F}^{*k}_{U,\gamma}(x + w) .$$

Using this inequality we get the following estimate:

$$\begin{aligned}
\mathrm{e}^{-\gamma w}\big(M_\gamma(x + w) - M_\gamma(x)\big) &\leq \mathrm{e}^{-\gamma w} M_\gamma(x + w) \\
&= \mathrm{e}^{-\gamma w} \sum_{k=1}^\infty p_k (\hat{m}_{F_U}(\gamma))^k \tilde{F}^{*k}_{U,\gamma}(x + w) \\
&\leq \mathrm{e}^{(-\gamma+\delta)w+\delta x} \sum_{k=1}^\infty p_k \big(\hat{m}_{F_U}(\gamma) \hat{m}_{\tilde{F}_{U,\gamma}}(-\delta)\big)^k .
\end{aligned}$$

On the other hand, by the definition of the associated distribution $\tilde{F}_{U,\gamma}$ (see also (2.3.7)), we have $\hat{m}_{F_U}(\gamma)\hat{m}_{\tilde{F}_{U,\gamma}}(-\delta) = \hat{m}_{F_U}(\gamma - \delta)$. Since $\gamma > \delta > 0$ we see that $\hat{m}_{F_U}(\gamma - \delta) < r$ and hence the remaining sum is bounded by a constant c, independently of w. We therefore have $\mathrm{e}^{-\gamma w}\big(M_\gamma(x+w) - M_\gamma(x)\big) \leq c\,\mathrm{e}^{-(\gamma-\delta)w+\delta x}$, where the bound tends to zero as $w \to \infty$. $\square$

We now formulate a general result that shows under what conditions the asymptotic behaviour of $\overline{F}_X(x)$ is basically exponential. One of these conditions is that the generating function of the claim number distribution has a *finite* radius of convergence. Note that unfortunately the Poisson distribution and a couple of other traditional claim number distributions escape the specific approach with associated distributions since their moment generating functions are entire.

Theorem 6.2.3 *Consider the compound distribution $F_X(x) = \sum_{k=0}^{\infty} p_k F_U^{*k}(x)$ and define $a_k = p_k \hat{m}_{F_U}^k(\gamma)$ and $b_k = (k+1)a_{k+1} - ka_k$. Now assume that*
(a) *the generating function $\hat{g}(z) = \sum_{k=0}^{\infty} p_k z^k$ has a finite radius of convergence $r_0 > 1$,*
(b) *there exists a value $\gamma \in (0, s_{F_U}^+)$ for which $\hat{m}_{F_U}(\gamma) = r_0$,*
(c) *$b_k \sim ck^\ell$ as $k \to \infty$, for some $\ell > -1$ and $c > 0$.*
Then, with the notation $\kappa = \hat{m}_{F_U}^{(1)}(\gamma)/r_0$,

$$1 - F_X(x) \sim \frac{c}{\kappa^{\ell+1}\gamma(1+\ell)} \mathrm{e}^{-\gamma x} x^\ell, \qquad x \to \infty. \tag{6.2.20}$$

Proof As shown in Example 3 of Section 6.2.1, we can apply Theorem 6.2.1 to the weighted renewal function $B(x) = \sum_{k=0}^{\infty} b_k \tilde{F}_{U,\gamma}^{*k}(x)$ to get

$$B(x) \sim \frac{c}{1+\ell}\Big(\frac{x}{\mu_{\tilde{F}_{U,\gamma}}}\Big)^{1+\ell}. \tag{6.2.21}$$

In particular, for all $y \in \mathbb{R}$, $B(x+y) \sim B(x)$ as $x \to \infty$. Furthermore, $b_k \geq 0$ for all k sufficiently large. By Theorem 6.2.2 we can then derive a Blackwell-type renewal theorem for the weighted renewal function $A(x) = \sum_{k=0}^{\infty} a_k \tilde{F}_{U,\gamma}^{*k}(x)$. Using the notation introduced in (6.2.17), this means that for all $v \in \mathbb{R}$

$$M_\gamma(x+v) - M_\gamma(x) \sim vB(x)/x, \qquad x \to \infty. \tag{6.2.22}$$

Note that $M_\gamma(x+v) - M_\gamma(x) \leq \sum_{k=1}^{\infty} a_k = \hat{g}(\hat{m}_{F_U}(\gamma))$, which is finite by assumptions (a) and (b). Since $s_{\tilde{F}_{U,\gamma}}^+ = s_{F_U}^+ - \gamma$ we have $B(x)\mathrm{e}^{(s_{F_U}^+ - \gamma - \varepsilon)x} \to 0$ for all $\varepsilon > 0$ and it follows that there is a constant $c > 0$ such that $x/B(x) \leq c\,\mathrm{e}^{-(s_{F_U}^+ - \gamma - \varepsilon)x}$. Thus we can apply the dominated convergence theorem to obtain from (6.2.22) that

$$\int_0^\infty \mathrm{e}^{-\gamma v} B(x)^{-1} x \big(M_\gamma(x+v) - M_\gamma(x)\big)\,\mathrm{d}v \to \int_0^\infty \mathrm{e}^{-\gamma v} v\,\mathrm{d}v, \qquad x \to \infty.$$

Thus, (6.2.18) and (6.2.21) give

$$\frac{\mathrm{e}^{\gamma x}}{\gamma}(1 - F_X(x)) \sim \frac{c}{(1+\ell)\gamma^2 x}\Big(\frac{x}{\mu_{\tilde{F}_{U,\gamma}}}\Big)^{1+\ell}, \qquad x \to \infty.$$

which leads to (6.2.20) since $\hat{m}_{\tilde{F}_{U,\gamma}}^{(1)}(0) = \hat{m}_{F_U}^{(1)}(\gamma)/\hat{m}_{F_U}(\gamma) = \kappa$ by (2.3.7). $\square$

Examples We illustrate the result of Theorem 6.2.3 by a few concrete examples of claim number distributions.

1. Recall from Section 4.3.1 that the *negative binomial distribution* $\mathrm{NB}(\alpha,p)$ is a popular claim number distribution. Thus, let us apply Theorem 6.2.3 to the case where

$$p_k = \binom{\alpha+k-1}{k}(1-p)^\alpha p^k .$$

Since the generating function $\hat{g}(z) = \sum_{k=0}^{\infty} p_k z^k$ is then given by the expression $\hat{g}(z) = \left((1-p)/(1-pz)\right)^\alpha$ we see that its radius of convergence is $r_0 = p^{-1}$. Hence define the quantity $\gamma \in (0, s_{F_U}^+)$ by the equation $\hat{m}_{F_U}(\gamma) = r_0$ provided that such a solution exists. Then we can identify the current situation with that of Theorem 6.2.3, where $a_k = (1-p)^\alpha\binom{\alpha+k-1}{k}$. Note that in this case $b_k = \alpha a_k$, as the reader can show by a simple calculation. Hence $b_k \sim (1-p)^\alpha(\Gamma(\alpha))^{-1}k^{\alpha-1}$. A straightforward application of Theorem 6.2.3 yields the estimate

$$1 - F_X(x) \sim \left(\frac{1}{\hat{m}_{F_U}^{(1)}(\gamma)}\right)^\alpha \frac{1}{\gamma\Gamma(\alpha)}\,\mathrm{e}^{-\gamma x}x^{\alpha-1} \tag{6.2.23}$$

where as before γ is the solution to $\hat{m}_{F_U}(\gamma) = p^{-1}$. We draw attention to the form of the right-hand side where we recognize the tail behaviour of a gamma distribution. As such (6.2.23) should be compared to (5.4.28) that has been obtained from the Beekman–Bowers approximation. Also note that the asymptotic expansion in (6.2.23) is of a different nature than approximations that are obtained from refined versions of the central limit theorem.

2. The *logarithmic distribution* $\mathrm{Log}(p)$ can be treated in the same spirit but the approach is technically simpler. Here $p_k = (-\log(1-p))^{-1}p^k/k$ for $k > 0$ where $0 < p < 1$. Since the radius of convergence of $\hat{g}(z)$ equals $r_0 = p^{-1}$ we look for a value γ that satisfies the equality $\hat{m}_{F_U}(\gamma) = p^{-1}$. Once this value has been chosen, we identify the quantity a_k in Theorem 6.2.3 by $a_k = c/k$ with $c = (-\log(1-p))^{-1}$. This implies that $b_k = c\delta_k(\{0\})$ and hence $B(x) = c$. The relation in (6.2.22) immediately yields $M_\gamma(x+v) - M_\gamma(x) \sim cv/x$. Hence, without recourse to Theorem 6.2.3, the relation in (6.2.18) immediately applies to give $1 - F_X(x) \sim (-\log(1-p)\gamma)^{-1}x^{-1}\,\mathrm{e}^{-\gamma x}$.

3. The *Engen distribution* $\mathrm{Eng}(\theta, a)$ can be found as a solution to the recursive system (4.3.1) starting at $k = 2$. Then the probabilities p_k are given by

$$p_k = \frac{\theta}{1-(1-a)^\theta}\,\frac{a^k\Gamma(k-\theta)}{k!\Gamma(1-\theta)}$$

where $0 < \theta < 1$ and $0 < a < 1$. The generating function $\hat{g}(z)$ has radius of convergence equal to a^{-1}. Again, take γ such that $\hat{m}_{F_U}(\gamma) = a^{-1}$. Then $a_k = c\Gamma(k-\theta)/k! \sim c\,k^{-\theta}$, where $c = \theta((1-(1-a)^\theta)\Gamma(1-\theta))^{-1}$. Further, $b_k = (1-\theta)a_k$ as follows from a simple calculation. Putting $\kappa = a\hat{m}_{F_U}^{(1)}(\gamma)$ we ultimately find the estimate $1 - F_X(x) \sim c\gamma^{-1}\kappa^{\theta-1}\,\mathrm{e}^{-\gamma x}x^{-\theta}$.

6.2.4 Lundberg-Type Bounds

The exponential estimates presented in Section 6.2.3 supplement the Lundberg bounds as derived in Section 4.5. We now show that similar upper bounds for the tail of the compound distribution $F_X(x) = \sum_{k=0}^{\infty} p_k F_U^{*k}(x)$ are possible whenever associated distributions of F_U are applicable. In connection with this we use the following *Berry–Esseen bound* for concentration functions.

Lemma 6.2.3 *Let $X_1, X_2, \ldots$ be independent and identically distributed random variables with $\mathbb{E}\, X^2 < \infty$. Then, for all $v \geq 0$, $n = 1, 2, \ldots$,*

$$\sup_{x \in \mathbb{R}} \mathbb{P}(x < X_1 + X_2 + \ldots + X_n \leq x + v) \leq cvn^{-1/2}\,, \tag{6.2.24}$$

where c is a positive constant independent of n and v.

The *proof* is omitted. It can be found, for example, in Section 3.2 of Petrov (1975).

Theorem 6.2.4 *Assume that $s_{F_U}^+ > 0$. Then, for all $s \in (0, s_{F_U}^+)$*

$$1 - F_X(x) \leq c\Big(1 + \frac{1}{s}\Big)\, \mathrm{e}^{-sx} \sum_{k=1}^{\infty} k^{-1/2} p_k (\hat{m}_{F_U}(s))^k\,. \tag{6.2.25}$$

Proof As in the proof of Lemma 6.2.2, formula (6.2.19) yields the equality

$$1 - F_X(x) = s\, \mathrm{e}^{-sx} \int_0^{\infty} \mathrm{e}^{-sv} \big(M_s(x+v) - M_s(x)\big)\, dv\,, \tag{6.2.26}$$

where in turn $M_s(y) = \sum_{k=1}^{\infty} a_k \tilde{F}_{U,s}^{*k}(y)$ and $a_k = p_k(\hat{m}_{F_U}(s))^k$. The k-fold convolution $\tilde{F}_{U,s}^{*k}$ can be interpreted as the distribution of a sum S_k of k independent copies of a random variable with distribution $\tilde{F}_{U,s}$. Thus, using the result of Lemma 6.2.3, the difference $\tilde{F}_{U,s}^{*k}(x+v) - \tilde{F}_{U,s}^{*k}(x)$ can be bounded by $cvk^{-1/2}$. Introducing this in formula (6.2.26) yields (6.2.25). □

The summation in (6.2.25) may be hard to handle. On the other hand one can still minimize the right-hand side of (6.2.25) over s satisfying $0 < s < s_{F_U}^+$.

Bibliographical Notes. The estimate in (6.2.11) has been found in Embrechts, Maejima and Omey (1984). The more specific form (6.2.23) can be found in Embrechts, Maejima and Teugels (1985). For the other cases and applications to stop-loss calculations, see Teugels and Willmot (1987). The proof of Lemma 6.2.2 has been inspired by Steinebach (1997).

6.3 RANDOM WALKS

We turn to the discussion of some basic properties of random walks on the real line $\mathbb{R}$. These processes are useful when computing ruin probabilities in the case where premiums are random or when extending bounds and asymptotic results as in Section 5.4 to the case of general inter-occurrence times.

Let $Y_1, Y_2, \ldots$ be a sequence of independent and identically distributed (not necessarily integer-valued) random variables with distribution F which can take both positive and negative values. The sequence $\{S_n,\ n \in \mathbb{N}\}$ with $S_0 = 0$ and $S_n = Y_1 + \ldots + Y_n$ for $n = 1, 2, \ldots$ is called a *random walk*. We assume that the first moment $\mathbb{E}\, Y$ exists and that Y is not concentrated at 0, i.e. $\mathbb{P}(Y = 0) < 1$.

6.3.1 Ladder Epochs

Look at the first entrance time of the random walk $\{S_n\}$ into the positive half-line $(0, \infty)$

$$\nu^+ = \min\{n > 0 :\ S_n > 0\}\,, \tag{6.3.1}$$

setting $\nu^+ = \infty$ if $S_n \le 0$ for all $n \in \mathbb{N}$, and call ν^+ the (first strong) *ascending ladder epoch* of $\{S_n\}$. Similarly we introduce the first entrance time to the nonpositive half-line $(-\infty, 0]$ by

$$\nu^- = \min\{n > 0 :\ S_n \le 0\}\,, \tag{6.3.2}$$

setting $\nu^- = \infty$ if $S_n > 0$ for all $n = 1, 2, \ldots$, and call ν^- the (first) *descending ladder epoch* of $\{S_n\}$. As we will see later, we need to know whether $\mathbb{E}\, Y$ is strictly positive, zero or strictly negative, as otherwise we cannot say whether ν^+ or ν^- are proper. In Figures 6.3.1 and 6.3.2 we depict the first ladder epochs ν^+ and ν^-. For each $k = 1, 2, \ldots$, the events

$$\{\nu^+ = k\} = \{S_1 \le 0, S_2 \le 0, \ldots, S_{k-1} \le 0, S_k > 0\} \tag{6.3.3}$$

and

$$\{\nu^- = k\} = \{S_1 > 0, S_2 > 0, \ldots, S_{k-1} > 0, S_k \le 0\} \tag{6.3.4}$$

are determined by the first k values of $\{S_n\}$. Note that this is a special case of the following, somewhat more general, property. Consider the σ-algebras $\mathcal{F}_0 = \{\emptyset, \Omega\}$ and $\mathcal{F}_k = \{\{\omega : (S_1(\omega), \ldots, S_k(\omega)) \in B\}, B \in \mathcal{B}(\mathbb{R}^k)\}$. Then, in view of (6.3.3) and (6.3.4), we have $\{\nu^+ = k\} \in \mathcal{F}_k$ and $\{\nu^- = k\} \in \mathcal{F}_k$ for $k \in \mathbb{N}$. This means that the ladder epochs ν^+ and ν^- are so-called *stopping times* with respect to the *filtration* $\{\mathcal{F}_n\}$ generated by $\{S_n\}$; see Chapter 9 for further details. From Corollary 9.1.1 proved there we have that, for each stopping time τ with respect to $\{\mathcal{F}_n\}$,

$$\mathbb{E}\, S_\tau = \mathbb{E}\, \tau \mathbb{E}\, Y \tag{6.3.5}$$

provided that $\mathbb{E}\,\tau < \infty$ and $\mathbb{E}\,|Y| < \infty$, which is known as *Wald's identity* for stopping times.

Actually, we can recursively define further ladder epochs. Define the sequence $\{\nu_n^+, n \in \mathbb{N}\}$ by

$$\nu_{n+1}^+ = \min\{j > \nu_n^+ : S_j > S_{\nu_n^+}\}, \tag{6.3.6}$$

where $\nu_0^+ = 0$ and $\nu_1^+ = \nu^+$and call ν_n^+ the n-th (strong ascending) *ladder epoch*. A priori, we cannot exclude the case that, from some random index on, all the ladder epochs are equal to ∞.

In a similar way, we recursively define the sequence $\{\nu_n^-, n \in \mathbb{N}\}$ of consecutive *descending ladder epochs* by $\nu_0^- = 0$, $\nu_1^- = \nu^-$ and

$$\nu_{n+1}^- = \min\{j > \nu_n^-, S_j \le S_{\nu_n^-}\}, \qquad n = 1, 2, \ldots. \tag{6.3.7}$$

Another interesting characteristic is the step ν at which the random walk $\{S_n\}$ has a local minimum for the last time before ν^-, i.e.

$$\nu = \max\Big\{n : 0 < n < \nu^-, S_n = \min_{0<j<\nu^-} S_j\Big\},$$

as depicted on Figure 6.3.1.

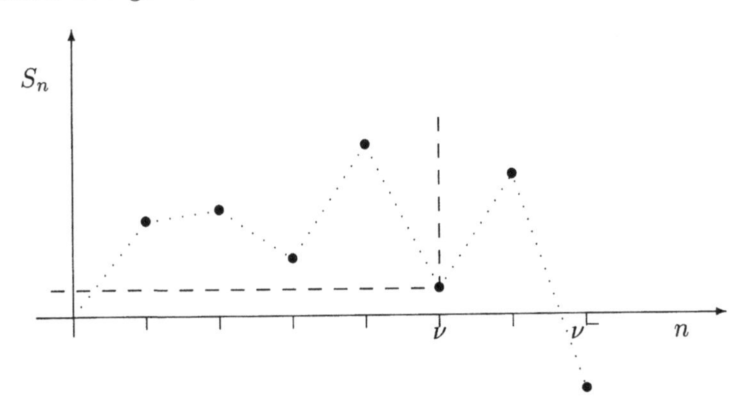

Figure 6.3.1 Last minimum before "ruin"

6.3.2 Random Walks with and without Drift

Depending on whether $\mathbb{E}\,Y$ is positive, zero or negative, we have three different kinds of evolution for the random walk $\{S_n\}$.

Theorem 6.3.1 (a) *If* $\mathbb{E}\,Y > 0$, *then* $\lim_{n\to\infty} S_n = \infty$.
(b) *If* $\mathbb{E}\,Y < 0$, *then* $\lim_{n\to\infty} S_n = -\infty$.
(c) *If* $\mathbb{E}\,Y = 0$, *then* $\limsup_{n\to\infty} S_n = \infty$ *and* $\liminf_{n\to\infty} S_n = -\infty$.

Proof From the strong law of large numbers we have $S_n/n \to \mathbb{E}Y$ with probability 1. This gives $S_n \to \infty$ with probability 1 if $\mathbb{E}Y > 0$, and $S_n \to -\infty$ with probability 1 if $\mathbb{E}Y < 0$. Thus, the statements (a) and (b) are proved. To prove (c) assume $\mathbb{E}Y = 0$ and define $N = \min\{n : S_n = \max_{j\ge 0} S_j\}$. We put $\eta = \infty$ in case $\max_{j\ge 0} S_j = \infty$. Then,

$$
\begin{aligned}
1 &\ge \mathbb{P}(N < \infty) = \sum_{n=0}^{\infty} \mathbb{P}(N = n) \\
&= \sum_{n=0}^{\infty} \mathbb{P}\left(\{S_j < S_n \text{ for all } 0 \le j \le n-1\} \cap \{S_j \le S_n \text{ for all } j > n\}\right) \\
&= \sum_{n=0}^{\infty} \mathbb{P}\big(\{\textstyle\sum_{k=j+1}^{n} Y_k > 0 \text{ for all } j = 0, 1, \ldots, n-1\} \\
&\qquad \cap \{\textstyle\sum_{k=n+1}^{j} Y_k \le 0 \text{ for all } j > n\}\big) \\
&= \sum_{n=0}^{\infty} \mathbb{P}\big(\textstyle\sum_{k=j+1}^{n} Y_k > 0 \text{ for all } j = 0, 1, \ldots, n-1\big) \\
&\qquad \times \mathbb{P}\big(\textstyle\sum_{k=n+1}^{j} Y_k \le 0 \text{ for all } j > n\big) \\
&= \sum_{n=0}^{\infty} \mathbb{P}\big(Y_n > 0, Y_n + Y_{n-1} > 0, \ldots, \textstyle\sum_{k=1}^{n} Y_k > 0\big) \\
&\qquad \times \mathbb{P}\big(S_j \le 0 \text{ for all } j \ge 0\big) \\
&= \sum_{n=0}^{\infty} \mathbb{P}\big(Y_1 > 0, Y_1 + Y_2 > 0, \ldots, \textstyle\sum_{k=1}^{n} Y_k > 0\big)\, \mathbb{P}(\nu^+ = \infty),
\end{aligned}
$$

where for the last equality we used the fact that $(Y_n, \ldots, Y_1) \stackrel{\mathrm{d}}{=} (Y_1, \ldots, Y_n)$; see Lemma 5.1.1. Thus,

$$
1 \ge \sum_{n=0}^{\infty} \mathbb{P}(\nu^- > n)\mathbb{P}(\nu^+ = \infty) = \mathbb{E}\,\nu^- \mathbb{P}(\nu^+ = \infty). \tag{6.3.8}
$$

Assume for the moment that $\mathbb{P}(\nu^+ = \infty) > 0$. Then it follows from (6.3.8) that $\mathbb{E}\,\nu^- = \sum_{n=0}^{\infty} \mathbb{P}(\nu^- > n) < \infty$. Thus, using Wald's identity (6.3.5), we have $\mathbb{E}\,S_{\nu^-} = \mathbb{E}\,\nu^- \mathbb{E}\,Y = 0$. Since $S_{\nu^-} \le 0$ by definition, we would get that $S_{\nu^-} = 0$ with probability 1. This leads to a contradiction because $\mathbb{P}(S_{\nu^-} < 0) \ge \mathbb{P}(Y_1 < 0) > 0$ if $\mathbb{E}\,Y_1 = 0$. Thus, $\mathbb{P}(\nu^+ < \infty) = 1$, i.e. the random variable S_{ν^+} is well-defined, and $S_{\nu^+} > 0$ by definition. Now consider the whole sequence $\{\nu_n^+,\ n \in \mathbb{N}\}$ of ladder epochs. Using the same argument as above we get that $\mathbb{P}(\nu_n^+ < \infty) = 1$ for all $n \in \mathbb{N}$ and that

$$
\{S_{\nu_{n+1}^+} - S_{\nu_n^+},\ n \in \mathbb{N}\} \tag{6.3.9}
$$

is a sequence of independent and identically distributed random variables which are strictly positive by definition. We leave the proof of this fact to the reader. In particular, $\mathbb{E}\, S_{\nu^+} > 0$. By the strong law of large numbers, we have

$$\frac{1}{n} S_{\nu_n^+} = \frac{1}{n} \sum_{k=0}^{n-1} (S_{\nu_{k+1}^+} - S_{\nu_k^+}) \to \mathbb{E}\, S_{\nu^+} > 0$$

for $n \to \infty$. Thus, $\lim_{n\to\infty} S_{\nu_n^+} = \infty$, i.e. $\limsup_{n\to\infty} S_n = \infty$. The proof that $\liminf_{n\to\infty} S_n = -\infty$ is analogous because we can consider the reflected random walk $\{-S_n\}$ with $\mathbb{E}\,(-Y) = 0$. $\square$

Theorem 6.3.1 motivates the use of the following terminology. We say that the random walk $\{S_n\}$

- has a *positive drift* provided that $\mathbb{E}\, Y > 0$,
- has a *negative drift* provided that $\mathbb{E}\, Y < 0$,
- is *without drift* or *oscillating* provided that $\mathbb{E}\, Y = 0$.

As already noticed in Section 5.1.2, the ladder epochs and, in particular, the maximum $M = \max\{0, S_1, S_2, \ldots\}$ of a random walk play an important role in the computation of ruin probabilities. Note that Theorem 6.3.1 implies that M is finite with probability 1 for a random walk with negative drift, and infinite otherwise.

6.3.3 Ladder Heights; Negative Drift

In this subsection we assume that the random walk $\{S_n\}$ has a negative drift, i.e. $\mathbb{E}\, Y < 0$. A basic characteristic of $\{S_n\}$ is then the first ascending ladder epoch ν^+. As one can expect, and we confirm this in Theorem 6.3.2, the distribution of the random variable ν^+ is defective under the assumption of a negative drift. The *overshoot* Y^+ above the zero level is defined by

$$Y^+ = \begin{cases} S_{\nu^+} & \text{if } \nu^+ < \infty, \\ \infty & \text{otherwise} \end{cases}$$

and is called the (first strong) *ascending ladder height*. A typical trajectory of the random walk $\{S_n\}$ which reflects this situation is presented in Figure 6.3.2.

More precisely, we have a result for $G^+(x) = \mathbb{P}(Y^+ \le x)$, the distribution function of Y^+ and $G^+(\infty) = \lim_{x\to\infty} G^+(x)$.

Theorem 6.3.2 *The following statements are equivalent:*
(a) $\mathbb{E}\, Y < 0$,
(b) *M is finite with probability* 1,
(c) $G^+(\infty) < 1$.

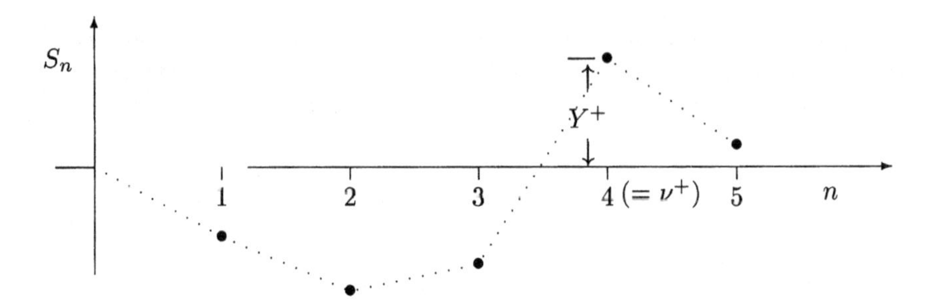

Figure 6.3.2 Ascending ladder height

The *proof* of this theorem is easy and is left to the reader.

Suppose that $\nu^+ < \infty$. We can then repeat the same argument as above, but now from the point (ν^+, Y^+), because of our assumption that the increments $Y_1, Y_2, \ldots$ of the random walk $\{S_n\}$ are independent and identically distributed. This means in particular, as illustrated in Figure 6.3.2, that we can define a new random walk $S_{\nu^++1} - S_{\nu^+}, S_{\nu^++2} - S_{\nu^+}, \ldots$ which can be proved to be an identically distributed copy of the original random walk $\{S_n\}$ and independent of $S_1, S_2, \ldots, S_{\nu^+}$. We leave it to the reader to show this. Iterating this procedure, we can recursively define the sequence $\{\nu_n^+\}$ of consecutive ladder epochs in the same way as this was done in (6.3.6). The random variable

$$Y_n^+ = \begin{cases} S_{\nu_n^+} - S_{\nu_{n-1}^+} & \text{if } \nu_n^+ < \infty, \\ \infty & \text{otherwise} \end{cases}$$

is called the n-th *ascending ladder height* of $\{S_n\}$. It is not difficult to show that the sequence $\{Y_1^+ + \ldots + Y_n^+, n = 0, 1, \ldots\}$ forms a terminating renewal process. Moreover, for the maximum $M = \max\{0, S_1, S_2, \ldots\}$ of $\{S_n\}$ we have (see also Figure 6.3.2)

$$M \stackrel{\text{d}}{=} \sum_{i=1}^{N} Y_i^+, \tag{6.3.10}$$

where $N = \max\{n : \nu_n^+ < \infty\}$ is the number of finite ladder epochs. Thus, with the notation $G_0(x) = G^+(x)/G^+(\infty)$, where $G_0(x)$ is a proper (i.e. nondefective) distribution function, we arrive at the following result, saying that M has a compound geometric distribution.

Theorem 6.3.3 *If* $\mathbb{E}Y < 0$, *then for all* $x \geq 0$ *and for* $p = G^+(\infty)$

$$\mathbb{P}(M \leq x) = (1-p)\sum_{k=0}^{\infty}(G^+)^{*k}(x) = \sum_{k=0}^{\infty}(1-p)p^k G_0^{*k}(x)\,. \tag{6.3.11}$$

Proof Recall that $\{Y_1^+ + \ldots + Y_n^+, n = 0,1,\ldots\}$ is a terminating renewal process and the distribution of $N = \max\{n : \nu_n^+ < \infty\}$ is geometric with parameter $p = G^+(\infty)$, i.e. $\mathbb{P}(N = k) = (1-p)p^k$ for $k = 0,1,\ldots$. Thus, using (6.3.10) and Theorem 6.1.1b, we have

$$\mathbb{P}(M \le x) = \mathbb{P}\Big(\sum_{i=1}^{N} Y_i^+ \le x\Big) = \sum_{k=0}^{\infty}(1-p)p^k G_0^{*k}(x) = \sum_{k=0}^{\infty}(1-p)(G^+)^{*k}(x)\,.$$

This completes the proof. □

Theorem 6.3.3 implies the following result for the ruin function $\psi(u) = \mathbb{P}(M > u)$ considered in Section 5.1.

Corollary 6.3.1 *For any $u \ge 0$, $\psi(u) = \sum_{k=1}^{\infty}(1-p)p^k \overline{G_0^{*k}}(u)$.*

Note that for the special case of integer-valued increments concentrated on the set $\{-1,0,1,\ldots\}$, the statement of Theorem 6.3.3 has already been proved in Theorem 5.1.1 where the distribution function $G_0(x)$ has been determined explicitly. In the general case, the calculation of $G_0(x)$ is more complicated. We will discuss this problem in Sections 6.4.2 and 6.4.3.

However, before doing so, we introduce the dual notions of descending ladder heights. Consider the descending ladder epoch ν^-. The *undershoot* Y^- below the zero level is defined by $Y^- = S_{\nu^-}$ and called the (first) *descending ladder height*. The n-th descending ladder height is defined by $Y_n^- = S_{\nu_n^-} - S_{\nu_{n-1}^-}$. Since $Y_1^-,\ldots,Y_n^-$ are independent and identically distributed copies of Y^-, it is clear that the sequence $\{-\sum_{i=1}^n Y_i^-,\ n \in \mathbb{N}\}$ is a nonterminating renewal process (in the case of the negative drift). Indeed, under our assumption on the negative drift it follows from Theorem 6.3.1 that all descending ladder epochs and heights are proper random variables.

Bibliographical Notes. The basic references for Section 6.3 are Feller (1971) and Chung (1974). The proof of Theorem 6.3.1 is from Resnick (1992).

6.4 THE WIENER–HOPF FACTORIZATION

6.4.1 General Representation Formulae

Define the ladder height distribution G^-, concentrated on $\mathbb{R}_-$, by

$$G^-(x) = \mathbb{P}(Y^- \le x)\,, \qquad x \in \mathbb{R}\,. \tag{6.4.1}$$

Thus G^- dualizes the ladder height distribution G^+ which is concentrated on $(0,\infty)$ and is given by

$$G^+(x) = \mathbb{P}(Y^+ \le x)\,, \qquad x \in \mathbb{R}\,. \tag{6.4.2}$$

Let H_0^- be the measure on $\mathbb{R}_-$ given by

$$H_0^-(B) = \sum_{k=0}^{\infty} (G^-)^{*k}(B)\,, \qquad B \in \mathcal{B}(\mathbb{R}_-)\,. \tag{6.4.3}$$

We also introduce as a dual measure H_0^+ on $\mathbb{R}_+$

$$H_0^+(B) = \sum_{k=0}^{\infty} (G^+)^{*k}(B)\,, \qquad B \in \mathcal{B}(\mathbb{R}_+)\,. \tag{6.4.4}$$

Note that the definition (6.4.3) of H_0^- is similar to that of the renewal measure H_0 considered in Section 6.1. Moreover, from (6.4.3) it follows that

$$H_0^- * G^- = H_0^- - \delta_0\,. \tag{6.4.5}$$

It turns out that H_0^- is equal to the so-called *pre-occupation measure* γ^- given by $\gamma^-(B) = \mathbb{E}\left(\sum_{i=0}^{\nu^+-1} \mathbb{I}(S_i \in B)\right)$ for $B \in \mathcal{B}(\mathbb{R})$, where obviously $\gamma^-(B) = 0$ for $B \subset (0,\infty)$.

Lemma 6.4.1 *For each $B \in \mathcal{B}(\mathbb{R})$ and $H_0^-(B) = H_0^-(B \cap \mathbb{R}_-)$ we have $H_0^-(B) = \gamma^-(B)$.*

Proof Note that $(G^-)^{*0}(B) = \mathbb{I}(0 \in B) = \mathbb{P}(S_0 \in B)$ and $(G^-)^{*n}(B) = \mathbb{P}(Y_1^- + \ldots + Y_n^- \in B) = \mathbb{P}(S_{\nu_n^-} \in B)$ for all $n = 1, 2, \ldots$. Thus

$$\begin{aligned}
H_0^-(B) &= \sum_{n=0}^{\infty} (G^-)^{*n}(B) = \mathbb{I}(0 \in B) + \sum_{n=1}^{\infty} \mathbb{P}(S_{\nu_n^-} \in B) \\
&= \mathbb{I}(0 \in B) + \mathbb{E}\Big(\sum_{n=1}^{\infty} \mathbb{I}(S_{\nu_n^-} \in B)\Big) \\
&= \mathbb{I}(0 \in B) + \mathbb{E}\Big(\sum_{n=1}^{\infty}\sum_{k=1}^{\infty} \mathbb{I}(\nu_n^- = k)\mathbb{I}(S_{\nu_n^-} \in B)\Big) \\
&= \mathbb{I}(0 \in B) + \sum_{k=1}^{\infty} \mathbb{E}\Big(\sum_{n=1}^{\infty} \mathbb{I}(\nu_n^- = k, S_{\nu_n^-} \in B)\Big).
\end{aligned}$$

On the other hand,

$$\begin{aligned}
&\mathbb{E}\Big(\sum_{n=1}^{\infty} \mathbb{I}(\nu_n^- = k, S_{\nu_n^-} \in B)\Big) = \mathbb{P}(S_k \le S_i,\ i = 0, \ldots, k-1,\ S_k \in B) \\
&= \mathbb{P}\Big(\sum_{i=1}^{k} Y_i \le 0, \sum_{i=2}^{k} Y_i \le 0, \ldots, Y_k \le 0, \sum_{i=1}^{k} Y_i \in B\Big) \\
&= \mathbb{P}\Big(Y_1 \le 0, \ldots, \sum_{i=1}^{k} Y_i \le 0, \sum_{i=1}^{k} Y_i \in B\Big),
\end{aligned}$$

where the last equality follows from Lemma 5.1.1. This gives

$$\begin{aligned} H_0^-(B) &= \mathbb{I}(0 \in B) + \sum_{k=1}^{\infty} \mathbb{P}(S_i \le 0, i = 0, \ldots, k, S_k \in B) \\ &= \mathbb{E}\Big(\sum_{k=0}^{\infty} \mathbb{I}(\nu^+ \ge k+1, S_k \in B)\Big). \end{aligned}$$

Thus, the proof is complete since

$$\mathbb{E}\Big(\sum_{k=0}^{\infty} \mathbb{I}(\nu^+ \ge k+1, S_k \in B)\Big) = \mathbb{E}\Big(\sum_{k=0}^{\nu^+-1} \mathbb{I}(S_k \in B)\Big).$$

□

Next we show that the distribution F of the increments $Y_1, Y_2, \ldots$ of the random walk $\{S_n\}$ can be expressed in terms of the ladder height distributions G^+ and G^-. This is the so-called *Wiener–Hopf factorization* of F, which is sometimes useful when computing the distribution of the maximum M of the random walk $\{S_n\}$.

Theorem 6.4.1 *The following relationship holds:*

$$F = G^+ + G^- - G^- * G^+. \tag{6.4.6}$$

Proof We first show that

$$\delta_0 + \gamma^- * F = \gamma^- + G^+. \tag{6.4.7}$$

Let $B \in \mathcal{B}(\mathbb{R})$ be an arbitrary Borel set. Then,

$$\sum_{n=0}^{\nu^+-1} \mathbb{I}(S_n \in B) + \mathbb{I}(S_{\nu^+} \in B) = \mathbb{I}(0 \in B) + \sum_{n=0}^{\nu^+-1} \mathbb{I}(S_{n+1} \in B). \tag{6.4.8}$$

Now $\mathbb{E}\big(\sum_{n=0}^{\nu^+-1} \mathbb{I}(S_{n+1} \in B)\big) = \mathbb{E}\big(\sum_{n=0}^{\infty} \mathbb{I}(\nu^+ > n, S_{n+1} \in B)\big)$ and, since the event $\{\nu^+ > n\}$ is independent of Y_{n+1}, the above equals

$$\sum_{n=0}^{\infty} \int_{-\infty}^{\infty} \mathbb{P}(\nu^+ > n, S_n \in B - y)\, F(\mathrm{d}y) = F * \gamma^-(B).$$

Taking the expected value of both sides of (6.4.8), we get (6.4.7). Convoluting both sides of (6.4.7) with G^- we obtain

$$G^- + G^- * \gamma^- * F = G^- * \gamma^- + G^- * G^+.$$

On the other hand, by (6.4.5) and Lemma 6.4.1, we have $\gamma^- * G^- = \gamma^- - \delta_0$. Thus, $G^- + (\gamma^- - \delta_0) * F = \gamma^- - \delta_0 + G^- * G^+$ and, equivalently, $F = G^- - G^- * G^+ + \gamma^- * F - \gamma^- + \delta_0$. Using (6.4.7) again, this gives (6.4.6). □

If we want to compute ruin probabilities, we need to determine the ladder height distribution G^+ that appears in Theorem 6.3.3. The Wiener–Hopf factorization (6.4.6) yields the following representation formula for G^+.

Corollary 6.4.1 *For* $B \in \mathcal{B}((0,\infty))$,

$$G^+(B) = F * H_0^-(B) = \int_{-\infty}^{0} F(B-y)\,\mathrm{d}H_0^-(y)\,, \tag{6.4.9}$$

while for $B \in \mathcal{B}(\mathbb{R}_-)$

$$G^-(B) = F * H_0^+(B) = \int_{0}^{\infty} F(B-y)\,\mathrm{d}H_0^+(y)\,. \tag{6.4.10}$$

Proof Convoluting both sides of the Wiener–Hopf factorization (6.4.6) by G^-, we obtain $F * G^- = G^+ * G^- + G^- * G^- - G^+ * (G^-)^{*2}$. Iterating this procedure we get $F * (G^-)^{*k} = G^+ * (G^-)^{*k} + (G^-)^{*(k+1)} - G^+ * (G^-)^{*(k+1)}$ for each $k = 1, 2, \ldots$. Summing over k from 0 to ∞ we obtain

$$F * \sum_{k=0}^{n} (G^-)^{*k}(B) = G^+(B) - G^+ * (G^-)^{*(n+1)}(B)\,,$$

for $B \in \mathcal{B}((0,\infty))$. This completes the proof, because $\lim_{n\to\infty}(G^-)^{*n}(B) = 0$ for all $B \in \mathcal{B}((0,\infty))$ and hence $\lim_{n\to\infty} G^+ * (G^-)^{*(n+1)}(B) = 0$. The proof of (6.4.10) is similar. □

6.4.2 An Analytical Factorization; Examples

We give two cases for which we can compute the distribution of the maximum $M = \max\{0, S_1, S_2, \ldots\}$ of the random walk $\{S_n\}$, using the Wiener–Hopf factorization (6.4.6). Recall that M is finite with probability 1 if and only if $\{S_n\}$ has a negative drift; see Theorem 6.3.2. Thus, throughout this section we suppose that $\mathbb{E}\,Y < 0$. From Theorem 6.3.3 we can easily draw the following general result.

Corollary 6.4.2 (a) *For* $s \le 0$,

$$\hat{m}_M(s) = \frac{1 - G^+(\infty)}{1 - \hat{m}_{G^+}(s)}\,. \tag{6.4.11}$$

(b) *If Y is integer-valued, then M is a nonnegative integer-valued random variable with generating function*

$$\hat{g}_M(s) = \frac{1 - G^+(\infty)}{1 - \hat{g}_{G^+}(s)}, \qquad -1 \le s \le 1. \tag{6.4.12}$$

The *proof* is obvious because (6.4.11) and (6.4.12) directly follow from (6.3.11) and from the product formulae (2.1.9) and (2.1.10). □

Recall that the moment generating functions of the distributions G^+, G^- and F can be defined as functions of a complex variable $z \in \mathbb{C}$. Since G^+ is concentrated on $(0, \infty)$, the moment generating function $\hat{m}_{G^+}(z)$ is well-defined on the half-plane $\Re(z) \le 0$. Analogously, since G^- is concentrated on $\mathbb{R}_-$, its moment generating function $\hat{m}_{G^-}(z)$ is well-defined on the half-plane $\Re(z) \ge 0$. The moment generating function $\hat{m}_F(z)$ is well-defined at least on the imaginary axis $\Re(z) = 0$ because each point on $\Re(z) = 0$ can be represented as $z = it$ for some real t and then $\hat{m}_F(z) = \int_{-\infty}^{\infty} \mathrm{e}^{itx}\, \mathrm{d}F(x)$, which is the characteristic function of F. An immediate consequence of the Wiener–Hopf factorization (6.4.6) is the following *analytical factorization* of the corresponding moment generating functions. For the generating function, see Corollary 6.4.4.

Corollary 6.4.3 *If for some $z \in \mathbb{C}$ all the moment generating functions $\hat{m}_F(z)$, $\hat{m}_{G^+}(z)$, $\hat{m}_{G^-}(z)$ exist, in particular if $\Re(z) = 0$, then*

$$1 - \hat{m}_F(z) = (1 - \hat{m}_{G^+}(z))(1 - \hat{m}_{G^-}(z)). \tag{6.4.13}$$

If Y is an integer-valued random variable with probability function $\{p_k\}$, i.e. $p_k = \mathbb{P}(Y = k)$, then both the ladder-height distributions G^+ and G^- are discrete: G^+ is concentrated on $\{1, 2, \ldots\}$ and G^- is concentrated on $\{0, -1, -2, \ldots\}$. Let $\{p_k^+\}$ and $\{p_k^-\}$ denote the probability function of G^+ and G^-, respectively. Let us consider the generating functions $\hat{g}_{G^+}(z)$, $\hat{g}_{G^-}(z)$ and $\hat{g}_F(z)$. Note that $\hat{g}_{G^+}(z)$ is well-defined in $|z| \le 1$, $\hat{g}_{G^-}(z)$ is well-defined in $|z| \ge 1$, and $\hat{g}_F(z)$ is well-defined at least on $|z| = 1$. In this case, from Euler's formula we have $z = \mathrm{e}^{it}$ for some $0 \le t < 2\pi$, and $\hat{g}_F(z) = \sum_{k=-\infty}^{\infty} p_k\, \mathrm{e}^{itk}$, which is the characteristic function of Y at t. Then (6.4.6) takes the form

$$p_k = p_k^+ + p_k^- - \sum_{j=-\infty}^{\infty} p_{k-j}^+ p_j^-, \qquad k \in \mathbb{Z}. \tag{6.4.14}$$

Analogous to the analytical factorization given in Corollary 6.4.3, we have the immediate consequence of (6.4.6).

Corollary 6.4.4 *For $|z| = 1$,*

$$1 - \hat{g}_F(z) = (1 - \hat{g}_{G^+}(z))(1 - \hat{g}_{G^-}(z)). \tag{6.4.15}$$

Proof Multiplying both sides of (6.4.14) by z^k ($|z|=1$) and summing over k from $-\infty$ to $+\infty$ we arrive at

$$\sum_{k=-\infty}^{\infty} p_k z^k = \sum_{k=-\infty}^{\infty} p_k z^k + \sum_{k=-\infty}^{\infty} p_k^- z^k - \sum_{k=-\infty}^{\infty}\Big(\sum_{j=-\infty}^{\infty} p_{k-j}^+ p_j^-\Big) z^k .$$

However

$$\sum_{k=-\infty}^{\infty}\Big(\sum_{j=-\infty}^{\infty} p_{k-j}^+ p_j^-\Big) z^k = \sum_{k=-\infty}^{\infty} p_j^- z^j \Big(\sum_{j=-\infty}^{\infty} p_{k-j}^+ z^{k-j}\Big) = \hat g_{G^-}(z)\hat g_{G^+}(z) .$$

This gives (6.4.15). □

In the remaining part of this section we derive the explicit factorization for the special case when Y is lattice and bounded either from above or from below. Subsequently, we will find $\hat g_{G^+}(z)$.

Suppose that Y is integer-valued and bounded from above, i.e.

$$p_b > 0\,, \quad \text{and} \quad p_{b+j} = 0\,, \quad j = 1,2,\ldots \tag{6.4.16}$$

for some integer $b > 0$. In this case G^+ is discrete and assumes values from the set $\{1,2,\ldots,b\}$. Thus, the generating function $\hat g_{G^+}(z)$ is a polynomial of degree b and is well-defined in the whole complex plane $\mathbb{C}$. Using the fundamental theorem of algebra we can write

$$1 - \hat g_{G^+}(z) = \text{const} \prod_{j=1}^{b} (z_j - z)\,, \tag{6.4.17}$$

where z_j ($j = 1,2,\ldots,b$) are the roots of equation $1 = \hat g_{G^+}(z)$. Note that $\hat g_{G^+}(0) = 0$ by setting $z = 0$ in (6.4.17). Hence

$$1 - \hat g_{G^+}(z) = \prod_{j=1}^{b} (1 - z/z_j)\,. \tag{6.4.18}$$

The left-hand side of (6.4.18) is equal to 1 for $z = 0$, i.e. $z_j \neq 0$ for $j = 1,2,\ldots,b$. Moreover, G^- is discrete and concentrated on the set $\{0,-1,-2,\ldots\}$ and $G^-(0) < 1$. Thus $1 - \hat g_{G^-}(z)$ is well-defined on $|z| \ge 1$ and hence $1 - \hat g_F(z)$ is well-defined on $|z| \ge 1$. Consequently, (6.4.15) holds on $|z| \ge 1$ too.

It is obvious that, if for a complex-valued function g, $|g(z)| < 1$ for some $z \in \mathbb{C}$, then z is not a root of the equation $g(z) = 1$. But then, $1 = \hat g_{G^+}(z)$ has no root in $|z| \le 1$, since $G^+(\infty) < 1$ by the fact that $\{S_n\}$ has negative drift. More explicitly, $|\hat g_{G^+}(z)| \le \sum_{j=1}^b p_j^+ |z|^j \le \sum_{j=1}^b p_j^+ = G^+(\infty) < 1$.

Lemma 6.4.2 *The equations* $1 = \hat{g}_{G^+}(z)$ *and* $1 = \hat{g}_F(z)$ *have exactly the same roots in* $|z| > 1$.

Proof Assume for the moment that there is a complex number z_0 such that $1 = \hat{g}_F(z_0)$ and $|z_0| > 1$, but $1 \neq \hat{g}_{G^+}(z_0)$. Then from the equation

$$(1 - \hat{g}_{G^+}(z))(1 - \hat{g}_{G^-}(z)) = 1 - \hat{g}_F(z), \qquad |z| \geq 1, \tag{6.4.19}$$

we would have that $1 = \hat{g}_{G^-}(z_0)$. However, since $|z_0| > 1$ we have $|\hat{g}_{G^-}(z_0)| < \sum_{j=-\infty}^{0} p_j^- = 1$ which means that z_0 cannot be a root of $1 = \hat{g}_{G^-}(z_0)$, i.e. each root of $1 = \hat{g}_F(z)$ with $|z| > 1$ is a root of $1 = \hat{g}_{G^+}(z)$. On the other hand, by (6.4.19) each root of $1 = \hat{g}_{G^+}(z)$ with $|z| > 1$ is a root of $1 = \hat{g}_F(z)$. □

From the above considerations, we get the following result.

Theorem 6.4.2 *Let* $z_1, \ldots, z_b$ *be the roots of* $1 = \hat{g}_F(z)$ *in* $|z| > 1$. *Then*

$$\hat{g}_{G^+}(z) = 1 - \prod_{j=1}^{b}(1 - z/z_j) \tag{6.4.20}$$

and for $|z| \leq 1$

$$\hat{g}_M(z) = \left(\prod_{j=1}^{b}(1 - z/z_j)\right)^{-1}(1 - \hat{g}_{G^+}(1)). \tag{6.4.21}$$

Proof Equation (6.4.20) follows from (6.4.18) and Lemma 6.4.2. Equation (6.4.21) follows from (6.4.12) because $G^+(\infty) = \hat{g}_{G^+}(1)$. □

Corollary 6.4.5 *If the roots* $z_1, z_2, \ldots, z_b$ *of the equation* $1 = \hat{g}_F(z)$ *in* $|z| > 1$ *are different, then by partial fraction decomposition we get*

$$\hat{g}_M(z) = (1 - \hat{g}_{G^+}(1))\sum_{j=1}^{b} c_j \frac{z_j}{z_j - z}, \qquad |z| \leq 1 \tag{6.4.22}$$

where $c_j = \prod_{k \neq j}((z_k - z_j)^{-1} z_k)$.

Proof Since $z_1, \ldots, z_b$ are different, we can write

$$\prod_{j=1}^{b} \frac{z_j}{z_j - z} = \sum_{j=1}^{b} c_j \frac{z_j}{z_j - z}. \tag{6.4.23}$$

Let $k \in \{1, \ldots, b\}$ be fixed. In order to determine c_j, it suffices to multiply both sides of (6.4.23) by $(z_k - z)$ to get

$$z_k \prod_{j \neq k} \frac{z_j}{z_j - z} = \sum_{j \neq k} c_j \frac{z_j(z_k - z)}{z_j - z} + c_k z_k.$$

Then for $z = z_k$ this gives $c_k = \prod_{j\neq k} z_j/(z_j - z_k)$. □

Notice that for each $j = 1, \ldots, b$, the function

$$\frac{z_j}{z_j - z} = \frac{1}{1 - (z/z_j)}, \qquad |z| \leq 1,$$

is the generating function of the sequence $\{z_j^{-k}\}, k \in \mathbb{N}$. Thus, the representation of $\hat{g}_M(z)$ by the linear combination of the moment generating functions given in (6.4.22) makes it possible to express the survival probability $1 - \psi(n) = \mathbb{P}(M \leq n)$ directly in terms of the roots $z_1, z_2, \ldots, z_b$ of $1 = \hat{g}_F(z)$ in $|z| > 1$. From (6.4.20) and (6.4.22), it follows that

$$\mathbb{P}(M = n) = \Big(\prod_{k=1}^{b} (1 - z_k^{-1})\Big) \sum_{j=1}^{b} c_j z_j^{-n}, \qquad n \in \mathbb{N},$$

and so

$$\psi(n) = \prod_{k=1}^{b} (1 - z_k^{-1}) \sum_{j=1}^{b} c_j \frac{z_j^{-(n+1)}}{1 - z_j^{-1}}.$$

We now suppose that Y is integer-valued and bounded from below, that is for some integer $b < 0$ we have $p_b > 0$ and $p_j = 0$ for all $j < b - 1$. In this case G^- is discrete and concentrated on $\{b, b+1, \ldots, 0\}$. The ladder height Y^- takes value b if $S_1 = Y_1 = b$. Therefore

$$p_b^- = \mathbb{P}(Y_1 = b) + \sum_{n=2}^{\infty} \mathbb{P}(S_1 > 0, \ldots, S_{n-1} > 0, S_n = b) = p_b,$$

because $\mathbb{P}(S_1 > 0, \ldots, S_{n-1} > 0, S_n = b) = 0$ for all $n \geq 2$. The generating function $\hat{g}_{G^-}$ is well-defined for $z \neq 0$ and the function $z^{|b|}\hat{g}_{G^-}(z)$ is a polynomial of degree $|b|$ defined on the whole complex plane $\mathbb{C}$. Therefore $1 = \hat{g}_{G^-}(z)$ has exactly $|b|$ roots. Since for $|z| > 1$

$$1 = \hat{g}_{G^-}(1) = \sum_{j=b}^{0} p_j^- > \sum_{j=b}^{0} p_j^- |z|^j \geq |\hat{g}_{G^-}(z)|,$$

all the roots are in $0 < |z| \leq 1$. On the other hand, G^+ is discrete and concentrated on $\{1, 2, \ldots\}$ and $1 - \hat{g}_{G^+}(z)$ is well-defined in $|z| \leq 1$. Similarly, $1 - \hat{g}_F(z)$ is well-defined for $0 < |z| \leq 1$. Therefore, equation (6.4.15) can be considered on $0 < |z| \leq 1$.

Lemma 6.4.3 *The equations $1 = \hat{g}_{G^-}(z)$ and $1 = \hat{g}_F(z)$ have exactly the same $|b|$ roots in $0 < |z| \leq 1$.*

Proof Since $|\hat{g}_{G^+}(z)| \leq \hat{g}_{G^+}(1) < 1$ for $|z| \leq 1$, we immediately get from

$$(1 - \hat{g}_{G^+}(z))(1 - \hat{g}_{G^-}(z)) = 1 - \hat{g}_F(z) \tag{6.4.24}$$

that the roots of $1 = \hat{g}_{G^-}(z)$ in $|z| \leq 1$ are roots of $1 = \hat{g}_F(z)$. Conversely, suppose that z is a root of $1 = \hat{g}_F(z)$, but $1 \neq \hat{g}_{G^-}(z)$. Then from (6.4.24) we would get that $1 = \hat{g}_{G^+}(z)$, which is not possible because $|\hat{g}_{G^+}(z)| < 1$ for $|z| \leq 1$. □

There is a single root $z = 1$ and it is the only root for which $|z| = 1$. The position of the other roots on $\mathbb{C}$ is discussed in the remark below. Readers who are not interested in the refined analytical details can pass immediately to Theorem 6.4.3. They however should accept the assumption that there exists a continuous function $h(z)$ in $0 < |z| \leq 1$ for which

$$1 - \hat{g}_F(z) = h(z) \prod_{j=1}^{|b|} (z - z_j). \tag{6.4.25}$$

Remark Let $z_1, z_2, \ldots, z_{|b|}$ be the roots of $1 = \hat{g}_F(z)$ in $|z| \leq 1$. To make a factorization like (6.4.25) the roots lying inside $0 < |z| \leq 1$ cause no problem since $1 - \hat{g}_F(z)$ is analytic there. So, let us first look at roots on the boundary $|z| = 1$. It is immediate that one root is 1. Let $z_1 = 1$. Since $1 - z^n = (1 - z)\sum_{k=0}^{n-1} z^k$ we can write

$$\begin{aligned} 1 - \hat{g}_F(z) &= \sum_{k=1}^{|b|} p_{-k}(1 - z^{-k}) + \sum_{k=1}^{\infty} p_k(1 - z^k) \\ &= -(1 - z)\sum_{k=1}^{|b|} \Big(\frac{\sum_{m=0}^{k-1} z^m}{z^k}\Big) p_{-k} + (1 - z)\sum_{n=0}^{\infty}\Big(\sum_{m=n}^{\infty} p_m\Big) z^n, \end{aligned}$$

provided that $\sum_{n=0}^{\infty}(\sum_{m=n}^{\infty} p_m) < \infty$. The latter condition is satisfied if the first moment of Y is finite. Therefore, there exists an analytic function $h_1(z)$ on $0 < |z| \leq 1$ such that $1 - \hat{g}_F(z) = (1 - z)h_1(z)$. Note that $z_1 = 1$ is a single root because otherwise $h_1(1) = 0$. However, $h_1(1) = \mathbb{E}\, Y < 0$, and hence the root $z_1 = 1$ is single. To continue the analysis, we make one more assumption in that $\mathbb{Z}$ is the minimal lattice on which F is concentrated, that is there is no $k = 2, 3, \ldots$ such that F_Y is concentrated on $\{kn,\ n \in \mathbb{Z}\}$. In this case, using Lemma 15.1.4 from Feller (1971), there are no other roots on $|z| = 1$. Now since $h_1(z)$ is analytic on $0 < |z| < 1$ and $h_1(z)$ has roots $z_2, \ldots, z_{|b|}$, we get formula (6.4.25).

We are now ready to state the following representation formula for $\hat{g}_{G^+}(z)$.

Theorem 6.4.3 *For* $0 < |z| \le 1$,

$$\hat{g}_{G^+}(z) = 1 - \frac{h(z) z^{|b|} \prod_{j=1}^{|b|} z_j}{(-1)^{|b|+1} p_b} \,. \tag{6.4.26}$$

Proof The fundamental theorem of algebra yields

$$z^{|b|}(\hat{g}_{G^-}(z) - 1) = c \prod_{j=1}^{|b|} (z - z_j) \,.$$

Letting $z = 0$ we see that $c = (-1)^{|b|} (\prod_{j=1}^{|b|} z_j)^{-1} p_b^-$ and hence, because $p_b = p_b^-$,

$$1 - \hat{g}_{G^-}(z) = \frac{p_b(-1)^{|b|+1}}{\prod_{j=1}^{|b|} z_j} \prod_{j=1}^{|b|} \Big(1 - \frac{z_j}{z}\Big) \,.$$

Now, (6.4.26) follows from (6.4.24) and (6.4.25). □

We conclude this section with a few comments on another form of the Wiener–Hopf factorization (6.4.6). Suppose that for some $\varepsilon > 0$ the function $\hat{g}_F(z)$ is well-defined in $1 - \varepsilon \le |z| \le 1 + \varepsilon$. Define the functions

$$d^+(z) = \exp\Big(-\sum_{n=1}^{\infty} \frac{\mathbb{E}\,[z^{S_n}; S_n > 0]}{n}\Big) \tag{6.4.27}$$

and

$$d^-(z) = \exp\Big(-\sum_{n=1}^{\infty} \frac{\mathbb{E}\,[z^{S_n}; S_n \le 0]}{n}\Big) . \tag{6.4.28}$$

It can be shown (see, for example, Prabhu (1980)) that, in $|z| < 1+\varepsilon$, $d_+(z)$ is analytic, bounded and bounded away from zero and that $d_-(z)$ has the same properties in $|z| > 1 - \varepsilon$. Moreover, $d^-(z) \to 1$ for $z \to 0$. We leave it to the reader to show that

$$1 - \hat{g}_F(z) = d^+(z) d^-(z) \,, \qquad 1 - \varepsilon \le |z| \le 1 + \varepsilon \,. \tag{6.4.29}$$

Hence,

$$1 - \hat{g}_{G^+}(z) = \exp\Big(-\sum_{n=1}^{\infty} \frac{\mathbb{E}\,[z^{S_n}; S_n > 0]}{n}\Big) \tag{6.4.30}$$

and

$$1 - \hat{g}_{G^-}(z) = \exp\Big(-\sum_{n=1}^{\infty} \frac{\mathbb{E}\,[z^{S_n}; S_n \le 0]}{n}\Big) , \tag{6.4.31}$$

since the factorization given in (6.4.29) is unique within the class of functions satisfying the same conditions as mentioned above for $d^+(z)$ and $d^-(z)$.

6.4.3 Ladder Height Distributions

In Theorem 6.3.3 we showed that the probability of ruin $\psi(u) = \mathbb{P}(\tau(u) < \infty)$ is closely related to the ladder height distribution G^+ of the random walk $\{S_n\}$ with $S_n = \sum_{i=1}^n (U_i - \beta T_i)$. In Sections 5.1.2 and 6.4.2 we determined G^+ for some cases when the increments $Y_n = U_n - \beta T_n$ were integer-valued. We now compute G^+, $p = G^+(\infty)$ and $G_0(u) = G^+(u)/G^+(\infty)$ for two further cases, i.e. the compound Poisson model with general claim size distribution, and the Sparre Andersen model with exponentially distributed claim sizes. We again assume that the drift of the random walk $\{S_n\}$ is negative, or equivalently that $\mathbb{E}\, U - \beta \mathbb{E}\, T < 0$.

We start with the compound Poisson model. We first prove a lemma of independent interest, which gives a simple expression for the pre-occupation measure $\gamma^- = H_0^-$ introduced in Section 6.4.1.

Lemma 6.4.4 *For the compound Poisson model,*

$$H_0^-((-x, 0]) = 1 + \lambda\beta^{-1}x\,, \qquad x > 0\,, \tag{6.4.32}$$

or, alternatively, $\mathrm{d}H_0^-(x) = \mathrm{d}\delta_0(x) + \lambda\beta^{-1}\,\mathrm{d}x$.

Proof From the definition (6.4.1) of the ladder height distribution G^-, we have

$$\begin{aligned} G^-(-x) &= \mathbb{P}(S_{\nu^-} \le -x) \\ &= \sum_{k=1}^\infty \mathbb{P}(\beta T_k \ge x + U_k + S_{k-1} \mid \{\beta T_k \ge U_k + S_{k-1} > 0\} \cap A_k)\mathbb{P}(\nu^- = k) \end{aligned}$$

where $A_k = \{S_{k-1} > 0, \ldots, S_1 > 0\}$. Apply Lemma 2.4.1 with $X = \beta T_k$, $W = U_k + S_{k-1}$ and $A = A_k$ to get that

$$\begin{aligned} G^-(-x) &= \sum_{k=1}^\infty \mathbb{P}(\beta T_k > x)\mathbb{P}(\nu^- = k) \\ &= \sum_{k=1}^\infty \mathrm{e}^{-\lambda\beta^{-1}x}\mathbb{P}(\nu^- = k) = \mathrm{e}^{-\lambda\beta^{-1}x}, \end{aligned}$$

because $\mathbb{P}(\nu^- < \infty) = 1$. The lack-of-memory property (2.4.7) implies that

$$G^-(x) = \begin{cases} \mathrm{e}^{\lambda\beta^{-1}x} & \text{if } x \le 0, \\ 1 & \text{if } x > 0. \end{cases} \tag{6.4.33}$$

From the definition (6.4.3) of H_0^- we get that $H_0^-((-x, 0))$ is the renewal function of a Poisson process with intensity $\lambda\beta^{-1}$. We finally use the first formula in (5.2.8) with some care as the direction of the renewal process has to be changed, and remember to add 1 for $(G^-)^{*0}$. □

Next, we derive an expression for the tail function $\overline{G^+}(x) = G^+(\infty)-G^+(x)$. It turns out that, in the compound Poisson model, the conditional ladder height distribution G_0 coincides with the integrated tail distribution F^s_U of claim sizes. The following result was already obtained as (5.3.11) and (5.3.18). But, for methodological reasons, we give a separate proof here.

Theorem 6.4.4 *For the compound Poisson model,*

$$\overline{G^+}(x) = \lambda\beta^{-1}\int_x^\infty \overline{F}_U(v)\,dv\,, \qquad x \ge 0\,. \tag{6.4.34}$$

Hence

$$p = \lambda\mu_U\beta^{-1} \tag{6.4.35}$$

and

$$\overline{G}_0(x) = \mu_U^{-1}\int_x^\infty \overline{F}_U(v)\,dv\,, \qquad x \ge 0\,. \tag{6.4.36}$$

Proof We have $\hat{m}_F(s) = \hat{m}_U(s)\lambda/(\lambda+\beta s)$ and, from (6.4.33), $\hat{m}_{G^-}(s) = \lambda/(\lambda+\beta s)$. Thus (6.4.13) implies

$$1-\hat{m}_{G^+}(s) = \frac{1-\hat{m}_U(s)\lambda/(\lambda+\beta s)}{1-\lambda/(\lambda+\beta s)}$$

for $-\lambda/\beta < s \le 0$. This is equivalent to (6.4.34). □

We turn to the Sparre Andersen model with general inter-occurrence time distribution but with exponentially distributed claim sizes. In particular, we derive the ladder height distribution G^+ and the probability p that the first ascending ladder epoch ν^+ is finite.

Theorem 6.4.5 *If the claim size distribution F_U is exponential with parameter $\delta > 0$, then G_0 is exponential with the same parameter δ and $\delta(1-p)$ is the unique positive root of*

$$\hat{m}_Y(s) = \frac{\delta}{\delta - s}\hat{l}_T(\beta s) = 1\,. \tag{6.4.37}$$

Proof As in the proof of Lemma 6.4.4, the ladder height distribution G^+ introduced in Section 6.3.3 is given by

$$\overline{G^+}(x) = \sum_{k=1}^\infty \mathbf{P}(S_k > x \mid S_k > 0, S_{k-1} \le 0, \ldots, S_1 \le 0)\mathbf{P}(\nu^+ = k)\,.$$

Apply Lemma 2.4.1 to

$$\mathbb{P}(U_k > x + \beta T_k - S_{k-1} \mid U_k > \beta T_k - S_{k-1} > 0, S_{k-1} \le 0, \ldots, S_1 \le 0)$$

to obtain $\overline{G^+}(x) = \sum_{k=1}^{\infty} \mathbb{P}(U_k > x)\mathbb{P}(\nu^+ = k) = \mathrm{e}^{-\delta x}\mathbb{P}(\nu^+ < \infty)$, or

$$\mathrm{d}G^+(x) = p\delta\,\mathrm{e}^{-\delta x}\,\mathrm{d}x \tag{6.4.38}$$

for all $x > 0$. From (6.4.13) we have

$$(1 - \hat{m}_Y(z)) = (1 - \hat{m}_{G^-}(z))\Big(1 - p\frac{\delta}{\delta - z}\Big)$$

for $0 \le \Re z < \delta$. If $\hat{m}_Y(z) = 1$ for some $z > 0$ then $\hat{m}_{G^-}(z) < 1$ and therefore $z = \delta(1-p)$. We still have to show that (6.4.37) has a positive root. Note that $\hat{m}_Y(0) = 1$ and $\hat{m}_Y^{(1)}(0) = \mathbb{E}[Y] < 0$. Moreover, $\hat{m}_Y(z) \to \infty$ as $z \uparrow \delta$ and $\hat{m}_Y(z)$ is continuous. Thus (6.4.37) has a positive solution. □

The following result is an obvious consequence of Theorems 6.4.4 and 6.4.5.

Corollary 6.4.6 *Consider the compound Poisson model with intensity λ and exponential claim size distribution $F_U = \mathrm{Exp}(\delta)$. Then $G_0 = \mathrm{Exp}(\delta)$ and $p = \lambda(\delta\beta)^{-1}$.*

Bibliographical Notes. Factorization theorems for random walks appear in many books and articles and in different forms. We refer, for example, to Chung (1974), Feller (1971), Prabhu (1980), and to Resnick (1992), Section 7.2. A probabilistic proof of the Wiener–Hopf factorization (6.4.6) has been given in Kennedy (1994). The exposition of Section 6.4.2 follows Asmussen (1987), Chapter 9.2. Theorems 6.4.4 and 6.4.5 are standard in the theory of random walks and can be found, for example, in Billingsley (1995), Feller (1968) and Resnick (1992).

6.5 RUIN PROBABILITIES: SPARRE ANDERSEN MODEL

6.5.1 Formulae of Pollaczek–Khinchin Type

Sometimes it is more convenient to consider the claim surplus process $\{S(t)\}$ with $S(t) = \sum_{i=1}^{N(t)} U_i - \beta t$ for $t \ge 0$ instead of the risk reserve process $\{R(t)\}$. The ruin function $\psi(u)$ is then given by $\psi(u) = \mathbb{P}(\tau(u) < \infty)$, where $\tau(u) = \min\{t : S(t) > u\}$ is the time of ruin for the initial risk reserve u. As already stated in Chapter 5, a fundamental question of risk theory is how to derive pleasing formulae for $\psi(u)$. However, most often this is impossible, as formulae turn out to be too complicated. As a result, various approximations are considered. From random walk theory, applied to the independent increments $Y_n = U_n - \beta T_n$, we already know that there is only one case that is interesting, namely when the coefficient $\rho = (\lambda \mathbb{E} U)/\beta$ is less than 1, as otherwise $\psi(u) \equiv 1$ (see Theorem 6.3.1). If $\rho < 1$, then the drift $\mathbb{E} U - \beta \mathbb{E} T$ of the random walk $\{S_n\}$ with $S_n = Y_1 + \ldots + Y_n$ is negative. In risk theory it is customary to express this condition in terms of the *relative safety loading* η, which is defined as

$$\eta = \frac{\beta \mathbb{E} T - \mathbb{E} U}{\mathbb{E} U} = \frac{1}{\rho} - 1 . \tag{6.5.1}$$

Obviously, $\eta > 0$ if and only if $\rho < 1$. The concept of relative safety loading comes from the following considerations. Consider a risk reserve process in the compound Poisson model,

$$R(t) = u + \lambda \mathbb{E} U t - \sum_{n=1}^{N(t)} U_n , \qquad t \geq 0 ,$$

where the premium rate $\beta = \lambda \mathbb{E} U$ is computed by the net premium principle. From random walk theory, we already know that the risk reserve process without drift will have unbounded large fluctuations as time goes on, and so ruin happens with probability 1. If we add a safety loading $\varepsilon \lambda \mathbb{E} U$ for some $\varepsilon > 0$, then ruin in the risk reserve process $\{R(t)\}$ with

$$R(t) = u + (1+\varepsilon) \lambda \mathbb{E} U t - \sum_{n=1}^{N(t)} U_n , \qquad t \geq 0 ,$$

will no longer occur with probability 1. Solving equation (6.5.1) for $\beta = (1+\varepsilon)\lambda \mathbb{E} U$, we have the relative safety loading $\eta = \varepsilon$.

In the sequel to this chapter, we always assume that $0 < \mathbb{E} T < \infty$, $0 < \mathbb{E} U < \infty$ and that the relative safety loading η is positive so that $\mathbb{E} U - \beta \mathbb{E} T < 0$. We know from Section 6.3.3 that the survival probability $1 - \psi(u)$ is given by the following formula of Pollaczek–Khinchin type.

Theorem 6.5.1 *For all* $u \geq 0$,

$$1 - \psi(u) = (1-p) \sum_{k=0}^{\infty} (G^+)^{*k}(u) = \sum_{k=0}^{\infty} (1-p) p^k G_0^{*k}(u) , \tag{6.5.2}$$

where G^+ *is the (defective) distribution of the ladder height of the random walk* $\{S_n\}$; $S_n = \sum_{i=1}^n (U_i - \beta T_i)$, $p = G^+(\infty)$ *and* $G_0(u) = G^+(u)/G^+(\infty)$.

Note that (6.5.2) implies

$$\psi(u) = (1-p)\sum_{k=1}^{\infty} p^k \overline{G_0^{*k}}(u)\,, \qquad u \geq 0\,. \tag{6.5.3}$$

After some simple algebraic manipulations, this reads

$$\psi(u) = \sum_{k=0}^{\infty} p^{k+1} \int_0^u \overline{G}_0(u-v)\,\mathrm{d}G_0^{*k}(v)\,, \qquad u \geq 0\,. \tag{6.5.4}$$

In the case of a compound Poisson model we know from Theorem 6.4.4 that G_0 is equal to the integrated tail distribution F_U^{s} of claim sizes. We rediscover the classical Pollaczek–Khinchin formula for the ruin probability $\psi(u)$ from Theorem 5.3.4.

Corollary 6.5.1 *The ruin function in the compound Poisson model is*

$$\psi(u) = \sum_{k=1}^{\infty}(1-\rho)\rho^k \overline{(F_U^{\mathrm{s}})^{*k}}(u), \tag{6.5.5}$$

which is the same as

$$\psi(u) = \sum_{k=0}^{\infty} \rho^{k+1} \int_0^u \overline{F_U^{\mathrm{s}}}(u-v)\,\mathrm{d}(F_U^{\mathrm{s}})^{*k}(v)\,. \tag{6.5.6}$$

The *proof* is immediate as it suffices to insert (6.4.35) and (6.4.36) into (6.5.3). In the same way (6.5.6) follows from (6.5.4).

Corollary 6.5.2 *In the Sparre Andersen model with exponential claim size distribution* $\mathrm{Exp}(\delta)$,

$$\psi(u) = (1-\gamma/\delta)\,\mathrm{e}^{-\gamma u} \tag{6.5.7}$$

for all $u \geq 0$, *where* γ *is the unique positive root of* (6.4.37).

Proof Theorem 6.4.5 and the representation formula (6.5.2) yield that $1-\psi(u)$ is the distribution function of a geometric compound with characteristics $(p, \mathrm{Exp}(\delta))$. It is not difficult to see that the zero-truncation of this compound distribution is the exponential distribution with parameter $\delta(1-p)$. Thus,

$$\psi(u)/\psi(0) = \mathrm{e}^{-\delta(1-p)u} \tag{6.5.8}$$

for all $u \geq 0$. From (6.5.2) we have that $\psi(0) = p$. Moreover, Theorem 6.4.5 implies that $p = 1 - \gamma\delta^{-1}$, where γ is the unique root of (6.4.37). This and (6.5.8) imply (6.5.7). □

Corollaries 6.5.1 and 6.5.2 yield the following result, which coincides with (5.3.8).

Corollary 6.5.3 *In the compound Poisson model with exponential claim size distribution* Exp(δ),

$$\psi(u) = \frac{\lambda}{\beta\delta}\,\mathrm{e}^{-(\delta-\lambda/\beta)u}\,, \qquad u \geq 0\,. \tag{6.5.9}$$

Again the *proof* is immediate as (6.5.8) and (6.4.35) imply (6.5.9), having in mind that $\psi(0) = p$.

We now determine the joint distribution of $(X^+(u), Y^+(u))$, where $X^+(u)$ is the surplus just before ruin time $\tau(u)$ and $Y^+(u)$ is the severity of ruin as defined in Section 5.1.4. More generally, we consider the multivariate ruin function

$$\psi(u,x,y) = \mathbf{P}(\tau(u) < \infty, X^+(u) \leq x, Y^+(u) > y)\,, \tag{6.5.10}$$

where $u, x, y \geq 0$. We derive a representation formula for $\psi(u,x,y)$, which generalizes the representation formula (6.5.2) for the (univariate) ruin function $\psi(u)$ and expresses $\psi(u,x,y)$ in terms of p, G_0 and $\psi(0,x,y)$. Here, $\psi(0,x,y)$ is obtained from the distribution of $(X^+(0), Y^+(0))$. Recall the pre-occupation measure $\gamma^- = H_0^+ = \sum_{k=0}^{\infty}(G^+)^{*k} = \sum_{k=0}^{\infty} p^k {G_0}^{*k}$ introduced in Section 6.4.1.

Theorem 6.5.2 *The multivariate ruin function* $\psi(u,x,y)$ *satisfies the integral equation*

$$\psi(u,x,y) = \psi(0, x-u, y+u) + p\int_0^u \psi(u-v,x,y)\,\mathrm{d}G_0(v) \tag{6.5.11}$$

for all $u, x, y \geq 0$*; its solution is*

$$\psi(u,x,y) = \int_0^u \psi(0, x-u+v, y+u-v)\,\mathrm{d}H_0^+(v)\,. \tag{6.5.12}$$

Proof First consider the event $\{\tau(u) < \infty, X^+(u) \leq x, Y^+(u) > y, Y^+(0) > u+y\}$, and denote $A = \{Y^+(0) > u+y\}$. Then, $A = A \cap \{\tau(u) = \tau(0)\}$ and, consequently,

$$\begin{aligned} \{\tau(u) < \infty\} \cap A &= \{\tau(0) < \infty\} \cap A\,, \\ \{Y^+(u) > y\} \cap A &= \{Y^+(0) > u+y\}\,, \\ \{X^+(u) \leq x\} \cap A &= \{X^+(0) \leq x-u\} \cap A\,. \end{aligned}$$

Using the law of total probability, we get from definition (6.5.10) of $\psi(u,x,y)$

$$\begin{aligned} \psi(u,x,y) &= \mathbf{P}(\tau(u) < \infty, X^+(u) \leq x, Y^+(u) > y, Y^+(0) > u+y) \\ &\quad + \mathbf{P}(\tau(u) < \infty, X^+(u) \leq x, Y^+(u) > y, Y^+(0) \leq u) \\ &= \mathbf{P}(\tau(0) < \infty, X^+(0) \leq x-u, Y^+(0) > y+u) \\ &\quad + \int_0^u \mathbf{P}(\tau(u-v) < \infty, X^+(u-v) \leq x, Y^+(u-v) > y)\,\mathrm{d}G^+(v)\,, \end{aligned}$$

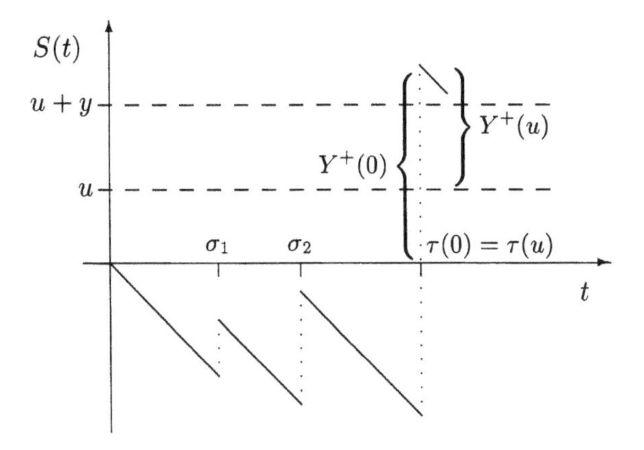

Figure 6.5.1 $Y^+(0) = Y^+(u) + u > u + y$

which proves (6.5.11). Note that, for $x, y \geq 0$ fixed, (6.5.11) is a defective renewal equation with respect to $g(u) = \psi(u, x, y)$. This gives (6.5.12) by applying Lemma 6.1.2 with $z(u) = \varphi(x - u, y + u)$ and $F(v) = pG_0(v) = G^+(v)$. □

Corollary 6.5.4 *For all* $u, y \geq 0$,

$$\psi(u, \infty, y) = p \int_0^u \overline{G}_0(y + u - v) \, \mathrm{d}H_0^+(v) \,. \tag{6.5.13}$$

Proof Since $\psi(0, \infty, y) = p\overline{G}_0(y)$, we obtain (6.5.13) from (6.5.12). □

Note that formulae (6.5.12) and (6.5.13) are extensions of (6.5.4), since $\psi(u) = \psi(u, \infty, 0)$. Furthermore, recall that in the case of the compound Poisson model, the characteristics p, G_0 and

$$\varphi(0, x, y) = \mathbb{P}(\tau(0) < \infty, X^+(0) > x, Y^+(0) > y) \tag{6.5.14}$$

can be easily expressed in terms of λ, β and F_U as shown in Sections 5.3.4 and 6.4.3. In particular, we have the representation formula (5.3.18):

$$\varphi(0, x, y) = \lambda \beta^{-1} \int_{x+y}^{\infty} (1 - F_U(v)) \, \mathrm{d}v \,, \qquad x, y \geq 0 \,. \tag{6.5.15}$$

Clearly, then $\psi(0, x, y) = \varphi(0, 0, y) - \varphi(0, x, y)$ can also be expressed by λ, β and F_U. Using (6.5.15), from (6.5.12) we immediately obtain the following formula for $\psi(u, x, y)$.

Theorem 6.5.3 *In the compound Poisson model,*

$$\psi(u,x,y) = \rho\int_0^u \left(F_U^{\mathrm{s}}((x-u+v)_+ + (y+u-v)) - F_U^{\mathrm{s}}(y+u-v)\right) \mathrm{d}H_0^+(v)\,, \tag{6.5.16}$$

for all $u, x, y \geq 0$, where $\rho = \lambda\beta^{-1}\mu_U$ and $H_0^+(v) = \sum_{k=0}^{\infty} \rho^k F_U^{\mathrm{s}\,*k}(v)$.

Note that the marginal ruin function $\psi(u,\infty,y)$ can be obtained directly from Corollary 6.5.4 and Theorem 6.4.4.

Corollary 6.5.5 *The probability $\psi(u,\infty,y)$ that, in the compound Poisson model, the overshoot $Y^+(u)$ at ruin time $\tau(u)$ exceeds y is given by*

$$\psi(u,\infty,y) = \sum_{k=0}^{\infty} \rho^{k+1} \int_0^u \overline{F_U^{\mathrm{s}}}(y+u-v)\,\mathrm{d}(F_U^{\mathrm{s}})^{*k}(v)\,. \tag{6.5.17}$$

The marginal ruin function $\psi(u,\infty,y)$ can also be obtained in the Sparre Andersen model with exponentially distributed claim sizes if one uses Corollary 6.5.4 and Theorem 6.4.5.

Corollary 6.5.6 *In the Sparre Andersen model with exponential claim size distribution* $\mathrm{Exp}(\delta)$

$$\psi(u,\infty,y) = \psi(u)\,\mathrm{e}^{-\delta y} = (1-\gamma/\delta)\,\mathrm{e}^{-(\gamma u+\delta y)} \tag{6.5.18}$$

for all $u, y \geq 0$, where γ is the unique positive root of (6.4.37).

Proof The formula (6.5.18) follows from (6.5.13) and (6.5.7), bearing in mind that G_0 is exponential with parameter δ. □

In order to determine the probability that, besides the overshoot $Y^+(u)$, the total maximal deficit $Z^+(u)$ after time $\tau(u)$ exceeds level z we define for $u, x, y, z \geq 0$:

$$\psi(u,x,y,z) = \mathbb{P}(\tau(u) < \infty, X^+(u) \leq x, Y^+(u) > y, Z^+(u) > z)\,.$$

Clearly, for $y \geq z$ we have $\psi(u,x,y,z) = \psi(u,x,y,y) = \psi(u,x,y)$. Using the same argument as in the proof of Theorem 6.5.2 we get the following defective renewal equation for $\psi(u,x,y,z)$. For all $u, x, y, z \geq 0$, we have

$$\psi(u,x,y,z) = \psi(0,x-u,y+u,z+u) + p\int_0^u \psi(u-v,x,y,z)\,\mathrm{d}G_0(v)\,. \tag{6.5.19}$$

The *proof* of (6.5.19) is left to the reader as an exercise. Hence, by Lemma 6.1.2

$$\psi(u,x,y,z) = \int_0^u \psi(0,x-u+v,y+u-v,z+u-v)\,\mathrm{d}H_0^+(v) \tag{6.5.20}$$

for all $u, x, y, z \geq 0$, where here and below $H_0^+(v) = \sum_{k=0}^{\infty} p^k G_0^{*k}(v)$. Moreover, since

$$\psi(0, \infty, y, z) = p\Big(\int_y^{\max\{y,z\}} \psi(z - v')\, \mathrm{d}G_0(v') + \overline{G}_0(\max\{y, z\})\Big),$$

(6.5.20) yields the following extension to (6.5.13). For all $u, y, z \geq 0$

$$\begin{aligned}\psi(u, \infty, y, z) \;&=\; p\int_0^u \int_{y+u-v}^{\max\{y,z\}+u-v} \psi(z + u - v - v')\, \mathrm{d}G_0(v') \\ &\qquad \times \overline{G}_0(\max\{y, z\} + u - v)\, \mathrm{d}H_0^+(v)\,.\end{aligned}$$

6.5.2 Lundberg Bounds

Formula (6.5.2) for $\psi(u)$ is of theoretical importance, as it can hardly be used for direct numerical computation. For the compound Poisson model, however, Theorem 5.4.1 gave useful two-sided bounds for $\psi(u)$ that were obtained by applying Theorem 4.5.1 to the compound geometric distribution in (6.5.3). Unfortunately, in general neither G_0 nor p is known.

In Theorem 6.5.4 below, we extend the result of Theorem 5.4.1 and derive a two-sided Lundberg bound for the ruin function $\psi(u)$ in the Sparre Andersen model with general distributions of inter-occurrence times and claim sizes. For the Sparre Andersen model, we have the geometric compound representation (6.5.3) for the ruin function $\psi(u)$. We can therefore try to prove a two-sided Lundberg bound by relying on Theorem 4.5.1 with $G = G_0$, in the same way as was done in Section 5.4.1 for the compound Poisson model. Unfortunately, the prefactors a_- and a_+ would then be expressed in terms of the unknown distribution G_0. To avoid this complication, we take a slightly different approach.

Consider the equation

$$\hat{m}_Y(s) = \hat{m}_U(s)\hat{l}_T(\beta s) = 1\,. \tag{6.5.21}$$

Clearly, $\hat{m}_Y(0) = 1$. This equation may have a second root. If such a root $s \neq 0$ exists, then it is unique and strictly positive. The solution to (6.5.21), if it exists, is called the *adjustment coefficient* and is denoted by γ. For the compound Poisson model, the solutions to (5.4.3) and (6.5.21) coincide. As we will see in the next Section 6.5.3, the adjustment coefficient γ in Theorem 6.5.4 satisfies $\int_0^\infty \mathrm{e}^{\gamma x}\, \mathrm{d}G_0(x) = p^{-1}$. This means that also in this more general case, γ coincides with the adjustment coefficient considered in Theorem 4.5.1. Let $x_0 = \sup\{x : F_Y(x) < 1\}$.

Theorem 6.5.4 *Suppose that there exists a positive solution γ to* (6.5.21). *Then*

$$b_-\, \mathrm{e}^{-\gamma u} \leq \psi(u) \leq b_+\, \mathrm{e}^{-\gamma u} \tag{6.5.22}$$

for all $u \ge 0$, *where*

$$b_- = \inf_{x\in[0,x_0)} \frac{e^{\gamma x}\overline{F}_Y(x)}{\int_x^\infty e^{\gamma y}\, dF_Y(y)}\,, \qquad b_+ = \sup_{x\in[0,x_0)} \frac{e^{\gamma x}\overline{F}_Y(x)}{\int_x^\infty e^{\gamma y}\, dF_Y(y)}\,. \tag{6.5.23}$$

Proof The proof is similar to that of Theorem 4.5.1. Suppose that $\overline{F}_0(x) = b_+ e^{-\gamma x}$ for all $x \ge 0$ and $\overline{F}_0(x) = 1$ otherwise. For $n \in \mathbb{N}$ define the sequence $\{F_n(x)\}$ of distribution functions by

$$F_{n+1}(x) = \begin{cases} \int_{-\infty}^{\infty} F_n(x-y)\, dF_Y(y) & \text{if } x \ge 0, \\ 0 & \text{if } x < 0. \end{cases} \tag{6.5.24}$$

Then, in view of (6.5.21), we have

$$\overline{F}_0(x) = b_+ e^{-\gamma x} \ge \frac{\overline{F}_Y(x)}{\int_x^\infty e^{\gamma y}\, dF_Y(y)} = \frac{\overline{F}_Y(x)}{1 - \int_{-\infty}^x e^{\gamma y}\, dF_Y(y)}\,.$$

From (6.5.24) we get $\overline{F}_1(x) = \overline{F}_Y(x) + b_+ e^{-\gamma x} \int_{-\infty}^x e^{\gamma y}\, dF_Y(y)$ for all $x \ge 0$ and $\overline{F}_1(x) = 1$ otherwise, and thus $\overline{F}_0(x) \ge \overline{F}_1(x)$. Use (6.5.24) recursively to obtain $\overline{F}_n(x) \ge \overline{F}_{n+1}(x)$ for all $x \in \mathbb{R}$ and $n \in \mathbb{N}$. For the upper bound in (6.5.22) it remains to show that

$$\lim_{n\to\infty} \overline{F}_n(u) = \psi(u)\,. \tag{6.5.25}$$

For this we consider a sequence $L_0, L_1, \ldots$ of nonnegative random variables, not necessarily integer-valued, fulfilling $L_{n+1} = (L_n + Y_{n+1})_+$ for $n = 1, 2, \ldots$, where L_0 is independent of the sequence $\{Y_n\}$ and distributed according to F_0. Then, F_n is the distribution of L_n and, in a similar manner as in Section 5.1.3, we can prove that

$$\begin{aligned} L_n &= \max\{0, Y_n, Y_{n-1}+Y_n, \ldots, Y_2+\ldots+Y_n, L_0+Y_1+\ldots+Y_n\} \\ &\overset{d}{=} \max\{0, Y_1, Y_1+Y_2, \ldots, Y_1+\ldots+Y_{n-1}, L_0+Y_1+\ldots+Y_n\}\,. \end{aligned}$$

Thus, repeating the proof of Theorem 5.1.2, we obtain (6.5.25) since, by the law of large numbers, $L_0 + Y_1 + \ldots + Y_n \to -\infty$ as $n \to \infty$. The lower bound in (6.5.22) can be derived similarly. □

A somewhat weaker though probably more useful bound is obtained if we express the prefactors in the two-sided Lundberg inequality (6.5.22) via the claim size distribution F_U. Thus we define further constants b_-^*, b_+^* by

$$b_-^* = \inf_{x\in[0,x_0')} \frac{e^{\gamma x}\overline{F}_U(x)}{\int_x^\infty e^{\gamma y}\, dF_U(y)}\,, \qquad b_+^* = \sup_{x\in[0,x_0')} \frac{e^{\gamma x}\overline{F}_U(x)}{\int_x^\infty e^{\gamma y}\, dF_U(y)}\,, \tag{6.5.26}$$

where γ is the solution to (6.5.21) and $x_0' = \sup\{x : F_U(x) < 1\}$. Note that

$$\frac{1}{b_-} = \sup_x \mathbb{E}\,(\mathrm{e}^{\gamma(Y-x)} \mid Y > x), \quad \frac{1}{b_-^*} = \sup_x \mathbb{E}\,(\mathrm{e}^{\gamma(U-x)} \mid U > x) \qquad (6.5.27)$$

and that $(b_+)^{-1}$ and $(b_+^*)^{-1}$ can be expressed in a similar way.

Theorem 6.5.5 *The constants b_-^*, b_-, b_+, b_+^* defined in* (6.5.23) *and* (6.5.26), *respectively, satisfy* $0 \le b_-^* \le b_- \le b_+ \le b_+^* \le 1$.

Proof By the law of total probability and our assumption that the random variables T and U are independent, (6.5.27) implies that

$$\begin{aligned}
(b_-)^{-1} &= \sup_x \mathbb{E}\,(\mathrm{e}^{\gamma(U-(\beta T+x))} \mid U > \beta T + x) \\
&= \sup_x \frac{\int_0^\infty \mathbb{E}\,(\mathrm{e}^{\gamma(U-(\beta t+x))}\mathbb{I}(U > \beta t + x))\,\mathrm{d}F_T(t)}{\mathbb{P}(U > \beta T + x)} \\
&= \sup_x \frac{\int_0^\infty \mathbb{E}\,(\mathrm{e}^{\gamma(U-(\beta t+x))} \mid U > \beta t + x)\mathbb{P}(U > \beta t + x)\,\mathrm{d}F_T(t)}{\mathbb{P}(U > \beta T + x)} \\
&\le \sup_x \frac{\int_0^\infty \sup_{s\ge 0}\mathbb{E}\,(\mathrm{e}^{\gamma(U-s)} \mid U > s)\mathbb{P}(U > \beta t + x)\,\mathrm{d}F_T(t)}{\mathbb{P}(U > \beta T + x)} \\
&= (b_-^*)^{-1} \sup_x \frac{\int_0^\infty \mathbb{P}(U > \beta t + x)\,\mathrm{d}F_T(t)}{\mathbb{P}(U > \beta T + x)} = (b_-^*)^{-1}.
\end{aligned}$$

Thus, $b_-^* < b_-$. The proof of $b_+ \le b_+^*$ is similar and $0 \le b_-^*, b_+^* \le 1$ directly follows from the definitions of b_-^*, b_+^* given in (6.5.26). □

It is easily seen that (5.4.3) and (6.5.21) coincide for the compound Poisson model. For the Sparre Andersen model with general inter-occurrence time distribution F_T and exponentially distributed claim sizes with parameter δ, the adjustment coefficient γ is the solution to $(\delta - s)^{-1}\delta\hat{l}_T(\beta s) = 1$. Moreover, $b_-^* = b_+^* = (\delta - \gamma)\delta^{-1}$ and we immediately obtain the expression (6.5.7) for $\psi(u)$ given in Corollary 6.5.2.

It is clear that the lower bound in (6.5.22) is only useful when the prefactor b_- is positive. A sufficient condition for this is given in the following result.

Theorem 6.5.6 *Suppose there exists a distribution F such that $\hat{m}_F(\gamma) < \infty$ and*

$$F_{U,x} \le_{\mathrm{st}} F \qquad (6.5.28)$$

for all $x \ge 0$, where $F_{U,x}(y) = \mathbb{P}(U - x < y \mid U > x)$ is the distribution function of the remaining claim size seen from level x and where $\gamma > 0$ is the solution to (6.5.21). *Then $0 < b_-^* \le b_-$.*

Proof We show that $(b^*_-)^{-1} < \infty$. From (6.5.27) we have

$$\begin{aligned}(b^*_-)^{-1} &= \sup_x \mathbb{E}\,(\mathrm{e}^{\gamma(U-x)} \mid U > x) = \sup_x \int_0^\infty \mathrm{e}^{\gamma y}\,\mathrm{d}F_{U,x}(y) \\ &\le \sup_x \int_0^\infty \mathrm{e}^{\gamma y}\,\mathrm{d}F(y) = \hat{m}_F(\gamma)\,,\end{aligned}$$

where in the last inequality we used (6.5.28) and the fact that the function $g(y) = \mathrm{e}^{\gamma y}$ is increasing. Taking into account the result of Theorem 6.5.5, this completes the proof since $\hat{m}_F(\gamma) < \infty$. □

Corollary 6.5.7 *Suppose that* (6.5.21) *has a positive solution and* F_U *is* IHR. *Then* $0 < b^*_- \le b_-$.

Proof From Theorem 2.4.2 we get that (6.5.28) is fulfilled for $F = F_U$. Thus, the assertion follows from Theorem 6.5.6. □

6.5.3 The Cramér–Lundberg Approximation

In this section we assume that the distribution F of Y is nonlattice and $\mathbb{E}Y < 0$. The reader should prove that then the ladder height distribution G^+ corresponding to F is nonlattice too. Furthermore, we assume that (6.5.21) has a positive solution γ. The following theorem deals with the asymptotic behaviour of $\psi(u)$ as u becomes unbounded large. It extends Theorem 5.4.2 of Section 5.4.2 from the compound Poisson model to the Sparre Andersen model.

Theorem 6.5.7 *For the Sparre Andersen model,*

$$\lim_{u\to\infty} \mathrm{e}^{\gamma u}\psi(u) = c \tag{6.5.29}$$

where the constant $c \ge 0$ *is finite and given by*

$$c = \frac{1 - G^+(\infty)}{\gamma \int_0^\infty v\mathrm{e}^{\gamma v}\,\mathrm{d}G^+(v)}\,. \tag{6.5.30}$$

Proof Recall that $\hat{m}_F(s) < \infty$ for $0 \le s \le \gamma$. Then, it is easily seen that $\hat{m}_{G^+}(s) < \infty$ for $0 \le s \le \gamma$. By Corollary 6.4.3,

$$1 - \hat{m}_F(s) = (1 - \hat{m}_{G^+}(s))(1 - \hat{m}_{G^-}(s))\,, \qquad 0 \le s \le \gamma \tag{6.5.31}$$

because $\hat{m}_{G^-}(s)$ is finite for all $s \ge 0$. Furthermore, $\hat{m}_{G^-}(\gamma) < 1$. Thus, $\hat{m}_F(\gamma) = 1$ and (6.5.31) give

$$\hat{m}_{G^+}(\gamma) = \int_0^\infty \mathrm{e}^{\gamma x}\,\mathrm{d}G^+(x) = 1\,. \tag{6.5.32}$$

But then $\psi(u)$ satisfies the defective renewal equation

$$\psi(u) = G^+(\infty) - G^+(u) + \int_0^u \psi(u-v)\,\mathrm{d}G^+(v)\,. \tag{6.5.33}$$

Multiplying both sides of (6.5.33) by $\mathrm{e}^{\gamma u}$, we have

$$\tilde{\psi}(u) = \mathrm{e}^{\gamma u}(G^+(\infty) - G^+(u)) + \int_0^u \tilde{\psi}(u-v)\,\mathrm{d}\tilde{G}^+(v) \tag{6.5.34}$$

for all $u \ge 0$, where $\tilde{\psi}(u) = \mathrm{e}^{\gamma u}\psi(u)$ and $\mathrm{d}\tilde{G}^+(x) = \mathrm{e}^{\gamma x}\mathrm{d}G^+(x)$. The distribution $\tilde{G}^+$ is nonlattice and in view of (6.5.32) nondefective, and therefore we can apply Theorem 6.1.11 to the renewal equation (6.5.34). For this we still have to show that the function $z(u) = \mathrm{e}^{\gamma u}(G^+(\infty) - G^+(u))$ is directly Riemann integrable. By Fubini's theorem we have

$$\begin{aligned}\int_0^\infty \mathrm{e}^{\gamma v}(G^+(\infty) - G^+(v))\,\mathrm{d}v &= \int_0^\infty \mathrm{e}^{\gamma v}\int_0^\infty \mathbb{I}(u > v)\,\mathrm{d}G^+(u)\,\mathrm{d}v \\ &= \int_0^\infty\int_0^\infty \mathrm{e}^{\gamma v}\mathbb{I}(u > v)\,\mathrm{d}v\,\mathrm{d}G^+(u) = \frac{1}{\gamma}(1 - G^+(\infty)) < \infty\,.\end{aligned}$$

Now, from Lemma 6.1.4 we derive that $z(u) = \mathrm{e}^{\gamma u}(G^+(\infty) - G^+(u))$ is directly Riemann integrable. Indeed, $z(u)$ is the product of the increasing function $z_1(u) = \mathrm{e}^{\gamma u}$ and the decreasing function $z_2(u) = G^+(\infty) - G^+(u)$ and condition (6.1.28) is obviously satisfied. By Theorem 6.1.11, (6.5.29) follows with

$$0 \le c = \frac{\int_0^\infty \mathrm{e}^{\gamma v}(G^+(\infty) - G^+(v))\,\mathrm{d}v}{\int_0^\infty v\mathrm{e}^{\gamma v}\,\mathrm{d}G^+(v)} = \frac{1 - G^+(\infty)}{\gamma\int_0^\infty v\mathrm{e}^{\gamma v}\,\mathrm{d}G^+(v)} < \infty\,. \qquad \square$$

Note that if $c > 0$ the asymptotic result obtained in Theorem 6.5.7 gives rise to the *Cramér–Lundberg approximation* $\psi_{\mathrm{app}}(u) = c\mathrm{e}^{-\gamma u}$ to the ruin function $\psi(u)$ when u is large.

Remark The constant c in Theorem 6.5.7 is positive if $\int_0^\infty v\mathrm{e}^{\gamma v}\,\mathrm{d}G^+(v) < \infty$. This condition holds if, for example, $\hat{m}_F(s) < \infty$ for $s < \gamma + \varepsilon$ for some $\varepsilon > 0$. Then $\hat{m}_F(s)$ is continuously differentiable in the interval $0 < s < \gamma + \varepsilon$ and hence from the Wiener–Hopf identity (6.5.31) the same property holds for $\hat{m}_{G^+}(s)$. Consequently, $\hat{m}^{(1)}_{G^+}(\gamma) = \int_0^\infty v\mathrm{e}^{\gamma v}\,\mathrm{d}G^+(v) < \infty$.

Feller (1971) gives a different proof of (6.5.32) which is based on a construction which is similar to the *change-of-measure technique* considered in Section 9.2. The idea is to introduce an *associated random walk* $\{\tilde{S}_n\}$ with $\tilde{S}_n = \sum_{i=1}^n \tilde{Y}_i$, where the distribution $\tilde{F}$ of the increments $\tilde{Y}_i$ of $\{\tilde{S}_n\}$ is given by $\mathrm{d}\tilde{F}(x) = \mathrm{e}^{\gamma x}\,\mathrm{d}F(x)$. Indeed, $\tilde{F}$ defined in this way is a (nondefective) distribution because γ solves (6.5.21), that is $\int_{-\infty}^{\infty} \mathrm{e}^{\gamma x}\,\mathrm{d}F(x) = 1$. To show the validity of (6.5.32) it is enough to prove that the random walk $\{\tilde{S}_n\}$ has a positive drift. This can be seen as follows. Since $\hat{m}_F(\gamma) < \infty$, the moment generating function $\hat{m}_F(s)$ exists in the interval $(0, \gamma]$. Moreover, $\hat{m}_F(s)$ is differentiable in $(0, \gamma)$ and

$$\frac{\mathrm{d}}{\mathrm{d}s}\hat{m}_F(s)\big|_{s=\gamma-0} = \int_{-\infty}^{\infty} x\mathrm{e}^{\gamma x}\,\mathrm{d}F_F(x) = \mathbb{E}\,\tilde{Y}. \tag{6.5.35}$$

Since $\hat{m}_F(0) = \hat{m}_F(\gamma) = 1$ and the function $\hat{m}_F(s)$ is strictly convex, we have $(\mathrm{d}/\mathrm{d}s)\hat{m}_F(s)|_{s=\gamma-0} > 0$. In view of this, the drift of $\{\tilde{S}_n\}$ is positive and as a consequence we get that the ladder height distribution $\tilde{G}^+$ of $\{\tilde{S}_n\}$ is nondefective. Furthermore, $\tilde{G}^+$ can be represented by

$$\mathrm{d}\tilde{G}^+(x) = \mathrm{e}^{\gamma x}\,\mathrm{d}G^+(x)\,, \qquad x \ge 0\,. \tag{6.5.36}$$

Namely, we have $1 - \tilde{G}^+(x) = \sum_{n=1}^{\infty} \mathbb{P}(\tilde{S}_1 \le 0, \ldots, \tilde{S}_{n-1} \le 0, \tilde{S}_n > x)$ and $1 - G^+(x) = \sum_{n=1}^{\infty} \mathbb{P}(S_1 \le 0, \ldots, S_{n-1} \le 0, S_n > x)$ for all $x \ge 0$. It is therefore sufficient to show that

$$\begin{aligned}\mathbb{P}(&\tilde{S}_1 \le 0, \ldots, \tilde{S}_{n-1} \le 0, \tilde{S}_n > x)\\ &= \int_x^{\infty} \mathrm{e}^{\gamma y}\mathbb{P}(S_1 \le 0, \ldots, S_{n-1} \le 0, S_n \in \mathrm{d}y)\,,\end{aligned}$$

for all $n = 1, 2, \ldots$ and $x \ge 0$. Putting $v_n = y_1 + \ldots + y_n$, we have

$$\begin{aligned}\mathbb{P}(\tilde{S}_1 &\le 0, \ldots, \tilde{S}_{n-1} \le 0, \tilde{S}_n > x)\\ &= \int_{-\infty}^{0}\int_{-\infty}^{-v_1}\cdots\int_{-\infty}^{-v_{n-2}}\int_{x-v_{n-1}}^{\infty} \mathbb{P}(\tilde{Y}_n \in \mathrm{d}y_n)\\ &\qquad\qquad \times \mathbb{P}(\tilde{Y}_{n-1} \in \mathrm{d}y_{n-1})\ldots\mathbb{P}(\tilde{Y}_2 \in \mathrm{d}y_2)\mathbb{P}(\tilde{Y}_1 \in \mathrm{d}y_1)\\ &= \int_{-\infty}^{0}\int_{-\infty}^{-v_1}\cdots\int_{-\infty}^{-v_{n-2}}\int_{x-v_{n-1}}^{\infty} \mathrm{e}^{\gamma v_n}\mathbb{P}(Y_n \in \mathrm{d}y_n)\\ &\qquad\qquad \times \mathbb{P}(Y_{n-1} \in \mathrm{d}y_{n-1})\ldots\mathbb{P}(Y_2 \in \mathrm{d}y_2)\mathbb{P}(Y_1 \in \mathrm{d}y_1)\\ &= \int_x^{\infty} \mathrm{e}^{\gamma y}\mathbb{P}(S_1 \le 0, \ldots, S_{n-1} \le 0, S_n \in \mathrm{d}y)\,.\end{aligned}$$

Hence we get (6.5.32) since the associated random walk $\{\tilde{S}_n\}$ has positive drift.

6.5.4 Compound Poisson Model with Aggregate Claims

In the compound Poisson model studied in Chapter 5 as well as in earlier sections of the present chapter, ruin could occur anytime whenever the risk reserve became negative. What happens if we are only able to inspect the value of the risk reserve at countably many, equally spaced time epochs $t = h, 2h, \ldots$ for some $h > 0$? To specify the problem, we consider the risk reserve process $\{R(t)\}$ given by $R(t) = u + \beta t - \sum_{i=1}^{N(t)} U_i = u + \beta t - X(t)$, where $\{X(t)\}$ is the compound Poisson process with the increments $X(t+h) - X(h) = \sum_{i=N(h)+1}^{N(t+h)} U_i$. We now say that ruin occurs if $R(kh) < 0$ for some $k = 1, 2, \ldots$. In terms of the claim surplus process $\{S(t)\}$ with $S(t) = X(t) - \beta t$, this can be written as $S(kh) > u$ for some $k = 1, 2, \ldots$. Since the compound Poisson process $\{X(t)\}$ has independent and stationary increments, the random variables $Y_k(h) = X(kh) - X((k-1)h) - \beta h$, $k = 1, 2, \ldots$, are independent and identically distributed. Hence, ruin occurs if the random walk $\{S(nh),\ n = 0, 1 \ldots\}$ with $S(nh) = \sum_{k=1}^{n} Y_k(h)$ crosses the level u. We call this model the *compound Poisson model with aggregate claims* as it is closely related to the risk model with discrete time considered in Section 5.1. However, now the aggregate claims do not necessarily take values in $\mathbb{N}$.

Another interpretation of a compound Poisson model with aggregate claims is that of a Sparre Andersen model with constant inter-occurrence times $T_n = h$, premium rate $\beta > 0$ and (individual) claim sizes $U_n(h) = X(nh) - X((n-1)h)$ having a compound Poisson distribution with characteristics $(\lambda h, F_U)$. For the initial reserve u, the ruin probability is then given by $\psi_h(u) = \mathbb{P}(\max_{n \geq 0} S(nh) > u)$, and $\psi_h(u)$, as a function of u, is called the *ruin function* of the compound Poisson model with aggregate claims. Below we derive a Lundberg bound and a Cramér–Lundberg approximation for this model. Note that in these results the adjustment coefficient γ is the same as for the ordinary compound Poisson model considered in Chapter 5.

Theorem 6.5.8 *In the compound Poisson model with aggregate claims there exist constants* $0 \leq b_-(h) \leq b_+(h) \leq 1$ *such that*

$$b_-(h) \mathrm{e}^{-\gamma u} \leq \psi_h(u) \leq b_+(h) \mathrm{e}^{-\gamma u}, \tag{6.5.37}$$

for all $u \geq 0$, *where the adjustment coefficient* γ *is the positive solution to* (5.4.3) *which is assumed to exist.*

Proof The inequalities (6.5.37) follow from Theorem 6.5.4 applied to the Sparre Andersen model with generic claim size $U(h) = \sum_{i=1}^{N(h)} U_i$, inter-occurrence time $T(h) = h$ and premium rate β. Indeed, the generic incremental random variable of the underlying random walk is $Y(h) = \sum_{i=1}^{N(h)} U_i - \beta h$. Applying (5.2.7), equation (6.5.21) for γ is then $\hat{m}_{Y(h)}(s) = \mathrm{e}^{\lambda h(\hat{m}_U(s)-1)-\beta h s} = 1$, and so $\lambda(\hat{m}_U(s) - 1) - \beta s = 0$, which coincides with (5.4.3). Thus, from (6.5.22) we obtain (6.5.37), where

$$b_-(h) = \inf_x \frac{\mathrm{e}^{\gamma x} \overline{F}_{Y(h)}(x)}{\int_x^\infty \mathrm{e}^{\gamma y}\, \mathrm{d}F_{Y(h)}(y)}, \qquad b_+(h) = \sup_x \frac{\mathrm{e}^{\gamma x} \overline{F}_{Y(h)}(x)}{\int_x^\infty \mathrm{e}^{\gamma y}\, \mathrm{d}F_{Y(h)}(y)}. \quad \square$$

Assume now that the distribution of U is nonlattice and (5.4.3) has a positive solution γ. We next derive a version of the Cramér–Lundberg approximation (6.5.29) for the compound Poisson model with aggregate claims.

Theorem 6.5.9 *There exists a positive and finite constant $c(h)$ such that*

$$\lim_{u\to\infty} \mathrm{e}^{\gamma u}\psi_h(u) = c(h)\,. \tag{6.5.38}$$

Proof We again consider the Sparre Andersen model with generic claim size $U(h) = \sum_{n=1}^{N(h)} U_n$, inter-occurrence time $T(h) = h$ and premium rate β. Applying Theorem 6.5.7 to this model and proceeding as in the proof of Theorem 6.5.8 we get (6.5.38). □

In general it is difficult to compare the constant $c(h)$ with the constant c that appears in the original Cramér–Lundberg approximation (5.4.10) for the compound Poisson model with permanent (time-continuous) inspection. Nevertheless, the following asymptotic result holds.

Theorem 6.5.10 *If $\hat{m}_U(\gamma+\varepsilon) < \infty$ for some $\varepsilon > 0$, then*

$$\lim_{h\to\infty} hc(h) = \frac{c}{\gamma\beta(1-\rho)} = \frac{c}{\eta\gamma\lambda\mu_U}\,, \tag{6.5.39}$$

where $\eta = \beta(\lambda\mu_U)^{-1} - 1$ is the relative safety loading of the compound Poisson model with permanent (time-continuous) inspection of risk reserve and $c = ((1-\rho)\beta)/(\lambda\hat{m}^{(1)}(\gamma) - \beta)$ is the constant appearing in the original Cramér–Lundberg approximation (5.4.10).

Proof The ascending and descending ladder heights of the random walk with generic increment $Y(h)$ are denoted by $Y^+(h)$ with distribution G_h^+ and by $Y^-(h)$ with distribution G_h^-, respectively. The distribution of $Y(h)$ is denoted by G_h. From $\hat{m}_U(\gamma+\varepsilon) < \infty$ we have $\hat{m}_{G_h}(\gamma+\varepsilon) < \infty$. Furthermore, this implies that $\hat{m}_{G_h^+}(\gamma+\varepsilon) < \infty$, as can be shown by the reader. Clearly, since $Y^-(h) \le 0$ we also have $\hat{m}_{G_h^-}(\gamma+\varepsilon) < \infty$. Now, from Corollary 6.4.3, we can derive that $1 - \hat{m}_{G_h}(s) = (1-\hat{m}_{G_h^+}(s))(1-\hat{m}_{G_h^-}(s))$ for $0 \le s \le \gamma+\varepsilon$. Differentiating this factorization identity at $s = \gamma$, one finds that

$$\begin{aligned}
-\int_0^\infty v\mathrm{e}^{\gamma v}\,\mathrm{d}G_h(v) &= -\int_0^\infty v\mathrm{e}^{\gamma v}\,\mathrm{d}G_h^+(v) - \int_0^\infty v\mathrm{e}^{\gamma v}\,\mathrm{d}G_h^-(v) \\
&\quad + \hat{m}_{G_h^+}(\gamma)\int_0^\infty v\mathrm{e}^{\gamma v}\,\mathrm{d}G_h^-(v) \\
&\quad + \hat{m}^-_{G_h}(\gamma)\int_0^\infty v\mathrm{e}^{\gamma v}\,\mathrm{d}G_h^+(v)\,.
\end{aligned}$$

Use $\hat{m}_{G^+}(\gamma) = 1$ (see (6.5.32)) to obtain

$$\int_0^\infty v e^{\gamma v} \, dG_h(v) = (1 - \hat{m}_{G_h^-}(\gamma)) \int_0^\infty v e^{\gamma v} \, dG_h^+(v) \, .$$

Hence, from (6.5.30) we obtain

$$c(h) = \frac{1 - G_h^+(\infty)}{\gamma \int_0^\infty v e^{\gamma v} \, dG_h^+(v)} = \frac{(1 - G_h^+(\infty))(1 - \hat{m}_{G_h^-}(\gamma))}{\gamma \hat{m}_{G_h}^{(1)}(\gamma)} \, . \qquad (6.5.40)$$

We now compute the derivative $\hat{m}_{G_h}^{(1)}(\gamma)$ of $\hat{m}_{G_h}(s) = \exp(s \sum_{i=1}^{N(h)} U_i - \beta h)$. By the result of Corollary 5.2.1 we have $\hat{m}_{G_h}(s) = \exp(h(\lambda(\hat{m}_U(s) - 1) - \beta s))$. As the definition of adjustment coefficient γ gives $\hat{m}_U(\gamma) - 1 = (\beta\gamma)/\lambda$, we get $\hat{m}_{G_h}^{(1)}(\gamma) = h(\lambda \hat{m}_U^{(1)}(\gamma) - \beta)$. Since in the original Poisson compound model $c = ((1-\rho)\beta)/(\lambda \hat{m}_U^{(1)}(\gamma) - \beta)$, this implies $\hat{m}_{G_h}^{(1)}(\gamma) = hc^{-1}(1-\rho)\beta$. Therefore by (6.5.40)

$$hc^{-1}\gamma(1-\rho)\beta c(h) = (1 - G_h^+(\infty))(1 - \hat{m}_{G_h^-}(\gamma)).$$

It remains to show that $\lim_{h\to\infty} G_h^+(\infty) = 0$ and $\lim_{h\to\infty} \hat{m}_{G_h^-}(\gamma) = 0$. By the strong law of large numbers we have $\lim_{h\to\infty} h^{-1} Y(h) = \lambda m_U - \beta < 0$. The formal proof is given in Theorem 10.3.4; see also Theorem 6.1.1. Hence

$$\lim_{h\to\infty} (1 - G_h^+(\infty)) = \lim_{h\to\infty} \mathbb{P}(S(h) < 0, S(2h) < 0, \ldots) = 1 \, .$$

We also have $Y^-(h) \le Y(h)$. But then $\lim_{h\to\infty} \mathbb{P}(Y^-(h) > x) = 0$ yields $\lim_{h\to\infty} \hat{m}_{G_h^-}(\gamma) = 0$. $\square$

6.5.5 Subexponential Claim Sizes

The Cramér–Lundberg approximation studied in Section 6.5.3 to the probability of ruin is valid for claim sizes having exponentially bounded or light-tailed distribution. To be more precise, the assumption that (6.5.21) has a positive solution γ means that the moment generating function $\hat{m}_{F_U}(s)$ is finite in a right neighbourhood of $s = 0$. Furthermore, for all $s > 0$ with $\hat{m}_{F_U}(s) < \infty$, the moment generating function $\hat{m}_{F_U^s}(s)$ of the integrated tail distribution F_U^s is

$$\hat{m}_{F_U^s}(s) = \frac{\hat{m}_{F_U}(s) - 1}{s \mu_{F_U}} \, .$$

Consequently, $\hat{m}_{F_U^s}(s)$ is finite in the same right neighbourhood of $s = 0$. When modelling large claims, one often uses claim size distributions F_U

like the Pareto or the lognormal distribution and that do not have this property. In the present section, we consider the Sparre Andersen model where the integrated tail distribution F^{s}_U of claim sizes belongs to the class $\mathcal{S}$ of subexponential distributions introduced in Section 2.5. We will show that the ruin function $\psi(u)$ has then the same asymptotic behaviour as the tail function $\overline{F^{\mathrm{s}}_U}(x)$. See Section 2.5.3 for sufficient conditions to have $F^{\mathrm{s}}_U \in \mathcal{S}$, in terms of the hazard rate function of F_U.

For heavy-tailed claim size distributions, the following result is an analogue to the Cramér–Lundberg approximation from Theorem 6.5.7. It extends Theorem 5.4.3 and shows that, for $\mathbb{E}\,U$ fixed, the asymptotics of the ruin function $\psi(u)$ depends on the claim size distribution $F_U(x)$ only through its behaviour for large values of x. Another interesting fact is that, in the case of a heavy-tailed claim size distribution, the asymptotic behaviour of $\psi(u)$ does not depend on the form of the inter-occurrence time distribution but only on its mean $\mathbb{E}\,T$.

Theorem 6.5.11 *If $F^{\mathrm{s}}_U \in \mathcal{S}$, then*

$$\lim_{u\to\infty} \frac{\psi(u)}{\overline{F^{\mathrm{s}}_U}(u)} = \frac{\mathbb{E}\,U}{\beta\mathbb{E}\,T - \mathbb{E}\,U}\,. \tag{6.5.41}$$

The *proof* of Theorem 6.5.11 will be partitioned into several steps. First we show the following auxiliary result for the integrated tail distribution $F^{\mathrm{s}}_{Y_+}$ of the generic increment $Y_+ = (U - \beta T)_+$. Recall that $Y_+ = \max\{0, U - \beta T\}$ and note that Y_+ is not the generic ladder height of a random walk, which we denote by Y^+.

Lemma 6.5.1 *If $F^{\mathrm{s}}_U \in \mathcal{S}$, then $F^{\mathrm{s}}_{Y_+} \in \mathcal{S}$ and*

$$\lim_{x\to\infty} \frac{\overline{F^{\mathrm{s}}_{Y_+}}(x)}{\overline{F^{\mathrm{s}}_U}(x)} = \frac{\mathbb{E}\,U}{\mathbb{E}\,Y_+}\,. \tag{6.5.42}$$

Proof Since $F_{Y_+}(x) = \int_0^\infty F_U(x+\beta t)\,\mathrm{d}F_T(t)$ for all $x \geq 0$, we have

$$\overline{F}_{Y_+}(x) = \int_0^\infty \overline{F}_U(x+\beta t)\,\mathrm{d}F_T(t)\,, \qquad x \geq 0\,.$$

Thus, by Fubini's theorem,

$$\begin{aligned}\overline{F^{\mathrm{s}}_{Y_+}}(x) &= \frac{1}{\mathbb{E}\,Y_+}\int_0^\infty \int_x^\infty \overline{F}_U(u+\beta t)\,\mathrm{d}u\,\mathrm{d}F_T(t)\\ &= \frac{\mathbb{E}\,U}{\mathbb{E}\,Y_+}\int_0^\infty \overline{F^{\mathrm{s}}_U}(x+\beta t)\,\mathrm{d}F_T(t)\,.\end{aligned}$$

Using now that $\overline{F^{\mathrm{s}}_U}(x+\beta t)/\overline{F^{\mathrm{s}}_U}(x) \le 1$ and that by Lemma 2.5.1,

$$\lim_{x\to\infty} \frac{\overline{F^{\mathrm{s}}_U}(x+\beta t)}{\overline{F^{\mathrm{s}}_U}(x)} = 1$$

for all $t \ge 0$, the dominated convergence theorem gives (6.5.42). The proof is completed because, by Lemma 2.5.4, $F^{\mathrm{s}}_U \in \mathcal{S}$ implies $F^{\mathrm{s}}_{Y_+} \in \mathcal{S}$. □

We are now in a position to prove that subexponentiality of the integrated tail distribution F^{s}_U of claim sizes implies subexponentiality of the conditional ladder height distribution G_0, where $G_0(x) = p^{-1}G^+(x)$ and $p = G^+(\infty)$.

Lemma 6.5.2 *If $F^{\mathrm{s}}_{Y_+} \in \mathcal{S}$, then $G_0 \in \mathcal{S}$ and*

$$\lim_{x\to\infty} \frac{\overline{F^{\mathrm{s}}_{Y_+}}(x)}{\overline{G}_0(x)} = \frac{p}{\mathbb{E}\, Y_+} \int_{-\infty}^{0} |t| \,\mathrm{d}G^-(t)\,. \tag{6.5.43}$$

Proof Note that

$$\begin{aligned}\int_{-\infty}^{0} F_Y(t)\,\mathrm{d}t \;&\le\; \int_{-\infty}^{0} \mathbb{P}(U-\beta T \le t)\,\mathrm{d}t \le \int_{-\infty}^{0} \mathbb{P}(-\beta T \le t)\,\mathrm{d}t\\ &=\; \int_{0}^{\infty} \mathbb{P}(T \ge t/\beta)\,\mathrm{d}t \le \beta \mathbb{E}\, T < \infty\,,\end{aligned}$$

and, by (6.4.6), $F_Y(t) = G^-(t) - \int_0^\infty G^-(t-y)\,\mathrm{d}G^+(y) \ge G^-(t)(1-p)$ for $t \le 0$. Thus,

$$\int_{-\infty}^{0} |t|\,\mathrm{d}G^-(t) = \int_{-\infty}^{0} G^-(t)\,\mathrm{d}t < \infty\,. \tag{6.5.44}$$

Also, by (6.4.6), we have for all $t \ge 0$

$$\overline{F}_{Y_+}(t) = \int_{-\infty}^{0} (G^+(t-y) - G^+(t))\,\mathrm{d}G^-(y)\,.$$

Integration of both sides of this equation from $x > 0$ to $a > x$ gives

$$\int_x^a \overline{F}_{Y_+}(t)\,\mathrm{d}t = \int_{-\infty}^{0}\int_x^a (G^+(t-y) - G^+(t))\,\mathrm{d}t\,\mathrm{d}G^-(y)\,, \tag{6.5.45}$$

where on the right-hand side the order of integration has been changed. Since

$$\begin{aligned}\int_x^a (G^+(t-y) - G^+(t))\,\mathrm{d}t &= \int_x^a (p - G^+(t))\,\mathrm{d}t + \int_x^a (G^+(t-y) - p)\,\mathrm{d}t\\ &= \int_x^{x-y} (p - G^+(t))\,\mathrm{d}t + \int_{x-y}^{a} (p - G^+(t))\,\mathrm{d}t + \int_{x-y}^{a-y} (G^+(t) - p)\,\mathrm{d}t\\ &= \int_x^{x-y} (p - G^+(t))\,\mathrm{d}t - \int_a^{a-y} (p - G^+(t))\,\mathrm{d}t\,,\end{aligned}$$

we get for all $y \le 0$

$$-y(G^+(a) - G^+(x-y)) \le \int_x^a (G^+(t-y) - G^+(t))\,dt \le -y(p - G^+(x)).$$

Substituting this into (6.5.45) and letting $a \to \infty$, we obtain, for $x > 0$,

$$\frac{\int_{-\infty}^0 |y|(p - G^+(x-y))\,dG^-(y)}{\mathbb{E}\,Y_+ \int_{-\infty}^0 |y|\,dG^-(y)} \le \frac{\overline{F^{\mathrm{s}}_{Y_+}}(x)}{\int_{-\infty}^0 |y|\,dG^-(y)} \le \frac{p - G^+(x)}{\mathbb{E}\,Y_+}, \quad (6.5.46)$$

where, in view of (6.5.44), the bounded convergence theorem is used for the lower bound. The upper bound in (6.5.46) gives, replacing x by $x + t$,

$$1 \le \int_{-\infty}^0 |y|\,dG^-(y)\,\frac{p - G^+(x+t)}{\mathbb{E}\,Y_+ \overline{F^{\mathrm{s}}_{Y_+}}(x+t)}, \qquad t, x > 0.$$

On the other hand, the lower bound in (6.5.46) yields

$$\frac{1}{\mathbb{E}\,Y_+}\int_{-t}^0 |y|\,dG^-(y)\,(p - G^+(x+t)) \le \overline{F^{\mathrm{s}}_{Y_+}}(x).$$

Thus,

$$1 \le \int_{-\infty}^0 |y|\,dG^-(y)\frac{p - G^+(x+t)}{\mathbb{E}\,Y_+ \overline{F^{\mathrm{s}}_{Y_+}}(x+t)} \le \frac{\int_{-\infty}^0 |y|\,dG^-(y)}{\int_{-t}^0 |y|\,dG^-(y)}\,\frac{\overline{F^{\mathrm{s}}_{Y_+}}(x)}{\overline{F^{\mathrm{s}}_{Y_+}}(x+t)}.$$

Since, by Lemma 2.5.1, $\lim_{x\to\infty} \overline{F^{\mathrm{s}}_{Y_+}}(x)/\overline{F^{\mathrm{s}}_{Y_+}}(x+t) = 1$, letting $x \to \infty$ and then $t \to \infty$ we get (6.5.43). Now, in view of Lemma 2.5.4, $G_0 \in \mathcal{S}$ follows. □

Proof of Theorem 6.5.11 Assume that $F^{\mathrm{s}}_U \in \mathcal{S}$. Then we have $G_0 \in \mathcal{S}$ by Lemmas 6.5.1 and 6.5.2. On the other hand we have by (6.5.3),

$$\frac{\psi(u)}{\overline{G}_0(u)} = (1-p)\sum_{k=1}^\infty p^k \frac{\overline{G_0^{*k}}(u)}{\overline{G}_0(u)}.$$

Use Theorem 2.5.4 to get $\lim_{u\to\infty}(\overline{G}_0(u))^{-1}\psi(u) = (1-p)\sum_{k=1}^\infty kp^k = p(1-p)^{-1}$. Thus, by (6.5.42) and (6.5.43),

$$\begin{aligned}\lim_{u\to\infty}\frac{\psi(u)}{\overline{F^{\mathrm{s}}_U}(u)} &= \lim_{u\to\infty}\left(\frac{\overline{F^{\mathrm{s}}_{Y_+}}(u)}{\overline{F^{\mathrm{s}}_U}(u)}\,\frac{\overline{G}_0(u)}{\overline{F^{\mathrm{s}}_{Y_+}}(u)}\,\frac{\psi(u)}{\overline{G}_0(u)}\right)\\ &= \frac{\mathbb{E}\,U}{\mathbb{E}\,Y_+}\,\frac{\mathbb{E}\,Y_+}{p\int_{-\infty}^0 |t|\,dG^-(t)}\,\frac{p}{1-p} = \frac{\mathbb{E}\,U}{(1-p)\int_{-\infty}^0 |t|\,dG^-(t)}.\end{aligned}$$

However, $(is)^{-1}(1-\hat{m}_{F_Y}(is)) = (1-\hat{m}_{G^+}(is))(is)^{-1}(1-\hat{m}_{G^-}(is))$ follows from rewriting (6.4.13). But then $\beta\mathbb{E}T - \mathbb{E}U = (1-p)\int_{-\infty}^{0} |t| \,\mathrm{d}G^-(t)$ as $s \to 0$. Hence,

$$\lim_{u\to\infty} \left(\overline{F^{\mathrm{s}}_U}(u)\right)^{-1} \psi(u) = \frac{\mathbb{E}U}{\beta\mathbb{E}T - \mathbb{E}U}. \qquad \square$$

Examples 1. We showed in Section 2.5.3 that the Weibull distribution $F = \mathrm{W}(r,c)$ with $0 < r < 1, c > 0$ belongs to $\mathcal{S}^*$. Furthermore, using Theorems 6.5.11 and 2.5.6 we have (for $F_U = \mathrm{W}(r,1)$)

$$\psi(u) \sim \frac{1}{r\beta\mathbb{E}T - \Gamma(1/r)} \int_{u^r}^{\infty} \mathrm{e}^{-y} y^{1/r-1}\,\mathrm{d}y\,, \qquad u \to \infty\,. \tag{6.5.47}$$

Note that the integral in (6.5.47) is the tail of an incomplete gamma function.

2. Let $F_U \in \mathcal{S}$ be the Pareto distribution with density

$$f_U(x) = \begin{cases} \alpha c^{\alpha} x^{-(\alpha+1)} & \text{if } x \geq c, \\ 0 & \text{if } x < c, \end{cases}$$

with $\alpha > 1, c > 0$. We leave it to the reader to show that then $\mu_U = \alpha c/(\alpha-1)$, $F^{\mathrm{s}}_U \in \mathcal{S}$ and $\psi(u) \sim c(\beta\mathbb{E}T(\alpha-1) - \alpha c)^{-1}(c/u)^{\alpha-1}$ as $u \to \infty$, where it suffices to prove that the condition of Corollary 2.5.1 is fulfilled and to use Theorem 6.5.11.

3. Let $F_U \in \mathcal{S}$ be the lognormal distribution $\mathrm{LN}(a,b)$ with $-\infty < a < \infty$, $b > 0$. If we show first that

$$\overline{F^{\mathrm{s}}_U}(x) \sim \frac{b^3 \exp(-b^2/2)}{\mathrm{e}^a\sqrt{2\pi}} \frac{x}{(\log x - a)^2} \exp\Big(-\frac{(\log x - a)^2}{2b^2}\Big)\,,$$

and then that the right-hand side belongs to $\mathcal{S}$, then we can conclude that $F^{\mathrm{s}}_U \in \mathcal{S}$. Now it is not difficult to show that

$$\psi(u) \sim c\frac{u}{(\log u - a)^2} \exp\Big(-\frac{(\log u - a)^2}{2b^2}\Big)\,, \qquad u \to \infty\,,$$

where $c = b^3(\sqrt{2\pi}(\beta\mathbb{E}T - \exp(a + b^2/2)))^{-1}$.

Bibliographical Notes. The surplus just before ruin and the severity of ruin were studied by many authors, mostly for the compound Poisson model; see the bibliographical notes to Section 5.3. Note however that results like formula (6.5.15) remain true even for much more general arrival processes with stationary increments. Using techniques of the theory of random point processes, such extensions to (6.5.15) have been derived, for example, in

Asmussen and Schmidt (1993, 1995) and Miyazawa and Schmidt (1993). Some of them will be discussed in Chapter 12. Theorem 6.5.4 was proved in the queueing setting in Kingman (1970) using monotonicity properties of the recursion defined in (6.5.24) and in Ross (1974) by martingale techniques; the proof presented in Section 6.5.2 is due to Kingman (1970); see also Stoyan (1983). The original proof of Theorem 6.5.7 given by H. Cramér is analytical, using Wiener–Hopf techniques and expansions of the resulting solutions. The approach via ladder heights, as presented in Section 6.5.3, is due to W. Feller. Theorem 6.5.10 is from Cramér (1955), p. 75. The exposition of Section 6.5.5 follows Embrechts and Veraverbeke (1982). Properties of subexponential distributions like those used in the proof of Theorem 6.5.11 can be found, for example, in Athreya and Ney (1972), Pakes (1975), Teugels (1975) and Veraverbeke (1977); see also Section 2.5. An extension of Theorem 6.5.11 to more general claim arrival processes is given in Chapter 12.

CHAPTER 7

Markov Chains

Throughout this book, a Markov chain is understood to be a stochastic process in discrete time possessing a certain conditional independence property. The state space may be finite, countably infinite or even more general. We begin with the simplest case of finitely many states. Note however that all definitions and statements presented in Section 7.1 remain valid for a countably infinite state space. Among the applications, we will pay special attention to the use of Markov chains to model bonus-malus systems in automobile insurance.

7.1 DEFINITION AND BASIC PROPERTIES

7.1.1 Initial Distribution and Transition Probabilities

Consider an evolution (of prices, premiums, exchange rates etc.) in discrete time on the finite state space $E = \{1, 2, \ldots, \ell\}$. Let $\alpha_i \in [0, 1]$ and interpret α_i as the probability that the evolution starts in state $i \in E$ at time 0. Furthermore, let $p_{ij} \in [0, 1]$ be interpreted as the probability that, in one step, the evolution moves from state i to state j. Since in each step we ultimately move somewhere we assume that

$$p_{ij} \geq 0\,, \qquad \sum_{j=1}^{\ell} p_{ij} = 1\,. \tag{7.1.1}$$

Each matrix $\boldsymbol{P} = (p_{ij})_{i,j=1,\ldots,\ell}$ fulfilling (7.1.1) is called a *stochastic matrix*. The future development of an evolution is often independent of its development in the past, provided that the present state of the evolution is given. We can formally define this conditional independence by introducing the following notion of a *homogeneous Markov chain*.

Definition 7.1.1 *A sequence $X_0, X_1, \ldots$ of E-valued random variables is called a homogeneous Markov chain if there exist a stochastic matrix $\boldsymbol{P} = (p_{ij})_{i,j \in E}$, called the (one step) transition matrix of $\{X_n\}$, and a probability function $\boldsymbol{\alpha} = (\alpha_1, \ldots, \alpha_\ell)$ on E, called the initial distribution (or* *the initial probability function) of* $\{X_n\}$*, such that for each* $n = 0, 1, \ldots$ *and* $i_0, i_1, \ldots, i_n \in E$,

$$\mathbb{P}(X_0 = i_0, X_1 = i_1, \ldots, X_n = i_n) = \alpha_{i_0} p_{i_0 i_1} \ldots p_{i_{n-1} i_n}. \tag{7.1.2}$$

Since throughout this chapter only the homogeneous case will be considered, we briefly speak of Markov chains omitting the term homogeneous. What is often most important is not the precise state of the Markov chain $\{X_n\}$ itself, but rather its distribution. In view of (7.1.2), the latter is uniquely determined by the probability function $\boldsymbol{\alpha}$ and the stochastic matrix $\boldsymbol{P}$.

Theorem 7.1.1 *Let* $\{X_n\}$ *be a sequence of E-valued random variables.* $\{X_n\}$ *is a Markov chain if and only if there exists a stochastic matrix* $\boldsymbol{P} = (p_{ij})$ *such that, for all* $n = 1, 2, \ldots$ *and* $i_0, i_1, \ldots, i_n \in E$,

$$\mathbb{P}(X_n = i_n \mid X_{n-1} = i_{n-1}, \ldots, X_0 = i_0) = p_{i_{n-1} i_n}, \tag{7.1.3}$$

whenever $\mathbb{P}(X_{n-1} = i_{n-1}, \ldots, X_0 = i_0) > 0$.

Proof If $\{X_n\}$ is a Markov chain, then (7.1.3) immediately follows from (7.1.2). Assume now that $\{X_n\}$ satisfies (7.1.3) for some stochastic matrix $\boldsymbol{P} = (p_{ij})$. Putting $\alpha_i = \mathbb{P}(X_0 = i)$ for all $i \in E$, we have

$$\mathbb{P}(X_0 = i_0, X_1 = i_1) = \begin{cases} 0 & \text{if } \alpha_{i_0} = 0, \\ \alpha_{i_0} p_{i_0 i_1} & \text{if } \alpha_{i_0} > 0, \end{cases}$$

i.e. (7.1.2) is proved for $n = 1$. Suppose that (7.1.2) holds for some $n = k - 1$. In this case we have $\mathbb{P}(X_0 = i_0, X_1 = i_1, \ldots, X_k = i_k) = 0$ if $\mathbb{P}(X_0 = i_0, X_1 = i_1, \ldots, X_{k-1} = i_{k-1}) = 0$, and

$$\begin{aligned} &\mathbb{P}(X_0 = i_0, X_1 = i_1, \ldots, X_k = i_k) \\ &= \mathbb{P}(X_0 = i_0, X_1 = i_1, \ldots, X_{k-1} = i_{k-1}) \\ &\quad \times \mathbb{P}(X_k = i_k \mid X_{k-1} = i_{k-1}, \ldots, X_0 = i_0) \\ &= \alpha_{i_0} p_{i_0 i_1} \ldots p_{i_{k-2} i_{i-1}} p_{i_{k-1} i_k} \end{aligned}$$

if $\mathbb{P}(X_0 = i_0, X_1 = i_1, \ldots, X_{k-1} = i_{k-1}) > 0$. Thus, (7.1.2) holds for every $n \in \mathbb{N}$ which completes the proof. □

Corollary 7.1.1 *If* $\{X_n\}$ *is a Markov chain, then*

$$\mathbb{P}(X_n = i_n \mid X_{n-1} = i_{n-1}, \ldots, X_0 = i_0) = \mathbb{P}(X_n = i_n \mid X_{n-1} = i_{n-1}), \tag{7.1.4}$$

whenever $\mathbb{P}(X_{n-1} = i_{n-1}, \ldots, X_0 = i_0) > 0$.

Proof It suffices to notice that $\mathbb{P}(X_{n-1} = i_{n-1}, \ldots, X_0 = i_0) > 0$ implies $\mathbb{P}(X_{n-1} = i_{n-1}) > 0$. Now, (7.1.2) yields

$$\begin{aligned}\frac{\mathbb{P}(X_n = i_n, X_{n-1} = i_{n-1})}{\mathbb{P}(X_{n-1} = i_{n-1})} &= \frac{\sum_{i_0,\ldots,i_{n-2}\in E} \alpha_{i_0} p_{i_0 i_1} \cdots p_{i_{n-2} i_{n-1}} p_{i_{n-1} i_n}}{\sum_{i_0,\ldots,i_{n-2}\in E} \alpha_{i_0} p_{i_0 i_1} \cdots p_{i_{n-2} i_{n-1}}} \\ &= p_{i_{n-1} i_n}.\end{aligned}$$

Using (7.1.3) this gives (7.1.4). □

The conditional independence property stated in Corollary 7.1.1 is called the *Markov property* of $\{X_n\}$. For $n \geq 1$ and $i, j \in E$ fixed, the product $p_{ii_1} p_{i_1 i_2} \cdots p_{i_{n-1} j}$ can be seen as the probability of the path $i \to i_1 \to \ldots \to i_{n-1} \to j$. Analogously, the sum

$$p_{ij}^{(n)} = \sum_{i_1,\ldots,i_{n-1}\in E} p_{ii_1} p_{i_1 i_2} \cdots p_{i_{n-1} j} \tag{7.1.5}$$

is interpreted as the probability of the transition from state i to state j in n steps. In particular, if $\{X_n\}$ is a Markov chain with $\mathbb{P}(X_0 = i) > 0$, then $p_{ij}^{(n)}$ is the n-step transition probability $p_{ij}^{(n)} = \mathbb{P}(X_n = j \mid X_0 = i)$ of $\{X_n\}$. In accordance with this, the matrix $\boldsymbol{P}^{(n)} = (p_{ij}^{(n)})_{i,j=1,\ldots,\ell}$ is called the *n-step transition matrix* corresponding to $\boldsymbol{P}$. We also set $\boldsymbol{P}^{(0)} = \boldsymbol{I}$, where $\boldsymbol{I}$ is the *identity matrix* whose entries are equal to 1 on the main diagonal, and 0 otherwise. From the following lemma we easily conclude that $\boldsymbol{P}^{(n)}$ is a stochastic matrix.

Lemma 7.1.1 *For all* $n, m = 0, 1, \ldots,$

$$\boldsymbol{P}^{(n)} = \boldsymbol{P}^n \tag{7.1.6}$$

and hence

$$\boldsymbol{P}^{(n+m)} = \boldsymbol{P}^{(n)} \boldsymbol{P}^{(m)}. \tag{7.1.7}$$

Proof Equation (7.1.6) is an immediate consequence of (7.1.5) and of the definition of matrix multiplication. □

The matrix identity (7.1.7) is usually called the *Chapman–Kolmogorov equation*. As an immediate consequence, we have the following useful result.

Corollary 7.1.2 *We have*

$$p_{ii}^{(n+m)} \geq p_{ij}^{(n)} p_{ji}^{(m)} \tag{7.1.8}$$

and

$$p_{ij}^{(r+n+m)} \geq p_{ik}^{(r)} p_{kk}^{(n)} p_{kj}^{(m)}. \tag{7.1.9}$$

Using Lemma 7.1.1, we get the following representation for the distribution of the state variable X_n. For convenience, we put $\mathbb{P}(X_n = j \mid X_0 = i) = 0$ if $\alpha_i = \mathbb{P}(X_0 = i) = 0$.

Theorem 7.1.2 *If $\{X_n\}$ is a Markov chain with transition matrix $\boldsymbol{P}$ and initial probability function $\boldsymbol{\alpha}$, then the distribution $\boldsymbol{\alpha}_n$ of X_n is given by*

$$\boldsymbol{\alpha}_n = \boldsymbol{\alpha}\boldsymbol{P}^n. \tag{7.1.10}$$

Proof We have $\mathbb{P}(X_n = j) = \sum_{i\in E} \mathbb{P}(X_n = j \mid X_0 = i)\alpha_i = \sum_{i\in E} \alpha_i p_{ij}^{(n)}$. It now suffices to use Lemma 7.1.1. □

Example Consider a random walk $\{S_n, n = 0, 1, \ldots\}$ from Chapter 5, where $S_0 = 0, S_n = Y_1 + \ldots + Y_n$ for $n \geq 1$ and $Y_1, Y_2, \ldots$ is a sequence of independent and identically distributed integer-valued random variables. Note that the $S_1, S_2, \ldots$ can be defined recursively by

$$S_n = S_{n-1} + Y_n\,. \tag{7.1.11}$$

By inspection it can be seen that the random walk $\{S_n, n = 0, 1, \ldots\}$ has the Markov property:

$$\mathbb{P}(S_n = i_n \mid S_{n-1} = i_{n-1}, \ldots, S_0 = i_0) = \mathbb{P}(S_n = i_n \mid S_{n-1} = i_{n-1}),$$

for $i_0 = 0, i_1, \ldots, i_n \in E$ and for all $n \geq 1$ provided that $\mathbb{P}(S_{n-1} = i_{n-1}, \ldots, S_1 = i_1) > 0$. In Section 7.1.3 we show that a Markov chain with state space $\mathbb{Z}$ is a natural extension of a random walk; instead of adding successive terms, a more general recursive scheme is considered which is useful for the simulation of a Markov chain with a given distribution.

7.1.2 Computation of the n-Step Transition Matrix

From (7.1.10) it is seen that the computation of the probability function $\boldsymbol{\alpha}_n$ of the state variable X_n is closely related to the computation of the n-th power of the transition matrix $\boldsymbol{P}$. In this section we discuss an algebraic method for computing $\boldsymbol{P}^n$ which makes use of the concept of eigenvalues and eigenvectors.

Assume that $\boldsymbol{A}$ is an arbitrary (not necessarily stochastic) $\ell \times \ell$ matrix, that $\boldsymbol{\phi}$ is an ℓ-dimensional vector with at least one component different from zero, and that θ is a real or complex number. A matrix of any dimension all of whose entries are 0 is denoted by $\boldsymbol{0}$. The *transposition* of any matrix $\boldsymbol{A} = (a_{ij})$ is denoted by $\boldsymbol{A}^\top$, i.e. $\boldsymbol{A}^\top = (a_{ji})$. If

$$\boldsymbol{A}\boldsymbol{\phi}^\top = \theta\boldsymbol{\phi}^\top, \tag{7.1.12}$$

then θ is said to be an *eigenvalue* of $\boldsymbol{A}$ and $\boldsymbol{\phi}$ is said to be a *right eigenvector* corresponding to θ. Writing (7.1.12) as $(\boldsymbol{A} - \theta\boldsymbol{I})\boldsymbol{\phi}^\top = \boldsymbol{0}$, from the theory of linear algebraic equations we get that the eigenvalues are exactly the solutions to the *characteristic equation*

$$\det(\boldsymbol{A} - \theta \boldsymbol{I}) = 0\,. \tag{7.1.13}$$

A nonzero vector $\boldsymbol{\psi}$ which is a solution to

$$\boldsymbol{\psi}\boldsymbol{A} = \theta\boldsymbol{\psi} \tag{7.1.14}$$

is called a *left eigenvector* corresponding to θ. It is easy to see that for each eigenvalue θ, a solution $\boldsymbol{\psi}$ to (7.1.14) always exists because (7.1.13) implies that $\det((\boldsymbol{A}-\theta\boldsymbol{I})^\top) = 0$, i.e. there exists a nonzero (column) vector $\boldsymbol{\psi}^\top$ such that $(\boldsymbol{A}-\theta\boldsymbol{I})^\top\boldsymbol{\psi}^\top = \boldsymbol{0}$, which is equivalent to (7.1.14).

Note that (7.1.13) is an algebraic equation of order ℓ, i.e. there are ℓ eigenvalues $\theta_1, \ldots, \theta_\ell$, which can be complex and some of them can coincide. We always assume that the eigenvalues $\theta_1, \ldots, \theta_\ell$ are numbered such that

$$|\theta_1| \geq |\theta_2| \geq \ldots \geq |\theta_\ell|\,.$$

Let $\boldsymbol{\Phi} = (\boldsymbol{\phi}_1^\top, \ldots, \boldsymbol{\phi}_\ell^\top)$ be an $\ell \times \ell$ matrix consisting of right (column) eigenvectors,

$$\boldsymbol{\Psi} = \begin{pmatrix} \boldsymbol{\psi}_1 \\ \vdots \\ \boldsymbol{\psi}_\ell \end{pmatrix}$$

an $\ell \times \ell$ matrix consisting of left eigenvectors $\boldsymbol{\psi}_1, \ldots, \boldsymbol{\psi}_\ell$, and $\boldsymbol{\theta} = (\theta_1, \ldots, \theta_\ell)$ the vector of eigenvalues. There results the equation

$$\boldsymbol{A}\boldsymbol{\Phi} = \boldsymbol{\Phi}\,\mathrm{diag}(\boldsymbol{\theta})\,, \tag{7.1.15}$$

where $\mathrm{diag}(\boldsymbol{\theta})$ denotes the diagonal matrix with diagonal elements $\theta_1, \ldots, \theta_\ell$ and all other elements equal to zero. We make a number of observations.

- If all eigenvectors $\boldsymbol{\phi}_1, \ldots, \boldsymbol{\phi}_\ell$ are linearly independent, and this is assumed to the end of the present section, then $\boldsymbol{\Phi}^{-1}$ exists. In this case, we can put $\boldsymbol{\Psi} = \boldsymbol{\Phi}^{-1}$.
- A direct consequence of (7.1.15) is $\boldsymbol{A} = \boldsymbol{\Phi}\,\mathrm{diag}(\boldsymbol{\theta})\boldsymbol{\Phi}^{-1} = \boldsymbol{\Phi}\,\mathrm{diag}(\boldsymbol{\theta})\boldsymbol{\Psi}$ and, consequently,

$$\boldsymbol{A}^n = \boldsymbol{\Phi}(\mathrm{diag}(\boldsymbol{\theta}))^n\boldsymbol{\Phi}^{-1} = \boldsymbol{\Phi}(\mathrm{diag}(\boldsymbol{\theta}))^n\boldsymbol{\Psi}. \tag{7.1.16}$$

- From (7.1.16) we get

$$\begin{aligned} \boldsymbol{A}^n &= (\boldsymbol{\phi}_1^\top, \ldots, \boldsymbol{\phi}_\ell^\top)(\mathrm{diag}(\boldsymbol{\theta}))^n \begin{pmatrix} \boldsymbol{\psi}_1 \\ \vdots \\ \boldsymbol{\psi}_\ell \end{pmatrix} \\ &= (\theta_1^n\boldsymbol{\phi}_1^\top, \ldots, \theta_\ell^n\boldsymbol{\phi}_\ell^\top) \begin{pmatrix} \boldsymbol{\psi}_1 \\ \vdots \\ \boldsymbol{\psi}_\ell \end{pmatrix}, \end{aligned}$$

which yields the *spectral representation* of $\boldsymbol{A}^n$, i.e.

$$\boldsymbol{A}^n = \sum_{i=1}^{\ell} \theta_i^n \boldsymbol{\phi}_i^\top \boldsymbol{\psi}_i \,. \tag{7.1.17}$$

Applying this procedure to the transition matrix $\boldsymbol{P}$, the spectral representation (7.1.17) gives us a basis for computing the n-step transition matrix $\boldsymbol{P}^n$. There of course remains the difficulty of computing the eigenvalues and the eigenvectors of $\boldsymbol{P}$. However, in many cases this can be done by means of standard software like MATLAB, MATHEMATICA or MAPLE. An important advantage of the method of spectral representation, however, is that the complexity of the numerical computations does not grow with n because, once the eigenvalues and eigenvectors of $\boldsymbol{P}$ are computed, it is easy to compute $\boldsymbol{P}^n$ by (7.1.17).

The crucial assumption for the validity of (7.1.17) is that the eigenvectors $\boldsymbol{\phi}_1, \ldots, \boldsymbol{\phi}_\ell$ are linearly independent. The following lemma gives a simple sufficient condition.

Lemma 7.1.2 *If the eigenvalues $\theta_1, \ldots, \theta_\ell$ are distinct, then $\boldsymbol{\phi}_1, \ldots, \boldsymbol{\phi}_\ell$ are linearly independent. Moreover, if the left eigenvectors $\boldsymbol{\psi}_1, \ldots, \boldsymbol{\psi}_\ell$ are defined via $\boldsymbol{\Psi} = \boldsymbol{\Phi}^{-1}$, then*

$$\boldsymbol{\psi}_i \boldsymbol{\phi}_j^\top = \begin{cases} 1 & \text{if } i = j, \\ 0 & \text{if } i \neq j. \end{cases} \tag{7.1.18}$$

Proof We show the asserted independence property by induction. Because the eigenvector $\boldsymbol{\phi}_1$ has at least one component different from 0, the only solution to $a_1 \boldsymbol{\phi}_1^\top = \boldsymbol{0}$ is $a_1 = 0$. Assume now that $\theta_1, \ldots, \theta_\ell$ are all distinct and that $\boldsymbol{\phi}_1, \ldots, \boldsymbol{\phi}_{k-1}$ are linearly independent for some $k \leq \ell$. In order to prove that also the eigenvectors $\boldsymbol{\phi}_1, \ldots, \boldsymbol{\phi}_k$ are linearly independent, we have to show that

$$\sum_{j=1}^{k} a_j \boldsymbol{\phi}_j^\top = \boldsymbol{0} \tag{7.1.19}$$

implies $a_1 = \ldots = a_k = 0$. If (7.1.19) holds, then

$$\boldsymbol{0} = \boldsymbol{A}\boldsymbol{0} = \sum_{j=1}^{k} a_j \boldsymbol{A} \boldsymbol{\phi}_j^\top = \sum_{j=1}^{k} a_j \theta_j \boldsymbol{\phi}_j^\top \,.$$

On the other hand, $\boldsymbol{0} = \theta_k \boldsymbol{0} = \theta_k \sum_{j=1}^{k} a_j \boldsymbol{\phi}_j^\top = \sum_{j=1}^{k} \theta_k a_j \boldsymbol{\phi}_j^\top$. This gives $\boldsymbol{0} = \sum_{j=1}^{k-1} (\theta_k - \theta_j) a_j \boldsymbol{\phi}_j^\top$ and, consequently,

$$(\theta_k - \theta_1)a_1 = (\theta_k - \theta_2)a_2 = \ldots = (\theta_k - \theta_{k-1})a_{k-1} = 0 \,.$$

Hence $a_1 = a_2 = \ldots = a_{k-1} = 0$, because $\theta_k \neq \theta_j$ for $1 \leq j \leq k-1$. This implies $a_k = 0$ by (7.1.19), and so (7.1.18) is a direct consequence of $\boldsymbol{\Psi} = \boldsymbol{\Phi}^{-1}$. □

We still need another application of the concept of eigenvalues which is related to the spectral representation (7.1.17) and which will be used in later sections.

Lemma 7.1.3 *Let $\boldsymbol{A}$ be an arbitrary $\ell \times \ell$ matrix. Then there exists a nonsingular matrix $\boldsymbol{C}$ such that $\boldsymbol{C}\boldsymbol{A}\boldsymbol{C}^{-1}$ is an upper triangular matrix with the eigenvalues of $\boldsymbol{A}$ along the main diagonal.*

Proof We use induction on ℓ and suppose that the lemma is true for each $(\ell-1) \times (\ell-1)$ matrix. Let θ be an eigenvalue of $\boldsymbol{A}$, and let $\boldsymbol{\phi}$ be the right eigenvector corresponding to θ, i.e. $\boldsymbol{A}\boldsymbol{\phi}^\top = \theta\boldsymbol{\phi}^\top$. Let $\boldsymbol{V}$ be any nonsingular matrix such that $\boldsymbol{\phi}^\top$ is the first column of $\boldsymbol{V}$. If $\boldsymbol{D}_{(1)}$ denotes the first column of the matrix $\boldsymbol{D}$, then

$$\begin{aligned}(\boldsymbol{V}^{-1}\boldsymbol{A}\boldsymbol{V})_{(1)} &= (\boldsymbol{V}^{-1}\boldsymbol{A})\,\boldsymbol{V}_{(1)} = \boldsymbol{V}^{-1}\boldsymbol{A}\boldsymbol{\phi}^\top = \theta\boldsymbol{V}^{-1}\boldsymbol{\phi}^\top \\ &= \theta\boldsymbol{V}^{-1}\boldsymbol{V}_{(1)} = \theta\left(\boldsymbol{V}^{-1}\boldsymbol{V}\right)_{(1)} = \theta\boldsymbol{e}_1^\top,\end{aligned}$$

where $\boldsymbol{e}_1 = (1, 0, \ldots, 0)$. Thus

$$\boldsymbol{V}^{-1}\boldsymbol{A}\boldsymbol{V} = \begin{pmatrix} \theta & \boldsymbol{B} \\ \boldsymbol{0} & \boldsymbol{A}' \end{pmatrix},$$

where $\boldsymbol{A}'$ and $\boldsymbol{B}$ are $(\ell-1) \times (\ell-1)$ and $1 \times \ell - 1$ matrices, respectively. By the induction hypothesis, there exists a nonsingular $(\ell-1) \times (\ell-1)$ matrix $\boldsymbol{W}$ such that $\boldsymbol{W}^{-1}\boldsymbol{A}'\boldsymbol{W}$ is upper triangular. Put $\boldsymbol{C} = \boldsymbol{V}\boldsymbol{W}'$, where

$$\boldsymbol{W}' = \begin{pmatrix} 1 & \boldsymbol{0} \\ \boldsymbol{0} & \boldsymbol{W} \end{pmatrix}.$$

Then $\boldsymbol{C}$ is nonsingular and $\boldsymbol{C}\boldsymbol{A}\boldsymbol{C}^{-1}$ is an upper triangular matrix. Since

$$\det\left(\boldsymbol{C}\boldsymbol{A}\boldsymbol{C}^{-1}\right) = \det(\boldsymbol{A}) = \prod_{i=1}^{\ell} \theta_i\,,$$

$\boldsymbol{C}\boldsymbol{A}\boldsymbol{C}^{-1}$ has the eigenvalues of $\boldsymbol{A}$ along the main diagonal. □

7.1.3 Recursive Stochastic Equations

Here we show that each sequence of random variables fulfilling a certain recursive stochastic equation is a Markov chain. However, it is also possible to show the reverse statement, namely that each Markov chain can be seen as a solution to a recursive stochastic equation.

Let $Y_1, Y_2, \ldots$ be independent and identically distributed integer-valued random variables. Let $X_0 : \Omega \to \mathbb{Z}$ be independent of $Y_1, Y_2, \ldots$ and let the random variables $X_1, X_2, \ldots$ be defined by the *recursive stochastic equation*

$$X_n = \phi(X_{n-1}, Y_n)\,, \tag{7.1.20}$$

where $\phi : \mathbb{Z} \times \mathbb{Z} \to \mathbb{Z}$ is an arbitrary function.

Theorem 7.1.3 *For all $n \geq 1$ and $i_0, i_1, \ldots, i_n \in \mathbb{Z}$,*

$$\mathbb{P}(X_n = i_n \mid X_{n-1} = i_{n-1}, \ldots, X_0 = i_0) = \mathbb{P}(X_n = i_n \mid X_{n-1} = i_{n-1})\,,$$

whenever $\mathbb{P}(X_{n-1} = i_{n-1}, \ldots, X_0 = i_0) > 0$.

Proof From (7.1.20) we get

$$\begin{aligned}
&\mathbb{P}(X_n = i_n \mid X_{n-1} = i_{n-1}, \ldots, X_0 = i_0) \\
&\quad= \mathbb{P}(\phi(i_{n-1}, Y_n) = i_n \mid X_{n-1} = i_{n-1}, \ldots, X_0 = i_0) \\
&\quad= \mathbb{P}(\phi(i_{n-1}, Y_n) = i_n) \qquad (7.1.21)\\
&\quad= \mathbb{P}(\phi(i_{n-1}, Y_n) = i_n \mid X_{n-1} = i_{n-1}) = \mathbb{P}(X_n = i_n \mid X_{n-1} = i_{n-1})\,,
\end{aligned}$$

where in (7.1.21) we used the fact that the random variables $X_0, \ldots, X_{n-1}$, defined by $Y_1, \ldots, Y_{n-1}$, are independent of $\phi(i_{n-1}, Y_n)$. □

From the proof of Theorem 7.1.3 we see that the conditional probability $p_{ij} = \mathbb{P}(X_n = j \mid X_{n-1} = i)$ is given by $p_{ij} = \mathbb{P}(\phi(i, Y_n) = j)$. Thus p_{ij} does not depend on n, because the Y_ns are identically distributed. Moreover, the joint probability $\mathbb{P}(X_0 = i_0, X_1 = i_1, \ldots, X_n = i_n)$ can be given by

$$\mathbb{P}(X_0 = i_0, X_1 = i_1, \ldots, X_n = i_n) = \alpha_{i_0} p_{i_0 i_1} \cdots p_{i_{n-1} i_n}\,, \tag{7.1.22}$$

where $\alpha_{i_0} = \mathbb{P}(X_0 = i_0)$. These properties of the stochastic process $\{X_n\}$ given by (7.1.20) are basic for the notion of a homogeneous Markov chain as introduced in Section 7.1.1.

We tackle the reverse problem. Suppose that $X_0, X_1, \ldots$ is a Markov chain on the finite set $E = \{1, 2, \ldots, \ell\}$, with initial distribution $\boldsymbol{\alpha} = (\alpha_1, \ldots, \alpha_\ell)$ and transition matrix $\boldsymbol{P} = (p_{ij})$. Starting from a recursive equation of type (7.1.20), we want to construct a Markov chain $\{X'_n\}$ with the same initial distribution and transition matrix, i.e. such that for each $n = 0, 1, \ldots$,

$$\mathbb{P}(X_0 = i_0, \ldots, X_n = i_n) = \mathbb{P}(X'_0 = i_0, \ldots, X'_n = i_n)\,, \tag{7.1.23}$$

for all $i_0, \ldots, i_n \in E$. Two stochastic processes $\{X_i\}$ and $\{X'_i\}$ for which (7.1.23) holds for all $n = 0, 1, \ldots$ are called *stochastically equivalent*.

Take $\{Z_n, n \in \mathbb{N}\}$ a sequence of independent random variables, uniformly distributed on $[0,1]$. We define an E-valued random variable X'_0 with probability function $\boldsymbol{\alpha} = (\alpha_1, \ldots, \alpha_\ell)$ by the statement

$$X'_0 = k \qquad \text{if and only if} \qquad Z_0 \in \Big(\sum_{i=1}^{k-1} \alpha_i, \sum_{i=1}^{k} \alpha_i\Big],$$

for all $k = 1, \ldots, \ell$, or explicitly by

$$X'_0 = \sum_{k=1}^{\ell} k \mathbb{I}\Big(\sum_{i=1}^{k-1} \alpha_i < Z_0 \le \sum_{i=1}^{k} \alpha_i\Big). \tag{7.1.24}$$

The random variables $X'_1, X'_2, \ldots$ are then defined recursively. Let the function $\phi : E \times [0,1] \to E$ be defined by

$$\phi(i, z) = \sum_{k=1}^{\ell} k \mathbb{I}\Big(\sum_{j=1}^{k-1} p_{ij} < z \le \sum_{j=1}^{k} p_{ij}\Big) \tag{7.1.25}$$

and put

$$X'_n = \phi(X'_{n-1}, Z_n). \tag{7.1.26}$$

It is easily seen that for the sequence $\{X'_n\}$ defined in (7.1.24)–(7.1.26), the joint probabilities $\mathbb{P}(X'_0 = i_0, X'_1 = i_1, \ldots, X'_n = i_n)$ are given by (7.1.2), i.e. $\{X'_n\}$ is a Markov chain with initial distribution $\boldsymbol{\alpha}$ and transition matrix $\boldsymbol{P}$.

The construction described above can be used to simulate a Markov chain with given initial distribution and transition matrix. Note that this construction remains valid for Markov chains with countably infinite state space.

7.1.4 Bonus-Malus Systems

As a first illustration of the use of Markov chains in insurance, we show how an *automobile insurance* problem can be modelled by the use of Markov chains. Up to minor modifications, most automobile insurances employ the following *bonus-malus system*. There is a finite number ℓ of classes (tariff groups) and the premium depends on the class to which the policy-holder belongs. Each year the class of a policy-holder is determined on the basis of the class of the previous year and on the number of reported claims during that year. If no claim has been reported, then the policy-holder gets a bonus expressed in the lowering to a class with a possibly lower premium. Depending on the number of reported claims, the policy-holder gets maluses, expressed by a shift to a higher class. Formally, we need the following ingredients:

- ℓ classes numbered by $1, \ldots, \ell$; we call class 1 *superbonus* and class ℓ *supermalus*; the annual premium depends on the number of the actual class and is computed from a given scale;
- a premium scale $\boldsymbol{b} = (b_1, b_2, \ldots, b_\ell)$, where we assume $b_1 \le b_2 \le \ldots \le b_\ell$;
- transition rules which say how the transfer from one class to another is determined once the number of claims is known; once k claims are reported, let

$$t_{ij}(k) = \begin{cases} 1 & \text{if the policy gets transferred from class } i \text{ to class } j, \\ 0 & \text{otherwise;} \end{cases}$$

- an initial class i_0 for a new policy holder entering the system.

Let $\boldsymbol{T}(k) = (t_{ij}(k))_{i,j=1,\ldots,\ell}$. Thus each $\boldsymbol{T}(k)$ is a 0–1 matrix having in each row exactly one 1. Suppose that for a policy-holder the numbers of yearly reported claims form an $\mathbb{N}$-valued sequence $Y_1, Y_2, \ldots$ of independent random variables with common probability function $\{q_k\}$. Denote by $X_0, X_1, \ldots$ the year-by-year classes for the policy-holder. Since we assume that the class for the next year is uniquely determined by the class of the preceding year and by the number of claims reported during that year, we can express $\{X_n\}$ by the recursive equation $X_n = \phi(X_{n-1}, Y_n)$, where $\phi(i,k) = j$ if and only if $t_{ij}(k) = 1$. Thus, in view of the results given in Section 7.1.3, $\{X_n\}$ is a Markov chain. The transition probability p_{ij} that the policy passes from class i to class j is $p_{ij} = \sum_{k=0}^{\infty} q_k t_{ij}(k)$. In practice, one usually assumes that the number of claims reported by the policy-holder follows a Poisson distribution with parameter λ, possibly depending on the policy-holder. Explicitly,

$$p_{ij} = p_{ij}(\lambda) = \sum_{k=0}^{\infty} \frac{\lambda^k}{k!} \mathrm{e}^{-\lambda} t_{ij}(k) . \tag{7.1.27}$$

For the analysis of bonus-malus systems it is interesting to study the following characteristics:

- the probability that in the n-th year the policy-holder is in class j;
- the expected accumulated (total) premium paid by the policy-holder over the period of n years.

At least two variants are possible for the computation of the accumulated premium: undiscounted or discounted premiums. We will discuss both of them later in Section 7.3.2. Another crucial issue is whether it is profitable for a policy-holder not to report small claims in order to avoid an increase in premium, a behaviour called *hunger for bonus*. Formally, we can define a strategy for the policy-holder by a vector $\boldsymbol{x} = (x_1, \ldots, x_\ell)$, where x_i is the *retention limit* for class i, i.e. the cost of any accident of amount less than x_i is borne by the policy-holder; the claims connected with higher costs are reported. The problem is to determine an optimal value of $\boldsymbol{x}$.

Example Table 7.1.1 presents as an example the German bonus-malus system. There are 18 bonus classes labelled from 1 to 18; new policies are placed in class 15. The bonus rules and the premium scale are given in the adjoining table. The transition probabilities p_{ij} in (7.1.27) can be given by Table 7.1.2 where $\{k\} = (\lambda^k/k!)e^{-\lambda}$.

Class	Premium scale	Class after one year (per no. of claims)				
		0	1	2	3	4, 5, ...
18	200	13	18	18	18	18
17	200	13	18	18	18	18
16	175	13	17	18	18	18
15	175	13	16	17	18	18
14	125	13	16	17	18	18
13	100	12	14	16	17	18
12	85	11	13	14	16	18
11	70	10	13	14	16	18
10	65	9	12	13	14	18
9	60	8	11	13	14	18
8	55	7	11	13	14	18
7	50	6	11	13	14	18
6	45	5	11	13	14	18
5	40	4	10	12	13	18
4	40	3	9	11	13	18
3	40	2	8	11	13	18
2	40	1	7	11	13	18
1	40	1	7	11	13	18

Table 7.1.1 German bonus-malus system

Bibliographical Notes. Further elementary properties of Markov chains with finitely or countably infinitely many states can be found in Berger (1993), Chung (1967), Feller (1968), Iosifescu (1980), Kemeny and Snell (1990) and Krengel (1991), for example. For more details on simulation of Markov chains, see Ross (1997b) and Winkler (1995), for example. The German bonus-malus system is discussed in Boos (1991), where the bonus-malus systems of further European countries are given as well. The Danish and Finnish bonus-malus systems are considered in Vepsäläinen (1972). Other references where bonus-malus systems are modelled by Markov chains are Dufresne (1984), Lemaire (1985,1995) and Loimaranta (1972). Strategies for the claim behaviour of a policy-holder have been investigated in Dellaert, Frenk and van Rijsoort (1993), where it is shown that it is optimal to claim

	1	2	3	4	5	6	7	8	9	10	11	12	13	14	15	16	17	18
1	{0}	.	.	.	.	.	{1}	.	.	.	{2}	.	{3}	.	.	.	.	{4, 5, .}
2	{0}	.	.	.	.	.	{1}	.	.	.	{2}	.	{3}	.	.	.	.	{4, 5, .}
3	.	{0}	.	.	.	.	.	{1}	.	.	{2}	.	{3}	.	.	.	.	{4, 5, .}
4	.	.	{0}	.	.	.	.	.	{1}	.	{2}	.	{3}	.	.	.	.	{4, 5, .}
5	.	.	.	{0}	.	.	.	.	.	{1}	.	{2}	{3}	.	.	.	.	{4, 5, .}
6	.	.	.	.	{0}	.	.	.	.	.	{1}	.	{2}	{3}	.	.	.	{4, 5, .}
7	.	.	.	.	.	{0}	.	.	.	.	{1}	.	{2}	{3}	.	.	.	{4, 5, .}
8	.	.	.	.	.	.	{0}	.	.	.	{1}	.	{2}	{3}	.	.	.	{4, 5, .}
9	.	.	.	.	.	.	.	{0}	.	.	{1}	.	{2}	{3}	.	.	.	{4, 5, .}
10	.	.	.	.	.	.	.	.	{0}	.	.	{1}	{2}	{3}	.	.	.	{4, 5, .}
11	.	.	.	.	.	.	.	.	.	{0}	.	.	{1}	{2}	.	{3}	.	{4, 5, .}
12	.	.	.	.	.	.	.	.	.	.	{0}	.	{1}	{2}	.	{3}	.	{4, 5, .}
13	.	.	.	.	.	.	.	.	.	.	.	{0}	.	{1}	.	{2}	{3}	{4, 5, .}
14	.	.	.	.	.	.	.	.	.	.	.	.	{0}	.	.	{1}	{2}	{3, 4, .}
15	.	.	.	.	.	.	.	.	.	.	.	.	{0}	.	.	{1}	{2}	{3, 4, .}
16	.	.	.	.	.	.	.	.	.	.	.	.	{0}	.	.	.	{1}	{2, 3, .}
17	.	.	.	.	.	.	.	.	.	.	.	.	{0}	.	.	.	.	{1, 2, .}
18	.	.	.	.	.	.	.	.	.	.	.	.	{0}	.	.	.	.	{1, 2, .}

Table 7.1.2 Transition probabilities for the German bonus-malus system

for damages only if its amount exceeds a certain retention limit. Further related results can be found, for instance, in Bonsdorff (1992), Islam and Consul (1992) and Szynal and Teugels (1993).

7.2 STATIONARY MARKOV CHAINS

7.2.1 Long-Run Behaviour

For large n, it may be difficult to compute the probability function $\boldsymbol{\alpha}_n = (\alpha_1^{(n)}, \ldots, \alpha_\ell^{(n)})$ of X_n using (7.1.10). One way out is to find conditions under which the $\boldsymbol{\alpha}_n$ converge to a limit, say $\boldsymbol{\pi} = \lim_{n\to\infty} \boldsymbol{\alpha}_n$, and then to use $\boldsymbol{\pi}$ as an approximation to $\boldsymbol{\alpha}_n$. This method is of practical importance because in many cases the computation of $\boldsymbol{\pi}$ is much easier than that of $\boldsymbol{\alpha}_n$. We begin with a simple example.

Example Let

$$\boldsymbol{P} = \begin{pmatrix} 1-p & p \\ p' & 1-p' \end{pmatrix}$$

with $0 < p, p' \le 1$. In this case, it is not difficult to show that the n-step transition matrix $\boldsymbol{P}^{(n)} = \boldsymbol{P}^n$ is given by

$$\boldsymbol{P}^n = \frac{1}{p+p'}\begin{pmatrix} p' & p \\ p' & p \end{pmatrix} + \frac{(1-p-p')^n}{p+p'}\begin{pmatrix} p & -p \\ -p' & p' \end{pmatrix}.$$

Assume that $p + p' < 2$; then we have

$$\lim_{n\to\infty} \boldsymbol{P}^n = \frac{1}{p+p'}\begin{pmatrix} p' & p \\ p' & p \end{pmatrix}$$

and, by (7.1.10),

$$\boldsymbol{\pi} = \lim_{n\to\infty} \boldsymbol{\alpha}_n = \Big(\frac{p'}{p+p'}, \frac{p}{p+p'}\Big). \tag{7.2.1}$$

However, if $p + p' = 2$, then

$$\boldsymbol{P}^n = \begin{cases} \boldsymbol{P} & \text{if } n \text{ is odd,} \\ \boldsymbol{I} & \text{if } n \text{ is even.} \end{cases}$$

Note that the limit distribution $\boldsymbol{\pi}$ in (7.2.1) does not depend on the choice of the initial distribution $\boldsymbol{\alpha} = \boldsymbol{\alpha}_0$. This invariance property of $\boldsymbol{\pi}$ is connected with the notion of *ergodicity of Markov chains.*

Definition 7.2.1 *A Markov chain $\{X_n\}$ with transition matrix $\boldsymbol{P} = (p_{ij})$ is said to be ergodic if*
(a) *the following limits exist for each $j \in E$:*

$$\pi_j = \lim_{n\to\infty} p_{ij}^{(n)}, \tag{7.2.2}$$

(b) *the π_j are strictly positive and independent of i and,*
(c) *(π_j) is a probability function, i.e. $\sum_{j\in E} \pi_j = 1$.*

The ergodicity of Markov chains will be characterized by a concept from matrix theory. An $\ell \times \ell$ matrix $\boldsymbol{A} = (a_{ij})$ is called *nonnegative* if all entries a_{ij} are nonnegative. A nonnegative matrix $\boldsymbol{A}$ is called *regular* if there exists some $n_0 \geq 1$ such that all entries of $\boldsymbol{A}^{n_0}$ are strictly positive.

Theorem 7.2.1 *A Markov chain $\{X_n\}$ with transition matrix $\boldsymbol{P}$ is ergodic if and only if $\boldsymbol{P}$ is regular.*

Proof We first show that the condition

$$\min_{i,j\in E} p_{ij}^{(n_0)} > 0 \tag{7.2.3}$$

is sufficient for ergodicity. Let $m_j^{(n)} = \min_{i\in E} p_{ij}^{(n)}$ and $M_j^{(n)} = \max_{i\in E} p_{ij}^{(n)}$. From (7.1.7) we have $p_{ij}^{(n+1)} = \sum_{k\in E} p_{ik} p_{kj}^{(n)}$ and, consequently,

$$m_j^{(n+1)} = \min_i p_{ij}^{(n+1)} = \min_i \sum_k p_{ik} p_{kj}^{(n)} \geq \min_i \sum_k p_{ik} \min_k p_{kj}^{(n)} = m_j^{(n)},$$

i.e. $m_j^{(n)} \le m_j^{(n+1)}$ for all $n \ge 1$. Analogously, we get $M_j^{(n)} \ge M_j^{(n+1)}$ for all $n \ge 1$. Thus, to prove (7.2.2), it suffices to show that

$$\lim_{n\to\infty} (M_j^{(n)} - m_j^{(n)}) = 0 \tag{7.2.4}$$

for each $j \in E$. Let $a = \min_{i,j\in E} p_{ij}^{(n_0)} > 0$. Then,

$$\begin{aligned} p_{ij}^{(n_0+n)} &= \sum_k p_{ik}^{(n_0)} p_{kj}^{(n)} = \sum_k (p_{ik}^{(n_0)} - a p_{jk}^{(n)}) p_{kj}^{(n)} + a \sum_k p_{jk}^{(n)} p_{kj}^{(n)} \\ &= \sum_k (p_{ik}^{(n_0)} - a p_{jk}^{(n)}) p_{kj}^{(n)} + a p_{jj}^{(2n)}. \end{aligned}$$

Since $p_{ik}^{(n_0)} - a p_{jk}^{(n)} \ge 0$, this gives

$$p_{ij}^{(n_0+n)} \ge m_j^{(n)} \sum_k (p_{ik}^{(n_0)} - a p_{jk}^{(n)}) + a p_{jj}^{(2n)} = m_j^{(n)} (1-a) + a p_{jj}^{(2n)}.$$

Thus, $m_j^{(n_0+n)} \ge m_j^{(n)}(1-a) + a p_{jj}^{(2n)}$. Analogously, we have $M_j^{(n_0+n)} \le M_j^{(n)}(1-a) + a p_{jj}^{(2n)}$. Consequently, $M_j^{(n_0+n)} - m_j^{(n_0+n)} \le (M_j^{(n)} - m_j^{(n)})(1-a)$, and by induction

$$M_j^{(kn_0+n)} - m_j^{(kn_0+n)} \le (M_j^{(n)} - m_j^{(n)})(1-a)^k, \tag{7.2.5}$$

for each $k \ge 1$. This means that there is a sequence $n_1, n_2, \ldots$ of natural numbers tending to infinity such that

$$\lim_{k\to\infty} (M_j^{(n_k)} - m_j^{(n_k)}) = 0 \tag{7.2.6}$$

for each $j \in E$. Since the differences $M_j^{(n)} - m_j^{(n)}$ are monotone in n, (7.2.6) holds for each sequence $n_1, n_2, \ldots$ of natural numbers tending to infinity, i.e. (7.2.4) is proved. The limits π_j are positive because

$$\pi_j = \lim_{n\to\infty} p_{ij}^{(n)} \ge \lim_{n\to\infty} m_j^{(n)} \ge m_j^{(n_0)} \ge a > 0.$$

Moreover, $\sum_{j\in E} \pi_j = \sum_{j\in E} \lim_{n\to\infty} p_{ij}^{(n)} = \lim_{n\to\infty} \sum_{j\in E} p_{ij}^{(n)} = 1$ because interchanging of limit and finite sum is always allowed. On the other hand, the necessity of (7.2.3) is an immediate consequence of $\min_{j\in E} \pi_j > 0$ and (7.2.2) having in mind that E is finite. □

As the limits $\pi_j = \lim_{n\to\infty} p_{ij}^{(n)}$ do not depend on i, (7.1.6) and (7.1.10) imply that $\lim_{n\to\infty} \boldsymbol{\alpha}_n = \boldsymbol{\alpha} \lim_{n\to\infty} \boldsymbol{P}^{(n)} = \boldsymbol{\pi}$. Note that one can prove an even stronger result than (7.2.2). Indeed, (7.2.5) gives

$$|p_{ij}^{(n)} - \pi_j| \le M_j^{(n)} - m_j^{(n)} \le (1-a)^{\lfloor n/n_0 \rfloor - 1}, \tag{7.2.7}$$

which is a geometric bound for the *rate of convergence* in (7.2.2).

Corollary 7.2.1 *If the Markov chain $\{X_n\}$ is ergodic, then $\boldsymbol{\pi} = (\pi_1, \ldots, \pi_\ell)$ is the unique probabilistic solution to the system of linear equations*

$$\pi_j = \sum_{i \in E} \pi_i p_{ij} \,, \qquad j \in E \,. \tag{7.2.8}$$

Proof From (7.1.7) and (7.2.2) we get, interchanging limit and summation,

$$\pi_j = \lim_{n\to\infty} p_{kj}^{(n)} = \lim_{n\to\infty} \sum_{i \in E} p_{ki}^{(n-1)} p_{ij} = \sum_{i \in E} \lim_{n\to\infty} p_{ki}^{(n-1)} p_{ij} = \sum_{i \in E} \pi_i p_{ij} \,.$$

Suppose now that there exists another probability function $\boldsymbol{\pi}' = (\pi'_1, \ldots, \pi'_\ell)$ such that $\pi'_j = \sum_{i \in E} \pi'_i p_{ij}$ for $j \in E$. By induction we can show that

$$\pi'_j = \sum_{i \in E} \pi'_i p_{ij}^{(n)}, \qquad j \in E \,, \tag{7.2.9}$$

for all $n = 1, 2, \ldots$. Thus, letting n tend to infinity in (7.2.9), we get

$$\pi'_j = \lim_{n\to\infty} \sum_{i \in E} \pi'_i p_{ij}^{(n)} = \sum_{i \in E} \pi'_i \lim_{n\to\infty} p_{ij}^{(n)} = \pi_j \,. \qquad \square$$

In matrix notation, (7.2.8) can be written as $\boldsymbol{\pi} = \boldsymbol{\pi}\boldsymbol{P}$. This equation is called the *balance equation* for $\boldsymbol{P}$. It yields a useful tool when computing the limiting probability function $\boldsymbol{\pi} = \lim_{n\to\infty} \boldsymbol{\alpha}_n$. In Section 7.2.4 we will discuss this problem in detail.

7.2.2 Application of the Perron–Frobenius Theorem

Let $\boldsymbol{A}$ be any nonnegative $\ell \times \ell$ matrix. Remember that the eigenvalues $\theta_1, \ldots, \theta_\ell$ of $\boldsymbol{A}$ are numbered so that $|\theta_1| \ge \ldots \ge |\theta_\ell|$. Let $\boldsymbol{e} = (1, \ldots, 1)$ be the ℓ-dimensional vector with all components equal to 1 and $\boldsymbol{E}$ the $\ell \times \ell$ matrix all of whose entries are 1, i.e. consisting of ℓ (row) vectors $\boldsymbol{e}$. Moreover, by $\boldsymbol{e}_i$ we denote the ℓ-dimensional (row) vector having zeros at all components with the exception of the i-th component, which is equal to 1, i.e.

$$\boldsymbol{e}_i = (\underbrace{0, \ldots, 0}_{i-1}, 1, 0, \ldots) \,.$$

Furthermore, let $\boldsymbol{P}$ be a regular stochastic $\ell \times \ell$ matrix, and $\boldsymbol{\pi}$ the probability function given by the limits (7.2.2) or, equivalently, by (7.2.8). By $\boldsymbol{\Pi}$ we denote the $\ell \times \ell$ matrix consisting of ℓ (row) vectors $\boldsymbol{\pi}$.

Besides the geometric bound (7.2.7) for the rate of convergence in (7.2.2), one can give further bounds using concepts of matrix algebra. These bounds are obtained from the following important result, called the *Perron–Frobenius theorem* for regular matrices.

Theorem 7.2.2 *If $\boldsymbol{A}$ is regular, then*
(a) $|\theta_1| > |\theta_i|$ *for* $i = 2, \ldots, \ell$;
(b) *the eigenvalue* θ_1 *is real and strictly positive;*
(c) *the right and left eigenvectors* $\boldsymbol{\phi}_1, \boldsymbol{\psi}_1$ *have all components strictly positive and are unique up to constant multiples.*

The *proof* of Theorem 7.2.2 goes beyond the scope of this book. It can be found, for example, in Chapter 1 of Seneta (1981). The eigenvalue θ_1 of a regular matrix $\boldsymbol{A}$ is called the *Perron–Frobenius eigenvalue.*

Corollary 7.2.2 *If $\boldsymbol{P}$ is a regular stochastic matrix, then*
(a) $\theta_1 = 1, \boldsymbol{\phi}_1 = \boldsymbol{e}$ *and* $\boldsymbol{\psi}_1 = \boldsymbol{\pi}$;
(b) $|\theta_i| < 1$ *for* $i = 2, \ldots, \ell$.

Proof By inspection we get that $\boldsymbol{P}\boldsymbol{e}^\top = \boldsymbol{e}^\top$, and from (7.2.8) we have $\boldsymbol{\pi}\boldsymbol{P} = \boldsymbol{\pi}$. Hence 1 is an eigenvalue of $\boldsymbol{P}$ and $\boldsymbol{e}, \boldsymbol{\pi}$ are right and left eigenvectors for this eigenvalue, respectively. Also, $\theta_1 = 1$, i.e. 1 is the eigenvalue with the largest modulus. Namely, let θ be some eigenvalue of $\boldsymbol{P}$, and $\boldsymbol{\phi} = (\phi_1, \ldots, \phi_\ell)$ the corresponding right eigenvector. Then, (7.1.12) gives $|\theta|\,|\phi_i| \le \sum_{j=1}^{\ell} p_{ij}|\phi_j| \le \max_{j\in E}|\phi_j|$ for each $i \in E$. Hence $|\theta| \le 1$. Thus, Theorem 7.2.2 gives that $|\theta_i| < 1$ for $i = 2, \ldots, \ell$. □

Moreover, if $\boldsymbol{P}$ is a regular stochastic matrix with distinct eigenvalues, then Theorem 7.2.2 leads to the following bound for the rate of convergence in (7.2.2).

Corollary 7.2.3 *If all eigenvalues $\theta_1, \ldots, \theta_\ell$ of $\boldsymbol{P}$ are distinct, then*

$$|p_{ij}^{(n)} - \pi_j| = O(|\theta_2|^n), \qquad n \to \infty. \tag{7.2.10}$$

Proof From Corollary 7.2.2 it follows that $\lim_{n\to\infty} \sum_{i=2}^{\ell} \theta_i^n \boldsymbol{\phi}_i^\top \boldsymbol{\psi}_i = 0$ since $|\theta_i| < 1$ for $i = 2, \ldots, \ell$. Moreover, also by Corollary 7.2.2, we have $\theta_1 = 1$ and $\boldsymbol{\phi}_1 = (1, \ldots, 1), \boldsymbol{\psi}_1 = \boldsymbol{\pi}$ for the right and left eigenvectors corresponding to θ_1. Using the spectral representation (7.1.17) of $\boldsymbol{P}^n$, we arrive at (7.2.10). □

There exists a slightly different variant of (7.2.10), which is still true when not all eigenvalues are distinct. This variant can be obtained from the theory of nonnegative matrices. By the algebraic multiplicity of an eigenvalue, one understands its multiplicity as a root of the characteristic equation (7.1.13).

Theorem 7.2.3 *Assume that $\boldsymbol{A}$ is regular. Further, assume that if $|\theta_2| = |\theta_3|$, the algebraic multiplicity m_2 of θ_2 is not smaller than that of θ_3, nor of any other eigenvalue having the same modulus as θ_2. Then,*
(a) *for* $\theta_2 \ne 0$ *and* $n \to \infty$,

$$\boldsymbol{A}^n = \theta_1^n \boldsymbol{\phi}_1^\top \boldsymbol{\psi}_1 + O(n^{m_2-1}|\theta_2|^n); \tag{7.2.11}$$

(b) *for* $\theta_2 = 0$ *and* $n \geq \ell - 1$,

$$\boldsymbol{A}^n = \theta_1^n \boldsymbol{\phi}_1^\top \boldsymbol{\psi}_1 . \tag{7.2.12}$$

The *proof* of this theorem can also be found in Chapter 1 of Seneta (1981). Moreover, proceeding in the same way as in the proof of Corollary 7.2.3 we arrive at the following result.

Corollary 7.2.4 *Assume that* $\boldsymbol{P}$ *is a regular stochastic matrix satisfying the conditions of Theorem* 7.2.3. *Then, for* $c > |\theta_2|$,

$$|p_{ij}^{(n)} - \pi_j| = O(c^n) , \qquad n \to \infty . \tag{7.2.13}$$

7.2.3 Irreducibility and Aperiodicity

In this section we study another type of ergodicity condition which is sometimes easier to verify than (7.2.3) thanks to its probabilistic interpretation. Let $\tau_j = \min\{n \geq 0 : X_n = j\}$ denote the step when $\{X_n\}$ is in state $j \in E$ for the first time; we put $\tau_j = \infty$ if $X_n \neq j$ for all $n \in \mathbb{N}$. For $i, j \in E$ we say that state j is *accessible* from i and write $i \to j$, if $\mathbb{P}(\tau_j < \infty \mid X_0 = i) > 0$.

Theorem 7.2.4 *State* j *is accessible from* i *if and only if* $p_{ij}^{(n)} > 0$ *for some* $n \geq 0$.

Proof Sufficiency is easy to see, because $\{X_n = j\} \subset \{\tau_j \leq n\} \subset \{\tau_j < \infty\}$ and, consequently, $0 < p_{ij}^{(n)} \leq \mathbb{P}(\tau_j < \infty \mid X_0 = i)$. Conversely, if $p_{ij}^{(n)} = 0$ for all $n \in \mathbb{N}$, then

$$\begin{aligned} \mathbb{P}(\tau_j < \infty \mid X_0 = i) &= \lim_{n\to\infty} \mathbb{P}(\tau_j < n \mid X_0 = i) \\ &= \lim_{n\to\infty} \mathbb{P}\Big(\bigcup_{k=0}^{n-1} \{X_k = j\} \;\Big|\; X_0 = i\Big) \\ &\leq \lim_{n\to\infty} \sum_{k=0}^{n-1} \mathbb{P}(X_k = j \mid X_0 = i) = \lim_{n\to\infty} \sum_{k=0}^{n-1} p_{ij}^{(k)} = 0 . \end{aligned}$$

□

Note that the relation of accessibility is transitive, i.e. $i \to k$ and $k \to j$ imply $i \to j$. This is an easy consequence of Theorem 7.2.4 and Corollary 7.1.2. If $i \to j$ and $j \to i$, then we say that states i and j *communicate*, and we write $i \leftrightarrow j$. Communication is an *equivalence relation* which means that

- $i \leftrightarrow i$ (reflexivity),
- $i \leftrightarrow j$ if and only if $j \leftrightarrow i$ (symmetry),

- $i \leftrightarrow k$ and $k \leftrightarrow j$ imply $i \leftrightarrow j$ (transitivity).

Consequently, the state space E can be partitioned into (disjoint and exhaustive) equivalence classes with respect to the relation of communication. A Markov chain $\{X_n\}$ or, equivalently, its transition matrix $\boldsymbol{P} = (p_{ij})$, is called *irreducible* if E consists of only one class, that is $i \leftrightarrow j$ for all $i, j \in E$.

Example It is easy to see that the matrices

$$\boldsymbol{P}_1 = \begin{pmatrix} 1/2 & 1/2 \\ 1/2 & 1/2 \end{pmatrix}, \qquad \boldsymbol{P}_2 = \begin{pmatrix} 1/2 & 1/2 \\ 1/4 & 3/4 \end{pmatrix}$$

are irreducible, whereas the 4×4 matrix $\boldsymbol{P}$ having the block structure

$$\boldsymbol{P} = \begin{pmatrix} \boldsymbol{P}_1 & 0 \\ 0 & \boldsymbol{P}_2 \end{pmatrix}$$

is not irreducible.

Besides irreducibility, there is still another property of the states which is important for ergodicity. Define the *period* $d_i = \gcd\{n \geq 1 : p_{ii}^{(n)} > 0\}$ of state i, where gcd means the greatest common divisor. We put $d_i = \infty$ if $p_{ii}^{(n)} = 0$ for all $n \geq 1$. A state $i \in E$ with $d_i = 1$ is called *aperiodic*. If all states are aperiodic, then the Markov chain $\{X_n\}$ or, equivalently, its transition matrix $\boldsymbol{P} = (p_{ij})$ is called *aperiodic*. The next theorem shows that the periods d_i, d_j coincide if i, j belong to the same equivalence class of communicating states. We will use the notation $i \to j[n]$ if $p_{ij}^{(n)} > 0$.

Theorem 7.2.5 *If states $i, j \in E$ communicate, then $d_i = d_j$.*

Proof If $j \to j[n]$, $i \to j[k]$ and $j \to i[m]$ for some $k, m, n \geq 1$, then using Corollary 7.1.2 $i \to i[k+m]$ and $i \to i[k+m+n]$. This means that d_i divides $k+m$ and $k+m+n$. Thus, d_i also divides $n = (k+m+n) - (k+m)$. Consequently, d_i is a common divisor of all n such that $p_{jj}^{(n)} > 0$, i.e. $d_i \leq d_j$. By symmetry we also get $d_j \leq d_i$. □

Corollary 7.2.5 *All states of an irreducible Markov chain have the same period.*

The remaining part of this section is devoted to the proof that condition (7.2.3) is fulfilled if and only if the Markov chain is irreducible and aperiodic. For this purpose we need an elementary result from number theory.

Lemma 7.2.1 *Let $k = 1, 2, \ldots$. Then, for some $n_0 \geq 1$,*

$$I = \{n_1 k + n_2(k+1);\ n_1, n_2 \in \mathbb{N}\} \supset \{n_0, n_0+1, n_0+2, \ldots\}.$$

Proof If $n \geq k^2$, then $n - k^2 = mk + d$ for some $m \in \mathbb{N}$ and $0 \leq d < k$. Thus $n = (k - d + m)k + d(k+1) \in I$, i.e. we can take $n_0 = k^2$. □

Theorem 7.2.6 *$\boldsymbol{P}$ is an irreducible and aperiodic stochastic matrix if and only if $\boldsymbol{P}$ is regular.*

Proof Assume that $\boldsymbol{P}$ is irreducible and aperiodic. Consider the set of integers $J(i) = \{n \geq 1 : p_{ii}^{(n)} > 0\}$ for each $i \in E$ and note that, by the aperiodicity of $\boldsymbol{P}$, the greatest common divisor of $J(i)$ is equal to one. Moreover, from Corollary 7.1.2, one gets that $n, m \in J(i)$ implies $n + m \in J(i)$. Next we show that $J(i)$ contains two consecutive numbers. For then, Lemma 7.2.1 implies

$$J(i) \supset \{n(i), n(i) + 1, \ldots\} \tag{7.2.14}$$

for some $n(i) \geq 1$. So, assume that $J(i)$ does not contain two consecutive numbers. Then there is a minimal difference $k \geq 2$ between any two integers of $J(i)$. Consequently, for some $m = 0, 1, \ldots$ and $d = 1, \ldots, k-1$, we have $n = mk + d \in J(i)$ because otherwise, for all $n \in J(i)$, we would have $n = mk$, in contradiction to our assumption that $\gcd(J(i)) = 1$. Let $n_1, n_1 + k \in J(i)$. We show that there exist $a, b \in \mathbb{N}$ such that the difference between $a(n_1 + k) \in J(i)$ and $n + bn_1 \in J(i)$ is strictly less than k. Namely, $a(n_1 + k) - n - bn_1 = (a - b)n_1 + (a - m)k - d$ and, if $a = b = m + 1$, then the difference is $k - d < k$. Therefore $J(i)$ contains two consecutive numbers and (7.2.14) holds for all $i \in E$. Now, from (7.1.9) and the irreducibility of $\boldsymbol{P}$, we also get that $J(ij) = \{n \geq 0 : p_{ij}^{(n)} > 0\} \supset \{n(ij), n(ij) + 1, \ldots\}$. Hence $\boldsymbol{P}$ is regular. The proof of the converse statement is left to the reader. □

7.2.4 Stationary Initial Distributions

In Corollary 7.2.1, we showed that the limit distribution $\boldsymbol{\pi} = \lim_{n\to\infty} \alpha_n$ of an ergodic Markov chain $\{X_n\}$ with transition matrix $\boldsymbol{P} = (p_{ij})$ satisfies the balance equation

$$\boldsymbol{\alpha} = \boldsymbol{\alpha}\boldsymbol{P}. \tag{7.2.15}$$

Moreover, under the assumption of ergodicity, $\boldsymbol{\pi}$ is the only probability solution to (7.2.15).

If we do not assume ergodicity, then (7.2.15) can have more than one probability solution. However, if the initial distribution $\boldsymbol{\alpha}_0$ of $\{X_n\}$ is equal to any probability solution to (7.2.15), then (7.2.15) implies that $\boldsymbol{\alpha}_1 = \boldsymbol{\alpha}_0\boldsymbol{P} = \boldsymbol{\alpha}_0$ and, by iteration, $\boldsymbol{\alpha}_k = \boldsymbol{\alpha}_0$ for all $k \geq 0$. Because of this invariance property, each probability solution $\boldsymbol{\alpha}$ to (7.2.15) is called a *stationary initial distribution* of $\{X_n\}$. Note that, besides the invariance property $\boldsymbol{\alpha}_0 = \boldsymbol{\alpha}_1 = \ldots$, a Markov chain $\{X_n\}$ with a stationary initial distribution possesses an even stronger invariance property. In this connection, it is worth considering the notion of a (strictly) *stationary sequence* of random variables.

Definition 7.2.2 *A sequence of E-valued random variables $X_0, X_1, \ldots$ is stationary if for all $k, n \in \mathbb{N}$ and $i_0, \ldots, i_n \in E$,*

$$\mathbb{P}(X_k = i_0, X_{k+1} = i_1, \ldots, X_{k+n} = i_n) = \mathbb{P}(X_0 = i_0, X_1 = i_1, \ldots, X_n = i_n).$$

Theorem 7.2.7 *A Markov chain $\{X_n\}$ with a stationary initial distribution is stationary in the above sense.*

The *proof* of this fact is left to the reader.

We finish this section with a brief discussion of three methods to solve the balance equation (7.2.15). The first two methods are called *direct methods*, in contrast to an iterative method that will be discussed later on.

Theorem 7.2.8 *Assume that the stochastic matrix $\boldsymbol{P}$ is regular. Then the matrix $\boldsymbol{I} - \boldsymbol{P} + \boldsymbol{E}$ is invertible and the solution to (7.2.15) is given by*

$$\boldsymbol{\pi} = \boldsymbol{e}(\boldsymbol{I} - \boldsymbol{P} + \boldsymbol{E})^{-1}. \tag{7.2.16}$$

Proof First we verify that $\boldsymbol{I} - \boldsymbol{P} + \boldsymbol{E}$ is invertible. We do this by showing that $(\boldsymbol{I} - \boldsymbol{P} + \boldsymbol{E})\boldsymbol{x}^\top = \boldsymbol{0}$ implies $\boldsymbol{x}^\top = \boldsymbol{0}$. From (7.2.15) we have $\boldsymbol{\pi}(\boldsymbol{I} - \boldsymbol{P}) = \boldsymbol{0}$. Thus, $(\boldsymbol{I} - \boldsymbol{P} + \boldsymbol{E})\boldsymbol{x}^\top = \boldsymbol{0}$ implies that $0 = \boldsymbol{\pi}(\boldsymbol{I} - \boldsymbol{P} + \boldsymbol{E})\boldsymbol{x}^\top = 0 + \boldsymbol{\pi}\boldsymbol{E}\boldsymbol{x}^\top$, i.e. $\boldsymbol{\pi}\boldsymbol{E}\boldsymbol{x}^\top = 0$. On the other hand, $\boldsymbol{\pi}\boldsymbol{E} = \boldsymbol{e}$. Thus, $\boldsymbol{e}\boldsymbol{x}^\top = 0$, which implies $\boldsymbol{E}\boldsymbol{x}^\top = \boldsymbol{0}$. Consequently, $(\boldsymbol{I} - \boldsymbol{P})\boldsymbol{x}^\top = \boldsymbol{0}$, which means that $\boldsymbol{P}\boldsymbol{x}^\top = \boldsymbol{x}^\top$. This implies for any $n \geq 1$ that $\boldsymbol{x}^\top = \boldsymbol{P}^n\boldsymbol{x}^\top$. From Theorem 7.2.1 we have $\boldsymbol{P}^n \to \boldsymbol{\Pi}$. Thus, as $n \to \infty$, $\boldsymbol{x}^\top = \boldsymbol{P}^n\boldsymbol{x}^\top \to \boldsymbol{\Pi}\boldsymbol{x}^\top$, i.e. $x_i = \sum_{j=1}^{\ell} \pi_j x_j$ for all $i = 1, \ldots, \ell$. Because the right-hand side of these equations does not depend on i, we have $\boldsymbol{x} = c\boldsymbol{e}$ for some $c \in \mathbb{R}$. Since we also have $0 = \boldsymbol{e}\boldsymbol{x}^\top = c\boldsymbol{e}\boldsymbol{e}^\top = c\ell$, we get $c = 0$. Thus, $\boldsymbol{I} - \boldsymbol{P} + \boldsymbol{E}$ is invertible. Furthermore, since $\boldsymbol{\pi}(\boldsymbol{I} - \boldsymbol{P}) = \boldsymbol{0}$, we have $\boldsymbol{\pi}(\boldsymbol{I} - \boldsymbol{P} + \boldsymbol{E}) = \boldsymbol{\pi}\boldsymbol{E} = \boldsymbol{e}$. This proves (7.2.16). □

If the number ℓ of states is small, the matrix $\boldsymbol{I} - \boldsymbol{P} + \boldsymbol{E}$ can easily be inverted. For larger ℓ, numerical methods have to be used like the Gaussian elimination algorithm. Another possibility for solving (7.2.15) is to transform this equation in a way slightly different from that used in the proof of Theorem 7.2.8. The inversion of this transformed version of (7.2.15) is facilitated by the next result.

Lemma 7.2.2 *Let $\boldsymbol{A}$ be an $\ell \times \ell$ matrix such that $\boldsymbol{A}^n \to \boldsymbol{0}$ as $n \to \infty$. Then $\boldsymbol{I} - \boldsymbol{A}$ is invertible and, for each $n = 1, 2, \ldots$,*

$$\boldsymbol{I} + \boldsymbol{A} + \ldots + \boldsymbol{A}^{n-1} = (\boldsymbol{I} - \boldsymbol{A})^{-1}(\boldsymbol{I} - \boldsymbol{A}^n). \tag{7.2.17}$$

Proof Note that

$$\begin{aligned}(\boldsymbol{I} - \boldsymbol{A})(\boldsymbol{I} + \boldsymbol{A} + \ldots + \boldsymbol{A}^{n-1}) &= \boldsymbol{I} + \boldsymbol{A} + \ldots + \boldsymbol{A}^{n-1} - \boldsymbol{A} - \ldots - \boldsymbol{A}^n \\ &= \boldsymbol{I} - \boldsymbol{A}^n.\end{aligned}$$

Since $\boldsymbol{A}^n \to \boldsymbol{0}$, the matrix $\boldsymbol{I}-\boldsymbol{A}^n$ is nonsingular for sufficiently large n. Hence $\det((\boldsymbol{I}-\boldsymbol{A})(\boldsymbol{I}+\boldsymbol{A}+\ldots+\boldsymbol{A}^{n-1})) = \det(\boldsymbol{I}-\boldsymbol{A})\det(\boldsymbol{I}+\boldsymbol{A}+\ldots+\boldsymbol{A}^{n-1}) \neq 0$, where det means determinant. Thus, $\boldsymbol{I}-\boldsymbol{A}$ is invertible and (7.2.17) follows. □

As $\boldsymbol{\pi}$ satisfies (7.2.15), so does $c\boldsymbol{\pi}$ for each $c \geq 0$. Hence we put $\hat{\pi}_\ell = 1$ and solve the transformed equation

$$\hat{\boldsymbol{\pi}}(\boldsymbol{I}-\hat{\boldsymbol{P}}) = \boldsymbol{b}\,, \tag{7.2.18}$$

where $\hat{\boldsymbol{P}} = (p_{ij})_{i,j=1,\ldots,\ell-1}$ and $\hat{\boldsymbol{\pi}} = (\hat{\pi}_1,\ldots,\hat{\pi}_{\ell-1}), \boldsymbol{b} = (p_{\ell 1},\ldots,p_{\ell,\ell-1})$. Then, the originally required solution is given by $\pi_i = \hat{\pi}_i/c$ for $i = 1,\ldots,\ell$, where $c = \hat{\pi}_1+\ldots+\hat{\pi}_\ell$. Note that the matrix $\boldsymbol{I}-\hat{\boldsymbol{P}}$ in (7.2.18) is invertible. This follows from the following result.

Lemma 7.2.3 *If the stochastic matrix $\boldsymbol{P}$ is regular, then $\hat{\boldsymbol{P}}^n \to 0$ as $n \to \infty$. Hence $\boldsymbol{I}-\hat{\boldsymbol{P}}$ is invertible and $(\boldsymbol{I}-\hat{\boldsymbol{P}})^{-1} = \sum_{n=0}^\infty \hat{\boldsymbol{P}}^n$.*

Proof In view of Lemma 7.2.2, it suffices to show that $\hat{\boldsymbol{P}}^n \to 0$. Since $\boldsymbol{P}$ is regular, there exists a natural number $n_0 \geq 1$ such that $\delta = \max_{i\in\hat{E}} \sum_{j\in\hat{E}} p_{ij}^{(n_0)} < 1$, where $\hat{E} = \{1,\ldots,\ell-1\}$. Note that

$$(\hat{\boldsymbol{P}}^n)_{ij} = \sum_{i_1,\ldots,i_{n-1}\in\hat{E}} p_{ii_1}p_{i_1i_2}\cdots p_{i_{n-1}j} \leq \sum_{i_1,\ldots,i_{n-1}\in E} p_{ii_1}p_{i_1i_2}\cdots p_{i_{n-1}j} = (\boldsymbol{P}^n)_{ij}$$

and consequently $0 \leq (\hat{\boldsymbol{P}}^n)_{ij} \leq (\boldsymbol{P}^n)_{ij} = p_{ij}^{(n)} < 1$ for all $n \geq n_0$; $i,j \in \hat{E}$. Thus, using the representation $n = kn_0 + m$ for some $k, m \in \mathbb{N}$ with $0 \leq m < n_0$, we have

$$\begin{aligned}
(\hat{\boldsymbol{P}}^n)_{ij} &= \sum_{i_1,\ldots,i_k\in\hat{E}} (\hat{\boldsymbol{P}}^{n_0})_{ii_1}(\hat{\boldsymbol{P}}^{n_0})_{i_1i_2}\cdots(\hat{\boldsymbol{P}}^{n_0})_{i_{k-1}i_k}(\hat{\boldsymbol{P}}^{m})_{i_kj} \\
&\leq \sum_{i_1,\ldots,i_k\in\hat{E}} p_{ii_1}^{(n_0)}p_{i_1i_2}^{(n_0)}\cdots p_{i_{k-1}i_k}^{(n_0)} \\
&= \sum_{i_1,\ldots,i_{k-1}\in\hat{E}} p_{ii_1}^{(n_0)}p_{i_1i_2}^{(n_0)}\cdots p_{i_{k-2}i_{k-1}}^{(n_0)}\Big(\sum_{i_k\in\hat{E}} p_{i_{k-1}i_k}^{(n_0)}\Big) \\
&\leq \delta \sum_{i_1,\ldots,i_{k-1}\in\hat{E}} p_{ii_1}^{(n_0)}p_{i_1i_2}^{(n_0)}\cdots p_{i_{k-2}i_{k-1}}^{(n_0)} \\
&\;\;\vdots \\
&\leq \delta^k\,,
\end{aligned}$$

provided that $n \geq n_0$. This gives $\lim_{n\to\infty}(\hat{\boldsymbol{P}}^n)_{ij} \leq \lim_{k\to\infty}\delta^k = 0$. □

Lemma 7.2.3 implies that one possible method for solving (7.2.18) is the Gaussian elimination algorithm. The computational effort of this method is proportional to ℓ^3. It requires that the whole coefficient matrix is stored, since this matrix must be updated at each step of the algorithm. In general, one cannot use sparse matrix storage techniques, since these procedures suffer from computer memory problems when ℓ gets large. Moreover, in some cases, the method tends to be numerically unstable due to the subtractions involved.

The following *iterative method* is well suited for sparse matrix computations. It can also be used for larger ℓ as the method works only with the original coefficient matrix. However, the rate of convergence may be slow. The coefficient matrix $I - \hat{\boldsymbol{P}}$ in (7.2.18) is invertible and the inverse $(\boldsymbol{I} - \hat{\boldsymbol{P}})^{-1}$ can be written in the form $(\boldsymbol{I} - \hat{\boldsymbol{P}})^{-1} = \sum_{n=0}^{\infty} \hat{\boldsymbol{P}}^n$. Hence, we have

$$\hat{\boldsymbol{\pi}} = \boldsymbol{b} \sum_{n=0}^{\infty} \hat{\boldsymbol{P}}^n , \tag{7.2.19}$$

which is the basis for computing $\hat{\boldsymbol{\pi}}$ iteratively. Start by defining $\boldsymbol{b}_0 = \boldsymbol{b}$; then put $\boldsymbol{b}_{n+1} = \boldsymbol{b}_n \hat{\boldsymbol{P}}$ for $n \geq 0$. As such, (7.2.19) can be written as

$$\hat{\boldsymbol{\pi}} = \sum_{n=0}^{\infty} \boldsymbol{b}_n \tag{7.2.20}$$

and, for some $n_0 = 1, 2, \ldots$, the quantity $\sum_{n=0}^{n_0} \boldsymbol{b}_n$ can be used as an approximation to $\hat{\boldsymbol{\pi}}$. In practice, one needs to estimate the error that occurs by using only a finite number of terms in (7.2.20).

Bibliographical Notes. More material on the long-run behaviour of Markov chains can be found in Berger (1993) and Chung (1967), for example. A detailed treatment of Perron–Frobenius-type theorems is given in Seneta (1981). Theorem 7.2.8 is taken from Resnick (1992). For further numerical aspects in solving the balance equation (7.2.15) we refer to Kulkarni (1995) and Tijms (1994).

7.3 MARKOV CHAINS WITH REWARDS

7.3.1 Interest and Discounting

We begin with the primary case of a deterministic interest and discounting. Suppose that $r > 0$ is the *interest rate* (or rate of return), i.e. investing one unit (of a currency) at time $k = 0$ we get $1 + r$ at $k = 1$. If we make investments of one unit at times $k = 0, 1, \ldots, n-1$, then the accumulated value y_n immediately before time n is $y_n = (1+r) + (1+r)^2 + \ldots + (1+r)^n$.

Conversely, we ask for the present value v at time $k = 0$ of the unit invested at $k = 1$. It is easily seen that v is the solution to $(1+r)v = 1$. The quantity $v = (1+r)^{-1}$ is called the *discount factor*. If investments of one currency unit are made at $k = 1, 2, \ldots, n$, then the discounted value at $k = 0$ is equal to $v + v^2 + \ldots + v^n$. We can also add the unit investment at 0 with discounted value 1. The accumulated discounted value is $1 + v + \ldots + v^n$ and converges to $(1-v)^{-1}$ as $n \to \infty$. This elementary model can be generalized in at least two ways.

- The interest rate can be different in each time interval, say r_k in $[k-1, k)$. Denote the corresponding discount factor by v_k. If investments are of unit value then the accumulated value y_n at time n is

$$\begin{aligned} y_n &= (1+r_n) + (1+r_n)(1+r_{n-1}) + \ldots + (1+r_n)\ldots(1+r_1) \\ &= (1+r_n)(1+y_{n-1}), \end{aligned}$$

  for $n = 1, 2, \ldots$ and $y_0 = 0$. The discounted accumulated value, including the investment at time 0, is given by $1 + v_1 + v_1v_2 + \ldots + v_1 \ldots v_n$, where $v_k = (1+r_k)^{-1}$.
- If the interest rates are constant and equal to r, but the value invested at k is z_k for $k = 0, 1, \ldots, n-1$, then the accumulated value y_n at time n is $y_n = \sum_{i=0}^{n-1} z_i(1+r)^{n-i}$. Moreover, if investments are made at $1, \ldots, n$, then the discounted value at time zero is $\sum_{k=1}^{n} z_k v^k$, where $v = (1+r)^{-1}$. If we also take into account the investment at time 0, then the discounted value at 0 is $\sum_{k=0}^{n} z_k v^k$.

7.3.2 Discounted and Undiscounted Rewards

Next we combine the concepts of interest and discounting with that of a Markov chain. Let the state space be $E = \{1, \ldots, \ell\}$ and consider a Markov chain $X_0, X_1, \ldots$ on E with transition matrix $\boldsymbol{P}$. We suppose that when the Markov chain $\{X_n\}$ is visiting state i, a fixed *reward* β_i is obtained, where β_i can be any real number. Let $\boldsymbol{\beta} = (\beta_1, \ldots, \beta_\ell)$ denote the vector of rewards. Note that the components of $\boldsymbol{\beta}$ can have other interpretations than rewards. For example, they can describe costs, or premiums as in the bonus-malus systems considered in Section 7.1.4.

The accumulated *discounted reward* R_n^{d} at 0 obtained from visits at times $0, \ldots, n-1$ is $R_n^{\mathrm{d}} = \sum_{k=0}^{n-1} v^k \beta_{X_k}$, where v is the constant discount factor. Since the state space is finite, this reward converges with probability 1 when $n \to \infty$ to the infinite-horizon discounted reward

$$R^{\mathrm{d}} = \sum_{k=0}^{\infty} v^k \beta_{X_k}. \tag{7.3.1}$$

We call this the *total discounted reward*. We can also consider the *undiscounted reward* $R_n^{\rm u}$ for visits at times $0, \ldots, n-1$ and given by $R_n^{\rm u} = \sum_{k=0}^{n-1} \beta_{X_k}$. If we want to emphasize that $\{X_n\}$ has the initial probability function $\boldsymbol{\alpha}$, then we denote the undiscounted reward for visits at times $0, \ldots, n-1$ by $R_n^{\rm u}(\boldsymbol{\alpha})$.

We are interested in computing the expected rewards $\mathbb{E}\, R_n^{\rm d}$, $\mathbb{E}\, R^{\rm d}$ and $\mathbb{E}\, R_n^{\rm u}$, where the existence of these expectations follows from the finiteness of E. For this purpose, remember that $\boldsymbol{e}_i$ denotes the ℓ-dimensional (row) vector having zeros at all components with the exception of the ith component, which is equal to 1.

Theorem 7.3.1 *Assume that the Markov chain $\{X_n\}$ starts at time $n = 0$ from state i_0, i.e. $X_0 = i_0$. Then,*

$$\mathbb{E}\, R_n^{\rm d} = \boldsymbol{e}_{i_0}(\boldsymbol{I} - v\boldsymbol{P})^{-1}(\boldsymbol{I} - v^n\boldsymbol{P}^n)\boldsymbol{\beta}^\top, \tag{7.3.2}$$

$$\mathbb{E}\, R^{\rm d} = \boldsymbol{e}_{i_0}(\boldsymbol{I} - v\boldsymbol{P})^{-1}\boldsymbol{\beta}^\top, \tag{7.3.3}$$

$$\mathbb{E}\, R_n^{\rm u} = \boldsymbol{e}_{i_0}\sum_{k=0}^{n-1}\boldsymbol{P}^k\boldsymbol{\beta}^\top. \tag{7.3.4}$$

Proof To show (7.3.2), note that

$$\begin{aligned}\mathbb{E}\, R_n^{\rm d} &= \mathbb{E}\sum_{k=0}^{n-1} v^k\beta_{X_k} = \sum_{k=0}^{n-1} v^k\mathbb{E}\,\beta_{X_k} \\ &= \sum_{k=0}^{n-1} v^k\boldsymbol{e}_{i_0}\boldsymbol{P}^k\boldsymbol{\beta}^\top = \boldsymbol{e}_{i_0}\Big(\sum_{k=0}^{n-1}(v\boldsymbol{P})^k\Big)\boldsymbol{\beta}^\top.\end{aligned}$$

Since $v^n\boldsymbol{P}^n \to \boldsymbol{0}$ as $n \to \infty$, Lemma 7.2.2 implies that $\boldsymbol{I} - v\boldsymbol{P}$ is invertible. Thus we get (7.3.2) from (7.2.17), and (7.3.3) immediately follows from (7.3.2). The proof of (7.3.4) is similar to the first part of the proof of (7.3.2). □

Note, however, that in (7.3.4) we cannot use the summation formula (7.2.17) since $\boldsymbol{I} - \boldsymbol{P}$ can be singular. In the rest of this section, we work out asymptotic formulae for the expectation and variance of the undiscounted reward $R_n^{\rm u}$ as $n \to \infty$.

Assume that the transition matrix $\boldsymbol{P}$ is regular. As usual, let $\boldsymbol{\pi}$ denote the uniquely determined stationary initial distribution corresponding to $\boldsymbol{P}$, and $\boldsymbol{\Pi}$ the $\ell \times \ell$ matrix consisting of ℓ vectors $\boldsymbol{\pi}$.

Lemma 7.3.1 *If $\boldsymbol{P}$ is regular, then*

$$(\boldsymbol{P} - \boldsymbol{\Pi})^n = \boldsymbol{P}^n - \boldsymbol{\Pi} \tag{7.3.5}$$

for $n \geq 1$, and

$$\lim_{n\to\infty}(\boldsymbol{P} - \boldsymbol{\Pi})^n = \boldsymbol{0}. \tag{7.3.6}$$

Proof Clearly, (7.3.5) is obvious for $n = 1$. Assume now that (7.3.5) holds for $n = k - 1$; $k > 1$. Then,

$$\begin{aligned} (\boldsymbol{P} - \boldsymbol{\Pi})^k &= (\boldsymbol{P} - \boldsymbol{\Pi})^{k-1}(\boldsymbol{P} - \boldsymbol{\Pi}) = (\boldsymbol{P}^{k-1} - \boldsymbol{\Pi})(\boldsymbol{P} - \boldsymbol{\Pi}) \\ &= \boldsymbol{P}^k - \boldsymbol{\Pi}\boldsymbol{P} - \boldsymbol{P}^{k-1}\boldsymbol{\Pi} + \boldsymbol{\Pi}^2 = \boldsymbol{P}^k - \boldsymbol{\Pi}, \end{aligned}$$

because $\boldsymbol{\Pi}\boldsymbol{P} = \boldsymbol{P}\boldsymbol{\Pi} = \boldsymbol{\Pi} = \boldsymbol{\Pi}^2$. Hence, by Theorem 7.2.1 we have $(\boldsymbol{P}-\boldsymbol{\Pi})^n \to \boldsymbol{0}$ as $n \to \infty$. □

Since $(\boldsymbol{P} - \boldsymbol{\Pi})^n \to \boldsymbol{0}$ as $n \to \infty$, Lemma 7.2.2 implies that the matrix $\boldsymbol{I} - (\boldsymbol{P} - \boldsymbol{\Pi})$ is invertible. The inverse $\boldsymbol{Z} = (\boldsymbol{I} - (\boldsymbol{P} - \boldsymbol{\Pi}))^{-1}$ is called the *fundamental matrix* of $\boldsymbol{P}$.

Lemma 7.3.2 *For the fundamental matrix $\boldsymbol{Z} = (\boldsymbol{I} - (\boldsymbol{P} - \boldsymbol{\Pi}))^{-1}$ of a regular stochastic matrix $\boldsymbol{P}$,*

$$\boldsymbol{Z} = \boldsymbol{I} + \sum_{k=1}^{\infty}(\boldsymbol{P}^k - \boldsymbol{\Pi}) \tag{7.3.7}$$

and, alternatively,

$$\boldsymbol{Z} = \boldsymbol{I} + \lim_{n\to\infty} \sum_{k=1}^{n-1} \frac{n-k}{n}(\boldsymbol{P}^k - \boldsymbol{\Pi}). \tag{7.3.8}$$

Proof Using Lemmas 7.2.2 and 7.3.1 we get (7.3.7). To prove (7.3.8), we note that

$$\sum_{k=1}^{n}(\boldsymbol{P}^k - \boldsymbol{\Pi}) - \sum_{k=1}^{n-1} \frac{n-k}{n}(\boldsymbol{P}^k - \boldsymbol{\Pi}) = \sum_{k=1}^{n} \frac{k}{n}(\boldsymbol{P}^k - \boldsymbol{\Pi}) = \frac{1}{n}\sum_{k=1}^{n} k(\boldsymbol{P} - \boldsymbol{\Pi})^k.$$

We show that the last expression tends to $\boldsymbol{0}$ as $n \to \infty$. Any matrix $\boldsymbol{A}$ satisfies the identity $(\boldsymbol{I} - \boldsymbol{A})\sum_{k=1}^{n} k\boldsymbol{A}^k = \sum_{k=1}^{n} \boldsymbol{A}^k - n\boldsymbol{A}^{n+1}$, and hence

$$\frac{1}{n}\sum_{k=1}^{n} k(\boldsymbol{P} - \boldsymbol{\Pi})^k = \frac{1}{n}\boldsymbol{Z}\sum_{k=1}^{n}(\boldsymbol{P} - \boldsymbol{\Pi})^k - \boldsymbol{Z}(\boldsymbol{P} - \boldsymbol{\Pi})^{n+1}.$$

Thus, $\lim_{n\to\infty} n^{-1}\sum_{k=1}^{n} k(\boldsymbol{P} - \boldsymbol{\Pi})^k = \boldsymbol{0}$. This gives (7.3.8). □

We turn to the expected undiscounted reward $\mathbb{E}\, R_n^{\mathrm{u}}$. We assume that the Markov chain $\{X_n\}$ has the regular transition matrix $\boldsymbol{P}$ and an arbitrary initial distribution $\boldsymbol{\alpha}$. Define the *stationary reward rate*, $\bar{\beta} = \boldsymbol{\pi}\boldsymbol{\beta}^\top$, i.e. the stationary expected reward per step, and $a = \boldsymbol{\alpha}\boldsymbol{Z}\boldsymbol{\beta}^\top - \bar{\beta}$, which sometimes is called the *excess reward* regarding the initial distribution $\boldsymbol{\alpha}$.

Theorem 7.3.2 *If $\boldsymbol{P}$ is regular, then the expected undiscounted reward $\mathbb{E}\, R_n^{\mathrm{u}}$ is*

$$\mathbb{E}\, R_n^{\mathrm{u}} = n\bar{\beta} + a + e_n\,, \tag{7.3.9}$$

where e_n is some remainder term with $e_n \to 0$ as $n \to \infty$.

Proof From (7.3.7) we get

$$\begin{aligned}\boldsymbol{\alpha Z \beta}^\top &= \boldsymbol{\alpha\beta}^\top + \boldsymbol{\alpha} \lim_{n\to\infty} \sum_{k=1}^{n-1} (\boldsymbol{P}^k - \boldsymbol{\Pi})\boldsymbol{\beta}^\top \\ &= \boldsymbol{\alpha\beta}^\top + \lim_{n\to\infty} \Big(\boldsymbol{\alpha}\Big(\sum_{k=1}^{n-1} \boldsymbol{P}^k\Big)\boldsymbol{\beta}^\top - (n-1)\boldsymbol{\alpha\Pi\beta}^\top\Big).\end{aligned}$$

Thus, for some $e_n \to 0$ as $n \to \infty$,

$$a = \boldsymbol{\alpha\beta}^\top + \boldsymbol{\alpha}\Big(\sum_{k=1}^{n-1} \boldsymbol{P}^k\Big)\boldsymbol{\beta}^\top - n\boldsymbol{\pi\beta}^\top - e_n = \mathbb{E}\, R_n^{\mathrm{u}} - n\boldsymbol{\pi\beta}^\top - e_n\,,$$

where we used a slight generalization of (7.3.4) in that, for an arbitrary initial distribution $\boldsymbol{\alpha}$, $\mathbb{E}\, R_n^{\mathrm{u}} = \boldsymbol{\alpha}\sum_{k=0}^{n-1} \boldsymbol{P}^k\boldsymbol{\beta}^\top$. This gives (7.3.9). □

An immediate consequence of Theorem 7.3.2 is that $\lim_{n\to\infty} n^{-1}\mathbb{E}\, R_n^{\mathrm{u}} = \bar\beta$. Moreover, the reader can show that $\lim_{n\to\infty} n^{-1} R_n^{\mathrm{u}} = \bar\beta$. Next we investigate the asymptotic behaviour of $\mathrm{Var}\, R_n^{\mathrm{u}}$ as $n \to \infty$.

Theorem 7.3.3 *Suppose $\boldsymbol{P}$ is regular. Then, for any initial distribution $\boldsymbol{\alpha}$,*

$$\lim_{n\to\infty} n^{-1}\mathrm{Var}\, R_n^{\mathrm{u}} = \bar\sigma_0^2 + 2\boldsymbol{\pi}\,\mathrm{diag}(\boldsymbol{\beta})(\boldsymbol{Z} - \boldsymbol{I})\boldsymbol{\beta}^\top, \qquad (7.3.10)$$

where $\bar\sigma_0^2 = \sum_{j=1}^{\ell} \pi_j(\beta_j - \bar\beta)^2$ and $\boldsymbol{Z}$ is the fundamental matrix of $\boldsymbol{P}$.

Proof We have

$$\begin{aligned}\mathrm{Var}\, R_n^{\mathrm{u}} &= \mathbb{E}\Big(\sum_{k=1}^{n} \beta_{X_k}\Big)^2 - \Big(\sum_{k=1}^{n} \mathbb{E}\,\beta_{X_k}\Big)^2 \\ &= \sum_{k=1}^{n} \mathbb{E}\,\beta_{X_k}^2 + 2\sum_{1\le k<l\le n} \mathbb{E}\,\beta_{X_k}\beta_{X_l} - \Big(\sum_{k=1}^{n} \mathbb{E}\,\beta_{X_k}\Big)^2.\end{aligned}$$

First we prove (7.3.10) under the additional assumption that the initial distribution $\boldsymbol{\alpha}$ is equal to the stationary distribution $\boldsymbol{\pi}$. Then

$$\Big(\sum_{k=1}^{n} \mathbb{E}\,\beta_{X_k}\Big)^2 = (n\bar\beta)^2\,, \qquad \sum_{k=1}^{n} \mathbb{E}\,\beta_{X_k}^2 = n\sum_{j=1}^{\ell} \pi_j\beta_j^2\,.$$

By the stationarity of $\{X_n\}$, we have

$$\sum_{1\le k<l\le n} \mathbb{E}\,\beta_{X_k}\beta_{X_l} = \sum_{k=1}^{n-1} (n-k)\mathbb{E}\,\beta_{X_0}\beta_{X_k} \qquad (7.3.11)$$

and $\mathbb{E}\,\beta_{X_0}\beta_{X_k} = \sum_{i=1}^{\ell}\sum_{j=1}^{\ell}\pi_i\beta_i p_{ij}^{(k)}\beta_j = \boldsymbol{\pi}\,\mathrm{diag}(\boldsymbol{\beta})\boldsymbol{P}^k\boldsymbol{\beta}^\top$. Hence,

$$\begin{aligned}\frac{1}{n}\mathrm{Var}\Big(\sum_{k=1}^{n}\beta_{X_k}\Big) &= \sum_{j=1}^{\ell}\pi_j\beta_j^2 + 2\boldsymbol{\pi}\,\mathrm{diag}(\boldsymbol{\beta})\sum_{k=1}^{n-1}\frac{n-k}{n}\boldsymbol{P}^k\boldsymbol{\beta}^\top - n\bar{\beta}^2\\ &= \bar{\sigma}_0^2 + 2\boldsymbol{\pi}\,\mathrm{diag}(\boldsymbol{\beta})\Big(\sum_{k=1}^{n-1}\frac{n-k}{n}\boldsymbol{P}^k\boldsymbol{\beta}^\top - \frac{n-1}{2}\boldsymbol{\Pi}\boldsymbol{\beta}^\top\Big)\\ &= \bar{\sigma}_0^2 + 2\boldsymbol{\pi}\,\mathrm{diag}(\boldsymbol{\beta})\Big(\sum_{k=1}^{n-1}\frac{n-k}{n}(\boldsymbol{P}^k - \boldsymbol{\Pi})\Big)\boldsymbol{\beta}^\top,\end{aligned}$$

where in the second equality we used that $\bar{\beta}^2 = \boldsymbol{\pi}\,\mathrm{diag}(\boldsymbol{\beta})\boldsymbol{\Pi}\boldsymbol{\beta}^\top$. Using (7.3.8), this gives (7.3.10) for $\boldsymbol{\alpha} = \boldsymbol{\pi}$. It is not difficult to show that $\lim_{n\to\infty} n^{-1}\left(\mathrm{Var}\left(R_n^{\mathrm{u}}(\boldsymbol{\alpha})\right) - \mathrm{Var}\left(R_n^{\mathrm{u}}(\boldsymbol{\pi})\right)\right) = 0$, where $R_n^{\mathrm{u}}(\boldsymbol{\alpha})$ is the undiscounted reward for visits at times $0,\ldots,n-1$ when $\{X_n\}$ has the initial distribution $\boldsymbol{\alpha}$. Thus, (7.3.10) holds for arbitrary $\boldsymbol{\alpha}$ as well. □

7.3.3 Efficiency of Bonus-Malus Systems

In this section we discuss one possible efficiency concept for bonus-malus systems. Let $\boldsymbol{P}(\lambda) = (p_{ij}(\lambda))_{i,j=1,\ldots,\ell}$ be the transition matrix of such a system given in (7.1.27), where $\lambda > 0$ is the rate of reported claims by the policyholder. Assume that $\boldsymbol{P}(\lambda)$ is regular. The main idea of introducing bonus-malus systems is to reduce the premium for good drivers while increasing it for bad ones. Remember that the system is modelled by the vector of premiums $\boldsymbol{\beta} = (\beta_1,\ldots,\beta_\ell)$, where $\boldsymbol{\beta} = \mathrm{const}\boldsymbol{b}$ and $\boldsymbol{b}$ is the underlying premium scale. Furthermore, in the undiscounted case we consider the stationary premium rate $\bar{\beta}(\lambda) = \boldsymbol{\pi}(\lambda)\boldsymbol{\beta}^\top$.

Suppose that consecutive claims are independent and identically distributed with mean $\mathbb{E}\,U$. Then the net premium is $\lambda\mathbb{E}\,U$. Suppose also that the scale of $\boldsymbol{\beta}$ is the same as for claim amounts. Ideally, we would like to have that $\bar{\beta}(\lambda) = \lambda\mathbb{E}\,U$. Typically, $\bar{\beta}(\lambda)$ is not linear in λ but, under some additional assumptions, the function $\bar{\beta}(\lambda)$ is continuous and increasing from β_1 to β_ℓ, as will be proved later in Theorem 7.4.6.

For simplicity, we take $\mathbb{E}\,U = 1$. The deviation of $\bar{\beta}(\lambda)$ from linearity can be measured as follows. If $\bar{\beta}(\lambda)$ were linear, then we would have

$$\log\bar{\beta}(\lambda) = \log c + \log\lambda \tag{7.3.12}$$

for some constant $c > 0$. Taking derivatives on both sides of (7.3.12) we get

$$\frac{\mathrm{d}\bar{\beta}(\lambda)}{\mathrm{d}\lambda}\Big/\bar{\beta}(\lambda) = 1/\lambda\,,$$

or, equivalently,

$$\frac{\mathrm{d}\bar{\beta}(\lambda)}{\mathrm{d}\lambda} \Big/ \frac{\bar{\beta}(\lambda)}{\lambda} = 1 .$$

Thus, the *efficiency* of a bonus-malus system might be measured by

$$\eta(\lambda) = \frac{\mathrm{d}\bar{\beta}(\lambda)}{\mathrm{d}\lambda} \Big/ \frac{\bar{\beta}(\lambda)}{\lambda} , \tag{7.3.13}$$

where the system is *perfectly efficient* for the arrival rate λ if

$$\eta(\lambda) = 1 . \tag{7.3.14}$$

As we have already noticed, $\eta(\lambda) \equiv 1$ $(\lambda > 0)$ is atypical. It is therefore desirable that $\eta(\lambda)$ takes values close to 1 for all λ restricted to a certain key interval (c_1, c_2) of interesting claim arrival rates. So,

$$\bar{\beta}(\lambda_0) = \lambda_0 \mathbf{E}\, U \quad (= \lambda_0) \tag{7.3.15}$$

for some $\lambda_0 \in (c_1, c_2)$. The upcoming representation formula for $\bar{\beta}(\lambda)$ implies that, if the function $\eta(\lambda)$ takes values close to 1 for all $\lambda \in (c_1, c_2)$, then $\bar{\beta}(\lambda)$ does not deviate too much from the expected risk λ for all $\lambda \in (c_1, c_2)$. There results that the average premium $\bar{\beta}(\lambda)$, paid by a policy-holder with any given claim arrival rate $\lambda \in (c_1, c_2)$, is nearly fair under the net premium calculation principle.

Theorem 7.3.4 *Let* λ_0 *be a solution to* (7.3.15). *Then*

$$\bar{\beta}(\lambda) = \lambda \exp\Big(\int_\lambda^{\lambda_0} (1 - \eta(x)) x^{-1} \, \mathrm{d}x \Big) . \tag{7.3.16}$$

Proof Definition (7.3.13) implies that

$$\begin{aligned} \int_\lambda^{\lambda_0} \eta(x) x^{-1} \, \mathrm{d}x &= \int_\lambda^{\lambda_0} (\log \bar{\beta}(x))' \, \mathrm{d}x \\ &= \log \bar{\beta}(\lambda_0) - \log \bar{\beta}(\lambda) = \log \lambda_0 - \log \bar{\beta}(\lambda) \\ &= \int_\lambda^{\lambda_0} x^{-1} \, \mathrm{d}x + \log \lambda - \log \bar{\beta}(\lambda) . \end{aligned}$$

Thus, $\log \bar{\beta}(\lambda) = \int_\lambda^{\lambda_0} (1 - \eta(x)) x^{-1} \, \mathrm{d}x + \log \lambda$ and (7.3.16) follows. □

In order to compute the efficiency $\eta(\lambda)$, we need to know the stationary distribution $\boldsymbol{\pi}(\lambda)$ corresponding to $\boldsymbol{P}(\lambda)$ and its derivative $\mathrm{d}\boldsymbol{\pi}(\lambda)/\,\mathrm{d}\lambda$. Methods for computing $\boldsymbol{\pi}(\lambda)$ have been discussed in Section 7.2.4. Moreover,

(7.2.15) implies that the vector $\mathrm{d}\boldsymbol{\pi}(\lambda)/\mathrm{d}\lambda = (\mathrm{d}\pi_1(\lambda)/\mathrm{d}\lambda, \ldots, \mathrm{d}\pi_\ell(\lambda)/\mathrm{d}\lambda)$ is the solution to the following system of linear algebraic equations

$$\frac{\mathrm{d}\boldsymbol{\pi}(\lambda)}{\mathrm{d}\lambda} = \frac{\mathrm{d}\boldsymbol{\pi}(\lambda)}{\mathrm{d}\lambda}\boldsymbol{P}(\lambda) + \boldsymbol{c}(\lambda), \tag{7.3.17}$$

where

$$\sum_{i=1}^{\ell} \frac{\mathrm{d}\pi_i(\lambda)}{\mathrm{d}\lambda} = 0 \tag{7.3.18}$$

and $\boldsymbol{c}(\lambda) = \boldsymbol{\pi}(\lambda)\,\mathrm{d}\boldsymbol{P}(\lambda)/\mathrm{d}\lambda$. These equations can be solved by the same methods as discussed in Section 7.2.4. For example, proceeding similarly as in the proof of Theorem 7.2.8, from (7.3.17) we have $\mathrm{d}\boldsymbol{\pi}(\lambda)/\mathrm{d}\lambda(\boldsymbol{I}-\boldsymbol{P}(\lambda)) = \boldsymbol{c}(\lambda)$ and, by using (7.3.18), $\mathrm{d}\boldsymbol{\pi}(\lambda)/\mathrm{d}\lambda(\boldsymbol{I}-\boldsymbol{P}(\lambda)+\boldsymbol{E}) = \boldsymbol{c}(\lambda)$. Thus, $\mathrm{d}\boldsymbol{\pi}(\lambda)/\mathrm{d}\lambda = \boldsymbol{c}(\lambda)(\boldsymbol{I}-\boldsymbol{P}(\lambda)+\boldsymbol{E})^{-1}$.

Bibliographical Notes. The concept of efficiency $\eta(\lambda)$ was introduced by Loimaranta (1972); see also Lemaire (1985), where in the case of discounting the notion of the efficiency $\eta^{\mathrm{d}}(\lambda)$ was proposed. A credibility theory for the evaluation of bonus-malus systems is given in Norberg (1976) and generalized in Borgan, Hoem and Norberg (1981).

7.4 MONOTONICITY AND STOCHASTIC ORDERING

7.4.1 Monotone Transition Matrices

In this section we assume that the state space E is countably infinite, say $E = \{1, 2, \ldots\}$, but finite state Markov chains are included as well. Let $\boldsymbol{\alpha} = (\alpha_1, \alpha_2, \ldots)$ and $\boldsymbol{\alpha}' = (\alpha'_1, \alpha'_2, \ldots)$ be two probability functions and $\boldsymbol{P}$ a stochastic matrix on E. We ask for conditions on $\boldsymbol{P}$ such that the following implication holds:

$$(\boldsymbol{\alpha} \le_{\mathrm{st}} \boldsymbol{\alpha}') \Rightarrow (\boldsymbol{\alpha P} \le_{\mathrm{st}} \boldsymbol{\alpha}'\boldsymbol{P}), \tag{7.4.1}$$

where the inequality $\boldsymbol{\alpha} \le_{\mathrm{st}} \boldsymbol{\alpha}'$ between two probability functions $\boldsymbol{\alpha}, \boldsymbol{\alpha}'$ on E means that

$$\sum_{i=k}^{\infty} \alpha_i \le \sum_{i=k}^{\infty} \alpha'_i \tag{7.4.2}$$

for all $k \ge 1$ and, equivalently, $\boldsymbol{\alpha f}^{\top} \le \boldsymbol{\alpha}'\boldsymbol{f}^{\top}$ for each increasing sequence $\boldsymbol{f} = \{f_1, f_2, \ldots\}$. If (7.4.1) is fulfilled then $\boldsymbol{P}$ is called a *stochastically monotone transition matrix*. In the following theorem we give four equivalent versions of this monotonicity property, where the probability function $\boldsymbol{p}_i = (p_{i1}, p_{i2}, \ldots)$ denotes the i-th row of $\boldsymbol{P}$.

Theorem 7.4.1 *Let $\boldsymbol{P}$ be a stochastic matrix on E. The following statements are equivalent:*
(a) *$\boldsymbol{P}$ is stochastically monotone,*
(b) $(i \le j) \Rightarrow (\boldsymbol{p}_i \le_{\rm st} \boldsymbol{p}_j)$,
(c) *for each increasing sequence $\boldsymbol{f} = (f_1, f_2, \ldots)$ of real numbers, the sequence $\boldsymbol{P}\boldsymbol{f}^\top$ is increasing,*
(d) $(\boldsymbol{\alpha} \le_{\rm st} \boldsymbol{\alpha}') \Rightarrow (\boldsymbol{\alpha}\boldsymbol{P}^n \le_{\rm st} \boldsymbol{\alpha}'\boldsymbol{P}^n$ *for all* $n = 1, 2, \ldots)$.

Proof (a)$\Rightarrow$(b) Take $\boldsymbol{\alpha} = \boldsymbol{e}_i$ and $\boldsymbol{\alpha}' = \boldsymbol{e}_j$.
(b)$\Rightarrow$(c) Note that

$$\boldsymbol{P}\boldsymbol{f}^\top = \begin{pmatrix} \boldsymbol{p}_1\boldsymbol{f}^\top \\ \boldsymbol{p}_2\boldsymbol{f}^\top \\ \vdots \end{pmatrix}$$

and that, in view of Theorem 3.2.1, $\boldsymbol{p}_i \le_{\rm st} \boldsymbol{p}_j$ means that $\boldsymbol{p}_i\boldsymbol{f}^\top \le \boldsymbol{p}_j\boldsymbol{f}^\top$ for each increasing sequence $\boldsymbol{f} = (f_1, f_2, \ldots)$.
(c)$\Rightarrow$(d) By induction we have that $\boldsymbol{P}^n\boldsymbol{f}^\top$ is an increasing sequence for all $n = 1, 2, \ldots$. Thus, $\boldsymbol{\alpha} \le_{\rm st} \boldsymbol{\alpha}'$ yields $(\boldsymbol{\alpha}\boldsymbol{P}^n)\boldsymbol{f}^\top = \boldsymbol{\alpha}(\boldsymbol{P}^n\boldsymbol{f}^\top) \le \boldsymbol{\alpha}'(\boldsymbol{P}^n\boldsymbol{f}^\top) = (\boldsymbol{\alpha}'\boldsymbol{P}^n)\boldsymbol{f}^\top$ for each increasing sequence $\boldsymbol{f} = (f_1, f_2, \ldots)$, i.e. $\boldsymbol{\alpha}\boldsymbol{P}^n \le_{\rm st} \boldsymbol{\alpha}'\boldsymbol{P}^n$.
(d)$\Rightarrow$(a) This step is obvious. □

Let $\phi : E \times [0,1] \to E$ be the function introduced in (7.1.25). The definition of ϕ, given in Section 7.1.3 for finite E, can easily be extended to a countably infinite state space. Explicitly,

$$\phi(i, z) = \sum_{m=1}^{\infty} m \mathbb{I}\Big(\sum_{k=1}^{m-1} p_{ik} < z \le \sum_{k=1}^{m} p_{ik}\Big) . \tag{7.4.3}$$

Here is a further equivalent version of (7.4.1), but now in terms of ϕ.

Corollary 7.4.1 *The transition matrix $\boldsymbol{P}$ is stochastically monotone if and only if $(i \le j) \Rightarrow (\phi(i, z) \le \phi(j, z)$ for all $z \in [0, 1])$.*

Proof Suppose that $\boldsymbol{P}$ is stochastically monotone. Then statement (b) of Theorem 7.4.1 gives $\sum_{k=1}^m p_{ik} \ge \sum_{k=1}^m p_{jk}$ for all $i \le j$ and $m \ge 1$. Thus if m_i and m_j are defined by

$$\sum_{k=1}^{m_i - 1} p_{ik} < z \le \sum_{k=1}^{m_i} p_{ik}, \qquad \sum_{k=1}^{m_j - 1} p_{jk} < z \le \sum_{k=1}^{m_j} p_{jk} ,$$

then $m_i \le m_j$. Hence $\phi(i, z) \le \phi(j, z)$ for $i \le j$. The proof of the reverse statement is analogous and is left to the reader. □

An immediate consequence of (d) in Theorem 7.4.1 is the following result. Let $X_0, X_1, \ldots$ and $X_0', X_1', \ldots$ be two Markov chains with the same stochastically monotone transition matrix $\boldsymbol{P}$ and with initial distribution $\boldsymbol{\alpha}$ and $\boldsymbol{\alpha}'$, respectively. If $\boldsymbol{\alpha} \le_{\rm st} \boldsymbol{\alpha}'$, then $X_n \le_{\rm st} X'_n$ for all $n \ge 0$; moreover, the following stronger statement on so-called *monotone coupling* of Markov chains is true.

Theorem 7.4.2 *Let $\boldsymbol{P}$ be stochastically monotone and $\boldsymbol{\alpha} \le_{\rm st} \boldsymbol{\alpha}'$. Then there exist a probability space $(\Omega, \mathcal{F}, \mathbb{P})$ and Markov chains $\{X_n\}, \{X'_n\}$ defined on $(\Omega, \mathcal{F}, \mathbb{P})$ having the same transition matrix $\boldsymbol{P}$ and the initial distribution $\boldsymbol{\alpha}$ and $\boldsymbol{\alpha}'$, respectively, such that for all $n = 0, 1, 2, \ldots$ and $\omega \in \Omega$,*

$$X_n(\omega) \le X'_n(\omega)\,. \tag{7.4.4}$$

Proof From probability theory (see Kolmogorov's extension theorem stated in Section 9.2.2) we know that one can construct a probability space $(\Omega', \mathcal{F}', \mathbb{P}')$ which carries a sequence $Z'_1, Z'_2, \ldots$ of independent and uniformly (on $[0,1]$) distributed random variables. Moreover, Theorem 3.1.2 implies that there exist a probability space $(\Omega'', \mathcal{F}'', \mathbb{P}'')$ and E-valued random variables X'' and Y'' with probability functions $\boldsymbol{\alpha}$ and $\boldsymbol{\alpha}'$, respectively, such that $X''(\omega) \le Y''(\omega)$ for all $\omega \in \Omega''$. Define now $\Omega = \Omega' \times \Omega'', \mathcal{F} = \mathcal{F}' \times \mathcal{F}'', \mathbb{P} = \mathbb{P}' \times \mathbb{P}''$, $Z_n(\omega) = Z'_n(\omega')$ with $\omega = (\omega', \omega'') \in \Omega' \times \Omega''$. Put $X_0(\omega) = X''(\omega'')$ and $X_{n+1}(\omega) = \phi(X_n(\omega), Z_n(\omega))$ for $n = 0, 1, \ldots$, where the function $\phi(i, z)$ defined in (7.4.3) is monotone in the variable i. Analogously, put $X'_0(\omega) = Y''(\omega'')$ and $X'_{n+1}(\omega) = \phi(X'_n(\omega), Z_n(\omega))$ for $n = 0, 1, \ldots$. Clearly, using arguments given in Section 7.1.3, we have that the sequences $\{X_n\}$ and $\{X'_n\}$ are Markov chains and (7.4.4) holds. $\square$

Examples 1. Note that all results given in this section can be analogously stated and proved for Markov chains with an arbitrary countably infinite state space E which is linearly ordered, in particular for $E = \mathbb{Z}$. As such, each random walk on $\mathbb{Z}$ is a Markov chain with stochastically monotone transition matrix. Indeed, from (7.1.11) one easily gets that in this case, for all $i \le j$ and $m \in \mathbb{Z}$,

$$\sum_{k=m}^{\infty} p_{ik} = \mathbb{P}(i + Y \ge m) \le \mathbb{P}(j + Y \ge m) = \sum_{k=m}^{\infty} p_{jk}\,.$$

2. In a completely analogous way one can show that the transition matrix of the following Markov chain is stochastically monotone. Let $\{Y_n, n \ge 1\}$ be a sequence of independent and identically distributed random variables with values in $\mathbb{N}$; $p_k = \mathbb{P}(Y_n = k)$. For some fixed $\ell, \ell' \ge 1$ with $\ell' < \ell$, let

$$X_n = \begin{cases} (X_{n-1} + 1 - Y_n)_+ & \text{if } 0 \le X_{n-1} \le \ell', \\ (X_{n-1} - Y_n)_+ & \text{if } \ell' < X_{n-1} \le \ell, \end{cases}$$

where X_0 is independent of $\{Y_n\}$ and takes values in $E = \{1, \ldots, \ell\}$. Then $\{X_n\}$ is a Markov chain with stochastically monotone transition matrix. The sequence $\{X_n\}$ can be interpreted as a discrete-time risk process with state-dependent increments where premiums are added to the portfolio only when the risk reserve process is below the critical level ℓ'. Furthermore, any downcrossing below the zero level is compensated immediately.

7.4.2 Comparison of Markov Chains

We continue our comparison of two Markov chains $\{X_n\}$ and $\{X'_n\}$ with state space $E = \{1, 2, \ldots\}$, initial probability functions $\boldsymbol{\alpha}, \boldsymbol{\alpha}'$ and transition matrices $\boldsymbol{P}, \boldsymbol{P}'$, respectively. We search for further conditions to have $X_n \le_{\rm st} X'_n$ for all $n = 0, 1, \ldots$. Clearly, $\boldsymbol{\alpha} \le_{\rm st} \boldsymbol{\alpha}'$ is necessary for this. Let ϕ and ϕ' be the functions induced by $\boldsymbol{P}$ and $\boldsymbol{P}'$ via (7.4.3). The proof of Corollary 7.4.1 indicates that a further condition

$$(i \le j) \Rightarrow (\phi(i, z) \le \phi'(j, z) \text{ for all } z \in [0, 1]) \tag{7.4.5}$$

is needed. The next theorem rewrites this condition in terms of $\boldsymbol{p}_i$ and $\boldsymbol{p}'_j$, the i-th and j-th row of $\boldsymbol{P}$ and $\boldsymbol{P}'$, respectively.

Theorem 7.4.3 *Condition* (7.4.5) *holds if and only if*

$$(i \le j) \Rightarrow (\boldsymbol{p}_i \le_{\rm st} \boldsymbol{p}'_j) \,. \tag{7.4.6}$$

As the *proof* is analogous to that of Corollary 7.4.1, we leave it to the reader.

If (7.4.6) holds for two stochastic matrices $\boldsymbol{P}$ and $\boldsymbol{P}'$, we say that $\boldsymbol{P}$ is *stochastically smaller* than $\boldsymbol{P}'$ and write $\boldsymbol{P} \le_{\rm st} \boldsymbol{P}'$. With this notation we are in a position to state the following extension to Theorem 7.4.2.

Theorem 7.4.4 *Let $\boldsymbol{\alpha} \le_{\rm st} \boldsymbol{\alpha}'$ and $\boldsymbol{P} \le_{\rm st} \boldsymbol{P}'$. Then there exist a probability space $(\Omega, \mathcal{F}, \mathbb{P})$ and Markov chains $\{X_n\}$, $\{X'_n\}$ defined on $(\Omega, \mathcal{F}, \mathbb{P})$ with initial probability function $\boldsymbol{\alpha}, \boldsymbol{\alpha}'$ and transition matrix $\boldsymbol{P}, \boldsymbol{P}'$, respectively, such that $X_n(\omega) \le X'_n(\omega)$ for all $\omega \in \Omega$ and $n = 0, 1, \ldots$.*

The *proof* is omitted as it is similar to the proof of Theorem 7.4.2 if the equivalence of $\boldsymbol{P} \le_{\rm st} \boldsymbol{P}'$ and (7.4.5) is taken into account.

7.4.3 Application to Bonus-Malus Systems

Consider a bonus-malus system as defined in Section 7.1.4. A bonus-malus system is called *regular* if the stochastic matrix $\boldsymbol{P}(\lambda) = (p_{ij}(\lambda))$ given in (7.1.27) is regular for some $\lambda > 0$. Note that the regularity of $\boldsymbol{P}(\lambda)$ for one specific $\lambda > 0$ implies the regularity of $\boldsymbol{P}(\lambda)$ for all $\lambda > 0$. Recall that the matrix $\boldsymbol{T}(k)$ $(k = 0, 1, \ldots)$ describes transitions of policies in the next year after k claims and that $\boldsymbol{t}_i(k)$ is the i-th row of $\boldsymbol{T}(k)$. Since in each row there is only one 1, each of the matrices $\boldsymbol{T}(k)$ is stochastic. Suppose that $\nu_i(k)$ denotes the position of 1 in the vector $\boldsymbol{t}_i(k)$. It is reasonable to postulate that for $i \le j$ we have $\nu_i(k) \le \nu_j(k)$ for all $k = 0, 1, \ldots$. In matrix notation, the assumption translates into the condition that

(A) for each $k = 0, 1, \ldots$, the matrix $\boldsymbol{T}(k)$ is stochastically monotone.

Similarly, it is natural to assume that a policy with more claims gets transferred to a worse class, that is for $k \le k'$ we have $\nu_i(k) \le \nu_i(k')$ for all $i = 0, 1, \ldots, \ell$. In matrix notation, this means that

(B) $\boldsymbol{T}(0) \le_{\rm st} \boldsymbol{T}(1) \le_{\rm st} \boldsymbol{T}(2) \le_{\rm st} \ldots$.

Note that, if postulate (B) holds, then there exists a natural number k_0 such that $\boldsymbol{T}(k) = \boldsymbol{T}(k_0)$ for all $k \ge k_0$. In real bonus-malus systems it is assumed that a policy-holder reaches class 1 if they do not report claims during sufficiently many successive years; similarly, they will get in class ℓ if they report at least one claim in a sufficiently long series of years. This leads to the assumption that

(C) for all $k = 1, 2, \ldots$, as $n \to \infty$,

$$\boldsymbol{T}^n(0) \to \begin{pmatrix} 1 & 0 & \ldots & 0 \\ 1 & 0 & \ldots & 0 \\ \vdots & \vdots & \ddots & \vdots \\ 1 & 0 & \ldots & 0 \end{pmatrix}, \qquad \boldsymbol{T}^n(k) \to \begin{pmatrix} 0 & 0 & \ldots & 1 \\ 0 & 0 & \ldots & 1 \\ \vdots & \vdots & \ddots & \vdots \\ 0 & 0 & \ldots & 1 \end{pmatrix}.$$

Theorem 7.4.5 *Let $Y(\lambda)$ be a Poisson distributed random variable with mean $\lambda > 0$. Then, for the stochastic matrix $\boldsymbol{P}(\lambda)$ defined in* (7.1.27), *the following statements hold:*
(a) $\boldsymbol{P}(\lambda) = \mathbb{E}\,\boldsymbol{T}(Y(\lambda))$;
(b) *$\boldsymbol{P}(\lambda)$ is stochastically monotone, provided that* (A) *is satisfied;*
(c) *$\boldsymbol{P}(\lambda) \le_{\rm st} \boldsymbol{P}(\lambda')$ for $\lambda \le \lambda'$, provided that* (B) *holds.*

Proof Statement (a) directly follows from the definition (7.1.27) of $\boldsymbol{P}(\lambda)$. Furthermore, note that (b) is an immediate consequence of the fact that $\boldsymbol{P}(\lambda)$ is a mixture of the monotone matrices $\boldsymbol{T}(0), \boldsymbol{T}(1), \ldots$. In order to prove (c) we use the fact that for the Poisson distributed random variables $Y(\lambda), Y(\lambda')$, we have $Y(\lambda) \le_{\rm st} Y(\lambda')$ whenever $\lambda \le \lambda'$, where the proof of this monotonicity property is left to the reader as an exercise. Thus, Theorem 3.2.1 implies that the random variables $Y(\lambda)$ and $Y(\lambda')$ can be defined on a common probability space $(\Omega, \mathcal{F}, \mathbb{P})$ such that $Y(\lambda, \omega) \le Y(\lambda', \omega)$ for almost all $\omega \in \Omega$. Now $\boldsymbol{P}(\lambda) \le_{\rm st} \boldsymbol{P}(\lambda')$ follows. $\square$

In the rest of this section we assume that postulates (A), (B) and (C) are fulfilled. Furthermore, we assume that the bonus-malus system is regular and that $\boldsymbol{\pi}(\lambda)$ is the stationary probability function for $\boldsymbol{P}(\lambda)$. We set $\boldsymbol{P}(0) = \boldsymbol{T}(0)$. Then, it is not difficult to show that the matrix function $\lambda \mapsto \boldsymbol{P}(\lambda)$ is continuous for $\lambda \geq 0$. Putting $\boldsymbol{\pi}(0) = \boldsymbol{e}_1$, we also have that the vector function $\lambda \mapsto \boldsymbol{\pi}(\lambda)$ is continuous for $\lambda \geq 0$. Note that the case $\lambda = 0$ has to be treated separately because $\boldsymbol{P}(0)$ is not regular.

Remember that $\bar{\beta}(\lambda) = \boldsymbol{\pi}(\lambda)\boldsymbol{\beta}^\top$ is the expected stationary premium per year in the undiscounted case while $\bar{\beta}^{\mathrm{d}}(\lambda) = \boldsymbol{e}_{i_0}(\lambda)\sum_{n=0}^{\infty} v^n \boldsymbol{P}(\lambda)^n \boldsymbol{\beta}^\top$ is the expected total discounted premium (at zero), where i_0 is the initial class of policy-holders entering the system.

Theorem 7.4.6 *The functions $\bar{\beta}(\lambda)$ and $\bar{\beta}^{\mathrm{d}}(\lambda)$ are continuous and increasing. Moreover,*

$$\lim_{\lambda\to 0} \bar{\beta}(\lambda) = \beta_1 , \qquad \lim_{\lambda\to\infty} \bar{\beta}(\lambda) = \beta_\ell , \tag{7.4.7}$$

and in the discounted case $1/(1-v)\beta_1 \leq \bar{\beta}^{\mathrm{d}}(\lambda) \leq 1/(1-v)\beta_\ell$.

Proof The proof of the continuity of $\bar{\beta}(\lambda)$ and $\bar{\beta}^{\mathrm{d}}(\lambda)$ uses the continuity of $\boldsymbol{P}(\lambda)$ and $\boldsymbol{\pi}(\lambda)$ and is left to the reader. The monotonicity of $\bar{\beta}(\lambda)$ and $\bar{\beta}^{\mathrm{d}}(\lambda)$ is obtained from statements (b) and (c) of Theorem 7.4.5, where we use the fact that $\boldsymbol{\pi}(\lambda) \leq_{\mathrm{st}} \boldsymbol{\pi}(\lambda')$ whenever $\lambda \leq \lambda'$. The limit behaviour of $\bar{\beta}(\lambda)$, that is (7.4.7), follows from postulates (B) and (C). The bounds for $\bar{\beta}^{\mathrm{d}}(\lambda)$ are obtained from

$$\begin{pmatrix} 1 & 0 & \ldots & 0 \\ 1 & 0 & \ldots & 0 \\ \vdots & \vdots & \ddots & \vdots \\ 1 & 0 & \ldots & 0 \end{pmatrix} \leq_{\mathrm{st}} \boldsymbol{P}(\lambda) \leq_{\mathrm{st}} \begin{pmatrix} 0 & 0 & \ldots & 1 \\ 0 & 0 & \ldots & 1 \\ \vdots & \vdots & \ddots & \vdots \\ 0 & 0 & \ldots & 1 \end{pmatrix} .$$

□

Bibliographical Notes. For stochastic monotonicity and comparability of Markov chains, see Stoyan (1983). Results for monotone and ordered Markov chains on abstract state spaces, in particular the monotone coupling of Markov chains, can be found in Lindvall (1992), for example.

7.5 AN ACTUARIAL APPLICATION OF BRANCHING PROCESSES

Assume that a very valuable item (like an airplane) has to be insured but that the value is so large that one single company can hardly underwrite an insurance policy for the item. Start out at time 0 with the first line insurer. As part of the necessary administrative paper work, this company starts negotiations with a number of second line insurance companies to which it sells part of the original policy. Each of these second line companies acts in a similar way, itself ending up with a number of third line companies, etc. The subsequent sequence of reinsurers makes up an (oversimplified) model of a *reinsurance chain.* In this section we are only interested in the expected number of companies involved in the coverage of the original policy in the n-th link or at any specific time point.

We assume that each company in the chain takes a random administrative time with distribution F before it simultaneously signs all its reinsurance contracts. The number of such reinsurance companies is assumed to be randomly distributed with a probability function $\{p_k,\ k \in \mathbb{N}\}$. Each company starts its own subsidiary chain, independently of the stochastic history of other companies acting prior or simultaneously with it. Also all of the administrative times are independent and follow the same distribution F.

The sequence of random variables $\{X_n\}$ counting the number of companies in each link of the chain is called a *Galton–Watson–Bienaymé branching process.* We do not need the knowledge of F when calculating the probability function of X_n. Let $R_{n,i}$ be the number of subsidiary companies of the i-th company in the n-th link. We assume that all the $R_{n,i}$ are independent and identically distributed with the same probability function $\{p_k,\ k \in \mathbb{N}\}$. We also write $R_{1,1} = R$. We leave it to the reader to show that $\{X_n\}$ is a Markov chain with state space $\mathbb{N}$, where

$$X_1 = 1\,, \qquad X_{n+1} = \sum_{i=1}^{X_n} R_{n,i}\,. \tag{7.5.1}$$

Let $\hat{g}_{X_n}(s)$ be the generating function of the number of companies X_n in the n-th link and $\hat{g}_R(s)$ the common generating function of the number of subsidiary companies of any company. Clearly, (7.5.1) immediately gives that $\hat{g}_{X_1}(s) = s$. Furthermore, the following formulae hold.

Theorem 7.5.1 *For $n = 1, 2, \ldots$ and $|s| < 1$,*

$$\hat{g}_{X_{n+1}}(s) = \hat{g}_{X_n}(\hat{g}_R(s)) = \hat{g}_R(\hat{g}_{X_n}(s)) \tag{7.5.2}$$

and consequently

$$\hat{g}_{X_{n+1}}(s) = \underbrace{\hat{g}_R(\hat{g}_R(\ldots \hat{g}_R}_{n \text{ times}}(s)\ldots))\,. \tag{7.5.3}$$

Proof In view of our assumptions, equation (7.5.1) implies that

$$\mathbb{E}\, s^{X_{n+1}} = \sum_{j=0}^{\infty} \mathbb{E}\left(s^{\sum_{i=1}^{X_n} R_{n,i}} \mid X_n = j\right) \mathbb{P}(X_n = j) = \hat{g}_{X_n}(\hat{g}_R(s))\,.$$

This gives the first equation in (7.5.2), and (7.5.3) follows by iteration. The second equation in (7.5.2) is now straightforward. □

The following corollary is an easy consequence of Theorem 7.5.1. We leave it to the reader to show this as an exercise.

Corollary 7.5.1 *For each* $n = 1, 2, \ldots$ *we have* $\mathbb{E}\, X_n = (\mu_R)^n$ *and*

$$\operatorname{Var} X_n = \begin{cases} \dfrac{\mu_R^{n-1}(\mu_R^n - 1)}{\mu_R - 1} \operatorname{Var} R & \text{if } \mu_R \neq 1, \\ n \operatorname{Var} R & \text{if } \mu_R = 1. \end{cases}$$

The process $\{X(t)\}$ counting the number of companies involved as a function of time is called an *age-dependent branching process* or a *Bellman–Harris process* and is much less trivial to analyse. We can formally introduce this process as follows. Let $X_{n,i}(t)$ be the number of subsidiary companies of the i-th one in the n-th line that are involved in the reinsurance chain, t units after signing its contract. Let $T = T_{1,1}$ be the administrative time for the first line company. We assume that T and all $X_{n,i}(t)$, $n = 2, 3, \ldots$, and $i = 1, 2, \ldots$, are independent. Carefully considering the generation tree of the reinsurance chain, we notice that for each company the stochastic mechanism is exactly the same as if the process took its start with the initial company. This yields

$$X(t) = \begin{cases} 1 & \text{if } t < T, \\ \sum_{i=1}^{R} X_{2,i}(t - T) & \text{if } t \geq T. \end{cases} \tag{7.5.4}$$

Let $\mu(t) = \mathbb{E}\, X(t)$ denote the expected number of companies involved at time t. Applying a conditioning on T we derive an integral equation for the function $\mu(t)$.

Lemma 7.5.1 *For all* $t \geq 0$,

$$\mu(t) = \overline{F}(t) + \mu_R \int_0^t \mu(t - v)\, \mathrm{d}F(v)\,. \tag{7.5.5}$$

Proof Starting from (7.5.4) and considering T as a kind of a renewal point, we condition both on the length T of the first administrative time and on the number R of its subsidiary companies. We begin by conditioning on T. Then

$$\begin{aligned} \mu(t) &= \mathbb{E}\,(\mathbb{E}\,(X(t) \mid T)) = \int_0^\infty \mathbb{E}\,(X(t) \mid T = v)\, \mathrm{d}F(v) \\ &= \int_0^t \mathbb{E}\,(X(t) \mid T = v)\, \mathrm{d}F(v) + \int_t^\infty \mathbb{E}\,(X(t) \mid T = v)\, \mathrm{d}F(v)\,. \end{aligned}$$

The second integral is easy since at time t only the first company is active and hence $\mathbb{E}\,(X(t) \mid T = v) = 1$ in this case. For the first integral we condition additionally on R. Then $\mathbb{E}\,(X(t) \mid T = v) = \sum_{j=0}^{\infty} p_j \mathbb{E}\,(X(t) \mid T = v,\ R = j)$ and we again rewrite the remaining conditional expectation. Since at time v

the first company underwrites j contracts, each one of these j descendants starts its own reinsurance chain. The total number of companies within that chain at time t consists of the sum of these j populations that had a time slot from v to t to deal with the administrative duties. Hence

$$\mathbb{E}\left(X(t) \mid T=v,\ R=j\right) = \mathbb{E}\left(\sum_{i=1}^{j} X_{2,i}(t-v) \,\Big|\, T=v,\ R=j\right)$$

and by our independence assumptions we are allowed to write

$$\mathbb{E}\left(X_{2,i}(t-v) \mid T=v,\ R=j\right) = \mu(t-v)\,.$$

Combining all of the above and using $\mu_R = \sum_{k=1}^{\infty} kp_k$ we get (7.5.5). □

Notice that the integral equation in Lemma 7.5.1 is more general than (6.1.18). Depending on the value of μ_R we have to treat three different cases in order to analyse the asymptotic behaviour of $\mu(t)$ as $t \to \infty$.

Theorem 7.5.2 *Assume that F is nondefective and nonlattice.*
(a) *If $\mu_R = 1$, then $\mu(t) = 1$ for all $t \geq 0$.*
(b) *If $\mu_R > 1$, then*

$$\lim_{t\to\infty} \frac{\mu(t)}{\mathrm{e}^{\gamma t}} = \frac{\mu_R - 1}{\gamma \mu_R^2 \,|\hat{l}_T^{(1)}(\gamma)|}\,, \tag{7.5.6}$$

where γ is the positive solution to equation $\hat{l}_T(\gamma) = \mu_R^{-1}$ (which always exists).
(c) *If $\mu_R < 1$ and if there exists a positive solution γ to $\hat{m}_T(\gamma) = \mu_R^{-1}$, then*

$$\lim_{t\to\infty} \frac{\mu(t)}{\mathrm{e}^{-\gamma t}} = \frac{\mu_R - 1}{\gamma \mu_R^2 \hat{m}_T^{(1)}(\gamma)}\,, \tag{7.5.7}$$

provided that $\hat{m}_T^{(1)}(\gamma) < \infty$. Otherwise the limit in (7.5.7) is zero.

Proof (a) Only in the case $\mu_R = 1$ is (7.5.5) a genuine renewal equation with a nondefective distribution F. It can be checked by inspection that $\mu(t) \equiv 1$ is a solution to (7.5.5) and the uniqueness follows from Lemma 6.1.2.
(b) The main problem with equation (7.5.5) is that we need to rewrite it in such a form that it becomes a genuine renewal equation. This can be done by using the concept of associated distribution introduced in Section 2.3. For this define the positive quantity γ by the equation $\hat{l}_T(\gamma) = \mu_R^{-1}$. Then the associated distribution $\tilde{F}_{-\gamma}$ is introduced by the integral

$$\tilde{F}_{-\gamma}(x) = \mu_R \int_0^x \mathrm{e}^{-\gamma y}\,\mathrm{d}F(y)\,;$$

see also (2.3.6). Multiplying (7.5.5) by $\mathrm{e}^{-\gamma t}$ we obtain

$$\mathrm{e}^{-\gamma t}\mu(t) = \mathrm{e}^{-\gamma t}\overline{F}(t) + \mu_R \int_0^t \mathrm{e}^{-\gamma(t-v)}\mu(t-v)\mathrm{e}^{-\gamma v}\,\mathrm{d}F(v)\,,$$

which leads to a renewal equation for the allied function $g(t) = \mathrm{e}^{-\gamma t}\mu(t)$. Thus, $g(t) = \mathrm{e}^{-\gamma t}\overline{F}(t) + \int_0^t g(t-v)\,\mathrm{d}\tilde{F}_{-\gamma}(v)$ and, using the notation $z(t) = \mathrm{e}^{-\gamma t}\overline{F}(t)$, we arrive at the renewal equation (6.1.18) with F replaced by $\tilde{F}_{-\gamma}$. We now apply the key renewal theorem (see Theorem 6.1.11). Notice that Lemma 6.1.4 can be applied with $z_1(x) = 1$, $z_2(x) = e^{-\gamma x}$ and next $z_1(x) = F(t)$, $z_2(x) = e^{-\gamma x}$, using the fact that the difference of two directly Riemann integrable functions is directly Riemann integrable. We then find that

$$\lim_{t\to\infty} g(t) = (\tilde{\mu}_{-\gamma})^{-1}\int_0^\infty z(x)\,\mathrm{d}x\,, \tag{7.5.8}$$

where $\tilde{\mu}_{-\gamma}$ is the mean of $\tilde{F}_{-\gamma}$ and hence

$$\tilde{\mu}_{-\gamma} = \mu_R\int_0^\infty y e^{-\gamma y}\,\mathrm{d}F(y) = -\mu_R \hat{l}_T^{(1)}(\gamma) = \mu_R\,|\hat{l}_T^{(1)}(\gamma)|\,.$$

For the numerator in (7.5.8) we have by integration by parts

$$\int_0^\infty z(x)\,\mathrm{d}x = \int_0^\infty \mathrm{e}^{-\gamma x}\overline{F}(x)\,\mathrm{d}x = \frac{1}{\gamma}(1-\hat{l}_T(\gamma)) = \frac{\mu_R - 1}{\gamma\mu_R}.$$

Thus we have proved part (b) for the case $\mu_R > 1$.
(c) Similarly as in the proof of Theorem 6.5.7 we can rewrite (7.5.5) to get

$$\mu(t) = \overline{F}(t) + \mu_R\int_0^t \mu(t-v)\,\mathrm{d}F(v) = \overline{F}(t) + \int_0^t \mu(t-v)\,\mathrm{d}F'(v)\,,$$

where now $F'(t) = \mu_R F(t)$ is a defective distribution function. However, since we assume that a positive solution γ to $\hat{m}_T(\gamma) = \mu_R^{-1}$ exists, we can define the associated distribution $\tilde{F}_\gamma(t) = \int_0^t \mathrm{e}^{\gamma v}\,\mathrm{d}F(v)(\hat{m}_T(\gamma))^{-1} = \int_0^t \mathrm{e}^{\gamma t}\,\mathrm{d}F'(t)$ as before and proceed in the same way as in the previous case. Rewrite (7.5.5) in the form

$$\mathrm{e}^{\gamma t}\mu(t) = \mathrm{e}^{\gamma t}\overline{F}(t) + \int_0^t \mathrm{e}^{\gamma(t-v)}\mu(t-v)\,\mathrm{d}\tilde{F}_\gamma(v) \tag{7.5.9}$$

and with $z(t) = \mathrm{e}^{\gamma t}\overline{F}(t)$ and $g(t) = \mathrm{e}^{\gamma t}\mu(t)$, the equation

$$g(t) = z(t) + \int_0^t g(t-v)\,\mathrm{d}\tilde{F}_\gamma(v)$$

is again a genuine renewal equation. Putting $z_1(x) = \mathrm{e}^{\gamma x}$ and $z_2(x) = \overline{F}(x)$, Lemma 6.1.4 implies that $z(t)$ is directly Riemann integrable. Thus, using Theorem 6.1.11, statement (c) follows. □

The result in part (b) of Theorem 7.5.2 shows that the average number of companies increases exponentially fast in time. The crucial quantity γ that measures the scale of increase is called the *Malthusian parameter* in demography. Note that its actual value is intrinsically dependent on all ingredients of the process. The case $\mu_T < 1$ is not very realistic in the insurance context; however it is included for completeness.

Bibliographical Notes. For more details on branching processes we refer to the books by Athreya and Ney (1972), Harris (1963) and Sevastyanov (1973).

CHAPTER 8

Continuous-Time Markov Models

In the previous chapter we studied sequences of random variables, called Markov chains, describing evolutions (of prices, premiums, exchange rates, etc.) in discrete time periods. It is sometimes convenient to have a model describing situations where states change at arbitrary time points. This is achieved by considering a collection of random variables $\{X(t),\ t \ge 0\}$, where the parameter t runs over the whole nonnegative half-line $\mathbb{R}_+$. For the time parameter t, other continuous sets like $[0,1]$, $\mathbb{R}$, etc. are also possible. Recall that such a nondenumerable collection of random variables is called a *stochastic process*. A continuous-time counterpart for the class of Markov chains considered in Chapter 7 are Markov processes in continuous time with a denumerable state space. In order to avoid technical difficulties we begin this chapter with the case of a finite state space $E = \{1, 2, \ldots, \ell\}$.

8.1 HOMOGENEOUS MARKOV PROCESSES

8.1.1 Matrix Transition Function

Markov chains in Section 7.1 were defined by a probability function $\boldsymbol{\alpha}$ and a one-step transition matrix $\boldsymbol{P}$, or equivalently by the probability function $\boldsymbol{\alpha}$ and the family of n-step transition matrices $\boldsymbol{P}^{(n)}$; $n = 1, 2, \ldots$. Recall that the $\boldsymbol{P}^{(n)}$ fulfil the Chapman–Kolmogorov equation (7.1.7). In continuous time we also consider a probability function $\boldsymbol{\alpha} = (\alpha_1, \alpha_2, \ldots, \alpha_\ell)$ and a family of stochastic matrices $\boldsymbol{P}(h) = (p_{ij}(h))_{i,j \in E}$, where $h \ge 0$. We assume that

$$\boldsymbol{P}(h_1 + h_2) = \boldsymbol{P}(h_1)\boldsymbol{P}(h_2) \tag{8.1.1}$$

for all $h_1, h_2 \ge 0$. The matrix identity (8.1.1) is called the (continuous-time) *Chapman–Kolmogorov equation.* We also assume continuity at zero, that is

$$\lim_{h \downarrow 0} \boldsymbol{P}(h) = \boldsymbol{P}(0) = \boldsymbol{I}\,. \tag{8.1.2}$$

We leave it to the reader to show as an exercise that then $\boldsymbol{P}(h)$ is uniformly continuous in $h \geq 0$. A family of stochastic matrices $\{\boldsymbol{P}(h),\ h > 0\}$ fulfilling (8.1.1) and (8.1.2) is called a *matrix transition function.*

Definition 8.1.1 *An E-valued stochastic process $\{X(t), t \geq 0\}$ is called a homogeneous Markov process if there exist a matrix transition function $\{\boldsymbol{P}(h), h \geq 0\}$ and a probability function $\boldsymbol{\alpha}$ on E such that*

$$\begin{aligned} &\mathbb{P}(X(0) = i_0, X(t_1) = i_1, \ldots, X(t_n) = i_n) \\ &= \alpha_{i_0} p_{i_0 i_1}(t_1) p_{i_1 i_2}(t_2 - t_1) \ldots p_{i_{n-1} i_n}(t_n - t_{n-1}), \end{aligned} \tag{8.1.3}$$

for all $n = 0, 1, \ldots,\ i_0, i_1, \ldots, i_n \in E,\ 0 \leq t_1 \leq \ldots \leq t_n$.

We interpret α_i as the probability that the evolution starts at time 0 in state $i \in E$, and $p_{ij}(h)$ as the probability that, in time h, the evolution moves from state i to state j. The probability function $\boldsymbol{\alpha} = (\alpha_1, \alpha_2, \ldots, \alpha_\ell)$ is called an *initial distribution.* In the sequel we will omit the phrase "homogeneous" if this does not lead to confusion. Note that, for each fixed $h \geq 0$, the matrix $\boldsymbol{P} = \boldsymbol{P}(h)$ is the transition matrix of the Markov chain $\{X_n, n \in \mathbb{N}\}$ with $X_n = X(nh)$. In accordance with this it is not surprising that continuous-time Markov processes have the following conditional independence property.

Theorem 8.1.1 *An E-valued stochastic process $\{X(t)\}$ is a Markov process if and only if there exists a matrix transition function $\{\boldsymbol{P}(h), h \geq 0\}$ such that, for all $n \geq 1$, $i_0, i_1, \ldots, i_n \in E$ and $0 \leq t_1 \leq \ldots \leq t_n$,*

$$\begin{aligned} &\mathbb{P}(X(t_n) = i_n \mid X(t_{n-1}) = i_{n-1}, \ldots, X(t_1) = i_1, X(0) = i_0) \\ &= p_{i_{n-1} i_n}(t_n - t_{n-1}), \end{aligned} \tag{8.1.4}$$

whenever $\mathbb{P}(X(t_{n-1}) = i_{n-1}, \ldots, X(t_1) = i_1, X(0) = i_0) > 0$.

The *proof* of Theorem 8.1.1 is analogous to the proof of Theorem 7.1.1. Moreover, analogous to Corollary 7.1.1, the following conditional independence property of continuous-time Markov processes is obtained.

Corollary 8.1.1 *If $\{X(t)\}$ is a Markov process, then*

$$\begin{aligned} &\mathbb{P}(X(t_n) = i_n \mid X(t_{n-1}) = i_{n-1}, \ldots, X(t_1) = i_1, X(0) = i_0) \\ &= \mathbb{P}(X(t_n) = i_n \mid X(t_{n-1}) = i_{n-1}), \end{aligned} \tag{8.1.5}$$

whenever $\mathbb{P}(X(t_{n-1}) = i_{n-1}, \ldots, X(t_1) = i_1, X(0) = i_0) > 0$.

We turn to the study of the main property of the *transition functions* $p_{ij}(h),\ h \geq 0$, i.e. the existence of the transition intensities. Let $\delta_{ij} = 1$ if $i = j$ and 0 otherwise.

Theorem 8.1.2 *If $\{\boldsymbol{P}(h), h \geq 0\}$ is a matrix transition function, then the following limits exist and are finite:*

$$q_{ij} = \lim_{h \downarrow 0} h^{-1}(p_{ij}(h) - \delta_{ij}). \tag{8.1.6}$$

Proof Without loss of generality we can assume that $\mathbb{P}(X(0) = i) > 0$ for all $i \in E$. First we show (8.1.6) for $i \neq j$. Define $p_{ii}^{j0}(h) = 1$ and

$$\begin{aligned} p_{ii}^{jv}(h) &= \mathbb{P}(X(vh) = i; X(kh) \neq j, 1 \leq k < v \mid X(0) = i), \\ f_{ij}^{v}(h) &= \mathbb{P}(X(vh) = j; X(kh) \neq j, 1 \leq k < v \mid X(0) = i). \end{aligned}$$

Then, by (8.1.1),

$$p_{ij}(nh) \geq \sum_{v=0}^{n-1} p_{ii}^{jv}(h) p_{ij}(h) p_{jj}((n-v-1)h) \tag{8.1.7}$$

and

$$p_{ii}(vh) = p_{ii}^{jv}(h) + \sum_{m=1}^{v-1} f_{ij}^{m}(h) p_{ji}((v-m)h). \tag{8.1.8}$$

Since $\sum_{m=1}^{v-1} f_{ij}^{m}(h) \leq 1$, (8.1.8) gives

$$p_{ii}^{jv}(h) \geq p_{ii}(vh) - \max_{1 \leq m < v} p_{ji}((v-m)h). \tag{8.1.9}$$

Now, by (8.1.2) we obtain that for all $\varepsilon > 0$ and $i, j \in E$ with $i \neq j$ there exists $h_0 > 0$ such that

$$\max_{0 \leq h \leq h_0} p_{ji}(h) < \varepsilon, \quad \min_{0 \leq h \leq h_0} p_{ii}(h) > 1 - \varepsilon, \quad \min_{0 \leq h \leq h_0} p_{jj}(h) > 1 - \varepsilon. \tag{8.1.10}$$

Hence if $nh < h_0$ and $v \leq n$, then (8.1.9) implies that $p_{ii}^{jv}(h) > 1 - 2\varepsilon$. Inserting this into (8.1.7) gives $p_{ij}(nh) \geq (1 - 2\varepsilon) \sum_{v=0}^{n-1} p_{ij}(h)(1 - \varepsilon) \geq (1 - 3\varepsilon) n p_{ij}(h)$ and, equivalently,

$$\frac{p_{ij}(nh)}{nh} \geq (1 - 3\varepsilon) \frac{p_{ij}(h)}{h} \tag{8.1.11}$$

if $nh < h_0$. Putting $a_{ij} = \liminf_{h \to 0} h^{-1} p_{ij}(h)$, this implies that $a_{ij} < \infty$. Indeed, if $a_{ij} = \infty$, we would find h arbitrarily small for which $p_{ij}(h)/h$ and, by (8.1.11), also $p_{ij}(nh)/nh$ would be arbitrarily large. On the other hand, choosing n such that $h_0/2 \leq nh < h_0$, (8.1.10) gives $(nh)^{-1} p_{ij}(nh) < (nh)^{-1} \varepsilon < h_0^{-1} 2\varepsilon$. Thus, $a_{ij} < \infty$ and it remains to show that

$$\limsup_{h \to 0} h^{-1} p_{ij}(h) \leq a_{ij}. \tag{8.1.12}$$

By the definition of a_{ij} there exists $h_1 < h_0$ such that $h_1^{-1}p_{ij}(h_1) < a_{ij} + \varepsilon$. Since $p_{ij}(h)$ is continuous, for all sufficiently small t_0 such that $h_1 + t_0 < h_0$ we have $p_{ij}(t)/t < a_{ij} + \varepsilon$ for $h_1 - t_0 < t < h_1 + t_0$. Now, by (8.1.11), for any $h < t_0$ we can find an integer n_h such that $h_1 - t_0 < n_h h < h_1 + t_0$ and $(1-3\varepsilon)h^{-1}p_{ij}(h) \le (n_h h)^{-1}p_{ij}(n_h h) < a_{ij} + \varepsilon$. Since $\varepsilon > 0$ is arbitrary, (8.1.12) follows. Thus, the existence of the limits q_{ij} in (8.1.6) is proved for $i \neq j$. Since the state space E is finite and the matrix $\boldsymbol{P}(h)$ is stochastic, we have

$$\lim_{h\downarrow 0} \frac{p_{ii}(h)-1}{h} = -\lim_{h\downarrow 0}\sum_{j\neq i}\frac{p_{ij}(h)}{h} = -\sum_{j\neq i}\lim_{h\downarrow 0}\frac{p_{ij}(h)}{h} = -\sum_{j\neq i} q_{ij}\,. \qquad (8.1.13)$$

This completes the proof. □

The matrix $\boldsymbol{Q} = (q_{ij})_{i,j=1,\ldots,\ell}$ is called the *intensity matrix* and its entries q_{ij} *transition intensities*. The matrix of transition intensities $\boldsymbol{Q}$ is sometimes called a q-matrix. In the case of a finite state space, $\boldsymbol{Q}$ is the generator of $\{\boldsymbol{P}(h), h \ge 0\}$ in the sense of the theory of transition semigroups. For a more general state space the concept of the generator requires a stronger definition; see Chapter 11.

Corollary 8.1.2 *For each $i \neq j$, $q_{ij} \ge 0$ and $q_{ii} \le 0$. Furthermore, $\boldsymbol{Q}\boldsymbol{e}^\top = \boldsymbol{0}$ or, equivalently, for each $i \in E$,*

$$\sum_{j\in E} q_{ij} = 0\,. \qquad (8.1.14)$$

Proof From definition (8.1.6) we immediately get $q_{ij} \ge 0$ and $q_{ii} \le 0$, for $i \neq j$. (8.1.14) follows from (8.1.13). □

Note that Definition 8.1.1 and Theorem 8.1.1 are completely analogous for Markov processes on a countably infinite state space, $E = \{1, 2, \ldots\}$ say. Also Theorem 8.1.2 remains true in a slightly modified form. In the proof above, the finiteness of the state space has *not* been used when showing the existence and finiteness of q_{ij} for $i \neq j$. In the case of a countably infinite state space, one still can show that the limits q_{ii} in (8.1.6) exist, but they may be infinite; see, for example, Karlin and Taylor (1981), Section 14.1. Moreover, instead of (8.1.13), one can only prove that $q_{ii} \ge -\sum_{j\neq i} q_{ij}$ for all $i \in E$. The case when equality prevails is of prime importance. A matrix transition function $\{\boldsymbol{P}(h), h \ge 0\}$, acting on a countably infinite state space E, is called *conservative* if

$$\sum_{j\neq i} q_{ij} = -q_{ii} < \infty \qquad (8.1.15)$$

for all $i \in E$. Most of the results that are stated and proved in the context of finite state Markov processes remain valid for conservative matrix transition functions on a countably infinite state space. However, the proofs are more involved.

A state $i \in E$ is called *absorbing* if $q_{ii} = 0$. The motivation for this terminology will be discussed in Section 8.1.3. The notion of an absorbing state plays an important role in the definition of the class of phase-type distributions; see Section 8.2. Alternatively, a state $i \in E$ is called *stable* if $0 \le -q_{ii} < \infty$, and *instantaneous* if $-q_{ii} = \infty$.

Example Let $E = \mathbb{N}$. The reader can show that any $\mathbb{N}$-valued stochastic process with independent and stationary increments, as defined in Section 5.2.1, is a Markov process. Any compound Poisson process with $\mathbb{N}$-valued claim sizes is Markov in the sense of Definition 8.1.1. In particular, if $\{X(t), t \ge 0\}$ is a Poisson process with intensity $\lambda > 0$, then $\alpha_0 = 1$ and

$$p_{ij}(h) = \begin{cases} \mathrm{e}^{-\lambda h} \dfrac{(\lambda h)^{j-i}}{(j-i)!} & \text{if } j \ge i, \\ 0 & \text{otherwise.} \end{cases} \tag{8.1.16}$$

This implies for $q_{ij} = p_{ij}^{(1)}(0+)$ that

$$q_{ij} = \begin{cases} \lambda & \text{if } j = i+1, \\ -\lambda & \text{if } j = i, \\ 0 & \text{otherwise.} \end{cases}$$

8.1.2 Kolmogorov Differential Equations

In this section we show that there is a one-to-one correspondence between matrix transition functions and their intensity matrices. In an extension to Theorem 8.1.2 we first show that the transition functions $p_{ij}(h)$ are differentiable for all $h \ge 0$.

Theorem 8.1.3 *For all $i, j \in E$ and $h \ge 0$, the transition functions $p_{ij}(h)$ are differentiable and satisfy the following system of differential equations:*

$$p_{ij}^{(1)}(h) = \sum_{k \in E} q_{ik} p_{kj}(h) \,. \tag{8.1.17}$$

Proof Let $h' > 0$. From the Chapman–Kolmogorov equation (8.1.1) we get

$$\begin{aligned} p_{ij}(h+h') - p_{ij}(h) &= \sum_{k \in E} p_{ik}(h') p_{kj}(h) - p_{ij}(h) \\ &= \sum_{k \ne i} p_{ik}(h') p_{kj}(h) + [p_{ii}(h') - 1] p_{ij}(h) \,. \end{aligned}$$

Similarly,

$$p_{ij}(h-h') - p_{ij}(h) = p_{ij}(h-h') - \sum_{k \in E} p_{ik}(h') p_{kj}(h-h')$$

$$= -\sum_{k\neq i} p_{ik}(h')p_{kj}(h-h') - [p_{ii}(h')-1]p_{ij}(h-h').$$

Dividing by h', letting $h' \downarrow 0$, and using the continuity of $p_{ij}(h)$ we obtain (8.1.17) because, by Theorem 8.1.2,

$$\lim_{h'\downarrow 0} \frac{1}{h'} \sum_{k\neq i} p_{ik}(h')p_{kj}(h) = \sum_{k\neq i} q_{ik}p_{kj}(h)$$

and

$$\lim_{h'\downarrow 0} \frac{1}{h'} \sum_{k\neq i} p_{ik}(h')p_{kj}(h-h') = \sum_{k\neq i} q_{ik}p_{kj}(h). \qquad \square$$

The differential equations in (8.1.17) are called the *Kolmogorov backward equations.* In matrix notation (8.1.17) takes the form

$$\boldsymbol{P}^{(1)}(h) = \boldsymbol{Q}\boldsymbol{P}(h) \tag{8.1.18}$$

for all $h \geq 0$. In the same way as (8.1.17) was proved, one can show that

$$\boldsymbol{P}^{(1)}(h) = \boldsymbol{P}(h)\boldsymbol{Q} \tag{8.1.19}$$

for all $h \geq 0$, which is the matrix notation of the *Kolmogorov forward equations.* The initial condition for both Kolmogorov equations is $\boldsymbol{P}(0) = \boldsymbol{I}$.

The solution to (8.1.18) and (8.1.19) needs concepts from matrix calculus. We assume that all matrices considered below have dimension $\ell \times \ell$ and that vectors have dimension $1 \times \ell$. The convergence of sequences of matrices and vectors is defined entry-wise. For example, if $\{\boldsymbol{A}_n\}$ is a sequence of matrices, $\boldsymbol{A}_n \to \boldsymbol{0}$ as $n \to \infty$ means that $(\boldsymbol{A}_n)_{ij} \to 0$ for all $i,j = 1,\ldots,\ell$. We introduce a norm implying the same topology as described above: for a vector $\boldsymbol{x} = (x_1,\ldots,x_\ell)$, we define $||\boldsymbol{x}|| = \sum_{i=1}^{\ell} |x_i|$ and, for an $\ell\times\ell$ matrix $\boldsymbol{A} = (a_{ij})$, we define $||\boldsymbol{A}|| = \sum_{i,j=1,\ldots,\ell} |a_{ij}|$. Note that for $h \in \mathbb{R}$ we have $||h\boldsymbol{A}|| = |h|\,||\boldsymbol{A}||$ and that $||\boldsymbol{A}|| = 0$ if and only if $\boldsymbol{A} = \boldsymbol{0}$. It is clear that $\boldsymbol{A}_n \to \boldsymbol{0}$ if and only if $||\boldsymbol{A}_n|| \to 0$. Furthermore, for $a \geq 0$ and arbitrary matrices $\boldsymbol{A}, \boldsymbol{B}$,

$$||\boldsymbol{A}+\boldsymbol{B}|| \leq ||\boldsymbol{A}|| + ||\boldsymbol{B}||, \quad ||\boldsymbol{A}\boldsymbol{B}|| \leq ||\boldsymbol{A}||\,||\boldsymbol{B}||, \quad ||a\boldsymbol{A}|| = a||\boldsymbol{A}||. \tag{8.1.20}$$

Lemma 8.1.1 *The series $\sum_{n=0}^{\infty}(h\boldsymbol{A})^n/(n!)$ converges uniformly with respect to $h \in [-h_0,h_0]$, for each $h_0 > 0$.*

Proof Let $h \in [-h_0,h_0]$ and $m \in \mathbb{N}$. By a generalized triangle inequality, deduced from (8.1.20), one has

$$\begin{aligned}\Big\|\sum_{n=0}^{\infty}\frac{(h\boldsymbol{A})^n}{n!} - \sum_{n=0}^{m}\frac{(h\boldsymbol{A})^n}{n!}\Big\| &= \Big\|\sum_{n=m+1}^{\infty}\frac{(h\boldsymbol{A})^n}{n!}\Big\| \leq \sum_{n=m+1}^{\infty}\frac{h^n||\boldsymbol{A}||^n}{n!}\\ &\leq \sum_{n=m+1}^{\infty}\frac{h_0^n||\boldsymbol{A}||^n}{n!} < \varepsilon\end{aligned}$$

for each sufficiently large m, uniformly in $h \in [-h_0, h_0]$. □

The series $\sum_{n=0}^{\infty}(h\boldsymbol{A})^n/(n!)$ is therefore a well-defined matrix function which is continuous with respect to h on the whole real line $\mathbb{R}$. We call this function the *matrix exponential function* and denote it by

$$\exp(h\boldsymbol{A}) = \boldsymbol{I} + h\boldsymbol{A} + \ldots + \frac{(h\boldsymbol{A})^n}{n!} + \ldots. \tag{8.1.21}$$

Let $\boldsymbol{A}(h)$ be a matrix function such that all entries are differentiable functions of h. We define the *matrix derivative* by

$$\boldsymbol{A}^{(1)}(h) = \lim_{h' \to 0} {h'}^{-1}(\boldsymbol{A}(h+h') - \boldsymbol{A}(h)).$$

Lemma 8.1.2 *The matrix exponential function* $\exp(h\boldsymbol{A})$ *is differentiable on the whole real line and*

$$\frac{\mathrm{d}\exp(h\boldsymbol{A})}{\mathrm{d}h} = \boldsymbol{A}\exp(h\boldsymbol{A}) = \exp(h\boldsymbol{A})\boldsymbol{A}. \tag{8.1.22}$$

Proof We have

$$\begin{aligned}\frac{\exp((h+h')\boldsymbol{A}) - \exp(h\boldsymbol{A})}{h'} &= \sum_{n=1}^{\infty} \frac{(h+h')^n - h^n}{h'} \frac{\boldsymbol{A}^n}{n!} \\ &= \sum_{n=1}^{\infty} n h^{n-1} \frac{\boldsymbol{A}^n}{n!} + h' \sum_{n=1}^{\infty} r_n(h,h') \frac{\boldsymbol{A}^n}{n!},\end{aligned}$$

where

$$0 \le r_n(h,h') \le n(n-1)(2h)^n \tag{8.1.23}$$

for $|h'| \le |h|$. The bound (8.1.23) is obtained by a Taylor expansion of the function $g(x) = (h+x)^n - h^n$. This gives $g(x) = xnh^{n-1} + \frac{x^2}{2}n(n-1)(h+\theta x)^{n-2}$, where $0 \le \theta \le 1$. Hence, letting $h' \to 0$, we get

$$\frac{\mathrm{d}\exp(h\boldsymbol{A})}{\mathrm{d}h} = \sum_{n=1}^{\infty} n h^{n-1} \frac{\boldsymbol{A}^n}{n!}.$$

Clearly

$$\sum_{n=1}^{\infty} n h^{n-1} \frac{\boldsymbol{A}^n}{n!} = \boldsymbol{A} \sum_{n=0}^{\infty} \frac{(h\boldsymbol{A})^n}{n!} = \sum_{n=0}^{\infty} \frac{(h\boldsymbol{A})^n}{n!} \boldsymbol{A}.$$ □

Arbitrary $\ell \times \ell$ matrices $\boldsymbol{A}, \boldsymbol{A}'$ are called commutative when $\boldsymbol{A}\boldsymbol{A}' = \boldsymbol{A}'\boldsymbol{A}$. In the next lemma we need that for such commutative matrices

$$\exp(\boldsymbol{A} + \boldsymbol{A}') = \exp(\boldsymbol{A})\exp(\boldsymbol{A}'). \tag{8.1.24}$$

The demonstration of this result is left as an exercise to the reader.

Lemma 8.1.3 *Let $\boldsymbol{Q}$ be an arbitrary $\ell \times \ell$ matrix such that $q_{ij} \geq 0$ for $i \neq j$, and $\boldsymbol{Q}\boldsymbol{e}^\top = \boldsymbol{0}$. Then the matrix exponential function $\{\exp(h\boldsymbol{Q}), h \geq 0\}$ is a matrix transition function which solves the Kolmogorov differential equations* (8.1.18) *and* (8.1.19).

Proof Note that for $\boldsymbol{Q} = \boldsymbol{0}$ the statement is obvious. Assume now that $\boldsymbol{Q} \neq \boldsymbol{0}$. We first check that $\exp(h\boldsymbol{Q})$ is a stochastic matrix for all $h \geq 0$. Since $\boldsymbol{Q}\boldsymbol{e}^\top = \boldsymbol{0}$, we have $\boldsymbol{Q}^n\boldsymbol{e}^\top = \boldsymbol{0}$ for all $n = 1, 2, \ldots$ and, moreover, $\exp(h\boldsymbol{Q})\boldsymbol{e}^\top = \boldsymbol{e}^\top$. To prove that all entries of $\exp(h\boldsymbol{Q})$ are nonnegative, let $a = \max\{-q_{ii} : i = 1, \ldots, \ell\}$. Then, $\tilde{\boldsymbol{P}}$ defined by

$$\tilde{\boldsymbol{P}} = a^{-1}\boldsymbol{Q} + \boldsymbol{I} \tag{8.1.25}$$

is a nonnegative matrix and since $\tilde{\boldsymbol{P}}\boldsymbol{e}^\top = a^{-1}\boldsymbol{Q}\boldsymbol{e}^\top + \boldsymbol{I}\boldsymbol{e}^\top = \boldsymbol{e}^\top$, the matrix $\tilde{\boldsymbol{P}}$ is stochastic. That the entries of $\exp(h\boldsymbol{Q})$ are nonnegative now follows from the representation

$$\exp(h\boldsymbol{Q}) = \exp(ah(\tilde{\boldsymbol{P}} - \boldsymbol{I})) = \sum_{n=0}^{\infty} \frac{(ah)^n(\tilde{\boldsymbol{P}})^n}{n!}\,\mathrm{e}^{-ah}. \tag{8.1.26}$$

In this equality, (8.1.24) has been used together with the fact that $\exp(-ah\boldsymbol{I}) = \mathrm{e}^{-ah}\boldsymbol{I}$. Furthermore, (8.1.24) implies that $\exp(h\boldsymbol{Q})$ fulfils the Chapman–Kolmogorov equation (8.1.1). Now, using (8.1.22) we see that $\boldsymbol{Q}$ is the intensity matrix of the matrix transition function $\exp(h\boldsymbol{Q})$, i.e. $\exp(h\boldsymbol{Q})$ is a solution to (8.1.18) and (8.1.19). □

We are equipped to state the main result of this section.

Theorem 8.1.4 *The matrix transition function $\{\boldsymbol{P}(h), h \geq 0\}$ can be represented by its intensity matrix $\boldsymbol{Q}$ via*

$$\boldsymbol{P}(h) = \exp(h\boldsymbol{Q}). \tag{8.1.27}$$

Proof By Lemma 8.1.3, $\{\boldsymbol{P}'(h)\} = \{\exp(h\boldsymbol{Q})\}$ is a solution to the Kolmogorov backward equation (8.1.18) and fulfils the initial condition $\boldsymbol{P}'(0) = \boldsymbol{I}$. From the theory of systems of ordinary linear differential equations we learn that such a solution is unique. Thus, $\boldsymbol{P}'(h) = \boldsymbol{P}(h)$ for each $h \geq 0$. □

If the eigenvalues $\theta_1, \ldots, \theta_\ell$ of $\boldsymbol{Q}$ are distinct, then the spectral representation (7.1.17) of $\boldsymbol{Q}^n$ can be used to determine the matrix exponential function $\boldsymbol{P}(h) = \exp(h\boldsymbol{Q})$. In this case we have

$$\boldsymbol{P}(h) = \sum_{i=1}^{\ell} \mathrm{e}^{h\theta_i}\boldsymbol{\phi}_i^\top\boldsymbol{\psi}_i\,, \tag{8.1.28}$$

where ϕ_i, ψ_i are the (right and left) eigenvectors corresponding to θ_i. The proof is analogous to that in Section 7.1.2 and is left to the reader.

Theorem 8.1.4 leads to another interesting conclusion. For each fixed pair $i, j \in E$, the function $h \mapsto p_{ij}(h)$ is either identically zero or everywhere positive in $(0, \infty)$, as can be easily shown by the reader.

Note that the assumption of a finite state space is essential for the result of Theorem 8.1.4. If the state space is infinite, $E = \{1, 2, \ldots\}$ say, the situation is much more complex. There may be many matrix transition functions corresponding to one intensity matrix.

8.1.3 An Algorithmic Approach

Our goal in this section is to construct a Markov process with state space $E = \{1, 2, \ldots, \ell\}$ and with a given intensity matrix $\boldsymbol{Q} = (q_{ij})_{i,j \in E}$. We first explain the construction and then show that the obtained process is indeed a Markov process with the preassigned intensity matrix $\boldsymbol{Q}$. The construction is realized in several steps and can be used for *simulation of Markov processes.*

Suppose the intensity matrix $\boldsymbol{Q}$ is given so that $q_{ij} \geq 0$ for $i \neq j$, and $\sum_{j=1}^{\ell} q_{ij} = 0$. Let $\boldsymbol{\alpha}$ be an initial distribution. Let $q(i) = \sum_{j \neq i} q_{ij}$ for all $i \in E$. We define a stochastic matrix $\boldsymbol{P}^\circ$ by setting

$$p^\circ_{ij} = \begin{cases} q_{ij}/q(i) & \text{if } i \neq j, \\ 0 & \text{if } i = j, \end{cases} \tag{8.1.29}$$

for all $i, j \in E$ with $q(i) > 0$. When $q(i) = 0$, the corresponding row of $\boldsymbol{P}^\circ$ is put equal to $\boldsymbol{e}_i$. From (8.1.6) and (8.1.29) follows that $\boldsymbol{P}^\circ = (p^\circ_{ij})$ is a stochastic matrix. Let $\{X_n, n \in \mathbb{N}\}$ be a Markov chain with initial distribution $\boldsymbol{\alpha}$ and transition matrix $\boldsymbol{P}^\circ$. Let $\{Z_n, n \in \mathbb{N}\}$ be a sequence of independent random variables with common exponential distribution Exp(1) and independent of $\{X_n\}$.

With respect to the Markov process $\{X(t), t \geq 0\}$ under construction, the random variables $X_0, X_1, \ldots$ will play the role of an *embedded Markov chain* which describes the state of $\{X(t)\}$ in the intervals between its jump epochs. The random variables $Z_0, Z_1, \ldots$ can be interpreted as unscaled *sojourn times* in these states. If $q(i) = 0$ for some $i \in E$, i.e. the state i is absorbing, then the sojourn time in this state is infinite. We construct $\{X(t), t > 0\}$ in the form $X(t) = \sum_{n=0}^{\infty} X_n \mathbb{I}(\sigma_n \leq t < \sigma_{n+1})$ as follows.

Step 1 Put $\sigma_0 = 0$ and $Z'_0 = Z_0/q(X_0)$. We interpret Z'_0 as the realized sojourn time in state X_0 which is chosen at time $\sigma_0 = 0$. Note that $\mathbb{P}(Z'_0 > x | X_0 = i) = \mathrm{e}^{-q(i)x}$ for all $i \in E$ with $\mathbb{P}(X_0 = i) > 0$; $x \geq 0$.

Step 2 Put $\sigma_1 = \sigma_0 + Z'_0$ and $X(t) = X_0$ for $\sigma_0 = 0 \leq t < \sigma_1$ which defines the trajectory of $\{X(t)\}$ until the first jump epoch σ_1.

Step 3 (analogous to Step 1) Put $Z'_1 = Z_1/q(X_1)$ which will be the sojourn time of $\{X(t)\}$ in state X_1 chosen at time σ_1; $\mathbb{P}(Z'_1 > x | X_1 = i) = \mathrm{e}^{-q(i)x}$ for $\mathbb{P}(X_1 = i) > 0$.

Step 4 Put $\sigma_2 = \sigma_1 + Z'_1$ and $X(t) = X_1$ for $\sigma_1 \le t < \sigma_2$.

$\vdots$

Step $2n-1$ Suppose that the $Z'_0, Z'_1, \ldots, Z'_{n-1}, \sigma_0, \sigma_1, \ldots, \sigma_n$ and $\{X(t), t \in [0, \sigma_n)\}$ are defined for some $n \ge 1$. Then put $Z'_n = Z_n/q(X_n)$.

Step $2n$ Put $\sigma_{n+1} = \sigma_n + Z'_n$ and $X(t) = X_n$ for $\sigma_n \le t < \sigma_{n+1}$.

In this way, we can define the sample paths of $\{X(t), t \ge 0\}$ on $\mathbb{R}_+$ because

$$\mathbb{P}(\lim_{n\to\infty} \sigma_n = \infty) = 1\,, \tag{8.1.30}$$

as can be shown by the reader.

Theorem 8.1.5 *The stochastic process $\{X(t), t \ge 0\}$ constructed above is a homogeneous Markov process.*

For a full *proof* of Theorem 8.1.5 we refer to Resnick (1992), pp. 378–379. Here we only remark that the sequence $\{(\sigma_n, X_n), n \in \mathbb{N}\}$ of states $X_0, X_1, \ldots$ and sojourn times $\sigma_1 - \sigma_0, \sigma_2 - \sigma_1, \ldots$ in these states has the following properties: (a) the times in between jumps $\sigma_1 - \sigma_0, \sigma_2 - \sigma_1, \ldots$ are conditionally independent and exponentially distributed provided that the $X_0, X_1, \ldots$ are given. Hence, for all $n \ge 1$, $i_0, \ldots, i_{n-1} \in E$ and $x_1, \ldots, x_n \ge 0$ we have

$$\begin{aligned}
&\mathbb{P}\Big(\bigcap_{m=1}^{n} \{\sigma_m - \sigma_{m-1} > x_m\} \;\Big|\; X_0 = i_0, \ldots, X_{n-1} = i_{n-1}\Big) \\
&= \mathbb{P}\Big(\bigcap_{m=1}^{n} \Big\{\frac{Z_{m-1}}{q(X_{m-1})} > x_m\Big\} \;\Big|\; X_0 = i_0, \ldots, X_{n-1} = i_{n-1}\Big) \\
&= \mathbb{P}\Big(\bigcap_{m=1}^{n} \Big\{\frac{Z_{m-1}}{q(i_{m-1})} > x_m\Big\} \;\Big|\; X_0 = i_0, \ldots, X_{n-1} = i_{n-1}\Big) \\
&= \mathbb{P}\Big(\bigcap_{m=1}^{n} \Big\{\frac{Z_{m-1}}{q(i_{m-1})} > x_m\Big\}\Big) = \prod_{m=1}^{n} \mathrm{e}^{-q(i_{m-1})x_m},
\end{aligned}$$

(b) the sequence $\{(\sigma_n, X_n), n \in \mathbb{N}\}$ is a *Markov renewal process*, i.e. for all $n \ge 1$, $i, j, i_0, \ldots, i_{n-1} \in E$ and $x, x_1, \ldots, x_n \ge 0$ we have

$$\mathbb{P}\Big(X_{n+1} = j, \sigma_{n+1} - \sigma_n > x \;\Big|\; \bigcap_{m=0}^{n-1} \{X_m = i_m\} \cap \{X_n = i\}$$

$$
\begin{aligned}
& \qquad \cap \bigcap_{m=1}^{n} \{\sigma_m - \sigma_{m-1} > x_m\}\Big) \\
= \;& \mathbb{P}\Big(X_{n+1} = j, \frac{Z_n}{q(i)} > x \;\Big|\; \bigcap_{m=0}^{n-1} \{X_m = i_m\} \cap \{X_n = i\} \\
& \qquad \cap \bigcap_{m=1}^{n} \{\frac{Z_{m-1}}{q(i_{m-1})} > x_m\}\Big) \\
= \;& \mathbb{P}\Big(X_{n+1} = j, \frac{Z_n}{q(i)} > x \;\Big|\; X_n = i\Big) \\
= \;& \mathbb{P}\left(X_{n+1} = j, \sigma_{n+1} - \sigma_n > x \mid X_n = i\right) \\
& \left[= \mathbb{P}\left(X_{n+1} = j \mid X_n = i\right) \mathbb{P}\Big(\frac{Z_n}{q(i)} > x\Big) = \frac{q_{ij}}{q(i)} \mathrm{e}^{-q(i)x}\right],
\end{aligned}
$$

provided that

$$
\mathbb{P}\Big(\bigcap_{m=0}^{n-1} \{X_m = i_m\} \cap \{X_n = i\} \cap \bigcap_{m=1}^{n} \{\sigma_m - \sigma_{m-1} > x_m\}\Big) > 0 .
$$

Note that the matrix $\boldsymbol{Q}$ is uniquely determined by the transition matrix $\boldsymbol{P}^\circ = (p^\circ_{ij})$ of the embedded Markov chain and by the vector of expected sojourn times $(1/q(1), \ldots, 1/q(\ell))$.

We now show that the Markov process $\{X(t), t \geq 0\}$ constructed above is the "right" one, i.e. its intensity matrix equals the preassigned matrix $\boldsymbol{Q}$. For that purpose, we need to show that the transition probabilities $p_{ij}(h)$ of this Markov process can be expressed in terms of the "local" characteristics $\{q(i)\}_{i\in E}$ and $\boldsymbol{P}^\circ = (p^\circ_{ij})_{i,j\in E}$.

Theorem 8.1.6 *For all $i, j \in E$ and $h \geq 0$,*

$$
p_{ij}(h) = \delta_{ij}\mathrm{e}^{-q(i)h} + \int_0^h q(i)\mathrm{e}^{-q(i)t} \sum_{k\neq i} p^\circ_{ik} p_{kj}(h-t)\,\mathrm{d}t . \tag{8.1.31}
$$

In particular, if $i \in E$ is an absorbing state, then $p_{ij}(h) = \delta_{ij}$.

Proof Without loss of generality we can assume that $\mathbb{P}(X(0) = i) > 0$. Consider the decomposition $p_{ij}(h) = I_{ij}(h) + I'_{ij}(h)$, where $I_{ij}(h) = \mathbb{P}(X(h) = j, \sigma_1 > h \mid X(0) = i)$ and $I'_{ij}(h) = \mathbb{P}(X(h) = j, \sigma_1 \leq h \mid X(0) = i)$. Then,

$$
I_{ij}(h) = \begin{cases} \mathbb{P}(Z'_0 > h \mid X_0 = i) = \mathrm{e}^{-q(i)h} & \text{if } i = j, \\ 0 & \text{if } i \neq j, \end{cases}
$$

and

$$
I'_{ij}(h) \;=\; \sum_{k\neq i} \int_0^h \frac{\mathbb{P}(X(h) = j, \sigma_1 \in \mathrm{d}t, X_1 = k, X_0 = i)}{\mathbb{P}(X_0 = i)}
$$

$$
\begin{aligned}
&= \sum_{k\neq i}\int_0^h \mathbb{P}(X(h)=j \mid X(t)=k)\mathbb{P}\Big(\frac{Z_0}{q(i)}\in \mathrm{d}t\Big)\mathbb{P}(X_1=k\mid X_0=i)\\
&= \int_0^h q(i)\mathrm{e}^{-q(i)t}\sum_{k\neq i} p^\circ_{ik}p_{kj}(h-t)\,\mathrm{d}t\,.
\end{aligned}
$$

If $i \in E$ is an absorbing state, then the process $\{X(t), t \geq 0\}$ constructed above stays in i once it gets there and hence $p_{ij}(h) = \delta_{ij}$ for all $h \geq 0$. □

Corollary 8.1.3 *For all* $i, j \in E$,

$$
p^{(1)}_{ij}(0+) = q_{ij} = \begin{cases} -q(i) & \text{if } i = j, \\ p^\circ_{ij}q(i) & \text{if } i \neq j. \end{cases} \tag{8.1.32}
$$

Proof Taking the first derivative in (8.1.31) with respect to h and letting $h \downarrow 0$, (8.1.32) follows. □

Comparing (8.1.6) and (8.1.32) we see that the intensity matrix of the Markov process $\{X(t), t \geq 0\}$ constructed in this section equals the preassigned matrix $\boldsymbol{Q}$.

Example A Markov process $\{X(t), t \geq 0\}$ with state space $E = \{1, \ldots, \ell\}$ is called a *birth-and-death process* if $p^\circ_{i,i-1} + p^\circ_{i,i+1} = 1$ for all $1 < i < \ell$ and $p^\circ_{12} = p^\circ_{\ell,\ell-1} = 1$. The products $p^\circ_{i,i+1}q(i)$ and $p^\circ_{i,i-1}q(i)$ are called *birth rate* and *death rate*, respectively. Indeed, for the Markov process constructed in this section, we showed in Corollary 8.1.3 that $p^\circ_{i,i+1}q(i)$ and $p^\circ_{i,i-1}q(i)$ are the transition intensities $q_{i,i+1}$ and $q_{i,i-1}$ for the transitions $i \to i+1$ and $i \to i-1$ in the sense of (8.1.6).

8.1.4 Monotonicity of Markov Processes

In this section we study monotonicity properties of Markov processes which are analogous to results given in Theorems 7.4.1 and 7.4.2 for (discrete-time) Markov chains.

We consider the finite state space $E = \{1, \ldots, \ell\}$ although all definitions and results given in this section can be formulated (usually under some extra conditions) for the case of a countable infinite state space too. Let $\boldsymbol{Q}$ be an intensity matrix on E. We say that $\boldsymbol{Q}$ and the underlying Markov process is *stochastically monotone* if

$$
\sum_{k\geq r} q_{ik} \leq \sum_{k\geq r} q_{jk}\,, \tag{8.1.33}
$$

for all $i, j, r \in E$ such that $i \leq j$, and $r \leq i$ or $r > j$.

The stochastic monotonicity of intensity matrices is easily linked to that of the corresponding transition matrices. Write $\boldsymbol{f}_r = (0, \ldots, 0, 1, \ldots, 1)$ for a vector with the first 1 at the r-th component and $\boldsymbol{e}_i = (0, \ldots, 0, 1, 0, \ldots, 0)$ as before. Condition (8.1.33) can be rewritten in the following way: for $r \le i \le j$ or $i \le j < r$,

$$\boldsymbol{e}_i \boldsymbol{Q} \boldsymbol{f}_r^\top \le \boldsymbol{e}_j \boldsymbol{Q} \boldsymbol{f}_r^\top . \tag{8.1.34}$$

Theorem 8.1.7 *The intensity matrix $\boldsymbol{Q}$ is stochastically monotone if and only if the transition matrix $\exp(h\boldsymbol{Q})$ is stochastically monotone for all $h \ge 0$.*

Proof Let $\exp(h\boldsymbol{Q})$ be a stochastically monotone matrix. Then

$$\boldsymbol{e}_i \exp(h\boldsymbol{Q}) \boldsymbol{f}_r^\top \le \boldsymbol{e}_j \exp(h\boldsymbol{Q}) \boldsymbol{f}_r^\top \tag{8.1.35}$$

for all $i \le j$. But $\boldsymbol{e}_i \boldsymbol{f}_r^\top = \boldsymbol{e}_j \boldsymbol{f}_r^\top$ for $r \le i \le j$ and $i \le j < r$. Subtract this equality from (8.1.35), divide by h, and let $h \to 0$ to find (8.1.33). The converse is left as an exercise. □

Let now $\boldsymbol{Q}$ be an arbitrary intensity matrix and choose $a > 0$ such that

$$a \ge \max_{i \in E} \{-q_{ii}\} . \tag{8.1.36}$$

Then $\boldsymbol{Q} + a\boldsymbol{I}$ is a nonnegative matrix and $\tilde{\boldsymbol{P}} = \boldsymbol{Q}/a + \boldsymbol{I}$ is a transition matrix. Thus for $\boldsymbol{P}(h) = \exp(h\boldsymbol{Q})$, we have

$$\boldsymbol{P}(h) = \sum_{n=0}^{\infty} \frac{(at)^n}{n!} \mathrm{e}^{-ah} (\tilde{\boldsymbol{P}})^n , \tag{8.1.37}$$

as in the proof of Lemma 8.1.3. From (8.1.37) we get the following useful representation of a Markov process $\{X(t)\}$ with intensity matrix $\boldsymbol{Q}$ and initial distribution $\boldsymbol{\alpha}$. Let $\{N(t)\}$ be a Poisson process with intensity a and let $\{X_n\}$ be a Markov chain with transition matrix $\tilde{\boldsymbol{P}}$ and initial distribution $\boldsymbol{\alpha}$. Furthermore, assume that $\{N(t)\}$ and $\{X_n\}$ are independent. Then it is easily shown that the stochastic process $\{X(t)\}$ with $X(t) = X_{N(t)}$ is a Markov process with intensity matrix $\boldsymbol{Q}$ and initial distribution $\boldsymbol{\alpha}$.

The representation of $\boldsymbol{P}(h)$ in (8.1.37) is called a *uniform representation*. It is also possible in the case of a countable infinite state space provided that (8.1.36) holds for some finite $a > 0$. Then, the Markov process is called *subordinated*. All results, derived in the rest of this section for a finite state space, extend to the case of a countably infinite state space.

Lemma 8.1.4 *The intensity matrix $\boldsymbol{Q}$ is stochastically monotone if and only if for some $a \ge 2\max_{i \in E}\{-q_{ii}\}$, the matrix $\boldsymbol{Q}/a + \boldsymbol{I}$ is a stochastically monotone transition matrix.*

Proof Assume first that $\boldsymbol{Q}$ is a stochastically monotone intensity matrix and that $a \geq 2\max_{i\in E}\{-q_{ii}\}$. Let $i, j, r \in E$ with $i \leq j$, and $r \leq i$ or $r > j$. Then, using (8.1.33) we have

$$\sum_{k\geq r} a^{-1} q_{ik} + \mathbb{I}(i \geq r) \leq \sum_{k\geq r} a^{-1} q_{jk} + \mathbb{I}(j \geq r) . \tag{8.1.38}$$

Furthermore, if $i < r \leq j$ then

$$\begin{aligned}\sum_{k\geq r} a^{-1} q_{ik} &= \sum_{k\geq i} a^{-1} q_{ik} - \sum_{k=i}^{r-1} a^{-1} q_{ik} \leq \sum_{k\geq i} a^{-1} q_{jk} - a^{-1} q_{ii} \\ &\leq \sum_{k\geq r} a^{-1} q_{jk} - a^{-1}(q_{ii} + q_{jj}) \leq \sum_{k\geq r} a^{-1} q_{jk} + 1 .\end{aligned}$$

Thus, using Theorem 7.4.1 we see that the matrix $\boldsymbol{Q}/a + \boldsymbol{I}$ is stochastically monotone. On the other hand, assuming that $\boldsymbol{Q}/a + \boldsymbol{I}$ is stochastically monotone for some $a \geq 2\max_{i\in E}\{-q_{ii}\}$, the monotonicity of $\boldsymbol{Q}$ immediately follows from (8.1.38). □

Theorem 8.1.8 *Let $\boldsymbol{Q}$ be a stochastically monotone intensity matrix and let $\boldsymbol{\alpha}$ and $\boldsymbol{\alpha}'$ be initial distributions on E such that $\boldsymbol{\alpha} \leq_{\rm st} \boldsymbol{\alpha}'$. Then there exist a probability space $(\Omega, \mathcal{F}, \mathbb{P})$ and Markov processes $\{X(t)\}$, $\{X'(t)\}$ defined on $(\Omega, \mathcal{F}, \mathbb{P})$, having the same intensity matrix $\boldsymbol{Q}$ and the initial distributions $\boldsymbol{\alpha}$ and $\boldsymbol{\alpha}'$, respectively, and such that for all $t \geq 0$,*

$$X(t) \leq X'(t) . \tag{8.1.39}$$

Proof By Lemma 8.1.4 and Theorem 7.4.2 there exist a probability space $(\Omega, \mathcal{F}, \mathbb{P})$ and Markov chains $\{X_n\}$, $\{X'_n\}$ defined on $(\Omega, \mathcal{F}, \mathbb{P})$, having the same transition matrix $\tilde{\boldsymbol{P}} = \boldsymbol{Q}/a + \boldsymbol{I}$ for some $a \geq 2\max_{i\in E}\{-q_{ii}\}$ and the initial distributions $\boldsymbol{\alpha}$ and $\boldsymbol{\alpha}'$, respectively, and such that with probability 1

$$X_n \leq X'_n \tag{8.1.40}$$

for all $n \in \mathbb{N}$. Assume now that $\{N(t)\}$ is a Poisson process with intensity a such that $a \geq 2\max_{i\in E}\{-q_{ii}\}$, and $\{N(t)\}$ is independent of $\{X_n\}$ and $\{X'_n\}$. Then by the uniform representation of Markov processes we can set $X(t) = X_{N(t)}$ and $X'(t) = X'_{N(t)}$. Clearly (8.1.40) implies (8.1.39). □

Example Let $\{X(t), t \geq 0\}$ be a birth-and-death process with state space $E = \{1, \ldots, \ell\}$ and intensity matrix $\boldsymbol{Q}$, that is $q_{i,i-1} + q_{ii} + q_{i,i+1} = 0$ for $i = 2, \ldots, \ell - 1$, and $q_{11} + q_{12} = 0$, $q_{\ell,\ell-1} + q_{\ell\ell} = 0$. We leave it to the reader to show that any birth-and-death process is a stochastically monotone Markov process.

8.1.5 Stationary Initial Distributions

We say that a probability function $\boldsymbol{\pi} = (\pi_1, \ldots, \pi_\ell)$ on $E = \{1, \ldots, \ell\}$ is a *stationary initial distribution* of a Markov process with matrix transition function $\{\boldsymbol{P}(h),\ h \geq 0\}$ if $\boldsymbol{\pi P}(h) = \boldsymbol{\pi}$ for all $h \geq 0$. Using the finiteness of the state space E we have $\mathbf{0} = \lim_{h\to 0} h^{-1}\boldsymbol{\pi}(\boldsymbol{P}(h) - \boldsymbol{I}) = \boldsymbol{\pi Q}$. Conversely, if $\boldsymbol{\pi Q} = \mathbf{0}$ then clearly $\boldsymbol{\pi Q}^n = \mathbf{0}$ and hence $\boldsymbol{\pi P}(h) = \boldsymbol{\pi} \sum_{n=0}^{\infty}(h\boldsymbol{Q})^n/n! = \boldsymbol{\pi}$ for $h \geq 0$.

As in the case of (discrete-time) Markov chains considered in Section 7.2, it is possible to give a characterization of stationary initial distributions as limit distributions. The situation is even easier for continuous-time Markov processes since only an irreducibility condition is needed. Notice that without any additional condition, we have $p_{ii}(h) > 0$ for all $i \in E$ and $h \geq 0$ as follows from (8.1.37). Furthermore, we call the Markov process $\{X(t)\}$ with matrix transition function $\{\boldsymbol{P}(h),\ h \geq 0\}$ *irreducible* if for all $i \neq j$, $p_{ij}(h) > 0$ for all $h > 0$. This is equivalent to the *irreducibility of the intensity matrix* $\boldsymbol{Q}$, which means that for each pair $i, j \in E$ with $i \neq j$ there exists a sequence $i_1, \ldots, i_n \in E$ $(i_k \neq i_l)$ such that $q_{ii_1} q_{i_1 i_2} \cdots q_{i_{n-1} j} > 0$. We recommend the reader to prove the equivalence of these two notions of irreducibility. Irreducibility implies that the stationary initial distribution $\boldsymbol{\pi}$ of $\{X(t)\}$ is uniquely determined and satisfies $\boldsymbol{\pi Q} = \mathbf{0}$.

Theorem 8.1.9 *If the Markov process* $\{X(t)\}$ *is irreducible, then for each* $i \in E$,

$$\lim_{t\to\infty} \mathbb{P}(X(t) = i) = \pi_i\,, \tag{8.1.41}$$

where $\boldsymbol{\pi} = (\pi_1, \ldots, \pi_\ell)$ *is the stationary initial distribution of* $\{X(t)\}$.

Proof Let $\{X(t)\}$ be irreducible, which means that the transition matrix $\boldsymbol{P}(h)$ is regular for each $h > 0$. Then, from Theorem 7.2.1 we get that $\lim_{n\to\infty} \boldsymbol{P}(nh) = \boldsymbol{\Pi}$ for each $h > 0$ where $\boldsymbol{\Pi}$ is the matrix with each row equal to $\boldsymbol{\pi}$. It follows from (8.1.1) and (8.1.2) that the matrix transition function $\{\boldsymbol{P}(h),\ h \geq 0\}$ is uniformly continuous. Thus, for each $\varepsilon > 0$ we can find a (small) number $h_0 > 0$ such that for all $t > 0$ sufficiently large, there is an $n \in \mathbb{N}$ for which $\|\boldsymbol{P}(t) - \boldsymbol{\Pi}\| \leq \|\boldsymbol{P}(t) - \boldsymbol{P}(nh_0)\| + \|\boldsymbol{P}(nh_0) - \boldsymbol{\Pi}\| \leq 2\varepsilon$. □

Bibliographical Notes. A coherent mathematical theory of Markov processes in continuous time was first introduced by Kolmogorov (1931). Important contributions to this class of stochastic processes were also made by W. Feller, W. Doeblin, J.L. Doob, P. Levy and others; see Feller (1971). More details on Markov processes with denumerable state space can be found, for example, in Chung (1967), Çinlar (1975), Karlin and Taylor (1981), Resnick (1992). Notice that by some authors, a Markov process with denumerable state space is called a continuous-time Markov chain. Standard references for the theory of systems of ordinary linear differential equations are books like Boyce and Di Prima (1969) and Simmons (1991). For stochastically monotone Markov processes, see Massey (1987) and Stoyan (1983).

8.2 PHASE-TYPE DISTRIBUTIONS

In this section we introduce the class of phase-type distributions which have a useful probabilistic interpretation and are convenient for numerical computations. We derive useful formulae for ruin functions and we show that an arbitrary distribution on $\mathbb{R}_+$ can be "approximated" by phase-type distributions. The probabilistic definition of phase-type distributions uses the theory of Markov processes. As a phase-type distribution can be characterized as a matrix exponential distribution, we need to recall some necessary concepts and results from matrix algebra.

8.2.1 Some Matrix Algebra and Calculus

Unless otherwise stated, we again assume that matrices have dimension $\ell \times \ell$ and that vectors have dimension $1 \times \ell$. By $\theta_i = \theta_i(\boldsymbol{A})$, $i = 1, \ldots, \ell$, we denote the eigenvalues of a matrix $\boldsymbol{A} = (a_{ij})_{i,j \in E}$; $E = \{1, \ldots, \ell\}$. As before we assume that $|\theta_1| \geq |\theta_2| \geq \ldots \geq |\theta_\ell|$. The following auxiliary result for linear transformations of matrices is easily proved.

Lemma 8.2.1 *If* $\boldsymbol{A}' = a\boldsymbol{A} + b\boldsymbol{I}$ *for some constants* $a, b \in \mathbb{R}$, *then*

$$\theta_i(\boldsymbol{A}') = a\theta_i(\boldsymbol{A}) + b\,, \qquad i = 1, \ldots, \ell\,. \tag{8.2.1}$$

We derive an upper bound for the "largest" eigenvalue θ_1 of a nonnegative matrix.

Lemma 8.2.2 *Let* $\boldsymbol{A}$ *be nonnegative. Then,*

$$|\theta_1| \leq \min\Big\{\max_{i \in E} \sum_{j=1}^{\ell} a_{ij}\,,\ \max_{j \in E} \sum_{i=1}^{\ell} a_{ij}\Big\}\,. \tag{8.2.2}$$

Proof We have $\theta_1 \phi_i = \sum_{j=1}^{\ell} a_{ij}\phi_j$, $i = 1, \ldots, \ell$, where $\boldsymbol{\phi} = (\phi_1, \ldots, \phi_\ell)$. This gives $|\theta_1||\phi_i| \leq \sum_{j=1}^{\ell} a_{ij} \max_{k \in E} |\phi_k|$, $i = 1, \ldots, \ell$, and consequently

$$|\theta_1| \max_{i \in E} |\phi_i| \leq \max_{i \in E} \sum_{j=1}^{\ell} a_{ij} \max_{k \in E} |\phi_k|\,.$$

Thus $|\theta_1| \leq \max_{i \in E} \sum_{j=1}^{\ell} a_{ij}$. The proof that $|\theta_1| \leq \max_{j \in E} \sum_{i=1}^{\ell} a_{ij}$ is similar because $\boldsymbol{A}^\top$ has the same eigenvalues as $\boldsymbol{A}$. $\square$

The bound given in Lemma 8.2.2 can be used to show that, for a certain class of matrices, the real parts of all eigenvalues are nonpositive. We say that $\boldsymbol{B} = (b_{ij})$ is a *subintensity matrix* if $b_{ij} \geq 0$ $(i \neq j)$ and $\sum_{j=1}^{\ell} b_{ij} \leq 0$, where for at least one $i \in E$, $\sum_{j=1}^{\ell} b_{ij} < 0$.

Theorem 8.2.1 *If $\boldsymbol{B}$ is a subintensity matrix, then $\theta_i(\boldsymbol{B}) = 0$ or $\Re(\theta_i(\boldsymbol{B})) < 0$, for each $i = 1, \ldots, \ell$.*

Proof Let $\boldsymbol{B}$ be a subintensity matrix and $c > \max_{i \in E}(-b_{ii})$. Then $\boldsymbol{B}' = \boldsymbol{B} + c\boldsymbol{I}$ is a nonnegative matrix. By Lemma 8.2.1, we have $\theta_i(\boldsymbol{B}) = \theta_i(\boldsymbol{B}') - c$. Furthermore, Lemma 8.2.2 yields $|\theta_1(\boldsymbol{B}')| \leq c$. Since $|\theta_1(\boldsymbol{B}')| \geq |\theta_2(\boldsymbol{B}')| \geq \ldots \geq |\theta_\ell(\boldsymbol{B}')|$, this completes the proof (see Figure 8.2.1). □

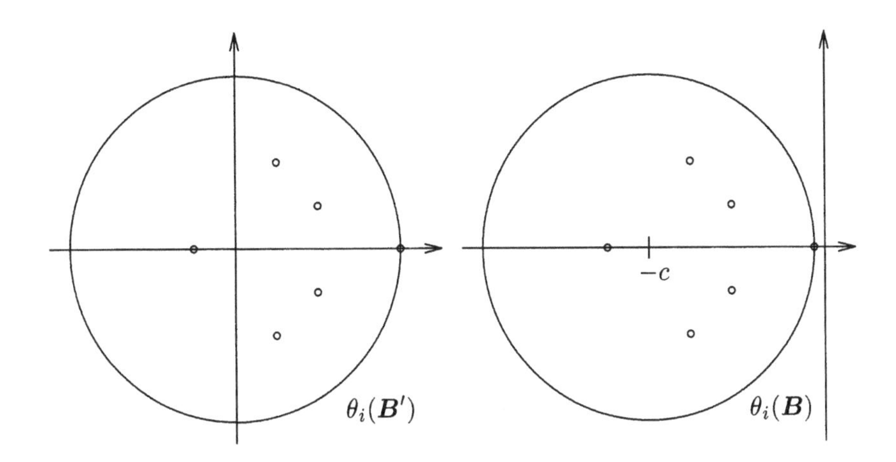

Figure 8.2.1 The eigenvalues of a nonsingular subintensity matrix

Corollary 8.2.1 *A subintensity matrix $\boldsymbol{B}$ is nonsingular if and only if $\Re(\theta_i(\boldsymbol{B})) < 0$ for each $i = 1, \ldots, \ell$.*

Proof It suffices to notice that 0 is not an eigenvalue of $\boldsymbol{B}$ if and only if $\boldsymbol{B}$ is nonsingular. Thus, the statement follows from Theorem 8.2.1. □

In the next theorem we give a representation formula for the matrix exponential function $\exp(t\boldsymbol{A})$ of an arbitrary $\ell \times \ell$ matrix $\boldsymbol{A}$.

Theorem 8.2.2 *Let $\theta_1, \ldots, \theta_\ell$ be the eigenvalues of $\boldsymbol{A}$. Then*

$$\exp(t\boldsymbol{A}) = a_1(t)\boldsymbol{A}_1 + \ldots + a_\ell(t)\boldsymbol{A}_\ell\,, \tag{8.2.3}$$

where $a_k(t)$, $\boldsymbol{A}_k$ are given recursively by $a_1(t) = \mathrm{e}^{\theta_1 t}$, $\boldsymbol{A}_1 = \boldsymbol{I}$ and $a_k(t) = \int_0^t \mathrm{e}^{\theta_k(t-x)} a_{k-1}(x)\,\mathrm{d}x$, $\boldsymbol{A}_k = (\boldsymbol{A} - \theta_1\boldsymbol{I}) \ldots (\boldsymbol{A} - \theta_{k-1}\boldsymbol{I})$ for $k = 2, \ldots, \ell$.

Proof Suppose first that $\boldsymbol{A} = \mathrm{diag}(\boldsymbol{\theta})$ with $\boldsymbol{\theta} = (\theta_1, \ldots, \theta_\ell)$ and that all eigenvalues $\theta_1, \ldots, \theta_\ell$ are distinct. Then, by the definition (8.1.21) of $\exp(t\boldsymbol{A})$, we have $\exp(t\boldsymbol{A}) = \mathrm{diag}(\mathrm{e}^{\theta_1 t}, \ldots, \mathrm{e}^{\theta_\ell t})$ and (8.2.3) is obvious for $\ell = 1$. Suppose that (8.2.3) holds for some $\ell = n-1$. Then, the right-hand side of (8.2.3) can be written as

$$\sum_{k=1}^{n} a_k(t)\boldsymbol{A}_k = \begin{pmatrix} \mathrm{diag}\left(\mathrm{e}^{\theta_1 t}, \ldots, \mathrm{e}^{\theta_{n-1} t}\right) & \boldsymbol{0} \\ \boldsymbol{0} & \sum_{k=1}^{n} a_k(t) \prod_{i=1}^{k-1}(\theta_n - \theta_i) \end{pmatrix}.$$

We have to show that, for all $n = 1, 2, \ldots$,

$$\sum_{k=1}^{n} a_k(t) \prod_{i=1}^{k-1} (\theta_n - \theta_i) = \mathrm{e}^{\theta_n t}. \tag{8.2.4}$$

For $n = 1$, (8.2.4) is obvious. Furthermore, for $k \geq 2$ we have

$$\begin{aligned} a_k(t) &= \int_0^t \int_0^{x_k} \cdots \int_0^{x_3} \mathrm{e}^{\theta_k(t-x_k)} \ldots \mathrm{e}^{\theta_2(x_3-x_2)} \mathrm{e}^{\theta_1 x_2}\, \mathrm{d}x_2 \ldots \mathrm{d}x_k \\ &= \mathrm{e}^{\theta_k t} \int_0^t \cdots \int_0^t \mathrm{I\!I}(0 < x_2 < \ldots < x_k) \\ &\qquad \times \mathrm{e}^{(\theta_{k-1}-\theta_k)x_k} \ldots \mathrm{e}^{(\theta_2-\theta_3)x_3} \mathrm{e}^{(\theta_1-\theta_2)x_2}\, \mathrm{d}x_2 \ldots \mathrm{d}x_k \\ &= \int_0^t \mathrm{e}^{\theta_1 x_2} \mathrm{e}^{\theta_k(t-x_2)} \Big(\int_0^{t-x_2} \cdots \int_0^{t-x_2} \mathrm{I\!I}(0 < x_3 < \ldots < x_k) \\ &\qquad \times \mathrm{e}^{(\theta_{k-1}-\theta_k)x_k} \ldots \mathrm{e}^{(\theta_2-\theta_3)x_3}\, \mathrm{d}x_3 \ldots \mathrm{d}x_k \Big)\, \mathrm{d}x_2 . \end{aligned}$$

Assuming that (8.2.4) holds for some $n = j-1$, this gives

$$\begin{aligned} \sum_{k=1}^{j} a_k(t) \prod_{i=1}^{k-1} (\theta_j - \theta_i) &= a_1(t) + (\theta_j - \theta_1) \sum_{k=2}^{j} a_k(t) \prod_{i=2}^{k-1} (\theta_j - \theta_i) \\ &= a_1(t) + (\theta_j - \theta_1) \int_0^t \mathrm{e}^{\theta_1 x_2} \mathrm{e}^{\theta_j (t-x_2)}\, \mathrm{d}x_2 \\ &= \mathrm{e}^{\theta_1 t} + \mathrm{e}^{\theta_j t}\left(1 - \mathrm{e}^{(\theta_1-\theta_j)t}\right) = \mathrm{e}^{\theta_j t}. \end{aligned}$$

Thus, (8.2.4) holds for all $n \geq 1$. Consequently, for $\boldsymbol{A} = \mathrm{diag}(\boldsymbol{\theta})$ and $\theta_1, \ldots, \theta_\ell$ distinct, (8.2.3) is true for all $\ell \geq 1$. Suppose now that $\boldsymbol{A}$ is an arbitrary (not necessarily diagonal) matrix with distinct eigenvalues $\theta_1, \ldots, \theta_\ell$. Then, it is not difficult to show that

$$\exp(t\boldsymbol{A}) = \boldsymbol{\Phi}\, \mathrm{diag}\left(\mathrm{e}^{\theta_1 t}, \ldots, \mathrm{e}^{\theta_\ell t}\right) \boldsymbol{\Phi}^{-1}, \tag{8.2.5}$$

where $\boldsymbol{\Phi} = (\boldsymbol{\phi}_1^\top, \ldots, \boldsymbol{\phi}_\ell^\top)$ is the $\ell \times \ell$ matrix consisting of right (column) eigenvectors of $\boldsymbol{A}$. Thus, from the first part of the proof we have

$$\begin{aligned}\exp(t\boldsymbol{A}) &= \boldsymbol{\Phi}\Big(\sum_{k=1}^{\ell} a_k(t) \prod_{i=1}^{k-1} (\mathrm{diag}(\boldsymbol{\theta}) - \theta_i \boldsymbol{I})\Big)\boldsymbol{\Phi}^{-1} \\ &= \sum_{k=1}^{\ell} a_k(t) \prod_{i=1}^{k-1} \big(\boldsymbol{\Phi}\,\mathrm{diag}(\boldsymbol{\theta})\boldsymbol{\Phi}^{-1} - \theta_i \boldsymbol{\Phi}\boldsymbol{\Phi}^{-1}\big) \\ &= \sum_{k=1}^{\ell} a_k(t) \prod_{i=1}^{k-1} (\boldsymbol{A} - \theta_i \boldsymbol{I}).\end{aligned}$$

If not all eigenvalues of $\boldsymbol{A}$ are distinct, we can proceed as follows. By Lemma 7.1.3,

$$\boldsymbol{A} = \boldsymbol{C}\boldsymbol{D}\boldsymbol{C}^{-1}, \tag{8.2.6}$$

where $\boldsymbol{D}$ is (upper) triangular and $\boldsymbol{C}$ nonsingular. Note that

$$\begin{aligned}\prod_{i=1}^{\ell}(\theta_i - s) &= \det(\boldsymbol{A} - s\boldsymbol{I}) = \det(\boldsymbol{C}\boldsymbol{D}\boldsymbol{C}^{-1} - s\boldsymbol{C}\boldsymbol{C}^{-1}) \\ &= \det(\boldsymbol{D} - s\boldsymbol{I}) = \prod_{i=1}^{\ell}(d_{ii} - s).\end{aligned}$$

Thus, the eigenvalues of $\boldsymbol{A}$ and $\boldsymbol{D}$ coincide, i.e. $\{d_{11}, \ldots, d_{\ell\ell}\} = \{\theta_1, \ldots, \theta_\ell\}$. Consider a triangular matrix $\boldsymbol{D}' = (d'_{ij})$ such that $d_{ij} = d'_{ij}$ for $i \neq j$ and, $\boldsymbol{D}'$ has distinct diagonal elements d'_{ii} with $|d'_{ii} - d_{ii}| < \varepsilon$ for all $i = 1, \ldots, \ell$ and some $\varepsilon > 0$. Then, the matrix $\boldsymbol{A}' = \boldsymbol{C}\boldsymbol{D}'\boldsymbol{C}^{-1}$ also has distinct eigenvalues and by (8.1.20)

$$||\boldsymbol{A} - \boldsymbol{A}'|| = ||\boldsymbol{C}(\boldsymbol{D} - \boldsymbol{D}')\boldsymbol{C}^{-1}|| \leq \ell ||\boldsymbol{C}||\, ||\boldsymbol{C}^{-1}||\varepsilon. \tag{8.2.7}$$

Moreover, it is not difficult to show that $||\exp(t\boldsymbol{A}') - \exp(t\boldsymbol{A})|| \to 0$ whenever $||\boldsymbol{A}' - \boldsymbol{A}|| \to 0$. This completes the proof since $\varepsilon > 0$ in (8.2.7) can be chosen arbitrarily small. □

If the eigenvalues $\theta_1, \ldots, \theta_\ell$ of $\boldsymbol{A}$ are distinct, then (8.2.5) immediately implies that $\lim_{t\to\infty} \mathrm{e}^{-\vartheta t} \exp(t\boldsymbol{A}) = \boldsymbol{0}$ for each $s > \max_{i\in E} \Re(\theta_i)$. Indeed, using (8.2.5) we have

$$\mathrm{e}^{-st} \exp(t\boldsymbol{A}) = \boldsymbol{\Phi}\,\mathrm{diag}\big(\mathrm{e}^{(\theta_1 - s)t}, \ldots, \mathrm{e}^{(\theta_\ell - s)t}\big)\boldsymbol{\Phi}^{-1} \to \boldsymbol{0}$$

since $\left|\mathrm{e}^{(\theta_i - s)t}\right| = \exp(-(s - \Re(\theta_i))t) \to 0$ as $t \to \infty$. If the eigenvalues $\theta_1, \ldots, \theta_\ell$ are not distinct, then the same exponential bound for the matrix exponential function $\exp(t\boldsymbol{A})$ can be obtained from Theorem 8.2.2, as is pointed out in the following result.

Corollary 8.2.2 *Let $\theta_1, \ldots, \theta_\ell$ be the eigenvalues of $\boldsymbol{A}$. Then, for each $s > \max_{i\in E} \Re(\theta_i)$,*

$$\lim_{t\to\infty} \mathrm{e}^{-st} \exp(t\boldsymbol{A}) = \boldsymbol{0}\,. \tag{8.2.8}$$

Proof In view of (8.2.3) it suffices to show that

$$\lim_{t\to\infty} \mathrm{e}^{-st} |a_k(t)| = 0\,, \tag{8.2.9}$$

for each $k = 1, \ldots, \ell$. Clearly (8.2.9) is true for $k = 1$. Suppose that (8.2.9) holds for some $k = n - 1 < \ell$. Then, for all $v \in (0, t)$,

$$\begin{aligned}\mathrm{e}^{-st}|a_n(t)| \;\; &\le \;\; \mathrm{e}^{(\Re(\theta_n)-s)t} \int_0^v |\mathrm{e}^{-\theta_n x} a_{n-1}(x)|\,\mathrm{d}x \\ &\quad + \int_v^t \mathrm{e}^{(\Re(\theta_n)-s)(t-x)} \mathrm{e}^{-sx} |a_{n-1}(x)|\,\mathrm{d}x\,,\end{aligned}$$

where the second integrand becomes arbitrarily small if v is sufficiently large. Indeed, because of our assumption that (8.2.9) holds for $k = n - 1$, for each $\varepsilon > 0$ there exists $v > 0$ such that $\mathrm{e}^{-sx}|a_{n-1}(x)| < \varepsilon$ for all $x > v$. Thus

$$\int_v^t \mathrm{e}^{(\Re(\theta_n)-s)(t-x)} \mathrm{e}^{-sx} |a_{n-1}(x)|\,\mathrm{d}x \le \varepsilon \int_v^t \mathrm{e}^{(\Re(\theta_n)-s)(t-x)}\,\mathrm{d}x \le \frac{\varepsilon}{s - \Re(\theta_n)}\,.$$

Since $\lim_{t\to\infty} \mathrm{e}^{(\Re(\theta_n)-s)t} \int_0^v |\mathrm{e}^{-\theta_n x} a_{n-1}(x)|\,\mathrm{d}x = 0$ for each fixed $v > 0$, the proof is complete. □

Suppose now that $\boldsymbol{A}(t) = (a_{ij}(t))$ is a matrix function where each entry $a_{ij}(t)$ is a function of t. If $\boldsymbol{A}(t)$ is differentiable as defined in Section 8.1.2, then $\boldsymbol{A}^{(1)}(t) = \mathrm{d}\boldsymbol{A}(t)/\mathrm{d}t$ has entries $a_{ij}^{(1)}(t)$. A differentiation rule for products of matrix functions is given in the following lemma.

Lemma 8.2.3 *If $\boldsymbol{A}(t)$ and $\boldsymbol{A}'(t)$ are two differentiable matrix functions, then*

$$\frac{\mathrm{d}}{\mathrm{d}t}\left(\boldsymbol{A}(t)\boldsymbol{A}'(t)\right) = \left(\frac{\mathrm{d}}{\mathrm{d}t}\boldsymbol{A}(t)\right)\boldsymbol{A}'(t) + \boldsymbol{A}(t)\frac{\mathrm{d}}{\mathrm{d}t}\boldsymbol{A}'(t)\,. \tag{8.2.10}$$

The *proof* of Lemma 8.2.3 is left to the reader as an exercise.

Conversely, by $\int_v^t \boldsymbol{A}(x)\,\mathrm{d}x$ we mean the matrix with entries $\int_v^t a_{ij}(x)\,\mathrm{d}x$ for $v < t$. In particular, for the matrix exponential function the following is true.

Lemma 8.2.4 (a) *If $\boldsymbol{A}$ is nonsingular, then*

$$\int_v^t \exp(x\boldsymbol{A})\,\mathrm{d}x = \boldsymbol{A}^{-1}(\exp(t\boldsymbol{A}) - \exp(v\boldsymbol{A}))\,. \tag{8.2.11}$$

(b) *If all eigenvalues of $\boldsymbol{A}$ have negative real parts, then*

$$\int_0^\infty \exp(x\boldsymbol{A})\,\mathrm{d}x = -\boldsymbol{A}^{-1}. \tag{8.2.12}$$

Proof Let the matrix function $\boldsymbol{F}(t)$ be differentiable. Then, $\int_v^t \boldsymbol{F}^{(1)}(x)\,\mathrm{d}x = \boldsymbol{F}(t) - \boldsymbol{F}(v)$ for $v < t$. Thus, setting $\boldsymbol{F}(x) = \boldsymbol{A}^{-1}\exp(x\boldsymbol{A})$, (8.2.11) follows from Lemma 8.1.2 because by (8.1.22) we have $(\mathrm{d}/\mathrm{d}x)\boldsymbol{A}^{-1}\exp(x\boldsymbol{A}) = \boldsymbol{A}^{-1}(\mathrm{d}/\mathrm{d}x)\exp(x\boldsymbol{A}) = \exp(x\boldsymbol{A})$. Now, (8.2.12) is an immediate consequence of (8.2.8) and (8.2.11). □

The following block operations on matrices will prove to be useful. Suppose we represent two $\ell \times \ell$ matrices $\boldsymbol{A}$ and $\boldsymbol{A}'$ by

$$\boldsymbol{A} = \begin{pmatrix} \boldsymbol{A}_{11} & \dots & \boldsymbol{A}_{1m} \\ \vdots & \ddots & \vdots \\ \boldsymbol{A}_{k1} & \dots & \boldsymbol{A}_{km} \end{pmatrix}, \qquad \boldsymbol{A}' = \begin{pmatrix} \boldsymbol{A}'_{11} & \dots & \boldsymbol{A}'_{1n} \\ \vdots & \ddots & \vdots \\ \boldsymbol{A}'_{m1} & \dots & \boldsymbol{A}'_{mn} \end{pmatrix},$$

where $1 \le k, m, n \le \ell$ and $\boldsymbol{A}_{ij}, \boldsymbol{A}'_{ij}$ are matrices such that the matrix multiplication $\boldsymbol{A}_{ir}\boldsymbol{A}'_{rj}$ is possible for all i, j, r. Then

$$\boldsymbol{A}\boldsymbol{A}' = \begin{pmatrix} \sum_{r=1}^m \boldsymbol{A}_{1r}\boldsymbol{A}'_{r1} & \dots & \sum_{r=1}^m \boldsymbol{A}_{1r}\boldsymbol{A}'_{rn} \\ \vdots & \ddots & \vdots \\ \sum_{r=1}^m \boldsymbol{A}_{kr}\boldsymbol{A}'_{r1} & \dots & \sum_{r=1}^m \boldsymbol{A}_{kr}\boldsymbol{A}'_{rn} \end{pmatrix}. \tag{8.2.13}$$

For example, if $\boldsymbol{A}_{ii}$ are $\ell_i \times \ell_i$ matrices with $\sum_{i=1}^k \ell_i = \ell$, $k = m$, and $\boldsymbol{A}_{ij} = \boldsymbol{0}$ $(i \ne j)$, then

$$\boldsymbol{A}^n = (\boldsymbol{A}_{ij})^n = \begin{pmatrix} \boldsymbol{A}_{11}^n & \boldsymbol{0} & \dots & \boldsymbol{0} \\ \boldsymbol{0} & \boldsymbol{A}_{22}^n & \dots & \boldsymbol{0} \\ \vdots & \vdots & \ddots & \vdots \\ \boldsymbol{0} & \boldsymbol{0} & \vdots & \boldsymbol{A}_{kk}^n \end{pmatrix}.$$

8.2.2 Absorption Time

We are ready to return to continuous-time Markov processes. We assume that none of the states $i \in E = \{1, 2, \ldots, \ell\}$ is absorbing. In Section 8.1.3 we showed that then the sample paths of a Markov process with the finite state space E can be chosen to be piecewise constant functions, where the distances between consecutive jump epochs are hyperexponentially distributed random variables.

We now extend the state space E by adding one new state, say 0, and which we assume to be absorbing. For the extended state space $E' = \{0, 1, \ldots, \ell\}$, we consider an intensity matrix $\boldsymbol{Q} = (q_{ij})_{i,j\in E'}$ written in the block form

$$\boldsymbol{Q} = \begin{pmatrix} 0 & \boldsymbol{0} \\ \boldsymbol{b}^\top & \boldsymbol{B} \end{pmatrix}, \tag{8.2.14}$$

where $\boldsymbol{b} = (b_1, \ldots, b_\ell)$ is an ℓ-dimensional vector with nonnegative components b_i, $\boldsymbol{B} = (b_{ij})$ is an $\ell \times \ell$ matrix with $b_{ij} \geq 0$ for $i \neq j$ and $b_{ii} \leq 0$ such that

$$\boldsymbol{b}^\top = -\boldsymbol{B}\boldsymbol{e}^\top. \tag{8.2.15}$$

Let $\{X(t), t \geq 0\}$ be the Markov process on E' with this intensity matrix $\boldsymbol{Q}$, constructed in the same way as shown in Section 8.1.3, where $X(t) = 0$ for all $t \geq \eta = \inf\{t' : X(t') = 0\}$. We do not want η to be infinite with probability 1. Hence $\boldsymbol{B}$ cannot be an intensity matrix (that is $b_i = 0$ for all $i = 1, \ldots, \ell$) because otherwise the Markov process $\{X(t)\}$ would never visit state 0 when being started in $E = \{1, \ldots, \ell\}$. For that reason $\boldsymbol{B}$ is assumed to be a subintensity matrix. According to Theorem 8.1.6, the transition probabilities of $\{X(t)\}$ are given by

$$p_{ij}(h) = \begin{cases} \delta_{ij} & \text{if } i = 0, \\ \delta_{ij} \mathrm{e}^{-q(i)h} + \int_0^h q(i) \mathrm{e}^{-q(i)t} \sum_{k \neq i} p^\circ_{ik} p_{kj}(h-t) \, \mathrm{d}t & \text{if } i \neq 0. \end{cases} \tag{8.2.16}$$

The random variable η is called the *absorption time* of $\{X(t)\}$. Its distribution is determined by the initial distribution $\boldsymbol{\alpha}' = (\alpha_0, \alpha_1, \ldots, \alpha_\ell)$ of $\{X(t)\}$ and by the subintensity matrix $\boldsymbol{B}$. Note that, instead of $\boldsymbol{\alpha}'$, it suffices to consider the (possibly defective) probability function $\boldsymbol{\alpha} = (\alpha_1, \ldots, \alpha_\ell)$ on E.

Definition 8.2.1 *The distribution of η is called a phase-type distribution with characteristics $(\boldsymbol{\alpha}, \boldsymbol{B})$. We denote this distribution by* $\mathrm{PH}(\boldsymbol{\alpha}, \boldsymbol{B})$.

Examples 1. Let $\ell = 1, \alpha_1 = 1$ and $\boldsymbol{B} = (-\lambda)$ for some $\lambda > 0$. Then $\mathrm{PH}(\boldsymbol{\alpha}, \boldsymbol{B})$ is the exponential distribution $\mathrm{Exp}(\lambda)$. Furthermore, for an arbitrary $\ell \geq 1$, for $\alpha_1 = 1$ and consequently $\alpha_2 = \ldots = \alpha_\ell = 0$, and for

$$\boldsymbol{B} = \begin{pmatrix} -\lambda & \lambda & 0 & \cdots & 0 & 0 \\ 0 & -\lambda & \lambda & \cdots & 0 & 0 \\ \vdots & \vdots & \vdots & & \vdots & \vdots \\ 0 & 0 & 0 & \cdots & 0 & -\lambda \end{pmatrix},$$

$\mathrm{PH}(\boldsymbol{\alpha}, \boldsymbol{B})$ is the Erlang distribution $\mathrm{Erl}(\ell, \lambda)$.

2. Let $\ell \geq 1$ and let $\boldsymbol{\alpha} = (\alpha_1, \ldots, \alpha_\ell)$ be an arbitrary (nondefective) probability function on E. If

$$\boldsymbol{B} = \begin{pmatrix} -\lambda_1 & 0 & 0 & \cdots & 0 & 0 \\ 0 & -\lambda_2 & 0 & \cdots & 0 & 0 \\ \vdots & \vdots & \vdots & & \vdots & \vdots \\ 0 & 0 & 0 & \cdots & 0 & -\lambda_\ell \end{pmatrix},$$

for some $\lambda_1, \ldots, \lambda_\ell > 0$, then $\mathrm{PH}(\boldsymbol{\alpha}, \boldsymbol{B})$ is the hyperexponential distribution $\sum_{k=1}^{\ell} \alpha_k \mathrm{Exp}(\lambda_k)$.

We next derive a formula for the tail of a phase-type distribution.

Theorem 8.2.3 *Consider the absorption time η with distribution* $\mathrm{PH}(\boldsymbol{\alpha}, \boldsymbol{B})$. *Then, for each $t \geq 0$,*

$$\mathbb{P}(\eta > t) = \boldsymbol{\alpha} \exp(t\boldsymbol{B})\boldsymbol{e}^{\top}. \tag{8.2.17}$$

Proof Clearly, $\{\eta > t\} = \{X(t) \neq 0\}$ and, consequently,

$$\mathbb{P}(\eta > t) = \mathbb{P}(X(t) \neq 0) = \sum_{i=1}^{\ell} \sum_{j=1}^{\ell} \alpha_i p_{ij}(t). \tag{8.2.18}$$

On the other hand, using formula (8.2.13) for block multiplication, we get

$$\exp(t\boldsymbol{Q}) = \begin{pmatrix} 1 & \boldsymbol{0} \\ \boldsymbol{e}^{\top} - \exp(t\boldsymbol{B})\boldsymbol{e}^{\top} & \exp(t\boldsymbol{B}) \end{pmatrix} \tag{8.2.19}$$

as can be shown by easy calculations. The proof is now completed by (8.1.27), (8.2.18) and (8.2.19). □

Remember that, with positive probability, $\{X(t), t \geq 0\}$ may never reach the absorbing state 0. The following result gives a necessary and sufficient condition for the finiteness of η.

Theorem 8.2.4 *The absorption time η is almost surely finite, i.e.*

$$\mathbb{P}(\eta < \infty) = 1, \tag{8.2.20}$$

for each (possibly defective) probability function $\boldsymbol{\alpha} = (\alpha_1, \ldots, \alpha_\ell)$ if and only if $\boldsymbol{B}$ is nonsingular.

Proof If $\boldsymbol{B}$ is nonsingular, then all eigenvalues of $\boldsymbol{B}$ have negative real parts as shown in Corollary 8.2.1. Thus by Corollary 8.2.2 (choosing $s = 0$) we have $\lim_{t\to\infty} \exp(t\boldsymbol{B}) = \boldsymbol{0}$ and hence, for each $\boldsymbol{\alpha}$,

$$\lim_{t\to\infty} \boldsymbol{\alpha} \exp(t\boldsymbol{B})\boldsymbol{e}^{\top} = 0. \tag{8.2.21}$$

Using (8.2.17) this gives (8.2.20). Conversely, let (8.2.21) hold for each probability function $\boldsymbol{\alpha}$ and suppose that $\boldsymbol{B}$ is singular. Then there exists a vector $\boldsymbol{x} = (x_1, \ldots, x_\ell) \neq \boldsymbol{0}$ with $\boldsymbol{B}\boldsymbol{x}^{\top} = \boldsymbol{0}$, and so $\boldsymbol{B}^n\boldsymbol{x}^{\top} = \boldsymbol{0}$. Consequently, $\exp(t\boldsymbol{B})\boldsymbol{x}^{\top} = \boldsymbol{x}^{\top}$ for all $t > 0$, and hence $\lim_{t\to\infty} \exp(t\boldsymbol{B}) = \boldsymbol{0}$ is not possible. Using similar arguments as in the proof of Lemma 8.1.3, it is not difficult to show that the matrix $\exp(t\boldsymbol{B})$ is nonnegative for each $t \geq 0$. Thus, $\limsup_{t\to\infty} (\exp(t\boldsymbol{B}))_{ij} > 0$ for some $i, j \in \{1, \ldots, \ell\}$ and (8.2.21) cannot hold for each probability function $\boldsymbol{\alpha}$, i.e. $\boldsymbol{B}$ must be nonsingular. □

We need the following auxiliary result.

Lemma 8.2.5 *Let $\boldsymbol{B}$ be a nonsingular subintensity matrix. Then $s\boldsymbol{I} - \boldsymbol{B}$ is nonsingular for each $s \geq 0$ and all entries of $(s\boldsymbol{I}-\boldsymbol{B})^{-1}$ are rational functions of $s \geq 0$. Furthermore, for all $s \geq 0$, $n \in \mathbb{N}$,*

$$\int_0^\infty \exp(t(-s\boldsymbol{I}+\boldsymbol{B}))\,\mathrm{d}t = (s\boldsymbol{I}-\boldsymbol{B})^{-1} \tag{8.2.22}$$

and

$$\frac{\mathrm{d}^n}{\mathrm{d}s^n}(s\boldsymbol{I}-\boldsymbol{B})^{-1} = (-1)^n n!(s\boldsymbol{I}-\boldsymbol{B})^{-n-1}. \tag{8.2.23}$$

Proof Let $s \geq 0$. By Corollary 8.2.1 all eigenvalues of $\boldsymbol{B}-s\boldsymbol{I}$ have negative real parts. Hence $s\boldsymbol{I}-\boldsymbol{B}$ is nonsingular. Furthermore, by Lemma 8.2.4, (8.2.22) follows. We now prove (8.2.23) by induction with respect to n. Using the differentiation rule (8.2.10) we have

$$\boldsymbol{0} = \frac{\mathrm{d}}{\mathrm{d}s}\boldsymbol{I} = \frac{\mathrm{d}}{\mathrm{d}s}\left((s\boldsymbol{I}-\boldsymbol{B})(s\boldsymbol{I}-\boldsymbol{B})^{-1}\right) = (s\boldsymbol{I}-\boldsymbol{B})^{-1} + (s\boldsymbol{I}-\boldsymbol{B})\frac{\mathrm{d}}{\mathrm{d}s}(s\boldsymbol{I}-\boldsymbol{B})^{-1}$$

and, consequently,

$$\frac{\mathrm{d}}{\mathrm{d}s}(s\boldsymbol{I}-\boldsymbol{B})^{-1} = -(s\boldsymbol{I}-\boldsymbol{B})^{-2}, \tag{8.2.24}$$

i.e. (8.2.23) holds for $n = 1$. Assume that (8.2.23) holds for $n = 1, 2, \ldots, k$. Then

$$\begin{aligned}\frac{\mathrm{d}^{k+1}}{\mathrm{d}s^{k+1}}(s\boldsymbol{I}-\boldsymbol{B})^{-1} &= \frac{\mathrm{d}}{\mathrm{d}s}\left(\frac{\mathrm{d}^k}{\mathrm{d}s^k}(s\boldsymbol{I}-\boldsymbol{B})^{-1}\right) = (-1)^k k!\frac{\mathrm{d}}{\mathrm{d}s}(s\boldsymbol{I}-\boldsymbol{B})^{-k-1}\\ &= (-1)^{k+1}(k+1)!(s\boldsymbol{I}-\boldsymbol{B})^{-(k+1)-1}\end{aligned}$$

because

$$\begin{aligned}\boldsymbol{0} &= \frac{\mathrm{d}}{\mathrm{d}s}\left((s\boldsymbol{I}-\boldsymbol{B})^{k+1}(s\boldsymbol{I}-\boldsymbol{B})^{-k-1}\right)\\ &= (k+1)(s\boldsymbol{I}-\boldsymbol{B})^{-1} + (s\boldsymbol{I}-\boldsymbol{B})^{k+1}\frac{\mathrm{d}}{\mathrm{d}s}(s\boldsymbol{I}-\boldsymbol{B})^{-k-1}.\end{aligned}$$

□

Theorem 8.2.5 *Assume that $\alpha_0 = 0$ and $\boldsymbol{B}$ is a nonsingular subintensity matrix. If F is the phase-type distribution with characteristics $(\boldsymbol{\alpha}, \boldsymbol{B})$, then F is continuous with*
(a) *density*

$$f(t) = \boldsymbol{\alpha}\exp(t\boldsymbol{B})\boldsymbol{b}^\top, \qquad t \geq 0, \tag{8.2.25}$$

(b) *Laplace–Stieltjes transform*

$$\hat{l}(s) = \boldsymbol{\alpha}(s\boldsymbol{I}-\boldsymbol{B})^{-1}\boldsymbol{b}^\top, \qquad s \geq 0, \tag{8.2.26}$$

(c) *n-th moment*

$$\mu^{(n)} = (-1)^n n! \boldsymbol{\alpha} \boldsymbol{B}^{-n} \boldsymbol{e}^\top, \qquad n \geq 1. \tag{8.2.27}$$

Proof Using Lemma 8.1.2, by (8.2.17) we have

$$f(t) = -\frac{\mathrm{d}}{\mathrm{d}t} \mathbb{P}(\eta > t) = \boldsymbol{\alpha} \exp(t\boldsymbol{B})(-\boldsymbol{B}\boldsymbol{e}^\top) = \boldsymbol{\alpha} \exp(t\boldsymbol{B})\boldsymbol{b}^\top,$$

where the last equality follows from (8.2.15). Thus, (a) is proved. In order to show (b) it suffices to note that, by (8.2.25),

$$\begin{aligned}
\hat{l}(s) &= \int_0^\infty \mathrm{e}^{-st} f(t)\,\mathrm{d}t = \int_0^\infty \mathrm{e}^{-st} \boldsymbol{\alpha} \exp(t\boldsymbol{B})\boldsymbol{b}^\top\,\mathrm{d}t \\
&= \int_0^\infty \boldsymbol{\alpha} \exp(-st\boldsymbol{I}) \exp(t\boldsymbol{B})\boldsymbol{b}^\top\,\mathrm{d}t = \boldsymbol{\alpha} \int_0^\infty \exp(t(-s\boldsymbol{I} + \boldsymbol{B}))\,\mathrm{d}t\boldsymbol{b}^\top \\
&= \boldsymbol{\alpha}(s\boldsymbol{I} - \boldsymbol{B})^{-1}\boldsymbol{b}^\top.
\end{aligned}$$

In this computation we used that $\mathrm{e}^{-st}\boldsymbol{I} = \exp(-st\boldsymbol{I})$ and

$$\exp(-st\boldsymbol{I}) \exp(t\boldsymbol{B}) = \exp(t(-s\boldsymbol{I} + \boldsymbol{B})),$$

where the last equality follows from Lemma 8.2.5. To show (c), we take the n-th derivative in (8.2.26). Then, (8.2.23) yields

$$\hat{l}^{(n)}(s) = \frac{\mathrm{d}^n}{\mathrm{d}s^n} \boldsymbol{\alpha}(s\boldsymbol{I} - \boldsymbol{B})^{-1}\boldsymbol{b}^\top = (-1)^n n! \boldsymbol{\alpha}(s\boldsymbol{I} - \boldsymbol{B})^{-n-1}\boldsymbol{b}^\top.$$

Putting $s = 0$ and using (2.1.6) and (8.2.15), this gives $\mu^{(n)} = (-1)^n \hat{l}^{(n)}(0) = (-1)^n n! \boldsymbol{\alpha} \boldsymbol{B}^{-n} \boldsymbol{e}^\top$. □

Note that a formula similar to (8.2.26) can be derived in the case when $\alpha_0 \geq 0$. Then,

$$\hat{l}(s) = \alpha_0 + \boldsymbol{\alpha}(s\boldsymbol{I} - \boldsymbol{B})^{-1}\boldsymbol{b}^\top, \qquad s \geq 0. \tag{8.2.28}$$

The *proof* of (8.2.28) is left to the reader.

8.2.3 Operations on Phase-Type Distributions

If it is convenient to indicate on which state space the Markov process $\{X(t)\}$ is defined, then we say that the phase-type distribution has the characteristics $(\boldsymbol{\alpha}, \boldsymbol{B}, E)$ and write $\mathrm{PH}(\boldsymbol{\alpha}, \boldsymbol{B}, E)$. Furthermore, we will always assume that $\boldsymbol{B}$ is nonsingular. Consider two phase-type distributions $\mathrm{PH}(\alpha_1, \boldsymbol{B}_1, E_1)$ and $\mathrm{PH}(\alpha_2, \boldsymbol{B}_2, E_2)$ simultaneously, where we take $E_1 = \{1, \ldots, \ell_1\}$ and $E_2 = \{\ell_1 + 1, \ldots, \ell_1 + \ell_2\}$. We prove that the family of phase-type distributions is closed under convolution and mixing.

Theorem 8.2.6 *The convolution of the phase-type distributions* $\mathrm{PH}(\boldsymbol{\alpha}_1, \boldsymbol{B}_1, E_1)$ *and* $\mathrm{PH}(\boldsymbol{\alpha}_2, \boldsymbol{B}_2, E_2)$ *is a phase-type distribution with characteristics* $(\boldsymbol{\alpha}, \boldsymbol{B}, E)$, *where* $E = E_1 \cup E_2$,

$$\alpha_i = \begin{cases} (\boldsymbol{\alpha}_1)_i & \textit{if } i \in E_1, \\ (\boldsymbol{\alpha}_1)_0(\boldsymbol{\alpha}_2)_i & \textit{if } i \in E_2, \end{cases} \tag{8.2.29}$$

and

$$\boldsymbol{B} = \begin{pmatrix} \boldsymbol{B}_1 & \boldsymbol{b}_1^\top \boldsymbol{\alpha}_2 \\ \boldsymbol{0} & \boldsymbol{B}_2 \end{pmatrix}, \tag{8.2.30}$$

with $(\boldsymbol{\alpha}_k)_i$, *the i-th component of* $\boldsymbol{\alpha}_k$, *and* $\boldsymbol{b}_k^\top = -\boldsymbol{B}_k \boldsymbol{e}^\top$, $k = 1, 2$.

Proof Let $\hat{l}(s), \hat{l}_1(s), \hat{l}_2(s)$ be the Laplace–Stieltjes transform of $\mathrm{PH}(\boldsymbol{\alpha}, \boldsymbol{B}, E)$, $\mathrm{PH}(\boldsymbol{\alpha}_1, \boldsymbol{B}_1, E_1)$, $\mathrm{PH}(\boldsymbol{\alpha}_2, \boldsymbol{B}_2, E_2)$, respectively. Then, it suffices to prove that $\hat{l}(s) = \hat{l}_1(s)\hat{l}_2(s)$ for $s \geq 0$. Note that

$$s\boldsymbol{I} - \boldsymbol{B} = \begin{pmatrix} s\boldsymbol{I} - \boldsymbol{B}_1 & -\boldsymbol{b}_1^\top \boldsymbol{\alpha}_2 \\ \boldsymbol{0} & s\boldsymbol{I} - \boldsymbol{B}_2 \end{pmatrix}.$$

We first show that the matrix $s\boldsymbol{I} - \boldsymbol{B}$ is invertible. This is equivalent to showing that there exists a matrix $\boldsymbol{A}$ for which

$$\begin{pmatrix} s\boldsymbol{I} - \boldsymbol{B}_1 & -\boldsymbol{b}_1^\top \boldsymbol{\alpha}_2 \\ \boldsymbol{0} & s\boldsymbol{I} - \boldsymbol{B}_2 \end{pmatrix} \times \begin{pmatrix} (s\boldsymbol{I} - \boldsymbol{B}_1)^{-1} & \boldsymbol{A} \\ \boldsymbol{0} & (s\boldsymbol{I} - \boldsymbol{B}_2)^{-1} \end{pmatrix} = \boldsymbol{I}.$$

In other words, $\boldsymbol{A}$ must satisfy $(s\boldsymbol{I} - \boldsymbol{B}_1)\boldsymbol{A} - \boldsymbol{b}_1^\top \boldsymbol{\alpha}_2 (s\boldsymbol{I} - \boldsymbol{B}_2)^{-1} = 0$. Thus,

$$(s\boldsymbol{I} - \boldsymbol{B})^{-1} = \begin{pmatrix} (s\boldsymbol{I} - \boldsymbol{B}_1)^{-1} & \boldsymbol{A} \\ \boldsymbol{0} & (s\boldsymbol{I} - \boldsymbol{B}_2)^{-1} \end{pmatrix},$$

where $\boldsymbol{A} = (s\boldsymbol{I} - \boldsymbol{B}_1)^{-1}\boldsymbol{b}_1^\top \boldsymbol{\alpha}_2 (s\boldsymbol{I} - \boldsymbol{B}_2)^{-1}$. Let $\tilde{\boldsymbol{b}}_1$ be the vector satisfying $\tilde{\boldsymbol{b}}_1^\top + (\boldsymbol{B}_1 +_{\mathbf{1}}^\top \boldsymbol{\alpha}_2)\boldsymbol{e}^\top = \boldsymbol{0}$, i.e. $\tilde{\boldsymbol{b}}_1^\top = (\boldsymbol{\alpha}_2)_0 \boldsymbol{b}_1^\top$. Now, using (8.2.26), for the Laplace–Stieltjes transform $\hat{l}(s)$ of $\mathrm{PH}(\boldsymbol{\alpha}, \boldsymbol{B}, E)$ we have

$$\begin{aligned} \hat{l}(s) &= (\boldsymbol{\alpha}_1)_0(\boldsymbol{\alpha}_2)_0 \\ &\qquad + (\boldsymbol{\alpha}_1, (\boldsymbol{\alpha}_1)_0\boldsymbol{\alpha}_2) \begin{pmatrix} (s\boldsymbol{I} - \boldsymbol{B}_1)^{-1} & \boldsymbol{A} \\ \boldsymbol{0} & (s\boldsymbol{I} - \boldsymbol{B}_2)^{-1} \end{pmatrix} (\tilde{\boldsymbol{b}}_1, \boldsymbol{b}_2)^\top \\ &= (\boldsymbol{\alpha}_1)_0(\boldsymbol{\alpha}_2)_0 + \boldsymbol{\alpha}_1 (s\boldsymbol{I} - \boldsymbol{B}_1)^{-1}\tilde{\boldsymbol{b}}_1^\top + \boldsymbol{\alpha}_1 \boldsymbol{A}\boldsymbol{b}_2^\top \\ &\qquad\qquad + (\boldsymbol{\alpha}_1)_0(\boldsymbol{\alpha}_2)(s\boldsymbol{I} - \boldsymbol{B}_2)^{-1}\boldsymbol{b}_2^\top \\ &= \left((\boldsymbol{\alpha}_1)_0 + \boldsymbol{\alpha}_1(s\boldsymbol{I} - \boldsymbol{B}_1)^{-1}\boldsymbol{b}_1^\top\right)\left((\boldsymbol{\alpha}_2)_0 + \boldsymbol{\alpha}_2(s\boldsymbol{I} - \boldsymbol{B}_2)^{-1}\boldsymbol{b}_2^\top\right) \\ &= \hat{l}_1(s)\hat{l}_2(s). \end{aligned}$$

□

It is clear how Theorem 8.2.6 can be generalized in order to show that the convolution $\mathrm{PH}(\boldsymbol{\alpha}_1, \boldsymbol{B}_1, E_1) * \ldots * \mathrm{PH}(\boldsymbol{\alpha}_n, \boldsymbol{B}_n, E_n)$ of n phase-type distributions is a phase-type distribution, $n \geq 2$. Its characteristics $(\boldsymbol{\alpha}, \boldsymbol{B}, E)$ are given by $E = \bigcup_{k=1}^{n} E_k$, $\alpha_i = \prod_{j=1}^{k-1} (\boldsymbol{\alpha}_j)_0 (\boldsymbol{\alpha}_k)_i$ if $i \in E_k$, where $k = 1, \ldots, n$, and

$$\boldsymbol{B} = \begin{pmatrix} \boldsymbol{B}_1 & \boldsymbol{b}_1^\top \boldsymbol{\alpha}_2 & \boldsymbol{0} & \cdots & \boldsymbol{0} \\ \boldsymbol{0} & \boldsymbol{B}_2 & \boldsymbol{b}_2^\top \boldsymbol{\alpha}_3 & \cdots & \boldsymbol{0} \\ \vdots & \vdots & \vdots & & \vdots \\ \boldsymbol{0} & \boldsymbol{0} & \boldsymbol{0} & \cdots & \boldsymbol{B}_n \end{pmatrix}.$$

Theorem 8.2.7 *Let $0 < p < 1$. The mixture $p\mathrm{PH}(\boldsymbol{\alpha}_1, \boldsymbol{B}_1, E_1) + (1 - p)\mathrm{PH}(\boldsymbol{\alpha}_2, \boldsymbol{B}_2, E_2)$ is a phase-type distribution with characteristics $(\boldsymbol{\alpha}, \boldsymbol{B}, E)$, where $E = E_1 \cup E_2$,*

$$\alpha_i = \begin{cases} p(\boldsymbol{\alpha}_1)_i & \text{if } i \in E_1, \\ (1-p)(\boldsymbol{\alpha}_2)_i & \text{if } i \in E_2, \end{cases} \qquad \boldsymbol{B} = \begin{pmatrix} \boldsymbol{B}_1 & \boldsymbol{0} \\ \boldsymbol{0} & \boldsymbol{B}_2 \end{pmatrix}. \tag{8.2.31}$$

As the *proof* of Theorem 8.2.7 is easy, we leave it to the reader.

Note that Theorem 8.2.7 implies that for $n \geq 2$ and for any probability function $(p_1, \ldots, p_n)$, the mixture $\sum_{k=1}^{n} p_k \mathrm{PH}(\boldsymbol{\alpha}_k, \boldsymbol{B}_k, E_k)$ is a phase-type distribution with characteristics $(\boldsymbol{\alpha}, \boldsymbol{B}, E)$, where $E = \bigcup_{k=1}^{n} E_k$, $\alpha_i = p_k(\boldsymbol{\alpha}_k)_i$ if $i \in E_k$ and

$$\boldsymbol{B} = \begin{pmatrix} \boldsymbol{B}_1 & \boldsymbol{0} & \boldsymbol{0} & \cdots & \boldsymbol{0} \\ \boldsymbol{0} & \boldsymbol{B}_2 & \boldsymbol{0} & \cdots & \boldsymbol{0} \\ \vdots & \vdots & \vdots & & \vdots \\ \boldsymbol{0} & \boldsymbol{0} & \boldsymbol{0} & \cdots & \boldsymbol{B}_n \end{pmatrix}.$$

We show now that the class of phase-type distributions is "dense" in the class of all distributions of nonnegative random variables. More concretely, for any distribution F on $\mathbb{R}_+$, there exists a sequence F_n of phase-type distributions such that

$$\lim_{n \to \infty} F_n(x) = F(x), \tag{8.2.32}$$

for each continuity point x of F. Recall that in Section 2.1.5 the sequence $\{F_n\}$ was then said to *converge weakly* to F, or $F_n \xrightarrow{\mathrm{d}} F$. We need the following lemma.

Lemma 8.2.6 *Let $\{X_n\}$ be a sequence of real-valued random variables, and $x \in \mathbb{R}$. If $\lim_{n\to\infty} \mathbb{E} X_n = x$ and $\lim_{n\to\infty} \mathbb{E} X_n^2 = x^2$, then $\lim_{n\to\infty} \mathbb{E} f(X_n) = f(x)$ for each bounded function $f : \mathbb{R} \to \mathbb{R}$ being continuous at x.*

Proof Without loss of generality we suppose that $\sup_y |f(y)| \le 1$. For each $\varepsilon > 0$ we choose $\delta > 0$ such that $|f(x) - f(y)| \le \varepsilon$ whenever $|x - y| \le \delta$. Then,

$$\begin{aligned}
|\mathbb{E}\,(f(X_n) - f(x))| &\le \mathbb{E}\,|f(X_n) - f(x)| \\
&\le \mathbb{E}\,[|f(X_n) - f(x)|; |X_n - x| \le \delta] + \mathbb{E}\,[|f(X_n) - f(x)|; |X_n - x| > \delta] \\
&\le \varepsilon + 2\mathbb{P}(|X_n - x| > \delta)\,.
\end{aligned}$$

From Chebyshev's inequality

$$\mathbb{P}(|X_n - x| > \delta) \le \frac{\mathbb{E}\,(X_n - x)^2}{\delta^2} = \frac{\mathbb{E}\,X_n^2 - x^2 + 2x(x - \mathbb{E}\,X_n)}{\delta^2}$$

and the proof is complete because $\varepsilon > 0$ is arbitrary and the numerator of the last expression tends to zero as $n \to 0$. □

Theorem 8.2.8 *The family of phase-type distributions is dense in the set of all distributions on* $\mathbb{R}_+$.

Proof Let F be an arbitrary distribution on $\mathbb{R}_+$ and define

$$F_n = F(0)\delta_0 + \sum_{k=1}^{\infty}(F(k/n) - F((k-1)/n))\mathrm{Erl}(k,n) \tag{8.2.33}$$

for $n \ge 1$. We first show that $\lim_{n\to\infty} F_n(x) = F(x)$ for each $x \ge 0$ with $F(x) = F(x-)$. Note that

$$F_n(x) = \sum_{k=0}^{\infty} F(k/n)\mathrm{e}^{-nx}\frac{(nx)^k}{k!} = \int_0^{\infty} F(t)\,\mathrm{d}G_{n,x}(t)\,, \tag{8.2.34}$$

where $G_{n,x} = \sum_{k=0}^{\infty} \mathrm{e}^{-nx}\frac{(nx)^k}{k!}\delta_{k/n}$. This can be seen by comparing the densities on $(0,\infty)$ of the distributions given in (8.2.33) and (8.2.34). Furthermore,

$$\int_0^{\infty} t\,\mathrm{d}G_{n,x}(t) = \sum_{k=0}^{\infty} \frac{k}{n}\frac{(nx)^k}{k!}\mathrm{e}^{-nx} = x\,,$$

and $\int_0^{\infty} t^2\,\mathrm{d}G_{n,x}(t) = n^{-1}x + x^2 \to x^2$ as $n \to \infty$. Thus, by Lemma 8.2.6, $\int_0^{\infty} F(t)\,\mathrm{d}G_{n,x}(t) = \mathbb{E}\,F(X_n) \to F(x)$, where X_n has distribution $G_{n,x}$. Now let $F_n' = (F(0) + \bar{F}(n))\delta_0 + \sum_{k=1}^{n^2}(F(k/n) - F((k-1)/n))\,\mathrm{Erl}(k,n)$. Since each Erlang distribution is phase-type (see Example 1 in Section 8.2.2), F_n' is a phase-type distribution by Theorem 8.2.7. This completes the proof because $\lim_{n\to\infty} |F_n(x) - F_n'(x)| = 0$ for each $x \ge 0$. □

One can easily show that the family of Erlang distributions $\{\mathrm{Erl}(n,n), n \geq 1\}$ possesses the following optimality property: among all phase-type distributions on a state space E with n elements, $\mathrm{Erl}(n,n)$ is the best approximation to δ_1 (in the sense of the L_2-norm). Furthermore, note that $\mathrm{Erl}(n,n)$ converges to δ_1 as $n \to \infty$.

Bibliographical Notes. More details on concepts and results from matrix algebra and calculus can be found for instance in Chatelin (1993), Graham (1981) and Wilkinson (1965). A comprehensive treatment of phase-type distributions is given in Neuts (1981); other references include Asmussen (1987), Latouche and Ramaswami (1998) and Neuts (1989). Theorem 8.2.8 is taken from Schassberger (1973).

8.3 RISK PROCESSES WITH PHASE-TYPE DISTRIBUTIONS

8.3.1 The Compound Poisson Model

In this section we consider the ruin function in the compound Poisson model with phase-type claim size distributions. As usual we denote the intensity of the Poisson arrival process by λ, and the claim size distribution by F_U. The ruin function $\psi(u)$ has been shown in Theorem 5.3.4 to satisfy the relation

$$1 - \psi(u) = \sum_{k=0}^{\infty} (1-\rho)\rho^k (F_U^{\mathrm{s}})^{*k}(u), \qquad u \geq 0 \tag{8.3.1}$$

(see also (6.5.2)), i.e. $1 - \psi(u)$ is the distribution function of the compound geometric distribution with characteristics (ρ, F_U^{s}), where $\rho = (\lambda \mathbb{E}\, U)/\beta < 1$ with β being the premium rate, and $F_U^{\mathrm{s}}(x) = (\mu_U)^{-1} \int_0^x \overline{F_U}(y)\, \mathrm{d}y$ for $x \geq 0$. For $F_U = \mathrm{PH}(\boldsymbol{\alpha}, \boldsymbol{B})$ we can derive a formula which is more suitable for numerical computations than (8.3.1).

We first show that by passing from F_U to F_U^{s} we do not leave the family of phase-type distributions.

Lemma 8.3.1 *Let* $\alpha_0 = 0$, $\boldsymbol{B}$ *a nonsingular subintensity matrix, and* F *a distribution on* $\mathbb{R}_+$. *If* F *is the phase-type distribution* $\mathrm{PH}(\boldsymbol{\alpha}, \boldsymbol{B})$, *then* F^{s} *is also phase-type and given by* $F^{\mathrm{s}} = \mathrm{PH}(\boldsymbol{\alpha}^{\mathrm{s}}, \boldsymbol{B})$, *where*

$$\boldsymbol{\alpha}^{\mathrm{s}} = -\mu_F^{-1} \boldsymbol{\alpha} \boldsymbol{B}^{-1}. \tag{8.3.2}$$

Proof Using (8.2.17) and (8.2.11), we get that

$$F^{\mathrm{s}}(x) = \mu_F^{-1} \int_0^x \overline{F}(y)\, \mathrm{d}y = \mu_F^{-1} \int_0^x \boldsymbol{\alpha} \exp(y\boldsymbol{B}) \boldsymbol{e}^{\top}\, \mathrm{d}y$$

$$
\begin{aligned}
&= \mu_F^{-1}\boldsymbol{\alpha}\int_0^x \exp(y\boldsymbol{B})\,\mathrm{d}y\,\boldsymbol{e}^\top = \mu_F^{-1}\boldsymbol{\alpha}\boldsymbol{B}^{-1}(\exp(x\boldsymbol{B})-\boldsymbol{I})\boldsymbol{e}^\top \\
&= -\mu_F^{-1}\boldsymbol{\alpha}\boldsymbol{B}^{-1}\boldsymbol{e}^\top - \mu_F^{-1}\boldsymbol{\alpha}\boldsymbol{B}^{-1}\exp(x\boldsymbol{B})\boldsymbol{e}^\top = 1-\boldsymbol{\alpha}^{\mathrm{s}}\exp(x\boldsymbol{B})\boldsymbol{e}^\top,
\end{aligned}
$$

where the last equality follows from (8.2.27). Thus, by (8.2.17), $F^{\mathrm{s}}(x)$ is the distribution function of $\mathrm{PH}(\boldsymbol{\alpha}^{\mathrm{s}}, \boldsymbol{B})$. □

The following useful theorem complements results for compound geometric distributions given in Section 4.5.1.

Lemma 8.3.2 *Let F, G be two distributions on $\mathbb{R}_+$. Assume that G is the phase-type distribution $\mathrm{PH}(\boldsymbol{\alpha}, \boldsymbol{B})$ with $\alpha_0 = 0$ and $\boldsymbol{B}$ nonsingular, and that F is the compound geometric distribution with characteristics (p, G), where $0 < p < 1$. Then $F = \mathrm{PH}(p\boldsymbol{\alpha}, \boldsymbol{B} + p\boldsymbol{b}^\top\boldsymbol{\alpha})$.*

Proof By Lemma 8.2.5 the matrix $s\boldsymbol{I}-\boldsymbol{B}$ is nonsingular for all $s \geq 0$. The series $\sum_{k=0}^\infty (s\boldsymbol{I}-\boldsymbol{B})^{-1}(p\boldsymbol{b}^\top\boldsymbol{\alpha}(s\boldsymbol{I}-\boldsymbol{B})^{-1})^k$ is a well-defined $\ell\times\ell$ matrix for each $s \geq 0$. Indeed,

$$
\begin{aligned}
&\sum_{k=n}^\infty \left\|(s\boldsymbol{I}-\boldsymbol{B})^{-1}(p\boldsymbol{b}^\top\boldsymbol{\alpha}(s\boldsymbol{I}-\boldsymbol{B})^{-1})^k\right\| \\
&= \sum_{k=n}^\infty \left\|p^k((s\boldsymbol{I}-\boldsymbol{B})^{-1}\boldsymbol{b}^\top)(\boldsymbol{\alpha}(s\boldsymbol{I}-\boldsymbol{B})^{-1}\boldsymbol{b}^\top)^{k-1}(\boldsymbol{\alpha}(s\boldsymbol{I}-\boldsymbol{B})^{-1})\right\| \\
&\leq \|(s\boldsymbol{I}-\boldsymbol{B})^{-1}\boldsymbol{b}^\top\boldsymbol{\alpha}(s\boldsymbol{I}-\boldsymbol{B})^{-1}\| \sum_{k=n}^\infty p^k(\boldsymbol{\alpha}(s\boldsymbol{I}-\boldsymbol{B})^{-1}\boldsymbol{b}^\top)^{k-1},
\end{aligned}
$$

where the above series is convergent since by (8.2.26), $0 \leq \boldsymbol{\alpha}(s\boldsymbol{I}-\boldsymbol{B})^{-1}\boldsymbol{b}^\top = \hat{l}_G(s) \leq 1$ for all $s \geq 0$. Now

$$
\begin{aligned}
&(s\boldsymbol{I}-\boldsymbol{B}-p\boldsymbol{b}^\top\boldsymbol{\alpha})\sum_{k=0}^\infty (s\boldsymbol{I}-\boldsymbol{B})^{-1}(p\boldsymbol{b}^\top\boldsymbol{\alpha}(s\boldsymbol{I}-\boldsymbol{B})^{-1})^k \\
&= (s\boldsymbol{I}-\boldsymbol{B})\sum_{k=0}^\infty (s\boldsymbol{I}-\boldsymbol{B})^{-1}(p\boldsymbol{b}^\top\boldsymbol{\alpha}(s\boldsymbol{I}-\boldsymbol{B})^{-1})^k \\
&\quad - p\boldsymbol{b}^\top\boldsymbol{\alpha}\sum_{k=0}^\infty (s\boldsymbol{I}-\boldsymbol{B})^{-1}(p\boldsymbol{b}^\top\boldsymbol{\alpha}(s\boldsymbol{I}-\boldsymbol{B})^{-1})^k \\
&= \sum_{k=0}^\infty (p\boldsymbol{b}^\top\boldsymbol{\alpha}(s\boldsymbol{I}-\boldsymbol{B})^{-1})^k - \sum_{k=1}^\infty (p\boldsymbol{b}^\top\boldsymbol{\alpha}(s\boldsymbol{I}-\boldsymbol{B})^{-1})^k = \boldsymbol{I}.
\end{aligned}
$$

Thus, $s\boldsymbol{I}-\boldsymbol{B}-p\boldsymbol{b}^\top\boldsymbol{\alpha}$ is invertible for each $s \geq 0$, and

$$
\sum_{k=1}^\infty (p\boldsymbol{\alpha}(s\boldsymbol{I}-\boldsymbol{B})^{-1}\boldsymbol{b}^\top)^k = p\boldsymbol{\alpha}\Big(\sum_{k=0}^\infty (s\boldsymbol{I}-\boldsymbol{B})^{-1}(p\boldsymbol{b}^\top\boldsymbol{\alpha}(s\boldsymbol{I}-\boldsymbol{B})^{-1})^k\Big)\boldsymbol{b}^\top
$$

$$= \quad p\boldsymbol{\alpha}(s\boldsymbol{I} - \boldsymbol{B} - p\boldsymbol{b}^\top\boldsymbol{\alpha})^{-1}\boldsymbol{b}^\top.$$

This is equivalent to

$$\sum_{k=0}^{\infty}(p\boldsymbol{\alpha}(s\boldsymbol{I} - \boldsymbol{B})^{-1}\boldsymbol{b}^\top)^k = 1 + p\boldsymbol{\alpha}(s\boldsymbol{I} - \boldsymbol{B} - p\boldsymbol{b}^\top\boldsymbol{\alpha})^{-1}\boldsymbol{b}^\top.$$

Hence

$$\frac{1-p}{1 - p\boldsymbol{\alpha}(s\boldsymbol{I} - \boldsymbol{B})^{-1}\boldsymbol{b}^\top} = (1-p) + p\boldsymbol{\alpha}(s\boldsymbol{I} - \boldsymbol{B} - p\boldsymbol{b}^\top\boldsymbol{\alpha})^{-1}(1-p)\boldsymbol{b}^\top. \quad (8.3.3)$$

By Theorem 4.2.1 and equation (8.2.26), the left-hand side of (8.3.3) is the Laplace–Stieltjes transform of the compound geometric distribution with characteristics $(p, \mathrm{PH}(\boldsymbol{\alpha}, \boldsymbol{B}))$; the right-hand side is the Laplace–Stieltjes transform of $\mathrm{PH}(p\boldsymbol{\alpha}, \boldsymbol{B} + p\boldsymbol{b}^\top\boldsymbol{\alpha})$, as follows from (8.2.28). The proof is completed by the one-to-one correspondence between distributions and their Laplace–Stieltjes transforms. □

The following probabilistic reasoning makes the statement of Lemma 8.3.2 intuitively clear. Let $X = \sum_{i=0}^{N} U_i$ have the distribution F_X, where $N, U_1, U_2, \ldots$ are independent random variables such that N is geometrically distributed with parameter p and $U_1, U_2, \ldots$ are distributed according to F_U. We show that X is distributed as the absorption time of a certain Markov process. For this purpose, we dissect the absorbing state 0 corresponding to the phase-type distributed random variables U_i into two "substates" 0_a and 0_b, where 0_b is no longer absorbing but *fictitious*. Assume that, given the state 0 is reached, the Markov process to be constructed takes substate 0_a with probability $(1-p)$ and substate 0_b with probability p. If 0_b is chosen, then a new state from $E = \{1, 2, \ldots, \ell\}$ is immediately chosen according to $\boldsymbol{\alpha}$, and the evolution is continued. In the spirit of the construction considered in Section 8.1.3, this leads to a Markov process on $E' = \{0_a, 1, 2, \ldots, \ell\}$ with initial distribution $\boldsymbol{\alpha}' = (1-p, p\boldsymbol{\alpha})$ and intensity matrix

$$\boldsymbol{Q} = \begin{pmatrix} 0 & \boldsymbol{0} \\ (1-p)\boldsymbol{b}^\top & \boldsymbol{B} + \boldsymbol{b}^\top p\boldsymbol{\alpha} \end{pmatrix}.$$

The geometric compound $X = \sum_{i=0}^{N} U_i$ is distributed as the absorption time of this Markov process since F_U is the distribution of the (stochastically independent) times between consecutive visits of the instantaneous state 0_b, and N is the number of visits in 0_b before the absorbing state 0_a is reached.

We are now ready to prove a numerically convenient representation formula for the ruin function $\psi(u)$ in terms of the matrix exponential function.

Theorem 8.3.1 *Assume that the claim sizes are distributed according to* $F_U = \mathrm{PH}(\boldsymbol{\alpha}, \boldsymbol{B})$ *with* $\alpha_0 = 0$, $\boldsymbol{B}$ *nonsingular and* $\rho = (\lambda \mathbb{E}\, U)/\beta < 1$. *Let* $\boldsymbol{\alpha}^{\mathrm{s}}$ *be defined by* (8.3.2). *Then, for all* $u \geq 0$,

$$\psi(u) = \rho \boldsymbol{\alpha}^{\mathrm{s}} \exp(u(\boldsymbol{B} + \rho \boldsymbol{b}^\top \boldsymbol{\alpha}^{\mathrm{s}}))\boldsymbol{e}^\top . \tag{8.3.4}$$

Proof Using (8.3.1), from Lemmas 8.3.1 and 8.3.2 we get that $1 - \psi(u)$ is the distribution function of $\mathrm{PH}(\rho\boldsymbol{\alpha}^{\mathrm{s}}, \boldsymbol{B} + \rho \boldsymbol{b}^\top \boldsymbol{\alpha}^{\mathrm{s}})$. Now, it is easily seen that formula (8.3.4) follows from (8.2.17). □

Also, for the multivariate ruin function $\psi(u, \infty, y)$ introduced in Section 5.1.4, a similar algorithmic formula can be given when the claim sizes are phase-type distributed. Recall that $\psi(u, \infty, y)$ denotes the probability that ruin occurs and that the overshoot (i.e. the deficit at the ruin epoch) is larger than y, where u is the initial risk reserve. Using the representation formula (6.5.17) for $\psi(u, \infty, y)$ we arrive at the following result.

Theorem 8.3.2 *Under the assumptions of Theorem* 8.3.1, *for all* $u, y \geq 0$,

$$\begin{aligned}\psi(u, \infty, y) \;=\; & \frac{\rho}{1-\rho}\Big((\boldsymbol{\alpha}^{\mathrm{s}} \exp(y\boldsymbol{B}), \boldsymbol{0}) \exp\Big(u\begin{pmatrix} \boldsymbol{B} & \rho \boldsymbol{b}^\top \boldsymbol{\alpha}^{\mathrm{s}} \\ \boldsymbol{0} & \boldsymbol{B} + \rho \boldsymbol{b}^\top \boldsymbol{\alpha}^{\mathrm{s}} \end{pmatrix}\Big)\boldsymbol{e}^\top \\ & - \rho \boldsymbol{\alpha}^{\mathrm{s}} \exp(u(\boldsymbol{B} + \rho \boldsymbol{b}^\top \boldsymbol{\alpha}^{\mathrm{s}}))\boldsymbol{e}^\top\Big) . \end{aligned} \tag{8.3.5}$$

Proof By G_1 we denote the phase-type distribution $\mathrm{PH}(\boldsymbol{\alpha}^{\mathrm{s}} \exp(y\boldsymbol{B}), \boldsymbol{B})$, and by G_2 the phase-type distribution $\mathrm{PH}(\rho\boldsymbol{\alpha}^{\mathrm{s}}, \boldsymbol{B} + \rho \boldsymbol{b}^\top \boldsymbol{\alpha}^{\mathrm{s}})$. By (8.2.17) and Lemma 8.3.1

$$\begin{aligned} F_U^{\mathrm{s}}(y + u - v) &= 1 - \boldsymbol{\alpha}^{\mathrm{s}} \exp((y + u - v)\boldsymbol{B})\boldsymbol{e}^\top \\ &= 1 - (\boldsymbol{\alpha}^{\mathrm{s}} \exp(y\boldsymbol{B})) \exp((u - v)\boldsymbol{B})\boldsymbol{e}^\top = G_1(u - v) , \end{aligned}$$

where (8.1.24) has been used in the second equality. From (6.5.17) we have

$$\psi(u, \infty, y) = \sum_{k=0}^{\infty} \rho^{k+1} (F_U^{\mathrm{s}})^{*k}(u) - \sum_{k=0}^{\infty} \rho^{k+1} \int_0^u F_U^{\mathrm{s}}(y + u - v) \, \mathrm{d}(F_U^{\mathrm{s}})^{*k}(v) .$$

Hence, by Lemma 8.3.2, $\psi(u, \infty, y) = (1 - \rho)^{-1}\rho(G_2(u) - G_1 * G_2(u))(u)$. Now, (8.2.17), (8.2.29), (8.2.30) and Theorem 8.2.6 give (8.3.5). (Note that $\boldsymbol{\alpha}_2 = \rho\boldsymbol{\alpha}^{\mathrm{s}}$ is defective.) □

8.3.2 Numerical Issues

In Theorem 8.2.8 we proved that the family of phase-type distributions forms a dense class of distributions on $\mathbb{R}_+$. Moreover, in Section 8.3.1 we showed that phase-type distributions lead to formulae involving matrix algebraic operations, like matrix multiplication and addition, inversion and matrix exponentiation. In this section we discuss a few numerical methods, helpful when computing the matrix exponential. We further show some numerical experiments in computing the ruin function $\psi(u)$ given in (8.3.4).

The most straightforward but, at the same time, most dubious method for numerical computation of matrix exponentials is the use of the *diagonalization method* which has already been considered in Chapter 7. This method is known for its numerical instability. Moreover, it requires that all eigenvalues of the considered matrix are distinct. Thus, in order to compute the right-hand side of (8.3.4) we first assume that the eigenvalues of the matrices $\boldsymbol{B}$ and $\boldsymbol{C} = \boldsymbol{B} + \rho \boldsymbol{b}^{\top} \boldsymbol{\alpha}^{\mathrm{s}}$ are distinct and represent these matrices by

$$\boldsymbol{B} = \boldsymbol{\Phi}(\boldsymbol{B})\mathrm{diag}(\boldsymbol{\theta}(\boldsymbol{B}))\boldsymbol{\Psi}(\boldsymbol{B}), \qquad \boldsymbol{C} = \boldsymbol{\Phi}(\boldsymbol{C})\mathrm{diag}(\boldsymbol{\theta}(\boldsymbol{C}))\boldsymbol{\Psi}(\boldsymbol{C}), \tag{8.3.6}$$

where $\boldsymbol{\Phi}(\boldsymbol{B}), \boldsymbol{\Phi}(\boldsymbol{C})$ and $\boldsymbol{\Psi}(\boldsymbol{B}), \boldsymbol{\Psi}(\boldsymbol{C})$ denote the $\ell \times \ell$ matrices consisting of right and left eigenvectors of $\boldsymbol{B}$, $\boldsymbol{C}$, respectively. Then we compute $\boldsymbol{B}^{-1} = \boldsymbol{\Phi}(\boldsymbol{B})\mathrm{diag}\big(\theta_1^{-1}(\boldsymbol{B}), \ldots, \theta_\ell^{-1}(\boldsymbol{B})\big)\boldsymbol{\Psi}(\boldsymbol{B})$ and also ρ, $\boldsymbol{\alpha}^{\mathrm{s}}$ and

$$\exp(u\boldsymbol{C}) = \boldsymbol{\Phi}(\boldsymbol{C})\mathrm{diag}\left(\exp(u\theta_1(\boldsymbol{C})), \ldots, \exp(u\theta_\ell(\boldsymbol{C}))\right)\boldsymbol{\Psi}_C\,. \tag{8.3.7}$$

This way we obtain all elements needed to compute the expression in (8.3.4).

Alternatively, $\exp(u\boldsymbol{C})$ can be computed by the following *uniformization method.* Put $\boldsymbol{C}' = u\boldsymbol{C}$ and note that $\max_{i,j\in E}(\boldsymbol{I} + a^{-1}\boldsymbol{C}')_{ij} \le 1$, where $a = \max\{|c'_{ij}| : i, j \in E\}$. Moreover, since all entries of $\boldsymbol{I} + a^{-1}\boldsymbol{C}'$ are nonnegative and all row sums of $\boldsymbol{I} + a^{-1}\boldsymbol{C}'$ are not greater than 1, we have

$$\boldsymbol{0} \le (\boldsymbol{I} + a^{-1}\boldsymbol{C}')(\boldsymbol{I} + a^{-1}\boldsymbol{C}') \le (\boldsymbol{I} + a^{-1}\boldsymbol{C}')\boldsymbol{E} \le \boldsymbol{E}\,,$$

where the inequalities are entry-wise. By induction, $\boldsymbol{0} \le (\boldsymbol{I} + a^{-1}\boldsymbol{C}')^k \le \boldsymbol{E}$ for each $k \in \mathbb{N}$. Thus,

$$\exp(\boldsymbol{C}') = \mathrm{e}^{-a}\exp\left(a(\boldsymbol{I} + a^{-1}\boldsymbol{C}')\right) = \sum_{k=0}^{n} \frac{a^k}{k!}\mathrm{e}^{-a}(\boldsymbol{I} + a^{-1}\boldsymbol{C}')^k + \boldsymbol{R}_n\,,$$

where for the remainder matrix $\boldsymbol{R}_n$ we have

$$0 \le \|\boldsymbol{R}_n\| \le \sum_{k=n+1}^{\infty} \frac{a^k}{k!}\mathrm{e}^{-a}. \tag{8.3.8}$$

This means that, for numerical purposes, the approximation

$$\exp(\boldsymbol{C}')_{\mathrm{app}} = \sum_{k=0}^{n} \frac{a^k}{k!}\mathrm{e}^{-a}(\boldsymbol{I} + a^{-1}\boldsymbol{C}')^k$$

can be used. It follows from (8.3.8) that the error of this approximation becomes arbitrarily small when n is sufficiently large.

Another way to compute the matrix exponential function $\exp(u\boldsymbol{C})$ is given by the *Runge–Kutta method.* This is based on the observation that, for each vector $\boldsymbol{x} = (x_1, \ldots, x_\ell)$, the (column) vector $\boldsymbol{f}(u) = \exp(u\boldsymbol{C})\boldsymbol{x}^\top$ satisfies the linear system of differential equations

$$\boldsymbol{f}^{(1)}(u) = \boldsymbol{C}\boldsymbol{f}(u)\,, \tag{8.3.9}$$

with initial condition $\boldsymbol{f}(0) = \boldsymbol{x}^\top$. A common algorithm for the computation of the solution to (8.3.9) is a standard fourth-order Runge–Kutta procedure. This method has the advantage that one computes the whole function $\exp(u\boldsymbol{C})\boldsymbol{x}^\top$ for all values u within some interval.

Still another approach for the computation of the ruin function $\psi(u)$ is based on the numerical inversion of the Laplace transform $\hat{c}(z) = \int_0^\infty \mathrm{e}^{-zu} c(u)\, du$, where $c(u) = \rho^{-1}\psi(u)$ and $\Re z > 0$, see Section 5.5. In the compound Poisson model with phase-type distributed claim sizes, $\hat{c}(z)$ can be given in closed form and, consequently, (5.5.14) can be applied to this case. Indeed, using Theorems 8.2.3 and 8.3.1 it is easily seen that $c(u)$ is the tail function of $F = \mathrm{PH}(\boldsymbol{\alpha}^{\mathrm{s}}, \boldsymbol{B} + \rho\boldsymbol{b}^\top\boldsymbol{\alpha}^{\mathrm{s}})$. By an integration by parts we find

$$\hat{c}(z) = z^{-1} + z^{-1}\int_0^\infty \mathrm{e}^{-zu}\,\mathrm{d}c(u) = z^{-1}(1 - \hat{l}_F(z))$$

and hence, by (8.2.26) we have

$$\hat{c}(z) = z^{-1}(1 - \boldsymbol{\alpha}^{\mathrm{s}}(z\boldsymbol{I} - \boldsymbol{B} - \rho\boldsymbol{b}^\top\boldsymbol{\alpha}^{\mathrm{s}})\boldsymbol{b}^\top)\,. \tag{8.3.10}$$

We can now use the approximation formula (5.5.14) when inverting the Laplace transform $\hat{c}(z)$ given in (8.3.10). The results of a numerical experiment for the example given below are included in Table 8.3.2 in the column called Euler.

Nowadays, the computation of the inverse matrix $\boldsymbol{B}^{-1}$ and the matrix exponential function $\exp(u\boldsymbol{C})$ can usually be done painlessly by standard software, as for example MATHEMATICA, MAPLE or MATLAB. For example, in the numerical experiment discussed below, the computation of $\exp(u\boldsymbol{C})$ by MAPLE led to the same values of $\psi(u)$ as those given in the first column of Table 8.3.2.

Example Let $\rho = 0.75$, and $F_U = \mathrm{PH}(\boldsymbol{\alpha}, \boldsymbol{B})$ with $\ell = 4$,

$$\boldsymbol{\alpha} = (0.9731, 0.0152, 0.0106, 0.0010),$$

$$\boldsymbol{B} = \begin{pmatrix} -28.648 & 28.532 & 0.089 & 0.027 \\ 0.102 & -8.255 & 8.063 & 0.086 \\ 0.133 & 0.107 & -5.807 & 5.296 \\ 0.100 & 0.102 & 0.111 & -2.176 \end{pmatrix}.$$

The eigenvalues of the subintensity matrices $\boldsymbol{B}, \boldsymbol{C}$ are distinct and their numerical values are given in Table 8.3.1. Furthermore, for $\mathbb{E}\,U = -\boldsymbol{\alpha}\boldsymbol{B}^{-1}\boldsymbol{e}^\top$ and $\boldsymbol{b} = (b_1, \ldots, b_4)$ we have

$$\mathbb{E}\,U = 0.888479\,, \qquad \boldsymbol{b} = (0.0, 0.004, 0.271, 1.863)\,.$$

Finally, using (8.3.4), the ruin function $\psi(u)$ has been computed by the diagonalization method and the results are presented in the first column of Table 8.3.2 called "Diagonalization".

i	$\theta_i(\boldsymbol{B})$	$\theta_i(\boldsymbol{C})$
1	−28.73487884	−28.736208960
2	−8.397648524	$-7.462913649 + i\,0.511062594$
3	−6.090958978	$-7.462913649 - i\,0.511062594$
4	−1.662513676	−0.379909177

Table 8.3.1 The eigenvalues

u	Diagonalization	Maple	Euler
0.1	0.728 043 6176	0.728 043 6171	0.728 041 3240
0.3	0.680 721 2139	0.680 721 2139	0.680 713 0932
0.5	0.632 869 6427	0.632 869 6429	0.632 853 8609
1.0	0.524 073 3050	0.524 073 3051	0.524 029 7168
2.0	0.358 447 3675	0.358 447 3678	0.358 310 3419
3.0	0.245 150 6038	0.245 150 6049	0.244 872 2425
4.0	0.167 664 2644	0.167 664 2654	0.167 202 5238
5.0	0.114 669 5343	0.114 669 5351	0.113 989 3001
6.0	0.078 425 1920	0.078 425 1924	0.077 498 0890

Table 8.3.2 The ruin function $\psi(u)$

Bibliographical Notes. Another proof of Theorem 8.3.1 can be found in Neuts (1981) with the interpretation that the ruin function $\psi(u)$ of the compound Poisson model can be seen as the tail function of the stationary waiting time distribution in the M/GI/1 queue. Theorem 8.3.2 extends related results which have been derived in Dickson (1992), Dickson and Waters (1992), Dufresne and Gerber (1988) and Gerber, Goovaerts and Kaas (1987), for example, for special phase-type distributions, in particular for hyperexponential and Erlang distributions. For Runge–Kutta procedures concerning the solution of linear systems of ordinary differential equations we refer to Press, Flannery, Teukolsky and Vetterling (1988). Further numerical methods to compute a matrix exponential function can be found in Moler and van Loan (1978). The method for computing the ruin function $\psi(u)$ by numerical inversion of its Laplace transform has been stated in Section 5.5. For further results concerning the numerical computation of the ruin function $\psi(u)$, see also Asmussen and Rolski (1991).

8.4 NONHOMOGENEOUS MARKOV PROCESSES

Life and pension insurance modelling require stochastic processes $\{X(t), t \ge 0\}$, for which the future evolution of the process after time t depends on the state $X(t) = x$ and also on time t. Therefore in this section we outline the theory of nonhomogeneous Markov processes. In order to gain more intuition we first consider an example. Let $T \ge 0$ be the lifetime of an insured. If T is exponentially distributed with parameter $\lambda > 0$, then the process $\{X(t),\ t \ge 0\}$ defined by

$$X(t) = \begin{cases} 1 & \text{if } t < T, \\ 2 & \text{if } t \ge T \end{cases} \tag{8.4.1}$$

is clearly a homogeneous Markov process with intensity matrix

$$\boldsymbol{Q} = \begin{pmatrix} 0 & 0 \\ \lambda & -\lambda \end{pmatrix}.$$

If $\mathbb{P}(T > x) = \exp\left(-\int_0^x m(v)\,\mathrm{d}v\right)$, where $m(t)$ is a hazard rate function, then the stochastic process $\{X(t)\}$ defined in (8.4.1) still fulfils the Markov property. The reader can show that, indeed, for all $n \ge 1$, $i_0, i_1, \ldots, i_n \in \{1,2\}$ and $0 \le t_1 \le \ldots \le t_n$,

$$\begin{aligned} &\mathbb{P}(X(t_n) = i_n \mid X(t_{n-1}) = i_{n-1}, \ldots, X(t_1) = i_1, X(0) = i_0) \\ &= \mathbb{P}(X(t_n) = i_n \mid X(t_{n-1}) = i_{n-1}), \end{aligned} \tag{8.4.2}$$

whenever $\mathbb{P}(X(t_{n-1}) = i_{n-1}, \ldots, X(t_1) = i_1, X(0) = i_0) > 0$. However, in general the transition probabilities $\mathbb{P}(X(t_n) = i_n \mid X(t_{n-1}) = i_{n-1})$ depend on the pair (t_{n-1}, t_n) and not just on the difference $t_n - t_{n-1}$, as was the case of a homogeneous Markov process. Throughout this section we consider Markov processes with the finite state space $E = \{1, \ldots, \ell\}$.

8.4.1 Definition and Basic Properties

The homogeneous Markov processes discussed in the preceding sections of this chapter form a special case of the following class of *nonhomogeneous Markov processes.* Consider a family of stochastic matrices $\boldsymbol{P}(t,t') = (p_{ij}(t,t'))_{i,j\in E}$ where $0 \le t \le t'$, fulfilling

- $\boldsymbol{P}(t,t) = \boldsymbol{I}$ for all $t \ge 0$, and
- for all $0 \le t \le v \le t'$,

$$\boldsymbol{P}(t,t') = \boldsymbol{P}(t,v)\boldsymbol{P}(v,t') \,. \tag{8.4.3}$$

Each family of such stochastic matrices $\{\boldsymbol{P}(t,t'), 0 \le t \le t'\}$ is said to be a (nonhomogeneous) *matrix transition function.* Referring to (7.1.7) and (8.1.1), the matrix identity (8.4.3) is also called the *Chapman–Kolmogorov equation.*

Definition 8.4.1 *An E-valued stochastic process $\{X(t), t \ge 0\}$ is called a nonhomogeneous Markov process if there exist a (nonhomogeneous) matrix transition function $\{\boldsymbol{P}(t,t'), 0 \le t \le t'\}$ and a probability function $\boldsymbol{\alpha} = (\alpha_1, \alpha_2, \ldots, \alpha_\ell)$ on E such that*

$$\begin{aligned}&\mathbb{P}(X(0) = i_0, X(t_1) = i_1, \ldots, X(t_n) = i_n) \\ &\quad = \alpha_{i_0} p_{i_0 i_1}(0,t_1) p_{i_1 i_2}(t_1,t_2) \ldots p_{i_{n-1} i_n}(t_{n-1},t_n)\,,\end{aligned} \tag{8.4.4}$$

for all $n = 0, 1, \ldots,$ $i_0, i_1, \ldots, i_n \in E$, $0 \le t_1 \le \ldots \le t_n$.

Similarly to the characterization in Theorem 8.1.1 for homogeneous Markov processes, we have the following result.

Theorem 8.4.1 *The E-valued stochastic process $\{X(t), t \ge 0\}$ is a nonhomogeneous Markov process if and only if there exists a matrix transition function $\{\boldsymbol{P}(t,t'),\ 0 \le t \le t'\}$ such that, for all $n \ge 1$, $i_0, i_1, \ldots, i_n \in E$ and $0 \le t_1 \le \ldots \le t_n$,*

$$\begin{aligned}&\mathbb{P}(X(t_n) = i_n \mid X(t_{n-1}) = i_{n-1}, \ldots, X(t_1) = i_1, X(0) = i_0) \\ &\quad = p_{i_{n-1},i_n}(t_{n-1},t_n)\,,\end{aligned} \tag{8.4.5}$$

whenever $\mathbb{P}(X(t_{n-1}) = i_{n-1}, \ldots, X(t_1) = i_1, X(0) = i_0) > 0$.

The *proof* is similar to that of Theorems 7.1.1 and 8.1.1.

In this section we assume that the matrix transition function $\{\boldsymbol{P}(t,t')\}$ is continuous at t for all $t \ge 0$, that is $\lim_{t'\downarrow 0} \boldsymbol{P}(0,t') = \boldsymbol{I}$ and

$$\lim_{t'\downarrow t} \boldsymbol{P}(t,t') = \lim_{t'\uparrow t} \boldsymbol{P}(t',t) = \boldsymbol{I} \tag{8.4.6}$$

for $t > 0$. We also assume that the limits

$$\boldsymbol{Q}(t) = \lim_{t'\downarrow t} \frac{\boldsymbol{P}(t,t') - \boldsymbol{I}}{t' - t} = \lim_{t'\uparrow t} \frac{\boldsymbol{P}(t',t) - \boldsymbol{I}}{t - t'} \tag{8.4.7}$$

exist for each $t > 0$ with the exception of a set of Lebesgue measure zero. Recall that all countable sets are of Lebesgue measure zero. On this set of exceptional points we put $\boldsymbol{Q}(t) = \mathbf{0}$. By inspection we check that, for $i \neq j$, $q_{ij}(t) \geq 0$, $q_{ii}(t) \leq 0$ and for all $i \in E$ and $t \geq 0$,

$$\sum_{j \in E} q_{ij}(t) = 0 \,. \tag{8.4.8}$$

Again $\{\boldsymbol{Q}(t), t \geq 0\}$ is called the *matrix intensity function* of $\{X(t)\}$.

In the nonhomogeneous case, problems can arise. We give an example that shows that the limits in (8.4.7) need not to exist for each $t > 0$. Consider the two-state Markov process defined in (8.4.1). Then, we have

$$p_{12}(t,t') = \mathbb{P}(T \leq t' \mid T > t) = 1 - \exp\Big(-\int_t^{t'} m(v)\,\mathrm{d}v\Big)\,, \qquad t' \geq t$$

and hence $q_{12}(t) = \lim_{t' \to t}(t'-t)^{-1} p_{12}(t,t') = m(t)$ requires that the hazard rate function $m(t)$ is continuous at t. Similarly we can prove that

$$\boldsymbol{Q}(t) = \begin{pmatrix} 0 & 0 \\ m(t) & -m(t) \end{pmatrix},$$

for each continuity point t of $m(t)$ and $\mathbf{0}$ otherwise.

Theorem 8.4.2 *For all $i, j \in E$, $0 \leq t < t'$ for which the limits in (8.4.7) exist, the partial derivatives $\partial/(\partial t) p_{ij}(t,t')$ and $\partial/(\partial t') p_{ij}(t,t')$ exist and satisfy the following systems of differential equations:*

$$\frac{\partial}{\partial t} p_{ij}(t,t') = -\sum_{k \in E} q_{ik}(t) p_{kj}(t,t') \tag{8.4.9}$$

and

$$\frac{\partial}{\partial t'} p_{ij}(t,t') = \sum_{k \in E} p_{ik}(t,t') q_{kj}(t') \,. \tag{8.4.10}$$

Proof Let $h > 0$ such that $t + h \leq t'$. From (8.4.3) we have

$$\begin{aligned} p_{ij}(t+h,t') - p_{ij}(t,t') &= p_{ij}(t+h,t') - \sum_{k \in E} p_{ik}(t,t+h) p_{kj}(t+h,t') \\ &= p_{ij}(t+h,t')(1 - p_{ii}(t,t+h)) - \sum_{k \neq i} p_{ik}(t,t+h) p_{kj}(t+h,t') \,. \end{aligned}$$

Using the continuity of $p_{ij}(t,t')$, we obtain

$$\lim_{h \downarrow 0} h^{-1} \left(p_{ij}(t+h,t') - p_{ij}(t,t')\right) = -\sum_{k \in E} q_{ik}(t) p_{kj}(t,t') \,.$$

In the same way,

$$\lim_{h\uparrow 0} h^{-1}\left(p_{ij}(t-h,t') - p_{ij}(t,t')\right) = -\sum_{k\in E} q_{ik}(t)p_{kj}(t,t')\,.$$

This gives (8.4.9). The proof of (8.4.10) is analogous. □

The terminology, used in the homogeneous case, becomes more transparent when the differential equations (8.4.9) and (8.4.10) are called the *Kolmogorov backward equations* and the *Kolmogorov forward equations*, respectively. In matrix notation, the equations take the form

$$\frac{\partial}{\partial t}\boldsymbol{P}(t,t') = -\boldsymbol{Q}(t)\boldsymbol{P}(t,t') \tag{8.4.11}$$

and

$$\frac{\partial}{\partial t'}\boldsymbol{P}(t,t') = \boldsymbol{P}(t,t')\boldsymbol{Q}(t') \tag{8.4.12}$$

with the boundary condition $\boldsymbol{P}(t,t) = \boldsymbol{I}$ for all $t \geq 0$. We can integrate the differential equations (8.4.11) and (8.4.12) to obtain the following result.

Theorem 8.4.3 *Suppose that $\{\boldsymbol{Q}(t), t \geq 0\}$ is measurable and that the function $\{\max_{1\leq i\leq \ell} |q_{ii}(t)|, t > 0\}$ is integrable on every finite interval in $\mathbb{R}_+$. Then, for all $0 \leq t < t'$, the matrix transition function $\{\boldsymbol{P}(t,t')\}$ satisfies the integral equations*

$$\boldsymbol{P}(t,t') = \boldsymbol{I} + \int_t^{t'} \boldsymbol{Q}(v)\boldsymbol{P}(v,t')\,\mathrm{d}v \tag{8.4.13}$$

and

$$\boldsymbol{P}(t,t') = \boldsymbol{I} + \int_t^{t'} \boldsymbol{P}(t,v)\boldsymbol{Q}(v)\,\mathrm{d}v\,. \tag{8.4.14}$$

Proof Let $t' \geq 0$ be fixed. Using the fact that, for $0 \leq t \leq t'$,

$$\boldsymbol{P}(t,t') = \boldsymbol{I} - \int_t^{t'} \frac{\partial}{\partial v}\boldsymbol{P}(v,t')\,\mathrm{d}v\,,$$

(8.4.13) is obtained from (8.4.11). The proof of (8.4.14) is analogous. □

Note that the matrix intensity function $\{\boldsymbol{Q}(t), t \geq 0\}$ fulfils the condition of Theorem 8.4.3 if, for instance, it is piecewise continuous and locally bounded. Relations (8.4.13) and (8.4.14) can be used to express the transition function $\{\boldsymbol{P}(t,t'), 0 \leq t < t'\}$ by the matrix intensity function $\{\boldsymbol{Q}(t), t \geq 0\}$, showing a one-to-one correspondence between the matrix transition function and the matrix intensity function of nonhomogeneous Markov processes with finite state space. We need the following auxiliary result, where we put $v_0 = t$.

Lemma 8.4.1 *Let $q : \mathbb{R}_+ \to \mathbb{R}$ be a measurable function which is integrable on every finite interval in $\mathbb{R}_+$. Then, for all $0 \le t \le t' < \infty$, $k = 1, 2, \ldots$,*

$$\int_t^{t'} \int_{v_1}^{t'} \cdots \int_{v_{k-1}}^{t'} q(v_1) \ldots q(v_k) \, dv_k \ldots dv_1 = \frac{(\int_t^{t'} q(v) \, dv)^k}{k!} \, . \tag{8.4.15}$$

Proof Note that

$$\begin{aligned}\Big(\int_t^{t'} q(v) \, dv\Big)^k &= \int_t^{t'} \cdots \int_t^{t'} q(v_1) \ldots q(v_n) \, dv_n \ldots dv_1 \\ &= \sum_{(i_1,\ldots,i_k)} \int_t^{t'} \cdots \int_t^{t'} \mathbb{I}(v_{i_1} \le v_{i_2} \le \ldots \le v_{i_k}) q(v_1) \ldots q(v_n) \, dv_n \ldots dv_1 \, ,\end{aligned}$$

where the summation is over all permutations $(i_1, \ldots, i_k)$ of $(1, \ldots, k)$. This in turn gives

$$\begin{aligned}&\Big(\int_t^{t'} q(v) \, dv\Big)^k \\ &= k! \int_t^{t'} \cdots \int_t^{t'} \mathbb{I}(v_1 \le v_2 \le \ldots \le v_k) q(v_1) \ldots q(v_n) \, dv_n \ldots dv_1 \, ,\end{aligned}$$

because all the summands in the above sum coincide. □

Theorem 8.4.4 *Under the assumptions of Theorem 8.4.3 we have, for $0 \le t \le t'$,*

$$\boldsymbol{P}(t, t') = \boldsymbol{I} + \sum_{n=1}^{\infty} \int_t^{t'} \int_{v_1}^{t'} \cdots \int_{v_{n-1}}^{t'} \boldsymbol{Q}(v_1) \ldots \boldsymbol{Q}(v_n) \, dv_n \ldots dv_1 \tag{8.4.16}$$

and alternatively

$$\boldsymbol{P}(t, t') = \boldsymbol{I} + \sum_{n=1}^{\infty} \int_t^{t'} \int_t^{v_1} \cdots \int_t^{v_{n-1}} \boldsymbol{Q}(v_1) \ldots \boldsymbol{Q}(v_n) \, dv_n \ldots dv_1 \, . \tag{8.4.17}$$

Proof Inserting (8.4.13) into the right-hand side of this equation yields

$$\begin{aligned}\boldsymbol{P}(t, t') &= \boldsymbol{I} + \int_t^{t'} \boldsymbol{Q}(v_1) \Big(\boldsymbol{I} + \int_{v_1}^{t'} \boldsymbol{Q}(v_2) \boldsymbol{P}(v_2, t') \, dv_2\Big) dv_1 \\ &= \boldsymbol{I} + \int_t^{t'} \boldsymbol{Q}(v) \, dv + \int_t^{t'} \int_{v_1}^{t'} \boldsymbol{Q}(v_1) \boldsymbol{Q}(v_2) \boldsymbol{P}(v_2, t') \, dv_2 dv_1 \, .\end{aligned}$$

By iteration we get, for arbitrary $k \geq 2$,

$$\begin{aligned} \boldsymbol{P}(t,t') &= \boldsymbol{I} + \int_t^{t'} \boldsymbol{Q}(v)\,\mathrm{d}v \\ &\quad + \sum_{n=2}^{k} \int_t^{t'} \int_{v_1}^{t'} \cdots \int_{v_{n-1}}^{t'} \boldsymbol{Q}(v_1)\ldots\boldsymbol{Q}(v_n)\,\mathrm{d}v_n\ldots\mathrm{d}v_1 \\ &\quad + \int_t^{t'} \int_{v_1}^{t'} \cdots \int_{v_{k-1}}^{t'} \boldsymbol{Q}(v_1)\ldots\boldsymbol{Q}(v_k)\boldsymbol{P}(v_k,t')\,\mathrm{d}v_k\ldots\mathrm{d}v_1 \,. \end{aligned}$$

Put $q(t) = \max_{1\leq i\leq \ell} |q_{ii}(t)|$. To complete the proof of (8.4.16), it suffices to observe that

$$\begin{aligned} &\left|\left(\int_t^{t'} \int_{v_1}^{t'} \cdots \int_{v_{k-1}}^{t'} \boldsymbol{Q}(v_1)\ldots\boldsymbol{Q}(v_k)\boldsymbol{P}(v_k,t')\,\mathrm{d}v_k\ldots\mathrm{d}v_1\right)_{ij}\right| \\ &\leq \ell^k \int_t^{t'} \int_{v_1}^{t'} \cdots \int_{v_{k-1}}^{t'} q(v_1)\ldots q(v_k)\,\mathrm{d}v_k\ldots\mathrm{d}v_1 \\ &= \frac{\ell(\int_t^{t'} q(v)\,\mathrm{d}v)^k}{k!} \to_{k\to\infty} 0 \,, \end{aligned}$$

where the last equality follows from Lemma 8.4.1. □

We still mention another property of the matrix transition function $\{\boldsymbol{P}(t,t')\}$. Under the assumptions of Theorem 8.4.3, the limit

$$\lim_{n\to\infty} \sum_{i=1}^{n} (\boldsymbol{P}(v_{i-1}^{(n)}, v_i^{(n)}) - \boldsymbol{I}) = \boldsymbol{A}(t,t') \tag{8.4.18}$$

exists for each sequence $\{(v_0^{(n)},\ldots,v_n^{(n)})\}$ such that $t = v_0^{(n)} \leq v_1^{(n)} \leq \ldots \leq v_n^{(n)} = t'$ and $\max_{1\leq i\leq n}\{v_i^{(n)} - v_{i-1}^{(n)}\} \to_{n\to\infty} 0$, where the limit $\boldsymbol{A}(t,t')$ does not depend on the particular choice of the sequence $\{(v_0^{(n)},\ldots,v_n^{(n)})\}$ of partitions of the interval $[t,t']$, and

$$\boldsymbol{A}(t,t') = \int_t^{t'} \boldsymbol{Q}(v)\,\mathrm{d}v \,. \tag{8.4.19}$$

8.4.2 Construction of Nonhomogeneous Markov Processes

Let $\boldsymbol{\alpha}$ be an initial distribution and $\{\boldsymbol{Q}(t), t \geq 0\}$ a measurable matrix intensity function. Assume that the function $\{q(t), t \geq 0\}$, where $q(t) = \max_{1\leq i\leq \ell} |q_{ii}(t)|$ is integrable on every finite interval in $\mathbb{R}_+$. Our goal now is to construct a nonhomogeneous Markov process with the state space $E =$

$\{1, 2, \ldots, \ell\}$, initial distribution $\boldsymbol{\alpha}$, and matrix intensity function $\{\boldsymbol{Q}(t), t \geq 0\}$. We outline a construction principle which is similar to that discussed in Section 8.1.3 for the homogeneous case. It can be used for *simulation of nonhomogeneous Markov processes* with a preassigned matrix intensity function.

We define the family of stochastic matrices $\{\boldsymbol{P}^\circ(t), t \geq 0\}$ setting

$$p^\circ_{ij}(t) = \begin{cases} (1 - \delta(i,j))q_{ij}(t)/q_i(t) & \text{if } q_i(t) > 0, \\ \delta_{ij} & \text{if } q_i(t) = 0, \end{cases} \tag{8.4.20}$$

where $q_i(t) = \sum_{j \neq i} q_{ij}(t)$. The nonhomogeneous Markov process $\{X(t), t \geq 0\}$ with initial distribution $\boldsymbol{\alpha}$ and intensity function $\{\boldsymbol{Q}(t), t \geq 0\}$, whose construction will be given below, has the form

$$X(t) = \sum_{n=0}^{\infty} X(\sigma_n)\mathbb{I}(\sigma_n \leq t < \sigma_{n+1}).$$

The jumps times σ_n of the process $\{X(t),\ t \geq 0\}$ and its states $X(\sigma_n)$ at the jump times are given by the following algorithm.

Step 1 Let $X(0)$ be an E-valued random variable with distribution $\boldsymbol{\alpha}$.

Step 2 If $X(0) = i_0$, then the sojourn time Z_0 in state i_0 of the process $\{X(t), t \geq 0\}$ has the conditional (possibly defective) distribution function

$$F^{[0]}_{i_0}(t) = 1 - \exp\Big(-\int_0^t q_{i_0}(v)\,\mathrm{d}v\Big), \tag{8.4.21}$$

where we put $\sigma_1 = Z_0$ and $X(t) = X(0)$ for $\sigma_0 = 0 \leq t < \sigma_1$.

Step 3 If $X(0) = i_0$ and $\sigma_1 = t_1 < \infty$, the process assumes state i_1 at time t_1 with probability $p^\circ_{i_0,i_1}(t_1)$, which gives the new state $X(\sigma_1)$ of $\{X(t)\}$ at jump time σ_1.

Step 4 (analogous to Step 2) If $\sigma_1 = t_1$ and $X(\sigma_1) = i_1$, then the sojourn time Z_1 in state i_1 has the conditional (possibly defective) distribution function $F^{[1]}_{i_1}(t) = 1 - \exp\big(-\int_{t_1}^{t_1+t} q(v, i_1)\,\mathrm{d}v\big)$, where we put $\sigma_2 = \sigma_1 + Z_1$ and $X(t) = X(\sigma_1)$ for $\sigma_1 = 0 \leq t < \sigma_2$.

Step 5 (analogous to Step 3) If $X(\sigma_1) = i_1$ and $\sigma_2 = t_2 < \infty$, the process jumps to state i_2 with probability $p^\circ_{i_1,i_2}(t_2)$, which gives the state $X(\sigma_2)$ of $\{X(t)\}$ at time σ_2.

$\vdots$

Following this construction, we define the sample paths of $\{X(t), t \geq 0\}$ on the whole nonnegative half-line $\mathbb{R}_+$, because

$$\mathbb{P}(\lim_{n\to\infty} \sigma_n = \infty) = 1, \tag{8.4.22}$$

as can be proved by the reader.

Theorem 8.4.5 *The stochastic process* $\{X(t), t \geq 0\}$ *constructed above is a nonhomogeneous Markov process.*

The *proof* is omitted. It can be found, for example, in Iosifescu and Tautu (1973).

Example A nonhomogeneous Markov process $\{X(t), t \geq 0\}$ with state space $E = \{1, \ldots, \ell\}$ is called a *nonhomogeneous birth-and-death process* if $p^\circ_{i,i-1}(t) + p^\circ_{i,i+1}(t) = 1$ and $p^\circ_{12}(t) = p^\circ_{\ell,\ell-1}(t) = 1$ for all $1 < i < \ell$, $t \geq 0$. The products $p^\circ_{i,i+1}(t)q(t,i)$ and $p^\circ_{i,i-1}(t)q(t,i)$ are called the *birth rate* and *death rate* in state i at time t, respectively.

We leave it to the reader to show that the Markov process $\{X(t), t \geq 0\}$ so constructed is the "right" one, i.e. its matrix intensity function equals the preassigned matrix intensity function $\{\boldsymbol{Q}(t), t \geq 0\}$.

8.4.3 Application to Life and Pension Insurance

We first review two basic economic factors: interest and discounting in continuous time. Suppose that the unit of time is one year and that the annual interest rate is r_1. If the interest were to be paid once per year, then the value of one monetary unit after k years would be equal to $(1 + r_1)^k$. Analogously, if the interest is paid n times per year and the annual interest rate is equal to r_n, then the value of one monetary unit after k payments of interest is equal to $(1+r_n/n)^k$. Now, letting k and n go to infinity in such a way that $k/n \to t$ and $r_n \to \delta$, then the value of one monetary unit at time t is equal to $\mathrm{e}^{\delta t}$. The value δ is called the *force of interest.* Thus, if δ is a force of interest and r is an annual interest rate which give the same value of one monetary unit after one year, then δ and r are related by $\delta = \log(1 + r)$. Conversely, the present value at time 0 of one monetary unit at time t is equal to $v(t) = \mathrm{e}^{-\delta t}$, which is called a *discount factor* in the case of continuous discounting.

Note that the above argument remains valid if a *time-dependent force of interest* $\delta(t)$ is considered. Assume that the function $\delta(t)$ is Riemann integrable and approximate it by a piecewise constant function which is equal to $\delta(j/n)$ for all $t \in [(j-1)/n, j/n)$. If k and n go to infinity so that $k/n \to t$, one can see as before that now the value of one monetary unit at time t is equal to $\exp(\int_0^t \delta(x)\,\mathrm{d}x)$. The subsequent discount factor is then $v(t) = \exp(-\int_0^t \delta(x)\,\mathrm{d}x)$. A formal proof of these facts is left to the reader.

Examples 1. We begin with the simplest *life insurance model*, considering a single life and only one cause of death. Suppose that the life time of an insured (after policy issue) is modelled by a random variable T with density function $f(t)$ and hazard rate function $m(t)$. In the introduction to the present Section 8.4 and just preceding Theorem 8.4.2, this model has been formulated in terms of a nonhomogeneous Markov process $\{X(t)\}$ with state space $E = \{1,2\}$. Assume that during his life the insured pays premiums at constant rate $\beta_1(t) \equiv \beta$ and that discounting is based on a constant force of interest $\delta > 0$. Assume further that the insurer provides a lump payment of amount $b_{12}(t) \equiv 1$ if the policy changes from state 1 (alive) to state 2 (dead). The *net prospective premium reserve* $\mu_1(t)$ at time t after issue is defined as the expected discounted value at that time of the subsequent benefits minus all future premiums payable until the policy changes from state 1 to state 2, i.e.

$$\mu_1(t) = \mathbb{E}\,(\mathrm{e}^{-\delta(T-t)} \mid T > t) - \beta\mathbb{E}\left(\int_0^{T-t} \mathrm{e}^{-\delta x}\,\mathrm{d}x \,\Big|\, T > t\right). \tag{8.4.23}$$

We leave it to the reader to show that the function $\mu_1(t)$ defined in (8.4.23) satisfies *Thiele's differential equation*

$$\frac{\mathrm{d}\mu_1(t)}{\mathrm{d}t} = \beta + \delta\mu_1(t) - m(t)(1 - \mu_1(t)). \tag{8.4.24}$$

If the equation $\mu_1(0) = 0$ can be solved for the net premium rate β, then (8.4.23) determines a premium calculation principle based on the reserve function $\mu_1(t)$. We can represent $\mu_1(t)$ in terms of the underlying nonhomogeneous Markov process $\{X(t)\}$ with two states $\{1,2\}$. Here, 2 is an absorbing state, $q_{12}(x) = m(x)$ and $p_{11}(t,x) = \exp(-\int_t^x m(y)\,\mathrm{d}y) = \overline{F}(x)/\overline{F}(t)$ for $x > t$. Note that the condition $T > t$ means that $X(t) = 1$. Thus, (8.4.23) can be rewritten in the form

$$\mu_1(t) = \frac{\int_t^\infty \mathrm{e}^{-\delta(x-t)} f(x)\,\mathrm{d}x}{\overline{F}(t)} - \beta\frac{\int_t^\infty (\int_0^{x-t} \mathrm{e}^{-\delta y}\,\mathrm{d}y) f(x)\,\mathrm{d}x}{\overline{F}(t)}$$

and consequently

$$\mu_1(t) = \int_t^\infty \mathrm{e}^{-\delta(x-t)} \left(p_{11}(t,x)q_{12}(x) - \beta p_{11}(t,x)\right) \mathrm{d}x\,. \tag{8.4.25}$$

The net prospective premium reserve $\mu_2(t)$ when the policy is in state 2 at time t is $\mu_2(t) \equiv 0$.

2. The above example from life insurance can be modified in the following way. Death occurs when the underlying process $\{X(t)\}$ passes from state 1 to state 2. Rather than paying one lump sum at the time of death, a *family income insurance* provides a continuous payment of one monetary unit per time unit, lasting from the instance of death till a (fixed) time w. As before, premiums are paid at a constant rate β but not longer than over a period w', during survival of the insured. The quantities w and w' are supposed to be

settled at the time the insurance contract is signed. In this case, the benefit rate function $b_2(t)$ in state 2 and the premium rate function $\beta_1(t)$ in state 1 are given by

$$b_2(x) = \begin{cases} 1 & \text{if } 0 \le x \le w, \\ 0 & \text{if } x > w, \end{cases} \qquad \beta_1(x) = \begin{cases} \beta & \text{if } 0 \le x \le w', \\ 0 & \text{if } x > w'. \end{cases}$$

Then

$$\begin{aligned} \mu_1(t) &= \mathbb{E}\left(\int_T^\infty e^{-\delta(x-t)} b_2(x)\, dx \,\middle|\, X(t) = 1\right) \\ &\quad - \mathbb{E}\left(\int_t^T e^{-\delta(x-t)} \beta_1(x)\, dx \,\middle|\, X(t) = 1\right) \end{aligned}$$

can be written in the form

$$\mu_1(t) = \int_t^\infty e^{-\delta(x-t)} \left(p_{12}(t,x)\, b_2(x) - p_{11}(t,x)\, \beta_1(x)\right) dx\,. \tag{8.4.26}$$

Furthermore

$$\mu_2(t) = \int_t^\infty e^{-\delta(x-t)} b_2(x)\, dx\,. \tag{8.4.27}$$

In particular, for $t \le w$,

$$\mu_2(t) = \int_t^w e^{-\delta(x-t)}\, dx = \delta^{-1}(1 - e^{-\delta(w-t)})\,. \tag{8.4.28}$$

3. Another example is the following insurance model with three states: 1-active, 2-disabled and 3-dead with possible transitions as depicted on Figure 8.4.1, where the matrix intensity function $\{\boldsymbol{Q}(t)\}$ is given by $q_{12}(t) =$

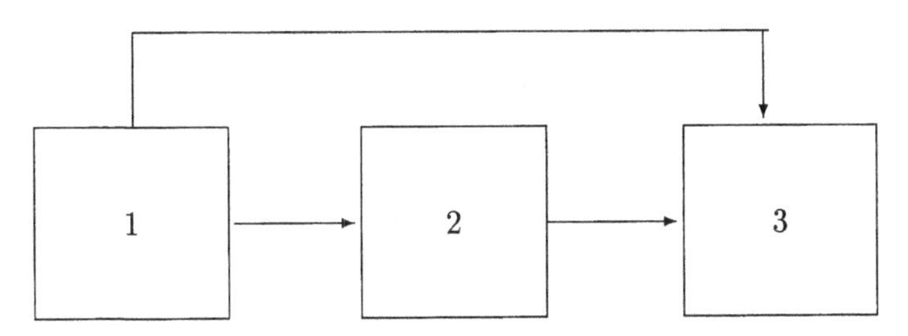

Figure 8.4.1 Transition graph

$a(t)$ and $q_{13}(t) = q_{23}(t) = b(t)$ for some nonnegative and continuous functions $a(t)$, $b(t)$. It is easily seen that the corresponding matrix transition function

$\{\boldsymbol{P}(t,t')\}$ has the entries

$$\begin{aligned}
p_{11}(t) &= \exp\Big(-\int_t^{t'} (a(y)+b(y))\,\mathrm{d}y\Big)\,,\\
p_{12}(t) &= \exp\Big(-\int_t^{t'} b(y)\,\mathrm{d}y\Big)\Big(1-\exp\Big(-\int_t^{t'} a(y)\,\mathrm{d}y\Big)\Big)\,,\\
p_{22}(t) &= \exp\Big(-\int_t^{t'} b(y)\,\mathrm{d}y\Big).
\end{aligned}$$

The above examples are special cases of a more general model which also includes a general life annuity payment, which is a kind of *pension insurance.* This general model allows two types of benefits: lump-type payments and continuous payments. Our concrete assumptions are the following:

- The state of the policy is described by a nonhomogeneous Markov process $\{X(t)\}$ with a finite state space $E=\{1,\ldots,\ell\}$; a subset E_{abs} of E describes the absorbing states; we assume that E_{abs} is nonempty and that the process $\{X(t)\}$ ultimately ends up in E_{abs}.
- The outgoing benefit for the transition $i\to j$ at time x (a lump payment) is $b_{ij}(x)$, and $\beta_j(x)$ and $b_j(x)$ are the premium rate and the annuity benefit rate in state j at time x, respectively; we will assume in this section that all the functions $\beta_j(x)$, $b_j(x)$ and $b_{ij}(t)$ are continuous and bounded.
- The discounting is based on a general time-dependent force of interest $\delta(t)$; we assume that $\delta(t)$ is bounded away from zero, i.e. $\delta(t)>\varepsilon>0$ for all $t\ge 0$ and some $\varepsilon>0$.

Let $N_{ij}(t)$ denote the number of transitions of the process $\{X(t)\}$ from state i to j by time t $(i\neq j)$. If all $q_{ij}(t)$ are bounded, the random variables $N_{ij}(t)$ are finite for all $t>0$. We therefore assume that all $q_{ij}(t)$ are bounded and moreover continuous. Thus the cumulative benefit up to time t is

$$B(t)=\int_0^t b_{X(y)}\,\mathrm{d}y+\sum_{i\neq j}\int_0^t b_{ij}(y)\,\mathrm{d}N_{ij}(y)$$

and the cumulative premium by time t is $\Pi(t)=\int_0^t \beta_{X(y)}\,\mathrm{d}y$. Further, the net prospective premium reserve $\mu_i(t)$ in state $i\in E$ at time $t\ge 0$ is

$$\mu_i(t)=\mathbb{E}\left(\int_t^\infty v(t,x)\,\mathrm{d}(B(x)-\Pi(x))\;\Big|\;X(t)=i\right),\tag{8.4.29}$$

where $v(t,x)=\exp(-\int_t^x \delta(y)\,\mathrm{d}y)$. If $i\in E_{\mathrm{abs}}$ is an absorbing state, one usually puts $\beta_i(x)\equiv 0$. In this case, (8.4.29) takes the form $\mu_i(t)=\int_t^\infty v(t,x)b_i(x)\,\mathrm{d}x$, generalizing (8.4.27). However, the following theorem shows that also for $i\in E\setminus E_{\mathrm{abs}}$ there is an alternative representation formula for $\mu_i(t)$ which generalizes (8.4.25) and (8.4.26).

Theorem 8.4.6 *For $i \in E$ and $t \geq 0$,*

$$\mu_i(t) = \sum_{j,k\in E,\, j\neq k} \int_t^\infty v(t,x) b_{jk}(x) p_{ij}(t,x) q_{jk}(x)\,\mathrm{d}x - \sum_{j\in E} \int_t^\infty v(t,x) p_{ij}(t,x)(\beta_j(x) - b_j(x))\,\mathrm{d}x\,. \qquad (8.4.30)$$

As the *proof* of Theorem 8.4.6 is based on martingale techniques, it will be given later in Chapter 11; see (11.2.12).

Notice that the premium reserve functions $\{\mu_i(t), i \in E, t \geq 0\}$ satisfy the following system of equations. Let $0 \leq t < t'$. Then,

$$\mu_i(t) = \int_t^{t'} v(t,x) \sum_{j\in E} p_{ij}(t,x)\Big((\beta_j(x) - b_j(x)) + \sum_{k\neq j} b_{jk}(x) q_{jk}(x)\Big)\,\mathrm{d}x + v(t,t') \sum_{j\in E} p_{ij}(t,t')\mu_j(t')\,, \qquad (8.4.31)$$

which is obtained from (8.4.30) by separating payments in $(t,t']$ from those in (t',∞), and by using the Chapman–Kolmogorov equations; see (8.4.3). Taking derivatives with respect to t on both sides of (8.4.31), we arrive at a generalization of Thiele's differential equation, see (8.4.24).

Theorem 8.4.7 *The premium reserve functions $\{\mu_i(t), i \in E, t \geq 0\}$ satisfy the following system of differential equations. For each $i \in E$,*

$$\frac{\mathrm{d}\mu_i(t)}{\mathrm{d}t} = (\delta(t) + q_i(t))\mu_i(t) - \sum_{j\neq i} q_{ij}(t)\mu_j(t) + (\beta_i(t) - b_i(t)) - \sum_{j\neq i} b_{ij}(t) q_{ij}(t)\,. \qquad (8.4.32)$$

The *proof* of this theorem is omitted. It can be found, for example, in Section 7.1 of Wolthuis (1994).

Example A married couple buys a combined life insurance and widow's pension policy specifying that premiums are to be paid at rate β as long as both husband and wife are alive; pensions are to be paid at rate b as long as the wife is widowed and a life insurance of amount c is paid immediately upon the death of the husband if the wife is already dead (as a benefit to their dependant). Assume that the force of interest is constant and equal to δ. The reader should write down the system of generalized Thiele's differential equations (8.4.32) for the net prospective premium reserves $\mu_i(t)$.

Note that in general the differential equations (8.4.32) do not help to calculate $\mu_i(t)$ because a boundary condition is needed. However, $\lim_{t\to\infty} \mu_i(t)$

does not even need to exist. However, if the contract has a finite expiration time w, then the boundary condition $\mu_i(w) = 0$ has to be satisfied. As there was no lump sum at w in our model, the boundedness of $b_i(t)$, $b_{ij}(t)$, $\beta_i(t)$ and $q_{ij}(t)$ imply $\lim_{t\uparrow w} \mu_i(t) = 0$. As no human being grows older than 200 years, we can safely assume that the contract expires after 200 years.

In general it is difficult to solve (8.4.32) analytically. However, the following *iteration procedure* leads to an approximate solution to (8.4.32). For each $t \geq 0$, let σ denote the next jump epoch after time t at which $\{X(t)\}$ changes its state. Put $\mu_i^0(t) \equiv 0$ and define $\mu_i^n(t)$ recursively by

$$\begin{aligned}\mu_i^n(t) &= \mathbb{E}\Big(\int_t^\sigma v(t,x)\sum_{j\in E} p_{ij}(t,x)\Big(\beta_j(x) - b_j(x) + \sum_{k\neq j} b_{jk}(x)q_{jk}(x)\Big)\,\mathrm{d}x \\ &\quad + v(t,\sigma)\sum_{\ell\in E} p_{i,\ell}(t,\sigma)\mu_\ell^{n-1}(\sigma)\;\Big|\; X(t) = i\Big)\,. \end{aligned} \tag{8.4.33}$$

Theorem 8.4.8 *For each $i \in E$ and $t \geq 0$,*

$$\lim_{n\to\infty} \mu_i^n(t) = \mu_i(t)\,. \tag{8.4.34}$$

Proof Denote by $\sigma_1 < \sigma_2 < \ldots$ the epochs after time t at which the state changes and let $\sigma_0 = t$. We show by induction that

$$\mu_i^n(t) = \mathbb{E}\Big(\int_t^{\sigma_n} v(t,x)\sum_{j\in E} p_{ij}(t,x)\Big(\beta_j(x) - b_j(x) + \sum_{k\neq j} b_{jk}(x)q_{jk}(x)\Big)\,\mathrm{d}x\Big)\,.$$

This clearly is true for $n = 0$. Assume the above equation holds for n. Then, using (8.4.3) and $v(t,\sigma_1)v(\sigma_1,x) = v(t,x)$ for $t \leq \sigma_1 \leq x$, we get

$$\begin{aligned}&\mu_i^{n+1}(t)\\ &= \mathbb{E}\Big(\int_t^{\sigma_1} v(t,x)\sum_{j\in E} p_{ij}(t,x)\Big(\beta_j(x) - b_j(x) + \sum_{k\neq j} b_{jk}(x)q_{jk}(x)\Big)\,\mathrm{d}x \\ &\qquad + v(t,\sigma_1)\sum_{\ell\in E} p_{i\ell}(t,\sigma_1)\mu_\ell^n(\sigma_1)\;\Big|\; X(t) = i\Big) \\ &= \mathbb{E}\Big(\int_t^{\sigma_1} v(t,x)\sum_{j\in E} p_{ij}(t,x)\Big(\beta_j(x) - b_j(x) + \sum_{k\neq j} b_{jk}(x)q_{jk}(x)\Big)\,\mathrm{d}x \\ &\qquad + v(t,\sigma_1)\sum_{\ell\in E} p_{i\ell}(t,\sigma_1)\mathbb{E}\Big(\int_{\sigma_1}^{\sigma_{n+1}} v(\sigma_1,x)\sum_{j\in E} p_{\ell j}(\sigma_1,x)\Big(\beta_j(x) - b_j(x) \\ &\qquad\qquad + \sum_{k\neq j} b_{jk}(x)q_{jk}(x)\Big)\,\mathrm{d}x\;\Big|\; X(\sigma_1) = \ell\Big)\;\Big|\; X(t) = i\Big) \\ &= \mathbb{E}\Big(\int_t^{\sigma_1} v(t,x)\sum_{j\in E} p_{ij}(t,x)\Big(\beta_j(x) - b_j(x) + \sum_{k\neq j} b_{jk}(x)q_{jk}(x)\Big)\,\mathrm{d}x\end{aligned}$$

$$
\begin{aligned}
& + \mathbb{E}\Big(\int_{\sigma_1}^{\sigma_{n+1}} v(t,\sigma_1)v(\sigma_1,x)\sum_{j\in E}\sum_{\ell\in E} p_{i\ell}(t,\sigma_1)p_{\ell j}(\sigma_1,x)\Big(\beta_j(x)-b_j(x) \\
&\qquad + \sum_{k\neq j} b_{jk}(x)q_{jk}(x)\Big)\,\mathrm{d}x \;\Big|\; X(\sigma_1)\Big) \;\Big|\; X(t)=i\Big) \\
= \;& \mathbb{E}\Big(\int_t^{\sigma_1} v(t,x)\sum_{j\in E} p_{ij}(t,x)\Big(\beta_j(x)-b_j(x)+\sum_{k\neq j} b_{jk}(x)q_{jk}(x)\Big)\,\mathrm{d}x \\
& + \int_{\sigma_1}^{\sigma_{n+1}} v(t,x)\sum_{j\in E} p_{ij}(t,x)\Big(\beta_j(x)-b_j(x) \\
&\qquad + \sum_{k\neq j} b_{jk}(x)q_{jk}(x)\Big)\,\mathrm{d}x \;\Big|\; X(t)=i\Big) \\
= \;& \mathbb{E}\Big(\int_t^{\sigma_{n+1}} v(t,x)\sum_{j\in E} p_{ij}(t,x)\Big(\beta_j(x)-b_j(x)+\sum_{k\neq j} b_{jk}(x)q_{jk}(x)\Big)\,\mathrm{d}x\Big)\,.
\end{aligned}
$$

Because $v(t,x) \to \infty$ as $x \to \infty$ and $\sigma_n \to \infty$ as $n \to \infty$, the assertion follows from (8.4.31) by bounded convergence. □

Bibliographical Notes. Further details and results on nonhomogeneous Markov processes with continuous transition functions, including the case of a general (infinite) state space, can be found, for example, in Iosifescu (1980) and Iosifescu and Tautu (1973). The speed of convergence of the series of *product integrals* in Theorem 8.4.4 has been investigated in Møller (1992). For nonhomogeneous Markov processes with discontinuous transition functions, the theory of product-integration and its application in insurance mathematics has been surveyed in Gill and Johansen (1990); see also Helbig and Milbrodt (1998). The original proof of (8.4.19) has been given in Dobrushin (1953). One of the first papers where nonhomogeneous Markov processes have been applied to problems of life insurance is Hoem (1969). Here, the central result is Thiele's differential equation (8.4.24) for the net prospective premium reserve and goes back to Thiele; see Hoem (1983). For generalizations in different directions, see, for example, Milbrodt and Stracke (1997), Møller (1993), Norberg (1991, 1992, 1995), Norberg and Møller (1996) and Ramlau-Hansen (1990). The exposition of Section 8.4.3 partially follows Wolthuis (1994), where further examples of the general person insurance model discussed in Section 8.4.3 (like widow's pension, disability annuity, AIDS models) can be found.

8.5 MIXED POISSON PROCESSES

In this section we give a thorough treatment of mixed Poisson processes which have found applications in insurance mathematics because of their flexibility. We show in particular that a mixed Poisson process can be represented as a pure birth process, a special type of nonhomogeneous Markov processes. However, mixed Poisson processes are interesting in their own right.

8.5.1 Definition and Elementary Properties

Consider a counting process $\{N(t),\ t \ge 0\}$. For example, the random variable $N(t)$ can be interpreted as the number of claims arriving within a portfolio up to time t. In this section we suppose that the distribution of the counting process $\{N(t)\}$ is given by a mixture of Poisson processes. This means that, conditioning on an extraneous random variable Λ, the process $\{N(t)\}$ behaves like a (homogeneous) Poisson process, as introduced in Section 5.2.1. Starting from $N(0) = 0$, a formal definition goes as follows.

Definition 8.5.1 *The counting process $\{N(t),\ t \ge 0\}$ is called a mixed Poisson process if there exists a positive random variable, the mixing random variable Λ with distribution function $F(x) = \mathbb{P}(\Lambda \le x)$ such that for each $n = 1, 2, \ldots$, for each sequence $\{k_r; r = 1, 2, \ldots, n\}$ of nonnegative integers, and for $0 < a_1 \le b_1 \le a_2 \le b_2 \le \ldots \le a_n \le b_n$,*

$$\mathbb{P}\Big(\bigcap_{r=1}^{n}\{N(b_r) - N(a_r) = k_r\}\Big) = \int_0^\infty \prod_{r=1}^{n} \frac{(\lambda(b_r - a_r))^{k_r}}{k_r!} \mathrm{e}^{-\lambda(b_r - a_r)}\, \mathrm{d}F(\lambda)\,. \tag{8.5.1}$$

From (8.5.1) it can be seen that a mixed Poisson process is a stochastic process with stationary increments. However, in general the increments are not independent. We return to this question in Lemma 8.5.2. However, it is already useful to rewrite (8.5.1) in a slightly different form.

Lemma 8.5.1 *Let $\{N(t),\ t \ge 0\}$ be a mixed Poisson process with mixing random variable Λ. Then, for each $n = 1, 2, \ldots$, for each sequence $\{k_r; r = 1, 2, \ldots, n\}$ of nonnegative integers, and for $0 \le a_1 \le b_1 \le a_2 \le b_2 \le \ldots \le a_n \le b_n$,*

$$\mathbb{P}\Big(\bigcap_{r=1}^{n}\{N(b_r) - N(a_r) = k_r\}\Big) = \prod_{r=1}^{n} \frac{(b_r - a_r)^{k_r}}{k_r!} (-1)^k \hat{l}_\Lambda^{(k)}\Big(\sum_{r=1}^{n}(b_r - a_r)\Big) \tag{8.5.2}$$

where $k = \sum_{r=1}^{n} k_r$, and $\hat{l}_\Lambda(s) = \mathbb{E}\,\mathrm{e}^{-\Lambda s}$ is the Laplace transform of Λ.

Proof The statement immediately follows from (8.5.1) since

$$\mathbb{P}\Big(\bigcap_{r=1}^{n}\{N(b_r)-N(a_r)=k_r\}\Big) = \int_0^\infty \prod_{r=1}^{n} \frac{(\lambda(b_r-a_r))^{k_r}}{k_r!} \mathrm{e}^{-\lambda(b_r-a_r)}\,\mathrm{d}F(\lambda)$$

$$= \prod_{r=1}^{n} \frac{(b_r-a_r)^{k_r}}{k_r!} \int_0^\infty \lambda^{k_\cdot} \exp\Big(-\lambda\sum_{r=1}^{n}(b_r-a_r)\Big)\,\mathrm{d}F(\lambda)$$

and $(\mathrm{d}^k/\mathrm{d}s^k)\int_0^\infty \mathrm{e}^{-\lambda s}\,\mathrm{d}F(\lambda) = (-1)^k \int_0^\infty \mathrm{e}^{-\lambda s}\lambda^k\,\mathrm{d}F(\lambda)$. □

Let $\{N(t)\}$ be a mixed Poisson process with mixing distribution function $F(\lambda)$. To facilitate the writing we use the notation $\alpha_i(t) = \mathbb{P}(N(t)=i)$. Note that (8.5.1) implies

$$\alpha_i(t) = \int_0^\infty \mathrm{e}^{-\lambda t}\frac{(\lambda t)^i}{i!}\,\mathrm{d}F(\lambda)\,, \qquad i \in \mathbb{N}\,. \tag{8.5.3}$$

Thus, for each fixed time point t, the random variable $N(t)$ has a mixed Poisson distribution as defined in Section 4.3.3. If F is degenerate at a fixed point λ, then we retrieve the Poisson random variable with intensity λ.

One remarkable property of mixed Poisson processes is that the probabilities $\{\alpha_0(t), t \ge 0\}$ for state 0 determine all the other probabilities $\{\alpha_i(t), t \ge 0\}$ for $i \ge 1$. Note that the function $\alpha_0(t)$ is differentiable infinitely often by (8.5.3) and Lebesgue's theorem on dominated convergence. Furthermore, (8.5.3) yields

$$\alpha_0^{(k)}(t) = (-1)^k \int_0^\infty \lambda^k \mathrm{e}^{-\lambda t}\mathrm{d}F(\lambda) = (-1)^k \frac{k!}{t^k}\alpha_k(t)\,, \qquad k \in \mathbb{N}\,. \tag{8.5.4}$$

We can of course transform the explicit formula (8.5.3) for $\alpha_i(t)$ into an equivalent expression for the generating function $\hat{g}_{N(t)}(s)$. This leads to the following result.

Lemma 8.5.2 *For* $|s| < 1$ *and* $t \ge 0$,

$$\hat{g}_{N(t)}(s) = \hat{l}_\Lambda(t(1-s))\,, \tag{8.5.5}$$

and hence

$$\mathbb{E}\,(N(t)(N(t)-1)\ldots(N(t)-r+1)) = t^r\mathbb{E}\,(\Lambda^r)\,, \tag{8.5.6}$$

for $r = 1,2,\ldots$. *In particular,*

$$\mathbb{E}\,N(t) = t\,\mathbb{E}\,\Lambda, \qquad \mathrm{Var}\,N(t) = t^2\,\mathrm{Var}\,\Lambda + t\,\mathbb{E}\,\Lambda \tag{8.5.7}$$

and for the index of dispersion

$$\mathrm{I}_{N(t)} = (\mathbb{E}\,N(t))^{-1}\mathrm{Var}\,N(t) = 1 + t\,\mathrm{I}_\Lambda\,. \tag{8.5.8}$$

Proof Applying Fubini's theorem we can write

$$\begin{aligned}\hat{g}_{N(t)}(s) &= \mathbb{E}\, s^{N(t)} = \sum_{n=0}^{\infty} \alpha_n(t) s^n = \int_0^\infty \mathrm{e}^{-\lambda t} \sum_{n=0}^{\infty} \frac{(\lambda t s)^n}{n!}\, \mathrm{d}F(\lambda) \\ &= \int_0^\infty \exp(-\lambda t(1-s))\, \mathrm{d}F(\lambda) = \hat{l}_\Lambda(t(1-s)).\end{aligned}$$

This shows (8.5.5). To evaluate the successive factorial moments of $N(t)$, take the rth derivative with respect to $s \in (0,1)$ of $\hat{g}_{N(t)}(s)$ to obtain $\hat{g}^{(r)}_{N(t)}(s) = \int_0^\infty (\lambda t)^r \exp(-\lambda t(1-s))\, \mathrm{d}F(\lambda)$ and hence as $s \uparrow 1$ we get, finite or not, $\mathbb{E}\,(N(t)(N(t)-1)\ldots(N(t)-r+1)) = \hat{g}^{(r)}_{N(t)}(1) = t^r \mathbb{E}\,(\Lambda^r)$. □

Lemma 8.5.2 implies that within the class of mixed Poisson processes, the Poisson process is the only one that is not overdispersed. It is also the one with the smallest variance function $\mathrm{Var}\, N(t)$, which on top of that is linear and not quadratic. We now generalize (8.5.4) to the case of bivariate distributions.

Lemma 8.5.3 *For $k_1, k_2 \in \mathbb{N}$ and $t, h \ge 0$,*

$$\mathbb{P}(N(t) = k_1, N(t+h) = k_1 + k_2) = \frac{t^{k_1}}{k_1!} \frac{h^{k_2}}{k_2!} (-1)^{k_1+k_2} \alpha_0^{(k_1+k_2)}(t+h) \quad (8.5.9)$$

and hence

$$\mathrm{Cov}\,(N(t), N(t+h) - N(t)) = th\, \mathrm{Var}\, \Lambda\,. \quad (8.5.10)$$

Proof We omit the proof of (8.5.9) since it is analogous to that of (8.5.4). Using (8.5.4) and (8.5.9) we have

$$\begin{aligned}&\mathbb{E}\,(N(t)(N(t+h) - N(t))) \\ &= \sum_{k_1,k_2 \in \mathbb{N}} k_1 k_2 \mathbb{P}(N(t) = k_1; N(t+h) - N(t) = k_2) \\ &= \sum_{k_1,k_2 \in \mathbb{N}} k_1 k_2 \frac{t^{k_1}}{k_1!} \frac{h^{k_2}}{k_2!} \int_0^\infty \lambda^{k_1+k_2} \mathrm{e}^{-\lambda(t+h)}\, \mathrm{d}F(\lambda) \\ &= \int_0^\infty \Big(\sum_{k_1=0}^\infty k_1 \frac{(t\lambda)^{k_1}}{k_1!}\Big)\Big(\sum_{k_2=0}^\infty k_2 \frac{(h\lambda)^{k_2}}{k_2!}\Big) \mathrm{e}^{-\lambda(t+h)}\, \mathrm{d}F(\lambda) \\ &= \int_0^\infty (t\lambda)(h\lambda) \mathrm{e}^{t\lambda} \mathrm{e}^{h\lambda} \mathrm{e}^{-\lambda(t+h)}\, \mathrm{d}F(\lambda) = th\, \mathbb{E}\,(\Lambda^2).\end{aligned}$$

Since $\{N(t)\}$ has stationary increments, this and (8.5.7) give (8.5.10). □

It follows from (8.5.10) that neighbouring increments of a mixed Poisson process with nondegenerate mixing distribution are positively correlated. This implies in particular that the increments are not independent, except in the case of a Poisson process. Thus, a large number of claim arrivals in a given time period has a tendency to trigger off a large number of claim arrivals in the next period as well.

The reader should see how the result of Lemma 8.5.3 can be generalized to the case of n-variate distributions.

8.5.2 Markov Processes with Infinite State Space

In order to represent a mixed Poisson process as a Markov process, we have to generalize the notion of a nonhomogeneous Markov process with a finite state space as it has been introduced in Section 8.4.1. Fortunately this extension to the case of a countably infinite state space is easy.

Consider the state space $E = \mathbb{N}$ and a family of matrices $\boldsymbol{P}(t,t') = (p_{ij}(t,t'))_{i,j\in\mathbb{N}}$, where $0 \le t \le t'$, fulfilling $\boldsymbol{P}(t,t) = \boldsymbol{I}$ for all $t \ge 0$, and

$$\boldsymbol{P}(t,t') = \boldsymbol{P}(t,v)\boldsymbol{P}(v,t'), \tag{8.5.11}$$

for all $0 \le t \le v \le t'$. Then, an $\mathbb{N}$-valued stochastic process $\{X(t),\ t \ge 0\}$ is called a *nonhomogeneous Markov process* if the conditions of Definition 8.4.1 are fulfilled. We further assume that the conditions formulated in (8.4.6),(8.4.7) and (8.4.8) hold, i.e. the *matrix intensity function* $\{\boldsymbol{Q}(t),\ t \ge 0\}$ is well-defined. Besides this we assume that $q_i(t) < \infty$ for all $t \ge 0$ and $i \in \mathbb{N}$, where $q_i(t) = \sum_{j\ne i} q_{ij}(t)$. This means that each state $i \in \mathbb{N}$ is *stable*. Under the above assumptions, the Kolmogorov differential equations (8.4.9)–(8.4.10) (and their matrix forms (8.4.11)–(8.4.12)) remain valid. However, the theory needed to solve these (infinite) systems of differential equations is more complicated. We omit the details and refer the reader to Chapter 2 in Bharucha–Reid (1960), for example. Instead we discuss a few examples of nonhomogeneous Markov processes with infinite state space. Note that for the probabilities $p_{0i}(0,t) = \alpha_i(t)$ for $t \ge 0$ and $i \in \mathbb{N}$, (8.4.10) implies

$$\alpha_i^{(1)}(t) = \sum_{k\ne i} \alpha_k(t) q_{ki}(t) - \alpha_i(t) q_i(t), \qquad i \in \mathbb{N}. \tag{8.5.12}$$

Examples 1. We first consider the case of a (nonhomogeneous) *pure birth process*, where $q_{ij}(t) \equiv 0$ for all $j \ne i, i+1$. In addition we take

$$q_{i,i+1}(t) = (b+t)^{-1}(a+i), \tag{8.5.13}$$

for all $i \in \mathbb{N}$ and $t \ge 0$, where $a, b \ge 0$ are some constants. The resulting process $\{X(t)\}$ is called a *Pólya process* with parameters a, b. Specifying the intensity functions q_{ki} and q_i in (8.5.12), we see that the probabilities $\alpha_i(t) = \mathbb{P}(X(t) = i \mid X(0) = 0)$ satisfy the system of differential equations:

$$\alpha_0^{(1)}(t) = -(b+t)^{-1} a\alpha_0(t), \tag{8.5.14}$$

and, for $i = 1, 2, \ldots$,

$$\alpha_i^{(1)}(t) = (b+t)^{-1}((a+i-1)\alpha_{i-1}(t) - (a+i)\alpha_i(t)). \tag{8.5.15}$$

Together with the initial condition

$$\alpha_i(0) = \begin{cases} 1 & \text{if } i = 0, \\ 0 & \text{otherwise,} \end{cases} \tag{8.5.16}$$

this system of differential equations has the solution

$$\alpha_i(t) = \binom{a+i-1}{i} \left(\frac{b}{t+b}\right)^a \left(\frac{t}{t+b}\right)^i \tag{8.5.17}$$

for all $i \in \mathbb{N}$ and $t \geq 0$. Note that this is the probability function of the negative binomial distribution $\mathrm{NB}(a, t/(t+b))$ and hence for each $t \geq 0$

$$\sum_{i=0}^{\infty} \alpha_i(t) = 1. \tag{8.5.18}$$

This means that a Pólya process does not explode in finite time. We leave it to the reader to derive (8.5.17) as an exercise, using the recursion formulae (8.5.14)–(8.5.16). Let us mention that a Pólya process can be approximated by a (homogeneous) Poisson process when $a, b \geq 0$ are large. The reader is invited to prove this.

2. The class of pure birth processes can be modified in the following way. Assume now that $q_{ij}(t) \equiv 0$ for all $j \neq i-1, i, i+1$. Then, $\{X(t)\}$ is called a (nonhomogeneous) *birth-and-death process*. If we additionally assume that

$$q_{ij}(t) = \begin{cases} \lambda(t)\, i & \text{if } j = i+1, \\ \mu(t)\, i & \text{if } j = i-1, \end{cases}$$

for some nonnegative functions $\lambda(t)$ and $\mu(t)$, then the probabilities $\alpha_i'(t) = \mathbb{P}(X(t) = i \mid X(0) = 1)$ are given by $\alpha_0'(t) = c_1(t)$ and

$$\alpha_i'(t) = (1 - c_1(t))(1 - c_2(t))(c_2(t))^{i-1}, \tag{8.5.19}$$

for all $i = 1, 2, \ldots$, where $c_1(t) = 1 - \mathrm{e}^{-g_1(t)}/g_2(t)$, $c_2(t) = 1 - (g_2(t))^{-1}$ and

$$g_1(t) = \int_0^t (\mu(v) - \lambda(v))\, \mathrm{d}v, \qquad g_2(t) = \mathrm{e}^{-g_1(t)}\left(1 + \int_0^t \mu(v)\mathrm{e}^{g_1(v)}\, \mathrm{d}v\right).$$

The *proof* of (8.5.19) is omitted. It can be found in Kendall (1948), for example. Note that (8.5.19) yields expressions for the expectation and the variance of $X(t)$:

$$\mathbb{E}\, X(t) = \mathrm{e}^{-g_1(t)}, \qquad \operatorname{Var} X(t) = \mathrm{e}^{-2g_1(t)} \int_0^t (\lambda(v) + \mu(v))\mathrm{e}^{g_1(v)}\, \mathrm{d}v.$$

Furthermore, the *probability of extinction* before time t is given by

$$\alpha_0'(t) = \frac{\int_0^t (\mu(v)) e^{g_1(v)} \, dv}{1 + \int_0^t (\mu(v)) e^{g_1(v)} \, dv} .$$

8.5.3 Mixed Poisson Processes as Pure Birth Processes

In this section we derive an intimate link between mixed Poisson processes and pure birth processes. In Example 1 of Section 8.5.2, the latter have been introduced as continuous-time but possibly nonhomogeneous Markov processes with state space $E = \mathbb{N}$. They are determined by the intensities or *instantaneous birth rates* $q_{i,i+1}(t) = -q_{ii}(t) = q_i(t)$ which are the only nonzero elements of the matrix intensity function $\{\boldsymbol{Q}(t)\}$. The Kolmogorov forward equations (8.4.10) are then

$$\begin{cases} \frac{\partial}{\partial t'} p_{i,j}(t,t') = -q_j(t') p_{i,j}(t,t') + q_{j-1}(t') p_{i,j-1}(t,t') & \text{if } i \neq j, \\ \frac{\partial}{\partial t'} p_{j,j}(t,t') = -q_j(t') p_{j,j}(t,t') & \text{if } i = j, \end{cases}$$

with boundary or initial condition $p_{i,j}(t,t) = \delta_i(j)$.

The link between mixed Poisson processes and pure birth processes is given by the following result.

Theorem 8.5.1 *Let $\{N(t), t \geq 0\}$ be a mixed Poisson process with mixing random variable Λ. Then, $\{N(t)\}$ is a pure birth process with intensities*

$$q_i(t) = -(\alpha_0^{(i)}(t))^{-1} \alpha_0^{(i+1)}(t) , \qquad i \in \mathbb{N} , \tag{8.5.20}$$

where $\alpha_0(t) = \hat{l}_\Lambda(t)$.

Proof We first show that the mixed Poisson process $\{N(t)\}$ is a Markov process. Notice that, by (8.5.2), we have $\mathbb{P}(N(t_n) = k_1 + \ldots + k_n, \ldots, N(0) = k_0) > 0$ for all $k_1, \ldots, k_{n+1} \in \mathbb{N}$ and $0 = t_0 < t_1 < \ldots < t_n$. Furthermore, for all $n \geq 1$, $k_1, \ldots, k_n \in \mathbb{N}$ and $0 = x_0 \leq x_1 \leq \cdots \leq x_n$,

$$\begin{aligned} &\mathbb{P}(N(x_1) = k_1, \ldots, N(x_n) = k_1 + \ldots + k_n) \\ &= \prod_{r=1}^{n} \frac{(x_r - x_{r-1})^{k_r}}{k_r!} (-1)^{k_1 + \ldots + k_r} \alpha_0^{(k_1 + \ldots + k_r)}(x_n) . \end{aligned}$$

Thus, we can write

$$\begin{aligned} &\mathbb{P}(N(t_{n+1}) = k_1 + \ldots + k_{n+1} \mid N(t_n) = k_1 + \ldots + k_n, \ldots, N(t_1) = k_1) \\ &= \frac{\prod_{r=1}^{n+1} (t_r - t_{r-1})^{k_r} (k_r!)^{-1} (-1)^{k_1 + \ldots + k_{n+1}} \alpha_0^{(k_1 + \ldots + k_{n+1})}(t_{n+1})}{\prod_{r=1}^{n} (t_r - t_{r-1})^{k_r} (k_r!)^{-1} (-1)^{k_1 + \ldots + k_n} \alpha_0^{(k_1 + \ldots + k_n)}(t_n)} \end{aligned}$$

$$
\begin{aligned}
&= \frac{(t_{n+1}-t_n)^{k_{n+1}}}{k_{n+1}!}(-1)^{k_{n+1}}\frac{\alpha_0^{(k_1+\ldots+k_{n+1})}(t_{n+1})}{\alpha_0^{(k_1+\ldots+k_n)}(t_n)} \\
&= \mathbb{P}(N(t_{n+1}) = k_1+\ldots+k_{n+1} \mid N(t_n) = k_1+\ldots+k_n)\,.
\end{aligned}
$$

Put $p_{ij}(t,t') = \mathbb{P}(N(t') = j \mid N(t) = i)$, where we assume that $0 \le t \le t'$. Then by the result of Lemma 8.5.3 we have, for $i \le j$ and $0 \le t \le t'$,

$$
p_{ij}(t,t') = \frac{\mathbb{P}(N(t)=i, N(t')=j)}{\mathbb{P}(N(t)=i)} = \frac{\frac{t^i}{i!}\frac{(t'-t)^{j-i}}{(j-i)!}(-1)^j\alpha_0^{(j)}(t')}{\frac{t^i}{i!}(-1)^i\alpha_0^{(i)}(t)}
$$

so that

$$
p_{ij}(t,t') = (-1)^{j+i}\frac{(t'-t)^{j-i}}{(j-i)!}\frac{\alpha_0^{(j)}(t')}{\alpha_0^{(i)}(t)}\,, \tag{8.5.21}
$$

for $i \le j$, and $p_{ij}(t,t') = 0$ otherwise. It can be seen that the matrices $\boldsymbol{P}(t,t') = (p_{ij}(t,t'))_{i,j\in\mathbb{N}}$ given by (8.5.21) satisfy $\boldsymbol{P}(t,t) = \boldsymbol{I}$ for all $t \ge 0$, and $p_{ij}(t,u) = \sum_{k=i}^{j} p_{ik}(t,t')p_{kj}(t',u)$ for all $0 \le t \le t' \le u$ and $i \le j$, i.e. condition (8.5.11) is satisfied. This shows that $\{N(t)\}$ is Markov. Furthermore, it follows from (8.5.21) that $q_{ij}(t) \equiv 0$ for all $j \ne i, i+1$ and that $q_i(t) = -q_{i,i+1}(t)$ is given by (8.5.20). □

One can show that the transition probabilities $p_{ij}(t,t')$ satisfy the Kolmogorov forward equations (8.4.10) with the prescribed intensities. For $i \ne j$, take first logarithms of both sides of (8.5.21) and then partial derivatives with respect to t'. This gives

$$
\frac{\frac{\partial}{\partial t'}p_{i,j}(t,t')}{p_{i,j}(t,t')} = \frac{j-i}{t'-t} + \frac{\alpha_0^{(j+1)}(t')}{\alpha_0^{(j)}(t')}\,.
$$

On the other hand we can also directly evaluate the right-hand side of the Kolmogorov forward equation. Then, by (8.5.20) and (8.5.21),

$$
\begin{aligned}
&-q_j(t') + q_{j-1}(t')\frac{p_{i,j-1}(t,t')}{p_{i,j}(t,t')} \\
&= \frac{\alpha_0^{(j+1)}(t')}{\alpha_0^{(j)}(t')} + \frac{\alpha_0^{(j)}(t')}{\alpha_0^{(j-1)}(t')}\,\frac{(-1)^{j-1+i}\frac{(t'-t)^{j-1-i}}{(j-1-i)!}\frac{\alpha_0^{(j-1)}(t')}{\alpha_0^{(i)}(t)}}{(-1)^{j+i}\frac{(t'-t)^{j-i}}{(j-i)!}\frac{\alpha_0^{(j)}(t')}{\alpha_0^{(i)}(t)}}\,,
\end{aligned}
$$

which coincides with the previous expression. For $i = j$ the calculation is even easier.

Combination of equations (8.5.4) and (8.5.21) expresses the transition probabilities $p_{ij}(t,t')$ in terms of the probabilities $\alpha_i(t) = \mathbb{P}(N(t)=i)$,

$$p_{ij}(t,t') = \binom{j}{i}\left(\frac{t}{t'}\right)^{i}\left(1-\frac{t}{t'}\right)^{j-i}\frac{\alpha_j(t')}{\alpha_i(t)}, \tag{8.5.22}$$

where $i \le j$ and $0 \le t \le t'$. Further, note that (8.5.20) can be written in the form $q_0(t) = -\alpha_0^{(1)}(t)$ and $q_{j+1}(t) = q_j(t) - q_j^{(1)}(t)/q_j(t)$ for $j \ge 0$. Alternatively, combine (8.5.4) with (8.5.20) to get another recursion formula

$$\alpha_j(t) = \frac{t}{j}q_{j-1}(t)\alpha_{j-1}(t), \qquad j = 1, 2, \ldots . \tag{8.5.23}$$

In many practical situations it suffices to solve the Kolmogorov forward equations for the case $i = 0$ and $s = 0$. We then arrive at

$$\begin{aligned} \alpha_j^{(1)}(t) &= -q_j(t)\alpha_j(t) + q_{j-1}(t)\alpha_{j-1}(t) && \text{if } j \neq 0, \\ \alpha_0^{(1)}(t) &= -q_0(t)\alpha_0(t) && \text{if } j = 0, \end{aligned}$$

with initial condition $\alpha_j(0) = \delta_0(j)$, and the intensities are given by (8.5.20). Once the functions $\{\alpha_j(t), j \in \mathbb{N}\}$ have been properly determined, the transition probabilities $p_{ij}(t,t')$ follow immediately from (8.5.22). That the resulting transition probabilities $p_{ij}(t,t')$ satisfy the general Kolmogorov forward equations should be proved by the reader.

A similar type of argument leads to the following result.

Corollary 8.5.1 *The mixed Poisson process $\{N(t)\}$ satisfies the binomial criterion, i.e. for $i \le j$ and $s < t$ the inverse transition probabilities $r_{ij}(t,t') = \mathbb{P}(N(t) = i \mid N(t') = j)$ are given by the binomial distribution*

$$r_{ij}(t,t') = \binom{j}{i}\left(\frac{t}{t'}\right)^{i}\left(1-\frac{t}{t'}\right)^{j-i}. \tag{8.5.24}$$

Proof Note that

$$r_{ij}(t,t') = \mathbb{P}(N(t') = j \mid N(t) = i)\,\frac{\mathbb{P}(N(t)=i)}{\mathbb{P}(N(t')=j)} = p_{ij}(t,t')\,\frac{\alpha_i(t)}{\alpha_j(t')}$$

and apply formulae (8.5.4) and (8.5.22). □

8.5.4 The Claim Arrival Epochs

We turn to the claim arrival epochs $\{\sigma_n\}$ defined by the mixed Poisson process $\{N(t)\}$. Recall that the arrival epoch σ_n of the nth claim satisfies the identity $\{N(t) \ge n\} = \{\sigma_n \le t\}$ for all $t \ge 0$. For the joint distribution of the arrival epochs $(\sigma_1, \ldots, \sigma_n)$ we have a generalization of the conditional uniformity property of homogeneous Poisson processes stated in Theorem 5.2.1.

Theorem 8.5.2 *The joint density function $f_{\sigma_1,\ldots,\sigma_n}(x_1,\ldots,x_n)$ of the random vector $(\sigma_1,\ldots,\sigma_n)$ is given by*

$$f_{\sigma_1,\ldots,\sigma_n}(x_1,\ldots,x_n) = \int_0^\infty \lambda^n \mathrm{e}^{-\lambda x_n}\,\mathrm{d}F(\lambda) = (-1)^n \alpha_0^{(n)}(x_n) = \frac{n!}{x_n^n}\alpha_n(x_n)\,, \tag{8.5.25}$$

for $0 < x_1 < \ldots < x_n$ while the density $f_{\sigma_1,\ldots,\sigma_n}(x_1,\ldots,x_n)$ is zero elsewhere.

Proof Let $b_0 = 0 \le a_1 \le b_1 \le \ldots \le a_n \le b_n = x_n$ be a sequence as in Lemma 8.5.1. Then

$$\begin{aligned}
&\mathbb{P}\Big(\bigcap_{r=1}^n \{a_r < \sigma_r \le b_r\}\Big)\\
&= \mathbb{P}\Big(\bigcap_{r=1}^n \{N(b_{r-1};a_r)=0\} \cap \bigcap_{r=1}^{n-1}\{N(a_r;b_r)=1\} \cap \{N(a_n;b_n)\ge 1\}\Big)\\
&= I_1 - I_2\,,
\end{aligned}$$

where

$$I_1 = \mathbb{P}\Big(\bigcap_{r=1}^n \{N(b_{r-1};a_r)=0\} \cap \bigcap_{r=1}^{n-1}\{N(a_r;b_r)=1\}\Big)$$

and

$$I_2 = \mathbb{P}\Big(\bigcap_{r=1}^n \{N(b_{r-1};a_r)=0\} \cap \bigcap_{r=1}^{n-1}\{N(a_r;b_r)=1\} \cap \{N(a_n;b_n)=0\}\Big)\,.$$

We apply the representation formula derived in Lemma 8.5.1 to both terms of the last difference. For the first term, in (8.5.2) we choose $k_r = 1$ when r is even, and $k_r = 0$ for r odd. This yields

$$I_1 = (-1)^{n-1}\alpha_0^{(n-1)}(a_n)\prod_{r=1}^n (b_r - a_r)\,.$$

For the second term we similarly find that $k_{2n} = 0$ and

$$I_2 = (-1)^{n-1}\alpha_0^{(n-1)}(b_n)\prod_{r=1}^n (b_r - a_r)\,.$$

Thus, we ultimately obtain

$$\begin{aligned}
\mathbb{P}\Big(\bigcap_{r=1}^n \{a_r < \sigma_n \le b_r\}\Big) &= \prod_{r=1}^n (b_r - a_r)(-1)^{n-1}\Big(\alpha_0^{(n-1)}(a_n) - \alpha_0^{(n-1)}(b_n)\Big)\\
&= \int_{a_1}^{b_1}\cdots\int_{a_n}^{b_n} (-1)^n \alpha_0^{(n)}(x_n)\,\mathrm{d}x_n\ldots\mathrm{d}x_1\,,
\end{aligned}$$

where the last equality follows from (8.5.4). □

From Theorem 8.5.2 a number of intriguing corollaries can be obtained.

Corollary 8.5.2 *The density $f_{\sigma_n}(x)$ of the arrival epoch σ_n is given by*

$$f_{\sigma_n}(x) = \frac{n}{x}\alpha_n(x) = \frac{(-1)^n}{(n-1)!} x^{n-1} \alpha_0^{(n)}(x), \qquad x > 0. \tag{8.5.26}$$

The latter result itself leads to a set of rather simple formulae linking characteristics of σ_n to the distribution of the mixing variable Λ.

Corollary 8.5.3 *For each $n \geq 1$,*

$$\hat{l}_{\sigma_n}(s) = \int_0^\infty \left(\frac{\lambda}{\lambda + s}\right)^n \mathrm{d}F(\lambda) \tag{8.5.27}$$

and

$$\mathbb{E}\,\sigma_n = n\,\mathbb{E}\,(\Lambda^{-1}), \qquad \mathbb{E}\,\sigma_n^2 = n(n+1)\,\mathbb{E}\,(\Lambda^{-2}). \tag{8.5.28}$$

Note that the expectations in (8.5.28) do not have to be finite. Furthermore, (8.5.28) shows that the expected waiting time for the arrival of the nth claim is proportional to n. The proportionality factor gets larger when the mixing distribution is more concentrated around small values of Λ.

The joint distribution of two consecutive claim arrival epochs illustrates a remarkable dependence structure in the mixed Poisson model.

Corollary 8.5.4 *The conditional densities $f_{\sigma_n|\sigma_1,\ldots,\sigma_{n-1}}(x_n \mid x_1,\ldots,x_{n-1})$ and $f_{\sigma_n|\sigma_{n-1}}(x_n \mid x_{n-1})$ of the sequence $\{\sigma_n; n \geq 1\}$ exist and are given by*

$$f_{\sigma_n|\sigma_1,\ldots,\sigma_{n-1}}(x_n \mid x_1,\ldots,x_{n-1}) = f_{\sigma_n|\sigma_{n-1}}(x_n \mid x_{n-1}) = -\frac{\alpha_0^{(n)}(x_n)}{\alpha_0^{(n-1)}(x_{n-1})}, \tag{8.5.29}$$

for $0 < x_1 < \ldots < x_{n-1} < x_n$, $n = 1, 2, \ldots$, and zero otherwise.

Proof Let $0 < x_1 < \ldots < x_n$. Then, using (8.5.25), we have

$$\begin{aligned} f_{\sigma_n|\sigma_1,\ldots,\sigma_{n-1}}(x_n \mid x_1,\ldots,x_{n-1}) &= \frac{f_{\sigma_1,\ldots,\sigma_{n-1},\sigma_n}(x_1,\ldots,x_{n-1},x_n)}{f_{\sigma_1,\ldots,\sigma_{n-1}}(x_1,\ldots,x_{n-1})} \\ &= \frac{n!\,\alpha_n(x_n)}{x_n^n}\,\frac{(x_{n-1})^{n-1}}{(n-1)!\alpha_{n-1}(x_{n-1})} = \frac{n}{x_n}\left(\frac{x_{n-1}}{x_n}\right)^{n-1}\frac{\alpha_n(x_n)}{\alpha_{n-1}(x_{n-1})}. \end{aligned}$$

Recall from (8.5.2) that

$$f_{\sigma_{n-1},\sigma_n}(x_{n-1},x_n) = \frac{n!}{x_n^n}\alpha_n(x_n)\frac{(x_{n-1})^{n-2}}{(n-2)!} = (-1)^n \frac{x_{n-1}^{n-2}}{(n-2)!}\,\alpha_0^{(n)}(x_n)$$

and

$$f_{\sigma_{n-1}}(x_{n-1}) = (-1)^{n-1} \frac{x_{n-1}^{n-2}}{(n-2)!} \alpha_0^{(n-1)}(x_{n-1}).$$

Now, with the help of (8.5.4), we obtain (8.5.29). □

From the result of Corollary 8.5.4 we realize that the sequence $\{\sigma_n, n \geq 1\}$ has the *Markov property*; see also Section 7.1.1.

8.5.5 The Inter-Occurrence Times

We turn to the times in between consecutive claim arrivals in the mixed Poisson process. The inter-occurrence times are defined by the relations $T_1 = \sigma_1$ and for $n \geq 2$ by $T_n = \sigma_n - \sigma_{n-1}$, and their joint density $f_{T_1,\ldots,T_n}(x_1,\ldots,x_n)$ has a simple form.

Theorem 8.5.3 *Let $\{N(t)\}$ be a mixed Poisson process. Then*

$$f_{T_1,\ldots,T_n}(x_1,\ldots,x_n) = (-1)^n \alpha_0^{(n)}(x_1 + \ldots + x_n) \tag{8.5.30}$$

for $x_1,\ldots,x_n > 0$.

Proof Use the chain rule for conditional densities to write

$$f_{T_1,\ldots,T_n}(x_1,\ldots,x_n) = \prod_{r=2}^{n} f_{T_r|T_1,\ldots,T_{r-1}}(x_r \mid x_1,\ldots,x_{r-1})\, f_{T_1}(x_1).$$

In order to determine the conditional densities $f_{T_r|T_1,\ldots,T_{r-1}}(x_r \mid x_1,\ldots,x_{r-1})$ notice that

$$\begin{aligned}
&\mathbb{P}(T_r \leq x_r \mid T_1 = x_1,\ldots,T_{r-1} = x_{r-1}) \\
&\quad = \mathbb{P}(\sigma_r \leq x_1 + \ldots + x_r \mid \sigma_1 = x_1,\ldots,\sigma_{r-1} = x_1 + \ldots + x_{r-1}) \\
&\quad = \mathbb{P}(\sigma_r \leq x_1 + \ldots + x_r \mid \sigma_{r-1} = x_1 + \ldots + x_{r-1}),
\end{aligned}$$

where in the second equation we used the Markov property of the sequence $\{\sigma_n\}$ of arrival epochs as derived in Corollary 8.5.4. Hence

$$f_{T_r|T_1,\ldots,T_{r-1}}(x_r \mid x_1,\ldots,x_{r-1}) = f_{\sigma_r|\sigma_{r-1}}(x_1 + \ldots + x_r \mid x_1 + \ldots + x_{r-1}).$$

Now (8.5.29) together with $f_{T_1}(x_1) = f_{\sigma_1}(x_1) = \alpha_0^{(1)}(x_1)$ yield the required result. □

One of the surprising consequences of Theorem 8.5.3 is that the inter-occurrence times are identically distributed but not independent. Moreover, (8.5.30) implies that for all permutations $(i_1,\ldots,i_n)$ of $(1,\ldots,n)$,

$$f_{T_1,T_2,\ldots,T_n}(x_1,x_2,\ldots,x_n) = f_{T_{i_1},T_{i_2},\ldots,T_{i_n}}(x_1,x_2,\ldots,x_n).$$

This means that $\{T_1, \ldots, T_n\}$ is an *exchangeable sequence* of random variables, since their joint distribution is invariant under permutations of the arguments. Restricting our attention to one or two inter-occurrence times, we immediately get the next consequences of Theorem 8.5.3.

Corollary 8.5.5 *Let $\{N(t)\}$ be a mixed Poisson process with mixing distribution function $F(\lambda) = \mathbb{P}(\Lambda \le \lambda)$. Then,*

$$f_{T_n}(x) = -\alpha_0^{(1)}(x) = \frac{\alpha_1(x)}{x} = \int_0^\infty \lambda \mathrm{e}^{-\lambda x}\, \mathrm{d}F(\lambda).$$

In particular $\mathbb{E}\, T_n = \mathbb{E}\,(\Lambda^{-1})$ while $\mathbb{E}\,(T_n^2) = 2\,\mathbb{E}\,(\Lambda^{-2})$. For the bivariate case,

$$f_{T_j,T_k}(x,y) = \alpha_0^{(2)}(x+y) = \int_0^\infty \lambda^2 \mathrm{e}^{-\lambda(x+y)}\, \mathrm{d}F(\lambda).$$

In particular $\mathbb{E}\,(T_j T_k) = \mathbb{E}\,(\Lambda^{-2})$ and $\mathrm{Cov}\,(T_j, T_k) = \mathrm{Var}\,(\Lambda^{-1})$.

These results entitle us to reformulate the conditional uniformity property of mixed Poisson processes from Theorem 8.5.2. The latter meant that a mixed Poisson process has the *order statistics property*, i.e. given $N(t) = n$, the claim arrival epochs $\sigma_1, \ldots, \sigma_n$ follow the same distribution as the sequence of order statistics from a uniform distribution on $[0, t]$. Using the notation $f_{\sigma_1,\ldots,\sigma_n | N(t)}(x_1, \ldots, x_n \mid n)$ for the density of the conditional distribution of $(\sigma_1, \ldots, \sigma_n)$ given that $N(t) = n$, we find the following result.

Theorem 8.5.4 *Let $\{N(t), t \ge 0\}$ be a mixed Poisson process. Then, for $0 < x_1 < \ldots < x_n < t$,*

$$\mathbb{P}(N(t) = n \mid \sigma_1 = x_1, \ldots, \sigma_n = x_n) = \frac{\alpha_0^{(n)}(t)}{\alpha_0^{(n)}(x_n)} \tag{8.5.31}$$

and

$$f_{\sigma_1,\ldots,\sigma_n | N(t)}(x_1, \ldots, x_n \mid n) = f_{\sigma_1,\ldots,\sigma_n | \sigma_{n+1}}(x_1, \ldots, x_n \mid t) = t^{-n} n!\,. \tag{8.5.32}$$

Proof As $\mathbb{P}(N(t) = n \mid \sigma_1 = x_1, \ldots, \sigma_n = x_n) = \mathbb{P}(T_{n+1} > t - x_n \mid \sigma_1 = x_1, \ldots, \sigma_n = x_n)$, (8.5.31) follows from Theorem 8.5.3. Furthermore,

$$\begin{aligned} & f_{\sigma_1,\ldots,\sigma_n | N(t)}(x_1, \ldots, x_n \mid n) \\ = & \frac{\mathbb{P}(N(t) = n \mid \sigma_1 = x_1, \ldots, \sigma_n = x_n)\, f_{\sigma_1,\ldots,\sigma_n}(x_1, \ldots, x_n)}{\mathbb{P}(N(t) = n)}. \end{aligned}$$

Replacing the factors in the numerator by the right-hand sides of (8.5.31) and (8.5.25), respectively, and applying (8.5.4) to the denominator, we finally get that $f_{\sigma_1,\ldots,\sigma_n|N(t)}(x_1,\ldots,x_n \mid n) = t^{-n}n!$ holds. In a similar fashion, applying Theorem 8.5.2 and Corollary 8.5.2 to the ratio $f_{\sigma_1,\ldots,\sigma_n|\sigma_{n+1}}(x_1,\ldots,x_n \mid t) = f_{\sigma_1,\ldots,\sigma_n,\sigma_{n+1}}(x_1,\ldots,x_n,t)/f_{\sigma_{n+1}}(t)$ we obtain (8.5.32). □

The following result is yet another phrasing of the same order statistics property.

Theorem 8.5.5 *Consider the sequence $\{\sigma_n, n \geq 1\}$ of arrival epochs of a mixed Poisson process. Then, for $0 < x_1 < \ldots < x_n < 1$,*

$$f_{\sigma_1/\sigma_{n+1},\sigma_2/\sigma_{n+1},\ldots,\sigma_n/\sigma_{n+1}}(x_1,x_2,\ldots,x_n) = n!\,. \tag{8.5.33}$$

The *proof* of Theorem 8.5.5 is based on standard transformation of random variables and is left to the reader.

8.5.6 Examples

We now give a few examples of mixed Poisson processes that have found their way into the actuarial literature.

1. The easiest example is the homogeneous Poisson process itself, which is characterized by the degenerate mixing distribution function $F(x) = \delta_0(x-\lambda)$, where λ is a positive constant. Since in this case $\alpha_i(t) = \mathrm{e}^{-\lambda t}(\lambda t)^i/i!$ for all $i \in \mathbb{N}$, we find that $q_i(t)$ in (8.5.20) is given by $q_i(t) = \lambda$.

2. Discrete mixtures of homogeneous Poisson processes form the next example. Suppose that there exists an increasing sequence $\{\lambda_n\}$ of positive values λ_n such that $F(x) = \sum_{n=0}^{\infty} a_n \delta_0(x-\lambda_n)$ for some sequence $\{a_n\}$ of weights with $a_n > 0$ and $\sum_n a_n = 1$. It is easily seen that $\alpha_i(t) = \sum_n a_n \mathrm{e}^{-\lambda_n t}(\lambda_n t)^i/i!$ holds. In the special case where $\lambda_n = n$ and $a_n = \mathrm{e}^{-\mu}\mu^n/n!$ for some $\mu > 0$, we have $\alpha_0(t) = \exp(-\mu(1-\mathrm{e}^{-t}))$. The corresponding mixed Poisson process is called *of Neyman type* A.

3. Another important example of a mixed Poisson process is obtained by choosing the gamma distribution $\Gamma(a,b)$ for the random variable Λ, i.e. $f(x) = \mathrm{d}F(x)/\mathrm{d}x = (b^a/\Gamma(a))\,\mathrm{e}^{-bx}x^{a-1}$, where a and b are positive constants. Some authors have coined this mixed Poisson process a *Pascal process*. It turns out, however, that the resulting mixed Poisson process is a *Pólya process* with parameters a, b, as has been defined in Example 1 of Section 8.5.2. A simple calculation reveals that the number $N(t)$ of claims arrivals up to time t is $\mathrm{NB}(a, t/(t+b))$ distributed, so that

$$\alpha_i(t) = \binom{a+i-1}{i}\left(\frac{b}{t+b}\right)^a\left(\frac{t}{t+b}\right)^i, \tag{8.5.34}$$

and this coincides with (8.5.17). For the generating function $\hat{g}_{N(t)}(s)$ we obtain

$$\hat{g}_{N(t)}(s) = \left(\frac{b}{b+t(1-s)}\right)^a. \tag{8.5.35}$$

Proofs of (8.5.34) and (8.5.35) are easy exercises. Furthermore, using (8.5.20) and (8.5.34), we arrive at the transition intensities

$$q_i(t) = \frac{i+1}{t}\frac{\alpha_{i+1}(t)}{\alpha_i(t)} = \frac{a+i}{b+t}, \qquad i \in \mathbb{N},$$

as in (8.5.13). Thanks to (8.5.22) we can even calculate the transition probabilities

$$p_{ij}(t,t') = \binom{a+j-1}{a+i-1}\left(\frac{t}{t'}\right)^i\left(\frac{t'-t}{t'+b}\right)^{j-i}, \qquad i \le j, t < t'.$$

4. The *Sichel process* is obtained by using a *generalized inverse Gaussian distribution* with density

$$f(x) = \begin{cases} \dfrac{\eta^{-\theta}x^{\theta-1}}{2K_\theta(\eta/\xi)}\exp\Big(-\dfrac{x^2+\eta^2}{2\xi x}\Big) & \text{if } x > 0, \\ 0 & \text{if } x \le 0, \end{cases}$$

and $\eta, \xi > 0$ and $\theta \in \mathbb{R}$ as mixing distribution. The function $K_\theta(x)$ is the modified Bessel function of the third kind defined in (2.2.2). For $\theta = -0.5$ we get the inverse Gaussian distribution, and the resulting mixed Poisson process is called an *inverse Gauss–Poisson process.* From the definition of the modified Bessel function of the third kind we can derive the relationship

$$\begin{aligned} &\int_0^\infty \exp(-bx - \frac{a}{x})x^{\theta-1}\,\mathrm{d}x \\ &= \int_0^\infty \exp\Big(-\sqrt{ab}\Big(\Big(\frac{b}{a}x\Big)^{1/2} + \Big(\frac{b}{a}x\Big)^{-1/2}\Big)\Big)x^{\theta-1}\,\mathrm{d}x \\ &= 2\left(\frac{a}{b}\right)^{\theta/2} K_\theta(2\sqrt{ab}) \end{aligned} \tag{8.5.36}$$

for $a, b > 0$. This integral representation gives us an expression for the Laplace–Stieltjes transform of Λ:

$$\begin{aligned} \hat{l}_\Lambda(s) &= \int_0^\infty \exp(-sx)\frac{\eta^{-\theta}x^{\theta-1}}{2K_\theta(\eta/\xi)}\exp\Big(-\frac{x^2+\eta^2}{2\xi x}\Big)\,\mathrm{d}x \\ &= (2\xi s+1)^{-\theta/2}\frac{K_\theta(\sqrt{2\xi s+1}\eta/\xi)}{K_\theta(\eta/\xi)}. \end{aligned}$$

By (8.5.5), the generating function of $N(t)$ is then

$$\hat{g}_{N(t)}(s) = (1+2\xi t(1-s))^{-\theta/2}\,\frac{K_\theta(\sqrt{1+2\xi t(1-s)}\eta/\xi)}{K_\theta(\eta/\xi)}.$$

Relation (8.5.36) leads to the explicit formula for the probabilities $\alpha_i(t)$:

$$\begin{aligned}\alpha_i(t) &= \int_0^\infty \exp(-t\lambda)\frac{(t\lambda)^i}{i!}\frac{\eta^{-\theta}\lambda^{\theta-1}}{2K_\theta(\theta/\xi)}\exp\Big(-\frac{\lambda^2+\eta^2}{2\xi x}\Big)\,\mathrm{d}\lambda \\ &= \frac{(\eta t)^i}{i!}(1+2\xi t)^{-(\theta+i)/2}\,\frac{K_{\theta+i}(\sqrt{1+2\xi t}\eta/\xi)}{K_\theta(\eta/\xi)}\,.\end{aligned}$$

Finally, (8.5.20) and the latter relation yield the following expression for the transition intensities $q_i(t)$:

$$q_i(t) = \frac{\eta}{\sqrt{1+2\xi t}}\,\frac{K_{\theta+i+1}(\sqrt{1+2\xi t}\eta/\xi)}{K_{\theta+i}(\sqrt{1+2\xi t}\eta/\xi)}\,.$$

5. As a final example we mention the *Delaporte process*, which is a mixed Poisson process with mixing distribution a *shifted gamma distribution* with density

$$f(x) = \begin{cases} \Gamma(a)^{-1}\eta^a(x-b)^{a-1}\exp\left(-\eta(x-b)\right) & \text{if } x > b,\\ 0 & \text{if } x \le b.\end{cases}$$

The Laplace–Stieltjes transform of Λ is $\hat{l}_\Lambda(s) = \mathrm{e}^{-bs}(\eta/(\eta+s))^a$ for $s \ge 0$ and hence, by formula (8.5.5), the generating function of $N(t)$ is

$$\hat{g}_{N(t)}(s) = \mathrm{e}^{-(1-s)tb}\Big(\frac{\eta}{\eta+(1-s)t}\Big)^a, \qquad |\, s\,| < 1\,.$$

From this, we immediately conclude that for all $t \ge 0$ the random variable $N(t)$ has the same distribution as $N_1(t) + N_2(t)$, where $N_1(t)$ and $N_2(t)$ are independent, $N_1(t)$ has the Poisson distribution $\mathrm{Poi}(bt)$ and $N_2(t)$ has the negative binomial distribution $\mathrm{NB}(a, t/(t+b))$. Moreover, each Delaporte process is the sum of a Poisson process and an independent Pólya process. We leave it to the reader to prove this. The probabilities $\alpha_i(t)$ are then given in terms of the confluent hypergeometric function $U(a,b;x)$, defined by its integral representation (2.2.5). Thus

$$\begin{aligned}\alpha_i(t) &= \int_b^\infty \mathrm{e}^{-\lambda t}\frac{(\lambda t)^i}{i!}\frac{\eta^a}{\Gamma(a)}(\lambda-b)^{a-1}\exp\left(-\eta(\lambda-b)\right)\,\mathrm{d}\lambda \\ &= \frac{t^i\eta^a b^{i+a}}{i!}\mathrm{e}^{-bt}\frac{1}{\Gamma(a)}\int_0^\infty (y+1)^i y^{a-1}\mathrm{e}^{-(\eta+t)by}\,\mathrm{d}y \\ &= \frac{(tb)^i(\eta b)^a}{i!}U(a, i+a+1; (\eta+t)b)\,.\end{aligned}$$

In particular, for $\alpha_0(t)$ we obtain a simple formula:

$$\begin{aligned}\alpha_0(t) &= \frac{(\eta b)^a\mathrm{e}^{-bt}}{\Gamma(a)}\int_0^\infty v^{a-1}\mathrm{e}^{-(\eta+t)bv}\,\mathrm{d}v \\ &= \frac{(\eta b)^a}{\Gamma(a)}\mathrm{e}^{-bt}((\eta+t)b)^{-a}\Gamma(a) = \mathrm{e}^{-bt}(\eta+t)^{-b}\,.\end{aligned}$$

Using now (8.5.20) we see that $q_0(t) = -\alpha_0^{(1)}(t)/\alpha_0(t) = b + a/(\eta + t)$.

Bibliographical Notes. A survey on nonhomogeneous continuous-time Markov processes with countable infinite state space can be found, for example, in Bharucha–Reid (1960). For further details on mixed Poisson processes we refer to Grandell (1997). The relationship between nonhomogeneous birth processes and mixed Poisson processes has been discussed, for example, in Lundberg (1964) and McFadden (1965). Mixed Poisson distributions whose mixing distribution is an inverse Gaussian distribution have been considered in Sichel (1971); see also Sichel (1974,1975). The Delaporte process was introduced in Delaporte (1960) as one of the first applications of mixed Poisson processes in an actuarial context; see also Delaporte (1965) and Willmot and Sundt (1989). Other special examples of mixed Poisson processes can be found in Albrecht (1984), Philipson (1960) and Willmot (1986).

CHAPTER 9

Martingale Techniques I

A variety of reasons can be given to study martingales. Not only do they constitute a large class of stochastic processes, but in recent years, insurance and financial mathematics have been prime fields of application of martingale techniques. In contrast to Markov models, the theory of martingales usually does not give tools for explicit computation of quantities of interest, like ruin probabilities. However, martingale techniques appear quite unexpectedly in proofs of various results. In particular, martingales turn out to be particularly useful when constructing bounds for ruin probabilities. Even more important is the backbone structure provided by martingale theory within the realm of financial mathematics.

The notions and results considered in the present chapter are basic for the modern theory of stochastic processes. Unfortunately, in contrast to most of the material given in the preceding chapters, we can no longer avoid using more advanced concepts of probability theory. More specifically, the notion of conditional expectation with respect to a σ-algebra will broadly be applied.

9.1 DISCRETE-TIME MARTINGALES

9.1.1 Fair Games

We start from a simple example that will help to understand the general martingale technique introduced later. A gambler wins or loses one currency unit in each game with equal probability and independently of the outcomes of other games. At most n_0 games can be played, but the gambler has the privilege of optional stopping before n_0. We can formalize his gains in terms of a random walk $\{S_n, n \in \mathbb{N}\}$. Put $S_0 = 0$ and $S_n = \sum_{i=1}^{n} Y_i$ and $Y_1, Y_2, \ldots$ as independent and identically distributed random variables assuming values -1 and 1 with probability $1/2$, respectively. Thus S_n is the gambler's gain after the n-th game. We now define a stopping rule bounded by n_0. Let $w_n : \mathbb{Z}^n \to \{0, 1\}$, $n = 1, 2, \ldots, n_0$, be a family of *test functions* with $w_{n_0} \equiv 1$. A *stopping rule* is a random variable τ taking values from $\{1, \ldots, n_0\}$

such that $\tau = n$ if and only if $w_i(Y_1, \ldots, Y_i) = 0$ for $i = 1, \ldots, n-1$ and $w_n(Y_1, \ldots, Y_n) = 1$; $n = 1, \ldots, n_0$. So, if after the n-th game the test function w_n shows 0, then the gambler continues to play, while if the outcome is 1, he quits. If the stopping rule τ is used, then the gambler leaves the game with the gain $S_\tau = \sum_{n=1}^{n_0} S_n \mathbb{I}(\tau = n)$. We show that

$$\mathbb{E}\, S_\tau = 0\,. \tag{9.1.1}$$

This equation is a special case of an optional sampling theorem in Section 9.1.6, stating that the game is fair. Note that the event $\tau = n$ is independent of $Y_{n+1}, \ldots, Y_{n_0}$ for $n = 1, \ldots, n_0 - 1$. Hence

$$\begin{aligned} \mathbb{E}\,[S_{n_0}; \tau = n] &= \mathbb{E}\,[S_n + Y_{n+1} + \ldots + Y_{n_0}; \tau = n] \\ &= \mathbb{E}\,[S_n; \tau = n] + \mathbb{E}\,[Y_{n+1} + \ldots + Y_{n_0}; \tau = n] \\ &= \mathbb{E}\,[S_n; \tau = n]\,, \end{aligned}$$

i.e.

$$\mathbb{E}\,[S_{n_0}; \tau = n] = \mathbb{E}\,[S_n; \tau = n]\,, \tag{9.1.2}$$

for $1 \le n \le n_0$ and, consequently,

$$\mathbb{E}\, S_\tau = \mathbb{E}\left(\sum_{n=1}^{n_0} S_n \mathbb{I}(\tau = n)\right) = \sum_{n=1}^{n_0} \mathbb{E}\,(S_{n_0} \mathbb{I}(\tau = n)) = \mathbb{E}\, S_{n_0} = 0\,.$$

Furthermore, note that a much stronger result than (9.1.2) is true in that

$$\mathbb{E}\,(S_{n+k} \mid Y_1, \ldots, Y_n) = S_n\,, \tag{9.1.3}$$

for all $k, n \in \mathbb{N}$. This equation illustrates in another way the fairness of the game. By $\mathbb{E}\,(S_{n+k} \mid Y_1, \ldots, Y_n)$, we mean the conditional expectation of S_{n+k} with respect to the sub-σ-algebra of $\mathcal{F}$ and which consists of the events $\{\omega : (Y_1(\omega), \ldots, Y_n(\omega)) \in B\}$, for all Borel sets $B \in \mathcal{B}(\mathbb{R}^n)$.

We continue with a slightly more general model from insurance. As in Chapter 5, consider a sequence $\{Z_n\}$ of aggregate claims over intervals of equal length, say $(n-1, n]$, $n = 1, 2, \ldots$. After the nth period the cumulative claim amount is $W_n = \sum_{i=1}^{n} Z_i$, where we set $W_0 = 0$. It turns out that we can decompose the random variables W_n into $W_n = S_n + V_n$, $n = 0, 1, \ldots$. Here the sequence $\{S_n, n \in \mathbb{N}\}$ fulfils a fairness property of the type (9.1.3) and hence no optimal bounded stopping rule for this sequence is induced by a family of test functions. Let us clarify this. Suppose that the risks $Z_1, Z_2, \ldots$ form a sequence of independent and identically distributed nonnegative random variables with finite mean $\mathbb{E}\, Z$. Put $Y_i = Z_i - \mathbb{E}\, Z$ and $S_n = \sum_{i=1}^{n} Y_i$, then $S_n = W_n - n\mathbb{E}\, Z$ and $V_n = n\mathbb{E}\, Z$. Furthermore, (9.1.3) holds, i.e. $\mathbb{E}\, Z$ is a fair premium applied to $\{Z_n\}$ for each interval $(n-1, n]$, $n = 1, 2, \ldots$. Note that

$\mathbb{E} Z_n$ is the net premium for the risk Z_n and so $\{n\mathbb{E} Z, n \in \mathbb{N}\}$ can be called the (cumulative) net premium process. Under our independence assumption, the sequence $\{V_n\}$ is deterministic. The more complex situation of dependent risks $Z_1, Z_2, \ldots$ will be studied in Section 9.1.8.

Equivalent formulations of (9.1.3) are $\mathbb{E}(S_{n+k} \mid S_1, \ldots, S_n) = S_n$ and

$$\mathbb{E}(S_{n+k} \mid \mathcal{F}_n^S) = S_n, \tag{9.1.4}$$

where $\mathcal{F}_n^S = \{\{\omega : (S_0(\omega), \ldots, S_n(\omega)) \in B\}, B \in \mathcal{B}(\mathbb{R}^{n+1})\}$. This version of the fairness property (9.1.3) can be seen as an introduction to the general theory of discrete-time martingales, based on filtrations.

9.1.2 Filtrations and Stopping Times

Suppose $\{X_n, \; n \in \mathbb{N}\}$ is an arbitrary sequence of real-valued random variables on $(\Omega, \mathcal{F}, \mathbb{P})$ with $\mathbb{E}|X_n| < \infty$ for each $n \in \mathbb{N}$. Inspired by the σ-algebras appearing in (9.1.4), the σ-algebra $\mathcal{F}_n^X$ containing the events $\{\omega : \; (X_0(\omega), \ldots, X_n(\omega)) \in B\}$, for all $B \in \mathcal{B}(\mathbb{R}^{n+1})$ is called the *history* of $\{X_n\}$ up to time n. We also say that $\mathcal{F}_n^X$ is generated by the random variables $X_0, \ldots, X_n$. The following statements are true: for all $n \in \mathbb{N}$,

- $\mathcal{F}_n^X \subset \mathcal{F}$,
- $\mathcal{F}_n^X \subset \mathcal{F}_{n+1}^X$,
- X_n is measurable with respect to $\mathcal{F}_n^X$.

The family of σ-algebras $\{\mathcal{F}_n^X, \; n \in \mathbb{N}\}$ is called the *history* of $\{X_n\}$. However, it is more common to say that $\{\mathcal{F}_n^X, \; n \in \mathbb{N}\}$ is the *filtration* generated by $\{X_n\}$. As such, it is a special case of the following definition.

Definition 9.1.1 *A family $\{\mathcal{F}_n, n \in \mathbb{N}\}$ of σ-algebras such that $\mathcal{F}_n \subset \mathcal{F}$ and $\mathcal{F}_n \subset \mathcal{F}_{n+1}$ for all $n \in \mathbb{N}$ is called a filtration. We say that the sequence $\{X_n, n \in \mathbb{N}\}$ is adapted to the filtration $\{\mathcal{F}_n\}$ if X_n is measurable with respect to $\mathcal{F}_n$ for all $n \in \mathbb{N}$.*

Example Consider the random walk $\{S_n\}$ from Section 9.1.1 which describes the evolution of the gambler's gain. In this case $\mathcal{F}_n^S$ is generated by the events $\{S_1 = i_1, \ldots, S_n = i_n\}$ with $i_1, \ldots, i_n \in \mathbb{Z}$ or, equivalently, by the events $\{Y_1 = i_1, \ldots, Y_n = i_n\}$ with $i_1, \ldots, i_n \in \{-1, 1\}$ because there is a one-to-one correspondence between the sequences $\{S_n, n \in \mathbb{N}\}$ and $\{Y_n, n = 1, 2, \ldots\}$, i.e. $\mathcal{F}_n^S = \mathcal{F}_n^Y$ for each $n \in \mathbb{N}$, and where $\mathcal{F}_0^S = \mathcal{F}_0^Y = \{\emptyset, \Omega\}$.

The stopping rule τ considered in Section 9.1.1 is an example of the important notion of a *stopping time*.

Definition 9.1.2 *A random variable τ taking values in $\mathbb{N} \cup \{\infty\}$ is said to be a stopping time with respect to a filtration $\{\mathcal{F}_n\}$ (or equivalently an $\{\mathcal{F}_n\}$-stopping time) if the event $\{\tau = n\}$ belongs to $\mathcal{F}_n$, for all $n \in \mathbb{N}$.*

Example Let $\{X_n\}$ be a sequence of real-valued random variables. Consider the *first entrance time* τ^B of $\{X_n\}$ to a Borel set $B \in \mathcal{B}(\mathbb{R})$, i.e

$$\tau^B = \begin{cases} \min\{n : X_n \in B\} & \text{if } X_n \in B \text{ for some } n \in \mathbb{N}, \\ \infty & \text{otherwise.} \end{cases}$$

The random variable τ^B is a stopping time with respect to $\{\mathcal{F}_n^X\}$ because

$$\{\tau^B = n\} = \{X_0 \notin B, \ldots, X_{n-1} \notin B, X_n \in B\} \in \mathcal{F}_n^X .$$

In the special case where $X_n = \sum_{j=1}^n Y_j$ is a random walk generated by a sequence $Y_1, Y_2, \ldots$, all (descending and ascending) ladder epochs ν_n^+, ν_n^- as defined in Section 6.3 are stopping times. The formal proof of this fact is left to the reader.

We say that a stopping time τ is *bounded* if there exists an $n_0 \in \mathbb{N}$ such that $\mathbb{P}(\tau \le n_0) = 1$. The following result is called *Komatsu's lemma*. It gives further motivation for the concept of martingales introduced in the next section.

Theorem 9.1.1 *Let $\{\mathcal{F}_n\}$ be a filtration and let $\{X_n\}$ be adapted to $\{\mathcal{F}_n\}$. Assume that for each bounded $\{\mathcal{F}_n\}$-stopping time τ,*

$$\mathbb{E}\, X_\tau = \mathbb{E}\, X_0 . \tag{9.1.5}$$

Then, for each $k \in \mathbb{N}$,

$$\mathbb{E}\,(X_{k+1} \mid \mathcal{F}_k) = X_k . \tag{9.1.6}$$

Proof Let $k \in \mathbb{N}$ and $A \in \mathcal{F}_k$ be fixed, and consider the random variable τ defined by

$$\tau(\omega) = \begin{cases} k & \text{if } \omega \in A, \\ k+1 & \text{if } \omega \notin A. \end{cases}$$

It is easy to see that τ is an $\{\mathcal{F}_n\}$-stopping time because

$$\{\tau \le n\} = \begin{cases} \emptyset & \text{if } n < k, \\ A & \text{if } n = k, \\ \Omega & \text{if } n \ge k+1, \end{cases}$$

and, consequently, $\{\tau = n\} \in \mathcal{F}_n$ for each $n \in \mathbb{N}$. Now, applying (9.1.5) consecutively to τ and to the bounded stopping time $\tau' = k+1$, we have

$$\mathbb{E}\,[X_k; A] + \mathbb{E}\,[X_{k+1}; A^c] = \mathbb{E}\, X_\tau = \mathbb{E}\, X_{k+1} = \mathbb{E}\,[X_{k+1}; A] + \mathbb{E}\,[X_{k+1}; A^c]$$

i.e. $\mathbb{E}\,[X_{k+1}; A] = \mathbb{E}\,[X_k; A]$. This gives (9.1.6) since $k \in \mathbb{N}$ and $A \in \mathcal{F}_k$ are arbitrary. $\square$

9.1.3 Martingales, Sub- and Supermartingales

Let $\{\mathcal{F}_n\}$ be a filtration and let $\{X_n\}$ be a sequence of random variables adapted to $\{\mathcal{F}_n\}$ such that $\mathbb{E}\,|X_n| < \infty$ for each $n \in \mathbb{N}$. Then $\{X_n\}$ is called a *martingale* with respect to $\{\mathcal{F}_n\}$ or an $\{\mathcal{F}_n\}$-martingale, if with probability 1

$$\mathbb{E}\,(X_{n+1} \mid \mathcal{F}_n) = X_n\,, \tag{9.1.7}$$

for all $n \in \mathbb{N}$. Similarly, $\{X_n\}$ is called a *submartingale* if

$$\mathbb{E}\,(X_{n+1} \mid \mathcal{F}_n) \geq X_n\,, \tag{9.1.8}$$

and a *supermartingale* if

$$\mathbb{E}\,(X_{n+1} \mid \mathcal{F}_n) \leq X_n\,, \tag{9.1.9}$$

for all $n \in \mathbb{N}$. Note that (9.1.7) implies

$$\mathbb{E}\,(X_{n+k} \mid \mathcal{F}_n) = X_n\,, \tag{9.1.10}$$

for all $k, n \in \mathbb{N}$. Indeed, repeatedly using (9.1.7) and basic properties of conditional expectation we have

$$\begin{aligned}
\mathbb{E}\,(X_{n+k} \mid \mathcal{F}_n) &= \mathbb{E}\,(\mathbb{E}\,(X_{n+k} \mid \mathcal{F}_{n+k-1}) \mid \mathcal{F}_n) = \mathbb{E}\,(X_{n+k-1} \mid \mathcal{F}_n) \\
&= \mathbb{E}\,(\mathbb{E}\,(X_{n+k-1} \mid \mathcal{F}_{n+k-2}) \mid \mathcal{F}_n) \\
&\;\vdots \\
&= \mathbb{E}\,(X_{n+1} \mid \mathcal{F}_n) = X_n\,.
\end{aligned}$$

Analogously, (9.1.8) and (9.1.9) imply

$$\mathbb{E}\,(X_{n+k} \mid \mathcal{F}_n) \geq X_n \tag{9.1.11}$$

and

$$\mathbb{E}\,(X_{n+k} \mid \mathcal{F}_n) \leq X_n \tag{9.1.12}$$

for all $k, n \in \mathbb{N}$. Taking expectations on both sides of (9.1.10)–(9.1.12) we get

- for a martingale, $\mathbb{E}\,X_n = \mathbb{E}\,X_0$ for all $n \in \mathbb{N}$,
- for a submartingale, $\mathbb{E}\,X_{n+k} \geq \mathbb{E}\,X_n$ for all $k, n \in \mathbb{N}$,
- for a supermartingale, $\mathbb{E}\,X_{n+k} \leq \mathbb{E}\,X_n$ for all $k, n \in \mathbb{N}$.

Let $\{X_n\}$ be a martingale with increments $Y_n = X_n - X_{n-1}$ having finite second moments $\mathbb{E}\,X_n^2 < \infty$. Then it is not difficult to show that

$$\mathbb{E}\,Y_n = 0, \qquad \mathrm{Cov}(Y_n, Y_{n+k}) = 0 \tag{9.1.13}$$

and consequently $\operatorname{Var} X_n = \sum_{i=0}^n \operatorname{Var} Y_i$.

Examples 1. As already mentioned in Section 9.1.1, every random walk $\{S_n\}$, $S_n = \sum_{i=1}^n Y_i$, with $\mathbb{E}\, Y = 0$ is a martingale with respect to $\{\mathcal{F}_n^S\} = \{\mathcal{F}_n^Y\}$ because $\mathbb{E}\,(S_{n+1} \mid \mathcal{F}_n^S) = \mathbb{E}\,(S_n \mid \mathcal{F}_n^S) + \mathbb{E}\,(Y_{n+1} \mid \mathcal{F}_n^S) = S_n$. We get a submartingale whenever $\mathbb{E}\, Y > 0$ and a supermartingale if $\mathbb{E}\, Y < 0$. For every random walk $\{S_n\}$ with $\mathbb{E}\,|Y| < \infty$, the sequence $\{X_n\}$, $X_n = S_n - n\mathbb{E}\, Y$, is an $\{\mathcal{F}_n^Y\}$-martingale.

2. Consider a random walk $\{S_n\}$, $S_n = \sum_{i=1}^n Y_i$, with $\mathbb{E}\, Y = 0$ and $\operatorname{Var} Y = \sigma^2 < \infty$. Then the sequence $\{X_n\}$, $X_n = S_n^2 - n\sigma^2$, is an $\{\mathcal{F}_n^Y\}$-martingale. The proof of this fact is similar to that given in Example 1.

3. Consider a random walk $\{S_n\}$, $S_n = \sum_{i=1}^n Y_i$, such that the moment generating function $\hat{m}_Y(s)$ is finite for some $s \in \mathbb{R}$. Then $\{X_n\}$ given by $X_n = \mathrm{e}^{sS_n}(\hat{m}_Y(s))^{-n}$ is an $\{\mathcal{F}_n^Y\}$-martingale. Indeed,

$$\begin{aligned}
\mathbb{E}\,(X_{n+1} \mid \mathcal{F}_n^Y) &= \frac{\mathbb{E}\,(\mathrm{e}^{sS_n}\mathrm{e}^{sY_{n+1}} \mid \mathcal{F}_n^Y)}{(\hat{m}_Y(s))^{n+1}} = \frac{\mathrm{e}^{sS_n}\mathbb{E}\,(\mathrm{e}^{sY_{n+1}} \mid \mathcal{F}_n^Y)}{(\hat{m}_Y(s))^{n+1}} \\
&= \frac{\mathrm{e}^{sS_n}\mathbb{E}\,\mathrm{e}^{sY_{n+1}}}{(\hat{m}_Y(s))^{n+1}} = \frac{\mathrm{e}^{sS_n}}{(\hat{m}_Y(s))^{n}} = X_n \,.
\end{aligned}$$

4. Consider a martingale $\{W_n, n \in \mathbb{N}\}$ with respect to a filtration $\{\mathcal{F}_n\}$ and a sequence $\{Z_n, n = 1, 2, \ldots\}$ of random variables such that Z_n is measurable with respect to $\mathcal{F}_{n-1}$ for each $n = 1, 2, \ldots$. Such a sequence $\{Z_n\}$ is said to be $\{\mathcal{F}_n\}$-*predictable*. If $\mathcal{F}_n = \mathcal{F}_n^W$, then the value of Z_n is determined by the values of $W_0, \ldots, W_{n-1}$. The sequence $\{X_n\}$ with $X_0 = 0$ and

$$X_n = \sum_{k=1}^n Z_k(W_k - W_{k-1}), \qquad n \in \mathbb{N}\,, \tag{9.1.14}$$

is a martingale with respect to $\{\mathcal{F}_n\}$ provided that the integrability condition $\mathbb{E}\,|Z_k(W_k - W_{k-1})| < \infty$ is fulfilled for all $k = 1, 2, \ldots$. Indeed,

$$\begin{aligned}
&\mathbb{E}\,(X_{n+1} \mid \mathcal{F}_n) \\
&= \sum_{k=1}^n \mathbb{E}\,(Z_k(W_k - W_{k-1}) \mid \mathcal{F}_n) + \mathbb{E}\,(Z_{n+1}(W_{n+1} - W_n) \mid \mathcal{F}_n) \\
&= \sum_{k=1}^n Z_k(W_k - W_{k-1}) + Z_{n+1}\mathbb{E}\,(W_{n+1} - W_n \mid \mathcal{F}_n) \\
&= \sum_{k=1}^n Z_k(W_k - W_{k-1}) = X_n \,.
\end{aligned}$$

Note that (9.1.14) is a discrete analogue to a stochastic integral of a predictable process with respect to a martingale.

5. Consider a homogeneous Markov chain $\{Z_n\}$ with finite state space $E = \{1, \ldots, \ell\}$ and transition matrix $\boldsymbol{P}$. If $\theta \neq 0$ is an eigenvalue of $\boldsymbol{P}$ and $\boldsymbol{\phi} = (\phi_1, \ldots, \phi_\ell)$ the corresponding right eigenvector, then $\{X_n\}$ with $X_n = \theta^{-n}\phi_{Z_n}$ is a martingale with respect to the filtration $\{\mathcal{F}_n^Z\}$. This can be seen as follows. From the Markov property (7.1.3) we have $\mathbb{E}(X_{n+1} \mid \mathcal{F}_n^Z) = \mathbb{E}(X_{n+1} \mid Z_n)$. Now, for all $i = 1, \ldots, \ell$, (7.1.12) implies

$$\mathbb{E}(X_{n+1} \mid Z_n = i) = \theta^{-n}\theta^{-1}\sum_{j=1}^{\ell} p_{ij}\phi_j = \theta^{-n}\phi_i\,,$$

from which we have $\mathbb{E}(X_{n+1} \mid Z_n) = \theta^{-n}\phi_{Z_n} = X_n$.

6. Suppose $Y_1, Y_2, \ldots$ are strictly positive, independent and identically distributed with $\mathbb{E}\,Y = 1$. Then the sequence $\{X_n\}$ given by

$$X_n = \begin{cases} 1 & \text{if } n = 0, \\ Y_1 Y_2 \ldots Y_n & \text{if } n \geq 1 \end{cases}$$

is a martingale with respect to the filtration $\{\mathcal{F}_n^Y\}$. Indeed, we have

$$\begin{aligned} \mathbb{E}(X_{n+1} \mid \mathcal{F}_n^Y) &= \mathbb{E}(Y_1 Y_2 \ldots Y_{n+1} \mid \mathcal{F}_n^Y) = Y_1 Y_2 \ldots Y_n \mathbb{E}(Y_{n+1} \mid \mathcal{F}_n^Y) \\ &= Y_1 Y_2 \ldots Y_n \mathbb{E}\,Y_{n+1} = X_n. \end{aligned}$$

7. Let f and $\tilde{f}$ be density functions on $\mathbb{R}$ such that $f \neq \tilde{f}$. For simplicity assume that the product $f(x)\tilde{f}(x) > 0$ for all $x \in \mathbb{R}$. Let $Y_1, Y_2, \ldots$ be a sequence of independent and identically distributed random variables, with the common density either f or $\tilde{f}$. The *likelihood ratio sequence* $\{X_n, n \in \mathbb{N}\}$ is then given by

$$X_n = \begin{cases} \displaystyle\prod_{k=1}^{n} \frac{\tilde{f}(Y_k)}{f(Y_k)} & \text{if } n \geq 1, \\ 1 & \text{if } n = 0. \end{cases}$$

We show that $\{X_n\}$ is an $\{\mathcal{F}_n^Y\}$-martingale if the Y_n have density f. Indeed,

$$\mathbb{E}(X_{n+1} \mid \mathcal{F}_n^Y) = \mathbb{E}\left(\prod_{k=1}^{n+1} \frac{\tilde{f}(Y_k)}{f(Y_k)} \mid \mathcal{F}_n^Y\right) = \prod_{k=1}^{n} \frac{\tilde{f}(Y_k)}{f(Y_k)} \mathbb{E}\left(\frac{\tilde{f}(Y_{n+1})}{f(Y_{n+1})}\right) = X_n\,,$$

because $\mathbb{E}(\tilde{f}(Y_{n+1})/f(Y_{n+1})) = \int_{-\infty}^{\infty} \tilde{f}(x)\,\mathrm{d}x = 1$. In the alternative situation that Y_n has density $\tilde{f}$, the additional assumption $\int_{-\infty}^{\infty} \tilde{f}^2(x)/f(x)\,\mathrm{d}x < \infty$ turns $\{X_n\}$ into a submartingale with respect to $\{\mathcal{F}_n^Y\}$. Indeed, in this case

$$\mathbb{E}\left(\frac{\tilde{f}(Y_{n+1})}{f(Y_{n+1})}\right) = \int_{-\infty}^{\infty} \frac{\left(\tilde{f}(x)\right)^2}{f(x)}\,\mathrm{d}x = \mathbb{E}\left(\frac{\tilde{f}(Z)}{f(Z)}\right)^2 \geq \left(\mathbb{E}\left(\frac{\tilde{f}(Z)}{f(Z)}\right)\right)^2 = 1\,,$$

where Z is a random variable with density f.

9.1.4 Life-Insurance Model with Multiple Decrements

In this section we give an application of discrete-time martingales to a general life insurance model. Let $J_0 : \Omega \to \{1, \ldots, \ell\}$ denote the (random) cause of decrement of an insured person and let $T_0 : \Omega \to \mathbb{N}$ denote the total lifetime of the insured (measured in years). Let b_{jk} be the payment at the end of the k-th year after policy issue, if decrement by cause j occurs during that year; $j = 1, \ldots, \ell$. Assume that the payments b_{jk} are not random but deterministic and that these payments are financed by annual premiums $\beta_0, \beta_1, \beta_2, \ldots$ which have to be paid by the insured at the beginning of each year.

For $i = 0, 1, \ldots,$ let the components of the random vector (J_i, T_i) be distributed as the (conditional) cause of decrement and the residual lifetime of an insured after policy issue at time i, respectively. Then $q_{ji}(m) = \mathbb{P}(J_i = j, T_i < m)$ denotes the probability that an insured will die of cause j within m years after time i. Note that $q_{ji}(m) = \mathbb{P}(J_0 = j, T_0 < i + m \mid T_0 \geq i)$ and $p_i(m) = 1 - \sum_{j=1}^{\ell} q_{ji}(m)$ is the probability that the insured survives at least m years after time i. By q_{ji} we denote the probability that the insured dies within one year after time i by cause j, i.e. $q_{ji} = q_{ji}(1)$. Then

$$\begin{aligned} \mathbb{P}(J_i = j, T_i = k) &= \mathbb{P}(J_i = j, T_0 = i + k \mid T_0 \geq i) \\ &= \mathbb{P}(T_0 \geq i + k \mid T_0 \geq i)\, \mathbb{P}(J_i = j, T_0 < i + k + 1 \mid T_0 \geq i + k) \\ &= p_i(k) q_{j,i+k} \,. \end{aligned}$$

We consider a constant annual discount factor v with $0 < v < 1$ and use the abbreviations $J = J_i$ and $T = T_i$. Then, at time i, the present value of the insured benefit is $b_{J,T+1} v^{T+1}$ and the present value of the insurer's overall loss X is given by

$$X = b_{J,T+1} v^{T+1} - \sum_{k=0}^{T} \beta_k v^k. \tag{9.1.15}$$

The annual premiums $\beta_0, \beta_1, \beta_2, \ldots$ are called *net premiums* if they satisfy the equation $\mathbb{E}\, X = 0$, which is equivalent to

$$\sum_{j=1}^{\ell} \sum_{k=0}^{\infty} b_{j,k+1} v^{k+1} p_i(k) q_{j,i+k} = \sum_{k=0}^{\infty} \beta_k v^k p_i(k).$$

Let μ_n denote the expectation of X with respect to the (conditional) probability measure $\mathbb{P}_n$, where $\mathbb{P}_n(A) = \mathbb{P}(A \mid T \geq n)$. Then we have $\mu_n = \mathbb{E}\,[X; T \geq n] / \mathbb{P}(T \geq n)$ and, by (9.1.15),

$$\mu_n = \sum_{j=1}^{\ell} \sum_{k=0}^{\infty} b_{j,n+k+1} v^{k+1} p_{i+n}(k) q_{j,i+n+k} - \sum_{k=0}^{\infty} \beta_{n+k} v^k p_{i+n}(k)\,. \tag{9.1.16}$$

In life insurance mathematics, the quantity μ_n is called the *net premium reserve* at time n after policy issue. It can be interpreted as the expectation of the difference between the present value of future benefit payments and the present value of future premiums at time n, provided that $T > n$.

Note that (9.1.16) gives

$$\mu_n + \beta_n = \mu_{n+1} v p_{i+n} + \sum_{j=1}^{\ell} b_{j,n+1} v q_{j,i+n} \,, \tag{9.1.17}$$

where $p_{i+n} = p_{i+n}(1)$. This recursion formula is useful for numerical computation of the net premium reserves μ_n. Moreover, it implies that the premium β_n can be decomposed into two components:

$$\beta_n = \mu_{n+1} v - \mu_n + \sum_{j=1}^{\ell} (b_{j,n+1} - \mu_{n+1}) v q_{j,i+n} = \beta_n^{\mathrm{s}} + \beta_n^{\mathrm{r}} \,, \tag{9.1.18}$$

where $\beta_n^{\mathrm{s}} = \mu_{n+1} v - \mu_n$ is the *savings premium* which increments the net premium reserve, and

$$\beta_n^{\mathrm{r}} = \sum_{j=1}^{\ell} (b_{j,n+1} - \mu_{n+1}) v q_{j,i+n} \tag{9.1.19}$$

is the *risk premium* which insures the net amount of risk for one year. Using this notation, the insurer's overall loss X given by (9.1.15) can be represented in the following form, provided that $\beta_0, \beta_1, \ldots$ are net premiums.

Lemma 9.1.1 *Assume that* $\mathbb{E} X = 0$. *Then*

$$X = \sum_{k=0}^{\infty} Y_k v^k \,, \tag{9.1.20}$$

where

$$Y_k = \begin{cases} 0 & \text{if } T \le k-1, \\ -\beta_k^{\mathrm{r}} + (b_{J,k+1} - \mu_{k+1}) v & \text{if } T = k, \\ -\beta_k^{\mathrm{r}} & \text{if } T \ge k+1. \end{cases}$$

Proof The decomposition (9.1.18) of β_n gives

$$\begin{aligned} X &= b_{J,T+1} v^{T+1} - \sum_{k=0}^{T} \big((\mu_{k+1} v - \mu_k) + \beta_k^{\mathrm{r}} \big) v^k \\ &= b_{J,T+1} v^{T+1} + \mu_0 - \mu_{T+1} v^{T+1} - \sum_{k=0}^{T} \beta_k^{\mathrm{r}} v^k . \end{aligned}$$

Thus, (9.1.20) follows since $\mu_0 = \mathbb{E}\,X = 0$. □

Note that the random variable Y_k appearing in (9.1.20) is the insurer's loss in year $k+1$, evaluated at time k. Moreover, it turns out that the partial sums $X_n = \sum_{k=0}^n Y_k$ form a martingale. Consider the filtration $\{\mathcal{F}_n\}$ with $\mathcal{F}_n$ the smallest σ-algebra containing the events $\{J = j, T = k\}$, for all $j = 1, \ldots, \ell$, $k = 0, 1, \ldots, n$.

Lemma 9.1.2 *If* $\mathbb{E}\,X = 0$, *then* $\{X_n\}$ *is a martingale with respect to* $\{\mathcal{F}_n\}$.

Proof Since X_n is $\mathcal{F}_n$-measurable, we have

$$\mathbb{E}\,(X_{n+1} \mid \mathcal{F}_n) = \mathbb{E}\,(X_n \mid \mathcal{F}_n) + \mathbb{E}\,(Y_{n+1} \mid \mathcal{F}_n) = X_n + \mathbb{E}\,(Y_{n+1} \mid \mathcal{F}_n).$$

Furthermore, $\mathbb{E}\,(Y_{n+1} \mid \mathcal{F}_n) = \mathbb{I}(T \geq n+1)\mathbb{E}\,(Y_{n+1} \mid T \geq n+1)$ and

$$\begin{aligned}
&\mathbb{E}\,(Y_{n+1} \mid T \geq n+1) \\
&\quad= \mathbb{E}\,\big((-\beta^{\mathrm{r}}_{n+1} + (b_{J,n+2} - \mu_{n+2})v)\mathbb{I}(T = n+1) \mid T \geq n+1\big) \\
&\qquad - \beta^{\mathrm{r}}_{n+1}\mathbb{P}(T \geq n+2 \mid T \geq n+1) \\
&\quad= \mathbb{E}\,\big((b_{J,n+2} - \mu_{n+2})v\mathbb{I}(T = n+1) \mid T \geq n+1\big) - \beta^{\mathrm{r}}_{n+1} \\
&\quad= 0
\end{aligned}$$

where the last equation follows from (9.1.19). Thus, $\mathbb{E}\,(X_{n+1} \mid \mathcal{F}_n) = X_n$. □

The next result is called *Hattendorff's theorem*. It shows how to compute the variance of the insurer's overall loss X. Interestingly, the yearly losses $Y_0, Y_1, \ldots$ are uncorrelated, but generally not independent.

Theorem 9.1.2 *Let* $\mathbb{E}\,X = 0$. *Then for arbitrary* $k, n = 0, 1, \ldots$,

$$\mathrm{Cov}\,(Y_k, Y_n) = \begin{cases} p_m(k)\Big(\sum\limits_{j=1}^{l}(b_{j,k+1} - \mu_{k+1})^2 v^2 q_{j,m+k} - (\beta^{\mathrm{r}}_k)^2\Big) & \text{if } n = k, \\ 0 & \text{if } n \neq k \end{cases} \tag{9.1.21}$$

and

$$\mathrm{Var}\,X = \sum_{k=0}^{\infty} v^{2k}\,\mathrm{Var}\,Y_k\,. \tag{9.1.22}$$

Proof It suffices to show that (9.1.21) is true because, by (9.1.20), equation (9.1.21) yields (9.1.22). However, we showed in Lemma 9.1.2 that $Y_n = X_n - X_{n-1}$, where $\{X_n\}$ is a martingale. Thus, the second part of (9.1.21) follows from (9.1.13). The first part of (9.1.21) is directly obtained from the definition of Y_k. □

9.1.5 Convergence Results

If the filtration $\{\mathcal{F}_n\}$ is not further specified, we simply speak of martingales, submartingales, supermartingales and stopping times without reference to the filtration. A useful tool is the following *submartingale convergence theorem.*

Theorem 9.1.3 *Let $\{X_n, n \geq 0\}$ be a submartingale and assume that*

$$\sup_{n\geq 0} \mathbb{E}\,(X_n)_+ < \infty\,. \tag{9.1.23}$$

Then there exists a random variable X_∞ such that, with probability 1,

$$\lim_{n\to\infty} X_n = X_\infty \tag{9.1.24}$$

and $\mathbb{E}\,|X_\infty| < \infty$. If, additionally,

$$\sup_{n\geq 0} \mathbb{E}\,X_n^2 < \infty\,, \tag{9.1.25}$$

then

$$\mathbb{E}\,X_\infty^2 < \infty\,, \qquad \lim_{n\to\infty} \mathbb{E}\,|X_n - X_\infty| = 0\,. \tag{9.1.26}$$

We first show an auxiliary result which will be used in the proof of Theorem 9.1.3. For arbitrary fixed real numbers $a, b \in \mathbb{R}$ with $a < b$, we consider the number of upcrossings of the interval (a, b) by the sample paths of $\{X_n\}$. Namely, we put

$$\begin{aligned}
\tau_0 &= 0\,,\\
\tau_1 &= \min\{n : n \geq 1, X_n \leq a\}\,,\\
\tau_2 &= \min\{n : n > \tau_1, X_n \geq b\}\,,\\
&\vdots\\
\tau_{2m-1} &= \min\{n : n > \tau_{2m-2}, X_n \leq a\}\,,\\
\tau_{2m} &= \min\{n : n > \tau_{2m-1}, X_n \geq b\}\,,\\
&\vdots
\end{aligned}$$

and call $U_n(a, b) = \max\{m : \tau_{2m} \leq n\}$ the number of upcrossings up to time n. With this notation we can derive the *upcrossing inequality* for discrete-time submartingales.

Lemma 9.1.3 *For each $n \geq 1$,*

$$\mathbb{E}\,U_n(a, b) \leq \frac{\mathbb{E}\,(X_n - a)_+}{b - a} \leq \frac{\mathbb{E}\,(X_n)_+ + |a|}{b - a}\,. \tag{9.1.27}$$

Proof Note that the number of upcrossings of (a,b) by $\{X_n\}$ is identical with the number of upcrossings of $(0, b-a)$ by $\{X'_n\}$, where $X'_n = (X_n - a)_+$. Furthermore, $\{X'_n\}$ is again a submartingale which follows from Jensen's inequality for conditional expectations. We leave it to the reader to show this. Thus, without loss of generality we can assume that $a = 0$ and that $\{X_n\}$ is nonnegative with $X_0 = 0$. Then, it remains to show that

$$\mathbb{E}\, U_n(0,b) \le b^{-1}\mathbb{E}\, X_n\,. \tag{9.1.28}$$

With the notation

$$\eta_i = \begin{cases} 1, & \text{if } \tau_m < i \le \tau_{m+1} \text{ and } m \text{ odd,} \\ 0, & \text{if } \tau_m < i \le \tau_{m+1} \text{ and } m \text{ even,} \end{cases}$$

we have $bU_n(0,b) \le \sum_{i=2}^n \eta_i(X_i - X_{i-1})$ and

$$\{\eta_i = 1\} = \bigcup_{m\in\mathbb{N},\text{ odd}} (\{\tau_m < i\} \setminus \{\tau_{m+1} < i\})\,.$$

Hence

$$\begin{aligned} b\mathbb{E}\, U_n(0,b) &\le \mathbb{E}\sum_{i=2}^n \eta_i(X_i - X_{i-1}) = \sum_{i=2}^n \int_{\{\eta_i=1\}} (X_i - X_{i-1})\, d\mathbb{P} \\ &= \sum_{i=2}^n \int_{\{\eta_i=1\}} \mathbb{E}\,(X_i - X_{i-1} \mid \mathcal{F}_{i-1})\, d\mathbb{P} \end{aligned}$$

since $\{\eta_i = 1\} \in \mathcal{F}_{i-1}$. Thus,

$$\begin{aligned} b\mathbb{E}\, U_n(0,b) &\le \sum_{i=2}^n \int_{\{\eta_i=1\}} (\mathbb{E}\,(X_i \mid \mathcal{F}_{i-1}) - X_{i-1})\, d\mathbb{P} \\ &\le \sum_{i=2}^n \int_{\Omega} (\mathbb{E}\,(X_i \mid \mathcal{F}_{i-1}) - X_{i-1})\, d\mathbb{P} = \mathbb{E}\, X_n - \mathbb{E}\, X_1 \le \mathbb{E}\, X_n\,, \end{aligned}$$

where in the last but one inequality we used that $\{X_n\}$ is a submartingale, that is $\mathbb{E}\,(X_i \mid \mathcal{F}_{i-1}) - X_{i-1} \ge 0$. □

Proof of Theorem 9.1.3 Note first that $X_n \to \infty$ is not possible because of (9.1.23). Let $A \subset \Omega$ be the set of those $\omega \in \Omega$ such that the limit $\lim_{n\to\infty} X_n(\omega)$ does not exist. Then, we have

$$A = \left\{\omega : \liminf_{j\to\infty} X_j(\omega) < \limsup_{j\to\infty} X_j(\omega)\right\} = \bigcup_{\{a,b\in\mathbb{Q}:a<b\}} A_{a,b}\,,$$

where $A_{a,b} = \left\{\liminf_{j\to\infty} X_j < a < b < \limsup_{j\to\infty} X_j\right\} \subset \{U_\infty(a,b) = \infty\}$. Thus, (9.1.23) and (9.1.27) imply that $\mathbb{P}(A_{a,b}) = 0$ and consequently $\mathbb{P}(A) =$

0, i.e. (9.1.24) is proved. Integrability of X_∞ follows from Fatou's lemma. In order to prove (9.1.26), we note that by Fatou's lemma

$$\mathbb{E}\, X_\infty^2 = \mathbb{E}\,(\liminf_{n\to\infty} X_n^2) \le \liminf_{n\to\infty} \mathbb{E}\, X_n^2 \le \sup_{n\ge 0} \mathbb{E}\, X_n^2 .$$

Thus, condition (9.1.25) implies $\mathbb{E}\, X_\infty^2 < \infty$. Furthermore, since (9.1.24) implies $\lim_{n\to\infty} \mathbb{P}(|X_n - X_\infty| > \varepsilon) = 0$ for each $\varepsilon > 0$, we have

$$\begin{aligned}
&\mathbb{E}\,|X_n - X_\infty| \\
&\quad = \mathbb{E}\,((\mathbb{I}(|X_n - X_\infty| \le \varepsilon) + \mathbb{I}(|X_n - X_\infty| > \varepsilon))|X_n - X_\infty|) \\
&\quad \le \varepsilon + \mathbb{E}\,(\mathbb{I}(|X_n - X_\infty| > \varepsilon)|X_n - X_\infty|) .
\end{aligned}$$

Now Schwartz's inequality gives for all sufficiently large $n \in \mathbb{N}$:

$$\begin{aligned}
\mathbb{E}\,|X_n - X_\infty| &\le \varepsilon + \mathbb{P}(|X_n - X_\infty| > \varepsilon)^{1/2}\Big(\sup_n \mathbb{E}\, X_n^2 \\
&\qquad + 2\sup_n (\mathbb{E}\, X_n^2)^{1/2} (\mathbb{E}\, X_\infty^2)^{1/2} + \mathbb{E}\, X_\infty^2\Big)^{1/2} \\
&\le 2\varepsilon .
\end{aligned}$$

This completes the proof of Theorem 9.1.3. □

9.1.6 Optional Sampling Theorems

The next few results are known as *optional sampling theorems* and can be seen as extensions of the fairness property (9.1.1).

Theorem 9.1.4 *Let $\{X_n\}$ be a martingale and τ a bounded stopping time. Then $\mathbb{E}\, X_\tau = \mathbb{E}\, X_0$.*

Proof Let τ be bounded by n_0. Then, by (9.1.10) we have $X_i = \mathbb{E}\,(X_{n_0} \mid \mathcal{F}_i)$ for $i \le n_0$ and consequently

$$\begin{aligned}
\mathbb{E}\, X_\tau &= \mathbb{E}\Big(\sum_{i=0}^{n_0} X_i \mathbb{I}(\tau = i)\Big) = \sum_{i=0}^{n_0} \mathbb{E}\,(\mathbb{E}\,(X_{n_0} \mid \mathcal{F}_i)\mathbb{I}(\tau = i)) \\
&= \sum_{i=0}^{n_0} \mathbb{E}\,(X_{n_0} \mathbb{I}(\tau = i)) = \mathbb{E}\, X_{n_0} .
\end{aligned}$$

□

Unless we make additional assumptions, the boundedness of stopping time τ is essential for the validity of $\mathbb{E}\, X_\tau = \mathbb{E}\, X_0$. With an appropriate finiteness condition, the equality also holds for τ not necessarily bounded.

Theorem 9.1.5 *Let $\{X_n\}$ be a martingale and τ a finite stopping time fulfilling*

$$\mathbb{E}|X_\tau| < \infty \tag{9.1.29}$$

and

$$\lim_{k\to\infty} \mathbb{E}[X_k; \tau > k] = 0 . \tag{9.1.30}$$

Then $\mathbb{E}X_\tau = \mathbb{E}X_0$.

Proof Note that $\tau_k = \min\{\tau, k\}$ is a bounded stopping time for each $k \in \mathbb{N}$. Thus, by Theorem 9.1.4, $\mathbb{E}X_0 = \mathbb{E}X_{\tau_k}$. Hence, using the dominated convergence theorem, (9.1.29) and (9.1.30) give

$$\begin{aligned}\mathbb{E}X_0 &= \lim_{k\to\infty}\mathbb{E}X_{\tau_k} = \lim_{k\to\infty}\mathbb{E}[X_\tau; \tau \le k] + \lim_{k\to\infty}\mathbb{E}[X_k; \tau > k] \\ &= \mathbb{E}X_\tau .\end{aligned}$$

□

We mention still another set of somewhat stronger conditions under which $\mathbb{E}X_\tau = \mathbb{E}X_0$ is true.

Theorem 9.1.6 *Let $\{X_n\}$ be a martingale and τ a stopping time fulfilling*

$$\mathbb{E}\tau < \infty \tag{9.1.31}$$

and, for some constant $c < \infty$,

$$\mathbb{E}(|X_{n+1} - X_n| \mid \mathcal{F}_n) \le c \qquad \text{a.s.} \tag{9.1.32}$$

for all $n \in \mathbb{N}$. Then $\mathbb{E}X_\tau = \mathbb{E}X_0$.

Proof In view of Theorem 9.1.5 it suffices to show that (9.1.29) and (9.1.30) are satisfied. Using the obvious identity $X_\tau = X_0 + \sum_{k=0}^{\infty}(X_{k+1} - X_k)\mathbb{I}(\tau > k)$, the triangle inequality and the monotone convergence theorem give

$$\begin{aligned}\mathbb{E}|X_\tau| &= \mathbb{E}\Big|X_0 + \sum_{k=0}^{\infty}(X_{k+1} - X_k)\mathbb{I}(\tau > k)\Big| \\ &\le \mathbb{E}|X_0| + \sum_{k=0}^{\infty}\mathbb{E}\left[|X_{k+1} - X_k|; \tau > k\right] \\ &= \mathbb{E}|X_0| + \sum_{k=0}^{\infty}\mathbb{E}\left[\mathbb{E}(|X_{k+1} - X_k| \mid \mathcal{F}_k); \tau > k\right] \\ &\le \mathbb{E}|X_0| + c\sum_{k=0}^{\infty}\mathbb{P}(\tau > k) = \mathbb{E}|X_0| + c\mathbb{E}\tau ,\end{aligned}$$

where (9.1.32) is used in the last inequality. Thus, by (9.1.31), (9.1.29) follows. Furthermore, using (9.1.32) repeatedly, we have

$$\begin{aligned} |\mathbb{E}[X_k;\tau > k]| &\le \mathbb{E}[|X_k|;\tau > k] \le \mathbb{E}[|X_0|;\tau > k] + c\mathbb{E}[k;\tau > k] \\ &\le \mathbb{E}[|X_0|;\tau > k] + c\mathbb{E}[\tau;\tau > k] \end{aligned}$$

and (9.1.30) follows from (9.1.31) and the dominated convergence theorem. □

Theorem 9.1.6 can be used to prove *Wald's identity* (6.3.5) for stopping times.

Corollary 9.1.1 *Consider a random walk* $\{S_n\}$ *with* $S_n = \sum_{i=1}^n Y_i$, *where* $Y_1, Y_2, \ldots$ *are independent and identically distributed random variables with* $\mathbb{E}|Y| < \infty$. *If* τ *is a stopping time with respect to the filtration* $\{\mathcal{F}_n^Y\}$ *and if* $\mathbb{E}\,\tau < \infty$, *then*

$$\mathbb{E}\,S_\tau = \mathbb{E}\,\tau\mathbb{E}\,Y. \tag{9.1.33}$$

Proof Applying Theorem 9.1.6 to the martingale $\{X_n\}$, $X_n = S_n - n\mathbb{E}\,Y$, we have to show that condition (9.1.32) is fulfilled. Note that

$$\begin{aligned} \mathbb{E}\,(|X_{n+1} - X_n| \mid \mathcal{F}_n^Y) &= \mathbb{E}\,(|Y_{n+1} - \mathbb{E}\,Y| \mid \mathcal{F}_n^Y) \\ &\le \mathbb{E}\,(|Y_{n+1}| \mid \mathcal{F}_n^Y) + |\mathbb{E}\,Y| \le 2\mathbb{E}\,|Y|, \end{aligned}$$

i.e. (9.1.32) holds. Consequently, Theorem 9.1.6 gives the equalities $0 = \mathbb{E}\,X_0 = \mathbb{E}\,X_\tau = \mathbb{E}\,S_\tau - \mathbb{E}\,\tau\mathbb{E}\,Y$. □

From the proofs of Theorems 9.1.4–9.1.6 it is easily seen that analogous results are also valid for sub- and supermartingales. If $\{X_n\}$ is a submartingale (supermartingale), then

$$\mathbb{E}\,X_\tau \ge (\le)\mathbb{E}\,X_0, \tag{9.1.34}$$

provided the conditions of one of the Theorems 9.1.4–9.1.6 are fulfilled. Moreover, Theorem 9.1.4 can be generalized in the following way.

Theorem 9.1.7 *Let* $\{X_n\}$ *be a submartingale and* τ *a stopping time such that* $\mathbb{P}(\tau \le n_0) = 1$ *for some* $n_0 \in \mathbb{N}$. *Then, for each* $x > 0$,

$$\mathbb{E}\,[X_\tau; X_\tau > x] \le \mathbb{E}\,[X_{n_0}; X_\tau > x]. \tag{9.1.35}$$

Proof We have

$$\begin{aligned} \mathbb{E}\,[X_\tau; X_\tau > x] &= \sum_{k=0}^{n_0} \mathbb{E}\,[X_k; X_k > x, \tau = k] \\ &\le \sum_{k=0}^{n_0} \mathbb{E}\,[X_{n_0}; X_k > x, \tau = k] = \mathbb{E}\,[X_{n_0}; X_\tau > x], \end{aligned}$$

where the last inequality follows from (9.1.11). □

Example The optional sampling theorem in the form of inequality (9.1.34) can be used to give another proof of a somewhat weaker version of the exponential bound derived in Theorem 4.5.2 for the tail function of compound distributions. Let $U_1, U_2, \ldots$ be nonnegative, independent and identically distributed with distribution F_U and let N be an $\mathbb{N}$-valued random variable with probability function $\{p_k\}$ which is independent of $U_1, U_2, \ldots$. Consider the compound distribution $F = \sum_{k=0}^{\infty} p_k F_U^{*k}$ of $\sum_{j=1}^{N} U_j$ and assume that, for some $0 < \theta < 1$,

$$\mathbb{P}(N > n+1 \mid N > n) \le \theta\,, \qquad n \in \mathbb{N}\,, \tag{9.1.36}$$

that is the probability function $\{p_k\}$ satisfies condition (4.5.13), i.e. $r_{n+1} \le \theta r_n$ for $n \ge 1$, where $r_n = \sum_{k=n}^{\infty} p_k$. Furthermore, assume that

$$\hat{m}_{F_U}(\gamma) = \theta^{-1} \tag{9.1.37}$$

has the solution $\gamma > 0$. We will show that

$$\bar{F}(x) \le \frac{1-p_0}{\theta} \mathrm{e}^{-\gamma x}, \qquad x \ge 0\,. \tag{9.1.38}$$

For each $n \in \mathbb{N}$, define

$$X_n = \begin{cases} \mathrm{e}^{\gamma S_{n+1}} & \text{if } N > n, \\ 0 & \text{if } N \le n, \end{cases} \tag{9.1.39}$$

where $S_n = U_1 + \ldots + U_n$. Then, $X_{n+1} = Z_{n+1} X_n$ for $n = 1, 2, \ldots$, where

$$Z_n = \begin{cases} \mathrm{e}^{\gamma U_{n+1}} & \text{if } N > n, \\ 0 & \text{if } N \le n. \end{cases}$$

Consider the filtration $\{\mathcal{F}_n\}$, where $\mathcal{F}_n$ is the σ-algebra generated by the random variables $\mathbb{I}(N = 0), \ldots, \mathbb{I}(N = n), U_1, \ldots, U_{n+1}$. Note that by (9.1.36) we have $\mathbb{P}(N > n+1 \mid \mathcal{F}_n) \le \theta$, and hence by (9.1.37),

$$\begin{aligned} &\mathbb{E}\,(Z_{n+1} \mid \mathcal{F}_n) = \mathbb{E}\,(\mathrm{e}^{\gamma U_{n+2}} \mathbb{I}(N > n+1) \mid \mathcal{F}_n) = \\ &= \mathbb{E}\,(\mathrm{e}^{\gamma U_{n+2}}) \mathbb{P}(N > n+1 \mid \mathcal{F}_n) = \theta^{-1} \mathbb{P}(N > n+1 \mid \mathcal{F}_n) \le 1\,. \end{aligned}$$

Thus $\mathbb{E}\,(X_{n+1} \mid \mathcal{F}_n) = \mathbb{E}\,(Z_{n+1} X_n \mid \mathcal{F}_n) = \mathbb{E}\,(Z_{n+1} \mid \mathcal{F}_n) X_n \le X_n$, that is $\{X_n, n \in \mathbb{N}\}$ is an $\{\mathcal{F}_n\}$-supermartingale. For each $x > 0$, consider the $\{\mathcal{F}_n\}$-stopping time $\tau = \min\{i : S_{i+1} > x\}$. We leave it to the reader to check that τ fulfils the conditions of Theorem 9.1.5. We then apply (9.1.34) to the supermartingale $\{X_n\}$ using the fact that

$$\begin{aligned} X_\tau &= \sum_{k=0}^{\infty} \mathbb{I}(\tau = k)X_k = \sum_{k=0}^{\infty} \mathbb{I}(\tau = k)\mathbb{I}(N > k)\mathrm{e}^{\gamma S_{k+1}} \\ &\geq \sum_{k=0}^{\infty} \mathbb{I}(\tau = k)\mathbb{I}(N > k)\mathrm{e}^{\gamma x} = \sum_{k=0}^{\infty} \mathbb{I}(\tau = k)\mathbb{I}\Big(\sum_{j=1}^{N} U_j > x\Big)\mathrm{e}^{\gamma x} \\ &= \mathbb{I}\Big(\sum_{j=1}^{N} U_j > x\Big)\mathrm{e}^{\gamma x}. \end{aligned}$$

Thus, (9.1.34) gives $(1 - p_0)\theta^{-1} = \mathbb{E}\,\big(\mathrm{e}^{\gamma U_1}\mathbb{I}(N > 0)\big) = \mathbb{E}\,X_0 \geq \mathbb{E}\,X_\tau \geq \mathrm{e}^{\gamma x}\mathbb{P}(\sum_{j=1}^{N} U_j > x)$, and hence the exponential bound (9.1.38).

9.1.7 Doob's Inequality

We now deal with *Doob's inequality* for sub- and supermartingales.

Theorem 9.1.8 (a) *If $\{X_n\}$ is a nonnegative submartingale, then*

$$\mathbb{P}\Big(\max_{0\leq k\leq n} X_k \geq x\Big) \leq \frac{\mathbb{E}\,X_n}{x}, \qquad x > 0, n \in \mathbb{N}. \tag{9.1.40}$$

(b) *If $\{X_n\}$ is a nonnegative supermartingale, then*

$$\mathbb{P}\Big(\max_{0\leq k\leq n} X_k \geq x\Big) \leq \frac{\mathbb{E}\,X_0}{x}, \qquad x > 0, n \in \mathbb{N}. \tag{9.1.41}$$

Proof Assume that $\{X_n\}$ is a submartingale. Let $A = \{\max_{0\leq k\leq n} X_k \geq x\}$. Then $A = A_0 \cup \ldots \cup A_n$ is the sum of the disjoint events

$$\begin{aligned} A_0 &= \{X_0 \geq x\} \in \mathcal{F}_0\,, \\ A_k &= \{X_0 < x, X_1 < x, \ldots, X_{k-1} < x, X_k \geq x\} \in \mathcal{F}_k\,, \qquad 1 \leq k \leq n\,. \end{aligned}$$

In view of the submartingale property, we have $\mathbb{E}\,[X_n; A_k] \geq \mathbb{E}\,[X_k; A_k] \geq x\mathbb{P}(A_k)$. Summing over $k = 0, \ldots, n$, we see that $X_k \geq 0$ implies $\mathbb{E}\,X_n \geq \mathbb{E}\,[X_n; A]$, and so statement (a) follows. The proof of statement (b) is analogous and is left to the reader. □

Remark Doob's inequality (9.1.41) can be used to give a simpler proof for the exponential bound (9.1.38). Define the filtration $\{\mathcal{F}_n\}$ and the supermartingale $\{X_n\}$ as in the example of Section 9.1.6. Then (9.1.41) gives

$$\begin{aligned} \mathbb{P}\Big(\sum_{j=1}^{N} U_j > x\Big) &= \mathbb{P}\Big(\max_{n\in\mathbb{N}}\{S_{n+1}\mathbb{I}(N > n)\} > x\Big) = \mathbb{P}\Big(\max_{n\in\mathbb{N}} X_n > \mathrm{e}^{\gamma x}\Big) \\ &= \lim_{m\to\infty} \mathbb{P}\Big(\max_{0\leq n\leq m} X_n > \mathrm{e}^{\gamma x}\Big) \leq \frac{\mathbb{E}\,X_0}{\mathrm{e}^{\gamma x}} = \frac{1 - p_0}{\theta}\mathrm{e}^{-\gamma x}. \end{aligned}$$

9.1.8 The Doob–Meyer Decomposition

We are now in a position to generalize the decomposition property for the sequence $\{W_n\}$ of cumulative claim amounts as discussed in Section 9.1.1. Note that $\{W_n\}$ is an $\{\mathcal{F}_n^W\}$-submartingale. We call a sequence $\{X_n, n \in \mathbb{N}\}$ of random variables *increasing* from zero if $X_0 = 0$ and $\mathbb{P}(\bigcap_{n=0}^{\infty}\{X_n \le X_{n+1}\}) = 1$. The reader should prove that each increasing sequence $\{X_n\}$ adapted to a filtration $\{\mathcal{F}_n\}$ is a submartingale. Two sequences $\{X_n\}$ and $\{X'_n\}$ are coined *indistinguishable* if $\mathbb{P}(\bigcap_{n=0}^{\infty}\{X_n = X'_n\}) = 1$. The following result is the well-known *Doob–Meyer decomposition* for submartingales.

Theorem 9.1.9 *Let $\{X_n\}$ be a submartingale with respect to a filtration $\{\mathcal{F}_n\}$. Then there exists an $\{\mathcal{F}_n\}$-martingale $\{M_n\}$ and an $\{\mathcal{F}_n\}$-predictable sequence $\{V_n\}$ which is increasing from zero and such that $X_n = X_0 + M_n + V_n$ for all $n \in \mathbb{N}$. This decomposition is unique modulo indistinguishability. Moreover, a version of $\{V_n\}$ is given by*

$$V_n = \sum_{k=1}^{n} \mathbb{E}\,(X_k - X_{k-1} \mid \mathcal{F}_{k-1}), \qquad n \ge 1, \tag{9.1.42}$$

which itself is called the compensator.

Proof Let V_n be the random variable given by (9.1.42) and define $M_n = X_n - X_0 - V_n$ for all $n = 1, 2, \ldots$; $M_0 = V_0 = 0$. Then

$$\begin{aligned}
\mathbb{E}\,(M_{n+1} \mid \mathcal{F}_n) &= \mathbb{E}\Big(X_{n+1} - X_0 - \sum_{k=1}^{n+1} \mathbb{E}\,(X_k - X_{k-1} \,\Big|\, \mathcal{F}_{k-1}) \mid \mathcal{F}_n\Big) \\
&= \mathbb{E}\,(X_{n+1} \mid \mathcal{F}_n) - X_0 - \sum_{k=1}^{n+1} \mathbb{E}\,(X_k - X_{k-1} \mid \mathcal{F}_{k-1}) \\
&= X_n - X_0 - \sum_{k=1}^{n} \mathbb{E}\,(X_k - X_{k-1} \mid \mathcal{F}_{k-1}) = M_n,
\end{aligned}$$

i.e. $\{M_n\}$ is an $\{\mathcal{F}_n\}$-martingale. By definition, (9.1.42) implies that V_n is measurable with respect to $\mathcal{F}_{n-1}$, for each $n = 1, 2, \ldots$. This means that $\{V_n\}$ is $\{\mathcal{F}_n\}$-predictable. Since $\{X_n\}$ is an $\{\mathcal{F}_n\}$-submartingale, we have $\mathbb{E}\,(X_{n+1} - X_n \mid \mathcal{F}_n) = \mathbb{E}\,(X_{n+1} \mid \mathcal{F}_n) - X_n \ge 0$ and consequently

$$V_{n+1} = \sum_{k=1}^{n+1} \mathbb{E}\,(X_k - X_{k-1} \mid \mathcal{F}_{k-1}) \ge \sum_{k=1}^{n} \mathbb{E}\,(X_k - X_{k-1} \mid \mathcal{F}_{k-1}) = V_n\,,$$

i.e. $\{V_n\}$ is increasing from zero. Suppose that there exists another decomposition $\{M'_n\}$, $\{V'_n\}$ of $\{X_n\}$ with the same properties. Then, $M'_n + V'_n = M_n + V_n$

for each $n \in \mathbb{N}$. This means that the $\{\mathcal{F}_n\}$-martingale $\{M_n - M'_n, n \in \mathbb{N}\}$ is $\{\mathcal{F}_n\}$-predictable. Hence

$$M_{n+1} - M'_{n+1} = \mathbb{E}(M_{n+1} - M'_{n+1} \mid \mathcal{F}_n) = M_n - M'_n,$$

which shows that $M_n - M'_n = 0$ for all $n \in \mathbb{N}$, since $M_0 - M'_0 = 0$. Consequently, $V_n - V'_n = 0$ for all $n \in \mathbb{N}$. □

The Doob–Meyer decomposition for submartingales, given in Theorem 9.1.9, can be used when defining the concept of a fair premium for the sequence $\{Z_n\}$ of aggregate claims considered in Section 9.1.1. However, we no longer assume that the risks $Z_1, Z_2, \ldots$ are independent. Consider the submartingale $\{W_n\}$, $W_n = \sum_{i=1}^n Z_i$, of cumulative claim amounts. Then by (9.1.42), the conditional increment $\mathbb{E}(Z_n \mid \mathcal{F}^Z_{n-1})$ is a fair premium to be paid for the aggregate claim over the interval $(n-1, n]$, in the sense that $\{M_n\}$ given by $M_n = \sum_{i=1}^n (Z_i - \mathbb{E}(Z_i \mid \mathcal{F}^Z_{i-1}))$ is a martingale and where $\mathcal{F}^Z_0 = \{\emptyset, \Omega\}$. The sequence $\{V_n\}$ with

$$V_n = \sum_{i=1}^n \mathbb{E}(Z_i \mid \mathcal{F}^Z_{i-1}) \qquad (9.1.43)$$

is called the (cumulative) *net premium process.*

Analogous to Theorem 9.1.9, a Doob–Meyer decomposition can also be proved for supermartingales. The reader can easily provide a proof if he uses the fact that $\{-X_n\}$ is a submartingale when $\{X_n\}$ is a supermartingale.

Bibliographical Notes. The introduction to martingale theory presented in this section is standard. For further details we refer to textbooks like Karr (1993) or Williams (1991). The exposition of Section 9.1.4 follows Gerber (1995). A continuous-time version of Hattendorff's theorem can be found, for example, in Wolthuis (1987). In risk theory, the usefulness of martingales was discovered in Gerber (1973); see also De Vylder (1977) and Gerber (1975). Scheike (1992) introduced the notion of the net premium process defined in (9.1.43). The idea of applying an optional sampling theorem for supermartingales to derive the exponential bound (9.1.38) for the tail function of compound distributions was also used in Gerber (1994). The proof of this result via Doob's inequality as mentioned in Section 9.1.7 seems to be new.

9.2 CHANGE OF THE PROBABILITY MEASURE

In this section we study concepts related to the likelihood ratio martingale which has been introduced in Example 7 of Section 9.1.3. Apart from examples, we show how to use the subsequent results in risk theory.

9.2.1 The Likelihood Ratio Martingale

We start from a sequence of random variables $Y_1, Y_2, \ldots$ defined on a measurable space $(\Omega, \mathcal{F})$, with filtration $\{\mathcal{F}_n^Y\}$. Put $\mathcal{F}_0^Y = \{\emptyset, \Omega\}$. For the underlying probability measure, we have to choose between $\mathbb{P}$ and $\tilde{\mathbb{P}}$. Under $\mathbb{P}$, the sequence $\{Y_n\}$ consists of independent and identically distributed random variables with common density $f(x)$, while under $\tilde{\mathbb{P}}$ the sequence $\{Y_n\}$ consists of independent and identically distributed random variables with common density $\tilde{f}(x)$. We assume that $f(x) > 0$ if and only if $\tilde{f}(x) > 0$. Define the *likelihood ratio function*

$$l(y_1, \ldots, y_n) = \begin{cases} \dfrac{\tilde{f}(y_1) \ldots \tilde{f}(y_n)}{f(y_1) \ldots f(y_n)} & \text{if } f(y_1) \ldots f(y_n) > 0, \\ 0 & \text{otherwise,} \end{cases}$$

and let

$$X_0 = 1, \qquad X_n = l(Y_1, \ldots, Y_n) \tag{9.2.1}$$

for $n \geq 1$. We have that

- the sequence $\{X_n\}$, considered on $(\Omega, \mathcal{F}, \mathbb{P})$, is an $\{\mathcal{F}_n^Y\}$-martingale,
- for all $A \in \mathcal{F}_n^Y$,

$$\tilde{\mathbb{P}}(A) = \int_A X_n(\omega) \mathbb{P}(\mathrm{d}\omega), \tag{9.2.2}$$

- $\mathbb{E} X_n = 1$.

The martingale property of $\{X_n\}$ was already noticed in Section 9.1.3. To prove (9.2.2), it suffices to consider events of the form $A = \{Y_1 \in B_1, \ldots, Y_n \in B_n\}$ which generate $\mathcal{F}_n^Y$ and where $B_1, \ldots, B_n \in \mathcal{B}(\mathbb{R})$. Then

$$\begin{aligned} \tilde{\mathbb{P}}(A) &= \int_{B_1} \cdots \int_{B_n} \tilde{f}(y_1) \ldots \tilde{f}(y_n) \, \mathrm{d}y_n \ldots \mathrm{d}y_1 \\ &= \int_{B_1} \cdots \int_{B_n} l(y_1, \ldots, y_n) f(y_1) \ldots f(y_n) \, \mathrm{d}y_n \ldots \mathrm{d}y_1 \\ &= \int_A X_n(\omega) \mathbb{P}(\mathrm{d}\omega) = \mathbb{E}\,[X_n; A]. \end{aligned}$$

Hence (9.2.2) holds and it is immediate that $\mathbb{E} X_n = \tilde{\mathbb{P}}(\Omega) = 1$.

The *likelihood ratio martingale* $\{X_n\}$ defined in (9.2.1) is a special case of the following model. We start from a sequence of random variables $Y_1, Y_2, \ldots$ defined on a measurable space $(\Omega, \mathcal{F})$, whose filtration $\{\mathcal{F}_n^Y\}$ is given. Again there are two candidates for the underlying probability measure $\mathbb{P}$ and $\tilde{\mathbb{P}}$. However, this time we assume nothing about the independence of $Y_1, Y_2, \ldots$.

In accordance with (9.2.2) we assume that, for each $n \in \mathbb{N}$, there exists an $\mathcal{F}_n^Y$-measurable, nonnegative random variable X_n such that

$$\tilde{\mathbb{P}}(A) = \int_A X_n(\omega)\mathbb{P}(\mathrm{d}\omega), \qquad A \in \mathcal{F}_n^Y. \tag{9.2.3}$$

Theorem 9.2.1 *On the probability space $(\Omega, \mathcal{F}, \mathbb{P})$ the sequence $\{X_n\}$ given by (9.2.3) is an $\{\mathcal{F}_n^Y\}$-martingale with mean $\mathbb{E}\, X_n = 1$.*

Proof Since $\mathcal{F}_n^Y \subset \mathcal{F}_{n+1}^Y$, it is clear that (9.2.3) implies $\mathbb{E}\,[X_{n+1}; A] = \mathbb{E}\,[X_n; A]$ for all $A \in \mathcal{F}_n^Y$. Together with the assumed $\mathcal{F}_n^Y$-measurability of X_n, this gives $\mathbb{E}\,(X_{n+1} \mid \mathcal{F}_n^Y) = X_n$ because, by the definition of the conditional expectation $\mathbb{E}\,(X_{n+1} \mid \mathcal{F}_n^Y)$, we have $\mathbb{E}\,[\mathbb{E}\,(X_{n+1} \mid \mathcal{F}_n^Y); A] = \mathbb{E}\,[X_{n+1}; A]$. Consequently, $\mathbb{E}\,[\mathbb{E}\,(X_{n+1} \mid \mathcal{F}_n^Y); A] = \mathbb{E}\,[X_n; A]$ for all $A \in \mathcal{F}_n^Y$. Furthermore, (9.2.3) obviously implies that $\mathbb{E}\, X_n = 1$ for all $n \in \mathbb{N}$. □

Note that we can rewrite (9.2.3) in terms of the restrictions of the probability measures $\mathbb{P}$ and $\tilde{\mathbb{P}}$ to the σ-algebra $\mathcal{F}_n^Y$, denoted by $\mathbb{P}_n$ and $\tilde{\mathbb{P}}_n$, respectively. Then (9.2.3) reads

$$\tilde{\mathbb{P}}_n(A) = \int_A X_n(\omega)\mathbb{P}_n(\mathrm{d}\omega), \qquad A \in \mathcal{F}_n^Y. \tag{9.2.4}$$

This assumption is justified by the Radon–Nikodym theorem, which says that (9.2.4) holds if and only if $\tilde{\mathbb{P}}_n(A) = 0$ whenever $\mathbb{P}_n(A) = 0$, for all $A \in \mathcal{F}_n^Y$. Thus, X_n is called the *Radon–Nikodym derivative* of $\tilde{\mathbb{P}}_n$ with respect to $\mathbb{P}_n$ and is denoted by $X_n(\omega) = (\mathrm{d}\tilde{\mathbb{P}}_n)/(\mathrm{d}\mathbb{P}_n)(\omega)$. In particular, if the densities $f_n(y_1, \ldots, y_n)$ and $\tilde{f}_n(y_1, \ldots, y_n)$ of the random vector $(Y_1, \ldots, Y_n)$ under $\mathbb{P}_n$ and $\tilde{\mathbb{P}}_n$ exist, respectively, then with $\mathbb{P}_n$-probability 1

$$X_n = \frac{\tilde{f}_n(Y_1, \ldots, Y_n)}{f_n(Y_1, \ldots, Y_n)}, \qquad n = 1, 2, \ldots . \tag{9.2.5}$$

9.2.2 Kolmogorov's Extension Theorem

We now consider the following converse question. Suppose we have a probability space $(\Omega, \mathcal{F}, \mathbb{P})$ with filtration $\{\mathcal{F}_n\}$ and where $\mathcal{F} = \mathcal{F}_\infty$, the smallest σ-algebra consisting of all events from $\bigcup_{n=0}^\infty \mathcal{F}_n$. Let $\{X_n\}$ be a sequence of nonnegative random variables on $(\Omega, \mathcal{F}, \mathbb{P})$ forming an $\{\mathcal{F}_n\}$-martingale with $\mathbb{E}\, X_n = 1$. By $\mathbb{P}_n$ we denote the restriction of $\mathbb{P}$ to $\mathcal{F}_n$ and, for each $n \in \mathbb{N}$, we define the set function $\tilde{\mathbb{P}}_n : \mathcal{F}_n \to [0, 1]$ by

$$\tilde{\mathbb{P}}_n(A) = \int_A X_n(\omega)\mathbb{P}_n(\mathrm{d}\omega) \quad \left(= \int_A X_n(\omega)\mathbb{P}(\mathrm{d}\omega)\right), \qquad A \in \mathcal{F}_n. \tag{9.2.6}$$

It is straightforward to check that $\tilde{\mathbb{P}}_n$ is a probability measure on $(\Omega, \mathcal{F}_n)$. The question is whether there exists a probability measure $\tilde{\mathbb{P}}$ on $(\Omega, \mathcal{F})$ such that the restriction of $\tilde{\mathbb{P}}$ to $\mathcal{F}_n$ is $\tilde{\mathbb{P}}_n$, for all $n \in \mathbb{N}$. An answer to this question is given by *Kolmogorov's extension theorem.* The discrete-time version of this theorem is stated for $\Omega = \mathbb{R}^\infty$, the set of all sequences $y_0, y_1, \ldots$ of real numbers. On this set Ω the σ-algebra $\mathcal{F}$ is defined in the following way. Let $\mathcal{G}_n$ be the family of those subsets of $\mathbb{R}^\infty$ which are finite unions of sets of the form $B_0 \times B_1 \times \ldots$, where $B_k \in \mathcal{B}(\mathbb{R})$ for all $k \le n$ and $B_k = \mathbb{R}$ for all $n > k$. Put $\mathcal{F} = \mathcal{B}(\mathbb{R}^\infty)$, where $\mathcal{B}(\mathbb{R}^\infty) = \sigma(\bigcup_{n=0}^\infty \mathcal{G}_n)$.

Theorem 9.2.2 *Suppose that, for each $n \ge 0$, there exists a probability measure P_n on $(\mathbb{R}^{n+1}, \mathcal{B}(\mathbb{R}^{n+1}))$ and suppose that the family $\{P_n\}$ fulfils the consistency condition*

$$P_{n+1}(B_0 \times \ldots \times B_n \times \mathbb{R}) = P_n(B_0 \times \ldots \times B_n)\,, \qquad n = 0, 1, \ldots . \tag{9.2.7}$$

Then there exists a uniquely determined probability measure, $\mathbb{P}$ say, on $(\mathbb{R}^\infty, \mathcal{B}(\mathbb{R}^\infty))$ such that for all $n = 0, 1, \ldots$ and $B_0, \ldots, B_n \in \mathcal{B}(\mathbb{R})$,

$$\mathbb{P}(B_0 \times \ldots \times B_n \times \mathbb{R}^\infty) = P_n(B_0 \times \ldots \times B_n)\,. \tag{9.2.8}$$

The *proof* of Theorem 9.2.2 is omitted and can be found, for example, in Shiryayev (1984). The probability space $(\Omega, \mathcal{F}, \mathbb{P}) = (\mathbb{R}^\infty, \mathcal{B}(\mathbb{R}^\infty), \mathbb{P})$ considered in Theorem 9.2.2 is called a *canonical probability space.*

Corollary 9.2.1 *Let $\{X_n, n \in \mathbb{N}\}$ be a nonnegative martingale on $(\mathbb{R}^\infty, \mathcal{B}(\mathbb{R}^\infty), \mathbb{P})$ with respect to the filtration $\{\mathcal{F}_n\}$ given by $\mathcal{F}_n = \sigma(\mathcal{G}_n)$. Assume that $\mathbb{E}\, X_n = 1$. Then there exists a uniquely determined probability measure $\tilde{\mathbb{P}}$ on $(\mathbb{R}^\infty, \mathcal{B}(\mathbb{R}^\infty))$ such that*

$$\tilde{\mathbb{P}}(A) = \tilde{\mathbb{P}}_n(A)\,, \tag{9.2.9}$$

for all $n \in \mathbb{N}, A \in \mathcal{F}_n$, where $\tilde{\mathbb{P}}_n$ is given by (9.2.6).

Proof For each $n = 0, 1, \ldots$, we put

$$P_n(B_0 \times \ldots \times B_n) = \tilde{\mathbb{P}}_n(B_0 \times \ldots \times B_n \times \mathbb{R}^\infty)\,, \quad B_0 \ldots, B_n \in \mathcal{B}(\mathbb{R})\,. \tag{9.2.10}$$

Then the family $\{P_n\}$ of probability measures defined in (9.2.10) fulfils (9.2.7) since $B_0 \times \ldots \times B_n \times \mathbb{R}^\infty \in \mathcal{F}_n$ and consequently

$$\begin{aligned} P_{n+1}(B_0 \times \ldots \times B_n \times \mathbb{R}) &= \int_{B_0 \times \ldots \times B_n \times \mathbb{R}^\infty} X_{n+1}(\omega) \mathbb{P}(\mathrm{d}\omega) \\ = \int_{B_0 \times \ldots \times B_n \times \mathbb{R}^\infty} \mathbb{E}\,(X_{n+1} \mid \mathcal{F}_n)(\omega) \mathbb{P}(\mathrm{d}\omega) &= \int_{B_0 \times \ldots \times B_n \times \mathbb{R}^\infty} X_n(\omega) \mathbb{P}(\mathrm{d}\omega) \\ = P_n(B_0 \times \ldots \times B_n)\,. \end{aligned}$$

In view of Theorem 9.2.2, this completes the proof. □

Remark In particular, Theorem 9.2.2 implies the existence of a "global" distribution $\mathbb{P}$ of a sequence $Y_0, Y_1, \ldots$ of independent and identically distributed random variables. Indeed, let F be the common distribution of the Y_n. Then, the family $\{P_n\}$ of probability measures given by

$$P_n(B_0 \times \ldots \times B_n) = \prod_{i=0}^{n} F(B_i), \qquad B_0, \ldots, B_n \in \mathcal{B}(\mathbb{R}), \tag{9.2.11}$$

satisfies the consistency condition (9.2.7). In an analogous fashion, Section 7.1.2 contains the notion of a Markov chain using a consistent family of finite-dimensional distributions. Again, in Sections 8.1.1 and 8.4.1 continuous-time Markov processes have been introduced in the same way.

9.2.3 Exponential Martingales for Random Walks

Let F be a distribution on $\mathbb{R}$ and let $\mathbb{P}$ be the probability measure on the measurable space $(\mathbb{R}^\infty, \mathcal{B}(\mathbb{R}^\infty))$ as given by (9.2.8) and (9.2.11). Furthermore, let $Y_0, Y_1, \ldots$ be a sequence of independent and identically distributed random variables on $(\mathbb{R}^\infty, \mathcal{B}(\mathbb{R}^\infty), \mathbb{P})$ with the common distribution F. Assume that $(\mathbb{R}^\infty, \mathcal{B}(\mathbb{R}^\infty), \mathbb{P})$ is the *canonical probability space* of $\{Y_n\}$, i.e. $Y_n(\omega) = y_n$ for all $n = 0, 1, \ldots$; $\omega = (y_0, y_1, \ldots)$. Assume that $\mathbb{E}\, Y < 0$ and consider the random walk $\{S_n\}$ with $S_n = \sum_{i=1}^{n} Y_i$. As shown in Example 1 of Section 9.1.3, $\{S_n\}$ is an $\{\mathcal{F}_n^Y\}$-supermartingale. Example 3 of Section 9.1.3 shows that the sequence $\{X_n, n \in \mathbb{N}\}$ with

$$X_n = \exp\Big(\gamma \sum_{i=1}^{n} Y_i\Big), \qquad n = 0, 1, \ldots, \tag{9.2.12}$$

is an $\{\mathcal{F}_n^Y\}$-martingale on $(\mathbb{R}^\infty, \mathcal{B}(\mathbb{R}^\infty), \mathbb{P})$ provided the equation

$$\hat{m}_F(s) = 1 \tag{9.2.13}$$

admits a positive solution γ. Corollary 9.2.1 now implies that there exists a well-defined probability measure $\tilde{\mathbb{P}}$ on $(\mathbb{R}^\infty, \mathcal{B}(\mathbb{R}^\infty))$ given by (9.2.6) and (9.2.9). Furthermore, the sequence $\{\tilde{X}_n, n \in \mathbb{N}\}$

$$\tilde{X}_n = \exp\Big(-\gamma \sum_{i=1}^{n} Y_i\Big), \qquad n = 0, 1, \ldots, \tag{9.2.14}$$

is an $\{\mathcal{F}_n^Y\}$-martingale on $(\mathbb{R}^\infty, \mathcal{B}(\mathbb{R}^\infty), \tilde{\mathbb{P}})$. This follows from the fact that, under $\tilde{\mathbb{P}}$, the random variables $Y_1, Y_2, \ldots$ are independent and identically distributed with distribution function $\tilde{F}(x)$, where $\tilde{F}(x) = F_\gamma(x)$ and

$$F_s(x) = \int_{-\infty}^{x} \frac{e^{sy}}{\hat{m}_F(s)}\, \mathrm{d}F(y) \tag{9.2.15}$$

is the associated distribution to F, for all $s \in \mathbb{R}$ such that $\hat{m}_F(s) < \infty$.

By $\tilde{\mathbb{E}}$ we denote the expectation taken with respect to $\tilde{\mathbb{P}}$. Note that $\tilde{\mathbb{E}}\, Y > 0$ because $\tilde{\mathbb{E}}\, Y = \int_{-\infty}^{\infty} x e^{\gamma x}\, dF(x) = \hat{m}_F^{(1)}(\gamma-)$ and $\hat{m}_F(s)$ is strictly increasing at γ. The random walk $\{S_n\}$ therefore tends to $-\infty$ under $\mathbb{P}$, but under $\tilde{\mathbb{P}}$ it tends to ∞; see also Theorem 6.3.1. The stopping time

$$\tau_d(u) = \inf\{n : S_n > u\} \tag{9.2.16}$$

is thus finite with $\tilde{\mathbb{P}}$-probability 1.

Since $\{\tilde{X}_n\}$ given by (9.2.14) is a martingale on $(\mathbb{R}^\infty, \mathcal{B}(\mathbb{R}^\infty), \tilde{\mathbb{P}})$, the change of measure $\mathbb{P} \mapsto \tilde{\mathbb{P}}$ defined in (9.2.6) and (9.2.9) can be iterated. For each $n \in \mathbb{N}$, let

$$\tilde{\tilde{\mathbb{P}}}_n(A) = \int_A \tilde{X}_n(\omega) \tilde{\mathbb{P}}(d\omega)\,, \qquad A \in \mathcal{F}_n^Y. \tag{9.2.17}$$

Then Corollary 9.2.1 implies that there exists a uniquely determined probability measure $\tilde{\tilde{\mathbb{P}}}$ on $(\mathbb{R}^\infty, \mathcal{B}(\mathbb{R}^\infty))$ such that $\tilde{\tilde{\mathbb{P}}}(A) = \tilde{\tilde{\mathbb{P}}}_n(A)$ for all $n \in \mathbb{N}, A \in \mathcal{F}_n^Y$. However, in view of (9.2.14) and (9.2.17), we have

$$\tilde{\tilde{\mathbb{P}}} = \mathbb{P}. \tag{9.2.18}$$

In what follows, we need a variant of the optional sampling theorems as stated in Section 9.1.6: if $\{X_n\}$ is a martingale and τ a stopping time, then $\mathbb{E}\,(X_n \mid \mathcal{F}_\tau) = X_{\tau \wedge n}$ for each $n \in \mathbb{N}$, where

$$\mathcal{F}_\tau = \{A : \{\tau = n\} \cap A \in \mathcal{F}_n, \text{ for all } n \in \mathbb{N}\} \tag{9.2.19}$$

is the σ-algebra consisting of all events prior to the stopping time τ. To show this, it suffices to note that, for $A \in \mathcal{F}_\tau$, we have

$$\mathbb{E}\,[X_n; A \cap \{\tau \le n\}] = \mathbb{E}\,[X_\tau; A \cap \{\tau \le n\}]\,. \tag{9.2.20}$$

We recommend the reader to prove this property as an exercise.

Theorem 9.2.3 *Let $\tau_d(u)$ be the stopping time given by* (9.2.16). *If $A \subset \{\tau_d(u) < \infty\}$ and $A \in \mathcal{F}_{\tau_d(u)}^Y$, then*

$$\mathbb{P}(A) = \tilde{\mathbb{E}}\left[\exp\Big(-\gamma \sum_{i=1}^{\tau_d(u)} Y_i\Big); A\right]. \tag{9.2.21}$$

Proof Let $n \in \mathbb{N}$ be fixed and consider the event $A \cap \{\tau_d(u) \le n\} \in \mathcal{F}_n^Y$. Then, by (9.2.17) and (9.2.18) we have

$$\mathbb{P}(A \cap \{\tau_d(u) \le n\}) = \tilde{\mathbb{E}}\left[\exp\Big(-\gamma \sum_{i=1}^{n} Y_i\Big); A \cap \{\tau_d(u) \le n\}\right].$$

Since $\{\exp(-\gamma\sum_{i=1}^n Y_i)\}$ is a martingale under $\tilde{\mathbb{P}}$, (9.2.20) gives

$$\tilde{\mathbb{E}}\left[\mathrm{e}^{-\gamma\sum_{i=1}^n Y_i}; A\cap\{\tau_{\mathrm{d}}(u)\le n\}\right] = \tilde{\mathbb{E}}\left[\mathrm{e}^{-\gamma\sum_{j=1}^{\tau_{\mathrm{d}}(u)} Y_j}; A\cap\{\tau_{\mathrm{d}}(u)\le n\}\right].$$

Letting n tend to ∞, the proof is completed by an appeal to the monotone convergence theorem. □

9.2.4 Finite-Horizon Ruin Probabilities

Consider the discrete-time risk process introduced in Section 5.1. Let $u \ge 0$ be the initial risk reserve, and $Y_1, Y_2, \ldots$ the net payouts encountered at epochs $1, 2, \ldots$ with distribution F. Assume that (9.2.13) has the solution $\gamma > 0$. Ruin occurs if the cumulative net payout exceeds the initial risk reserve at some epoch $n = 1, 2, \ldots$. We want to compute the *ruin function* $\psi(u) = \mathbb{P}(\tau_{\mathrm{d}}(u) < \infty)$, where $\tau_{\mathrm{d}}(u)$ is the ruin time defined in (9.2.16). If we restrict ourselves to a finite time horizon, then we ask for the ruin until time n and we need to compute the *finite-horizon ruin function* $\psi(u;n) = \mathbb{P}(\tau_{\mathrm{d}}(u) \le n)$. We can apply Theorem 9.2.3 to the events $A = \{\tau_{\mathrm{d}}(u) < \infty\}$ and $A = \{\tau_{\mathrm{d}}(u) \le n\}$ to obtain representations for the ruin functions $\psi(u)$ and $\psi(u;n)$, respectively.

Theorem 9.2.4 *For $u \ge 0$ and $n = 1, 2, \ldots$,*

$$\psi(u) = \mathrm{e}^{-\gamma u}\tilde{\mathbb{E}}\left(\mathrm{e}^{-\gamma\left(\sum_{i=1}^{\tau_{\mathrm{d}}(u)} Y_i - u\right)}\right) \tag{9.2.22}$$

$$\psi(u;n) = \mathrm{e}^{-\gamma u}\tilde{\mathbb{E}}\left[\mathrm{e}^{-\gamma\left(\sum_{i=1}^{\tau_{\mathrm{d}}(u)} Y_i - u\right)}; \{\tau_{\mathrm{d}}(u)\le n\}\right]. \tag{9.2.23}$$

Proof In view of Theorem 9.2.3, we have to comment only on formula (9.2.22). Since under $\tilde{\mathbb{P}}$ the event $\{\tau_{\mathrm{d}}(u) < \infty\}$ has probability 1, Theorem 9.2.3 gives

$$\begin{aligned}\psi(u) &= \mathrm{e}^{-\gamma u}\tilde{\mathbb{E}}\left[\mathrm{e}^{-\gamma\left(\sum_{i=1}^{\tau_{\mathrm{d}}(u)} Y_i - u\right)}; \{\tau_{\mathrm{d}}(u) < \infty\}\right]\\ &= \mathrm{e}^{-\gamma u}\tilde{\mathbb{E}}\left[\mathrm{e}^{-\gamma\left(\sum_{i=1}^{\tau_{\mathrm{d}}(u)} Y_i - u\right)}\right].\end{aligned}$$

□

Note that Theorem 9.2.4 can be generalized in the following way. Let $s \in \mathbb{R}$ be such that $\hat{m}_F(s) < \infty$ and put $X_n = \exp(-s\sum_{i=1}^n Y_i + n\log\hat{m}_F(s))$. Then $\{X_n\}$ is an $\{\mathcal{F}_n^Y\}$-martingale on $(\mathbb{R}^\infty, \mathcal{B}(\mathbb{R}^\infty), \mathbb{P}^{(s)})$, where $\mathbb{P}^{(s)}$ is the probability measure under which $Y_1, Y_2, \ldots$ are independent and identically distributed with the common distribution function $F_s(x)$ given by (9.2.15). Moreover, for each $A \in \mathcal{F}_{\tau_{\mathrm{d}}(u)}^Y$ such that $A \subset \{\tau_{\mathrm{d}}(u) < \infty\}$, we have

$$\mathbb{P}(A) = \mathbb{E}^{(s)}\left[\exp\Big(-s\sum_{i=1}^{\tau_{\mathrm{d}}(u)} Y_i + \tau_{\mathrm{d}}(u)\log\hat{m}_F(s)\Big); A\right] \tag{9.2.24}$$

provided that $\mathbb{P}^{(s)}(\tau_{\rm d}(u) < \infty) = 1$, where $\mathbb{E}^{(s)}$ denotes expectation with respect to $\mathbb{P}^{(s)}$. Proofs of these properties can be provided by arguments similar to those used in Section 9.2.3. Details are left to the reader.

The computation of the finite-horizon ruin probabilities is a notoriously difficult problem, even in the compound Poisson model. Contrarily, exponential bounds for the ruin probability $\mathbb{P}(\tau_{\rm d}(u) < n)$ arc easier. Let $\chi(s) = \log \hat{m}_F(s)$ and take $\gamma > 0$ the solution to $\chi(\gamma) = 0$. If $\chi(s_0) < \infty$ for some $s_0 > \gamma$, then $\chi(s)$ is differentiable in $(0, s_0)$. We leave it to the reader to show that $\chi(s)$ is convex in $\mathbb{R}$, even strictly convex in $(0, s_0)$. Let $x > 0$, and $s = s_x$ the solution to $\chi^{(1)}(s) = x^{-1}$.

Theorem 9.2.5 *Let $x > 0$. If $\chi^{(1)}(\gamma) < x^{-1}$, then $x\chi(s_x) - s_x < 0$ and*

$$\mathbb{P}(\tau_{\rm d}(u) \le xu) \le \exp\left((x\chi(s_x) - s_x)u\right) \tag{9.2.25}$$

for all $u \ge 0$. If $x^{-1} \le \chi^{(1)}(\gamma)$, then $s_x > 0$ and, for all $u \ge 0$,

$$\mathbb{P}(\tau_{\rm d}(u) < \infty) - \mathbb{P}(\tau_{\rm d}(u) \le xu) \le {\rm e}^{-s_x u}. \tag{9.2.26}$$

Proof Suppose that $\chi^{(1)}(\gamma) < x^{-1}$. Since $\hat{m}_F(s)$ is strictly convex in $(0, s_0)$, it follows that $\hat{m}_F^{(1)}(s)$ is strictly increasing in $(0, s_0)$ and hence $\chi(s_x) > 0$ for $s_x > \gamma$. Now, from (9.2.24) we get

$$\begin{aligned} \mathbb{P}(\tau_{\rm d}(u) \le xu) &= \mathbb{E}^{(s_x)}\left[\exp(\tau_{\rm d}(u)\chi(s_x) - s_x S_{\tau_{\rm d}(u)}); \tau_{\rm d}(u) \le xu\right] \\ &\le \exp\left((x\chi(s_x) - s_x)u\right). \end{aligned}$$

For $0 < x^{-1} \le \chi^{(1)}(\gamma)$, we have $0 < s_x \le \gamma$ and consequently $\chi(s_x) \le 0$. Thus, (9.2.24) gives

$$\begin{aligned} &\mathbb{P}(\tau_{\rm d}(u) < \infty) - \mathbb{P}(\tau_{\rm d}(u) \le xu) \\ &\quad = \mathbb{E}^{(s_x)}\left[\exp(\tau_{\rm d}(u)\chi(s_x) - s_x S_{\tau_{\rm d}(u)}); xu < \tau_{\rm d}(u) < \infty\right] \le {\rm e}^{-s_x u}. \end{aligned}$$

□

9.2.5 Simulation of Ruin Probabilities

Consider the discrete-time risk process generated by a random walk, as in Section 9.2.3. We collect a few remarks on how to use the change-of-measure theory presented above in the approximation to the ruin probability $\psi(u; n)$ via simulation. The simplest approach is to simulate l independent replications of the random walk until ruin occurs. In each replication we stop the experiment at n, unless ruin occurs before n, in which case we stop at the ruin epoch. As an estimator $\bar{\psi}$ for $\psi(u; n)$ we take the ratio $\bar{\psi} = L/l$ of the number L of replications in which ruin occurred until n, over the total number l of runs. Since L has the binomial distribution $\mathrm{Bin}(l,\psi(u;n))$,

$$\mathbb{E}\,(L/l)=\psi(u;n)\,,\qquad \mathrm{Var}\,(L/l)=l^{-1}\psi(u;n)(1-\psi(u;n))\,. \tag{9.2.27}$$

In mathematical statistics, an estimator fulfilling the first equality in (9.2.27) is called *unbiased*. However, we cannot expect $\bar{\psi}$ to be a good estimator since typically $\psi(u;n)$ is very small. To reach a prescribed relative accuracy a with probability $1-p$, we must run a certain minimum number l of replications such that the relative error $|\bar{\psi}-\psi(u;n)|/\psi(u;n)$ satisfies

$$\mathbb{P}\Big(\frac{|\bar{\psi}-\psi(u;n)|}{\psi(u;n)}>a\Big)=p\,. \tag{9.2.28}$$

Let ε_p be determined from $\mathbb{P}\,(|Z|>\varepsilon_p)=p$, where Z has the standard normal distribution $\mathrm{N}(0,1)$. The central limit theorem gives

$$\lim_{l\to\infty}\mathbb{P}\Big(\frac{|\bar{\psi}-\psi(u;n)|}{\sqrt{l^{-1}\psi(u;n)(1-\psi(u;n))}}>\varepsilon_p\Big)=p\,. \tag{9.2.29}$$

From the 2-sigma law of normal distributions, the value of p is close to 0.05 if $\varepsilon_p=2$. Hence, for all sufficiently large l, the probability

$$\mathbb{P}\Big(\frac{|\bar{\psi}-\psi(u;n)|}{\psi(u;n)}>2\frac{\sqrt{l^{-1}\psi(u;n)(1-\psi(u;n))}}{\psi(u;n)}\Big)$$

is close to 0.05 and consequently $a_{\mathrm{app}}=2\sqrt{l^{-1}\psi(u;n)(1-\psi(u;n))}/\psi(u;n)$. Thus, for all sufficiently large $u\ge 0$,

$$l_{\mathrm{app}}=\frac{4(1-\psi(u;n))}{a^2\psi(u;n)}\ge\frac{2}{a^2}[\psi(u)]^{-1}\sim\frac{2}{a^2c}\mathrm{e}^{\gamma u}$$

where in the last relation the Cramér–Lundberg estimate (6.5.29) has been used. This shows that the number of replications l has to be at least proportional to $\mathrm{e}^{\gamma u}$ and so the number of required replications grows very fast with u.

Let $\mathbb{P}^{(s)}$ be the probability measure on $(\mathbb{R}^\infty,\mathcal{B}(\mathbb{R}^\infty))$ defined analogously to $\tilde{\mathbb{P}}$, but by the martingale $\{\exp(s\sum_{i=1}^n Y_i)\,,\,n\in\mathbb{N}\}$ as considered in Example 3 of Section 9.1.3 and in Section 9.2.4. For the special case when $s=\gamma$, we have $\mathbb{P}^{(\gamma)}=\tilde{\mathbb{P}}$. In an attempt to lower the number of replications while keeping a given precision, we simulate the random walk under the probability measure $\mathbb{P}^{(s)}$ for some properly chosen s for which $\hat{m}_F(s)<\infty$. We then use (9.2.24) to estimate the ruin probability $\psi(u;n)$. In the particular case $s=\gamma$, we proceed as follows. Relation (9.2.23) shows that it suffices to estimate the expectation $\tilde{\mathbb{E}}\left[\exp(-\gamma\sum_{i=1}^{\tau_{\mathrm{d}}(u)}Y_i);\tau_{\mathrm{d}}(u)\le n\right]$. The crucial point is that, under $\tilde{\mathbb{P}}$, the stopping time $\tau_{\mathrm{d}}(u)$ is finite with probability 1; see Section 9.2.3. For each simulation run we compute a realization of the random variable

$$Z = \exp\Big(-\gamma \sum_{i-1}^{\tau_{\mathrm{d}}(u)} Y_i\Big) \mathbb{I}(\tau_{\mathrm{d}}(u) \le n) . \qquad (9.2.30)$$

Thus for l independent replications we obtain the values $z_1, z_2, \ldots, z_l$ computed from (9.2.30). We now use $\bar{\psi}_\gamma(z_1, \ldots, z_l) = \sum_{i=1}^{l} z_i/l$ as an estimate for $\psi(u;n)$. It is clear that $\tilde{\mathbb{E}}\,\bar{\psi}_\gamma = \psi(u;n)$, showing that the estimator $\bar{\psi}_\gamma$ is again unbiased. If we denote by $\tilde{\mathbb{D}}^2$ the variance with respect to $\tilde{\mathbb{P}}$, then $\tilde{\mathbb{D}}^2 \bar{\psi}_\gamma < \operatorname{Var} \bar{\psi}$ because

$$\begin{aligned} l\, \tilde{\mathbb{D}}^2 \bar{\psi}_\gamma &= \tilde{\mathbb{E}}\,(Z^2) - (\psi(u;n))^2 \\ &= \tilde{\mathbb{E}}\left[\mathrm{e}^{-2\gamma \sum_{i=1}^{\tau_{\mathrm{d}}(u)} Y_i}; \tau_{\mathrm{d}}(u) \le n\right] - (\psi(u;n))^2 \\ &= \mathbb{E}\left[\mathrm{e}^{-\gamma \sum_{i=1}^{\tau_{\mathrm{d}}(u)} Y_i}; \tau_{\mathrm{d}}(u) \le n\right] - (\psi(u;n))^2 \\ &< \psi(u;n) - (\psi(u;n))^2 , \end{aligned}$$

where the third equality follows from the definition of $\tilde{\mathbb{E}}$. Thus, simulation under $\tilde{\mathbb{P}}$ leads to an estimator for $\psi(u;n)$, with a *reduced variance.*

Bibliographical Notes. The family of distribution functions F_s defined in (9.2.15) generates a family of associated distributions. A detailed account of such families was given in Section 2.3; see also Asmussen (1987). Formulae like (9.2.23) appear in Asmussen (1982), Siegmund (1975) and von Bahr (1974). In the theory of Monte Carlo simulations, the proper choice of an underlying probability measure is called *importance sampling.* The early ideas for solving such problems go back to Siegmund (1976), who considered the simulation of probabilities occurring in sequential tests. In Lehtonen and Nyrhinen (1992a) importance sampling is studied for random walks; see also Lehtonen and Nyrhinen (1992b). For further papers discussing importance sampling in connection with stochastic simulation, see, for example, Asmussen and Rubinstein (1995) and Glynn and Iglehart (1989). More details on Monte Carlo simulations can be found, for example, in Crane and Lemoine (1977), Fishman (1996) and Ross (1997a).

CHAPTER 10

Martingale Techniques II

10.1 CONTINUOUS-TIME MARTINGALES

The theory of continuous-time martingales is deeper and often requires lengthy proofs that will not always be presented in full detail and generality. The aim of this section is to outline and discuss some selected aspects of continuous-time martingales and to study their applications in risk theory and other branches of insurance and financial mathematics.

10.1.1 Stochastic Processes and Filtrations

Under the notion of a *stochastic process* we understand a collection of random variables $\{X(t),\ t \in \mathcal{T}\}$ on a common probability space $(\Omega, \mathcal{F}, \mathbb{P})$. Here $\mathcal{T}$ is an ordered space of parameters. Typically in this book $\mathcal{T} \subset \mathbb{R}$ and in particular $\mathcal{T} = \mathbb{N}, \mathbb{Z}, \mathbb{R}_+$ or $\mathcal{T} = \mathbb{R}$. However, in a few places we will feel the need for more general parameter spaces like sets of stopping times or families of subsets. Formally, a stochastic process is a mapping $X : \mathcal{T} \times \Omega \to \mathbb{R}$, but in general we do not require the measurability of this mapping. If $\mathcal{T}$ is a subset of $\mathbb{R}$ and X is measurable with respect to the product-σ-algebra $\mathcal{B}(\mathcal{T}) \otimes \mathcal{F}$, then we say the stochastic process $\{X(t), t \in \mathcal{T}\}$ is *measurable.*

In this section we always assume that $\mathcal{T} = \mathbb{R}_+$. Then the set $\mathcal{T}$ of parameters plays the role of time and so we speak about *continuous-time stochastic processes.* For each fixed $\omega \in \Omega$, the function $t \mapsto X(t, \omega)$ is called a *sample path* or *trajectory*; however, we usually drop the dependence on $\omega \in \Omega$. In general, sample paths can be quite irregular. We will mostly deal with processes having sample paths belonging to one of the following two spaces:

- the space of continuous functions $g : \mathbb{R}_+ \to \mathbb{R}$ denoted by $C(\mathbb{R}_+)$,
- the space of right-continuous functions $g : \mathbb{R}_+ \to \mathbb{R}$ with left-hand limits denoted by $D(\mathbb{R}_+)$.

Note that the process $\{X(t)\}$ is measurable if the sample paths of $\{X(t)\}$ are from $D(\mathbb{R}_+)$; see, for example, Lemma 2.1.1 in Last and Brandt (1995). In this book we say that a stochastic process with sample paths from $D(\mathbb{R}_+)$

is *càdlàg*, which is the abbreviated French name of this property. In the literature, one usually makes the somewhat weaker assumption that only almost all sample paths are from $D(\mathbb{R}_+)$. However, in most cases relevant in insurance mathematics, one can consider a canonical probability space. In particular, one can restrict Ω to the set $\Omega_0 \subset \Omega$ such that $\Omega_0 = \{\omega : X(\cdot, \omega) \in D(\mathbb{R}_+)\}$.

Let $t \geq 0$. By the *history of* $\{X(t)\}$ *up to time* t we mean the smallest σ-algebra $\mathcal{F}_t^X$ containing the events $\{\omega : (X(t_1, \omega), \ldots, X(t_n, \omega)) \in B\}$ for all Borel sets $B \in \mathcal{B}(\mathbb{R}^n)$, for all $n = 1, 2, \ldots$ and arbitrary sequences $t_1, t_2, \ldots, t_n$ with $0 \leq t_1 \leq t_2 \leq \ldots \leq t_n \leq t$. Note that for $0 \leq t \leq t'$

- $\mathcal{F}_t^X \subset \mathcal{F}$,
- $\mathcal{F}_t^X \subset \mathcal{F}_{t'}^X$,
- $X(t)$ is measurable with respect to $\mathcal{F}_t^X$.

The family of σ-algebras $\{\mathcal{F}_t^X\}$ is called the *history* of the process $\{X(t)\}$. Similarly to the discrete-time case (see Section 9.1.2) we also say that $\{\mathcal{F}_t^X\}$ is the *filtration* generated by $\{X(t)\}$. This is a special case of the following definition. An arbitrary family $\{\mathcal{F}_t, t \in \mathcal{T}\}$ of σ-algebras such that $\mathcal{T} \subset \mathbb{R}$ and $\mathcal{F}_t \subset \mathcal{F}$, $\mathcal{F}_t \subset \mathcal{F}_{t'}$ for all $t, t' \in \mathcal{T}$ with $t \leq t'$ is called a *filtration*. We say that the process $\{X(t), t \in \mathcal{T}\}$ is *adapted* to the filtration $\{\mathcal{F}_t, t \in \mathcal{T}\}$ if $X(t)$ is measurable with respect to $\mathcal{F}_t$, for all $t \in \mathcal{T}$.

10.1.2 Stopping Times

A random variable τ taking values in $\mathbb{R}_+ \cup \{\infty\}$ is said to be a *stopping time* with respect to a filtration $\{\mathcal{F}_t,\ t \geq 0\}$ (or equivalently an $\{\mathcal{F}_t\}$-stopping time) if the event $\{\tau \leq t\}$ belongs to $\mathcal{F}_t$, for all $t \geq 0$. We define $\mathcal{F}_{t+} = \bigcap_{\epsilon > 0} \mathcal{F}_{t+\epsilon}$. Note that $\mathcal{F}_{t+}$ is a σ-algebra because the intersection of any family of σ-algebras is a σ-algebra. If $\mathcal{F}_{t+} = \mathcal{F}_t$ for all $t \in \mathbb{R}_+$, we say that the filtration $\{\mathcal{F}_t,\ t \geq 0\}$ is *right-continuous*. In this case we have the following equivalent definition of a stopping time.

Lemma 10.1.1 *The random variable τ is an $\{\mathcal{F}_{t+}\}$-stopping time if and only if $\{\tau < t\} \in \mathcal{F}_t$ for all $t \geq 0$. In particular, if $\{\mathcal{F}_t\}$ is a right-continuous filtration, then τ is an $\{\mathcal{F}_t\}$-stopping time if and only if $\{\tau < t\} \in \mathcal{F}_t$ for all $t \geq 0$.*

Proof If τ is an $\{\mathcal{F}_{t+}\}$-stopping time, then $\{\tau < t\} \in \mathcal{F}_t$ since $\{\tau < t\} = \bigcup_{n=1}^{\infty} \{\tau \leq t - n^{-1}\} \in \mathcal{F}_t$. Conversely suppose that the random variable τ has the property that $\{\tau < t\} \in \mathcal{F}_t$ for all $t \geq 0$. Then $\{\tau \leq t\} = \bigcap_{n=1}^{\infty} \{\tau < t + n^{-1}\} \in \mathcal{F}_{t+}$. The second part of the statement is now obvious. □

Throughout the present section we assume that the stochastic process $\{X(t), t \geq 0\}$ is càdlàg. Let $B \in \mathcal{B}(\mathbb{R})$ and define the *first entrance time*

τ^B of $\{X(t)\}$ to the set B by

$$\tau^B = \begin{cases} \inf\{t : X(t) \in B\} & \text{if } X(t) \in B \text{ for some } t \geq 0, \\ \infty & \text{otherwise.} \end{cases}$$

In contrast to the discrete-time case considered in Section 9.1.2, the question whether τ^B is a stopping time is not obvious. A positive answer can only be given under additional assumptions, for example on B or on the filtration $\{\mathcal{F}_t\}$. We now discuss this problem in more detail for sets of the form $B = (u, \infty)$ and $B = [u, \infty)$, where $u \in \mathbb{R}$. Let $\tau(u) = \inf\{t \geq 0 : X(t) > u\}$ denote the first entrance time of $\{X(t)\}$ to the open interval (u, ∞), where we put $\inf \emptyset = \infty$ as usual. For the interval $[u, \infty)$ it is more convenient to consider the *modified first entrance time*

$$\tau^*(u) = \inf\{t \geq 0 : X(t - 0) \geq u \text{ or } X(t) \geq u\} . \tag{10.1.1}$$

Theorem 10.1.1 *Let $u \in \mathbb{R}$. If the process $\{X(t)\}$ is adapted to a filtration $\{\mathcal{F}_t\}$, then $\tau(u)$ is an $\{\mathcal{F}_{t+}\}$-stopping time and $\tau^*(u)$ is an $\{\mathcal{F}_t\}$-stopping time. In particular, if $\{\mathcal{F}_t\}$ is right-continuous then $\tau(u)$ is an $\{\mathcal{F}_t\}$-stopping time too.*

Proof Since the trajectories of $\{X(t)\}$ belong to $D(\mathbb{R}_+)$, we have

$$\{\tau(u) < t\} = \bigcup_{q \in \mathbb{Q}_t} \{X(q) > u\} \in \mathcal{F}_t \tag{10.1.2}$$

for each $t \geq 0$, where $\mathbb{Q}_t$ is the set of all rational numbers in $[0, t)$. Hence $\tau(u)$ is an $\{\mathcal{F}_{t+}\}$-stopping time by the result of Lemma 10.1.1. Furthermore, $\{\tau^*(u) \leq t\} = \bigcap_{n \in \mathbb{N}} \{\tau(u - n^{-1}) < t\} \cup \{X(t) \geq u\}$. Thus, (10.1.2) implies that $\{\tau^*(u) \leq t\} \in \mathcal{F}_t$ for each $t \geq 0$. $\square$

Remarks 1. The proof of Theorem 10.1.1 can easily be extended in order to show that the first entrance time τ^B to an arbitrary open set B is an $\{\mathcal{F}_{t+}\}$-stopping time. Moreover, it turns out that τ^B is a stopping time for each Borel set $B \in \mathcal{B}(\mathbb{R}_+)$ provided that some additional conditions are fulfilled. In connection with this we need the following concept. We say that the probability space $(\Omega, \mathcal{F}, \mathbb{P})$ is *complete* if for each subset $A \subset \Omega$ for which an event $A' \in \mathcal{F}$ exists with $A \subset A'$ and $\mathbb{P}(A') = 0$, we have $A \in \mathcal{F}$. We now say that the filtration $\{\mathcal{F}_t, t \geq 0\}$ is *complete* if the probability space is complete and $\{A \in \mathcal{F} : \mathbb{P}(A) = 0\} \subset \mathcal{F}_0$. If the filtration $\{\mathcal{F}_t\}$ is right-continuous and complete, $\{\mathcal{F}_t\}$ is said to fulfil the *usual conditions*. Furthermore, if $\{\mathcal{F}_t\}$ fulfils the usual conditions and if $\{X(t)\}$ is adapted to $\{\mathcal{F}_t\}$, then τ^B is an $\{\mathcal{F}_t\}$-stopping time for each $B \in \mathcal{B}(\mathbb{R}_+)$. A proof of this statement can be found, for example, in Dellacherie (1972), p. 51. We mention, however, that in some cases it can be difficult to show that a given filtration is right-continuous.

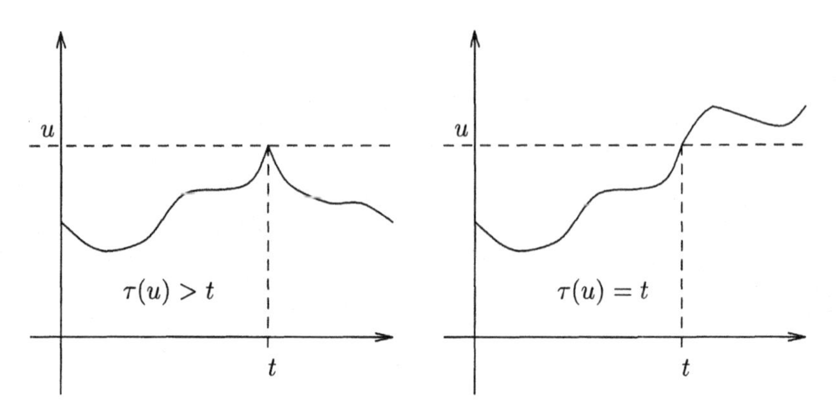

Figure 10.1.1 Two paths coinciding till time t

2. Theorem 10.1.1 indicates that the first entrance time $\tau(u)$ is not always a stopping time, unless the considered filtration is right-continuous. An example where this problem appears can easily be found if the underlying probability space is large enough. Consider the process $\{X(t)\}$ on the canonical probability space $(\Omega, \mathcal{F}, \mathbb{P})$ with $\Omega = D(\mathbb{R}_+)$ and $\mathcal{F} = \mathcal{B}(D(\mathbb{R}_+))$. Then, $\{\tau(u) \le t\} \notin \mathcal{F}_t^X$ for each $t > 0$, i.e. $\tau(u)$ is not a stopping time with respect to the history $\{\mathcal{F}_t^X\}$ of $\{X(t)\}$. Indeed, the two sample paths given in Figure 10.1.1 show that from the knowledge of the process $\{X(t)\}$ up to time t it is not possible to recognize whether $\tau(u) \le t$ or $\tau(u) > t$.

3. An important characteristic of the claim surplus process $\{S(t)\}$ introduced in Section 5.1.4 is the time of ruin for a given initial risk reserve $u \ge 0$, i.e. the first entrance time of $\{S(t)\}$ to the open interval (u, ∞). However, in this case the measurability problem mentioned above does not appear if we consider $\{S(t)\}$ on its canonical probability space. This means that we restrict $\Omega = D(\mathbb{R}_+)$ to the set $\Omega_0 \subset \Omega$ of those functions from Ω which have only finitely many jumps in each bounded interval and which decrease linearly between the jumps; see also Figure 5.1.1. We leave it to the reader to show as an exercise that on this smaller probability space the ruin time $\tau(u) = \min\{t \ge 0 : S(t) > u\}$ is an $\{\mathcal{F}_t^S\}$-stopping time.

10.1.3 Martingales, Sub- and Supermartingales

Suppose that the stochastic process $\{X(t),\ t \in \mathcal{T}\}$ is adapted to $\{\mathcal{F}_t,\ t \in \mathcal{T}\}$ and that $\mathbb{E}\,|X(t)| < \infty$ for all $t \in \mathcal{T}$. We say that $\{X(t)\}$ is an $\{\mathcal{F}_t\}$-*martingale* if with probability 1

$$\mathbb{E}\,(X(t+h) \mid \mathcal{F}_t) = X(t)\,, \qquad (10.1.3)$$

for all $t, t+h \in \mathcal{T}$ with $h \geq 0$. Similarly, $\{X(t)\}$ is called a *submartingale* if

$$\mathbb{E}\,(X(t+h) \mid \mathcal{F}_t) \geq X(t)\,, \tag{10.1.4}$$

and a *supermartingale* if

$$\mathbb{E}\,(X(t+h) \mid \mathcal{F}_t) \leq X(t)\,, \tag{10.1.5}$$

for all $t, t+h \in \mathcal{T}$ with $h \geq 0$.

The definition (10.1.3) for continuous-time martingales entails a property for the discrete version. If $\{X(t), t \geq 0\}$ is an $\{\mathcal{F}_t\}$-martingale and $0 \leq t_0 < t_1 < \ldots$, then $\{X(t_n), n \in \mathbb{N}\}$ is an $\{\mathcal{F}_{t_n}\}$-martingale in discrete time, i.e. (9.1.7) holds with $X_n = X(t_n)$ and $\mathcal{F}_n = \mathcal{F}_{t_n}$.

Examples 1. Consider the cumulative arrival process $\{X(t), t \geq 0\}$ introduced in Section 5.2.2 for the compound Poisson model with characteristics (λ, F_U), where $X(t) = \sum_{i=1}^{N(t)} U_i$. If $\mathbb{E}\,U < \infty$, then the process $\{X'(t), t \geq 0\}$ with $X'(t) = X(t) - t\lambda\mathbb{E}\,U$ is a martingale with respect to the filtration $\{\mathcal{F}_t^X\}$. This is a special case of the next example.

2. Let $\{X(t),\ t \geq 0\}$ be a process with stationary and independent increments. If $\mathbb{E}\,|X(1)| < \infty$, then the process $\{X'(t), t \geq 0\}$ with $X'(t) = X(t) - t\mathbb{E}\,X(1)$ is a martingale with respect to the filtration $\{\mathcal{F}_t^X\}$. We leave the proof of this fact to the reader. We only remark that it suffices to show that $\mathbb{E}\,(X'(t+h) \mid X'(t_1), \ldots, X'(t_n), X'(t)) = X'(t)$ whenever $0 \leq t_1 < t_2 < \ldots < t_n < t < t+h$.

3. Consider the claim surplus process $\{S(t), t \geq 0\}$ with $S(t) = \sum_{i=1}^{N(t)} U_i - \beta t$ for the compound Poisson model, with arrival rate λ, premium rate β and claim size distribution F_U. Note that, by formula (5.2.7) in Corollary 5.2.1, we have

$$\mathbb{E}\,\mathrm{e}^{sS(t)} = \mathrm{e}^{tg(s)}\,, \qquad s \in \mathbb{R}\,, \tag{10.1.6}$$

where $g(s) = \lambda(\hat{m}_U(s) - 1) - \beta s$. Now we use this result to show that the process $\{X(t), t \geq 0\}$ with $X(t) = \mathrm{e}^{sS(t) - g(s)t}$ is a martingale with respect to the filtration $\{\mathcal{F}_t^S\}$, where $s \in \mathbb{R}$ is fixed. For $t, h \geq 0$ we have

$$\begin{aligned}
\mathbb{E}\,(X(t+h) \mid \mathcal{F}_t^S) &= \mathbb{E}\,(\mathrm{e}^{sS(t+h) - g(s)(t+h)} \mid \mathcal{F}_t^S) \\
&= \mathrm{e}^{sS(t) - g(s)t}\,\mathbb{E}\,(\mathrm{e}^{s(S(t+h) - S(t)) - g(s)h} \mid \mathcal{F}_t^S) \\
&= \mathrm{e}^{sS(t) - g(s)t}\,\mathbb{E}\,(\mathrm{e}^{sS(h) - g(s)h})\,,
\end{aligned}$$

where the last equation follows from the fact that $\{S(t)\}$ has independent and stationary increments, and which is known from Corollary 5.2.1. From (10.1.6) we have $\mathbb{E}\,(\mathrm{e}^{sS(h) - g(s)h}) = 1$ and consequently $\mathbb{E}\,(X(t+h) \mid \mathcal{F}_t^S) = X(t)$.

4. The current example indicates the close relationship between martingales and the concept of the infinitesimal generator in the theory of Markov

processes; see also Section 11.1.3. Consider a continuous-time homogeneous Markov process $\{Z(t),\ t \ge 0\}$ with finite state space $E = \{1, 2, \ldots, \ell\}$ and intensity matrix $\boldsymbol{Q}$. Then, for each vector $\boldsymbol{b} = (b_1, \ldots, b_\ell) \in \mathbb{R}^\ell$, the process $\{X(t), t \ge 0\}$ with

$$X(t) = b_{Z(t)} - b_{Z(0)} - \int_0^t (\boldsymbol{Q}\boldsymbol{b}^\top)_{Z(v)}\,\mathrm{d}v\,, \qquad t \ge 0 \tag{10.1.7}$$

is an $\{\mathcal{F}_t^Z\}$-martingale, where the integral in (10.1.7) is defined pathwise. In order to demonstrate this fact, write for $t, h \ge 0$

$$\begin{aligned}\mathbb{E}\,(X(t+h) \mid \mathcal{F}_t^Z) &= \mathbb{E}\,(X(t+h) \mid Z(t))\\ &= X(t) + \mathbb{E}\left(b_{Z(t+h)} - b_{Z(t)} - \int_t^{t+h} (\boldsymbol{Q}\boldsymbol{b}^\top)_{Z(v)}\,\mathrm{d}v \;\Big|\; Z(t)\right).\end{aligned}$$

Because $\{Z(t)\}$ is homogeneous we have

$$\begin{aligned}&\mathbb{E}\left(b_{Z(t+h)} - b_{Z(t)} - \int_t^{t+h} (\boldsymbol{Q}\boldsymbol{b}^\top)_{Z(v)}\,\mathrm{d}v \;\Big|\; Z(t) = i\right)\\ &\quad = \mathbb{E}\left(b_{Z(h)} - b_{Z(0)} - \int_0^h (\boldsymbol{Q}\boldsymbol{b}^\top)_{Z(v)}\,\mathrm{d}v \;\Big|\; Z(0) = i\right).\end{aligned}$$

Thus it suffices to show that, for all $i \in E$,

$$\mathbb{E}\,(b_{Z(h)} \mid Z(0) = i) - b_i = \int_0^h \mathbb{E}\,((\boldsymbol{Q}\boldsymbol{b}^\top)_{Z(v)} \mid Z(0) = i)\,\mathrm{d}v \tag{10.1.8}$$

since $\mathbb{E}\left(\int_0^h (\boldsymbol{Q}\boldsymbol{b}^\top)_{Z(v)}\,\mathrm{d}v \mid Z(0) = i\right) = \int_0^h \mathbb{E}\,((\boldsymbol{Q}\boldsymbol{b}^\top)_{Z(v)} \mid Z(0) = i)\,\mathrm{d}v$. However, recalling from Theorem 8.1.4 that the matrix transition function $\{\boldsymbol{P}(v), v \ge 0\}$ of $\{Z(t), t \ge 0\}$ is given by $\boldsymbol{P}(v) = \exp(\boldsymbol{Q}v)$, we have

$$\mathbb{E}\,\left(b_{Z(h)} \mid Z(0) = i\right) = \boldsymbol{e}_i \exp(\boldsymbol{Q}h)\boldsymbol{b}^\top \tag{10.1.9}$$

and

$$\mathbb{E}\left((\boldsymbol{Q}\boldsymbol{b}^\top)_{Z(v)} \mid Z(0) = i\right) = \boldsymbol{e}_i \exp(\boldsymbol{Q}v)\boldsymbol{Q}\boldsymbol{b}^\top, \tag{10.1.10}$$

where $\boldsymbol{e}_i$ is the ℓ-dimensional (row) vector with all components equal to 0 but the i-th equal to 1. Using (10.1.9) and (10.1.10) we see that (10.1.8) is equivalent to

$$\boldsymbol{e}_i \exp(\boldsymbol{Q}h)\boldsymbol{b}^\top - b_i = \int_0^h \boldsymbol{e}_i \exp(\boldsymbol{Q}v)\boldsymbol{Q}\boldsymbol{b}^\top\,\mathrm{d}v\,. \tag{10.1.11}$$

The latter can be verified by differentiation and by using Lemma 8.1.2. So far, we have shown that the process $\{X(t), t \ge 0\}$ given by (10.1.7) is an

$\{\mathcal{F}_t^Z\}$-martingale. Note that (10.1.7) is a special case of *Dynkin's formula* for Markov processes with general state space; see also (11.1.17). This martingale technique yields a versatile approach to the concept of infinitesimal generators for Markov processes. For example, the following converse statement is true. Suppose for the moment that $\boldsymbol{Q}'$ is an arbitrary $\ell \times \ell$ matrix, i.e. not necessarily the intensity matrix of the Markov process $\{Z(t)\}$. Moreover, assume that the process $\{X'(t)\}$ with

$$X'(t) = b_{Z(t)} - b_i - \int_0^t (\boldsymbol{Q}'\boldsymbol{b}^\top)_{Z(v)}\,\mathrm{d}v\,, \qquad t \geq 0 \tag{10.1.12}$$

is an $\{\mathcal{F}_t^Z\}$-martingale for each vector $\boldsymbol{b} \in \mathbb{R}^\ell$ and for each initial state $Z(0) = i$ of $\{Z(t)\}$, Then, analogously to (10.1.11), we have

$$\boldsymbol{e}_i\boldsymbol{P}(h)\boldsymbol{b}^\top - b_i = \int_0^h \boldsymbol{e}_i \exp(\boldsymbol{Q}v)\boldsymbol{Q}'\boldsymbol{b}^\top\,\mathrm{d}v\,. \tag{10.1.13}$$

On the other hand, using Theorem 8.1.2 we see that

$$\lim_{h\to 0}\frac{\boldsymbol{e}_i\boldsymbol{P}(h)\boldsymbol{b}^\top - \boldsymbol{e}_i\boldsymbol{b}^\top}{h} = \boldsymbol{e}_i\boldsymbol{Q}\boldsymbol{b}^\top$$

for all $i = 1, \ldots, \ell$ and $\boldsymbol{b} \in \mathbb{R}^\ell$. This means that $\boldsymbol{Q}'$ must be equal to the intensity matrix $\boldsymbol{Q}$ of $\{Z(t)\}$. In Section 11.1.3 we will return to questions of this type in a more general setting.

5. This example shows how stochastic integrals with respect to martingales can be used to create new martingales. In an attempt to avoid technical difficulties, we make rather restrictive assumptions on the (deterministic) integrated function f and on the martingale $\{X(t)\}$ with respect to which we integrate. Let the process $\{X(t), t \geq 0\}$ be càdlàg and such that, with probability 1, the trajectories of $\{X(t)\}$ have locally bounded variation. Then for each continuous function $f : \mathbb{R}_+ \to \mathbb{R}$ and $t \geq 0$ the stochastic integral $X_f(t) = \int_0^t f(v)\,\mathrm{d}X(v)$ is defined pathwise as a Riemann–Stieltjes integral, i.e. for each $\omega \in \Omega$ and $t, h \geq 0$,

$$\begin{aligned} X_f(\omega, t+h) - X_f(\omega, t) &= \int_t^{t+h} f(v)\,\mathrm{d}X(\omega, v) \\ &= \lim_{n\to\infty} \sum_{\lfloor nt+1\rfloor < i \leq \lfloor n(t+h)\rfloor} f\Big(\frac{i-1}{n}\Big)\Big(X\big(\omega, \frac{i}{n}\big) - X\big(\omega, \frac{i-1}{n}\big)\Big)\,. \end{aligned}$$

Note that the integrated process $\{X_f(t),\ t \geq 0\}$ is an $\{\mathcal{F}_t^X\}$-martingale whenever $\{X(t)\}$ is an $\{\mathcal{F}_t^X\}$-martingale. Indeed, the random variable $X_f(t)$

is $\mathcal{F}_t^X$-measurable, since it is the limit of $\mathcal{F}_t^X$-measurable random variables. Furthermore, for $t, h \le 0$ we have

$$\mathbb{E}\left(X_f(t+h) \mid \mathcal{F}_t^X\right) = \int_0^t f(v)\,\mathrm{d}X(v) + \mathbb{E}\left(\int_t^{t+h} f(v)\,\mathrm{d}X(v) \,\Big|\, \mathcal{F}_t^X\right).$$

If $\{X(t)\}$ is a martingale, then

$$\mathbb{E}\Big(\sum_{\lfloor nt+1\rfloor < i \le \lfloor n(t+h)\rfloor} f\big(\frac{i-1}{n}\big)\Big(X\big(\frac{i}{n}\big) - X\big(\frac{i-1}{n}\big)\Big) \,\Big|\, \mathcal{F}_t^X\Big) = 0\,.$$

The reader should prove that for fixed $t, h \ge 0$ the sequence $\{Z_n\}$ with

$$Z_n = \sum_{\lfloor nt+1\rfloor < i \le \lfloor n(t+h)\rfloor} f\big(\frac{i-1}{n}\big)\Big(X\big(\frac{i}{n}\big) - X\big(\frac{i-1}{n}\big)\Big), \qquad n = 1, 2, \ldots$$

is *uniformly integrable*, that is

$$\lim_{x\to\infty}\Big(\sup_{n\in\mathbb{N}} \mathbb{E}\left[|Z_n|; \{|Z_n| > x\}\right]\Big) = 0\,. \tag{10.1.14}$$

We can conclude that in this case $\mathbb{E}\left(\int_t^{t+h} f(v)\,\mathrm{d}X(v) \mid \mathcal{F}_t^X\right) = 0$ and consequently $\mathbb{E}\left(X_f(t+h) \mid \mathcal{F}_t^X\right) = X_f(t)$. For f which are random and for $\{X(t)\}$ with trajectories of unbounded variation, the theory of stochastic integrals is much more complicated; see Section 13.1.1.

6. For two $\{\mathcal{F}_t\}$-martingales $\{X(t)\}$ and $\{Y(t)\}$, the process $\{X(t)+Y(t)\}$ is also an $\{\mathcal{F}_t\}$-martingale. The proof of this fact is left to the reader.

7. If the random variable Z is measurable with respect to $\mathcal{F}_0$ for some filtration $\{\mathcal{F}_t, t \ge 0\}$, and if $\mathbb{E}\,|Z| < \infty$, then the process $\{Y(t)\}$ defined by $Y(t) \equiv Z$ is an $\{\mathcal{F}_t\}$-martingale. Moreover if $\{X(t)\}$ is another $\{\mathcal{F}_t\}$-martingale, then the process $\{ZX(t),\ t \ge 0\}$ is an $\{\mathcal{F}_t\}$-martingale, provided $\mathbb{E}\,(ZX(0)) = 0$. We leave the proofs of these simple properties to the reader.

10.1.4 Brownian Motion and Related Processes

In the risk model introduced in Section 5.1.4 we assumed that the premium income is a linear function of time. This reflects the situation that pricing of insurance products is sometimes evaluated on a basis where no interest is taken into account. However, if the return of the company's investments is included into the balance, then a deterministic (linear) income process is no longer an appropriate model, as the return is affected over time by random changes of market values of assets and inflation. For that reason, one also considers risk models with a stochastic income process $\{X(t)\}$. In Section 13.2 we discuss this question in detail assuming that $\{X(t)\}$ is a diffusion process. An important special case is given when the (scheduled) deterministic income βt up to time t is perturbed by a Brownian motion $\{W(t), t \ge 0\}$, i.e. $X(t) = \beta t + W(t)$, where the stochastic process $\{W(t),\ t \ge 0\}$ is called a *σ^2-Brownian motion* if

- $W(0) = 0$,
- $\{W(t)\}$ has stationary and independent increments,
- $W(t)$ is $\mathrm{N}(0, \sigma^2 t)$-distributed for each $t \ge 0$,
- $\{W(t)\}$ has all sample paths from $C(\mathbb{R}_+)$.

It is a nontrivial question to show that the notion of Brownian motion is not empty, that is to show that there exists a stochastic process satisfying these four conditions. The positive answer to this question was given by Norbert Wiener in 1923. Note that the condition $W(0) = 0$ is merely a normalization rather than a basic requirement. It was shown later that the paths of the Brownian motion have curious properties. For example, with probability 1, all the paths are nowhere differentiable, or they pass through 0 infinitely often in every neighbourhood of zero. Besides this, all the sample paths have unbounded variation on the interval $[0, t]$ for each $t > 0$. If $\sigma = 1$, then $\{W(t)\}$ is called a *standard Brownian motion*. We say that $\{X(t)\}$ is a *σ^2-Brownian motion with drift* if $X(t) = W(t) + \mu t$ for some $\mu \in \mathbb{R}$, where $\{W(t)\}$ is a σ^2-Brownian motion. For short, we say that $\{X(t)\}$ is a (μ, σ^2)-Brownian motion.

The following result shows that the σ^2-Brownian motion is a martingale and gives two further martingales related to Brownian motion.

Theorem 10.1.2 *Let $\{W(t)\}$ be a σ^2-Brownian motion. The following processes are martingales with respect to the filtration $\{\mathcal{F}_t^W\}$:*
(a) $\{W(t)\}$,
(b) $\{W^2(t) - \sigma^2 t\}$,
(c) $\{\exp(sW(t) - \sigma^2 s^2 t/2)\}$ *for each fixed* $s \in \mathbb{R}$.

Proof Statement (a) directly follows from the property of processes with stationary and independent increments mentioned in Example 2 of Section 10.1.3. To show (b), note that for $t, h \ge 0$ we have

$$
\begin{aligned}
\mathbb{E}\,(W^2(t+h) - \sigma^2(t+h) \mid \mathcal{F}_t^W) &= \mathbb{E}\,((W(t+h) - W(t))^2 \\
&\quad + 2(W(t+h) - W(t))W(t) + W^2(t) \mid \mathcal{F}_t^W) - \sigma^2(t+h) \\
&= W^2(t) - \sigma^2 t\,.
\end{aligned}
$$

Similarly, for $t, h \ge 0$,

$$
\mathbb{E}\left(\mathrm{e}^{sW(t+h) - \sigma^2 s^2 (t+h)/2} \mid \mathcal{F}_t^W\right) = \mathbb{E}\left(\mathrm{e}^{s(W(t+h) - W(t))}\right) \mathrm{e}^{sW(t) - \sigma^2 s^2 (t+h)/2}.
$$

Since $\mathbb{E}\left(\mathrm{e}^{s(W(t+h) - W(t))}\right) = \mathrm{e}^{\sigma^2 s^2 h/2}$ we get statement (c). □

Let $\{X(t),\ t \ge 0\}$ be a stochastic process which is càdlàg and adapted to a filtration $\{\mathcal{F}_t\}$. We say that $\{X(t),\ t \ge 0\}$ is a *homogeneous Markov process* with respect to $\{\mathcal{F}_t\}$ if, with probability 1,

$$\mathbb{P}(X(t+h) \in B \mid \mathcal{F}_t) = \mathbb{P}(X(t+h) \in B \mid X(t))$$

for all $t, h \ge 0$ and $B \in \mathcal{B}(\mathbb{R})$. The case that $\{X(t)\}$ is Markov with respect to its history $\{\mathcal{F}_t^X\}$ is discussed in more detail in Chapter 11. Furthermore, we say that $\{X(t),\ t \ge 0\}$ is a *strong Markov process* if, with probability 1,

$$\mathbb{P}(X(\tau+h) \in B \mid \mathcal{F}_\tau) = \mathbb{P}(X(\tau+h) \in B \mid X(\tau))$$

on $\{\tau < \infty\}$ for each $\{\mathcal{F}_t\}$-stopping time τ. It can be proved that most processes with independent stationary increments (including Brownian motion and the claim surplus process in the Poisson compound model) are strong Markov processes with respect to their history; see Breiman (1992).

For many results on martingales given in the literature, a right-continuous filtration is required. For example, such an assumption is needed in order to prove that the first entrance time to an open set is a stopping time; see Section 10.1.2. It turns out that if $\{X(t),\ t \ge 0\}$ is a strong Markov process with respect to a complete filtration $\{\mathcal{F}_t\}$, then this complete filtration is right-continuous; see Proposition 7.7 of Karatzas and Shreve (1991). However, for our purposes the notion of a complete filtration is not very useful; see, for example, the remark at the end of Section 10.2.6.

Another result, due to Brémaud (1981), Theorem A2T26, states that if $\{X(t),\ t \ge 0\}$ is a stochastic process defined on a probability space $(\Omega, \mathcal{F}, \mathbb{P})$ such that, for all $t \ge 0$ and all $\omega \in \Omega$, there exists a strictly positive real number $\varepsilon(t,\omega)$ for which

$$X(t+h,\omega) = X(t,\omega) \qquad \text{if } h \in [0, \varepsilon(t,\omega)), \tag{10.1.15}$$

then the history $\{\mathcal{F}_t^X,\ t \ge 0\}$ is right-continuous. A stochastic process with property (10.1.15) is called a *pure jump process.*

10.1.5 Uniform Integrability

The introduction on continuous-time martingales closes with a discussion of some further related results. We extend the definition of uniform integrability that has been mentioned in (10.1.14) for sequences of random variables. The concept can be generalized to any collection of random variables, that is to a stochastic process $\{X(t),\ t \in \mathcal{T}\}$ with a general space $\mathcal{T}$ of parameters. We say that $\{X(t),\ t \in \mathcal{T}\}$ is *uniformly integrable* if $\mathbb{E}\,|X(t)| < \infty$ for all $t \in \mathcal{T}$ and if

$$\lim_{x\to\infty} \Big(\sup_{t\in\mathcal{T}} \mathbb{E}\,(|X(t)|; |X(t)| > x)\Big) = 0\,. \tag{10.1.16}$$

The following is a characterization of uniform integrability.

Theorem 10.1.3 *The stochastic process* $\{X(t), t \in \mathcal{T}\}$ *is uniformly integrable if and only if the following two conditions hold:*
(a) $\sup_{t\in\mathcal{T}} \mathbb{E}\,|X(t)| < \infty$,
(b) *for every* $\varepsilon > 0$ *there exists a* $\delta > 0$ *such that for every* $t \in \mathcal{T}$ *and* $A \in \mathcal{F}$, $\mathbb{P}(A) < \delta$ *implies* $\mathbb{E}\,[|X(t)|; A] < \varepsilon$.

Proof To show sufficiency of conditions (a) and (b) it suffices to prove that (10.1.16) holds. Note that $x\mathbb{P}(|X(t)| > x) \le \mathbb{E}\,|X(t)|$ for all $t \in \mathcal{T}$ and $x > 0$. Hence $\sup_{t\in\mathcal{T}} \mathbb{P}(|X(t)| > x) \le x^{-1} \sup_{t\in\mathcal{T}} \mathbb{E}\,|X(t)| < \delta$ for each $\delta > 0$ provided that $x > 0$ is large enough. Thus, for each $\varepsilon > 0$,

$$\limsup_{x\to\infty}\Big(\sup_{t\in\mathcal{T}} \mathbb{E}\,(|X(t)|; |X(t)| > x)\Big) \le \varepsilon\,.$$

This gives (10.1.16), since $\varepsilon > 0$ can be chosen arbitrarily small. Assume now that $\{X(t), t \in \mathcal{T}\}$ is uniformly integrable. Then for each x large enough,

$$\sup_{t\in\mathcal{T}} \mathbb{E}\,|X(t)| \le \sup_{t\in\mathcal{T}} \{\mathbb{E}\,[|X(t)|; |X(t)| > x] + x\} < \infty\,,$$

i.e. condition (a) is fulfilled. Now, for $\varepsilon > 0$ given, choose $x > 0$ such that

$$\sup_{t\in\mathcal{T}} \mathbb{E}\,[|X(t)|; |X(t)| > x] < \frac{\varepsilon}{2}\,.$$

Then, for each $\delta > 0$ such that $\delta < x^{-1}\varepsilon/2$ and for each $A \in \mathcal{F}$ with $\mathbb{P}(A) < \delta$, we have $\mathbb{E}\,[|X(t)|; A] \le \mathbb{E}\,[|X(t)|; |X(t)| > x] + x\mathbb{P}(A) \le \varepsilon$. This shows that condition (b) is fulfilled. $\square$

The relevance of uniform integrability in the realm of convergence of random variables becomes clear in the next theorem.

Theorem 10.1.4 *Let* $t_0, t_1, \ldots$ *be an arbitrary sequence of parameters* $t_n \in \mathcal{T}$ *and consider the sequence* $X, X_0, X_1, \ldots$ *of random variables with* $X_n = X(t_n)$ *and* $\mathbb{E}\,|X_n| < \infty$ *for each* $n \in \mathbb{N}$. *If* $\lim_{n\to\infty} X_n = X$ *with probability* 1, *then the following statements are equivalent:*
(a) $\{X_n,\ n \in \mathbb{N}\}$ *is uniformly integrable,*
(b) $\mathbb{E}\,|X| < \infty$ *and* $\lim_{n\to\infty} \mathbb{E}\,|X_n - X| = 0$.

Proof Suppose that $\lim_{n\to\infty} X_n = X$ with probability 1 and note that then $\lim_{n\to\infty} \mathbb{P}(|X_n - X| > \varepsilon) = 0$ for each $\varepsilon > 0$. Furthermore, for each $\varepsilon > 0$ we have

$$\begin{aligned}
&\mathbb{E}\,|X_n - X| \\
&\quad\le \mathbb{E}\,[|X_n - X|; |X_n - X| \le \varepsilon] + \mathbb{E}\,[|X_n - X|; |X_n - X| > \varepsilon] \\
&\quad\le \varepsilon + \mathbb{E}\,[|X_n|; |X_n - X| > \varepsilon] + \mathbb{E}\,[|X|; |X_n - X| > \varepsilon]\,.
\end{aligned}$$

Let $\{X_n,\ n \in \mathbb{N}\}$ be uniformly integrable. Then the second term of the last expression tends to zero by Theorem 10.1.3 choosing $A = \{|X_n - X| > \varepsilon\}$. The third term also tends to zero since $\sup_{n\in\mathbb{N}} \mathbb{E}\,|X_n| < \infty$ by Theorem 10.1.3 and therefore $\mathbb{E}\,X < \infty$ by Fatou's lemma. Thus, (b) holds because $\varepsilon > 0$ can be chosen arbitrarily small. Assume now that (b) holds. Then

$$\sup_{n\in\mathbb{N}} \mathbb{E}\,|X_n| \le \sup_{n\in\mathbb{N}} \mathbb{E}\,|X_n - X| + \mathbb{E}\,|X| < \infty. \tag{10.1.17}$$

Fix $\varepsilon > 0$ and choose $n_0 \in \mathbb{N}$ such that $\mathbb{E}\,[|X_n - X|; A] \le \mathbb{E}\,|X_n - X| \le \varepsilon/2$ for all $n \ge n_0$ and $A \in \mathcal{F}$. Now, choose $\delta > 0$ such that $\mathbb{P}(A) < \delta$ implies

$$\sup_{0\le n\le n_0} \mathbb{E}\,[|X_n - X|; A] + \mathbb{E}\,[|X|; A] \le \frac{\varepsilon}{2}\,.$$

Thus, for each $\varepsilon > 0$ there exists $\delta > 0$ such that

$$\sup_{n\in\mathbb{N}} \mathbb{E}\,[|X_n|; A] \le \sup_{n\in\mathbb{N}} \mathbb{E}\,[|X_n - X|; A] + \mathbb{E}\,[|X|; A] \le \varepsilon$$

whenever $\mathbb{P}(A) < \delta$. By Theorem 10.1.3, this and (10.1.17) imply that $\{X_n, n \in \mathbb{N}\}$ is uniformly integrable. □

Corollary 10.1.1 *Let $X, X_0, X_1, \ldots$ be random variables such that $\mathbb{E}\,|X_n| < \infty$ for each $n \in \mathbb{N}$ and $\lim_{n\to\infty} X_n = X$ with probability 1. If $\{X_n, n \in \mathbb{N}\}$ is uniformly integrable, then $\mathbb{E}\,|X| < \infty$ and $\lim_{n\to\infty} \mathbb{E}\,X_n = \mathbb{E}\,X$.*

Proof Since $|\mathbb{E}\,X_n - \mathbb{E}\,X| \le \mathbb{E}\,|X_n - X|$, the statement is an immediate consequence of Theorem 10.1.4. □

Bibliographical Notes. The general concepts of stochastic processes are studied in many books, such as Chung (1982), Dellacherie (1972) and Meyer (1966). The notion of a martingale was introduced to probability theory by Ville (1939) and developed by Doob (1953). For continuous-time martingales we refer to the books mentioned above and to Dellacherie and Meyer (1982), Elliot (1982) and Liptser and Shiryaev (1977). The first quantitative results for Brownian motion are due to Bachelier (1900), Einstein (1905) and Thiele (1880). A rigorous mathematical treatment of this class of stochastic processes began from Wiener (1923,1924). Basic properties of Brownian motion are given, for example, in Billingsley (1995), Breiman (1992), Ito and McKean (1974) and Karatzas and Shreve (1991). The characterization of uniform integrability stated in Section 10.1.5 follows the approach of Karr (1993) and Williams (1991).

10.2 SOME FUNDAMENTAL RESULTS

In this section, we state and prove some fundamental results for continuous-time martingales that will prove to be highly useful when studying more specific martingales that appear in insurance and financial mathematics. The main idea in the proofs of all these results is to draw on the discrete-time theory by considering a continuous-time martingale at a sequence of discrete time instants.

Let $\{\mathcal{F}_t\}$ be an arbitrary but fixed filtration. Unless otherwise stated, stochastic processes are always assumed to be adapted to this filtration and stopping times refer to $\{\mathcal{F}_t\}$. If not needed, we will not refer to the filtration explicitly. Contrary to common practice in continuous-time martingales, sample paths of stochastic processes considered in this section are càdlàg, not just right-continuous functions.

10.2.1 Doob's Inequality

The following result is a counterpart of *Doob's inequality* (9.1.40) given in Section 9.1.7 for discrete-time submartingales. Armed with this inequality, one can easily derive exponential (Lundberg-type) upper bounds for infinite-horizon ruin probabilities in the compound Poisson model.

Theorem 10.2.1 *Let $\{X(t)\}$ be a submartingale. Then for each $x > 0$ and $t \ge 0$,*

$$\mathbb{P}\Big(\sup_{0\le v\le t} X(v) \ge x\Big) \le \frac{\mathbb{E}\,(X(t))_+}{x}\,. \tag{10.2.1}$$

Proof Without loss of generality, we assume that for each $t \ge 0$ the random variable $X(t)$ is nonnegative. Indeed, for $x \ge 0$ we have

$$\mathbb{P}\Big(\sup_{0\le v\le t} X(v) \ge x\Big) = \mathbb{P}\Big(\sup_{0\le v\le t} (X(v))_+ \ge x\Big)$$

and the stochastic process $\{X'(t), t \ge 0\}$ with $X'(t) = (X(t))_+$ is again a submartingale. This follows easily from Jensen's inequality for conditional expectations. Let B be a finite subset of $[0,t]$ such that $0 \in B$ and $t \in B$. Then, (9.1.40) gives the inequality

$$x\mathbb{P}\Big(\max_{v\in B} X(v) \ge x\Big) \le \mathbb{E}\,X(t)\,. \tag{10.2.2}$$

Considering an increasing sequence $B_1, B_2, \ldots$ of finite sets with union $([0,t) \cap \mathbb{Q}) \cup \{t\}$, we can replace the set B in (10.2.2) by this union. The right-continuity of $\{X(t)\}$ then implies (10.2.1). $\square$

Examples 1. For $s \in \mathbb{R}$ fixed, consider the martingale $\{e^{sS(t)-tg(s)}, t \ge 0\}$ of Example 3 in Section 10.1.3. Here $\{S(t)\}$ is the claim surplus process for the compound Poisson model with arrival rate λ, premium rate β and claim size distribution F_U, while $g(s) = t^{-1} \log \mathbb{E}\, \mathrm{e}^{sS(t)}$. Choose $s = \gamma > 0$ such that $g(\gamma) = 0$, i.e. γ is the adjustment coefficient for this model. In Section 10.1.3 it was shown that $\{\mathbb{E}\, \mathrm{e}^{\gamma S(t)}\}$ is an $\{\mathcal{F}_t^S\}$-martingale. Since $\mathbb{E}\, \mathrm{e}^{\gamma S(t)} = 1$ and $\{\sup_{0 \le v \le t} S(v) \ge x\} = \{\sup_{0 \le v \le t} \mathrm{e}^{\gamma S(v)} \ge \mathrm{e}^{\gamma x}\}$, the bound

$$\mathbb{P}\Big(\sup_{0 \le v \le t} S(v) \ge x\Big) \le \mathrm{e}^{-\gamma x},$$

for all $x \ge 0$ follows from Doob's inequality (10.2.1). Let $t \to \infty$ to get $\mathbb{P}(\sup_{v \ge 0} S(v) \ge x) \le \mathrm{e}^{-\gamma x}$. Note that a stronger version of this *Lundberg inequality* has already been derived in Corollary 5.4.1 but using a more complex argument.

2. Let $\{W(t)\}$ be a standard Brownian motion and consider the $(-\mu, 1)$-Brownian motion $\{X(t)\}$ with negative drift, where $X(t) = W(t) - \mu t$; $\mu > 0$. In Theorem 10.1.2 we showed that, for $s \in \mathbb{R}$ fixed, $\{\mathrm{e}^{s(X(t)+\mu t) - s^2 t/2}, t \ge 0\}$ is an $\{\mathcal{F}_t^W\}$-martingale. Putting $s = 2\mu$ we see that $\{\mathrm{e}^{2\mu X(t)}\}$ is a martingale and since $\sup_{0 \le v \le t} X(v) \ge x$ is equivalent to $\sup_{0 \le v \le t} \mathrm{e}^{2\mu X(v)} \ge \mathrm{e}^{2\mu x}$, again Doob's inequality (10.2.1) yields a bound

$$\mathbb{P}\Big(\sup_{t \ge 0} X(t) \ge x\Big) \le \mathrm{e}^{-2\mu x} \tag{10.2.3}$$

for all $x \ge 0$. Later, in Section 10.3.1, we will prove that even equality holds in (10.2.3), so that $\mathbb{P}(\sup_{t \ge 0} X(t) \ge x) = \mathrm{e}^{-2\mu x}$ for all $x \ge 0$.

10.2.2 Convergence Results

The next theorem is usually called the *submartingale convergence theorem* and is a consequence of Theorem 9.1.3.

Theorem 10.2.2 *Let $\{X(t), t \ge 0\}$ be a submartingale and assume that*

$$\sup_{t \ge 0} \mathbb{E}\, (X(t))_+ < \infty\,. \tag{10.2.4}$$

Then there exists a random variable $X(\infty)$ such that, with probability 1,

$$\lim_{t \to \infty} X(t) = X(\infty) \tag{10.2.5}$$

and $\mathbb{E}\, |X(\infty)| < \infty$. If, additionally,

$$\sup_{t \ge 0} \mathbb{E}\, X^2(t) < \infty\,, \tag{10.2.6}$$

then

$$\mathbb{E}\, X^2(\infty) < \infty\,, \qquad \lim_{t \to \infty} \mathbb{E}\, |X(t) - X(\infty)| = 0\,. \tag{10.2.7}$$

Proof Let $\{t_k\}$ be a strictly increasing sequence of positive numbers such that $t_k \to \infty$. Because $\{X_k\}$ with $X_k = X(t_k)$ is a submartingale in discrete time, Theorem 9.1.3 guarantees the existence of an integrable random variable $X(\infty)$ such that $\lim_{k\to\infty} X(t_k) \to X(\infty)$. Let now $\{t'_k\}$ be any other increasing sequence such that $t'_k \to \infty$. By the same Theorem 9.1.3 there exists a random variable Y such that $\lim_{k\to\infty} X(t'_k) \to Y$. Let $\{v_k\}$ be the increasing sequence such that $\{t_k\} \cup \{t'_k\} = \{v_k\}$. Because $\lim_{k\to\infty} X(v_k)$ exists we must have $Y = X(\infty)$. But this proves $\lim_{t\to\infty} X(t) \to X(\infty)$. The rest of the theorem follows from Theorem 9.1.3. □

10.2.3 Optional Sampling Theorems

Consider an arbitrary filtration $\{\mathcal{F}_t\}$, a stochastic process $\{X(t)\}$ and a stopping time τ. Define $X_\tau : \Omega \to \mathbb{R}$ by

$$X_\tau(\omega) = \begin{cases} X(\tau(\omega), \omega) & \text{if } \tau(\omega) < \infty, \\ X & \text{if } \tau(\omega) = \infty, \end{cases} \tag{10.2.8}$$

where X is a certain random variable. We put $X = X(\infty) = \lim_{t\to\infty} X(t)$ if this limit is well-defined as, for example, in the submartingale case of Theorem 10.2.2. It is generally not obvious that X_τ is a random variable. However, for processes with right-continuous sample paths, it is easy to give a positive answer to this question.

We first construct a standard discrete approximation to the stopping time τ. For $n = 1, 2, \ldots$, define the random variable $\tau^{(n)}$ by

$$\tau^{(n)} = \begin{cases} \dfrac{k+1}{2^n} & \text{if} \quad k2^{-n} < \tau \le (k+1)2^{-n} \text{ for some } k = 0, 1, \ldots, \\ \infty & \text{if} \quad \tau = \infty. \end{cases} \tag{10.2.9}$$

Then for $k2^{-n} \le t < (k+1)2^{-n}$ we have

$$\{\tau^{(n)} \le t\} = \{\tau^{(n)} \le k2^{-n}\} = \{\tau \le k2^{-n}\} \in \mathcal{F}_{k2^{-n}} \subset \mathcal{F}_t \tag{10.2.10}$$

and hence the random variables $\tau^{(n)}$ are stopping times. Moreover, with probability 1, we have $\tau^{(1)} \ge \tau^{(2)} \ge \ldots \downarrow \tau$.

Theorem 10.2.3 *If $\tau < \infty$ or $X(\infty)$ exists then X_τ is a random variable.*

Proof Since $\{X(t)\}$ is assumed to be càdlàg, we have $X_\tau = \lim_{n\to\infty} X_{\tau^{(n)}}$. Thus it suffices to prove that $X_{\tau^{(n)}}$ is measurable. However, for each Borel set $B \in \mathcal{B}(\mathbb{R})$,

$$\begin{aligned} \{X_{\tau^{(n)}} \in B\} &= \{X(\infty) \in B, \tau^{(n)} = \infty\} \\ &\quad \cup \bigcup_{k=0}^{\infty} \Big\{X\Big(\frac{k+1}{2^n}\Big) \in B\Big\} \cap \Big\{\tau^{(n)} = \frac{k+1}{2^n}\Big\} \in \mathcal{F}. \end{aligned}$$ □

Let $t \geq 0$ be fixed. If τ is a stopping time, then also the random variable $t \wedge \tau = \min\{t, \tau\}$ is a stopping time, as the reader can easily show. The following theorem is concerned with the "stopped" version $\{X(t \wedge \tau), t \geq 0\}$ of the stochastic process $\{X(t)\}$. We first deal with an auxiliary result.

Lemma 10.2.1 *If $\{X(t)\}$ is a submartingale, then the sequence $X(\tau^{(1)} \wedge t), X(\tau^{(2)} \wedge t), \ldots$ is uniformly integrable for each $t \geq 0$.*

Proof Let $Y = \sup_{n \geq 1} X(\tau^{(n)} \wedge t)$. Note that $\mathbb{P}(|Y| < \infty) = 1$. Indeed, apply Doob's inequality (10.2.1) to the submartingale $\{|X(t)|, t \geq 0\}$ to find $\mathbb{P}(|Y| \geq x) \leq \mathbb{P}(\sup_{0 \leq v \leq t} |X(v)| \geq x) \leq x^{-1}\mathbb{E}\,|X(t)|$ and, consequently, $\lim_{x \to \infty} \mathbb{P}(|Y| \geq x) = 0$. Since

$$\mathbb{E}\,|X(\tau^{(n)} \wedge t)| \leq \sum_{\{k:(k+1)2^{-n} < t\}} \mathbb{E}\left|X\left(\frac{k+1}{2^n} \wedge t\right)\right| + \mathbb{E}\,|X(t)| < \infty\,,$$

it remains to show that (10.1.16) is fulfilled. By the result of Theorem 9.1.7,

$$\begin{aligned}
&\sup_{n \geq 1} \mathbb{E}\,[|X(\tau^{(n)} \wedge t)|; X(\tau^{(n)} \wedge t) > x] \\
&\quad \leq \sup_{n \geq 1} \mathbb{E}\,[|X(t)|; X(\tau^{(n)} \wedge t) > x] \leq \sup_{n \geq 1} \mathbb{E}\,[|X(t)|\,; Y \geq x] \\
&\quad = \mathbb{E}\,[|X(t)|; Y \geq x]\,.
\end{aligned}$$

Furthermore, $\lim_{x \to \infty} \mathbb{E}\,[|X(t)|; Y \geq x] = 0$ since $\mathbb{E}\,|X(t)| < \infty$ and $|Y| < \infty$ with probability 1. □

Theorem 10.2.4 *Let $\{X(t)\}$ be a martingale and τ a stopping time. Then also the stochastic process $\{X(\tau \wedge t), t \geq 0\}$ is a martingale.*

Proof We have to prove that

$$\mathbb{E}\,|X(\tau \wedge t)| < \infty \tag{10.2.11}$$

for each $t \geq 0$ and that

$$\mathbb{E}\,[X(\tau \wedge t); A] = \mathbb{E}\,[X(\tau \wedge v); A] \tag{10.2.12}$$

for all $0 \leq v < t$ and $A \in \mathcal{F}_v$. As in the proof of Theorem 9.1.7, we can show that

$$\mathbb{E}\,[X(\tau^{(n)} \wedge t); A] = \mathbb{E}\,[X(\tau^{(n)} \wedge v); A]\,. \tag{10.2.13}$$

On the other hand, by Lemma 10.2.1 the sequences $\{X(\tau^{(n)} \wedge v),\ n \geq 1\}$ and $\{X(\tau^{(n)} \wedge t),\ n \geq 1\}$ are uniformly integrable, and since $\{X(t)\}$ is càdlàg we have $\lim_{n \to \infty} X(\tau^{(n)} \wedge t) = X(\tau \wedge t)$ and $\lim_{n \to \infty} X(\tau^{(n)} \wedge v) = X(\tau \wedge v)$. By Corollary 10.1.1, both (10.2.11) and (10.2.12) hold. □

We remark that Theorem 10.2.4 on *stopped martingales* can easily be extended to submartingales. It suffices to replace (10.2.13) by

$$\mathbb{E}\,[X(\tau^{(n)} \wedge t); A] \geq \mathbb{E}\,[X(\tau^{(n)} \wedge v); A]$$

to conclude that for all $0 \leq v < t$ and $A \in \mathcal{F}_v$

$$\mathbb{E}\,[X(\tau \wedge t); A] \geq \mathbb{E}\,[X(\tau \wedge v); A]\,. \tag{10.2.14}$$

We turn to continuous-time versions of the *optional sampling theorems* as derived in Section 9.1.6 for discrete-time martingales. In the same way as in Section 9.2.3 the reader can show that the family of events $\mathcal{F}_\tau$ consisting of those $A \in \mathcal{F}$ for which $A \cap \{\tau \leq t\} \in \mathcal{F}_t$ for every $t \geq 0$ is a σ-algebra. We call $\mathcal{F}_\tau$ the σ-algebra of *events prior to the stopping time* τ. Note that formally there is a difference between this definition and the definition of $\mathcal{F}_\tau$ given in (9.2.19) for the discrete-time case. However, in the latter case we can use both approaches.

Theorem 10.2.5 *Let $\{X(t)\}$ be a martingale and τ an arbitrary stopping time. Then, for each $t \geq 0$,*

$$\mathbb{E}\,(X(t) \mid \mathcal{F}_\tau) = X(\tau \wedge t)\,. \tag{10.2.15}$$

Proof Let $\tau^{(n)}$ be given by (10.2.9). It then follows that $\mathbb{E}\,(X(t) \mid \mathcal{F}_{\tau^{(n)}}) = X(\tau^n \wedge t)$, as can be shown by the reader. Now, since $\mathcal{F}_\tau \subset \mathcal{F}_{\tau^{(n)}}$, we have

$$\mathbb{E}\,(X(t) \mid \mathcal{F}_\tau) = \mathbb{E}\,(\mathbb{E}\,(X(t) \mid \mathcal{F}_{\tau^{(n)}}) \mid \mathcal{F}_\tau) = \mathbb{E}\,(X(\tau^{(n)} \wedge t) \mid \mathcal{F}_\tau)\,.$$

By Lemma 10.2.1, the sequence $\{X(\tau^{(n)} \wedge t), n \geq 1\}$ is uniformly integrable. But $\{X(t)\}$ is càdlàg and so Corollary 10.1.1 shows that

$$\mathbb{E}\,(X(t) \mid \mathcal{F}_\tau) = \lim_{n\to\infty} \mathbb{E}\,(X(\tau^{(n)} \wedge t) \mid \mathcal{F}_\tau) = \mathbb{E}\,(X(\tau \wedge t) \mid \mathcal{F}_\tau) = X(\tau \wedge t)\,,$$

which completes the proof. □

Theorem 10.2.6 *If $\{X(t)\}$ is a martingale and τ a bounded stopping time, then*

$$\mathbb{E}\,X(\tau) = \mathbb{E}\,X(0)\,. \tag{10.2.16}$$

Proof Putting $A = \Omega$ and $s = 0$, (10.2.16) is an immediate consequence of (10.2.12) if we choose $t \geq 0$ such that $\mathbb{P}(\tau \leq t) = 1$. □

The following result is a continuous-time counterpart of Theorem 9.1.5.

Theorem 10.2.7 *Let $\{X(t)\}$ be a martingale and τ a finite stopping time such that $\mathbb{E}\,|X(\tau)| < \infty$ and $\lim_{t\to\infty} \mathbb{E}\,[|X(t)|; \tau > t] = 0$. Then*

$$\mathbb{E}\,X(\tau) = \mathbb{E}\,X(0)\,. \tag{10.2.17}$$

Proof Apply Theorem 10.2.6 to the bounded stopping time $\tau \wedge t$, where $t \geq 0$ to obtain $\mathbb{E}\, X(\tau \wedge t) = \mathbb{E}\, X(0)$. By the dominated convergence theorem,

$$\begin{aligned} \mathbb{E}\, X(0) &= \lim_{t\to\infty} \mathbb{E}\, X(\tau \wedge t) \\ &= \lim_{t\to\infty} \mathbb{E}\,[X(\tau); \tau \leq t] + \lim_{t\to\infty} \mathbb{E}\,[X(t); \tau > t] = \mathbb{E}\, X(\tau)\,. \end{aligned}$$ □

10.2.4 The Doob–Meyer Decomposition

Our next result provides a continuous-time analogue to the *Doob–Meyer decomposition* which was derived in Section 9.1.8 for discrete-time submartingales. We employ the concept of a uniformly integrable family of random variables from Section 10.1.5, with $\mathcal{T}$ the family of all $\{\mathcal{F}_t\}$-stopping times. We say that the submartingale $\{X(t)\}$ belongs to the *class DL* if for each $t \geq 0$, the family $\{X(t \wedge \tau), \tau \in \mathcal{T}\}$ is uniformly integrable. Examples of such submartingales can be found as follows. Let $\{X(t)\}$ be a right-continuous submartingale. Then $\{X(t)\}$ belongs to the class DL if $\{X(t)\}$ is a martingale or if $\{X(t)\}$ is bounded from below. The reader is invited to show this as an exercise.

The Doob–Meyer decomposition stated below is unique up to indistinguishability where we say that the stochastic processes $\{X(t), t \in \mathcal{T}\}$ and $\{X'(t), t \in \mathcal{T}\}$ are *indistinguishable* if they are defined on the same probability space $(\Omega, \mathcal{F}, \mathbb{P})$ and if there exists $A \in \mathcal{F}$ such that $\{\omega : X(t,\omega) \neq X'(t,\omega) \text{ for some } t \in \mathcal{T}\} \subset A$ and $\mathbb{P}(A) = 0$.

The aim of the Doob–Meyer decomposition is to represent the submartingale $\{X(t)\}$ in the following way: $X(t) = X(0) + M(t) + A(t)$ for all $t \geq 0$ where

- $\{M(t)\}$ is a martingale with respect to the filtration $\{\mathcal{F}_t\}$,
- $\{A(t)\}$ is an *increasing process*, that is, for each $\omega \in \Omega$, the sample path $\{A(\omega, t), t \geq 0\}$ is a nondecreasing function and $A(0) = 0$, where as usual we assume that $\{A(t)\}$ is adapted to $\{\mathcal{F}_t\}$.

Note that every right-continuous nondecreasing function $a : \mathbb{R}_+ \to \mathbb{R}_+$ with $a(0) = 0$ determines a Borel measure m_a on $\mathcal{B}(\mathbb{R}_+)$ by $m_a([0,t]) = a(t)$ for all $t \geq 0$. For each measurable function $g : \mathbb{R}_+ \to \mathbb{R}$ we define the integral $\int_0^t g(v)\,\mathrm{d}a(v)$ by

$$\int_0^t g(v)\,\mathrm{d}a(v) = \int_{[0,t]} g(v)\, m_a(\mathrm{d}v)$$

provided that the integral on the right-hand side exists.

Theorem 10.2.8 *Let $\{\mathcal{F}_t\}$ be right-continuous, and let $\{X(t)\}$ be a submartingale of class DL. Then:*

(a) *there exists an increasing process* $\{A(t)\}$ *and a martingale* $\{M(t)\}$ *fulfilling for all* $t \geq 0$

$$X(t) = X(0) + M(t) + A(t) \tag{10.2.18}$$

and, for all nonnegative right-continuous martingales $\{Y(t)\}$ *and* $\tau \in \mathcal{T}$,

$$\mathbb{E} \int_0^{t\wedge\tau} Y(v-)\, \mathrm{d}A(v) = \mathbb{E} \int_0^{t\wedge\tau} Y(v)\, \mathrm{d}A(v) = \mathbb{E}\left(Y(t\wedge\tau)A(t\wedge\tau)\right), \tag{10.2.19}$$

(b) *up to indistinguishability, there exists only one right-continuous increasing process* $\{A(t), t \geq 0\}$ *which satisfies* (10.2.18) *and* (10.2.19).

The *proof* of Theorem 10.2.8 goes beyond the scope of this book. It can be found, for example, in Ikeda and Watanabe (1989), Theorem 6.12.

10.2.5 Kolmogorov's Extension Theorem

In this section we discuss a variant of *Kolmogorov's extension theorem* for the continuous-time case. As a canonical model we take $\Omega = \mathbb{R}^{[0,\infty)}$ to be the set of all functions $\omega : \mathbb{R}_+ \to \mathbb{R}$. A filtration on Ω is introduced as follows. Let $A_{t_1,\ldots,t_n}(B_1,\ldots,B_n) = \{\omega \in \Omega : \omega(t_1) \in B_1,\ldots,\omega(t_n) \in B_n\}$, where $t_1 \leq \ldots \leq t_n$, $B_1,\ldots,B_n \in \mathcal{B}(\mathbb{R})$. We call $A_{t_1,\ldots,t_n}(B_1,\ldots,B_n)$ a *cylindrical set*. Define now $\mathcal{F}_t$ to be the smallest σ-algebra containing all cylindrical sets $A_{t_1,\ldots,t_n}(B_1,\ldots,B_n)$ such that $t_1 \leq \ldots \leq t_n \leq t$. Let the σ-algebra $\mathcal{F}$ of all events in Ω be given by $\mathcal{F} = \sigma(\bigcup_{t\geq 0} \mathcal{F}_t)$.

Theorem 10.2.9 *Suppose that, for each* $n \geq 1$ *and for all* $t_1 \leq \ldots \leq t_n$, *there is a probability measure* $P_{t_1,\ldots,t_n}$ *on* $(\mathbb{R}^n, \mathcal{B}(\mathbb{R}^n))$ *for which the family* $\{P_{t_1,\ldots,t_n}\}$ *satisfies the following consistency condition:*

$$P_{t_1,\ldots,t_n,t_{n+1}}(B_1 \times \ldots \times B_n \times \mathbb{R}) = P_{t_1,\ldots,t_n}(B_1 \times \ldots \times B_n), \tag{10.2.20}$$

for all $n \geq 1$, $t_1 \leq \ldots \leq t_n$ *and* $B_1,\ldots,B_n \in \mathcal{B}(\mathbb{R})$. *Then there exists a uniquely determined probability measure,* $\mathbb{P}$ *say, on* $(\Omega, \mathcal{F})$ *such that*

$$\mathbb{P}(A_{t_1,\ldots,t_n}(B_1,\ldots,B_n)) = P_{t_1,\ldots,t_n}(B_1 \times \ldots \times B_n), \tag{10.2.21}$$

for all $n \geq 1$, $t_1 \leq \ldots \leq t_n$ *and* $B_1,\ldots,B_n \in \mathcal{B}(\mathbb{R})$.

The *proof* of Theorem 10.2.9 is omitted and can be found, for example, in Shiryayev (1984). The probability space $(\Omega, \mathcal{F}, \mathbb{P})$ with the filtration $\{\mathcal{F}_t\}$ and the stochastic process $X(t,\omega) = \omega(t)$ considered in Theorem 10.2.9 is called a *canonical probability space* for $\{X(t)\}$. In the following corollary we assume that $\{X(t)\}$ and $\{\mathcal{F}_t\}$ are given by this canonical model.

Corollary 10.2.1 *Let $\{X(t),\ t \geq 0\}$ be a nonnegative $\{\mathcal{F}_t\}$-martingale such that $\mathbb{E}\, X(0) = 1$ and, for each $t \geq 0$, let $\tilde{\mathbb{P}}_t$ be the probability measure on $\mathcal{F}_t$ defined by $\tilde{\mathbb{P}}_t(A) = \mathbb{E}\,[X(t); A]$. Then there exists a unique probability measure $\tilde{\mathbb{P}}$ on $(\Omega, \mathcal{F})$ such that, for all $A \in \mathcal{F}_t$ and $t \geq 0$,*

$$\tilde{\mathbb{P}}(A) = \tilde{\mathbb{P}}_t(A)\,. \tag{10.2.22}$$

The *proof* is similar to the proof of Corollary 9.2.1 and is therefore omitted.

10.2.6 Change of the Probability Measure

The change of the probability measure considered in Section 9.2 turned out to be useful when studying ruin probabilities of discrete-time risk processes as shown in Section 9.2.4. Here is a continuous-time version of this concept.

Consider an arbitrary probability space $(\Omega, \mathcal{F}, \mathbb{P})$ and an arbitrary filtration $\{\mathcal{F}_t\}$ on it. Put $\mathcal{F}_\infty = \sigma(\cup_{t\geq 0}\mathcal{F}_t)$. Let $\{M(t)\}$ be a positive $\{\mathcal{F}_t\}$-martingale. Without loss of generality we assume that $\mathbb{E}\, M(0) = 1$. Let $t \geq 0$ be fixed. Then, as in Corollary 10.2.1, we can define a new probability measure $\tilde{\mathbb{P}}_t$ on the σ-algebra $\mathcal{F}_t$ by

$$\tilde{\mathbb{P}}_t(A) = \mathbb{E}\,[M(t); A]\,, \qquad A \in \mathcal{F}_t\,. \tag{10.2.23}$$

Lemma 10.2.2 (a) *If $t, h \geq 0$, then $\tilde{\mathbb{P}}_t = \tilde{\mathbb{P}}_{t+h}$ on $\mathcal{F}_t$. In particular, for $A \in \mathcal{F}_t$*

$$\mathbb{E}\,[M(t); A] = \mathbb{E}\,[M(t+h); A]\,.$$

(b) *Assume that there exists a probability measure $\tilde{\mathbb{P}}$ on $\mathcal{F}_\infty$ such that $\tilde{\mathbb{P}} = \tilde{\mathbb{P}}_t$ on $\mathcal{F}_t$ for all $t \geq 0$. Let τ be a stopping time and $A \subset \{\tau < \infty\}$ such that $A \in \mathcal{F}_\tau$. Then*

$$\tilde{\mathbb{P}}(A) = \mathbb{E}\,[M(\tau); A]\,. \tag{10.2.24}$$

Proof (a) By conditioning on $\mathcal{F}_t$ we have

$$\begin{aligned} \mathbb{E}\,[M(t+h); A] &= \mathbb{E}\,(\mathbb{E}\,[M(t+h); A] \mid \mathcal{F}_t) = \mathbb{E}\,[\mathbb{E}\,(M(t+h) \mid \mathcal{F}_t); A] \\ &= \mathbb{E}\,[M(t); A]\,. \end{aligned}$$

(b) Recall that $A \cap \{\tau \leq t\} \in \mathcal{F}_t$. Thus, by (a),

$$\tilde{\mathbb{P}}(A \cap \{\tau \leq t\}) = \mathbb{E}\,[M(t); A \cap \{\tau \leq t\}]\,.$$

Hence, from Theorem 10.2.5 we conclude that

$$\begin{aligned} &\tilde{\mathbb{P}}(A \cap \{\tau \leq t\}) = \mathbb{E}\,(\mathbb{E}\,[M(t); A \cap \{\tau \leq t\}] \mid \mathcal{F}_\tau) \\ &\quad = \mathbb{E}\,[\mathbb{E}\,(M(t) \mid \mathcal{F}_\tau); A \cap \{\tau \leq t\}] = \mathbb{E}\,[M(\tau \wedge t); A \cap \{\tau \leq t\}] \\ &\quad = \mathbb{E}\,[M(\tau); A \cap \{\tau \leq t\}]\,. \end{aligned}$$

Thus, the assertion follows from the monotone convergence theorem. □

Later on, we will need the conditional expectation $\tilde{\mathbb{E}}_t(Y \mid \mathcal{G})$ under $\tilde{\mathbb{P}}_t$, where Y is $\mathcal{F}_\tau$-measurable, $\mathcal{G} \subset \mathcal{F}_\tau$ and τ is either deterministic or a stopping time bounded by t. If τ is unbounded we consider the conditional expectation $\tilde{\mathbb{E}}(Y \mid \mathcal{G})$ under $\tilde{\mathbb{P}}$, where we have to assume that the measure $\tilde{\mathbb{P}}$ on $\mathcal{F}_\infty$ exists. For clarity of notation we sometimes write $\tilde{\mathbb{E}}_\infty$ instead of $\tilde{\mathbb{E}}$, and $\tilde{\mathbb{P}}_\infty$ instead of $\tilde{\mathbb{P}}$.

Theorem 10.2.10 *Let $t \le \infty$. Consider the random variable M_τ defined in (10.2.8). Let Y be an $\mathcal{F}_\tau$-measurable random variable such that Y is integrable under $\tilde{\mathbb{P}}_t$. If $\tau \le t$, then*

$$\tilde{\mathbb{E}}_t(Y \mid \mathcal{G}) = \frac{\mathbb{E}(M_\tau Y \mid \mathcal{G})}{\mathbb{E}(M_\tau \mid \mathcal{G})}. \tag{10.2.25}$$

In particular, for $t, h \ge 0$,

$$\tilde{\mathbb{E}}_{t+h}(Y \mid \mathcal{F}_t) = \frac{\mathbb{E}(M(t+h) Y \mid \mathcal{F}_t)}{M(t)}. \tag{10.2.26}$$

Proof Let Z be a bounded $\mathcal{G}$-measurable random variable. Note that (10.2.24) implies

$$\begin{aligned}
\tilde{\mathbb{E}}_t(YZ) &= \mathbb{E}(M_\tau YZ) = \mathbb{E}(\mathbb{E}(M_\tau YZ \mid \mathcal{G})) \\
&= \mathbb{E}(\mathbb{E}(M_\tau Y \mid \mathcal{G})Z) = \mathbb{E}\Big(\frac{\mathbb{E}(M_\tau Y \mid \mathcal{G})}{\mathbb{E}(M_\tau \mid \mathcal{G})}\mathbb{E}(M_\tau \mid \mathcal{G})Z\Big) \\
&= \mathbb{E}\Big(\mathbb{E}\Big(\frac{\mathbb{E}(M_\tau Y \mid \mathcal{G})}{\mathbb{E}(M_\tau \mid \mathcal{G})} M_\tau Z \,\Big|\, \mathcal{G}\Big)\Big) = \mathbb{E}\Big(\frac{\mathbb{E}(M_\tau Y \mid \mathcal{G})}{\mathbb{E}(M_\tau \mid \mathcal{G})} M_\tau Z\Big) \\
&= \tilde{\mathbb{E}}_t\Big(\frac{\mathbb{E}(M_\tau Y \mid \mathcal{G})}{\mathbb{E}(M_\tau \mid \mathcal{G})} Z\Big),
\end{aligned}$$

where we used the fact that $\mathbb{E}(M_\tau Y \mid \mathcal{G})Z/\mathbb{E}(M_\tau \mid \mathcal{G})$ is $\mathcal{G}$-measurable. □

Most of the results derived in Section 9.2.3 hold in the continuous-time case as well. For instance, for each $t \le \infty$, $\{M(v)^{-1}, 0 \le v < t\}$ is a martingale under $\tilde{\mathbb{P}}_t$. Indeed, (10.2.26) implies

$$\tilde{\mathbb{E}}_t(M(v+h)^{-1} \mid \mathcal{F}_v) = \frac{\mathbb{E}(M(v+h)M(v+h)^{-1} \mid \mathcal{F}_v)}{M(v)} = M(v)^{-1} \tag{10.2.27}$$

for all $t \ge v + h \ge v$. Moreover, (10.2.24) implies that

$$\tilde{\mathbb{E}}_t(M(0)^{-1}) = \mathbb{E}(M(0)^{-1} M(0)) = 1$$

and the martingale $\{M(v)^{-1}, 0 \le v < t\}$ can be used to change the measure $\tilde{\mathbb{P}}_t$. It then readily follows that, on $\mathcal{F}_t$,

$$\widetilde{\left(\tilde{\mathbb{P}}_t\right)}_t = \mathbb{P}_t\,, \qquad t \le \infty\,. \tag{10.2.28}$$

Let $\{S(t)\}$ be a stochastic process, as for example the claim surplus process introduced in Section 5.1.4 and assume that $\tilde{\mathbb{P}}$ exists. We then interpret $\{S(t)\}$ as a generalized surplus process. Assume that there exists an $\mathbb{R}^d$-valued process $\{T(t)\}$ for which $\{M(t)\}$ given by $M(t) = \exp\{s[S(t) - g(T(t))]\}$ is a martingale for some $s \in \mathbb{R}$ and for some measurable function $g : \mathbb{R}^d \to \mathbb{R}$; $d \ge 1$. Without loss of generality, we assume that g is chosen in such a way that $\mathbb{E}\, M(0) = 1$. Furthermore, assume that the probability space on which the processes $\{S(t)\}$ and $\{T(t)\}$ are defined is small enough such that the first entrance time $\tau(u) = \inf\{t \ge 0 : S(t) > u\}$ is a stopping time and $S_{\tau(u)}$, $T_{\tau(u)}$ are well-defined random variables. Consider the "ruin probabilities" $\psi(u;x) = \mathbb{P}(\tau(u) \le x)$ and $\psi(u) = \mathbb{P}(\tau(u) < \infty)$. By changing the measure $\tilde{\mathbb{P}}$ and using (10.2.28), Lemma 10.2.2b applied to $\tilde{\mathbb{P}}$ and to the "changed measure" $\widetilde{\left(\tilde{\mathbb{P}}\right)} = \mathbb{P}$ gives

$$\psi(u;x) = \tilde{\mathbb{E}}\left[\mathrm{e}^{-sS(\tau(u))+su+g(T(\tau(u)))}; \tau(u) \le x\right]\mathrm{e}^{-su}. \tag{10.2.29}$$

Furthermore,

$$\psi(u) = \tilde{\mathbb{E}}\left[\mathrm{e}^{-sS(\tau(u))+su+g(T(\tau(u)))}; \tau(u) \le \infty\right]\mathrm{e}^{-su}. \tag{10.2.30}$$

The latter formula is in particular useful if s is chosen in such a way that ruin occurs almost surely under the measure $\tilde{\mathbb{P}}$, in which case

$$\psi(u) = \tilde{\mathbb{E}}\left(\mathrm{e}^{-sS(\tau(u))+su+g(T(\tau(u)))}\right)\mathrm{e}^{-su}. \tag{10.2.31}$$

If $s > 0$, an upper bound for the ruin probability $\psi(u)$ follows easily:

$$\psi(u) < \tilde{\mathbb{E}}\left(\mathrm{e}^{g(T(\tau(u)))}\right)\mathrm{e}^{-su}\,.$$

However, in most cases the hard problem will be to find an estimate for $\tilde{\mathbb{E}}\,\mathrm{e}^{g(T(\tau(u)))}$.

Remarks 1. If $\tau(u)$ is not a stopping time, we have to replace $\tau(u)$ by the modified first entrance time $\tau^*(u) = \inf\{t \ge 0 : S(t-) \ge u \text{ or } S(t) \ge u\}$ of $\{S(t)\}$ to the interval $[u, \infty)$. Recall that $\tau^*(u)$ is always a stopping time by Theorem 10.1.1. Also ruin probabilities are modified to $\psi^*(u;x) = \mathbb{P}(\tau^*(u) \le x)$ and $\psi^*(u) = \mathbb{P}(\tau^*(u) < \infty)$. In many cases these modified ruin probabilities are equal to the ruin probabilities $\psi(u;x)$ and $\psi(u)$ related with the "usual" ruin time $\tau(u)$. In particular, when the claim surplus process

$\{S(t)\}$ is that of the compound Poisson model it is not difficult to show that $\mathbb{P}(\tau^*(u) \le x) = \mathbb{P}(\tau(u) \le x)$ and consequently $\mathbb{P}(\tau^*(u) < \infty) = \mathbb{P}(\tau(u) < \infty)$ for all $u, x \ge 0$.

2. Using Theorem 10.2.9 with Corollary 10.2.1, one can give sufficient conditions that ensure that the probability measure $\tilde{\mathbb{P}}$ considered in this section exists. The following situation will repeatedly occur in Chapters 11 and 12. Assume that the stochastic process $\{(S(t), T(t)), t \ge 0\}$ is defined on its canonical probability space $(\Omega, \mathcal{F}, \mathbb{P})$, where Ω is a certain (Borel) subset of the set of all right-continuous functions $\omega : \mathbb{R}_+ \to \mathbb{R}^{d+1}$ with left-hand limits and $\mathcal{F} = \mathcal{B}(\Omega)$ is the Borel σ-algebra on Ω. Furthermore, assume that $\{\mathcal{F}_t\}$ is the (uncompleted) history of $\{(S(t), T(t))\}$. Then, $\mathcal{F} = \mathcal{F}_\infty = \sigma\big(\bigcup_{t\ge 0} \mathcal{F}_t\big)$. Theorem 10.2.9 ensures that the probability measures $\tilde{\mathbb{P}}_t$ can be extended to a probability measure $\tilde{\mathbb{P}}$ on $\mathcal{F}_\infty$.

3. Note that this argument does not work if the filtration $\{\mathcal{F}_t\}$ is complete. We will observe in Theorem 11.3.1 that in fact $\mathbb{P}$ and $\tilde{\mathbb{P}}$ may be singular on $\mathcal{F}_\infty$, i.e. there exists a set $A \in \mathcal{F}_\infty$ such that $\mathbb{P}(A) = 1 = 1 - \tilde{\mathbb{P}}(A)$. If $\mathcal{F}_0$ is complete then $A \in \mathcal{F}_0$ and $\tilde{\mathbb{P}}(A) = \mathbb{P}(A) = 1$ because $\mathbb{P}$ and $\tilde{\mathbb{P}}$ are equivalent probability measures on $\mathcal{F}_0$. Hence the $\tilde{\mathbb{P}}_t$ cannot be extended to $\mathcal{F}_\infty$.

Bibliographical Notes. For some of the results presented in this section, the proofs are only sketched and the reader can find them in a number of textbooks, for example, in Dellacherie and Meyer (1982), Ethier and Kurtz (1986) and Karatzas and Shreve (1991).

10.3 RUIN PROBABILITIES AND MARTINGALES

That the computation of finite-horizon ruin probabilities is a difficult task was already illustrated in Section 5.6, and in Section 9.2.4 for the discrete-time risk process. In Theorem 9.2.5 we were able to derive a bound on the ruin probability $\mathbb{P}(\tau_{\mathrm{d}}(u) \le x)$. We now use an optional sampling theorem for continuous-time martingales to prove an analogous result for the finite-horizon ruin probability $\mathbb{P}(\tau(u) \le x)$ in the compound Poisson model. The martingale approach will also lead to similar bounds for level-crossing probabilities of additive processes. As a special case, we show that the supremum of Brownian motion with negative drift is exponentially distributed.

10.3.1 Ruin Probabilities for Additive Processes

In this section we suppose that the stochastic process $\{X(t), t \ge 0\}$ satisfies the following conditions:

- $X(0) = 0$,

- $\{X(t), t \geq 0\}$ has independent and stationary increments,
- there exists some $s_0 > 0$ such that $\mathbb{E}\, e^{s_0 X(t)} < \infty$ for all $t \geq 0$,
- $\mathbb{E}\, X(t) = -\mu t$ for some $\mu > 0$ and all $t \geq 0$,
- $a_s(t) = \mathbb{E}\, e^{sX(t)}$ is a right-continuous function at 0, for each $s \in [0, s_0)$ and some $s_0 > 0$.

The above conditions define the class of *additive processes* with negative drift. We say that $\{X'(t), t \in \mathcal{T}\}$ is a *version* of the stochastic process $\{X(t), t \in \mathcal{T}\}$ if both processes are defined on the same probability space $(\Omega, \mathcal{F}, \mathbb{P})$ and $\mathbb{P}(X(t) \neq X'(t)) = 0$ for each $t \in \mathcal{T}$. From this definition we immediately conclude that each version $\{X'(t)\}$ of $\{X(t)\}$ has the same finite-dimensional distributions as $\{X(t)\}$. It is known (see, for example, Breiman (1992), Chapter 14.4) that an additive process $\{X(t)\}$ always has a version which is càdlàg. In the same way one can show that there is a version of $\{X(t)\}$ with left-continuous sample paths having right-hand limits. That $a_s(t)$ is a continuous function of the variable $t \geq 0$ can be proved by the reader.

Note that both the claim surplus process in the compound Poisson model with $\rho < 1$ and the $(-\mu, \sigma^2)$-Brownian motion with negative drift fulfil the five conditions stated above and so are additive processes.

Lemma 10.3.1 *There exists a function* $g : \mathbb{R}_+ \to \mathbb{R}$ *such that for all* $t \in \mathbb{R}_+$, $s \in [0, s_0)$,

$$\mathbb{E}\, e^{sX(t)} = e^{tg(s)}. \tag{10.3.1}$$

Proof Recall that $a_s(t) = \mathbb{E}\, e^{sX(t)}$ is a continuous function of the variable t. Since $\{X(t)\}$ has independent and stationary increments, the function $a_s(t)$ fulfils the functional equation $a_s(t+h) = a_s(t)a_s(h)$ for all $t, h \geq 0$. As the only continuous solution to this equation is $a_s(t) = e^{tg(s)}$ for some constant $g(s)$ the lemma is proved. □

For the claim surplus process in the compound Poisson model $g(s) = \lambda(\hat{m}_U(s) - 1) - \beta s$, by formula (5.2.1) in Corollary 5.2.1. For the $(-\mu, \sigma^2)$-Brownian motion with negative drift we have $\mathbb{E}\, e^{sX(t)} = \exp(t(-\mu s + (\sigma)^2/2))$ and consequently $g(s) = -\mu s + (\sigma s)^2/2$.

Lemma 10.3.2 *The following process* $\{M(t)\}$ *is an* $\{\mathcal{F}_t^X\}$*-martingale, where*

$$M(t) = e^{sX(t) - tg(s)}, \qquad t \geq 0. \tag{10.3.2}$$

The *proof* of Lemma 10.3.2 is analogous to the proof given in Example 3 of Section 10.1.3 and is therefore omitted.

Consider the modified first entrance time $\tau^*(u)$ of $\{X(t)\}$ to the set $[u, \infty)$ – see (10.1.1) for definition – and let $\tau(u) = \inf\{t \geq 0 : X(t) > u\}$. We showed in Theorem 10.1.1 that $\tau^*(u)$ is a stopping time. Recall also that for all $u, x \geq 0$

$$\mathbb{P}(\tau^*(u) \leq x) = \mathbb{P}(\tau(u) \leq x) \tag{10.3.3}$$

and in particular

$$\mathbb{P}(\tau^*(u) < \infty) = \mathbb{P}(\tau(u) < \infty) \tag{10.3.4}$$

if $\{X(t)\}$ is the claim surplus process in the compound Poisson model; see Remark 1 in Section 10.2.6. We leave it to the reader to show that (10.3.3) and (10.3.4) also hold if $\{X(t)\}$ is the $(-\mu, \sigma^2)$-Brownian motion.

Now assume that $\{X(t)\}$ is càdlàg and apply Theorem 10.2.6 to the stopped martingale $\{M(t \wedge \tau^*), t \geq 0\}$, where $\{M(t)\}$ is given by (10.3.2) and $\tau^* = \tau^*(u)$. Then (10.2.16) gives

$$\begin{aligned} 1 &= M(0) = \mathbb{E}\, M(t \wedge \tau^*) \\ &= \mathbb{E}\left[M(t \wedge \tau^*); \tau^* \leq t\right] + \mathbb{E}\left[M(t \wedge \tau^*); \tau^* > t\right]. \end{aligned} \tag{10.3.5}$$

From (10.3.5) we can directly draw a few interesting results. Using (10.3.3) the first of them gives explicit formulae for the ruin probability $\psi(u) = \mathbb{P}(\tau < \infty)$ and for the Laplace–Stieltjes transform of $\tau = \tau(u)$ when $\{X(t)\}$ is the $(-\mu, \sigma^2)$-Brownian motion with negative drift. Put

$$\gamma = \sup\{s > 0,\ g(s) \leq 0\}. \tag{10.3.6}$$

Then we have $\gamma = 2\mu/\sigma^2$ and $g(\gamma) = 0$. Note that, according to the terminology of Chapter 6, γ is called the *adjustment coefficient* if there exists $s_0 > 0$ such that $g(s_0) = 0$. Clearly, in this case we have $\gamma = s_0$.

Theorem 10.3.1 *Let $\{X(t)\}$ be the $(-\mu, \sigma^2)$-Brownian motion with negative drift. Then,*

$$\psi(u) = \mathrm{e}^{-\gamma u} \tag{10.3.7}$$

and

$$\mathbb{E}\left[\mathrm{e}^{-s\tau}; \tau < \infty\right] = \exp\Big(-\frac{u}{\sigma^2}\big(\mu + \sqrt{\mu^2 + 2\sigma^2 s}\big)\Big), \quad s \geq 0. \tag{10.3.8}$$

Proof Note that $\{X(t)\}$ has continuous trajectories and that consequently $X(\tau) = u$. Then, applying (10.3.5) to the martingale $\{M(t)\}$ given by (10.3.2) and using (10.3.3), we obtain

$$1 = M(0) = \mathbb{E}\left[\mathrm{e}^{su - \tau(\sigma^2 s^2/2 - \mu s)}; \tau < t\right] + \mathbb{E}\left[\mathrm{e}^{sX(t) - t(\sigma^2 s^2/2 - \mu s)}; \tau > t\right]. \tag{10.3.9}$$

Since $\lim_{t\to\infty} X(t) = -\infty$, we have $\mathrm{e}^{sX(t) - t(\sigma^2 s^2/2 - \mu s)}\mathbb{I}(\tau > t) \to 0$ with probability 1, for $0 < s \leq \gamma$. Moreover,

$$\mathrm{e}^{sX(t) - t(\sigma^2 s^2/2 - \mu s)}\mathbb{I}(\tau > t) \leq \mathrm{e}^{su - t(\sigma^2 s^2/2 - \mu s)}\mathbb{I}(\tau > t),$$

which is an integrable bound, and hence by the dominated convergence theorem, $\lim_{t\to\infty} \mathbb{E}\left[e^{sX(t)-t(\sigma^2 s^2/2-\mu s)}; \tau > t\right] = 0$. Thus, letting $t \to \infty$, (10.3.9) implies

$$e^{-su} = \mathbb{E}\left[e^{-\tau(\sigma^2 s^2/2-\mu s)}; \tau < \infty\right]. \tag{10.3.10}$$

Now setting $s = \gamma$ we get (10.3.7). It remains to show (10.3.8). Turn once more to equation (10.3.10) and take $s = g^{-1}(s')$, where $0 \le s' = g(s) = -\mu s + (\sigma s)^2/2$. Then the inverse function $g^{-1}(s')$ of $g(s)$ is well-defined in the interval $(\mu/\sigma^2, \infty)$, where $g^{-1}(s') = \sigma^{-2}\big(\mu + \sqrt{\mu^2 + 2\sigma^2 s'}\big)$. □

10.3.2 Finite-Horizon Ruin Probabilities

We begin with a formula for the finite-horizon ruin probabilities of a Brownian motion with drift.

Theorem 10.3.2 *Let $u \ge 0$ be fixed. Let $\{X(t)\}$ be the $(-\mu, \sigma^2)$-Brownian motion with negative drift and let $\tau(u) = \inf\{t \ge 0,\ X(t) > u\}$. Then, for all $x \ge 0$,*

$$\mathbb{P}(\tau(u) > x) = \Phi\Big(\frac{u - \mu x}{\sigma x^{1/2}}\Big) - \exp\Big(\frac{2\mu}{\sigma^2} u\Big)\Phi\Big(\frac{-u - \mu x}{\sigma x^{1/2}}\Big) \tag{10.3.11}$$

and

$$\frac{d}{dx}\mathbb{P}(\tau(u) \le x) = \frac{u}{\sigma\sqrt{2\pi x^3}} \exp\Big(-\frac{(u - \mu x)^2}{2\sigma^2 x}\Big), \tag{10.3.12}$$

where $\Phi(x)$ denotes the distribution function of the standard normal distribution.

Proof In view of (10.3.8) and the uniqueness property of Laplace transforms it suffices to show that

$$\int_0^\infty e^{-sx} \frac{u}{\sigma\sqrt{2\pi x^3}} \exp\Big(-\frac{(u - \mu x)^2}{2\sigma^2 x}\Big)\, dx = \exp\Big(-\frac{u}{\sigma^2}\big(\mu + \sqrt{\mu^2 + 2\sigma^2 s}\big)\Big).$$

This can be done by using, for example, the fact that the Laplace transform of

$$h(x) = \frac{k}{2\sqrt{\pi x^3}} \exp\Big(-\frac{k^2}{4x}\Big)$$

is given by

$$\int_0^\infty e^{-sx} h(x)\, dx = \exp(-k\sqrt{s}), \qquad s > 0,$$

see the table of Laplace transforms in Korn and Korn (1968), for example. □

In the proof of Theorem 10.3.1 the continuity of the trajectories of $\{X(t)\}$ has been crucial. Within the class of additive processes this is only possible for

$\{X(t)\}$ being a Brownian motion with negative drift. However, if we are using (10.3.5) for additive processes with jumps, then we can get upper bounds of the Lundberg-type for the probability $\mathbb{P}(\tau^* \le x)$. Equation (10.3.5) implies that

$$\begin{aligned} 1 &\ge \mathbb{E}\,(M(x \wedge \tau^*) \mid \tau^* \le x)\mathbb{P}(\tau^* \le x) \\ &= \mathbb{E}\,(\mathrm{e}^{sX(\tau^*)-\tau^* g(s)} \mid \tau^* \le x)\mathbb{P}(\tau^* \le x) \end{aligned}$$

and since $X(\tau^*) \ge u$, we have $1 \ge \mathbb{E}\,(\mathrm{e}^{su-\tau^* g(s)} \mid \tau^* \le x)\mathbb{P}(\tau^* \le x)$. Hence

$$\mathbb{P}(\tau^* \le x) \le \frac{\mathrm{e}^{-su}}{\mathbb{E}\,(\mathrm{e}^{-\tau^* g(s)} \mid \tau^* \le x)} \le \mathrm{e}^{-su} \sup_{0 \le v \le x} \mathrm{e}^{vg(s)} . \tag{10.3.13}$$

Let $x \to \infty$, then

$$\psi^*(u) = \mathbb{P}(\tau^* < \infty) \le \mathrm{e}^{-su} \sup_{v \ge 0} \mathrm{e}^{vg(s)} . \tag{10.3.14}$$

Note that (10.3.14) implies that, with probability 1,

$$\sup_{t \ge 0} X(t) < \infty . \tag{10.3.15}$$

We try to make the bound in (10.3.14) as sharp as possible, at least in the asymptotic sense. We therefore choose s as large as possible under the restriction that $\sup_{v\ge 0} \mathrm{e}^{vg(s)} < \infty$, i.e. we choose $s = \gamma$, where γ is defined in (10.3.6). But this then yields the Lundberg bound $\psi^*(u) \le \mathrm{e}^{-\gamma u}$ for all $u \ge 0$.

We now investigate the *finite-horizon ruin function* $\psi(u;x) = \mathbb{P}(\tau(u) \le x)$ in the compound Poisson model. So, the claim surplus process $\{X(t)\}$ is given by $X(t) = S(t) = \sum_{i=1}^{N(t)} U_i - \beta t$, where λ is the arrival rate, β the premium rate and F_U the claim size distribution. Using (10.3.3), from (10.3.13) we have $\psi(u;x) \le \mathrm{e}^{-su} \sup_{0\le v\le x} \mathrm{e}^{vg(s)}$, where in this case $g(s) = \lambda(\hat{m}_U(s) - 1) - \beta s$. For $y \ge 0$, consider the function $f_y(s) = s - yg(s)$ and let

$$\gamma_y = \sup_{s \ge 0} \min\{s, f_y(s)\} . \tag{10.3.16}$$

If $\hat{m}_U^{(1)}(\gamma) < \infty$ let $y_0 = (g^{(1)}(\gamma))^{-1} = (\lambda \hat{m}_U(\gamma) - \beta)^{-1}$. We call y_0 the critical value. The following results are called *finite-horizon Lundberg inequalities.*

Theorem 10.3.3 *For all $u, y \ge 0$,*

$$\psi(u; yu) \le \mathrm{e}^{-\gamma_y u} \tag{10.3.17}$$

and

$$\psi(u) - \psi(u; yu) \le \mathrm{e}^{\gamma^y u} , \tag{10.3.18}$$

where $\gamma^y = \sup_{0 \le s \le \gamma} f_y(s)$. Moreover, if $\hat{m}_U(s) < \infty$ for some $s > \gamma$ then $\gamma_{y_0} = \gamma^{y_0} = \gamma$. If $y < y_0$ then $\gamma_y > \gamma$ and $\gamma^y = \gamma$. If $y > y_0$ then $\gamma_y = \gamma$ and $\gamma^y > \gamma$.

Proof Suppose first that $\gamma < s$, i.e. $g(s) \geq 0$. Setting $x = yu$, by (10.3.13) we have $\psi(u; uy) \leq e^{-su}e^{yug(s)} = e^{-u(s-yg(s))}$ for all $u \geq 0$. If $s \leq \gamma$, then $g(s) \leq 0$ and by (10.3.13) we have $\psi(u; uy) \leq e^{-su}$ for all $u \geq 0$. Hence $\psi(u; yu) \leq e^{-u\min\{s, s-yg(s)\}}$ for all $u \geq 0$ which yields (10.3.17). In order to show (10.3.18) we need a finer estimation procedure. Let $x' > x$ and consider the bounded stopping time $x' \wedge \tau^*$. Then, as with (10.3.5), we get

$$\begin{aligned} 1 &= \mathbb{E}\, M(x' \wedge \tau^*) \geq \mathbb{E}\,(M(x' \wedge \tau^*) \mid x < \tau^* \leq x')\, \mathbb{P}(x < \tau^* \leq x') \\ &= \mathbb{E}\,(M(\tau^*) \mid x < \tau^* \leq x')\, \mathbb{P}(x < \tau^* \leq x'). \end{aligned}$$

Thus $1 \geq e^{su}\mathbb{E}\,(e^{-\tau^* g(s)} \mid x < \tau^* \leq x')\mathbb{P}(x < \tau^* \leq x')$ and hence, by (10.3.3), $\mathbb{P}(x < \tau \leq x') \leq e^{-su}\sup_{x < v \leq x'} e^{vg(s)}$. Let $x' \to \infty$ to get

$$\psi(u) - \psi(u; x) = \mathbb{P}(x < \tau < \infty) \leq e^{-su} \sup_{v > x} e^{vg(s)}.$$

For $x = yu$ this gives

$$\psi(u) - \psi(u; yu) \leq e^{-su}\sup\{e^{xug(s)} : x \geq y\} = \begin{cases} e^{-f_y(s)u} & \text{if } g(s) \leq 0, \\ \infty & \text{if } g(s) > 0. \end{cases}$$

Since $g(s) \leq 0$ if and only if $s \leq \gamma$ we obtain (10.3.18). Assume now that there is an $s > \gamma$ such that $\hat{m}_U(s) < \infty$. Then $\hat{m}_U^{(1)}(\gamma) < \infty$. Note that $f_y^{(1)}(s) = 1 - yg^{(1)}(s)$ and $f_y^{(2)}(s) = -yg^{(2)}(s)$, showing that $f_y(s)$ is a concave function. Moreover, $f_y(0) = 0$ and $f_y(\gamma) = \gamma$. Thus it follows that $\gamma_y \geq \gamma$ and $\gamma^y \geq \gamma$. If $y = y_0$ then $f_{y_0}^{(1)}(\gamma) = 0$, i.e. $f_{y_0}(s) \leq f_{y_0}(\gamma) = \gamma$ and $\gamma_{y_0} = \gamma^{y_0} = \gamma$ follows. If $y < y_0$ then $f_y^{(1)}(\gamma) > 0$. Thus there exists $s > \gamma$ such that $f_y(s) > \gamma$. This gives $\gamma_y > \gamma$. If $y > y_0$ then $f_y^{(1)}(\gamma) < 0$. Thus there exists $s < \gamma$ for which $f_y(s) > \gamma$. This gives $\gamma^y > \gamma$. □

Corollary 10.3.1 *Assume that there is an $s > \gamma$ such that $\hat{m}_U(s) < \infty$. Then $u^{-1}\tau(u) \to y_0$ in probability on the set $\{\tau(u) < \infty\}$, as $u \to \infty$.*

Proof Let $\varepsilon > 0$ such that $\hat{m}_U(\gamma_{y_0-\varepsilon}) < \infty$ and $\hat{m}_U(\gamma_{y_0+\varepsilon}) < \infty$. This will be the case for ε small enough. Recall that $\min\{\gamma_{y_0-\varepsilon}, \gamma_{y_0+\varepsilon}\} > \gamma$. Then

$$\begin{aligned} &\mathbb{P}\Big(\Big|\frac{\tau(u)}{u} - y_0\Big| > \varepsilon \;\Big|\; \tau(u) < \infty\Big) \\ &= \frac{\psi(u; (y_0 - \varepsilon)u) + \psi(u) - \psi(u; (y_0 + \varepsilon)u)}{\psi(u)} \\ &\leq \frac{e^{-\gamma_{y_0-\varepsilon}u} + e^{-\gamma_{y_0+\varepsilon}u}}{\psi(u)} = \frac{e^{-(\gamma_{y_0-\varepsilon}-\gamma)u} + e^{-(\gamma_{y_0+\varepsilon}-\gamma)u}}{\psi(u)e^{\gamma u}}, \end{aligned}$$

where we used (10.3.17) and (10.3.18). But the latter expression tends to 0 as $u \to \infty$ by (5.4.4). □

10.3.3 Law of Large Numbers for Additive Processes

We need the notion of a reversed martingale. Let $\mathcal{T} \subset \mathbb{R}$ be an arbitrary but fixed subset of the real line, and consider a family of σ-algebras $\{\mathcal{F}_t^*, t \in \mathcal{T}\}$ such that

- $\mathcal{F}_t^* \subset \mathcal{F}$ for all $t \in \mathcal{T}$,
- $\mathcal{F}_{t+h}^* \subset \mathcal{F}_t^*$ for all $t, h \in \mathcal{T}$.

We call $\{\mathcal{F}_t^*, t \in \mathcal{T}\}$ a *reversed filtration*. Let $\{X^*(t),\ t \in \mathcal{T}\}$ be a stochastic process with left-continuous trajectories, adapted to $\{\mathcal{F}_t^*\}$ and such that $\mathbb{E}\,|X^*(t)| < \infty$ for each $t \in \mathcal{T}$. We say that $\{X^*(t),\ t \in \mathcal{T}\}$ is a *reversed martingale* if, for all $t, h \in \mathcal{T}$,

$$\mathbb{E}\,\left(X^*(t) \mid \mathcal{F}_{t+h}^*\right) = X^*(t+h)\,. \tag{10.3.19}$$

Similarly, $\{X^*(t)\}$ is a *reversed submartingale* if in (10.3.19) we replace $=$ by $\geq$. Note that $\{|X^*(t)|\}$ is a reversed submartingale if $\{X^*(t)\}$ is a reversed martingale.

All results of this section are valid for right-continuous processes. However, as we will later apply an optional sampling theorem in reversed time, the subsequent formulations are for left-continuous processes.

Lemma 10.3.3 *Let $t_0 \geq 0$ be fixed and let $\{X^*(t), t \geq t_0\}$ be a reversed martingale. Then there exists a random variable $X^*(\infty)$ such that, with probability* 1,

$$\lim_{t\to\infty} X^*(t) = X^*(\infty)\,. \tag{10.3.20}$$

Proof Analysing the proof of Theorem 10.2.2 we see that it suffices to show the following. For all $t_0 \leq a \leq b$ and increasing sequences $\{t'_k\}$ converging to infinity with $t'_0 = t_0$ one has

$$\mathbb{P}\,(U_\infty(a,b) = \infty) = 0 \tag{10.3.21}$$

where $U_\infty(a,b) = \lim_{n\to\infty} U_n(a,b)$ and $U_n(a,b)$ denotes the number of upcrossings of (a,b) by the reversed martingale $\{X^*(t'_j), j = 0,\ldots,n\}$. Indeed, if we apply Lemma 9.1.3 to the martingale $\{X_j, j = 0,\ldots,n\}$ with $X_j = X^*(n-j)$, then (9.1.27) implies

$$\mathbb{E}\,U_n(a,b) \leq \frac{\mathbb{E}\,X^*(t_0) + |a|}{b-a} < \infty\,,$$

that is $\mathbb{E}\,U_\infty(a,b) = \lim_{n\to\infty} \mathbb{E}\,U_n(a,b) < \infty$ and therefore (10.3.21) holds. This proves the lemma. $\square$

In the rest of this section we deal with the process $\{X^*(t),\ t > 0\}$ defined by

$$X^*(t) = X(t)/t\,, \tag{10.3.22}$$

where $\{X(t), t \geq 0\}$ is assumed to be an additive process with left-continuous trajectories. Furthermore, the reversed filtration $\{\mathcal{F}_t^*\}$ is chosen by taking $\mathcal{F}_t^*$ to be the smallest σ-algebra of subsets of Ω containing the events $\{\omega : (X(t_1,\omega),\ldots,X(t_n,\omega)) \in B\}$ for all Borel sets $B \in \mathcal{B}(\mathbb{R}^n)$, for all $n = 1,2,\ldots$ and arbitrary sequences $t_1, t_2, \ldots, t_n$ with $t \leq t_1 \leq t_2 \leq \ldots$.

Lemma 10.3.4 *Let $\{X(t), t \geq 0\}$ be a stochastic process with left-continuous trajectories. Assume that $\{X(t)\}$ has stationary and independent increments such that $\mathbb{E}\,|X(1)| < \infty$. Then*

$$\lim_{n\to\infty} \mathbb{E}\,(X(t_n) \mid X(t+h)) = \mathbb{E}\,(X(t) \mid X(t+h)), \tag{10.3.23}$$

for all $t, h \geq 0$ and for each sequence $t_1, t_2, \ldots$ of nonnegative real numbers such that $t_n \leq t$ and $t_n \uparrow t$.

Proof Use the fact that $\{X'(t), t \geq 0\}$ with $X'(t) = X(t) - t\mathbb{E}\,X(1)$ is a martingale. Then, proceeding similarly as in the proof of Lemma 10.2.1, it is not difficult to show that the sequence $X(t_1), X(t_2), \ldots$ is uniformly integrable. Now, (10.3.23) follows from a well-known convergence theorem for conditional expectations; see, for example, Liptser and Shiryayev (1977), p.16. $\square$

Lemma 10.3.5 *If $0 < v \leq t$, then*

$$\mathbb{E}\left(\frac{X(v)}{v} \,\Big|\, X(t)\right) = \frac{X(t)}{t}\,. \tag{10.3.24}$$

Proof Let $B \in \mathcal{B}(\mathbb{R}_+)$ be a Borel set and suppose that $v = \frac{m}{n}t$ for some $m, n \in \mathbb{N}$ with $m \leq n$. Note that the random variables

$$X_k = X\Big(\frac{kt}{n}\Big) - X\Big(\frac{(k-1)t}{n}\Big), \qquad k = 1,\ldots,n\,,$$

are independent and identically distributed. Thus,

$$\begin{aligned}
&\mathbb{E}\left(\frac{X(v)}{v}\mathbb{I}(X(t) \in B)\right)\\
&= \frac{n}{mt}\sum_{i=1}^{m}\mathbb{E}\left(X_i\mathbb{I}\Big(\sum_{k=1}^{n}X_k \in B\Big)\right) = \frac{n}{t}\mathbb{E}\left(X_1\mathbb{I}\Big(\sum_{i=1}^{n}X_i \in B\Big)\right)\\
&= \frac{1}{t}\sum_{i=1}^{n}\mathbb{E}\left(X_i\mathbb{I}\Big(\sum_{k=1}^{n}X_k \in B\Big)\right) = \mathbb{E}\left(\frac{X(t)}{t}\mathbb{I}(X(t) \in B)\right).
\end{aligned}$$

This shows (10.3.24) for $v = mt/n$. Now for arbitrary $v < t$, let $v_k = (m_k/n_k)t \uparrow v$. Since the trajectories of $\{X(t)\}$ are left-continuous with probability 1, we have $X(v_k) \to X(v)$. By Lemma 10.3.4 we also have $\mathbb{E}\,(X(v_k) \mid X(t)) \to \mathbb{E}\,(X(v) \mid X(t))$. Thus, equation (10.3.24) holds for all $v < t$. $\square$

Lemma 10.3.6 *The process* $\{X^*(t),\ t>0\}$ *defined in* (10.3.22) *is a reversed martingale.*

Proof We first show that for $0<v<t=t_1\le t_2\le\ldots\le t_n$,

$$\mathbb{E}\left(\frac{X(v)}{v}\;\Big|\;X(t_1),\ldots,X(t_n)\right)=\mathbb{E}\left(\frac{X(v)}{v}\;\Big|\;X(t)\right).$$

Indeed, since the process $\{X(t)\}$ has independent increments, the random vector $(X(v),X(t_1))$ is independent of $X(t_2)-X(t_1),\ldots,X(t_n)-X(t_{n-1})$ and hence

$$\begin{aligned}
&\mathbb{E}\left(\frac{X(v)}{v}\;\Big|\;X(t_1),\ldots,X(t_n)\right)\\
&= \mathbb{E}\left(\frac{X(v)}{v}\;\Big|\;X(t_1),X(t_2)-X(t_1),\ldots,X(t_n)-X(t_{n-1})\right)\\
&= \mathbb{E}\left(\frac{X(v)}{v}\;\Big|\;X(t_1)\right)=\mathbb{E}\left(\frac{X(v)}{v}\;\Big|\;X(t)\right).
\end{aligned}$$

Now the statement follows from Lemma 10.3.5. □

Summarizing the results of Lemmas 10.3.3 to 10.3.6, we arrive at the following strong *law of large numbers* for additive processes.

Theorem 10.3.4 *Let* $\{X(t),t\ge 0\}$ *be an additive process with left-continuous trajectories. Then, with probability* 1,

$$\lim_{t\to\infty}\frac{X(t)}{t}=-\mu\,. \tag{10.3.25}$$

Proof From Lemmas 10.3.3 and 10.3.6 we have

$$\lim_{t\to\infty}\frac{X(t)}{t}=X^*(\infty)\,, \tag{10.3.26}$$

for some random variable $X^*(\infty)$. However, applying the usual law of large numbers to the sequence $X(1),X(2)-X(1),\ldots$ of independent and identically distributed random variables, we have $\lim_{n\to\infty}n^{-1}X(n)=-\mu$ with probability 1. This and (10.3.26) gives (10.3.25). □

10.3.4 An Identity for Finite-Horizon Ruin Probabilities

The aim of this section is to derive an identity for finite-horizon ruin probabilities in terms of the aggregate claim amount. We consider the claim surplus process $\{S(t)\}$ with $S(t)=\sum_{i=1}^{N(t)}U_i-\beta t$ in the compound Poisson model with arrival rate λ, premium rate β and claim size distribution F_U. As usually, let $\tau(u) = \min\{t : S(t) > u\}$ be the ruin time and, as in Section 10.3.2, consider the finite-horizon ruin function $\psi(u;x) = \mathbb{P}(\tau(u) \le x)$. Let $\{Y(t), t \ge 0\}$ with

$$Y(t) = \sum_{i=1}^{N(t-0)} U_i, \qquad t \ge 0, \tag{10.3.27}$$

denote the left-continuous version of $\{\sum_{i=1}^{N(t)} U_i; t \ge 0\}$ and consider the last time $\tau^0 = \sup\{v : v \le x,\ S(v) \ge u\}$ in the interval $[0,x]$ where the claim surplus is above level u. We put $\tau^0 = 0$ if $S(v) < u$ for all $v \in [0,x]$. With these notations we can formulate the following representation formula for $\psi(u;x)$; see also Theorem 5.6.2.

Theorem 10.3.5 *For all $u \ge 0$ and $x > 0$,*

$$1 - \psi(u;x) = \mathbb{E}\left(1 - \frac{Y(x)}{u+\beta x}\right)_+ + \mathbb{E}\left[\int_{\tau^0}^{x} \frac{Y(v)}{v}\frac{u}{(u+\beta v)^2}\,\mathrm{d}v;\ S(x) \le u\right]. \tag{10.3.28}$$

In particular, for $u = 0$

$$1 - \psi(0;x) = \mathbb{E}\left(1 - \frac{Y(x)}{\beta x}\right)_+. \tag{10.3.29}$$

The *proof* of Theorem 10.3.5 relies on the notion of a reversed martingale as introduced in Section 10.3.3. Suppose first that $u > 0$ and put

$$X^*(t) = \frac{Y(t)}{u+\beta t} + \int_t^x \frac{Y(v)}{v}\frac{u}{(u+\beta v)^2}\,\mathrm{d}v, \qquad 0 < t \le x. \tag{10.3.30}$$

Let $\{\mathcal{F}_t^X,\ 0 < t \le x\}$ be the filtration generated by the (time-reversed) process $\{X(t),\ 0 \le t < x\}$, where

$$X(t) = X^*(x-t), \tag{10.3.31}$$

i.e. the σ-algebra $\mathcal{F}_t^X$ is generated by the events $\{X^*(t_1) \in B_1, \ldots, X^*(t_n) \in B_n\}$, where $x - t \le t_1 < \ldots < t_n \le x$, $B_1, \ldots, B_n$ are Borel sets and $n = 1, 2, \ldots$. In this case the process $\{X(t), t \in [0,x)\}$ is càdlàg.

Lemma 10.3.7 *Let $u > 0$. Then the process $\{X(t), t \in [0,x)\}$ given by* (10.3.30) *and* (10.3.31) *is an $\{\mathcal{F}_t^X\}$-martingale.*

Proof By Lemma 10.3.6, $\{Y(t)/t,\ 0 < t \le x\}$ is a reversed martingale. By Example 5 in Section 10.1.3 applied to the continuous function $f(v) = v/(u+\beta v)$ we see that $\{\int_t^x v(u+\beta v)^{-1}\,\mathrm{d}(Y(v)/v),\ 0 < t \le x\}$ is a reversed martingale. But then by Examples 6 and 7 in Section 10.1.3 also $\{Y(x)(u+\beta x)^{-1} - \int_t^x v(u+\beta v)^{-1}\,\mathrm{d}(Y(v)/v),\ 0 < t \le x\}$ is a reversed martingale. This proves the lemma, since by an integration by parts we obtain that

$$\frac{Y(x)}{u+\beta x}-\int_t^x \frac{v}{u+\beta v}\,\mathrm{d}\Big(\frac{Y(v)}{v}\Big)=\frac{Y(t)}{u+\beta t}+\int_t^x \frac{Y(v)}{v}\frac{u}{(u+\beta v)^2}\,\mathrm{d}v\,. \qquad \square$$

Proof of Theorem 10.3.5. Consider the process $\{\mathbb{I}(S(x-)\le u)X(t), 0\le t<x\}$ which is an $\{\mathcal{F}_t^X\}$-martingale by the results of Lemma 10.3.7 and Example 7 in Section 10.1.3. Note that $x-\tau^0$ can be seen as the modified first entrance time of this process to the set $[u,\infty)$. Since $x-\tau^0$ is an $\{\mathcal{F}_t^X\}$-stopping time by the result of Theorem 10.1.1, using Theorem 10.2.6 we have

$$\begin{aligned}&\mathbb{E}\Big[\frac{Y(x)}{u+\beta x};Y(x)\le u+\beta x\Big]\\ &\quad= \mathbb{E}\Big[\frac{Y(\tau^0)}{u+\beta\tau^0}+\int_{\tau^0}^x \frac{Y(v)}{v}\frac{u}{(u+\beta v)^2}\,\mathrm{d}v; S(x-)\le u\Big]\,. \qquad (10.3.32)\end{aligned}$$

On the other hand, we have

$$\mathbb{P}(\tau(u)>x)=\mathbb{P}(S(x)\le u,\tau^0=0)=\mathbb{P}(S(x)\le u)-\mathbb{P}(S(x)\le u,\tau^0>0)$$

and, since $Y(\tau^0)=u+\beta\tau^0$ for $\tau^0>0$,

$$\begin{aligned}\mathbb{E}\Big[\frac{Y(\tau^0)}{u+\beta\tau^0};\ S(x)\le u\Big] &= \mathbb{E}\Big[\frac{Y(\tau^0)}{u+\beta\tau^0};S(x)\le u,\tau^0>0\Big]\\ &= \mathbb{P}(S(x)\le u,\tau^0>0)\,.\end{aligned}$$

Thus, for $u>0$, (10.3.28) follows from (10.3.32). The verification of formula (10.3.29) by letting $u\downarrow 0$ in (10.3.28) is left to the reader. $\square$

Corollary 10.3.2 *For all $u\ge 0$ and $x>0$,*

$$\psi(u;x)\le 1-\mathbb{E}\Big(1-\frac{Y(x)}{u+\beta x}\Big)_+=\mathbb{E}\Big(\min\Big\{1,\frac{Y(x)}{u+\beta x}\Big\}\Big)\,. \qquad (10.3.33)$$

Proof The inequality (10.3.33) immediately follows from (10.3.28). $\square$

Bibliographical Notes. Section 10.3.1 on additive processes is in the spirit of Grandell (1991a). Inequality (10.3.17) was first proved in Arfwedson (1955), and (10.3.18) in Cramér (1955). A stronger version of Corollary 10.3.1 can be found in Segerdahl (1955). The martingale proof of (10.3.17) goes back to Gerber (1973) and (10.3.18) to Grandell (1991a); see also Grandell (1991b). Further bounds for finite-horizon ruin probabilities can be found in Centeno (1997). Formula (10.3.29) for the finite-horizon ruin function $\psi(0, x)$ is attributed to H. Cramér. Formula (10.3.28) for the finite-horizon ruin function $\psi(u, x)$ with an arbitrary initial risk reserve $u \geq 0$ has been derived in Delbaen and Haezendonck (1985); see also Aven and Jensen (1997) and Schmidli (1996) for some ramifications of the proof. In Møller (1996), martingale techniques have been used for analysing prospective events in risk theory, in cases with random time horizon.

CHAPTER 11

Piecewise Deterministic Markov Processes

We now extend the concept of Markov processes from the case of denumerable many states to a possibly uncountable state space. Special emphasis is put on piecewise deterministic Markov processes (PDMP). Of course, the denumerable state space remains a special case. Typical illustrations from insurance are included in this and the next chapter.

We assume that the stochastic processes $\{X(t), t \geq 0\}$ considered in the present chapter are càdlàg, i.e. their sample paths belong to the set $D(\mathbb{R}_+)$ of right-continuous functions $g : \mathbb{R}_+ \to E$ with left-hand limits, where E denotes the state space of $\{X(t)\}$.

11.1 MARKOV PROCESSES WITH CONTINUOUS STATE SPACE

In order to avoid technical difficulties, we only consider the case of a finite-dimensional state space E. More precisely, we assume that $E = \mathbb{R}^d$ for some $d \geq 1$ or that E consists of possibly disconnected components in $\mathbb{R}^d$, as given in Section 11.2. Let $\mathcal{B}(E)$ denote the σ-algebra of Borel sets in E, and $M(E)$ the family of all real-valued measurable functions on E. Further, let $M_\mathrm{b}(E) \subset M(E)$ be the subfamily of all bounded functions from $M(E)$ with, for each $g \in M_\mathrm{b}(E)$, its *supremum norm* $\|g\| = \sup_{x\in E} |g(x)|$.

11.1.1 Transition Kernels

In Chapter 8, we defined Markov processes on a denumerable state space E by a probability function and a family of stochastic matrices, describing the probability of being in a finite set of states. Now, in the case of an uncountable state space E, we have to consider more general subsets of E.

Let $\mathcal{P}(E)$ denote the set of all probability measures on $\mathcal{B}(E)$. A function

$P : \mathbb{R}_+ \times E \times \mathcal{B}(E) \to [0,1]$ is said to be a *transition kernel* if the following four conditions are fulfilled for all $h, h_1, h_2 \geq 0$, $x \in E$, $B \in \mathcal{B}(E)$:

$$P(h, x, \cdot) \in \mathcal{P}(E)\,, \tag{11.1.1}$$

$$P(0, x, \{x\}) = 1\,, \tag{11.1.2}$$

$$P(\cdot, \cdot, B) \in M(\mathbb{R}_+ \times E)\,, \tag{11.1.3}$$

$$P(h_1 + h_2, x, B) = \int_E P(h_2, y, B) P(h_1, x, \mathrm{d}y)\,. \tag{11.1.4}$$

Condition (11.1.4) corresponds to the Chapman–Kolmogorov equation (8.1.1). At this moment we refrain from requiring a continuity property of P, that would correspond to (8.1.2).

Definition 11.1.1 *An E-valued stochastic process $\{X(t), t \geq 0\}$ is called a (homogeneous) Markov process if there exists a transition kernel P and a probability measure $\alpha \in \mathcal{P}(E)$ such that*

$$\begin{aligned}
&\mathbb{P}(X(0) \in B_0, X(t_1) \in B_1, \ldots, X(t_n) \in B_n) \\
&= \int_{B_0} \int_{B_1} \cdots \int_{B_n} P(t_n - t_{n-1}, x_{n-1}, \mathrm{d}x_n) \ldots P(t_1, x_0, \mathrm{d}x_1) \alpha(\mathrm{d}x_0)\,,
\end{aligned}$$

for all $n = 0, 1, \ldots$, $B_0, B_1, \ldots, B_n \in \mathcal{B}(E)$, $t_0 = 0 \leq t_1 \leq \ldots \leq t_n$.

The probability measure α is called an *initial distribution* and we interpret $P(h, x, B)$ as the probability that, in time h, the stochastic process $\{X(t)\}$ moves from state x to a state in B.

In Sections 7.1.3 and 8.1.3 we were able to construct a Markov chain and a Markov process with finite state space from the initial distribution and from the transition probabilities and the transition intensities, respectively. For Markov processes with continuous state space such a general construction principle is not always possible. However, using a continuous-time version of Kolmogorov's extension theorem (see Theorem 10.2.9) one can show that, in the case $E = \mathbb{R}^d$, there exists a Markov process $\{X(t)\}$ such that α is its initial distribution and P its transition kernel, whatever the pair (α, P). Note that this existence theorem remains valid if E is a complete separable metric space (see, for example, Ethier and Kurtz (1986)).

Analogously to Theorem 8.1.1, we have the following conditional independence property.

Theorem 11.1.1 *Let $\{X(t), t \geq 0\}$ be an E-valued stochastic process. Then, $\{X(t)\}$ is a Markov process if and only if there exists a transition kernel $P = \{P(h, x, B)\}$ such that for all $t, h \geq 0$, $B \in \mathcal{B}(E)$*

$$\mathbb{P}(X(t+h) \in B \mid \mathcal{F}_t^X) = P(h, X(t), B)\,, \tag{11.1.5}$$

or equivalently for all $t, h \geq 0$, $g \in M_{\mathrm{b}}(E)$

$$\mathbb{E}\,(g(X(t+h)) \mid \mathcal{F}_t^X) = \int_E g(y) P(h, X(t), \mathrm{d}y)\,. \tag{11.1.6}$$

The *proof* of Theorem 11.1.1 is omitted. The sufficiency part follows immediately from Definition 11.1.1. A proof of the necessity part can be found in Chapter 4 of Ethier and Kurtz (1986).

In Section 10.1.4, we introduced the notion of a Markov process with respect to an arbitrary filtration. Theorem 11.1.1 shows that each Markov process of that type is a Markov process in the sense of Definition 11.1.1. Consistent with the concept in Section 10.1.4, we call the $\{X(t)\}$ a *strong Markov process* with respect to its history $\{\mathcal{F}_t^X\}$, if with probability 1

$$\mathbb{P}(X(\tau+h) \in B \mid \mathcal{F}_\tau^X) = P(h, X(\tau), B) \tag{11.1.7}$$

on $\{\tau < \infty\}$, for all $h \geq 0$, $B \in \mathcal{B}(E)$ and for each $\{\mathcal{F}_t^X\}$-stopping time τ.

11.1.2 The Infinitesimal Generator

Our main goal will be to construct martingales from a given Markov process. We will do this by subtracting the infinitesimal drift from the process, which generalizes the idea from Example 4 of Section 10.1.3. For this purpose we define the infinitesimal generator of a transition kernel which is a generalization of the notion of an intensity matrix introduced in Section 8.1.1 for Markov processes with discrete state space. Let $\{\boldsymbol{T}(h), h \geq 0\}$ be a family of mappings from $M_{\mathrm{b}}(E)$ to $M_{\mathrm{b}}(E)$. Then, $\{\boldsymbol{T}(h)\}$ is called a *contraction semigroup* on $M_{\mathrm{b}}(E)$ if

$$\boldsymbol{T}(0) = \boldsymbol{I}\,, \tag{11.1.8}$$

$$\boldsymbol{T}(h_1 + h_2) = \boldsymbol{T}(h_1)\boldsymbol{T}(h_2)\,, \tag{11.1.9}$$

$$\|\boldsymbol{T}(h)g\| \leq \|g\| \tag{11.1.10}$$

for all $h, h_1, h_2 \geq 0$ and $g \in M_{\mathrm{b}}(E)$, where $\boldsymbol{I}$ denotes the identity mapping.

Lemma 11.1.1 *Assume that* $\{X(t)\}$ *is an* E*-valued Markov process with transition kernel* $P = \{P(h, x, B)\}$. *Let*

$$\boldsymbol{T}(h)g(x) = \int_E g(y) P(h, x, \mathrm{d}y) = \mathbb{E}\,(g(X(h)) \mid X(0) = x) \tag{11.1.11}$$

for any $g \in M_{\mathrm{b}}(E)$. *Then* $\{\boldsymbol{T}(h)\}$ *is a contraction semigroup on* $M_{\mathrm{b}}(E)$.

Proof First note that $\boldsymbol{T}(0)g(x) = g(x)$ by (11.1.2) for all $x \in E$, $g \in M_{\mathrm{b}}(E)$, i.e. $\boldsymbol{T}(0) = \boldsymbol{I}$. For $h_1, h_2 \geq 0$ we have by (11.1.4)

$$\begin{aligned}\boldsymbol{T}(h_1 + h_2)g(x) &= \int g(y)P(h_1 + h_2, x, \mathrm{d}y) \\ &= \int g(y) \int P(h_2, z, \mathrm{d}y)P(h_1, x, \mathrm{d}z) \\ &= \int \boldsymbol{T}(h_2)g(z)P(h_1, x, \mathrm{d}z) = \boldsymbol{T}(h_1)\boldsymbol{T}(h_2)g(x)\end{aligned}$$

and (11.1.9) follows. Note that $|\boldsymbol{T}(h)g(x)| \leq \int \|g\| P(h, x, \mathrm{d}y) = \|g\|$, where we used (11.1.1) in the equality. This gives (11.1.10). □

Consider a contraction semigroup $\{\boldsymbol{T}(h)\}$ and define

$$\boldsymbol{A}g = \lim_{h \downarrow 0} h^{-1}(\boldsymbol{T}(h)g - g) \tag{11.1.12}$$

for each function $g \in M_{\mathrm{b}}(E)$ for which this limit exists in the supremum norm and belongs to $M_{\mathrm{b}}(E)$. Let $\mathcal{D}(\boldsymbol{A}) \subset M_{\mathrm{b}}(E)$ denote the set of all functions from $M_{\mathrm{b}}(E)$ which have these two properties. Then, the mapping $\boldsymbol{A} : \mathcal{D}(\boldsymbol{A}) \to M_{\mathrm{b}}(E)$ given by (11.1.12) is called the *infinitesimal generator* of $\{\boldsymbol{T}(h)\}$. The set $\mathcal{D}(\boldsymbol{A})$ is called the *domain* of $\boldsymbol{A}$. For the semigroup given in (11.1.11) we find

$$\boldsymbol{A}g(x) = \lim_{h \downarrow 0} h^{-1}\mathbb{E}\left(g(X(h)) - g(x) \mid X(0) = x\right) \tag{11.1.13}$$

for all functions $g \in \mathcal{D}(\boldsymbol{A})$.

For a mapping $\boldsymbol{B} : (a, b) \to M_{\mathrm{b}}(E)$, where $(a, b) \subset \mathbb{R}$ is an arbitrary open interval, we define the notions of the derivative and the Riemann integral in the usual way, considering convergence with respect to the supremum norm. We leave it to the reader to check that such an integral exists for right-continuous semigroups. In particular, let $\{\boldsymbol{T}(h), h \geq 0\}$ be the contraction semigroup given in (11.1.11). If $g \in M_{\mathrm{b}}(E)$ such that $\lim_{h \downarrow 0} \boldsymbol{T}(h)g = g$ then it is not difficult to show that the mapping $h \mapsto \boldsymbol{T}(h)g$ is right-continuous and the Riemann integral $\int_0^t \boldsymbol{T}(v + h)g \, \mathrm{d}v$ exists for all $t, h \geq 0$.

The next theorem collects a number of important results for contraction semigroups.

Theorem 11.1.2 *Let $\{\boldsymbol{T}(h)\}$ be a contraction semigroup and $\boldsymbol{A}$ its infinitesimal generator. Then the following statements hold:*

(a) *If $g \in M_{\mathrm{b}}(E)$ such that the mapping $h \mapsto \boldsymbol{T}(h)g$ is right-continuous at $h = 0$, then for $t \geq 0$, $\int_0^t \boldsymbol{T}(v)g \, \mathrm{d}v \in \mathcal{D}(\boldsymbol{A})$ and*

$$\boldsymbol{T}(t)g - g = \boldsymbol{A} \int_0^t \boldsymbol{T}(v)g \, \mathrm{d}v \, . \tag{11.1.14}$$

(b) *If* $g \in \mathcal{D}(\boldsymbol{A})$ *and* $t \geq 0$, *then* $\boldsymbol{T}(t)g \in \mathcal{D}(\boldsymbol{A})$ *and*

$$\frac{\mathrm{d}^+}{\mathrm{d}t}\boldsymbol{T}(t)g = \boldsymbol{A}\boldsymbol{T}(t)g = \boldsymbol{T}(t)\boldsymbol{A}g\,, \tag{11.1.15}$$

where $\mathrm{d}^+/\mathrm{d}t$ *denotes the derivative from the right.*
(c) *If* $g \in \mathcal{D}(\boldsymbol{A})$ *and* $t \geq 0$, *then* $\int_0^t \boldsymbol{T}(v)g\,\mathrm{d}v \in \mathcal{D}(\boldsymbol{A})$ *and*

$$\boldsymbol{T}(t)g - g = \boldsymbol{A}\int_0^t \boldsymbol{T}(v)g\,\mathrm{d}v = \int_0^t \boldsymbol{A}\boldsymbol{T}(v)g\,\mathrm{d}v = \int_0^t \boldsymbol{T}(v)\boldsymbol{A}g\,\mathrm{d}v\,. \tag{11.1.16}$$

Proof (a) Since $\{\boldsymbol{T}(h)\}$ is a contraction semigroup, it is not difficult to show that the mapping $v \mapsto \boldsymbol{T}(v)g$ is right-continuous for all $v \geq 0$. Moreover, the Riemann integral $\int_0^t \boldsymbol{T}(v+h)g\,\mathrm{d}v$ exists for all $t, h \geq 0$. Let $t_i^n = ti/n$. Then $\lim_{n\to\infty}(t/n)\sum_{i=1}^n \boldsymbol{T}(t_i^n)g = \int_0^t \boldsymbol{T}(v)g\,\mathrm{d}v$. Since

$$\begin{aligned}
&\boldsymbol{T}(h)\int_0^t \boldsymbol{T}(v)g\,\mathrm{d}v \\
&= \boldsymbol{T}(h)\Big(\int_0^t \boldsymbol{T}(v)g\,\mathrm{d}v - \frac{t}{n}\sum_{i=1}^n \boldsymbol{T}(t_i^n)g\Big) + \frac{t}{n}\sum_{i=1}^n \boldsymbol{T}(h)\boldsymbol{T}(t_i^n)g\,,
\end{aligned}$$

we have $\boldsymbol{T}(h)\int_0^t \boldsymbol{T}(v)g\,\mathrm{d}v = \int_0^t \boldsymbol{T}(h)\boldsymbol{T}(v)g\,\mathrm{d}v = \int_0^t \boldsymbol{T}(v+h)g\,\mathrm{d}v$ which follows from the contraction property (11.1.10) of $\{\boldsymbol{T}(h)\}$. Thus,

$$\begin{aligned}
\frac{1}{h}(\boldsymbol{T}(h) - \boldsymbol{I})\int_0^t \boldsymbol{T}(v)g\,\mathrm{d}v &= \frac{1}{h}\int_0^t (\boldsymbol{T}(v+h)g - \boldsymbol{T}(v)g)\,\mathrm{d}v \\
&= \frac{1}{h}\int_t^{t+h} \boldsymbol{T}(v)g\,\mathrm{d}v - \frac{1}{h}\int_0^h \boldsymbol{T}(v)g\,\mathrm{d}v\,.
\end{aligned}$$

The right-continuity of $v \mapsto \boldsymbol{T}(v)g$ implies that the right-hand side of the last equation tends to $\boldsymbol{T}(t)g - g$ as $h \downarrow 0$.
(b) We have $h^{-1}(\boldsymbol{T}(h)\boldsymbol{T}(t)g - \boldsymbol{T}(t)g) = \boldsymbol{T}(t)h^{-1}(\boldsymbol{T}(h)g - g)$. Thus, (11.1.10) yields $\boldsymbol{T}(t)g \in \mathcal{D}(\boldsymbol{A})$ and $\boldsymbol{A}\boldsymbol{T}(t)g = \boldsymbol{T}(t)\boldsymbol{A}g$. Moreover

$$h^{-1}(\boldsymbol{T}(h+t)g - \boldsymbol{T}(t)g) = h^{-1}(\boldsymbol{T}(h) - \boldsymbol{I})\boldsymbol{T}(t)g\,,$$

which gives the right-hand derivative in (11.1.15).
(c) The first part follows from (a) by noting that $g \in \mathcal{D}(\boldsymbol{A})$ implies $\boldsymbol{T}(h)g \to g$ as $h \downarrow 0$. The second part follows from (b) and from the fact that

$$\int_0^t \frac{\mathrm{d}^+}{\mathrm{d}v}(\boldsymbol{T}(v)g)\,\mathrm{d}v = \boldsymbol{T}(t)g - \boldsymbol{T}(0)g\,,$$

which can be proved as in the case of Riemann integrals of real-valued functions. □

11.1.3 Dynkin's Formula

There is a close relationship between martingales and the infinitesimal generator as defined in (11.1.12). In particular, using Theorem 11.1.2 we can easily construct a class of martingales. This leads to the following generalization of *Dynkin's formula*; see (10.1.7).

Theorem 11.1.3 *Assume that $\{X(t)\}$ is an E-valued Markov process with transition kernel $P = \{P(h, x, B)\}$. Let $\{\boldsymbol{T}(h)\}$ denote the semigroup defined in* (11.1.11) *and let $\boldsymbol{A}$ be its generator. Then, for each $g \in \mathcal{D}(\boldsymbol{A})$ the stochastic process $\{M(t), t \geq 0\}$ is an $\{\mathcal{F}_t^X\}$-martingale, where*

$$M(t) = g(X(t)) - g(X(0)) - \int_0^t \boldsymbol{A}g(X(v))\, \mathrm{d}v\,. \tag{11.1.17}$$

Proof Recall that for each $g \in \mathcal{D}(\boldsymbol{A})$, we have $\boldsymbol{A}g \in M_{\mathrm{b}}(E)$ and therefore $\boldsymbol{A}g$ is measurable. Since $\{X(t)\}$ is càdlàg, the function $\boldsymbol{A}g(X(\cdot, \omega))$ is measurable as well. Thus, the Lebesgue integral $\int_0^t \boldsymbol{A}g(X(v, \omega))\, \mathrm{d}v$ is well-defined for each $\omega \in \Omega$ because $\boldsymbol{A}g$ is bounded. The assertion is now an easy consequence of Theorem 11.1.2. For $t, h \geq 0$ we have

$$\begin{aligned}
&\mathbb{E}\,(M(t+h) \mid \mathcal{F}_t^X) + g(X(0)) \\
&= \mathbb{E}\Big(g(X(t+h)) - \int_t^{t+h} \boldsymbol{A}g(X(v))\, \mathrm{d}v \,\Big|\, \mathcal{F}_t^X\Big) - \int_0^t \boldsymbol{A}g(X(v))\, \mathrm{d}v \\
&= \int g(y)P(h, X(t), \mathrm{d}y) - \int_t^{t+h} \int \boldsymbol{A}g(y)P(v-t, X(t), \mathrm{d}y)\, \mathrm{d}v \\
&\qquad\qquad - \int_0^t \boldsymbol{A}g(X(v))\, \mathrm{d}v \\
&= \boldsymbol{T}(h)g(X(t)) - \int_0^h \boldsymbol{T}(v)\boldsymbol{A}g(X(t))\, \mathrm{d}v - \int_0^t \boldsymbol{A}g(X(v))\, \mathrm{d}v \\
&= g(X(t)) - \int_0^t \boldsymbol{A}g(X(v))\, \mathrm{d}v = M(t) + g(X(0))\,,
\end{aligned}$$

where (11.1.16) has been used in the last but one equality. □

Examples 1. Let $\{N(t)\}$ be a Poisson process with intensity λ and let the process $\{X(t)\}$ be defined by $X(t) = N(t) - ct$ for some $c > 0$. Then $\{X(t)\}$ is a Markov process since it has stationary and independent increments; $E = \mathbb{R}$. Furthermore, Theorem 5.2.1 implies that the transition kernel $P = \{P(h, x, B)\}$ of $\{X(t)\}$ is given by

$$P(h, x, B) = \sum_{k=0}^{\infty} \frac{(\lambda h)^k}{k!} \mathrm{e}^{-\lambda h}\, \mathbb{I}(x + k - ch \in B)\,. \tag{11.1.18}$$

Consider the semigroup $\{\boldsymbol{T}(h)\}$ given in (11.1.11). Then, (11.1.18) implies that $\boldsymbol{T}(h)g(x) = \sum_{k=0}^{\infty}((\lambda h)^k/k!)\mathrm{e}^{-\lambda h}g(x+k-ch)$ for $x \in \mathbb{R}$ and $g \in M_{\mathrm{b}}(\mathbb{R})$. It follows readily that $\mathcal{D}(\boldsymbol{A})$ consists of all continuous bounded measurable functions which are differentiable from the left with a bounded left derivative and

$$\boldsymbol{A}g(x) = \lambda(g(x+1) - g(x)) - cg^{(1)}(x)\,, \tag{11.1.19}$$

where we use the same notation for the left derivative as for the derivative itself. By Theorem 11.1.3, the process $\{M(t)\}$ with

$$M(t) = g(X(t)) - \int_0^t [\lambda(g(X(v)+1) - g(X(v))) - cg^{(1)}(X(v))]\,\mathrm{d}v\,,$$

is a martingale for each $g \in \mathcal{D}(\boldsymbol{A})$. In particular, if $\boldsymbol{A}g = 0$ then $g(X(t))$ is a martingale. Therefore it is interesting to solve the equation

$$\lambda(g(x+1) - g(x)) - cg^{(1)}(x) = 0\,. \tag{11.1.20}$$

Let us try the function $g(x) = \mathrm{e}^{sx}$ for some $s \in \mathbb{R}$. Then the condition

$$\lambda(\mathrm{e}^s - 1) - cs = 0 \tag{11.1.21}$$

has to be fulfilled. Since the function $g'(s) = \lambda(\mathrm{e}^s - 1) - cs$ is convex, (11.1.21) admits a (nontrivial) solution $s \neq 0$ if and only if $\lambda \neq c$. Let s_0 denote this solution. If $\lambda = c$ then $g(x) = x$ is a solution to (11.1.20). But Theorem 11.1.3 cannot be applied because the functions $g(x) = \mathrm{e}^{sx}$ with $s \neq 0$ and $g(x) = x$ are unbounded. However, from Examples 2 and 3 of Section 10.1.3, we see that the processes $\{\exp(s_0X(t))\}$ in the case $\lambda \neq c$ and $\{X(t)\}$ if $\lambda = c$ are $\{\mathcal{F}_t^X\}$-martingales.

2. Assume $\{W(t)\}$ is a standard Brownian motion. Let g be a bounded, twice continuously differentiable function such that $g^{(2)} \in M_{\mathrm{b}}(\mathbb{R})$. Then it is not difficult to see that $g \in \mathcal{D}(\boldsymbol{A})$ and $\boldsymbol{A}g = \frac{1}{2}g^{(2)}$, where $\boldsymbol{A}$ is the infinitesimal generator of $\{W(t)\}$. To show this, one can use the fact that $g(x+y) = g(x) + yg^{(1)}(x) + \frac{1}{2}y^2g^{(2)}(x) + y^2r(y)$, where $r(y)$ is a continuous bounded function converging to 0 as $y \to 0$.

11.1.4 The Full Generator

We generalize the notion of the infinitesimal generator, keeping the same symbol $\boldsymbol{A}$ as before since no confusion is possible.

The example considered in Section 11.1.3 shows that it would be desirable to also allow unbounded functions g in $\boldsymbol{A}g$. An auxiliary definition is that of a *multivalued linear operator*. This is simply a set $\boldsymbol{A} \subset \{(g,\tilde{g}) : g, \tilde{g} \in M(E)\}$ such that, if $(g_i, \tilde{g}_i) \in \boldsymbol{A}$ for $i \in \{1,2\}$ then also $(ag_1 + bg_2, a\tilde{g}_1 + b\tilde{g}_2) \in \boldsymbol{A}$ for all $a, b \in \mathbb{R}$. The set $\mathcal{D}(\boldsymbol{A}) = \{g \in M(E) : (g,\tilde{g}) \in \boldsymbol{A}$ for some $\tilde{g} \in M(E)\}$

is called the *domain* of the operator $\boldsymbol{A}$. The multivalued operator $\boldsymbol{A}$ that consists of all pairs $(g, \tilde{g}) \in M(E) \times M(E)$ for which

$$\left\{g(X(t)) - g(X(0)) - \int_0^t \tilde{g}(X(v))\,\mathrm{d}v\,, \quad t \geq 0\right\} \tag{11.1.22}$$

becomes an $\{\mathcal{F}_t^X\}$-martingale, is called the *full generator* of the Markov process $\{X(t)\}$.

Sometimes one requires that the process in (11.1.22) is only a local martingale. In that case, the set consisting of all pairs $(g, \tilde{g}) \in M(E) \times M(E)$ with this weaker martingale property is called the *extended generator* of $\{X(t)\}$. The concept of an extended generator will be studied in more detail in Chapter 13.

Theorem 11.1.3 implies that the domain of the infinitesimal generator of a Markov process is always contained in the domain of its full generator. In the sequel, the generator will always mean the full generator. In what follows, we give criteria for a function g to be in the domain $\mathcal{D}(\boldsymbol{A})$ of a generator $\boldsymbol{A}$. We derive a formula showing how to obtain a function $\tilde{g}$ such that $(g, \tilde{g}) \in \boldsymbol{A}$. The resulting martingale will then be used to determine ruin probabilities for risk models, that are more complex than those already studied in this book.

Trying to simplify the notation, we will write $\boldsymbol{A}g$ if we mean a function $\tilde{g}$ such that $(g, \tilde{g}) \in \boldsymbol{A}$. The reader should keep in mind that this $\tilde{g}$ is only one version of all functions $\tilde{g}$ for which $(g, \tilde{g}) \in \boldsymbol{A}$.

Bibliographical Notes. For a broader introduction to Markov processes with general state space and their infinitesimal and full generators, see Ethier and Kurtz (1986) and Rogers and Williams (1994).

11.2 CONSTRUCTION AND PROPERTIES OF PDMP

The compound Poisson and the Sparre Andersen model are prime examples for stochastic processes having sample paths that are deterministic between claim arrival epochs. In the compound Poisson model, both the risk reserve process $\{R(t)\}$ and the claim surplus process $\{S(t)\}$ are even Markov processes. This is no longer true in the Sparre Andersen model. However, $\{R(t)\}$ and $\{S(t)\}$ can be easily "Markovized" by considering the "age" of the actual inter-occurrence time at time t as a "supplementary variable". Another possible supplementary variable is the (forward) residual inter-occurrence time up to the next claim arrival epoch. Even more general risk models can be forced within this Markovian framework, as will be seen later in this and the next chapter. These models include the following situations (see also Figure 11.2.1):

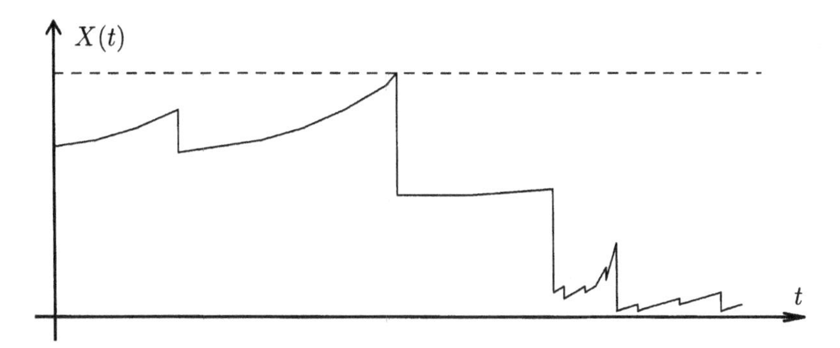

Figure 11.2.1 Possible modifications of the risk reserve process

- nonlinear deterministic paths between claim arrival epochs,
- additional jumps caused by reaching a boundary,
- claim arrival intensity and distribution of claim sizes depend on the actual value of the claim surplus or of a stochastic environmental process.

The goal is to describe the evolution of such models by a Markov process $\{X(t)\}$ whose trajectories have countably many jump epochs. The jump epochs and also the jump sizes are random in general. But, between the jump epochs, the trajectories are governed by a deterministic rule.

We again assume that the state space E for the piecewise deterministic Markov process $\{X(t)\}$ to be constructed can be identified with a subset of some Euclidean space $\mathbb{R}^d$. More specifically, let I be an arbitrary finite non-empty set and let $\{d_\nu, \nu \in I\}$ be a family of natural numbers. For each $\nu \in I$, let C_ν be an open subset of $\mathbb{R}^{d_\nu}$. Put $E = \{(\nu, z) : \nu \in I,\ z \in C_\nu\}$ and, as before, denote by $\mathcal{B}(E)$ the σ-algebra of Borel sets of E. Thus I is the set of possible different external states of the process. For instance, in life insurance one could choose I = {"healthy", "sick", "dead"}. C_ν is the state space of the process if the external state is ν. This allows us to consider different state spaces in different external states. For simplicity, we only consider finite sets I even though the theory extends to infinite but countable I. Then $d = \sum_\nu d_\nu = \infty$, i.e. E is an infinite dimensional state space. In what follows, we use the notation $X(t) = (J(t), Z(t))$, where $\{J(t)\}$ describes the external states of $\{X(t)\}$ and $\{Z(t)\}$ indicates the evolution of the external component.

11.2.1 Behaviour between Jumps

Between jumps, the process $\{X(t)\}$ follows a deterministic path, while the external component $J(t)$ is fixed, $J(t) = \nu$, say. Starting at some point

$z \in C_\nu$, the development of the deterministic path is complete specified by its velocities at all points of C_ν, i.e. through an appropriate function $\boldsymbol{c}_\nu = (c_1, \ldots, c_{d_\nu}) : C_\nu \to \mathbb{R}^{d_\nu}$, called a *vector field.* If a sufficiently smooth vector field is given, then for every $z \in C_\nu$ there exists a path $\varphi_\nu(t, z)$, called an *integral curve*, such that $\varphi_\nu(0, z) = z$ and $(\mathrm{d}/\mathrm{d}t)\varphi_\nu(t, z) = \boldsymbol{c}_\nu(\varphi_\nu(t, z))$. We need to assume that the function $\boldsymbol{c}_\nu$ satisfies enough regularity conditions to ensure the uniqueness of all integral curves, regardless of initial conditions. Sometimes it is convenient to describe a vector field as a differential operator $\boldsymbol{X}$ given by $\boldsymbol{X}g(z) = \sum_{i=1}^{d_\nu} c_i(z)\,(\partial g/\partial z_i)(z)$ acting on differentiable functions g. If g is continuously differentiable, then for $z(t) = \varphi_\nu(t, z)$ we have $(\mathrm{d}/\mathrm{d}t)g(z(t)) = \sum_{i=1}^{d_\nu} c_i(z(t))(\partial g/\partial z_i)(z(t))$. In other words, the integral curve $\{\varphi_\nu(t, z), t < t^*(\nu, z)\}$, where

$$t^*(\nu, z) = \sup\{t > 0 : \varphi_\nu(t, z) \text{ exists and } \varphi_\nu(t, z) \in C_\nu\}\,,$$

is the solution to the differential equation

$$\frac{\mathrm{d}}{\mathrm{d}t} g(\varphi_\nu(t, z)) = (\boldsymbol{X}g)(\varphi_\nu(t, z)), \qquad \varphi_\nu(0, z) = z\,. \tag{11.2.1}$$

Denote by ∂C_ν the boundary of C_ν and let

$$\begin{aligned} \partial^* C_\nu &= \{\tilde{z} \in \partial C_\nu : \tilde{z} = \varphi_\nu(t, z) \text{ for some } (t, z) \in \mathbb{R}^+ \times C_\nu\}\,, \\ \Gamma &= \{(\nu, z) \in \partial E : \nu \in I, z \in \partial^* C_\nu\}\,. \end{aligned}$$

We will assume that $\varphi_\nu(t^*(\nu, z), z) \in \Gamma$ if $t^*(\nu, z) < \infty$. The set Γ is called the *active boundary* of E. More transparently, Γ is the set of those boundary points of E, that can be reached from E via integral curves within finite time and $t^*(\nu, z)$ is the time needed to reach the boundary from the point (ν, z). The condition $\varphi_\nu(t^*(\nu, z), z) \in \Gamma$ if $t^*(\nu, z) < \infty$ ensures that the integral curves cannot "disappear" inside E.

Examples 1. For the compound Poisson model, consider the risk reserve process $\{R(t)\}$ defined in (5.1.14). Then the deterministic paths between claim arrival epochs have the form $\varphi(t, z) = z + \beta t$; $\beta > 0$. Let $g : \mathbb{R} \to \mathbb{R}$ be differentiable. Since $(\boldsymbol{X}g)(z + \beta t) = (\mathrm{d}/\mathrm{d}t)g(z + \beta t) = \beta g^{(1)}(z + \beta t)$, the operator $\boldsymbol{X}$ has the form $\boldsymbol{X} = \beta \mathrm{d}/\mathrm{d}z$. Another choice is to use the integral curves $\varphi'(t, (z, h)) = (z + \beta t, h + t)$ where the state vector (z, h) describes the actual risk reserve z and the time parameter h. For a differentiable function $g' : \mathbb{R} \times \mathbb{R} \to \mathbb{R}$ we have

$$\begin{aligned} (\boldsymbol{X}'g')(z + \beta t, h + t) &= \frac{\mathrm{d}}{\mathrm{d}t} g'(z + \beta t, h + t) \\ &= \beta \frac{\partial g'}{\partial z}(z + \beta t, h + t) + \frac{\partial g'}{\partial h}(z + \beta t, h + t)\,. \end{aligned}$$

Thus we formally write that $\boldsymbol{X}'$ takes the form $\boldsymbol{X} = \beta(\partial/\partial z) + (\partial/\partial h)$. A more general case will be discussed in the next example.

2. Let $\{X(t)\}$ be a PDMP with full generator $\boldsymbol{A}$; see Section 11.2.2 for further details on the construction of $\{X(t)\}$. It is easy to incorporate an explicit time-dependence in the model. Note that the process $\{X'(t)\}$, where $X'(t) = (X(t), t)$ is also a PDMP, now acting on $E' = E \times \mathbb{R}$. Let $z' = (z_1, \ldots, z_{\nu_d}, h) \in C_\nu \times \mathbb{R}$ and let $z' \mapsto g'(z') \in \mathbb{R}$ be a differentiable function. For each $h \geq 0$, denote the function $C_\nu \ni z \mapsto g'(z, h)$ by g_h. Then

$$\begin{aligned}
(\boldsymbol{X}'g')(\varphi_\nu(t,z),t) &= \frac{\mathrm{d}}{\mathrm{d}t} g'(\varphi_\nu(t,z),t) \\
&= \sum_{i=1}^{d_\nu} c_i(\varphi_\nu(t,z)) \frac{\partial g'}{\partial z_i}(\varphi_\nu(t,z),t) + \frac{\partial g'}{\partial h}(\varphi_\nu(t,z),t) \\
&= (\boldsymbol{X} g_t)(\varphi_\nu(t,z)) + \frac{\partial g'}{\partial h}(\varphi_\nu(t,z),t) \,. \qquad (11.2.2)
\end{aligned}$$

Hence formally we write $\boldsymbol{X}' = \boldsymbol{X} + (\partial/\partial h)$, for the differential operator $\boldsymbol{X}'$ acting as in (11.2.2). We add a remark. Let $\boldsymbol{A}'$ denote the full generator of $\{X'(t)\}$ and suppose that $g' \in \mathcal{D}(\boldsymbol{A}')$. Then, $g_h \in \mathcal{D}(\boldsymbol{A})$ for each $h \geq 0$, where $g_h(z) = g'(z,h)$. Moreover, a representative of $(\boldsymbol{A}'g')(z,h)$ is given by

$$(\boldsymbol{A}'g')(z,h) = (\boldsymbol{A}g_h)(z) + \frac{\partial g'}{\partial h}(z,h) \,, \qquad (11.2.3)$$

as can be shown by the reader.

11.2.2 The Jump Mechanism

To fully define a PDMP on $(E, \mathcal{B}(E))$, we need more than a family of vector fields $\{\boldsymbol{c}_\nu, \nu \in I\}$. We also require a *jump intensity*, i.e. a measurable function $\lambda : E \to \mathbb{R}_+$, and a *transition kernel* $Q : (E \cup \Gamma) \times \mathcal{B}(E) \to [0,1]$, i.e. $Q(x, \cdot)$ is a probability measure for all $x \in E \cup \Gamma$ and $Q(\cdot, B)$ is measurable for all $B \in \mathcal{B}(E)$. Note that in actuarial terminology, the jump intensity λ can be interpreted as a "force of transition", whereas $Q(x, \cdot)$ is the "after jump" distribution of a jump from state x (if $x \in E$) or from the boundary point x (if $x \in \Gamma$).

We construct a stochastic process $\{X(t)\}$ with (deterministic) initial state $x_0 = (\nu_0, z_0) \in E$. Let

$$F_1(t) = 1 - \exp\Big(- \int_0^t \lambda(\nu_0, \varphi_{\nu_0}(v, z_0))\, \mathrm{d}v \Big)\, \mathbb{I}(t < t^*(x_0)) \,.$$

Further, let σ_1 be a nonnegative random variable with distribution function $F_1(t)$, and (N_1, Z_1) an E-valued random vector with conditional distribution

$\mathbb{P}((N_1, Z_1) \in \cdot \mid \sigma_1) = Q(\varphi_{\nu_0}(\sigma_1, z_0), \cdot)$. Let $X(t) = (\nu_0, \varphi_{\nu_0}(t, z_0))$ if $0 \le t < \sigma_1$. Assume now the process $\{X(t)\}$ is constructed up to time $\sigma_{k-1} - 0$. Let

$$F_k(t) = 1 - \exp\Big(-\int_0^t \lambda(N_{k-1}, \varphi_{N_{k-1}}(v, Z_{k-1}))\,\mathrm{d}v\Big)\,\mathbb{I}(t < t^*(N_{k-1}, Z_{k-1}))\,,$$

and let $\sigma_k > \sigma_{k-1}$ be a random variable such that $\mathbb{P}(\sigma_k \le \sigma_{k-1} + t \mid \mathcal{F}^X_{\sigma_{k-1}}) = F_k(t)$. Furthermore, let (N_k, Z_k) be an E-valued random vector with conditional distribution

$$\mathbb{P}((N_k, Z_k) \in \cdot \mid \mathcal{F}^X_{\sigma_{k-1}}, \sigma_k) = Q(\varphi_{N_{k-1}}(\sigma_k - \sigma_{k-1}, Z_{k-1}), \cdot)\,.$$

If $\sigma_{k-1} \le t < \sigma_k$, let $X(t) = (N_{k-1}, \varphi_{N_{k-1}}(t - \sigma_{k-1}, Z_{k-1}))$. Denote by $\{N(t), t \ge 0\}$ the counting process given by $N(t) = \sum_{i=1}^{\infty} \mathbb{I}(\sigma_i \le t)$. We assume for all $t \in \mathbb{R}_+$ that

$$\mathbb{E}\, N(t) < \infty\,, \tag{11.2.4}$$

so that $\lim_{k\to\infty} \sigma_k = \infty$ and the random variables $X(t)$ are well-defined for each $t \ge 0$ by the above construction. To construct a process with a random initial state $X(0)$, sample with respect to an arbitrary initial distribution α.

We leave it to the reader to show that (11.2.4) holds if (a) the jump intensity $\lambda(x)$ is bounded and if one of the following conditions is fulfilled: (b_1) $t^*(x) = \infty$ for each $x \in E$, i.e. $\Gamma = \emptyset$, meaning that there are no active boundary points, or (b_2) for some $\varepsilon > 0$ we have $Q(x, B_\varepsilon) = 1$ for all $x \in \Gamma$, where $B_\varepsilon = \{x \in E : t^*(x) \ge \varepsilon\}$, i.e. the minimal distance between consecutive boundary hitting times is not smaller than ε.

Theorem 11.2.1 *The stochastic process $\{X(t), t \ge 0\}$ defined above is a strong Markov process with respect to its history $\{\mathcal{F}^X_t\}$.*

The *proof* of Theorem 11.2.1 is omitted since it goes beyond the scope of this book. It can be found in Davis (1984), p. 364. Since the behaviour of the trajectories of $\{X(t)\}$ between the jump epochs is governed by a deterministic rule, one says that $\{X(t)\}$ is a *piecewise deterministic Markov process* (PDMP).

11.2.3 The Generator of a PDMP

Our next step is to construct martingales associated with a PDMP $\{X(t)\}$. According to the definition of the full generator given in Section 11.1.4, we have to find a function in the domain $\mathcal{D}(\boldsymbol{A})$ of the generator $\boldsymbol{A}$ of $\{X(t)\}$. This raises two problems:

- to find conditions for a measurable real-valued function g on E to belong to $\mathcal{D}(\boldsymbol{A})$ and

- to determine a function $\boldsymbol{A}g$ such that $(g, \boldsymbol{A}g) \in \boldsymbol{A}$.

These problems are generally hard to solve. Fortunately, solutions are possible if we restrict ourselves to a subset of $\mathcal{D}(\boldsymbol{A})$, amply sufficient for the insurance setup. Example 1 in Section 11.1.3 shows a clear way on how to find a function $\boldsymbol{A}g$ for which $(g, \boldsymbol{A}g) \in \boldsymbol{A}$. For $\{N(t)\}$ a Poisson process, the process $\{X(t)\}$ with $X(t) = N(t) - ct$ is easily shown to be a PDMP. Moreover, in Section 11.1.3 we showed that $(g, \boldsymbol{A}g) \in \boldsymbol{A}$ if both the function g is bounded, measurable and differentiable from the left with a bounded derivative and if $\boldsymbol{A}g$ is given by (11.1.19). The following theorem yields an extension which is crucial for many results derived further in the present chapter. Recall that a function $g(y)$ is called *absolutely continuous* if there exists a Lebesgue integrable function $f(y)$ such that $g(y) = g(y_0) + \int_{y_0}^{y} f(z)\,\mathrm{d}z$. As usual in analysis, we use the identity $\int_{y_0}^{y} f(z)\,\mathrm{d}z = -\int_{y}^{y_0} f(z)\,\mathrm{d}z$.

Theorem 11.2.2 *Let $\{X(t)\}$ be a* PDMP *and let $g^* : E \cup \Gamma \to \mathbb{R}$ be a measurable function satisfying the following conditions:*
(a) *for each $(\nu, z) \in E$, the function $t \mapsto g^*(\nu, \varphi_\nu(t, z))$ is absolutely continuous on $(0, t^*(\nu, z))$,*
(b) *for each x on the boundary Γ*

$$g^*(x) = \int_E g^*(y)Q(x, \mathrm{d}y)\,, \tag{11.2.5}$$

(c) *for each $t \geq 0$,*

$$\mathbb{E}\Big(\sum_{i:\sigma_i \leq t} |g^*(X(\sigma_i)) - g^*(X(\sigma_i-))|\Big) < \infty\,. \tag{11.2.6}$$

Then $g \in \mathcal{D}(\boldsymbol{A})$, where g denotes the restriction of g^* to E, and $(g, \boldsymbol{A}g) \in \boldsymbol{A}$, where $\boldsymbol{A}g$ is given by

$$(\boldsymbol{A}g)(x) = (\boldsymbol{X}g)(x) + \lambda(x)\int_E (g(y) - g(x))Q(x, \mathrm{d}y)\,. \tag{11.2.7}$$

Proof Inserting (11.1.22) into (10.1.3), we have to show that

$$\mathbb{E}\Big(g(X(t+h)) - g(X(t)) - \int_t^{t+h} (\boldsymbol{A}g)(X(v))\,\mathrm{d}v \;\Big|\; \mathcal{F}_t^X\Big) = 0 \tag{11.2.8}$$

for all $t, h \geq 0$. The above condition only makes sense if the random variable on the left-hand side of (11.2.8) is absolutely integrable. Our first attention goes to the verification of this integrability property. By Theorem 11.2.1, the left-hand side of (11.2.8) can be written in the form

$$\mathbb{E}\Big(g(X(t+h)) - g(X(t)) - \int_t^{t+h} \Big((\boldsymbol{X}g)(X(v)) + \lambda(X(v))$$

$$\times \int_E (g(y) - g(X(v)))Q(X(v), \mathrm{d}y)\Big) \,\mathrm{d}v \;\Big|\; \mathcal{F}_t^X\Big)$$

$$= \mathbb{E}\Big(g(X(t+h)) - g(X(t)) - \int_t^{t+h} \Big((\boldsymbol{X}g)(X(v)) + \lambda(X(v)) \times \int_E (g(y) - g(X(v)))Q(X(v), \mathrm{d}y)\Big) \,\mathrm{d}v \;\Big|\; X(t)\Big).$$

We now condition on $X(t) = x \in E$. Then

$$\mathbb{E}\Big(g(X(t+h)) - g(X(t)) - \int_t^{t+h} \Big((\boldsymbol{X}g)(X(v)) + \lambda(X(v)) \times \int_E (g(y) - g(X(v)))Q(X(v), \mathrm{d}y)\Big) \,\mathrm{d}v \;\Big|\; X(t) = x\Big)$$

$$= \mathbb{E}\Big(g(X(h)) - g(X(0)) - \int_0^{h} \Big((\boldsymbol{X}g)(X(v)) + \lambda(X(v)) \times \int_E (g(y) - g(X(v)))Q(X(v), \mathrm{d}y)\Big) \,\mathrm{d}v \;\Big|\; X(0) = x\Big).$$

It therefore suffices to show that for each $t \ge 0$

$$\mathbb{E}\Big(g(X(t)) - g(X(0)) - \int_0^{t} \Big((\boldsymbol{X}g)(X(v)) + \lambda(X(v)) \int_E (g(y) - g(X(v)))Q(X(v), \mathrm{d}y)\Big) \,\mathrm{d}v\Big) = 0\,.$$

If no jump takes place at time v, then any integral curve of the differential operator $\boldsymbol{X}$ satisfying condition (a) yields $(\boldsymbol{X}g)(X(v)) = (\mathrm{d}/\mathrm{d}v)g(X(v))$. Therefore

$$g(X(t)) - g(X(0)) - \int_0^t (\boldsymbol{X}g)(X(v))\,\mathrm{d}v = \sum_{\sigma_i \le t} g(X(\sigma_i)) - g^*(X(\sigma_i - 0))\,,$$

and we need to show that

$$\mathbb{E}\Big(\sum_{\sigma_i \le t} (g(X(\sigma_i)) - g^*(X(\sigma_i - 0))) - \int_0^t \lambda(X(v)) \int_E (g(y) - g(X(v)))Q(X(v), \mathrm{d}y)\,\mathrm{d}v\Big) = 0\,. \quad (11.2.9)$$

For notational convenience, assume $X(0) = x$, where $x = (\nu, z) \in E$ is an arbitrary (deterministic) initial state of $\{X(t)\}$. Let us first consider the expression $\mathbb{E}\,(g(X(\sigma_1 \wedge t)) - g^*(X((\sigma_1 \wedge t) - 0)))$. Condition (11.2.5) implies that $\mathbb{E}\,[g(X(\sigma_1 \wedge t)) - g^*(X((\sigma_1 \wedge t) - 0)); \sigma_1 \geq t \wedge t^*(x)] = 0$. From the construction of the PDMP $\{X(t)\}$ we have

$$
\begin{aligned}
&\mathbb{E}\,[g(X(\sigma_1)) - g^*(X(\sigma_1 - 0)); \sigma_1 < t \wedge t^*(x)] \\
&\quad = \int_0^{t\wedge t^*(x)} \int_E (g(y) - g(\nu, \varphi_\nu(v,z)))Q(\nu, \varphi_\nu(v,z), \mathrm{d}y)\, \lambda(\nu, \varphi_\nu(v,z)) \overline{F}_1(v)\, \mathrm{d}v\,,
\end{aligned}
$$

where $\overline{F}_1(v) = \int_v^\infty \lambda(\nu, \varphi_\nu(w,z))\overline{F}_1(w)\,\mathrm{d}w$ thanks to the definition of F_1 in Section 11.2.2. We split the above expression into two parts to obtain

$$
\begin{aligned}
&\int_0^{t\wedge t^*(x)} \int_E (g(y) - g(\nu, \varphi_\nu(v,z)))Q(\nu, \varphi_\nu(v,z), \mathrm{d}y)\lambda(\nu, \varphi_\nu(v,z)) \\
&\qquad\qquad \times \int_v^{t\wedge t^*(x)} \lambda(\nu, \varphi_\nu(w,z))\overline{F}_1(w)\,\mathrm{d}w\,\mathrm{d}v \\
&\quad = \int_0^{t\wedge t^*(x)} \int_0^w \lambda(\nu, \varphi_\nu(v,z)) \int_E (g(y) - g(\nu, \varphi_\nu(v,z)))Q(\nu, \varphi_\nu(v,z))\,\mathrm{d}y\,\mathrm{d}v \\
&\qquad\qquad \times \lambda(\varphi_\nu(w,z))\overline{F}_1(w)\,\mathrm{d}w \\
&\quad = \mathbb{E}\Big[\int_0^{\sigma_1} \lambda(X(v)) \int_E (g(y) - g(X(v)))Q(X(v), \mathrm{d}y)\,\mathrm{d}v; \sigma_1 < t \wedge t^*(x)\Big]\,,
\end{aligned}
$$

and

$$
\begin{aligned}
&\int_0^{t\wedge t^*(x)} \int_E (g(y) - g(\varphi_\nu(v,z)))Q(\varphi_\nu(v,z), \mathrm{d}y)\lambda(\varphi_\nu(v,z))\overline{F}_1(t \wedge t^*(x))\,\mathrm{d}v \\
&\quad = \mathbb{E}\Big[\int_0^{\sigma_1 \wedge t} \lambda(X(v)) \int_E (g(y) - g(X(v)))Q(X(v), \mathrm{d}y)\,\mathrm{d}v; \sigma_1 \geq t \wedge t^*(x)\Big]\,.
\end{aligned}
$$

Thus we find

$$
\begin{aligned}
&\mathbb{E}\,(g(X(\sigma_1 \wedge t)) - g^*(X(\sigma_1 \wedge t) - 0)) \\
&\quad = \mathbb{E}\Big(\int_0^{t\wedge\sigma_1} \lambda(X(v)) \int_E (g(y) - g(X(v)))Q(X(v), \mathrm{d}y)\,\mathrm{d}v\Big)\,.
\end{aligned}
$$

As the last expression is also valid for a random initial state, we can drop the assumption $X(0) = x$. The reader should show that, for each $k \geq 1$, the jump epoch σ_k is an $\{\mathcal{F}_t^X\}$-stopping time. Moreover, from the strong Markov property of $\{X(t)\}$ in Theorem 11.2.1, we conclude that

$$
\begin{aligned}
&\mathbb{E}\,(g(X(\sigma_k \wedge t)) - g^*(X(\sigma_k \wedge t) - 0)) \\
&= \mathbb{E}\,(\mathbb{E}\,(g(X(\sigma_k \wedge t)) - g^*(X(\sigma_k \wedge t) - 0) \mid \mathcal{F}_{\sigma_{k-1}})) \\
&= \mathbb{E}\Big(\mathbb{E}\Big(\int_{t\wedge\sigma_{k-1}}^{t\wedge\sigma_k} \lambda(X(v)) \int_E (g(y) - g(X(v)))Q(X(v), \mathrm{d}y)\,\mathrm{d}v \;\Big|\; \mathcal{F}_{\sigma_{k-1}}\Big)\Big) \\
&= \mathbb{E}\Big(\int_{t\wedge\sigma_{k-1}}^{t\wedge\sigma_k} \lambda(X(v)) \int_E (g(y) - g(X(v)))Q(X(v), \mathrm{d}y)\,\mathrm{d}v\Big)\,.
\end{aligned}
$$

This yields

$$
\begin{aligned}
&\mathbb{E}\Big(\sum_{i=0}^{N(t)\wedge n} g(X(\sigma_i)) - g^*(X(\sigma_i - 0))\Big) \\
&\quad = \mathbb{E}\Big(\int_0^{t\wedge\sigma_n} \lambda(X(v)) \int_E (g(y) - g(X(v)))Q(X(v), \mathrm{d}y)\,\mathrm{d}v\Big)\,.
\end{aligned}
$$

Analogously, it follows that

$$
\begin{aligned}
&\mathbb{E}\Big(\sum_{i=0}^{N(t)\wedge n} |g(X(\sigma_i)) - g^*(X(\sigma_i - 0))|\Big) \\
&\quad = \mathbb{E}\Big(\int_0^{t\wedge\sigma_n} \lambda(X(v)) \int_E |g(y) - g(X(v))|Q(X(v), \mathrm{d}y)\,\mathrm{d}v\Big)\,,
\end{aligned}
$$

and by the monotone convergence theorem

$$
\begin{aligned}
&\mathbb{E}\Big(\sum_{i=0}^{N(t)} |g(X(\sigma_i)) - g^*(X(\sigma_i - 0))|\Big) \\
&\quad = \mathbb{E}\Big(\int_0^{t} \lambda(X(v)) \int_E |g(y) - g(X(v))|Q(X(v), \mathrm{d}y)\,\mathrm{d}v\Big)\,.
\end{aligned}
$$

In particular, by (11.2.6), the random variable on the left-hand side of (11.2.9) is absolutely integrable. Hence (11.2.9) is valid and the assertion of the theorem has been proved. □

Note that in the proof of Theorem 11.2.2 we could not use the infinitesimal generator and (11.1.17) because we did not want to assume that g is bounded. Also, the finiteness condition in (11.2.6) is fulfilled when g is bounded.

In forthcoming applications of Theorem 11.2.2, we will frequently use the following method. Let $\{X(t)\}$ be a PDMP with generator $\boldsymbol{A}$. We look for a measurable function g^* satisfying the conditions of Theorem 11.2.2 and for which $\boldsymbol{A}g = 0$. The process $\{g(X(t)) - g(X(0)) : \ t \geq 0\}$ is then an $\{\mathcal{F}_t^X\}$-martingale; see (11.1.22). For the extended PDMP $\{X'(t)\}$, we have the following easy consequence of Theorem 11.2.2.

Corollary 11.2.1 *Let $\{X(t)\}$ be a* PDMP *on the state space E with active boundary Γ. Consider the* PDMP *$\{X'(t)\}$ with $X'(t) = (X(t), t)$, acting on the state space $E' = E \times \mathbb{R}$ with the active boundary $\Gamma' = \Gamma \times \mathbb{R}$. Let $g^* : E' \cup \Gamma' \to \mathbb{R}$ be a measurable function which satisfies the conditions of Theorem* 11.2.2 *with respect to the extended* PDMP *$\{X'(t)\}$. Then $g \in \mathcal{D}(\boldsymbol{A}')$, where g denotes the restriction of g^* to E', and $(g, \boldsymbol{A}'g) \in \boldsymbol{A}'$, where $\boldsymbol{A}'g$ is given by*

$$(\boldsymbol{A}'g)(x,t) = (\boldsymbol{X}g_t)(x) + \frac{\partial g}{\partial t}(x,t) + \lambda(x)\int_E (g_t(y) - g_t(x))Q(x,\mathrm{d}y)\,. \quad (11.2.10)$$

Proof The jump intensity $\lambda(x,t)$ and the transition kernel $Q((x,t),\mathrm{d}(y,t))$ of the extended PDMP $\{X'(t)\}$ are given by

$$\lambda(x,t) \equiv \lambda(x)\,, \qquad Q((x,t),\mathrm{d}(y,t)) \equiv Q(x,\mathrm{d}y)\,, \quad (11.2.11)$$

where $\lambda(x)$ and $Q(x,\mathrm{d}y)$ are the corresponding characteristics of $\{X(t)\}$. By using (11.2.2), (11.2.10) follows from Theorem 11.2.2. □

Example Here is an application of Theorem 11.2.2 to nonhomogeneous Markov processes. In Section 8.4 we considered a class of nonhomogeneous Markov processes $\{X(t)\}$ with state space $\{1,\ldots,\ell\}$ and measurable matrix intensity function $\boldsymbol{Q}(t)$. We took the function $\max_{1\le i\le \ell}|q_{ii}(t)|$ to be integrable on every finite interval in $\mathbb{R}_+$. Under these assumptions, the number of all transitions of the process $\{X(t)\}$ in any finite interval is bounded and integrable. A crucial observation is that the extended process $\{(X(t),t)\}$ is a homogeneous Markov process and moreover a PDMP with $I = \{1,\ldots,\ell\}$, $C_\nu = \mathbb{R}$, $\lambda(\nu,t) = -\sum_{j\neq\nu} q_{\nu j}(t)$, $Q((\nu,t),(\nu',t)) = q_{\nu\nu'}(t)/\lambda(\nu,t)$, $(\nu' \neq \nu)$. From Theorem 11.2.2 we get the following extension of Dynkin's formula derived in Section 10.1.3 for the homogeneous case; see (10.1.7). Let $g : I \times \mathbb{R}_+ \to \mathbb{R}_+$ be such that for each $i \in I$ the function $g(i,t)$ is absolutely continuous with respect to t. Then the process $\{M(t)\}$ with

$$\begin{aligned} M(t) &= g(X(t),t) - g(X(0),0) \\ &\quad - \int_0^t \Big(\frac{\partial g}{\partial t}(X(v),v) + \sum_{j\neq X(v)} (g(j,v) - g(X(v),v))\, q_{X(v),j}(v)\Big)\,\mathrm{d}v \\ &= g(X(t),t) - g(X(0),0) \end{aligned}$$

$$-\int_0^t \Big(\frac{\partial g}{\partial t}(X(v),v) + (\boldsymbol{Q}(v)(\boldsymbol{g}(v))^\top)_{X(v)}\Big)\,dv$$

is an $\{\mathcal{F}_t^X\}$-martingale, where $\boldsymbol{g}(t) = (g(1,t),\ldots,g(\ell,t))$. Further martingales of the same type can be obtained as follows. Let $i,j \in I$ be fixed with $i \neq j$ and consider the number $N_{ij}(t)$ of transitions of the nonhomogeneous Markov process $\{X(t)\}$ from i to j by time t; see also Section 8.4.3. The process $\{X'(t)\}$ defined by $X'(t) = (X(t), N_{ij}(t), t)$ is again a PDMP whose characteristics can be easily specified. Consider the function $g' : I \times \mathbb{N} \times \mathbb{R}_+ \to \mathbb{R}$ given by $g'(i,n,t) = n$ which satisfies the conditions of Theorem 11.2.2. The process $\{M_{ij}(t)\}$ defined by $M_{ij}(t) = N_{ij}(t) - \int_0^t \mathbb{I}(X(v) = i)q_{ij}(v)\,dv$ is an $\{\mathcal{F}_t^X\}$-martingale. Hence, for any locally integrable function $a : \mathbb{R}_+ \to \mathbb{R}_+$

$$\mathbb{E}\Big(\int_t^{t'} a(v)\,dN_{ij}(v) \;\Big|\; X(t) = i\Big) = \int_t^{t'} a(v)p_{ij}(t,v)q_{ij}(v)\,dv\,. \qquad (11.2.12)$$

In various applications, particularly in optimization problems, expectations of the form $\mathbb{E}\,(\int_0^{t_0} \exp(-\int_0^t \kappa(X(v),v)\,dv)\,\gamma(X(t),t)\,dt)$ have to be computed for some fixed time horizon t_0, where $\Delta(t) = \exp(-\int_0^t \kappa(X(v),v)\,dv)$ can be interpreted as a discounting factor and $\gamma(X(t),t)$ as a cost function. For this purpose we need the next result.

Theorem 11.2.3 *Let $\{X(t)\}$ be a* PDMP *with state space E and generator $\boldsymbol{A}$. For $t_0 > 0$ fixed, let $\kappa : E\times[0,t_0] \to \mathbb{R}_+$ and $\gamma : E\times[0,t_0] \to \mathbb{R}$ be measurable functions. Furthermore, consider the measurable functions $g : E \times \mathbb{R} \to \mathbb{R}$ and $g_{\rm ter} : E \to \mathbb{R}$, where the latter function models terminal costs. Suppose that*

(a) *$g(x,t)$ fulfils the conditions of Corollary* 11.2.1,

(b) *$g(x,t_0) = g_{\rm ter}(x)$ for all $x \in E$,*

(c) $\dfrac{\partial g}{\partial t}(x,t) + (\boldsymbol{A}g_t)(x) - \kappa(x,t)g(x,t) + \gamma(x,t) = 0$ *for all $t \le t_0$.* $\qquad (11.2.13)$

Then

$$g(X(0),0) = \mathbb{E}\Big(\int_0^{t_0} \exp\Big(-\int_0^t \kappa(X(v),v)\,dv\Big)\,\gamma(X(t),t)\,dt + \exp\Big(-\int_0^{t_0} \kappa(X(v),v)\,dv\Big)\,g_{\rm ter}(X(t_0))\Big)\,. \qquad (11.2.14)$$

Proof From (11.2.3) we have that the generator $\boldsymbol{A}'$ of the extended PDMP $\{X'(t)\}$, where $X'(t) = (X(t),t)$ is given by $(\boldsymbol{A}'g)(x,t) = (\boldsymbol{A}g_t)(x) + (\partial/\partial t)g(x,t)$, where $g_t(x) = g(x,t)$. Thus Corollary 11.2.1 implies that the process $\{M(t), t \ge 0\}$ with

$$M(t) = g(X(t),t) - g(X(0),0) - \int_0^t \Big(\frac{\partial g}{\partial v}(X(v),v) + (\boldsymbol{A}g_v)(X(v))\Big)\,dv$$

is a martingale. Note that $\{M(t)\}$ is pathwise absolutely continuous by condition (a) of Theorem 11.2.2. Thus, in differential terms we can write $\mathrm{d}g(X(t),t) = \mathrm{d}M(t) + ((\partial g/\partial t)(X(t),t) + (\boldsymbol{A}g_t)(X(t)))\,\mathrm{d}t$. Then, for the function $\Delta(t) = \exp(-\int_0^t \kappa(X(v),v)\,\mathrm{d}v)$ we have

$$\begin{aligned}
\mathrm{d}(\Delta(t)g(X(t),t)) &= \Delta(t)(\mathrm{d}g(X(t),t) - \kappa(X(t),t)\,g(X(t),t)\,\mathrm{d}t) \\
&= \Delta(t)\Big(\mathrm{d}M(t) + \Big(\frac{\partial g}{\partial t}(X(t),t) + (\boldsymbol{A}g_t)(X(t)) - \kappa(X(t),t)g(X(t),t)\Big)\,\mathrm{d}t\Big) \\
&= -\Delta(t)\,(\gamma(X(t),t)\,\mathrm{d}t - \mathrm{d}M(t))\,,
\end{aligned}$$

where we used (11.2.13) in the last equality. This yields

$$\mathbb{E}\,(\Delta(t_0)g(X(t_0),t_0) - g(X(0),0)) = -\mathbb{E}\int_0^{t_0} \Delta(t)\gamma(X(t),t)\,\mathrm{d}t\,,$$

see Example 5 in Section 10.1.3. The assertion follows. □

11.2.4 An Application to Health Insurance

Suppose that an insurance contract guarantees the insured an annuity payment at constant rate during sickness. However, there is an elimination period of length y_0, by which we mean that the annuity payment only starts when the insured is sick for longer than y_0 units of time. Suppose that the insured can be in any one of three states: healthy (1), sick (2), dead (3). Transitions from state 1 are assumed to depend only on the age t of the insured, those from state 2 also depend on the duration of sickness at time t, while state 3 is absorbing. Define the stochastic process $\{X(t), t \ge 0\}$ by

$$X(t) = \begin{cases} (1,t) & \text{if the insured is healthy at } t, \\ (2,(t,y)) & \text{if the insured is sick at } t \text{ for a period } y, \\ (3,t) & \text{if the insured is dead at } t. \end{cases} \qquad (11.2.15)$$

We assume that $\{X(t)\}$ is a PDMP, with $I = \{1,2,3\}$, $C_1 = \mathbb{R}$, $C_2 = \mathbb{R}^2$ and $C_3 = \mathbb{R}$. The active boundary Γ is the empty set. The transitions from one state to another are governed by the following transition intensities. Let $\lambda_{1i}(t)$ denote the transition intensity from state $(1,t)$ to state $(2,(t,0))$ if $i=2$ and to state $(3,t)$ if $i=3$. Further, $\lambda_{2,i}(t,y)$ is the transition intensity from state $(2,(t,y))$ to state (i,t), where y denotes the duration of sickness at time t. With the notation introduced in Section 11.2.2, we have

$$\lambda_{1i}(t) = \begin{cases} \lambda(1,t)\,Q((1,t),(2,(t,0))), & \text{if } i=2, \\ \lambda(1,t)\,Q((1,t),(3,t)), & \text{if } i=3, \end{cases}$$

$$\lambda_{2i}(t,y) = \begin{cases} \lambda(2,(t,y))\,Q((2,(t,y)),(1,t)), & \text{if } i=1, \\ \lambda(2,(t,y))\,Q((2,(t,y)),(3,t)), & \text{if } i=3. \end{cases}$$

Lemma 11.2.1 *The generator* $\boldsymbol{A}$ *of the PDMP* $\{X(t)\}$ *is given by*

$$
\begin{aligned}
(\boldsymbol{A}g)(1,t) &= \frac{\partial g}{\partial t}(1,t) + \lambda_{12}(t)g(2,t,0) + \lambda_{13}(t)g(3,t) \\
&\quad - (\lambda_{12}(t) + \lambda_{13}(t))g(1,t)\,,
\end{aligned}
$$

$$
\begin{aligned}
(\boldsymbol{A}g)(2,t,y) &= \frac{\partial g}{\partial t}(2,t,y) + \frac{\partial g}{\partial y}(2,t,y) + \lambda_{23}(t,y)g(3,t) \\
&\quad + \lambda_{21}(t,y)g(1,t) - (\lambda_{21}(t,y) + \lambda_{23}(t))g(2,t,y)\,,
\end{aligned}
$$

$$
(\boldsymbol{A}g)(3,t) = \frac{\partial g}{\partial t}(3,t)\,.
$$

Proof Due to the special structure of the PDMP $\{X(t)\}$ defined in (11.2.15), the statement immediately follows from (11.2.7). □

Suppose that, if the insured is in state 2, the insurance company pays an annuity at rate $a(t,y) = \mathbb{I}(y \ge y_0)$ after the elimination period of length y_0. Thus, we use here the annuity rate as a monetary unit. Let $0 \le t < t_0$. We are interested in the expected payment in the interval $[t, t_0]$ knowing the state at time t. As before, the external state component of $X(t)$ is denoted by $J(t)$. Clearly, if $J(t) = 3$ then there is no payment after t and therefore we have the expected payment $\mu_3(t) \equiv 0$. We want to know the expected payments $\mu_1(t)$ and $\mu_2(t,y)$ defined by

$$
\mu_1(t) = \mathbf{E}\left(\int_t^{t_0} \mathbb{I}(J(v) = 2, Y(v) \ge y_0)\,\mathrm{d}v \;\Big|\; J(t) = 1\right),
$$

$$
\mu_2(t,y) = \mathbf{E}\left(\int_t^{t_0} \mathbb{I}(J(v) = 2, Y(v) \ge y_0)\,\mathrm{d}v \;\Big|\; J(t) = 2, Y(t) = y\right),
$$

where $Y(t)$ denotes the duration of sickness at time t.

The current model is essentially different from the life insurance model in Section 8.4.3. There, the nonhomogeneous Markov process $\{X(t)\}$ could be extended to a homogeneous Markov process by simply adding the time t, i.e. by replacing $X(t)$ by $X'(t) = (X(t), t)$. To get a homogeneous Markov process, we needed the additional state variable $Y(s)$ to deal with the elimination period y_0 during which the insured does not get any annuity payment even though he is in state 2. Also, the current model discards the economic context originally present in the discounting.

The derivation of the next result is similar to that of Theorem 11.2.3. However, we do not directly apply Theorem 11.2.3 to the presently studied PDMP $\{X(t)\}$ since the time variable t is already included into $X(t)$.

Theorem 11.2.4 *Assume that the functions $g_1(t)$ and $g_2(t,y)$ solve the following system of partial differential equations:*

$$\frac{\partial g_1}{\partial t}(t) + \lambda_{12}(t)g_2(t,0) - (\lambda_{12}(t) + \lambda_{13}(t))g_1(t) = 0\,, \tag{11.2.16}$$

$$\begin{aligned}&\frac{\partial g_2}{\partial t}(t,y) + \frac{\partial g_2}{\partial y}(t,y) + \lambda_{21}(t,y)g_1(t)\\ &\qquad - (\lambda_{21}(t,y) + \lambda_{23}(t,y))g_2(t,y) + \mathbb{I}(y \ge y_0) = 0\,,\end{aligned} \tag{11.2.17}$$

with boundary conditions $g_1(t_0) = g_2(t_0,y) = 0$. If for $x = (\nu, z)$ the function

$$g(x) = \begin{cases} g_1(t) & \text{if } \nu = 1,\ z = t \le t_0 \\ g_2(t,y) & \text{if } \nu = 2,\ z = (t,y) \text{ and } t \le t_0 \\ 0 & \text{otherwise} \end{cases}$$

satisfies conditions (a)–(c) *of Theorem* 11.2.2, *then*

$$\mathbb{E}\, g(X(t)) = \mathbb{E}\left(\int_t^{t_0} \mathbb{I}(J(v) = 2, Y(v) \ge y_0)\, \mathrm{d}v\right). \tag{11.2.18}$$

Proof Theorem 11.2.2 implies that the process $\{M(t), t \ge 0\}$ with

$$M(t) = g(X(t)) - g(X(0)) - \int_0^t (\boldsymbol{A}g)(X(v))\, \mathrm{d}v$$

is a martingale. Thus, in differential terms we have $\mathrm{d}g(X(t)) = \mathrm{d}M(t) + (\boldsymbol{A}g)(X(t))\,\mathrm{d}t$. Using Lemma 11.2.1 and the system of partial differential equations (11.2.16)–(11.2.17), this leads to $\mathrm{d}g(X(t)) = \mathrm{d}M(t) - \mathbb{I}(J(t) = 2, Y(t) \ge y_0)\,\mathrm{d}t$. Integrate this equation over the interval $[t, t_0]$ and use the fact that $\{M(t)\}$ is a martingale to obtain

$$\mathbb{E}\,(g(X(t_0)) - g(X(t))) = -\mathbb{E}\left(\int_t^{t_0} \mathbb{I}(J(v) = 2, Y(v) \ge y_0)\, \mathrm{d}v\right).$$

By the boundary conditions we have $\mathbb{E}\, g(X(t_0)) = 0$, and hence (11.2.18) follows. This finishes the proof of the theorem. □

Corollary 11.2.2 *If the function g considered in Theorem* 11.2.4 *satisfies the conditions of Theorem* 11.2.2 *for every (degenerate) initial distribution of the* PDMP $\{X(t+s), s \ge 0\}$, *then $g_1(t) = \mu_1(t)$ and $g_2(t,y) = \mu_2(t,y)$.*

Proof Conditioning on $X(t)$, the statement follows from Theorem 11.2.4. □

Theorem 11.2.4 and Corollary 11.2.2 show that we have to determine the functions $g_1(t)$ and $g_2(t,y)$ if we want to compute the conditional expected payments $\mu_1(t)$ and $\mu_2(t,y)$ or the unconditional expected payment considered in (11.2.18). This then requires the solution to the system of partial differential equations (11.2.16)–(11.2.17). Under some regularity conditions on the transition intensities $\lambda_{ij}(t)$ and $\lambda_{ij}(t,y)$, this system admits a unique solution which is bounded on compact sets. For instance, this is the case if $\lambda_{ij}(t,y)$ are uniformly continuous on bounded intervals; see, for example, Chapter 1 of Forsyth (1906).

In the rest of this section we assume that these regularity conditions hold as well as the conditions of Corollary 11.2.2. The following approximation technique can then be invoked when solving the equations (11.2.16)–(11.2.17) numerically.

Consider the sequence $\{\sigma_n\}$ of all consecutive instants when the external component $\{J(t)\}$ jumps and let $N(t)$ denote the number of these jumps in the interval $(0,t]$. Start with $g_1^0(t) = g_2^0(t,y) = 0$ for all $t \in [0,t_0]$ and define the functions $g_1^n(t)$ and $g_2^n(t,y)$ recursively by

$$g_1^{n+1}(t) = \mathbb{E}\,(g_2^n(\sigma \wedge t_0, 0)\,\mathbb{I}(J(\sigma)=2) \mid J(t)=1)\,, \tag{11.2.19}$$

$$g_2^{n+1}(t,y) = \mathbb{E}\,((\sigma \vee y_0 - y \vee y_0) + g_1^n(\sigma \wedge t_0)\,\mathbb{I}(J(\sigma)=1) \mid J(t)=2, Y(t)=y)\,. \tag{11.2.20}$$

Here, $\sigma = \sigma_{N(t)+1}$ denotes the next jump epoch of the external state component after t. It is easily seen by induction that $g_1^n(t_0) = g_2^n(t_0,y) = 0$ for all $n \geq 1$.

Theorem 11.2.5 *Let $g_1(t)$ and $g_2(t,y)$ be the solutions to* (11.2.16)–(11.2.17). *Then*

$$\lim_{n\to\infty} g_1^n(t) = g_1(t)\,, \qquad \lim_{n\to\infty} g_2^n(t,y) = g_2(t,y)\,. \tag{11.2.21}$$

Proof The reader should prove that the functions $g_1^n(t)$ and $g_2^n(t,y)$, defined in (11.2.19) and (11.2.20), can be represented in the form

$$g_1^n(t) = \mathbb{E}\Big(\int_t^{\sigma_n \wedge t_0} \mathbb{I}(J(v)=2, Y(v) \geq y_0)\,\mathrm{d}v \;\Big|\; J(t)=1, N(t)=0\Big), \tag{11.2.22}$$

$$g_2^n(t,y) = \mathbb{E}\Big(\int_t^{\sigma_n \wedge t_0} \mathbb{I}(J(v)=2, Y(v) \geq y_0)\,\mathrm{d}v \;\Big|\; J(t)=2, Y(t)=y, N(t)=0\Big)\,. \tag{11.2.23}$$

By Corollary 11.2.2, the statement of the theorem then follows from the fact that $\{X(t)\}$ is Markov and that $\lim_{n\to\infty} \sigma_n = \inf\{t : J(t)=3\}$. □

In a simulation of the PDMP $\{X(t)\}$, estimates of $g_1^n(t)$ and $g_2^n(t,y)$ can be directly computed from (11.2.19) and (11.2.20). An alternative is to use the following result.

Theorem 11.2.6 *The functions $g_1^n(t)$ and $g_2^n(t,y)$ defined in* (11.2.19) *and* (11.2.20) *solve the following system of differential equations:*

$$\frac{\mathrm{d}}{\mathrm{d}t} g_1^{n+1}(t) + \lambda_{12}(t) g_2^n(t) - (\lambda_{12}(t) + \lambda_{13}(t)) g_1^{n+1}(t) = 0\,, \tag{11.2.24}$$

$$\frac{\partial g_2^{n+1}}{\partial t}(t,y) + \frac{\partial g_2^{n+1}}{\partial y}(t,y) + \lambda_{21}(t,y) g_1^n(t) - (\lambda_{21}(t,y) + \lambda_{23}(t,y)) g_2^{n+1}(t,y) + \mathbb{I}(y \geq y_0) = 0\,, \tag{11.2.25}$$

with the boundary conditions $g_1^{n+1}(t_0) = g_2^{n+1}(t_0, y) = 0$.

The *proof* of this theorem is left to the reader. Notice that the system (11.2.24)–(11.2.25) has a simpler structure than the original system (11.2.16)–(11.2.17) since the equations (11.2.24)–(11.2.25) can be solved separately.

Bibliographical Notes. The general class of PDMP was introduced in Davis (1984), where Theorem 11.2.3 was proved for the first time. For (partially) more specific classes of Markov processes with piecewise deterministic paths and for other related results, see also Dassios and Embrechts (1989), Davis (1993), Embrechts and Schmidli (1994), Franken, König, Arndt and Schmidt (1982), Gnedenko and Kovalenko (1989), Miyazawa, Schassberger and Schmidt (1995) and Schassberger (1978). Notice in particular that Dassios and Embrechts (1989, p. 211) used Theorem 11.2.3 to determine an optimal dividend barrier in the compound Poisson model. The representation of the permanent health insurance model considered in Section 11.2.4 in terms of a PDMP is due to Davis and Vellekoop (1995). They also showed that the solution to the system (11.2.16)–(11.2.17) can be found numerically. For an application of PDMP to disability insurance, see Möller and Zwiesler (1996).

11.3 THE COMPOUND POISSON MODEL REVISITED

We first reconsider the compound Poisson model of Section 5.3 and demonstrate on simple examples how the method of PDMP works in the context of risk theory. In this section, the net profit condition (5.3.2) is taken for granted and we look for new expressions for characteristics of the time of ruin. The extension to the usual economic environment is considered in Section 11.4, while in Chapter 12 a stochastic environmental process controls the risk model.

11.3.1 Exponential Martingales via PDMP

We again apply the method of PDMP to get martingales related to the risk reserve process in the compound Poisson model, paving the road for more general risk models considered in Chapter 12.

Let $\{R(t)\}$ with $R(t) = u + \beta t - \sum_{i=1}^{N(t)} U_i$ be the risk reserve process in the compound Poisson model considered in Section 5.3. It is easy to see that $\{(R(t), t)\}$ is a PDMP with state space $E = \mathbb{R}^2$. The set I of external states consists of only one element and is therefore omitted. The characteristics of the PDMP $\{(R(t), t)\}$ are given by

$$\boldsymbol{X}g(y,t) = \Big(\beta\frac{\partial g}{\partial y} + \frac{\partial g}{\partial t}\Big)(y,t)$$

and $\lambda(y,t) = \lambda$, $Q((y,t), B_1 \times B_2) = \mathbb{I}(t \in B_2)F_U(y - B_1)$, where $y - B = \{y - v : v \in B\}$; $B_i \in \mathcal{B}(\mathbb{R})$ for $i = 1, 2$. The active boundary Γ is empty.

We are interested in functions $g \in \mathcal{D}(\boldsymbol{A})$ or more specifically functions that satisfy the conditions of Theorem 11.2.2. Condition (a) is fulfilled if and only if g is absolutely continuous. Condition (11.2.5) becomes trivial because $\Gamma = \emptyset$. Suppose now that g satisfies (11.2.6). Then, by Theorem 11.2.2,

$$\boldsymbol{A}g(y,t) = \Big(\beta\frac{\partial g}{\partial y} + \frac{\partial g}{\partial t}\Big)(y,t) + \lambda\Big(\int_0^\infty g(y-v,t)\,\mathrm{d}F_U(v) - g(y,t)\Big).$$

By Theorem 11.1.3, we need to solve the equation $\boldsymbol{A}g = 0$ if we want to find a martingale of the form $\{g(R(t), t), t \ge 0\}$.

If we would be interested in a martingale $\{g(R(t))\}$ that does not explicitly depend on time, then the equation to solve would be

$$\beta g^{(1)}(y) + \lambda\Big(\int_0^\infty g(y-v)\,\mathrm{d}F_U(v) - g(y)\Big) = 0\,. \tag{11.3.1}$$

The latter equation is similar to (5.3.3). Indeed, if we put $g(y) = 0$ for all $y < 0$ and $g(y) = \overline{\psi}(y)$ for $y \ge 0$, then we recover (5.3.3). But the only absolutely continuous function $g(y)$ satisfying (11.3.1) with boundary condition $g(y) = 0$ for $y < 0$ is the function $g(y) \equiv 0$, as can be easily shown by the reader.

We now try a function of the form $g(y,t) = \exp(-sy - \theta t)$, where we assume that $\hat{m}_U(s) < \infty$. We will see in a moment that θ has to depend on s, i.e. $\theta = \theta(s)$. Then the equation $\boldsymbol{A}g = 0$ yields

$$-\beta s g(y,t) - \theta(s)g(y,t) + \lambda\Big(\int_0^\infty \mathrm{e}^{-s(y-v)}\,\mathrm{d}F_U(v)\mathrm{e}^{-\theta(s)t} - g(y,t)\Big) = 0$$

or, because $g(y,t) > 0$, equivalently $-\beta s - \theta(s) + \lambda(\hat{m}_U(s) - 1) = 0$. Hence

$$\theta(s) = \lambda(\hat{m}_U(s) - 1) - \beta s\,. \tag{11.3.2}$$

Notice that this function was already considered in Section 5.4.1 in connection with the adjustment coefficient in the compound Poisson model. Moreover, the stochastic process $\{M(t)\}$ with

$$M(t) = \exp(-sR(t) - (\lambda(\hat{m}_U(s) - 1) - \beta s)t) \tag{11.3.3}$$

has already been obtained in Example 3 of Section 10.1.3. Trying to apply Theorem 11.2.2 in an attempt to get a martingale, it still remains to verify that (11.2.6) holds for $g(y,t) = \exp(-sy - \theta(s)t)$, where $\theta(s)$ is given by (11.3.2). This can be seen from the following computation:

$$
\begin{aligned}
&\mathbb{E}\Big(\sum_{i=1}^{N(t)}(\mathrm{e}^{-sR(\sigma_i)} - \mathrm{e}^{-sR(\sigma_i-0)})\mathrm{e}^{-\theta(s)\sigma_i}\Big) \\
&\quad\le\ \mathbb{E}\Big(\sum_{i=1}^{N(t)}\mathrm{e}^{-sR(\sigma_i)}\Big)\max\{1,\mathrm{e}^{-\theta(s)t}\} \\
&\quad\le\ \mathbb{E}\Big(\sum_{i=1}^{N(t)}\exp\big(s\sum_{j=1}^{i}U_j\big)\Big)\max\{1,\mathrm{e}^{-\theta(s)t}\} \\
&\quad\le\ \mathbb{E}\Big(N(t)\exp\big(s\sum_{j=1}^{N(t)}U_j\big)\Big)\max\{1,\mathrm{e}^{-\theta(s)t}\} \\
&\quad=\ \mathbb{E}\Big(N(t)\mathbb{E}\Big(\exp\big(s\sum_{j=1}^{N(t)}U_j\big)\mid N(t)\Big)\Big)\max\{1,\mathrm{e}^{-\theta(s)t}\} \\
&\quad=\ \mathbb{E}\,(N(t)(\hat{m}_U(s))^{N(t)})\max\{1,\mathrm{e}^{-\theta(s)t}\} \\
&\quad=\ \sum_{n=1}^{\infty}n(\hat{m}_U(s))^n\frac{(\lambda t)^n}{n!}\mathrm{e}^{-\lambda t}\max\{1,\mathrm{e}^{-\theta(s)t}\} \\
&\quad=\ \lambda\hat{m}_U(s)t\mathrm{e}^{\lambda t(\hat{m}_U(s)-1)}\max\{1,\mathrm{e}^{-\theta(s)t}\} < \infty\,.
\end{aligned}
$$

11.3.2 Change of the Probability Measure

In this section we consider the canonical probability space $(\Omega,\mathcal{F},\mathbb{P})$ of the risk reserve process $\{R(t), t\ge 0\}$ in the compound Poisson model, where Ω is the (Borel) subset of $D(\mathbb{R}_+)$ consisting of all possible sample paths of $\{R(t)\}$, and $\mathcal{F} = \mathcal{B}(\Omega)$. Let $\{\mathcal{F}_t\}$ be the (uncompleted) history of $\{R(t)\}$.

Recall that by T_n, U_n we denote the interoccurrence times and claim sizes, respectively, where in the compound Poisson model $\{T_n\}$ is a sequence of independent random variables with common exponential distribution $\mathrm{Exp}(\lambda)$, independent of the sequence $\{U_n\}$. The sequences T_n and U_n are considered as random variables on the canonical probability space $(\Omega,\mathcal{F},\mathbb{P})$ of $\{R(t)\}$. It is rather obvious that for each $n\in\mathbb{N}$, the claim arrival epoch $\sigma_n = T_1+\ldots+T_n$ is an $\{\mathcal{F}_t\}$-stopping time.

Let $s\in\mathbb{R}$ be fixed such that $\hat{m}_U(s) < \infty$. Then, for the martingale $\{M(t), t\ge 0\}$ given in (11.3.3) we consider the family of probability measures

$\{\mathbb{P}_t^{(s)}, t \ge 0\}$ defined as in (10.2.23), that is

$$\mathbb{P}_t^{(s)}(A) = \mathbb{E}\,[M(t); A]\,, \qquad A \in \mathcal{F}_t\,. \tag{11.3.4}$$

From Kolmogorov's extension theorem, see Remark 2 in Section 10.2.6, we get that there exists a "global" probability measure $\mathbb{P}^{(s)}$ on $(\Omega, \mathcal{F})$ such that the restriction of $\mathbb{P}^{(s)}$ to $\mathcal{F}_t$ is $\mathbb{P}_t^{(s)}$. Let $\mathbb{E}^{(s)}$ denote the expectation under $\mathbb{P}^{(s)}$.

Lemma 11.3.1 *For all $t \ge 0$ and $A \in \mathcal{F}_t$,*

$$\mathbb{P}^{(s)}(A) = \mathbb{E}\,[\mathrm{e}^{-s(R(t)-u)-\theta(s)t}; A] \tag{11.3.5}$$

and

$$\mathbb{P}(A) = \mathbb{E}^{(s)}[\mathrm{e}^{s(R(t)-u)+\theta(s)t}; A]\,, \tag{11.3.6}$$

where $\theta(s)$ is given by (11.3.2). *Moreover, if τ is an $\{\mathcal{F}_t\}$-stopping time and $A \subset \{\tau < \infty\}$ such that $A \in \mathcal{F}_\tau$, then*

$$\mathbb{P}^{(s)}(A) = \mathbb{E}\,[\mathrm{e}^{-s(R(\tau)-u)-\theta(s)\tau}; A] \tag{11.3.7}$$

and

$$\mathbb{P}(A) = \mathbb{E}^{(s)}[\mathrm{e}^{s(R(\tau)-u)+\theta(s)\tau}; A]\,. \tag{11.3.8}$$

Proof Observe that (11.3.5) is an immediate consequence of the defining equations (11.3.3) and (11.3.4). Furthermore, using (10.2.28) we get (11.3.6). Formula (11.3.7) follows from Lemma 10.2.2b. To show the validity of (11.3.8) we use (10.2.28) and apply Lemma 10.2.2b to the martingale $\{M^{-1}(t), t \ge 0\}$ given on the probability space $(\Omega, \mathcal{F}, \mathbb{P}^{(s)})$. □

The change of measure technique stated in Lemma 11.3.1, combined with the method of PDMP, is a powerful tool when investigating ruin probabilities. In the present chapter we illustrate this for the compound Poisson model, leaving further examples to Chapter 12. We show first that, under the measure $\mathbb{P}^{(s)}$, the process $\{R(t)\}$ remains a risk reserve process in a compound Poisson model. For convenience, we denote the original probability measure by $\mathbb{P}^{(0)} = \mathbb{P}$.

Theorem 11.3.1 *Let $s \in \mathbb{R}$ such that $\hat{m}_U(s) < \infty$. Consider the probability space $(\Omega, \mathcal{F}, \mathbb{P}^{(s)})$. Then, the following statements are true:*
(a) *Under the measure $\mathbb{P}^{(s)}$, the process $\{R(t)\}$ is the risk reserve process in the compound Poisson model with premium rate β, claim arrival intensity $\lambda^{(s)} = \lambda\hat{m}_U(s)$ and claim size distribution $F_U^{(s)}(x) = \int_0^x \mathrm{e}^{sy}\,\mathrm{d}F_U(y)/\hat{m}_U(s)$. In particular,*

$$\mathbb{E}^{(s)}R(1) - u = -\theta^{(1)}(s)\,. \tag{11.3.9}$$

(b) *If $s' \in \mathbb{R}$ such that $\hat{m}_U(s') < \infty$ and $s \neq s'$, then $\mathbb{P}^{(s)}$ and $\mathbb{P}^{(s')}$ are singular on $\mathcal{F}$.*

Proof (a) Since the set of trajectories of the risk reserve process $\{R(t)\}$ is the same under the measures $\mathbb{P}^{(0)}$ and $\mathbb{P}^{(s)}$, it is clear that the premium rate is β under both measures. Let $n \in \mathbb{N}$ be fixed and let $B_i, B_i' \in \mathcal{B}(\mathbb{R})$, $1 \le i \le n$. Notice that $\sigma_n = \sum_{i=1}^n T_i$ is an $\{\mathcal{F}_t\}$-stopping time which is finite under $\mathbb{P}^{(0)}$. Thus, from (11.3.7) we get that

$$\begin{aligned}
&\mathbb{P}^{(s)}\Big(\bigcap_{i=1}^n \{T_i \in B_i, U_i \in B_i'\}\Big)\\
&= \mathbb{E}^{(0)}\Big[\exp\Big(s\sum_{i=1}^n U_i\Big)\exp\Big(-\lambda(\hat{m}_U(s)-1)\sum_{i=1}^n T_i\Big); \bigcap_{i=1}^n \{T_i \in B_i, U_i \in B_i'\}\Big]\\
&= \prod_{i=1}^n (\mathbb{E}^{(0)}[\mathrm{e}^{sU_i}; U_i \in B_i']\,\mathbb{E}^{(0)}[\mathrm{e}^{-\lambda(\hat{m}_U(s)-1)T_i}; T_i \in B_i])\\
&= \prod_{i=1}^n \Big(\int_{B_i'} \mathrm{e}^{sy}\,\mathrm{d}F_U(y)/\hat{m}_U(s)\int_{B_i} \lambda\hat{m}_U(s)\mathrm{e}^{-\lambda\hat{m}_U(s)v}\,\mathrm{d}v\Big),
\end{aligned}$$

and the first part of assertion (a) follows. The expression $\mathbb{E}^{(s)}R(1) - u = \beta - (\lambda\hat{m}_U(s))\,(\hat{m}_U^{(1)}(s)/\hat{m}_U(s)) = -\theta^{(1)}(s)$ is now obtained from (5.2.8) because $\mathbb{E}^{(s)}U_i = \int_0^\infty y\mathrm{e}^{sy}\,\mathrm{d}F_U(y)/\hat{m}_U(s) = \hat{m}_U^{(1)}(s)/\hat{m}_U(s)$. In order to prove (b), observe that from (11.3.9) and from the law of large numbers for additive processes (see Theorem 10.3.4) it follows that

$$\mathbb{P}^{(s)}\big(\lim_{t\to\infty} t^{-1}(R(t)-u) = \lim_{t\to\infty} t^{-1}(R(t-)-u) = -\theta^{(1)}(s)\big) = 1,$$

where $\{R(t-)\}$ is the left-continuous version of $\{R(t)\}$. Thus, the measures $\mathbb{P}^{(s)}$ and $\mathbb{P}^{(s')}$ are singular unless $\theta^{(1)}(s) = \theta^{(1)}(s')$. However, the latter can only happen if $s = s'$ since $\theta(s)$ is strictly convex, which means that $\theta^{(1)}(s)$ is strictly increasing for all $s > 0$ where $\theta(s)$ is finite. □

Using Lemma 11.3.1 one can easily show that the ruin probabilities $\psi(u)$ and $\psi(u;x)$ can be expressed under the new measure $\mathbb{P}^{(s)}$.

Theorem 11.3.2 *For each $s \in \mathbb{R}$ such that $\hat{m}_U(s) < \infty$,*

$$\psi(u) = \mathbb{E}^{(s)}[\mathrm{e}^{sR(\tau(u))+\theta(s)\tau(u)}; \tau(u) < \infty]\mathrm{e}^{-su} \tag{11.3.10}$$

and

$$\psi(u;x) = \mathbb{E}^{(s)}[\mathrm{e}^{sR(\tau(u))+\theta(s)\tau(u)}; \tau(u) \le x]\mathrm{e}^{-su}. \tag{11.3.11}$$

The *proof* of this theorem is left to the reader.

Formula (11.3.11) constitutes a continuous analogue to (9.2.23) and is useful to simulate ruin probabilities. Indeed, if the event $\{\tau(u) \le x\}$ has a large probability under $\mathbb{P}^{(s)}$, the right-hand side of (11.3.11) can be simulated efficiently and yields an estimator for $\psi(u;x)$, see also Section 9.2.5. Equivalent representations for $\psi(u)$ and $\psi(u;x)$ in terms of the claim surplus process were derived in (10.2.29) and (10.2.30).

11.3.3 Cramér–Lundberg Approximation

In this section we show how (11.3.10) can be used to recover the Cramér–Lundberg approximation $\psi_{\rm app}(u)$ to $\psi(u)$ given in (5.4.16). Assume that the adjustment coefficient $\gamma > 0$ exists and $\hat{m}_U^{(1)}(\gamma) < \infty$. First note that (11.3.10) is not very useful for $s \ne \gamma$ because then the joint distribution of $R(\tau(u))$ and $\tau(u)$ is needed. However, we can get rid of the term involving $\tau(u)$ by choosing an s such that $\theta(s) = 0$, that is $s = \gamma$. Since the function $\theta(s)$ defined in (11.3.2) is convex, we have $\theta^{(1)}(\gamma) > 0$. From Theorem 11.3.1 we know that $\mathbb{E}^{(\gamma)} R(1) - u = -\theta^{(1)}(\gamma) < 0$ and thus $\mathbb{P}^{(\gamma)}(\tau(u) < \infty) = 1$. Using (11.3.10) this gives

$$\psi(u) = {\rm e}^{-\gamma u}\mathbb{E}^{(\gamma)} \exp(\gamma R(\tau(u)))\,. \tag{11.3.12}$$

Comparing (5.4.10) and (11.3.12) we still have to show that

$$\lim_{u\to\infty} \mathbb{E}^{(\gamma)} \exp(\gamma R(\tau(u))) = (\beta - \lambda\mu)(\lambda \hat{m}_U^{(1)}(\gamma) - \beta)^{-1}.$$

Let $g(u) = \mathbb{E}^{(\gamma)} \exp(\gamma R(\tau(u)))$, $\tau_- = \inf\{t \ge 0 : R(t) < u\}$ and $F(y) = \mathbb{P}^{(\gamma)}(u - R(\tau_-) \le y)$. Notice that $\tau_- \stackrel{\rm d}{=} \tau(0)$ and that F is the ladder-height distribution at the first descending ladder epoch τ_- of $\{R(t)\}$. Moreover,

$$\begin{aligned} g(u) &= \mathbb{E}^{(\gamma)}\Big(\mathbb{E}^{(\gamma)}({\rm e}^{\gamma R(\tau(u))} \mid \tau_-, R(\tau_-))\Big) \\ &= \mathbb{E}^{(\gamma)}\Big(\mathbb{E}^{(\gamma)}({\rm e}^{\gamma R(\tau(u))} \mid \tau_-, R(\tau_-))\mathbb{I}(R(\tau_-) \ge 0)\Big) \\ &\qquad + \mathbb{E}^{(\gamma)}\Big(\mathbb{E}^{(\gamma)}({\rm e}^{\gamma R(\tau(u))} \mid \tau_-, R(\tau_-))\mathbb{I}(R(\tau_-) < 0)\Big) \\ &= \int_0^u g(u-y)\,{\rm d}F(y) + \int_u^\infty {\rm e}^{\gamma(u-y)}\,{\rm d}F(y)\,. \end{aligned}$$

Thus, the function $g(u)$ satisfies the renewal equation

$$g(u) = z(u) + \int_0^u g(u-y)\,{\rm d}F(y)\,,$$

where $z(u) = \int_u^\infty {\rm e}^{\gamma(u-y)}\,{\rm d}F(y) = {\rm e}^{\gamma u}\int_u^\infty {\rm e}^{-\gamma y}\,{\rm d}F(y)$. Furthermore, we have

$$\int_0^\infty z(u)\,{\rm d}u = \int_0^\infty\int_u^\infty {\rm e}^{\gamma(u-y)}\,{\rm d}F(y)\,{\rm d}u = \int_0^\infty\int_0^y {\rm e}^{\gamma(u-y)}\,{\rm d}u\,{\rm d}F(y)$$

$$
\begin{aligned}
&= \frac{1}{\gamma}(1 - \mathbb{E}^{(\gamma)} \mathrm{e}^{\gamma R(\tau(0))}) \\
&= \frac{1}{\gamma}(1 - \mathbb{E}^{(0)}[\mathrm{e}^{-\gamma R(\tau(0))} \mathrm{e}^{\gamma R(\tau(0))}; \tau(0) < \infty]) \\
&= \frac{1}{\gamma}(1 - \mathbb{P}^{(0)}(\tau(0) < \infty)) = \frac{1}{\gamma}\Big(1 - \frac{\lambda\mu}{\beta}\Big) = \frac{\beta - \lambda\mu}{\gamma\beta},
\end{aligned}
$$

where we used (5.3.11) in the last but one inequality. Thus, by Lemma 5.4.2, $z(u)$ is directly Riemann integrable. In order to apply Theorem 6.1.11 we finally need the expectation of the ladder height distribution F:

$$
\begin{aligned}
\mathbb{E}^{(\gamma)}(-R(\tau(0))) &= \mathbb{E}^{(0)}[-R(\tau(0))\mathrm{e}^{-\gamma R(\tau(0))}; \tau(0) < \infty] \\
&= \int_0^\infty y \mathrm{e}^{\gamma y} \frac{\lambda}{\beta} \overline{F}_U(y)\,\mathrm{d}y = \frac{\lambda}{\beta}\int_0^\infty \int_0^v y\mathrm{e}^{\gamma y}\,\mathrm{d}y\,\mathrm{d}F_U(v) \\
&= \frac{\lambda}{\gamma^2\beta}(\gamma \hat{m}_U^{(1)}(\gamma) - (\hat{m}_U(\gamma) - 1)) = \frac{\lambda \hat{m}_U^{(1)}(\gamma) - \beta}{\gamma\beta},
\end{aligned}
$$

where we used (5.3.18) in the second equality and the definition of γ (see (5.4.3)) in the last equality. Thus, using Theorem 6.1.11 we obtain (5.4.10).

11.3.4 A Stopped Risk Reserve Process

A rather useful way of modelling is to stop the risk reserve process $\{R(t)\}$ at the time of ruin $\tau(u)$ and to let it jump to a cemetery state. In other words, we consider the PDMP $\{(X(t), t), t \geq 0\}$, where

$$
X(t) = \begin{cases} (1, R(t)) & \text{if } t \leq \tau(u), \\ (0, R(\tau(u))) & \text{if } t > \tau(u). \end{cases} \tag{11.3.13}
$$

Here we have $I = \{0, 1\}$, where 0 means that the process is in the cemetery state, $C_0 = (-\infty, 0) \times \mathbb{R}$ and $C_1 = \mathbb{R}^2$. Since the external state is uniquely determined by the continuous component of $X(t)$, we can and will simplify the notation by omitting the external state. The vector field of $\{(X(t), t)\}$ is given by $\boldsymbol{X} g(y,t) = \beta \mathbb{I}(y \geq 0)(\partial g)(\partial y)(y,t) + (\partial g)(\partial t)(y,t)$. The jump intensity is $\lambda(y,t) = \mathbb{I}(y \geq 0)\lambda$ and the transition kernel is $Q((y,t), B_1 \times B_2) = \mathbb{I}(t \in B_2)F_U(y - B_1)$ for $B_1, B_2 \in \mathcal{B}(\mathbb{R})$.

Notice that the introduction of the cemetery state within the stopped risk reserve process entitles us to apply Theorem 11.2.2 to an essentially broader class of test functions. Indeed, now a function $g(y,t)$ fulfils condition (a) of Theorem 11.2.2 if and only if it is absolutely continuous in y on $[0, \infty)$ and absolutely continuous in t. However, what is important, there is no need to assume that $g(y,t)$ is continuous in y at $y = 0$, in contrast to the situation discussed in Section 11.3.1. Furthermore, the active boundary Γ is empty and therefore condition (11.2.5) becomes trivial. Now, assuming (11.2.6) from Theorem 11.2.2, the generator of the PDMP $\{(X(t),t)\}$ is given by

$$\boldsymbol{A}g(y,t)=\frac{\partial g}{\partial t}(y,t)+\mathbb{I}(y\ge 0)\Big(\beta\frac{\partial g}{\partial y}(y,t)+\lambda\Big(\int_0^\infty g(y-v,t)\,\mathrm{d}F_U(v)-g(y,t)\Big)\Big). \tag{11.3.14}$$

We show how this representation formula for the generator of the PDMP $\{(X(t),t)\}$ can be used to determine the probability $\overline{\psi}(u)=\mathbb{P}(\tau(u)=\infty)$ of survival in infinite time. We will indeed prove that, under some conditions, the survival function $\overline{\psi}(u)$ in the compound Poisson model is the *only* solution to the integro-differential equation considered in Theorem 5.3.1.

Theorem 11.3.3 *Let* $\{(X(t),t)\}$ *be the PDMP defined in* (11.3.13) *with generator* $\boldsymbol{A}$ *given in* (11.3.14). *Let* $g(y,t)$ *be a function which satisfies the conditions of Theorem* 11.2.2 *for this PDMP. Then the following statements are true:*
(a) *If* $g(y)=g(y,t)$ *does not depend on* t*, then the only solution* $g(y)$ *(up to a multiplicative constant) to* $\boldsymbol{A}g=0$ *such that* $g(0)>0$ *and fulfilling the boundary condition* $g(y)=0$ *on* $(-\infty,0)$ *is the survival function* $\overline{\psi}(y)$.
(b) *Let* $x>0$ *be fixed. Let* $g(y,t)$ *solve* $\boldsymbol{A}g=0$ *in* $\mathbb{R}\times[0,x]$ *with boundary condition* $g(y,x)=\mathbb{I}(y\ge 0)$. *Then* $g(y,0)=\mathbb{P}(\tau(y)>x)$.

Proof (a) Assume that $g(y)=g(y,t)$ does not depend on t and that $g(y)=0$ for $y<0$. In view of (11.3.14), equation $\boldsymbol{A}g=0$ reads then for $y>0$

$$\beta g^{(1)}(y)+\lambda\Big(\int_0^y g(y-z)\,\mathrm{d}F_U(z)-g(y)\Big)=0\,. \tag{11.3.15}$$

From Theorem 5.3.1 we see that $g(y)=\overline{\psi}(y)$ is a solution to (11.3.15). Suppose that there is another solution $g(y)$ to (11.3.15). Since $g(0)>0$ and $g(y)$ has to be absolutely continuous on $[0,\infty)$, we have $\lim_{y\downarrow 0}g(y)=g(0)>0=\lim_{y\downarrow 0}\int_0^y g(y-v)\,\mathrm{d}F_U(v)$. Thus, (11.3.15) implies that $g^{(1)}(y)>0$ for all sufficiently small $y>0$, i.e. $g(y)$ is strictly increasing in a right neighbourhood of the origin. Let $y_0=\inf\{y>0:g^{(1)}(y)\le 0\}$ and suppose $y_0<\infty$. The continuity of $g(y)$ and (11.3.15) imply that $g^{(1)}(y_0)\le 0$. Then, using (11.3.15) we have $0\ge\beta\lambda^{-1}g^{(1)}(y_0)=g(y_0)-\int_0^{y_0}g(y_0-v)\,\mathrm{d}F_U(v)$ and consequently

$$0<g(y_0)\le\int_0^{y_0}g(y_0-v)\,\mathrm{d}F_U(v)<\int_0^{y_0}g(y_0)\,\mathrm{d}F_U(v)\le g(y_0)\,,$$

which is a contradiction. Thus, $g(y)$ is strictly increasing on the whole nonnegative halfline $\mathbb{R}_+=[0,\infty)$. Recall that by the net profit condition (5.3.2), $R(t)\to\infty$ as $t\to\infty$. Thus, the possibly infinite random variable $R(\tau(u))=\lim_{t\to\infty}R(\tau(u)\wedge t)$ is well-defined. Since $\tau(u)$ is a stopping time and since the stochastic process $\{g(R(t)),t\ge 0\}$ is a martingale by

Theorem 11.2.2, we get from Theorem 10.2.4 that $\{g(R(\tau(u) \wedge t)), t \geq 0\}$ is a martingale too. Using Theorem 10.2.5, we have $g(u) = \mathbb{E}\, g(R(t)) = \mathbb{E}\, g(R(\tau(u) \wedge t))$ for each $t \geq 0$. Thus, Theorem 10.2.2 implies that $g(R(\tau(u)))$ is integrable. Since $\mathbb{P}(\tau(u) = \infty) > 0$ the latter is only possible if g is bounded. Moreover, since g is increasing, the limit $g(\infty) = \lim_{y\to\infty} g(y)$ exists and is finite. But then by the dominated convergence theorem

$$g(u) = \lim_{t\to\infty} \mathbb{E}\, g(R(\tau(u) \wedge t)) = \mathbb{E}\, g(R(\tau(u))) = g(\infty)\mathbb{P}(\tau(u) = \infty)\,,$$

where the last equality follows from the fact that $g(R(\tau(u))) = 0$, whenever $\tau(u) < \infty$. This finishes the proof of the first part of the theorem.
(b) We are now interested in $\mathbb{P}(\tau(u) > x)$, where we assume that $x > 0$ is fixed. The equation to solve is then

$$\frac{\partial}{\partial t} g(y,t) + \beta \frac{\partial}{\partial y} g(y,t) + \lambda\Big(\int_0^y g(y-z,t)\, \mathrm{d}F_U(z) - g(y,t)\Big) = 0\,. \quad (11.3.16)$$

As in the proof of part (a) it follows that $\{g(R(\tau(u) \wedge t), t), t \geq 0\}$ is a martingale. By Theorem 10.2.4 the process $\{g(R(\tau(u) \wedge x \wedge t), x \wedge t), t \geq 0\}$ is a martingale too. Thus, using the boundary condition $g(y,x) = \mathbb{I}(y \geq 0)$ in the second inequality, we finally get that $g(u,0) = \mathbb{E}\, g(R(\tau(u) \wedge x), x) = \mathbb{P}(\tau(u) > x)$. □

11.3.5 Characteristics of the Ruin Time

In this section we study several characteristics of the ruin time $\tau(u)$, in particular the conditional expectation of $\tau(u)$ provided $\tau(u)$ is finite, and the Laplace–Stieltjes transform $\hat{l}_{\tau(u)}(s) = \mathbb{E} \exp(-s\tau(u))$, $s \geq 0$. Let $\eta_w(u) = \mathbb{E} \int_0^{\tau(u)} \mathrm{e}^{-wv}\, \mathrm{d}v$ and notice that then

$$\eta_w(u) = w^{-1}(1 - \hat{l}_{\tau(u)}(w))\,, \qquad u \geq 0,\, w > 0\,. \quad (11.3.17)$$

We first show how Theorem 11.2.3 can be applied to determine the Laplace transform $\hat{L}_{\eta_w}(s) = \int_0^\infty \eta_w(u)\, \mathrm{e}^{-su}\, \mathrm{d}u$ of $\eta_w(u)$. The idea of using Theorem 11.2.3 is suggested by the fact that $\eta_w(u)$ can be represented in the form of equation (11.2.14). Namely, from the definition of $\eta_w(u)$ we immediately get

$$\eta_w(u) = \mathbb{E} \int_0^\infty \mathrm{e}^{-wv} \mathbb{I}(\tau(u) > v)\, \mathrm{d}v\,. \quad (11.3.18)$$

Since the risk reserve process $\{R(t)\}$ in the compound Poisson model has independent and stationary increments, the reader can rewrite (11.3.18) to obtain

$$\eta_w(u) = \mathbb{E}\Big(\int_0^{t_0} e^{-wv}\mathbb{I}(\tau(u) > v)\,dv + e^{-wt_0}\mathbb{I}(\tau(u) > t_0)\eta_w(R(t_0))\Big) \tag{11.3.19}$$

for each $t_0 > 0$. Now, putting

$$g(y,s) = g(y) = \mathbb{I}(y \geq 0)\eta_w(y)\,, \tag{11.3.20}$$

$\kappa(y,v) = w$, $\gamma(y,v) = \mathbb{I}(y \geq 0)$ and $g_{\rm ter}(y) = \mathbb{I}(y \geq 0)\eta_w(y)$, we see that (11.3.19) is identical with (11.2.14). Notice, however, that the function $g(y,s)$ defined in (11.3.20) does not fulfil the conditions of Theorem 11.2.3 with respect to the PDMP $\{R(t)\}$ since $g(y,s)$ is not continuous at $y = 0$. Using the procedure of Section 11.3.4, we can get rid of this problem by replacing $\{R(t)\}$ by the stopped risk reserve process defined in (11.3.13). Before we state Theorem 11.3.4, we formulate a technical lemma.

Lemma 11.3.2 *There exists a function $g : \mathbb{R} \to \mathbb{R}$, which is absolutely continuous, bounded, positive and increasing on $[0,\infty)$ satisfying the integro-differential equation*

$$\beta g^{(1)}(u) + \lambda\Big(\int_0^u g(u-v)\,dF_U(v) - g(u)\Big) - wg(u) + 1 = 0\,, \quad u \geq 0\,. \tag{11.3.21}$$

The *proof* of this lemma is omitted but can be found in Schmidli (1992).

Theorem 11.3.4 *For all $s, w > 0$,*

$$\hat{L}_{\eta_w}(s) = \frac{s_w^{-1} - s^{-1}}{\beta s - \lambda(1 - \hat{l}_U(s)) - w}\,, \tag{11.3.22}$$

where s_w is the unique positive solution to the equation

$$\beta s - \lambda(1 - \hat{l}_U(s)) - w = 0\,. \tag{11.3.23}$$

Proof Consider the stopped risk reserve process defined in (11.3.13). Because $\eta_w(u)$ can be represented in the form (11.2.14) corresponding to this PDMP, we are looking for a function $g(u,s) = g(u)$ for which $g(u) = 0$ for $u < 0$ and that satisfies the conditions of Theorem 11.2.3. Specifically, (11.2.13) and (11.3.14) imply that $g(u)$ should be a solution to equation (11.3.21). Since $\eta_w(u)$ is absolutely continuous, bounded, positive and increasing on $[0,\infty)$, it suffices to look for a function $g(u)$ which shares these properties and satisfies (11.3.21). The existence of such a function is guaranteed by Lemma 11.3.2. Multiply (11.3.21) by e^{-sy} and integrate over $(0,\infty)$ to find that for $\hat{L}_g(s) = \int_0^\infty g(u)e^{-su}\,du$,

$$\beta(-g(0) + s\hat{L}_g(s)) - \lambda\hat{L}_g(s)(1 - \hat{l}_U(s)) - w\hat{L}_g(s) + s^{-1} = 0\,.$$

See also the proof of Theorem 5.3.3. This gives

$$\hat{L}_g(s) = \frac{\beta g(0) - s^{-1}}{\beta s - \lambda(1 - \hat{l}_U(s)) - w} \,. \tag{11.3.24}$$

Since $g(u)$ should be bounded, $\hat{L}_g(s)$ has to exist for all $s > 0$. The denominator in (11.3.24), however, has a unique positive root s_w since the function $\theta_w(s) = \beta s - \lambda(1 - \hat{l}_U(s)) - w$ is convex, negative at 0 and converges to ∞ as $s \to \infty$. Hence $\beta g(0) = s_w^{-1}$ because otherwise $\hat{L}_g(s_w)$ would be infinite. With the choice $\kappa(x,t) = w$, $\gamma(x,t) = \mathbb{I}(x \ge 0)$ and $g_{\mathrm{ter}}(x) = \mathbb{I}(x \ge 0)g(x)$, $g(u)$ also fulfils conditions (a) and (b) of Theorem 11.2.3, because $g(u)$ is bounded and absolutely continuous on $[0, \infty)$. Thus, by the result of Theorem 11.2.3, we get that

$$g(u) = \mathbb{E}\Big(\int_0^{\tau(u)\wedge t_0} \mathrm{e}^{-ws}\,\mathrm{d}s + \mathrm{e}^{-wt_0}\mathbb{I}(\tau(u) > t_0)g(R(t_0))\Big) \tag{11.3.25}$$

for each $t_0 > 0$. Notice that, by letting $t_0 \to \infty$, the second term in (11.3.25) disappears because g is bounded. This gives $g(u) = \eta_w(u)$. Now, (11.3.22) follows from (11.3.24). □

Using the result of Theorem 11.3.4 we can express the conditional expected ruin time $\mathbb{E}(\tau(u) \mid \tau(u) < \infty)$ in terms of the ruin function $\psi(u) = \mathbb{P}(\tau(u) < \infty)$.

Theorem 11.3.5 *Assume that $\mu_U^{(2)} < \infty$. Then for each $u \ge 0$*

$$\mathbb{E}[\tau(u); \tau(u) < \infty] = \frac{1}{\beta - \lambda\mu}\Big(\frac{\lambda\mu_U^{(2)}}{2(\beta - \lambda\mu)}\overline{\psi}(u) - \int_0^u \psi(u-y)\overline{\psi}(y)\,\mathrm{d}y\Big). \tag{11.3.26}$$

Proof Let $w > 0$ and put $\tau(u)\mathrm{e}^{-w\tau(u)} = 0$ if $\tau(u) = \infty$. By the monotone convergence theorem, we have

$$\begin{aligned}\int_0^\infty \mathbb{E}[\tau(u); \tau(u) < \infty]\mathrm{e}^{-su}\,\mathrm{d}u &= \int_0^\infty \mathbb{E}(\lim_{w\downarrow 0} \tau(u)\mathrm{e}^{-w\tau(u)})\mathrm{e}^{-su}\,\mathrm{d}u \\ &= \lim_{w\downarrow 0}\int_0^\infty \mathbb{E}(\tau(u)\mathrm{e}^{-w\tau(u)})\mathrm{e}^{-su}\,\mathrm{d}u\,.\end{aligned}$$

$\mathbb{E}\,\mathrm{e}^{-w\tau(u)}$ is an analytic function for $w > 0$. In particular, $\mathbb{E}(\tau(u)\mathrm{e}^{-w\tau(u)}) = -(\partial)(\partial w)\mathbb{E}\,\mathrm{e}^{-w\tau(u)}$ and

$$\int_0^\infty \frac{\partial}{\partial w}\mathbb{E}\,\mathrm{e}^{-w\tau(u)}\,\mathrm{e}^{-su}\,\mathrm{d}u = \frac{\partial}{\partial w}\int_0^\infty \mathbb{E}\,\mathrm{e}^{-w\tau(u)}\,\mathrm{e}^{-su}\,\mathrm{d}u\,.$$

Thus, using (11.3.17) or equivalently $\hat{l}_{\tau(u)}(w) = 1 - w\eta_w(u)$, we have

$$\begin{aligned}
&\int_0^\infty \mathbb{E}\,[\tau(u); \tau(u) < \infty] \mathrm{e}^{-su}\,\mathrm{d}u \\
&= -\lim_{w\downarrow 0} \frac{\partial}{\partial w} \int_0^\infty (1 - w\eta_w(u))\mathrm{e}^{-su}\,\mathrm{d}u \\
&= -\lim_{w\downarrow 0} \frac{\partial}{\partial w}\left(s^{-1} - wL_{\eta_w}(s)\right) = \lim_{w\downarrow 0}\Big(L_{\eta_w}(s) + w\frac{\partial}{\partial w}L_{\eta_w}(s)\Big) \\
&= \lim_{w\downarrow 0}\Big(\frac{s_w^{-1} - s^{-1} - ws_w^{(1)}s_w^{-2}}{\beta s - \lambda(1 - \hat{l}_U(s)) - w} + \frac{w(s_w^{-1} - s^{-1})}{(\beta s - \lambda(1 - \hat{l}_U(s)) - w)^2}\Big),
\end{aligned}$$

where we used Theorem 11.3.4 and the implicit function theorem in the last equality. Indeed, from the implicit function theorem (see Theorems 17.1.1 and 17.4.1 in Hille (1966)) we get that the function s_w defined by (11.3.23) is twice continuously differentiable and $\beta s_w^{(1)} + \lambda\hat{l}_U^{(1)}(s_w)s_w^{(1)} - 1 = 0$, which gives $s_w^{(1)} = (\beta + \lambda\hat{l}_U^{(1)}(s_w))^{-1}$. Moreover $\lim_{w\downarrow 0} s_w = s_0 = 0$. By L'Hospital's rule

$$\lim_{w\downarrow 0} ws_w^{-1} = \lim_{w\downarrow 0}\frac{1}{s_w^{(1)}} = \beta + \lambda\hat{l}_U^{(1)}(0) = \beta - \lambda\mu_U$$

and consequently

$$\lim_{w\downarrow 0}\frac{w(s_w^{-1} - s^{-1})}{(\beta s - \lambda(1 - \hat{l}_U(s)) - w)^2} = \frac{\beta - \lambda\mu_U}{(\beta s - \lambda(1 - \hat{l}_U(s)))^2}\,.$$

Using L'Hospital's rule again, we get

$$\begin{aligned}
\lim_{w\downarrow 0}\frac{s_w - ws_w^{(1)}}{s_w^2} &= \lim_{w\downarrow 0} -\frac{ws_w^{(2)}}{2s_w s_w^{(1)}} = \lim_{w\downarrow 0}\frac{w\lambda\hat{l}_U^{(2)}(s_w)}{2s_w s_w^{(1)}(\beta + \lambda\hat{l}_U^{(1)}(s_w))^2} \\
&= \frac{\lambda\mu_U^{(2)}}{2(\beta - \lambda\mu_U)}\,.
\end{aligned}$$

Putting the above together we find that

$$\begin{aligned}
\int_0^\infty \mathbb{E}\,[\tau(u); \tau(u) < \infty]\mathrm{e}^{-su}\,\mathrm{d}u &= \frac{1}{\beta - \lambda\mu_U}\,\frac{\beta - \lambda\mu_U}{\beta s - \lambda(1 - \hat{l}_U(s))} \\
&\quad\times\Big(\frac{\lambda\mu_U^{(2)}}{2(\beta - \lambda\mu_U)} - \Big(s^{-1} - \frac{\beta - \lambda\mu_U}{\beta s - \lambda(1 - \hat{l}_U(s))}\Big)\Big).
\end{aligned}$$

Using (5.3.13) and (5.3.14), this gives

$$\int_0^\infty \mathbb{E}\,[\tau(u); \tau(u) < \infty]\mathrm{e}^{-su}\,\mathrm{d}u = \frac{1}{\beta - \lambda\mu_U}\,L_{\overline{\psi}}(s)\Big(\frac{\lambda\mu_U^{(2)}}{2(\beta - \lambda\mu_U)} - L_\psi(s)\Big)\,.$$

Since the right-hand side of the last equation is the Laplace transform of the right-hand side of (11.3.26), this proves the theorem. □

Corollary 11.3.1 *If* $\mu_U^{(2)} < \infty$, *then*

$$\mathbb{E}\,(\tau(0) \mid \tau(0) < \infty) = \frac{\mu_U^{(2)}}{2\mu_U(\beta - \lambda\mu_U)}\,. \tag{11.3.27}$$

Proof It suffices to put $u = 0$ in (11.3.26) since (5.3.11) gives (11.3.27). □

Corollary 11.3.2 *If* $\mu_U^{(2)} < \infty$, *then for each* $x > 0$

$$\mathbb{P}(x < \tau(u) < \infty) \le \frac{\lambda\mu_U^{(2)}}{2x(\beta - \lambda\mu_U)^2}\,. \tag{11.3.28}$$

Proof From Markov's inequality we have

$$\mathbb{P}(x < \tau(u) < \infty) = \mathbb{P}(\tau(u)\mathbb{I}(\tau(u) < \infty) > x) \le x^{-1}\mathbb{E}\,(\tau(u)\mathbb{I}(\tau(u) < \infty))\,.$$

Now the assertion is an immediate consequence of Theorem 11.3.5. □

We remark that Theorem 11.3.5 and (5.3.8) imply that for the compound Poisson model with exponential claim size distribution $\text{Exp}(\delta)$, we have $\mathbb{E}\,(\tau(u) \mid \tau(u) < \infty) = (\beta + \lambda u)(\beta(\beta\delta - \lambda))^{-1}$ for each $u \ge 0$.

Bibliographical Notes. The risk reserve process in the compound Poisson model was first treated as a PDMP in Dassios and Embrechts (1989). The results for the ruin time presented in Section 11.3.5, in particular Theorem 11.3.5 and Corollary 11.3.2, can be found in Schmidli (1996).

11.4 COMPOUND POISSON MODEL IN AN ECONOMIC ENVIRONMENT

11.4.1 Interest and Discounting

When studying economic phenomena, the effects of interest and inflation have to be taken into account. By interest we mean that the capital increases in time due to investments as in money markets or riskless bonds. In Sections 7.3 and 9.1.4 we considered the case of discrete-time interest or discounting. Then, in Section 8.4.3, we took up the idea of an instantaneous interest rate and showed how this can be obtained as a limit of the corresponding operations in discrete time. In the present section we consider the risk reserve process in the compound Poisson model in a continuous-time economic environment.

We first introduce the necessary notation. In case of inflation, a monetary unit at time 0 has the value $e^{-I(t)}$ at time t, where $I : \mathbb{R}_+ \to \mathbb{R}_+$ is a certain function with $I(0) = 0$. We call $e^{-I(t)}$ a *discounting factor*. If $I(t)$ is absolutely continuous, that is $I(t) = \int_0^t i(v)\,dv$ for some function $i : \mathbb{R}_+ \to \mathbb{R}_+$, then $i(t)$ is called the (instantaneous) *inflation rate* at time t. Interest is modelled by $e^{B(t)}$ as the value at time t of a monetary unit invested at time 0, where $B : \mathbb{R}_+ \to \mathbb{R}_+$ is an increasing function with $B(0) = 0$. If $B(t)$ is absolutely continuous, then $B(t) = \int_0^t \delta(v)\,dv$ for some function $\delta : \mathbb{R}_+ \to \mathbb{R}_+$, and $\delta(t)$ is called the *force of interest* at time t or *spot rate*; see also Section 8.4.3.

As a first illustration of the effects of inflation and interest, we treat a simple but useful example. Suppose that we have a risk reserve process with initial risk reserve $X(0)$ at time $t = 0$ and with income rate $x(t) \in \mathbb{R}$ at time $t \geq 0$ which can be deterministic or random. If we would ignore the economic environment, then the risk reserve $X(t)$ at time t would be given by $X(t) = X(0) + \int_0^t x(v)\,dv$ for $t \geq 0$. However in the model with interest the income in the interval $[v, v + dv)$ yields the risk reserve $x(v)\exp(B(t) - B(v))\,dv$ at time t. In this case the risk reserve $X(t)$ at time t is given by $X(t) = X(0)e^{B(t)} + \int_0^t x(v)e^{B(t)-B(v)}\,dv$. If we want to consider inflation and interest jointly, then we need to introduce the *economic factor* $e(t) = e^{I(t)-B(t)}$.

11.4.2 A Discounted Risk Reserve Process

In prior chapters, we assumed that the effects of interest and inflation were cancelling out. We now introduce a more general class of risk reserve processes where this is no longer valid. Denote by $e^{I(t)}$ the inflated monetary unit at time t, and by $e^{B(t)}$ the value at time t of a unit invested at time 0. Suppose that the claim sizes have to be adjusted to inflation. Then the aggregate claim amount process $\{X(t)\}$ is given by $X(t) = \sum_{i=1}^{N(t)} U_i e^{I(\sigma_i)}$ for $t \geq 0$, where σ_i denotes the arrival epoch of the ith claim, and U_i its size at time 0. Keeping track of inflation, the premium rate also has to increase with inflation, i.e. the premium rate at time t is assumed to be $\beta e^{I(t)}$. This leads to the following risk reserve process $\{R'(t)\}$ with

$$R'(t) = u + \int_0^t \beta e^{I(v)}\,dv - \sum_{i=1}^{N(t)} U_i e^{I(\sigma_i)}.$$

Usually the insurer has to invest the surplus. Then the resulting risk reserve process $\{R''(t)\}$ is given by

$$R''(t) = ue^{B(t)} + \int_0^t \beta e^{I(v)} e^{B(t)-B(v)}\,dv - \sum_{i=1}^{N(t)} U_i e^{I(\sigma_i)} e^{B(t)-B(\sigma_i)}.$$

As the process $\{R''(t)\}$ is a bit clumsy to analyse, we consider its *discounted version* $\{R(t)\}$ given by

$$R(t) = R''(t)\mathrm{e}^{-B(t)} = u + \int_0^t \beta e(v)\,\mathrm{d}v - \sum_{i=1}^{N(t)} U_i e(\sigma_i), \qquad (11.4.1)$$

where $e(v) = \mathrm{e}^{I(v)-B(v)}$. Notice that the event of ruin is the same for both processes $\{R(t)\}$ and $\{R''(t)\}$.

Assume as before that the claim arrival process is a compound Poisson process. All stochastic processes considered in this section are defined on the canonical probability space of the claim arrival process.

Recall that the function $e(t) = \mathrm{e}^{I(t)-B(t)}$ is deterministic. In a nondeterministic economic environment, $\{e(t)\}$ is a stochastic process. We then take $\{e(t)\}$ independent of $\{N(t)\}$ and $\{U_i\}$ and condition on $\{e(t)\}$. As a last constraint we assume that $e(t)$ is continuous.

The process $\{X(t)\}$ with $X(t) = (R(t), t)$ is a PDMP. Here, $I = \{1\}$, $C_1 = \mathbb{R} \times \mathbb{R}$, $\lambda(y,t) = \lambda$ and $Q((y,t), B_1 \times B_2) = \mathbb{I}(t \in B_2) F_U(y - B_1)$, where the external state is omitted. Since the deterministic paths between claim arrival epochs have the form $\varphi(t,(y,h)) = (y + \int_h^{t+h} \beta e(v)\,\mathrm{d}v, t+h)$ we have, for a differentiable function $g(y,h)$,

$$(\boldsymbol{X}g)(\varphi(t,(y,h))) = \frac{\partial g}{\partial h}(\varphi(t,(y,h))) + \beta e(t+h)\frac{\partial g}{\partial y}(\varphi(t,(y,h))). \qquad (11.4.2)$$

Hence we find the following auxiliary martingale.

Lemma 11.4.1 *Let $s \in \mathbb{R}$ be fixed and assume that $\hat{m}_U(se(t)) < \infty$ for all $t \geq 0$. Then the process $\{M(t)\}$ with*

$$M(t) = \exp\Big(-sR(t) - \int_0^t \theta(se(v))\,\mathrm{d}v\Big) \qquad (11.4.3)$$

is a martingale, where the function $\theta(s)$ is given by (11.3.2).

Proof Consider the PDMP $\{(R(t),t)\}$. Then, (11.4.2) implies that the equation $\boldsymbol{A}g = 0$ for a function g fulfilling the conditions of Theorem 11.2.2 is

$$\frac{\partial g}{\partial t}(y,t) + \beta e(t)\frac{\partial g}{\partial y}(y,t) + \lambda\Big(\int_0^\infty g(y - e(t)v, t)\,\mathrm{d}F_U(v) - g(y,t)\Big) = 0\,.$$

Trying a function of the form $g(y,t) = a(t)\mathrm{e}^{-sy}$, where $a(t)$ is positive and differentiable, yields the equation

$$a^{(1)}(t)\mathrm{e}^{-sy} - \beta s e(t) g(y,t) + \lambda\Big(\int_0^\infty \mathrm{e}^{se(t)v} g(y,t)\,\mathrm{d}F_U(v) - g(y,t)\Big) = 0\,,$$

or equivalently $a^{(1)}(t)+\theta(se(t))a(t)=0$. The latter equation has the solution $a(t) = a(0)\exp(-\int_0^t \theta(se(v))\,dv)$, where we can assume that $a(0) = 1$. Condition (a) of Theorem 11.2.2 is fulfilled since $g(y,t) = a(t)e^{-sy}$ is absolutely continuous, (11.2.5) is trivial since $\Gamma = \emptyset$, and (11.2.6) follows as in the case of the martingale given in (11.3.3). By Theorem 11.2.2, the process given by (11.4.3) is a martingale. □

The ruin probability $\psi(u) = \mathbb{P}(\inf_{t\geq 0} R(t) < 0)$ can be estimated using the result of the next theorem.

Theorem 11.4.1 *The following statements are true:*
(a) *If $s \geq 0$ is fixed and $\hat{m}_U(se(t)) < \infty$ for all $t \geq 0$, then*

$$\psi(u) \leq e^{-su} \sup_{t\geq 0} \exp\Big(\int_0^t \theta(se(v))\,dv\Big). \tag{11.4.4}$$

(b) *Let $\gamma = \sup\{s \geq 0 : \sup_{t\geq 0} \int_0^t \theta(se(v))\,dv < \infty\}$. Then for all $\varepsilon > 0$*

$$\lim_{u\to\infty} \psi(u)e^{(\gamma-\varepsilon)u} = 0. \tag{11.4.5}$$

Proof Using Lemma 11.4.1 and Theorem 10.2.4 we get that the stopped process $\{M(\tau(u)\wedge t), t \geq 0\}$ is a martingale where $\tau(u) = \inf\{t \geq 0 : R(t) < 0\}$. Thus, for each $x \geq 0$

$$\begin{aligned}
e^{-su} &= \mathbb{E}\exp\Big(-sR(\tau(u)\wedge x) - \int_0^{\tau(u)\wedge x} \theta(se(v))\,dv\Big)\\
&\geq \mathbb{E}\Big[\exp\Big(-sR(\tau(u)) - \int_0^{\tau(u)} \theta(se(v))\,dv\Big); \tau(u) \leq x\Big]\\
&\geq \mathbb{E}\Big[\exp\Big(-\int_0^{\tau(u)} \theta(se(v))\,dv\Big); \tau(u) \leq x\Big]\\
&\geq \inf_{0\leq t\leq x} \exp\Big(-\int_0^t \theta(se(v))\,dv\Big)\mathbb{P}(\tau(u) \leq x),
\end{aligned}$$

where we used in the second inequality that $R(\tau(u)) < 0$. Let $x \to \infty$ to obtain (a). Further, (b) is trivial if $\gamma = 0$. Now let $0 < \varepsilon < \gamma$ and choose $s = \gamma - \varepsilon/2$. Then $c = \sup_{t\geq 0} \exp\Big(\int_0^t \theta(se(v))\,dv\Big) < \infty$ and $\psi(u) \leq ce^{-su}$. Thus $\lim_{u\to\infty} \psi(u)e^{(\gamma-\varepsilon)u} \leq \lim_{u\to\infty} ce^{-\varepsilon u/2} = 0$. □

11.4.3 The Adjustment Coefficient

As before, the *adjustment coefficient* (or *Lundberg exponent*) for a ruin function $\psi(u)$ is a strictly positive number γ for which (11.4.5) and

$$\lim_{u\to\infty} \psi(u)e^{(\gamma+\varepsilon)u} = \infty \tag{11.4.6}$$

hold for all $\varepsilon > 0$. We leave it to the reader to show that this definition is equivalent to the notion of the *Lyapunov constant* γ defined for a ruin function $\psi(u)$ by

$$\lim_{u\to\infty} \frac{-\log\psi(u)}{u} = \gamma\,. \tag{11.4.7}$$

The constant γ in Theorem 11.4.1b can only be called an adjustment coefficient, if we prove that (11.4.6) is true for all $\varepsilon > 0$. However, this requires additional assumptions on more specific economic factors as have been discussed in Section 11.4.4.

Remarks 1. For the compound Poisson model without economical environment as well as for the Sparre Andersen model, Theorems 5.4.2 and 6.5.7 show that, under appropriate conditions, $\lim_{u\to\infty}\psi(u)\exp(\gamma u) = c$ holds for a certain positive finite constant c. Hence (11.4.5) and (11.4.6) are clearly satisfied. In these two special cases the adjustment coefficient γ turns out to be the solution to equations (5.4.3) and (6.5.21), respectively.

2. The converse statement is, however, false. It does not follow from (11.4.5) and (11.4.6) that the limit $\lim_{u\to\infty}\psi(u)\exp(\gamma u)$ is positive. We illustrate this by considering a compound Poisson model without economical environment, i.e. $e(t) = 1$ for all $t \ge 0$. Consider the function given in (11.3.2) and suppose that for some $\gamma > 0$

$$\theta(s) < 0 \text{ if } s \le \gamma, \qquad \theta(s) = \infty \text{ if } s > \gamma, \tag{11.4.8}$$

and that $\theta^{(1)}(\gamma) > 0$. Then by Theorem 11.3.1 we have $\mathbb{E}^{(\gamma)}R(1) - u < 0$. Under $\mathbb{P}^{(\gamma)}$ the ruin time $\tau(u)$ is therefore finite and $\tau(u) \to \infty$ as $u \to \infty$. Thus by (11.3.10) we have $\lim_{u\to\infty}\psi(u)\exp(\gamma u) = 0$, from which (11.4.5) immediately follows. To check that (11.4.6) is also true, we use the integral equation (5.3.9) to obtain

$$\frac{\beta}{\lambda}\psi(u)\mathrm{e}^{(\gamma+\varepsilon)u} \ge \mathrm{e}^{(\gamma+\varepsilon)u}\int_u^\infty \overline{F}_U(x)\,\mathrm{d}x, \qquad u > 0\,. \tag{11.4.9}$$

However, if $\hat{m}_U(\gamma+\varepsilon) = \infty$ for all $\varepsilon > 0$, then the right-hand side of (11.4.9) tends to infinity as $u \to \infty$; see Lemma 2.3.1. Using similar considerations the reader can show that we can even drop the assumption $\theta^{(1)}(\gamma) > 0$.

3. A similar situation holds if $\theta(\gamma) = 0$ for some $\gamma > 0$, but $\theta^{(1)}(\gamma) = \infty$. In this case we have $\lim_{u\to\infty}\psi(u)\exp(\gamma u) = 0$ by Theorem 5.4.2.

4. There exist models for the claim arrival process where the limit $\psi(u)\mathrm{e}^{\gamma u}$ as $u \to \infty$ does not exist or is infinite even if γ is the adjustment coefficient.

11.4.4 Decreasing Economic Factor

We return to the discounted risk reserve process $\{R(t)\}$ in the compound Poisson model in an economic environment introduced in (11.4.1). We explicitly deal with two special cases where the function $e(t)$ is decreasing and where the constant γ, occurring in Theorem 11.4.1, is the adjustment coefficient. To show the latter, we only need to prove that in both cases (11.4.6) also holds.

First we consider the case of *discounting at constant rate* $\delta > 0$, that is $e(t) = \mathrm{e}^{-\delta t}$. Assume that $s^+ > 0$ and that $\beta > \lambda\mu_U$, where $s^+ = \sup\{s \geq 0 : \hat{m}_U(s) < \infty\}$, i.e. there exists an $s > 0$ such that $\theta(s) < 0$. Furthermore, $\theta(s\mathrm{e}^{-\delta v}) < 0$ for all v large enough. Thus $\sup_{t\geq 0} \int_0^t \theta(s\mathrm{e}^{-\delta v})\,\mathrm{d}v < \infty$ if $s < s^+$ and $\sup_{t\geq 0} \int_0^t \theta(s\mathrm{e}^{-\delta v})\,\mathrm{d}v = \infty$ if $s > s^+$. Hence it follows that $\gamma = s^+$. On the other hand,

$$\begin{aligned}
\psi(u) &= \mathbf{P}\Big(\bigcup_{t\geq 0}\Big\{\sum_{i=1}^{N(t)} U_i \mathrm{e}^{-\delta\sigma_i} > u + \frac{\beta}{\delta}(1-\mathrm{e}^{-\delta t})\Big\}\Big)\\
&\geq \mathbf{P}\Big(\bigcup_{t\geq 0}\Big\{\sum_{i=1}^{N(t)} U_i \mathrm{e}^{-\delta\sigma_i} > u + \frac{\beta}{\delta}\Big\}\Big) = \mathbf{P}\Big(\sum_{i=1}^{\infty} U_i \mathrm{e}^{-\delta\sigma_i} > u + \frac{\beta}{\delta}\Big).
\end{aligned}$$

If $s > s^+$, then it is clear that $\mathbf{E}\exp\left(s\sum_{i=1}^{\infty} U_i\mathrm{e}^{-\delta\sigma_i}\right) = \infty$ and by Lemma 2.3.1 also $\lim_{u\to\infty}\mathbf{P}(\sum_{i=1}^{\infty} U_i\mathrm{e}^{-\delta\sigma_i} > u + \beta/\delta)\mathrm{e}^{su} = \infty$. Thus, we have $\lim_{u\to\infty}\psi(u)\mathrm{e}^{su} = \infty$ for all $s > s^+$. This means that for $e(t) = \mathrm{e}^{-\delta t}$ the constant γ occurring in Theorem 11.4.1 is the adjustment coefficient.

We next consider a more general economic factor. Suppose that $e(t)$ is decreasing and denote by $e(\infty)$ the limit of $e(t)$ as $t \to \infty$. As before assume that $s^+ > 0$ and $\theta(s) < 0$ for some $s > 0$. Then $\gamma = \sup\{s \in [0, s^+] : \theta(se(\infty)) < 0\}$. In particular, if $e(\infty) = 0$ then $\gamma = s^+$ and, as in the special case that $e(t) = \mathrm{e}^{-\delta t}$ for some $\delta > 0$ which we discussed before, we get that $\lim_{u\to\infty}\psi(u)\mathrm{e}^{su} = \infty$ for $s > s^+$.

Assume now that $\gamma < s^+$ and let $\gamma < s < s^+$. Then, in the same way as in the proof of Theorem 11.4.1, we get

$$\begin{aligned}
\mathrm{e}^{-su} &= \mathbf{E}\exp\Big(-sR(\tau(u)\wedge x) - \int_0^{\tau(u)\wedge x}\theta(se(v))\,\mathrm{d}v\Big)\\
&= \mathbf{E}\Big[\exp\Big(-sR(\tau(u)) - \int_0^{\tau(u)}\theta(se(v))\,\mathrm{d}v\Big); \tau(u) \leq x\Big]\\
&\quad + \mathbf{E}\Big[\exp\Big(-sR(x) - \int_0^{x}\theta(se(v))\,\mathrm{d}v\Big); \tau(u) > x\Big].
\end{aligned}$$

The random variable in the second expectation is bounded by 1. Moreover, this random variable tends to 0 with probability 1 as $x \to \infty$ since, for the terms in the exponent, we have $\int_0^x \theta(se(v))\,\mathrm{d}v \to \infty$ and $R(x) \to \infty$. By the dominated convergence theorem we get

$$\mathrm{e}^{-su} = \mathbb{E}\left[\exp\Big(-sR(\tau(u)) - \int_0^{\tau(u)} \theta(se(s))\,\mathrm{d}s\Big); \tau(u) < \infty\right],$$

and therefore

$$\psi(u) = \frac{\mathrm{e}^{-su}}{\mathbb{E}\left(\exp(-sR(\tau(u)) - \int_0^{\tau(u)} \theta(se(v))\,\mathrm{d}v) \mid \tau(u) < \infty\right)}. \tag{11.4.10}$$

Note that, using $e(t) \le e(0) = 1$, we have for all $u, y \ge 0$

$$\begin{aligned}
\mathbb{E}\left(\mathrm{e}^{-sR(\tau(u))} \mid \tau(u), R(\tau(u)-0) = y\right) &= \frac{\int_{y/e(\tau(u))}^{\infty} \mathrm{e}^{s(ve(\tau(u))-y)}\,\mathrm{d}F_U(v)}{\overline{F}_U(y/e(\tau(u)))} \\
&\le \frac{\int_{y/e(\tau(u))}^{\infty} \mathrm{e}^{s(v-y/e(\tau(u)))}\,\mathrm{d}F_U(v)}{\overline{F}_U(y/e(\tau(u)))} \le \sup_{x\ge 0} \frac{\int_x^{\infty} \mathrm{e}^{sv}\,\mathrm{d}F_U(v)}{\mathrm{e}^{sx}\overline{F}_U(x)},
\end{aligned}$$

and therefore

$$\begin{aligned}
\psi(u)\mathrm{e}^{su} &\ge \inf_{x\ge 0} \frac{\mathrm{e}^{sx}\overline{F}_U(x)}{\int_x^{\infty} \mathrm{e}^{sv}\,\mathrm{d}F_U(v)} \frac{1}{\mathbb{E}\left(\exp(-\int_0^{\tau(u)} \theta(se(v))\,\mathrm{d}v) \mid \tau(u) < \infty\right)} \\
&\ge \inf_{x\ge 0} \frac{\mathrm{e}^{sx}\overline{F}_U(x)}{\int_x^{\infty} \mathrm{e}^{sv}\,\mathrm{d}F_U(v)} \frac{1}{\mathbb{E}\left(\exp(-\theta(se(\infty))\tau(u)) \mid \tau(u) < \infty\right)},
\end{aligned}$$

where the last inequality follows from the fact that $\theta(se(\infty)) \ge 0$ for $s > \gamma$, $e(v) \ge e(\infty)$ for all $v \ge 0$, and consequently $0 \le \theta(se(\infty)) \le \theta(se(v))$. Note that the infimum in this bound for $\psi(u)\mathrm{e}^{su}$ is positive. Furthermore, $\lim_{u\to\infty} \mathbb{E}\,(\tau(u) \mid \tau(u) < \infty) = \infty$, as can be shown as an exercise. It therefore follows that $\lim_{u\to\infty} \psi(u)\mathrm{e}^{su} = \infty$ and γ is the adjustment coefficient.

Bibliographical Notes. An early paper on the compound Poisson model within an economic environment is Gerber (1971). The martingale given by (11.4.3) has been introduced by Delbaen and Haezendonck (1987). In Schmidli (1994) the concept of PDMP has been applied to this model. Bounds and asymptotic approximations to ruin probabilities in the case of a constant interest force are obtained in Boogaert and Crijns (1987), Boogaert, Haezendonck and Delbaen (1988), Gerber (1971), Harrison (1977a), Klüppelberg and Stadtmüller (1998), Segerdahl (1942, 1954), Sundt and Teugels (1995, 1997) and Vittal and Vasudevan (1987). In the literature, further classes of modified risk reserve processes with a compound Poisson claim arrival process have been studied. For the case where the premium rate is a function of the current reserve, see, for example, Asmussen and Bladt (1996), Davidson (1969), Harrison (1977a), Petersen (1990) and Taylor (1980). It is clear that also in this case the risk reserve process can be described as a PDMP. For risk processes where the limit $\lim_{u\to\infty}\psi(u)e^{\gamma u}$ does not exist or is infinite even if γ is the adjustment coefficient, see, for example, Embrechts and Schmidli (1994). Asmussen and Nielsen (1995) considered the notion of a local adjustment coefficient for ruin probabilities in the compound Poisson model with state-dependent premiums. The probability of ruin under different types of dividend barriers is studied in Boogaert, Delbaen and Haezendonck (1988), Dickson (1991), Dickson and Gray (1984), di Lorenzo and Sibillo (1994), Gerber (1979, 1981) and Vittal and Vasudevan (1987). The equivalence between characteristics of storage and risk processes with state-dependent release and premium rates, respectively, has been discussed in Asmussen and Petersen (1988) and Harrison and Resnick (1978), for example. For relationships between reliability and risk models, see Aven and Jensen (1998).

11.5 EXPONENTIAL MARTINGALES: THE SPARRE ANDERSEN MODEL

In this section we show how to change non-Markov risk processes into Markov processes by adding supplementary components to the process. We start with the previously discussed Sparre Andersen model. The Markovization of stochastic processes in other risk models will be studied in Chapter 12. In the case of the Sparre Andersen model we consider two types of supplementary components and obtain the corresponding martingales. The first approach is natural, but needs an additional assumption on the distribution of inter-occurrence times. The second type of supplementary components easily leads to simple martingales. This preferred approach is considered in Section 11.5.3.

Assume that the claim counting process $\{N(t)\}$ is a renewal process where the inter-occurrence times are denoted by T_n, and the claim arrival epochs by $\sigma_n = \sum_{i=1}^{n} T_i$, where $\sigma_0 = 0$.

The (continuous-time) risk reserve process $\{R(t)\}$ is not a Markov process, unless the inter-occurrence times are exponentially distributed. But the reader can show that considering the process $\{R(t)\}$ only at the claim arrival epochs σ_n yields a (discrete-time) Markov process. Furthermore, the behaviour of $\{R(t)\}$ is piecewise deterministic in the intervals between claim arrival epochs. For the Markovization of $\{R(t)\}$ we therefore have to add information on the neighbouring claim arrival epochs. The two possibilities mentioned above consist of either adding information on the arrival epoch of the last claim or to include information on the arrival epoch of the next claim. We only discuss the martingales resulting from these two ways of Markovization. To prove Lundberg bounds or the Cramér–Lundberg approximation, one can proceed as in Section 11.3.

11.5.1 An Integral Equation

We begin with an auxiliary result on the solution to an integral equation. The latter is similar to the equation (6.5.21) that defines the adjustment coefficient in the Sparre Andersen model.

Lemma 11.5.1 *Let $s \in \mathbb{R}$ such that $\hat{m}_U(s) < \infty$. Then, the following statements are true:*
(a) *There exists at most one solution $\theta = \theta(s)$ to the equation*

$$\hat{m}_U(s)\hat{l}_T(\theta + \beta s) = 1\,. \tag{11.5.1}$$

More specifically, if $s \geq 0$ then there exists a unique solution $\theta(s)$ to (11.5.1).
(b) *Let $s^+ = \sup\{s \geq 0 : \hat{m}_{F_T}(s) < \infty\} > 0$. Then, the function $s \mapsto \theta(s)$ is strictly convex on $[0, s^+)$ provided not both U_n and T_n are deterministic. In any case, $\theta^{(1)}(0) = \lambda\mu_U - \beta$, where $\lambda = \mu_T^{-1}$.*

Proof (a) The function $\hat{l}_T(s)$ is monotone. Thus, (11.5.1) admits at most one solution. Assume $s \geq 0$. Because the function $\hat{l}_T : [0, \infty) \to (0, 1]$ is one-to-one and $\hat{m}_U(s) \geq 1$, there exists a unique solution $z(s)$ to $\hat{l}_T(z) = (\hat{m}_U(s))^{-1}$. Thus $\theta(s) = z(s) - \beta s$ is the unique solution to (11.5.1).
(b) By the implicit function theorem, see Hille (1966), $\theta(s)$ is differentiable on $[0, s^+)$ and has the derivative

$$\theta^{(1)}(s) = -\frac{\hat{m}_U^{(1)}(s)\hat{l}_T(\theta(s) + \beta s)}{\hat{m}_U(s)\hat{l}_T^{(1)}(\theta(s) + \beta s)} - \beta\,. \tag{11.5.2}$$

For $s = 0$, this gives $\theta^{(1)}(0) = \lambda\mu_U - \beta$. Moreover, it follows that $\theta(s)$ is infinitely often differentiable. Let us rewrite (11.5.1) in the form $\log\hat{m}_U(s) + \log\hat{l}_T(\theta(s) + \beta s) = 0$. Differentiating twice, we find

$$\begin{aligned}(\log\hat{m}_U(s))^{(2)} + (\theta^{(1)}(s) + \beta)^2(\log\hat{l}_T(v))^{(2)}\big|_{v=\theta(s)+\beta s}& \\ +\theta^{(2)}(s)(\log\hat{l}_T(v))^{(1)}\big|_{v=\theta(s)+\beta s} &= 0\,.\end{aligned}$$

We first observe that $\theta(s) + \beta s \geq 0$ for $s \in [0, s^+)$ and $(\log\hat{l}_T(v))^{(1)} = \hat{l}_T^{(1)}(v)/\hat{l}_T(v) < 0$ for $v \geq 0$. The second derivative of $\log\hat{m}_U(s)$ can be represented as the variance of an associated distribution, see Lemma 2.3.2, and is therefore nonnegative and strictly positive provided U_n is not deterministic. Note that $(\log\hat{l}_T(v))^{(2)} = (\log\hat{m}_T(-v))^{(2)} \geq 0$. Thus, in the same way as before, we can conclude that $(\log\hat{l}_T(v))^{(2)}\big|_{v=\theta(s)+\beta s} > 0$ if T_n is not deterministic. From (11.5.2) it follows that $\theta^{(1)}(s) + \beta \neq 0$. If we assume that at least one of the random variables U_n and T_n is not deterministic we find that $\theta^{(2)}(s) > 0$. □

11.5.2 Backward Markovization Technique

We now consider the *backward Markovization technique*. Let $T'(t) = t - \sigma_{N(t)}$ be the time elapsed since the last claim arrival. Often $T'(t)$ is called the *age* of the inter-occurrence time at time t. It is not difficult to see that the process $\{X(t)\}$ with $X(t) = (R(t), T'(t), t)$ is a PDMP. The only problem is to find the jump intensity $\lambda(x)$. Observe that $\lambda(x)$ only depends on the component w of $x = (y, w, t)$ and so write $\lambda(w)$ instead. From the construction of the PDMP $\{X(t)\}$, it follows that

$$\overline{F}_T(t) = \exp(-\int_0^t \lambda(w)\,\mathrm{d}w)\,; \tag{11.5.3}$$

see Section 11.2.2. This is only possible if the distribution of the inter-occurrence times T_n is absolutely continuous. We therefore have to assume that F_T is absolutely continuous with density f_T. Differentiating (11.5.3) we obtain

$$\lambda(w) = \lambda(y, w, t) = f_T(w)/\overline{F}_T(w)\,. \tag{11.5.4}$$

To determine the other characteristics of the PDMP $\{X(t)\}$ is left to the reader. We find the following martingale.

Theorem 11.5.1 *Let $s \in \mathbb{R}$ such that $\hat{m}_U(s) < \infty$ and the solution $\theta(s)$ to (11.5.1) exists. Then, the stochastic process $\{M(t), t \geq 0\}$ with*

$$M(t) = \hat{m}_U(s)\frac{\mathrm{e}^{(\theta(s)+\beta s)T'(t)}}{\overline{F}_T(T'(t))}\int_{T'(t)}^{\infty} \mathrm{e}^{-(\theta(s)+\beta s)v} f_T(v)\,\mathrm{d}v\,\mathrm{e}^{-sR(t)}\mathrm{e}^{-\theta(s)t} \tag{11.5.5}$$

is a martingale.

Proof With regard to Theorem 11.2.2 we have to solve the partial differential equation

$$\begin{aligned}\frac{\partial}{\partial t}g(y,w,t) + \beta\frac{\partial}{\partial y}g(y,w,t) + \frac{\partial}{\partial w}g(y,w,t)&\\ +\frac{f_T(w)}{\overline{F}_T(w)}\Big(\int_0^\infty g(y-v,0,t)\,\mathrm{d}F_U(v) - g(y,w,t)\Big) &= 0\,,\end{aligned} \tag{11.5.6}$$

where we used the representation formula (11.5.4) for the jump intensity $\lambda(w)$. The general solution to this equation is hard to find. Motivated by the results obtained in Section 11.3.1 for the compound Poisson model, we try a function of the form $g(y, w, t) = a(w)\exp(-sy - \theta t)$. Then

$$-\theta a(w) - \beta s a(w) + a^{(1)}(w) + (\overline{F}_T(w))^{-1} f_T(w)(a(0)\hat{m}_U(s) - a(w)) = 0\,,$$

which has the solution

$$a(w) = (\overline{F}_T(w))^{-1}\mathrm{e}^{(\theta+\beta s)w}\Big(a(0)\hat{m}_U(s)\int_w^\infty \mathrm{e}^{-(\theta+\beta s)v} f_T(v)\,\mathrm{d}v + c\Big)$$

for some constant $c \in \mathbb{R}$. It is possible (but not trivial) to show that condition (11.2.6) cannot be fulfilled if $c \neq 0$, hence choose $c = 0$. Letting $w = 0$ yields the equation

$$a(0) = a(0)\hat{m}_U(s)\hat{l}_T(\theta + \beta s),$$

which shows that (11.5.1) has to be fulfilled. We can assume $a(0) = 1$. Thus

$$g(y, w, t) = \hat{m}_U(s)\frac{\mathrm{e}^{(\theta(s)+\beta s)w}}{\overline{F}_T(w)}\int_w^\infty \mathrm{e}^{-(\theta(s)+\beta s)v} f_T(v)\,\mathrm{d}v\,\mathrm{e}^{-sy-\theta(s)t}. \quad (11.5.7)$$

The verification that this function g satisfies the conditions of Theorem 11.2.2 is left to the reader. □

11.5.3 Forward Markovization Technique

Next to the approach discussed in Section 11.5.2, we consider the alternative *forward Markovization technique.* Let $T(t) = \sigma_{N(t)+1} - t$ be the time remaining to the next claim arrival and which is also called the *excess* of the inter-occurrence time at time t; see Section 6.1.2. The use of the stochastic process $\{T(t)\}$ seems to be rather strange because $\{T(t)\}$ is not measurable with respect to the natural filtration of $\{R(t)\}$ unless the inter-occurrence times are deterministic. The natural filtration of the PDMP $\{X(t)\}$ with $X(t) = (R(t), T(t), t)$ is therefore different from the natural filtration of $\{R(t)\}$, and the process $\{T(t)\}$ is not observable in reality. We will, however, see that the approach considered in the present section is much simpler than that of Section 11.5.2.

First note that $\lambda(y, w, t) = 0$ because claims can only occur when $T(t)$ reaches the boundary 0. The active boundary consists of all points $\Gamma = \{(y, 0, t) : (y, t) \in \mathbb{R}^2\}$.

According to Theorem 11.2.2, we will arrive at a martingale of the form $\{g(R(t), T(t), t), t \geq 0\}$ if the partial differential equation

$$\frac{\partial}{\partial t}g(y, w, t) + \beta\frac{\partial}{\partial y}g(y, w, t) - \frac{\partial}{\partial w}g(y, w, t) = 0 \quad (11.5.8)$$

is satisfied together with the boundary condition (11.2.5)

$$g(y, 0, t) = \int_0^\infty\int_0^\infty g(y - v, w, t)\,\mathrm{d}F_T(w)\,\mathrm{d}F_U(v). \quad (11.5.9)$$

Note that the differential equation (11.5.8) is much simpler than (11.5.6). But we get the additional integral equation (11.5.9).

Theorem 11.5.2 *Let $s \in \mathbb{R}$ such that $\hat{m}_U(s) < \infty$ and the solution $\theta(s)$ to* (11.5.1) *exist. Then the stochastic process $\{M(t), t \geq 0\}$ with*

$$M(t) = \mathrm{e}^{-(\theta(s)+\beta s)T(t)}\mathrm{e}^{-sR(t)}\mathrm{e}^{-\theta(s)t} \quad (11.5.10)$$

is an $\{\mathcal{F}_t^X\}$*-martingale.*

Proof We solve (11.5.8) and (11.5.9) by a function of the form $g(y,w,t) = b(w)\mathrm{e}^{-sy}\mathrm{e}^{-\theta t}$. Plugging this function $g(y,w,t)$ into (11.5.8) yields $-\theta b(w) - \beta s b(w) - g'^{(1)}(w) = 0$ or equivalently $b(w) = b(0)\mathrm{e}^{-(\theta+\beta s)w}$. Note that $b(0) = 0$ would imply $g(y,w,t) = 0$, and hence we put $b(0) = 1$. Substitution of this solution to (11.5.8) into (11.5.9) shows that (11.5.1) is satisfied. The remaining conditions of Theorem 11.2.2 can easily be verified by the reader. □

Theorem 11.5.2 amply shows the advantages of the forward Markovization technique. First of all, no condition on the distribution of inter-occurrence times is needed. Secondly, it is much easier to arrive at the martingale in (11.5.10) than at that of Section 11.5.2.

Bibliographical Notes. The Sparre Andersen model was first investigated by means of PDMP in Dassios (1987); see also Dassios and Embrechts (1989) and Embrechts, Grandell and Schmidli (1993).

CHAPTER 12

Point Processes

In earlier chapters of this book we introduced four classes of claim arrival processes in continuous time: homogeneous Poisson processes and compound Poisson processes in Section 5.2, renewal point processes in Section 6.1, and mixed Poisson processes in Section 8.5. For each one of these claim arrival processes, at least one of the following stationarity properties holds: the claim counting process $\{N(t), t \geq 0\}$ has stationary increments or the sequence $\{T_n, n \geq 1\}$ of inter-occurrence times is stationary. The point processes considered in Section 12.1 provide a general model for claim arrival processes with such stationarity properties. In Section 12.1.4 we generalize these models further by considering marked point processes, giving us the possibility to include the claim sizes into the model as well. In Section 12.2 we extend the class of homogeneous Poisson processes in a different direction by introducing a notion of nonhomogeneity. The corresponding claim counting process $\{N(t)\}$ has independent but not necessarily stationary increments. Nonhomogeneous Poisson processes are an appropriate tool to define Cox processes, a wider class of claim arrival processes that can be seen as a mixture of nonhomogeneous Poisson processes. For this reason Cox processes are often called doubly stochastic Poisson processes. We also discuss other constructions of new point processes obtained by several kinds of compounding, in particular superposition and clustering. This general point-process approach is later combined with techniques from piecewise deterministic Markov processes and subexponential distributions in order to study ruin probabilities.

In this chapter, a point process is usually understood to be a two-sided sequence $\{\sigma_n, n \in \mathbb{Z}\}$ of random variables, where $\ldots \leq \sigma_{-1} \leq \sigma_0 \leq 0 < \sigma_1 \leq \ldots$. Furthermore, we consider the two-sided infinite sequence $\{T_n, n \in \mathbb{Z}\}$ of inter-occurrence times $T_n = \sigma_n - \sigma_{n-1}$ where we interpret $\{T_n, n \leq 0\}$ and $\{T_n, n > 0\}$ as the sequence of past and future inter-occurrence times, respectively. We also consider the *random counting measure* $\{N(B), B \in \mathcal{B}(\mathbb{R})\}$ with $N(B) = \sum_{i \in \mathbb{Z}} \mathbb{I}(\sigma_i \in B)$. Notice that there is a one-to-one correspondence between $\{\sigma_n\}$ and $\{N(B)\}$. The elements of the sequence $\{\sigma_n\}$ are called claim arrival epochs but we will also speak of claim arrival points, or briefly of points of the point process $\{\sigma_n\}$.

The introduction of point processes and their corresponding random counting measures on the whole real line rather than on the more usual nonnegative halfline, is particularly useful when studying stationarity properties of point processes.

12.1 STATIONARY POINT PROCESSES

In this section we study point processes having general stationarity properties. Hitherto we have usually assumed that the sequence of inter-occurrence times $\{T_n\}$ consists of independent and identically distributed random variables. However, we met two exceptions: the delayed renewal point process considered in Section 6.1.2, where the distribution of T_1 could be different from the distribution of $T_2, T_3, \ldots$, and the mixed Poisson process considered in Section 8.5.5, where the inter-occurrence times $T_1, T_2, \ldots$ were exchangeable. It turns out that these two point-process models can be embedded into a general stationary framework.

12.1.1 Definition and Elementary Properties

We will discuss two kinds of stationarity. In the first case we assume that $\{T_n\}$ is *stationary*, i.e. for all $n \geq 1$ and $k \in \mathbb{Z}$ the (joint) distribution of the random vector $(T_{1+k}, \ldots, T_{n+k})$ does not depend on k. We also assume that $0 < \mathbb{E}T < \infty$. For simplicity we suppose that the T_n are positive random variables, that is $\mathbb{P}(T = 0) = 0$.

If there is a point at 0, then we have $\sigma_n = \sum_{i=1}^{n} T_i$ for $n > 0$, $\sigma_0 = 0$, and $\sigma_n = -\sum_{i=n+1}^{0} T_i$ for $n < 0$. If $\{T_n\}$ is stationary and if there is a claim arrival at the origin, we call $\{\sigma_n, n \in \mathbb{Z}\}$ a *Palm-stationary point process.*

However, from the insurer's point of view, it might sometimes be more convenient to put the origin of the time axis differently, so that this choice is in a sense "independent from data". The origin is then put "at random" between two consecutive claim arrival epochs. This alternative model of stationarity of a claim arrival process is defined with the help of the random counting measure $\{N(B)\}$ corresponding to $\{\sigma_n\}$. Assume that the distribution of $\{N(B)\}$ is invariant under (deterministic) time shifting, that is, for $B+t = \{v+t,\ v \in B\}$,

$$\{N(B+t), B \in \mathcal{B}(\mathbb{R})\} \stackrel{\mathrm{d}}{=} \{N(B), B \in \mathcal{B}(\mathbb{R})\} \tag{12.1.1}$$

for all $t \in \mathbb{R}$. This means in particular that for all $n \geq 1$, $t \in \mathbb{R}$ and bounded $B_1, \ldots, B_n \in \mathcal{B}(\mathbb{R})$, the distribution of the random vector $(N(B_1 + t), \ldots, N(B_n + t))$ is independent of t. Then $\{N(B)\}$ is called a *time-stationary counting measure* and the sequence $\{\sigma_n\}$ corresponding to

$\{N(B)\}$ is called a *time-stationary point process.* In this case, the counting process $\{N(t), t \geq 0\}$ with $N(t) = \sum_{i=1}^{\infty} \mathbb{I}(\sigma_i \leq t)$ has stationary increments. We assume that the expectation $\lambda = \mathbb{E}\, N((0,1])$ is positive and finite and we call λ the *intensity* of $\{\sigma_n\}$. We leave it to the reader to show that, for each time-stationary counting measure $\{N(B)\}$,

$$\mathbb{E}\, N(B) = \lambda |B| \tag{12.1.2}$$

for all bounded $B \in \mathcal{B}(\mathbb{R})$, where $|B|$ denotes the Lebesgue measure of B. In particular, for each $t \in \mathbb{R}$ we have $\mathbb{P}(N(\{t\}) > 0) = 0$.

Examples 1. *Homogeneous Poisson process.* When we introduced the continuous-time risk model in Section 5.1.4 we assumed for convenience that $\sigma_0 = 0$. However, in the definition of the counting process $\{N(t), t \geq 0\}$ given in (5.1.13) we did not take this claim arrival into consideration. In the case of a homogeneous Poisson process, see Section 5.2.1, we assumed that the sequence $\{T_n, n \geq 1\}$ of inter-occurrence times consists of identically (exponentially) distributed random variables. Furthermore, we showed in Theorem 5.2.1 that the corresponding counting process $\{N(t), t \geq 0\}$ has stationary increments. This is the reason that a homogeneous Poisson point process $\{\sigma_n, n \geq 1\}$ on the positive halfline, where $\sigma_n = \sum_{i=1}^{n} T_i$, can easily be extended to a time-stationary model on the whole real line. It suffices to assume that $-\sigma_0, \sigma_0 - \sigma_{-1}, \sigma_{-1} - \sigma_{-2}, \ldots$ is a sequence of independent and identically (exponentially) distributed random variables which is independent of $\{\sigma_n, n \geq 1\}$. Anyhow, at the same time the sequence $\{T_n, n \geq 1\}$ consists of identically distributed random variables and therefore can be seen as the restriction to the positive halfline of the Palm version of the time-stationary Poisson model.

2. *Mixed Poisson process.* By the result of Theorem 8.5.3, the mixed Poisson process introduced in Section 8.5.1 is a Palm model. However, as in the case of a homogeneous Poisson process (see Example 1 above), we obtain a time-stationary model if we cancel the point at the origin. The fact that by simply adding a point at the origin one can pass from a time-stationary model to a Palm model is characteristic for (mixed) Poisson processes. We return to this later in Section 12.2.3; see Theorem 12.2.7.

3. *Renewal process.* Here we have to distinguish more carefully between the time-stationary model and the Palm model, unless we consider the special case of a homogeneous Poisson process. By Theorem 6.1.8 we arrive at a time-stationary model if the distribution of T_1 is chosen according to (6.1.15). On the other hand, T_1 must have the same distribution as $T_2, T_3, \ldots$ to get a Palm model. Consider now a Palm renewal point process on $\mathbb{R}$. In this case, $\{T_n,\ n \in \mathbb{Z}\}$ is a sequence of independent and identically distributed random variables. We denote the common distribution function by $F(x)$. To obtain a time-stationary renewal point process on $\mathbb{R}$ we choose σ_0, σ_1 to have a joint

distribution defined by $\mathbb{P}(-\sigma_0 > x, \sigma_1 > y) = \overline{F^s}(x+y)$ for $x, y \geq 0$. We choose $\sigma_0 - \sigma_{-1}, \sigma_{-1} - \sigma_{-2}, \ldots$ and $\sigma_2 - \sigma_1, \sigma_3 - \sigma_2, \ldots$ to be independent and identically distributed and also independent of (σ_0, σ_1). It is left to the reader to show that the corresponding counting measure $\{N(B), B \in \mathcal{B}(\mathbb{R})\}$ is time-stationary.

12.1.2 Palm Distributions and Campbell's Formula

There is a one-to-one correspondence between time-stationary and Palm models of point processes. To show this, it is convenient to use the representation of point processes as counting measures and to consider them on a canonical probability space.

In this section Ω is the set of all integer-valued measures $\omega : \mathcal{B}(\mathbb{R}) \to \mathbb{N} \cup \{\infty\}$ such that $\omega(B) < \infty$ for all bounded $B \in \mathcal{B}(\mathbb{R})$. Furthermore, let $\mathcal{F}$ be the smallest σ-algebra of subsets of Ω that contains all events of the form $\{\omega : \omega(B) = j\}$, where $B \in \mathcal{B}(\mathbb{R})$, $j \in \mathbb{N}$. The *canonical representation* of a point process is then given by the triple $(\Omega, \mathcal{F}, \mathbb{P})$, where $\mathbb{P}$ is a probability measure on $\mathcal{F}$. Thus, we identify a point process with its distribution $\mathbb{P}$ (on the canonical probability space). In what follows we only consider *simple* point processes, that is $\mathbb{P}(\omega : \omega(\{t\}) > 1 \text{ for some } t \in \mathbb{R}) = 0$. Furthermore, in accordance with definition (12.1.1), we say that $\mathbb{P}$ is *time-stationary* if

$$\mathbb{P}(A) = \mathbb{P}(\boldsymbol{T}_x A) \tag{12.1.3}$$

for all $A \in \mathcal{F}$, $x \in \mathbb{R}$ where the shift operator $\boldsymbol{T}_x : \Omega \to \Omega$ is defined by $(\boldsymbol{T}_x\omega)(B) = \omega(B+x)$. It is left to the reader to show that for each time-stationary distribution $\mathbb{P}$ we have

$$\mathbb{P}(\{\omega : \omega(\mathbb{R}_-) = \omega(\mathbb{R}_+) = \infty\} \cup \{\omega : \omega(\mathbb{R}) = 0\}) = 1\,. \tag{12.1.4}$$

Thus, if

$$\mathbb{P}(\omega : \omega(\mathbb{R}) = 0) = 0\,, \tag{12.1.5}$$

then there are infinitely many points on both halflines.

In the rest of this section we take (12.1.3) and (12.1.5) under the probability measure $\mathbb{P}$ for granted. Moreover, the intensity $\lambda = \int \omega((0,1])\mathbb{P}(\mathrm{d}\omega)$ is taken to be positive and finite. For brevity, we call $\mathbb{P}$ a *stationary distribution.* The following *symmetry property* of stationary distributions is often useful.

Lemma 12.1.1 *Let $g : \Omega \times \mathbb{R}^2 \to \mathbb{R}_+$ be a measurable function. Then*

$$\int_{\mathbb{R}}\int_{\Omega}\int_{\mathbb{R}} g(\boldsymbol{T}_x\omega, x, y)\omega(\mathrm{d}x)\mathbb{P}(\mathrm{d}\omega)\,\mathrm{d}y = \int_{\mathbb{R}}\int_{\Omega}\int_{\mathbb{R}} g(\boldsymbol{T}_x\omega, y, x)\omega(\mathrm{d}x)\mathbb{P}(\mathrm{d}\omega)\,\mathrm{d}y. \tag{12.1.6}$$

Proof From the stationarity of $\mathbb{P}$ we have

$$\begin{aligned}\iiint g(\boldsymbol{T}_x\omega, x, y)\omega(\mathrm{d}x)\mathbb{P}(\mathrm{d}\omega)\,\mathrm{d}y &= \iiint g(\boldsymbol{T}_{x+y}\omega, x, y)(\boldsymbol{T}_y\omega)(\mathrm{d}x)\mathbb{P}(\mathrm{d}\omega)\,\mathrm{d}y \\ &= \iiint g(\boldsymbol{T}_x\omega, x-y, y)\omega(\mathrm{d}x)\mathbb{P}(\mathrm{d}\omega)\,\mathrm{d}y\,.\end{aligned}$$

In the same way we get

$$\iiint g(\boldsymbol{T}_x\omega, y, x)\omega(\mathrm{d}x)\mathbb{P}(\mathrm{d}\omega)\,\mathrm{d}y = \iiint g(\boldsymbol{T}_x\omega, y, x-y)\omega(\mathrm{d}x)\mathbb{P}(\mathrm{d}\omega)\,\mathrm{d}y\,.$$

This gives (12.1.6) since for all $x \in \mathbb{R}$ and $\omega \in \Omega$

$$\int f(\boldsymbol{T}_x\omega, y, x-y)\,\mathrm{d}y = \int f(\boldsymbol{T}_x\omega, x-y, y)\,\mathrm{d}y$$

which follows from the substitution $y' = x - y$. □

We are now ready to introduce the Palm-stationary model corresponding to a time-stationary point process on the canonical probability space. For each $B \in \mathcal{B}(\mathbb{R})$ such that $0 < |B| < \infty$ we define the probability measure $\mathbb{P}^0 : \mathcal{F} \to [0,1]$ by

$$\mathbb{P}^0(A) = \frac{1}{\lambda|B|}\int_\Omega\int_B \mathbb{I}(\boldsymbol{T}_x\omega \in A)\,\omega(\mathrm{d}x)\mathbb{P}(\mathrm{d}\omega)\,, \qquad A \in \mathcal{F}\,. \tag{12.1.7}$$

Furthermore, using Lemma 12.1.1, we can show that the value $\mathbb{P}^0(A)$ does not depend on the choice of the set B. Indeed, for $g(\omega, x, y) = \mathbb{I}(\omega \in A)\mathbb{I}(x \in B)\mathbb{I}(y \in (0,1])$, equation (12.1.6) gives

$$\int_\Omega\int_B \mathbb{I}(\boldsymbol{T}_x\omega \in A)\,\omega(\mathrm{d}x)\mathbb{P}(\mathrm{d}\omega) = |B|\int_\Omega\int_{(0,1]} \mathbb{I}(\boldsymbol{T}_x\omega \in A)\,\omega(\mathrm{d}x)\mathbb{P}(\mathrm{d}\omega)\,.$$

The probability measure $\mathbb{P}^0$ defined in (12.1.7) is called the *Palm distribution* corresponding to the time-stationary distribution $\mathbb{P}$. It is not difficult to see that, under $\mathbb{P}^0$, with probability 1 there is a point at the origin, that is $\mathbb{P}^0(\Omega^0) = 1$, where $\Omega^0 = \{\omega : \omega(\mathbb{R}_-) = \omega(\mathbb{R}_+) = \infty, \omega(\{0\}) > 0\}$. Indeed, this immediately follows from (12.1.7) if we put $A = \Omega^0$ in (12.1.7) and use (12.1.4) and (12.1.5). However, in some cases it is not very convenient to have this point at the origin. Besides $\mathbb{P}^0$ one can still consider another type of a Palm distribution given by the probability measure $\mathbb{P}^! : \mathcal{F} \to [0,1]$ with

$$\mathbb{P}^!(A) = \frac{1}{\lambda|B|}\int_\Omega\int_B \mathbb{I}(\boldsymbol{T}_x\omega - \delta_0 \in A)\,\omega(\mathrm{d}x)\mathbb{P}(\mathrm{d}\omega)\,, \qquad A \in \mathcal{F}\,, \tag{12.1.8}$$

which is called the *reduced Palm distribution.*

We now show that under the Palm distribution $\mathbb{P}^0$, the inter-occurrence times $T_n = \sigma_n - \sigma_{n-1}$ form a stationary sequence. Denote by $\boldsymbol{S} : \Omega^0 \to \Omega^0$ the pointwise shift defined by $\boldsymbol{S}\omega = \boldsymbol{T}_{\sigma_1(\omega)}\omega$; $\mathcal{F}^0 = \mathcal{F} \cap \Omega^0$.

Theorem 12.1.1 *For each $A \in \mathcal{F}^0$,*

$$\mathbb{P}^0(A) = \mathbb{P}^0(\boldsymbol{S}A). \tag{12.1.9}$$

Proof Let $t > 0$ and $B = (0, t]$. Then, (12.1.7) gives

$$|\mathbb{P}^0(A) - \mathbb{P}^0(\boldsymbol{S}A)| \le \frac{1}{\lambda t}\int \Big|\sum_{i=1}^{N(t)} \left(\mathbb{I}\left(\boldsymbol{T}_{\sigma_i(\omega)}\omega \in A\right) - \mathbb{I}\left(\boldsymbol{T}_{\sigma_{i+1}(\omega)}\omega \in A\right)\right)\Big| \,\mathbb{P}(\mathrm{d}\omega) \le \frac{2}{\lambda t}.$$

Thus, (12.1.9) follows since t can be taken arbitrarily large. □

The following relationship between $\mathbb{P}$ and $\mathbb{P}^0$ is called *Campbell's formula.* It is rather useful when determining characteristics of functionals of stationary point processes, in particular when computing the ruin probability $\psi(0)$ in the case of an arbitrary time-stationary claim arrival process; see Section 12.1.5.

Theorem 12.1.2 *For each measurable function $g : \Omega \times \mathbb{R} \to \mathbb{R}_+$,*

$$\int_\Omega \int_{\mathbb{R}} g(\omega, x)\,\mathrm{d}x\,\mathbb{P}^0(\mathrm{d}\omega) = \frac{1}{\lambda}\int_\Omega \int_{\mathbb{R}} g(\boldsymbol{T}_x\omega, x)\,\omega(\mathrm{d}x)\,\mathbb{P}(\mathrm{d}\omega). \tag{12.1.10}$$

Proof We write (12.1.7) in the form

$$\iint \mathbb{I}(\omega \in A, x \in B)\,\mathrm{d}x\,\mathbb{P}^0(\mathrm{d}\omega) = \frac{1}{\lambda}\iint \mathbb{I}(\boldsymbol{T}_x\omega \in A, x \in B)\,\omega(\mathrm{d}x)\,\mathbb{P}(\mathrm{d}\omega).$$

This shows (12.1.10) for functions of the form $g(\omega, x) = \mathbb{I}(\omega \in A, x \in B)$. Thus, (12.1.10) also holds for linear combinations of such functions and, by the monotone class theorem, for each measurable $g : \Omega \times \mathbb{R} \to \mathbb{R}_+$. □

In some cases a *dual version* of Campbell's formula (12.1.10) is even more convenient.

Corollary 12.1.1 *For each measurable function $g : \Omega \times \mathbb{R} \to \mathbb{R}_+$,*

$$\int_\Omega \int_{\mathbb{R}} g(\boldsymbol{T}_{-x}\omega, x)\,\mathrm{d}x\,\mathbb{P}^0(\mathrm{d}\omega) = \frac{1}{\lambda}\int_\Omega \int_{\mathbb{R}} g(\omega, x)\,\omega(\mathrm{d}x)\,\mathbb{P}(\mathrm{d}\omega). \tag{12.1.11}$$

Proof The result follows from (12.1.10) applied to the function $g'(\omega, x) = g(\boldsymbol{T}_{-x}\omega, x)$. □

This in turn leads to the following *inversion formula* which expresses $\mathbb{P}$ in terms of $\mathbb{P}^0$.

Corollary 12.1.2 *Let $g : \Omega \times \mathbb{R} \to \mathbb{R}_+$ be a measurable function with*

$$\sum_n g(\omega, \sigma_n(\omega)) = 1 \tag{12.1.12}$$

if $\omega(\mathbb{R}) > 0$, and $g(\omega, x) \equiv 0$ if $\omega(\mathbb{R}) = 0$. Then, for each $A \in \mathcal{F}$

$$\mathbb{P}(A) = \lambda \int_\Omega \int_{\mathbb{R}} \mathbb{I}(\boldsymbol{T}_{-x}\omega \in A) g(\boldsymbol{T}_{-x}\omega, x) \, \mathrm{d}x \, \mathbb{P}^0(\mathrm{d}\omega) \,. \tag{12.1.13}$$

Proof Suppose g fulfils (12.1.12) and put $g'(\omega, x) = \mathbb{I}(\omega \in A) g(\omega, x)$. Then, (12.1.13) immediately follows from (12.1.11). $\square$

Remarks 1. To give an example fulfilling (12.1.12), consider the function

$$g(\omega, x) = \begin{cases} 1 & \text{if } x = \sigma_1(\omega), \\ 0 & \text{otherwise.} \end{cases}$$

In this case, (12.1.13) takes the form

$$\mathbb{P}(A) = \lambda \int_\Omega \int_0^{-\sigma_{-1}(\omega)} \mathbb{I}(\boldsymbol{T}_{-x}\omega \in A) \, \mathrm{d}x \, \mathbb{P}^0(\mathrm{d}\omega) \,. \tag{12.1.14}$$

Thus, using (12.1.9) we get that

$$\mathbb{P}(A) = \lambda \int_\Omega \int_0^{\sigma_1(\omega)} \mathbb{I}(\boldsymbol{T}_x\omega \in A) \, \mathrm{d}x \, \mathbb{P}^0(\mathrm{d}\omega) \tag{12.1.15}$$

and in particular, for $A = \Omega$,

$$\mathbb{E}^0 \sigma_1 = \lambda^{-1} \tag{12.1.16}$$

where $\mathbb{E}^0$ denotes the expectation under $\mathbb{P}^0$. This shows that, starting from the Palm model $\mathbb{P}^0$ with stationary inter-point distances and with a point at the origin, the time-stationary model $\mathbb{P}$ given by (12.1.15) is obtained by putting the origin "at random" between $\sigma_0 = 0$ and the next claim arrival epoch σ_1.

2. The formulae (12.1.7) and (12.1.15) constitute a one-to-one relationship between time-stationary and Palm distributions of point processes. To see this, start from an $\boldsymbol{S}$-invariant distribution Q on $\mathcal{F}^0$ with $0 < \int \sigma_1(\omega) \, Q(\mathrm{d}\omega) < \infty$ and put

$$\mathbb{P}(A) = \frac{\int_\Omega \int_0^{\sigma_1(\omega)} \mathbb{I}(\boldsymbol{T}_x\omega \in A) \, \mathrm{d}x \, Q(\mathrm{d}\omega)}{\int_\Omega \sigma_1(\omega) \, Q(\mathrm{d}\omega)} \,. \tag{12.1.17}$$

Defining $\mathbb{P}^0$ by (12.1.7), we have

$$\mathbb{P}^0 = Q \,. \tag{12.1.18}$$

Notice that by stating (12.1.18) we implicitly assume that the distribution $\mathbb{P}$ defined in (12.1.17) is $\boldsymbol{T}$-invariant. We leave it to the reader to prove this.

12.1.3 Ergodic Theorems

The right-hand side of (12.1.7) can be interpreted as a ratio of two intensities. The integral in (12.1.7) is the *partial intensity* of those points in $B = (0,1]$ from whose perspective the shifted counting measure $\boldsymbol{T}_x\omega$ has property A. This partial intensity is divided by the total intensity λ of *all* points in the unit interval $(0,1]$. Thus, taking $\mathbb{P}$ as the basic model, $\mathbb{P}^0(A)$ can be seen as the pointwise averaged relative frequency of points with property A. This interpretation of $\mathbb{P}^0$ can be stated more precisely if the stationary distribution $\mathbb{P}$ is ergodic. One possible definition of ergodicity is given by the following *nondecomposability* property: $\mathbb{P}$ is *ergodic* if each representation

$$\mathbb{P} = p\mathbb{P}' + (1-p)\mathbb{P}'', \qquad 0 \le p \le 1 \tag{12.1.19}$$

of $\mathbb{P}$ as a mixture of stationary distributions $\mathbb{P}', \mathbb{P}''$ on $\mathcal{F}$ must be trivial in the sense that either $\mathbb{P}' = \mathbb{P}''$ or $p(1-p) = 0$.

The following basic results of ergodic theory are useful. For their proofs we refer to Breiman (1992), Krengel (1985) and Tempelman (1992). Let $(\Omega, \mathcal{F}, \mathbb{P})$ be an arbitrary probability space and $\boldsymbol{S} : \Omega \to \Omega$ a measure-preserving mapping, i.e. $\mathbb{P}(\boldsymbol{S}^{-1}A) = \mathbb{P}(A)$ for all $A \in \mathcal{F}$, where $\boldsymbol{S}^{-1}A = \{\omega \in \Omega : \boldsymbol{S}\omega \in A\}$.

Theorem 12.1.3 *Let $g : \Omega \to \mathbb{R}_+$ be measurable such that $\int g(\omega)\mathbb{P}(d\omega) < \infty$. Then the limit*

$$\bar{g}(\omega) = \lim_{n\to\infty} \frac{1}{n} \sum_{k=1}^{n} g(\boldsymbol{S}^k\omega) \tag{12.1.20}$$

exists and the limit function $\bar{g} : \Omega \to \mathbb{R}_+$ is invariant with respect to $\boldsymbol{S}$, i.e. $\bar{g}(\boldsymbol{S}\omega) = \bar{g}(\omega)$ for all $\omega \in \Omega$. Moreover, $\int \bar{g}(\omega)\mathbb{P}(d\omega) < \infty$ and

$$\lim_{n\to\infty} \int \Big| \frac{1}{n} \sum_{k=1}^{n} g(\boldsymbol{S}^k\omega) - \bar{g}(\omega) \Big| \mathbb{P}(d\omega) = 0\,. \tag{12.1.21}$$

The first part of Theorem 12.1.3 is called the *individual ergodic theorem*, whereas the second part is called the *statistical ergodic theorem*. We also observe that there are several equivalent definitions of ergodicity, as summarized in the following lemma. In its statement, $\mathcal{I} \subset \mathcal{F}$ denotes the sub-σ-algebra of invariant events, i.e. $\boldsymbol{S}A = A$ if and only if $A \in \mathcal{I}$.

Lemma 12.1.2 *The following statements are equivalent.*
(a) *Each representation of $\mathbb{P}$ as a mixture of $\boldsymbol{S}$-invariant probability measures is trivial, that is,* (12.1.19) *implies that either $\mathbb{P}' = \mathbb{P}''$ or $p(1-p) = 0$.*
(b) $\mathbb{P}(A)(1 - \mathbb{P}(A)) = 0$ *for each $A \in \mathcal{I}$.*
(c) $\mathbb{E}(g \mid \mathcal{I}) = \mathbb{E}\, g$ *for each measurable function $g : \Omega \to \mathbb{R}_+$ with $\mathbb{E}\, g < \infty$.*

An immediate consequence of Theorem 12.1.3 and Lemma 12.1.2 is the following result. If $\mathbb{P}$ is ergodic, then for the limit function $\bar{g}$ in (12.1.20) we have

$$\bar{g}(\omega) = \mathbb{E}\, g\,. \tag{12.1.22}$$

Furthermore, the following continuous-time analogues to Theorem 12.1.3 and Lemma 12.1.2 are useful. Instead of a single measure-preserving mapping we then consider a whole family $\boldsymbol{T} = \{\boldsymbol{T}_x, x \in \mathbb{R}\}$ of such mappings $\boldsymbol{T}_x : \Omega \to \Omega$, where

- $\mathbb{P}(A) = \mathbb{P}(\boldsymbol{T}_x A)$ for all $x \in \mathbb{R}$, $A \in \mathcal{F}$,
- $\boldsymbol{T}_x \boldsymbol{T}_y = \boldsymbol{T}_{x+y}$ for all $x, y \in \mathbb{R}$,
- $\{(\omega, x) : \boldsymbol{T}_x \omega \in A\} \in \mathcal{F} \otimes \mathcal{B}(\mathbb{R})$ for all $A \in \mathcal{F}$.

The quadruple $(\Omega, \mathcal{F}, \mathbb{P}, \boldsymbol{T})$ is called a *dynamical system* in continuous time. Let $\mathcal{I} \subset \mathcal{F}$ denote the sub-σ-algebra of $\boldsymbol{T}$-invariant sets, i.e. $A \in \mathcal{I}$ if $\boldsymbol{T}_x A = A$ for all $x \in \mathbb{R}$.

Theorem 12.1.4 *Let $g : \Omega \to \mathbb{R}_+$ be measurable such that $\int g(\omega)\mathbb{P}(\mathrm{d}\omega) < \infty$. Then the limit*

$$\bar{g}(\omega) = \lim_{t\to\infty} \frac{1}{t} \int_0^t g(\boldsymbol{T}_x \omega)\,\mathrm{d}x \tag{12.1.23}$$

exists and

$$\bar{g} = \mathbb{E}\,(g \mid \mathcal{I})\,. \tag{12.1.24}$$

Moreover, $\int \bar{g}(\omega)\mathbb{P}(\mathrm{d}\omega) < \infty$ and

$$\lim_{t\to\infty} \int \left| \frac{1}{t} \int_0^t g(\boldsymbol{T}_x \omega)\,\mathrm{d}x - \bar{g}(\omega) \right| \mathbb{P}(\mathrm{d}\omega) = 0\,. \tag{12.1.25}$$

Lemma 12.1.3 *Each representation of $\mathbb{P}$ as a mixture of $\boldsymbol{T}$-invariant probability measures is trivial if and only if one of the conditions* (b) *or* (c) *of Lemma* 12.1.2 *holds.*

We now return to the interpretation of the Palm probability $\mathbb{P}^0(A)$ as the relative frequency of points with property A.

Theorem 12.1.5 *Let $\mathbb{P}$ be a time-stationary distribution on the canonical point-process space. Assume that $\mathbb{P}$ is ergodic. Then*

$$\lim_{t\to\infty} t^{-1}\omega((0,t]) = \lambda \tag{12.1.26}$$

and for each $A \in \mathcal{F}$

$$\lim_{t\to\infty} \frac{1}{\omega((0,t])} \int_{(0,t]} \mathbb{I}(\boldsymbol{T}_x \omega \in A)\,\omega(\mathrm{d}x) = \mathbb{P}^0(A)\,. \tag{12.1.27}$$

Proof Using the inequalities

$$\int_0^{t-1} (\boldsymbol{T}_x\omega)((0,1])\,\mathrm{d}x \le \omega((0,t]) \le \int_{-1}^{t} (\boldsymbol{T}_x\omega)((0,1])\,\mathrm{d}x\,,$$

the existence of the limit in (12.1.26) is obtained from (12.1.23) for $g(\omega) = \omega((0,1])$. Taking into account that by Lemma 12.1.3 this limit is constant, (12.1.24) implies (12.1.26). Using a similar argument, the limit in (12.1.27) follows from Theorem 12.1.4 and Lemma 12.1.3. □

Theorem 12.1.5 provides the motivation to say that, in the ergodic case, $\mathbb{P}^0(A)$ is the probability of the event A seen from the *typical point* of a time-stationary point process with distribution $\mathbb{P}$. The next result shows that, in order to have ergodicity of a point process, it is immaterial whether we work with the time-stationary model or with the Palm model.

Theorem 12.1.6 *The time-stationary distribution* $\mathbb{P}$ *is* $\boldsymbol{T}$*-ergodic if and only if the corresponding Palm distribution* $\mathbb{P}^0$ *is* $\boldsymbol{S}$*-ergodic.*

Proof Assume first that $\mathbb{P}$ is not ergodic. Thus there is a nontrivial representation of $\mathbb{P}$ as a mixture $\mathbb{P} = p\mathbb{P}' + (1-p)\mathbb{P}''$ of two stationary distributions $\mathbb{P}'$, $\mathbb{P}''$ with intensities λ', λ'', respectively. We leave it to the reader to show that then

$$\mathbb{P}^0 = \frac{p\lambda'}{p\lambda' + (1-p)\lambda''}(\mathbb{P}')^0 + \frac{(1-p)\lambda''}{p\lambda' + (1-p)\lambda''}(\mathbb{P}'')^0\,.$$

This means that $\mathbb{P}^0$ can also be represented as a nontrivial mixture (of Palm distributions). Thus, $\mathbb{P}^0$ cannot be ergodic. Conversely, if $\mathbb{P}^0$ is not ergodic, then $\mathbb{P}^0$ can be represented as a nontrivial mixture of the form

$$\mathbb{P}^0 = pQ' + (1-p)Q''\,, \qquad 0 < p < 1 \tag{12.1.28}$$

for two $\boldsymbol{S}$-invariant probability measures Q' and Q'' such that $0 < \int \sigma_1\,\mathrm{d}Q' < \infty$, $0 < \int \sigma_1\,\mathrm{d}Q'' < \infty$. The construction given in (12.1.17) leads to a $\boldsymbol{T}$-invariant probability measure. Hence, if we apply the transformation (12.1.15) to both sides of (12.1.28) we see that $\mathbb{P}$ can be represented as a nontrivial mixture of $\boldsymbol{T}$-invariant measures. □

Corollary 12.1.3 *A Palm renewal point process is* $\boldsymbol{S}$*-ergodic. Moreover, a time-stationary renewal point process is* $\boldsymbol{T}$*-ergodic.*

Proof The 0–1 law of Kolmogorov implies that a Palm renewal point process fulfils condition (b) of Lemma 12.1.2. Hence, such a process is $\boldsymbol{S}$-ergodic. Ergodicity of the corresponding time-stationary renewal point process now follows from Theorem 12.1.6. □

Notice that a mixed Poisson process is not ergodic unless the mixing distribution is concentrated at a single point. This immediately follows from Lemma 12.1.3 and from the defining equation (8.5.1) of mixed Poisson processes. Further examples of ergodic point processes will be discussed in Section 12.2.

We conclude this section with a property on events, invariant under time shifts $\boldsymbol{T}_x$.

Theorem 12.1.7 *Let $\mathbb{P}$ be a $\boldsymbol{T}$-ergodic distribution on the canonical point-process space and let $A \in \mathcal{I}$ be a $\boldsymbol{T}$-invariant set. Then, $\mathbb{P}(A) = 1$ implies $\mathbb{P}^0(A) = 1$.*

Proof Since A is $\boldsymbol{T}$-invariant, the assertion follows from (12.1.7) and

$$\int_B \mathbb{I}(\boldsymbol{T}_x\omega \in A)\,\omega(\mathrm{d}x) = \omega(B)\mathbb{I}(\omega \in A)\,. \qquad \square$$

Since the event $\{\omega : \lim_{t\to\infty} t^{-1}\omega((0,t]) \text{ exists}\}$ is $\boldsymbol{T}$-invariant, Theorem 12.1.7 entails that, in the ergodic case, the law of large numbers (12.1.26) also holds under $\mathbb{P}^0$. By Corollary 12.1.3, this can be seen as a generalization of the law of large numbers which had been derived in Theorem 6.1.1a for (nondelayed) renewal point processes. Other interesting results, whose proofs are based on Theorem 6.1.1a, remain valid in the general ergodic framework. As a specific example we mention the law of small numbers stated in Theorem 6.1.3.

12.1.4 Marked Point Processes

Marked point processes are useful when we want the model to include other information about the claims like their size or type. We first generalize the canonical probability space introduced in Section 12.1.2. Let K be a complete separable metric space, for example $K = \mathbb{R}^d$, and let $\mathcal{K}$ be the σ-algebra of Borel sets in K. Let Ω_K denote the set of all integer-valued measures $\omega : \mathcal{B}(\mathbb{R})\otimes\mathcal{K} \to \mathbb{N}\cup\{\infty\}$ such that $\omega(B\times K) < \infty$ for all bounded $B \in \mathcal{B}(\mathbb{R})$. Furthermore, let $\mathcal{F}_K$ be the smallest σ-algebra of subsets of Ω_K containing all events of the form $\{\omega : \omega(B \times C) = j\}$, where $B \in \mathcal{B}(\mathbb{R})$, $C \in \mathcal{K}$, $j \in \mathbb{N}$. Note that there is a one-to-one correspondence between the counting measures $\omega \in \Omega_K$ and the set of sequences $\{(\sigma_n(\omega), X_n(\omega)), n \in \mathbb{Z}\}$, where the mark X_n of σ_n is a random variable with values in K.

The *canonical representation* of a marked point process is given by the triple $(\Omega_K, \mathcal{F}_K, \mathbb{P})$, where $\mathbb{P}$ is some probability measure on $\mathcal{F}_K$. As in Section 12.1.2, we only consider *simple* marked point processes, for which $\mathbb{P}(\omega : \omega(\{t\} \times K) > 1 \text{ for some } t \in \mathbb{R}) = 0$. We call $\mathbb{P}$ *time-stationary* if (12.1.3) holds for all $A \in \mathcal{F}_K$, $x \in \mathbb{R}$, where now the shift operator $\boldsymbol{T}_x : \Omega_K \to \Omega_K$ is defined by $(\boldsymbol{T}_x\omega)(B \times C) = \omega((B + x) \times C)$, i.e. the marks are not changed under the time shift $\boldsymbol{T}_x$.

In what follows we assume that $\mathbb{P}$ is a time-stationary distribution on $\mathcal{F}_K$. All notions and results stated in Sections 12.1.2 and 12.1.3 can be transferred into the framework of marked point processes. In particular, for each $C \in \mathcal{K}$ we consider the intensity $\lambda(C) = \int \omega((0,1] \times C)\mathbb{P}(\mathrm{d}\omega)$ of points with a mark from C. If $\lambda(C) > 0$, then we assume $\mathbb{P}(\omega : \omega(\mathbb{R} \times C) = 0) = 0$. In the same way as in (12.1.7), for each $C \in \mathcal{K}$ with $\lambda(C) > 0$, we introduce the (conditional) *Palm distribution* $\mathbb{P}_C$ by

$$\mathbb{P}_C(A) = \frac{1}{\lambda(C)} \int_{\Omega_K} \int_{(0,1]} \mathbb{I}(\boldsymbol{T}_x\omega \in A)\, \omega(\mathrm{d}x \times C)\mathbb{P}(\mathrm{d}\omega)\,, \quad A \in \mathcal{F}_K\,. \tag{12.1.29}$$

Then, $\mathbb{P}_C(\Omega_K^C) = 1$, where

$$\Omega_C^K = \{\omega : \omega(\mathbb{R}_- \times C) = \omega(\mathbb{R}_+ \times C) = \infty, \omega(\{0\} \times C) > 0\}$$

and $\mathbb{P}_C(A) = \mathbb{P}_C(\boldsymbol{S}_C A)$, $A \in \mathcal{F}_K^C$, where $\mathcal{F}_K^C = \mathcal{F}_K \cap \Omega_K^C$, $\boldsymbol{S}_C : \Omega_K^C \to \Omega_K^C$ with $\boldsymbol{S}_C\omega = \boldsymbol{T}_{\sigma_C(\omega)}\omega$ and $\sigma_C(\omega) = \min\{t > 0 : \omega(\{t\} \times C) > 0\}$. Furthermore, for each $C \in \mathcal{K}$ with $\lambda(C) > 0$, we introduce the (conditional) *Palm mark distribution* D_C by

$$D_C(C') = (\lambda(C))^{-1}\lambda(C')\,, \qquad C' \in \mathcal{K}_C \tag{12.1.30}$$

where $\mathcal{K}_C = \mathcal{K} \cap C$. If $C = K$, then we use the notation $D^0 = D_K$ and $\mathbb{P}^0 = \mathbb{P}_K$. It is clear that D_C can be used to establish a relationship between Palm distributions taken with respect to different mark sets.

Theorem 12.1.8 *Let $C, C' \in \mathcal{K}$ such that $\lambda(C), \lambda(C') > 0$ and $C' \subset C$. Then,*

$$\mathbb{P}_{C'}(A) = (D_C(C'))^{-1}\mathbb{P}_C(A \cap \{X_0 \in C'\})\,, \qquad A \in \mathcal{F}_K\,, \tag{12.1.31}$$

and, in particular,

$$D_C(C') = \mathbb{P}_C(X_0 \in C')\,. \tag{12.1.32}$$

Proof From (12.1.29) and (12.1.30) we get

$$\begin{aligned}
\mathbb{P}_{C'}(A) &= \frac{1}{\lambda(C')} \iint \mathbb{I}(\boldsymbol{T}_x\omega \in A)\, \omega(\mathrm{d}x \times C')\mathbb{P}(\mathrm{d}\omega) \\
&= \frac{\lambda(C)}{\lambda(C')} \frac{1}{\lambda(C)} \iint \mathbb{I}(\boldsymbol{T}_x\omega \in A, X_0(\boldsymbol{T}_x\omega) \in C')\, \omega(\mathrm{d}x \times C)\mathbb{P}(\mathrm{d}\omega) \\
&= \frac{\mathbb{P}_C(A \cap \{X_0 \in C'\})}{D_C(C')}\,.
\end{aligned}$$

□

Analogously to (12.1.10) and (12.1.11), for each $C \in \mathcal{K}$ such that $\lambda(C) > 0$ and for each measurable function $g : \Omega_K \times \mathbb{R} \to \mathbb{R}_+$, we have the *Campbell formulae*

$$\int_{\Omega_K} \int_{\mathbb{R}} g(\omega, x)\, \mathrm{d}x\, \mathbf{P}_C(\mathrm{d}\omega) = \frac{1}{\lambda(C)} \int_{\Omega_K} \int_{\mathbb{R}} g(\boldsymbol{T}_x\omega, x)\, \omega(\mathrm{d}x \times C)\, \mathbf{P}(\mathrm{d}\omega)\,, \tag{12.1.33}$$

and

$$\int_{\Omega_K} \int_{\mathbb{R}} g(\boldsymbol{T}_{-x}\omega, x)\, \mathrm{d}x\, \mathbf{P}_C(\mathrm{d}\omega) = \frac{1}{\lambda(C)} \int_{\Omega_K} \int_{\mathbb{R}} g(\omega, x)\, \omega(\mathrm{d}x \times C)\, \mathbf{P}(\mathrm{d}\omega)\,. \tag{12.1.34}$$

Moreover, analogously to (12.1.14) and (12.1.15), we get the *inversion formulae*

$$\mathbf{P}(A) = \lambda(C) \int_{\Omega_K} \int_0^{-\sigma'_C(\omega)} \mathbb{I}(\boldsymbol{T}_{-x}\omega \in A)\, \mathrm{d}x\, \mathbf{P}_C(\mathrm{d}\omega), \tag{12.1.35}$$

where $\sigma'_C(\omega) = \max\{t < 0 : \omega(\{t\} \times C) > 0\}$, and

$$\mathbf{P}(A) = \lambda(C) \int_{\Omega_K} \int_0^{\sigma_C(\omega)} \mathbb{I}(\boldsymbol{T}_x\omega \in A)\, \mathrm{d}x\, \mathbf{P}_C(\mathrm{d}\omega) \tag{12.1.36}$$

and in particular

$$\int_{\Omega_K} \sigma_C(\omega) \mathbf{P}_C(\mathrm{d}\omega) = (\lambda(C))^{-1}\,. \tag{12.1.37}$$

The notion of an ergodic marked point process is introduced in the same way as in Section 12.1.3 for (nonmarked) point processes.

Examples 1. The *compound Poisson process* $\{(\sigma_n, U_n), n \geq 1\}$, introduced in Section 5.2.2, can be seen as the restriction to the nonnegative halfline of a time-stationary marked point process $\{(\sigma_n, X_n)\}$ with $X_n = U_n$ and mark space $K = \mathbb{R}_+$. It is a special case of an *independently marked point process* $\{(\sigma_n, X_n)\}$, where one assumes that the sequences $\{\sigma_n\}$ and $\{X_n\}$ are independent and that $\{X_n\}$ consists of independent random variables with a common distribution D. But the (nonmarked) point process $\{\sigma_n\}$ itself can be arbitrary. If the point process $\{\sigma_n\}$ is also stationary, then the Palm mark distribution D_K defined in (12.1.30) coincides with D. The proof of this is left to the reader.

2. The *mixed Poisson process*, introduced in Section 8.5, can also be seen as a marked point process. Each point σ_n can be "marked" by a nonnegative random variable X_n, say, indicating to which mixing component the point σ_n belongs. In this case, $\{(\sigma_n, X_n)\}$ is *not* independently marked unless the distribution of the X_n is concentrated at a single point.

12.1.5 Ruin Probabilities in the Time-Stationary Model

In this section we consider the claim surplus process $\{S(t), t \geq 0\}$ introduced in Section 5.1.4. We suppose that $\{S(t)\}$ is given on the canonical probability space of the (extended) claim arrival process $\{(\sigma_n, X_n), n \in \mathbb{Z}\}$. Here $X_n = (U_n, V_n)$, where $U_n \geq 0$ denotes the size of the claim arriving at time σ_n and $V_n \geq 0$ is the type of this claim. The mark space is then $K = (\mathbb{R}_+)^2$. We assume $\{(\sigma_n, X_n)\}$ to be a $\boldsymbol{T}$-ergodic time-stationary marked point process. We also assume that the *net profit condition*

$$\beta > \lambda\mu^0 \tag{12.1.38}$$

is fulfilled, where $\mu^0 = \mathbb{E}^0 U$ denotes the expected claim size under the Palm mark distribution $D^0 = D_{(\mathbb{R}_+)^2}$ given in (12.1.32), i.e. $\mu^0 = \int x \mathbb{P}^0(U_0 \in \mathrm{d}x)$. By Theorem 12.1.4 and Lemma 12.1.3, we then have $\lim_{t\to\infty} S(t) = -\infty$ and consequently $\lim_{u\to\infty} \psi(u) = 0$ for the ruin function

$$\psi(u) = \mathbb{P}(\tau(u) < \infty) = \mathbb{P}\Big(\sup_{t\geq 0}\Big\{\sum_{n=1}^{N(t)} U_n - \beta t\Big\} > u\Big), \tag{12.1.39}$$

where $\tau(u)$ is the ruin time. Take the initial reserve u to be 0 and consider the ruin probability

$$\varphi(x, y, C) = \mathbb{P}(\tau < \infty, X^+ > x, Y^+ > y, V^+ \in C). \tag{12.1.40}$$

Here $\tau = \tau(0)$, $X^+ = X^+(0)$ is the surplus prior to τ, $Y^+ = Y^+(0)$ is the severity of ruin, $V^+ = V_{\nu^+}$ is the type of the claim that triggers ruin and $\nu^+ = \min\{n > 0 : \sum_{i=1}^n U_i - \beta\sigma_n > 0\}$. In the next theorem we state a surprisingly simple formula for the ruin probability $\varphi(x, y, C)$. It shows that $\varphi(x, y, C)$ does not depend on the distribution of $\{\sigma_n\}$ provided that λ is fixed and that $\{(\sigma_n, X_n)\}$ is independently marked. This is in agreement with the results given in (5.3.18) and in Theorems 6.4.4 and 6.5.15, where we obtained the same type of formulae for ladder height distributions in the compound Poisson model.

Theorem 12.1.9 *For all $x, y \geq 0$ and $C \in \mathcal{B}(\mathbb{R}_+)$,*

$$\varphi(x, y, C) = \lambda\beta^{-1} \int_{x+y}^{\infty} \mathbb{P}^0(U_0 \geq v, V_0 \in C)\,\mathrm{d}v. \tag{12.1.41}$$

Proof First note that the probability $\varphi(x, y, C)$ and the integral on the right-hand side of (12.1.41) do not change if we rescale the time axis by the factor β considering the new claim arrival process $\{(\beta\sigma_n, X_n)\}$ with intensity $\lambda\beta^{-1}$ instead of $\{(\sigma_n, X_n)\}$. Thus, without loss of generality we can assume $\beta = 1$. Introduce the notations

$$g(\omega, v) = \mathbb{I}(X^+ > x, Y^+ > y, V^+ \in C)\mathbb{I}(\tau = v),$$

and $\omega_v = \boldsymbol{T}_{-v}\omega$. We then have

$$\begin{aligned}
\mathbb{P}(\tau < \infty, X^+ > x, Y^+ > y, V^+ \in C) &= \int_{\Omega_K} \sum_{i=-\infty}^{\infty} g(\omega, \sigma_i(\omega))\mathbb{P}(\mathrm{d}\omega) \\
&= \int_{\Omega_K}\int_{\mathbb{R}} g(\omega, v)\, \omega(\mathrm{d}v \times K)\mathbb{P}(\mathrm{d}\omega) = \lambda \int_{\Omega_K}\int_{\mathbb{R}} g(\omega_v, v)\, \mathrm{d}v\, \mathbb{P}^0(\mathrm{d}\omega) \\
&= \lambda \int_0^\infty \mathbb{P}^0(\omega : X^+(\omega_v) > x, Y^+(\omega_v) > y, V^+(\omega_v) \in C, \tau(\omega_v) = v)\, \mathrm{d}v\,,
\end{aligned}$$

where we used Campbell's formula (12.1.34) in the last but one equality. Trying to evaluate the integrand in the last expression, it is convenient to introduce an auxiliary stochastic process $\{S^*(t), t \geq 0\}$ on the Palm probability space $(\Omega_K, \mathcal{F}_K, \mathbb{P}^0)$. Assume that $\{S^*(t)\}$ makes an upward jump of size U_{-n} at time $-\sigma_{-n}$ and moves down linearly at unit rate between the jumps. Assume further that $S^*(0) = U_0$. This gives for $B_1, B_2, C \in \mathcal{B}(\mathbb{R}_+)$,

$$\begin{aligned}
&\mathbb{P}^0(\omega : X^+(\omega_v) \in B_1, Y^+(\omega_v) \in B_2, V^+(\omega_v) \in C, \tau(\omega_v) = v) \\
&= \mathbb{P}^0(U_0 - S^*(v) \in B_1, S^*(v) \in B_2, V_0 \in C, S^*(v) \leq S^*(v-t)\ \forall t \in (0,v)) \\
&= \mathbb{P}^0(\{U_0 - S^*(v) \in B_1, S^*(v) \in B_2, V_0 \in C\} \cap A_v)\,,
\end{aligned}$$

where $A_v = \{S^*(v) \leq S^*(t)\ \forall t \in (0,v)\}$ is the event that $\{S^*(t)\}$ has a relative minimum at v. Consider the random measure $\{M^*(B), B \in \mathcal{B}(\mathbb{R})\}$ given by

$$M^*(B) = \int_0^\infty \mathbb{I}(S^*(v) \in B)\mathbb{I}(A_v)\, \mathrm{d}v\,.$$

Since $S^*(0) = U_0$, the support of $\{M^*(B)\}$ has right endpoint U_0. Moreover $\mathbb{P}^0$ is $\boldsymbol{S}$-ergodic by Theorem 12.1.6, and $\mathbb{E}^0 U_n < \mathbb{E}^0(\sigma_{n+1} - \sigma_n)$ by (12.1.37) and (12.1.38). Henceforth, we get from Theorem 12.1.4 and Lemma 12.1.3 that $\lim_{v\to\infty} S^*(v) = -\infty$. Thus, the left endpoint of the support of $\{M^*(B)\}$ is $-\infty$ and consequently $\{M^*(B)\}$ is the Lebesgue measure on $(-\infty, U_0]$. Putting the above together, we have

$$\begin{aligned}
\varphi(x,y,C) &= \lambda \int_0^\infty \mathbb{P}^0(\{U_0 - S^*(v) > x, S^*(v) > y, V_0 \in C\} \cap A_v)\, \mathrm{d}v \\
&= \lambda \mathbb{E}^0\Big(\mathbb{I}(V_0 \in C) \int_0^\infty \mathbb{I}(\{y < S^*(v) < U_0 - x\} \cap A_v)\, \mathrm{d}v\Big) \\
&= \lambda \mathbb{E}^0\Big(\mathbb{I}(V_0 \in C) \int_0^\infty \mathbb{I}(y < z < U_0 - x)\, \mathrm{d}z\Big) \\
&= \lambda \int_y^\infty \mathbb{P}^0(V_0 \in C, U_0 > x + z)\, \mathrm{d}z = \lambda \int_{x+y}^\infty \mathbb{P}^0(V_0 \in C, U_0 > z)\, \mathrm{d}z\,. \qquad \square
\end{aligned}$$

Corollary 12.1.4 *Let $\rho = \lambda\beta^{-1}\mu^0$ and let $U^+ = X^+ + Y^+$ denote the size of the claim that triggers ruin. Then the following statements hold:*
(a) *The time-stationary ruin probability $\psi(0)$ for the initial reserve $u = 0$ is given by $\psi(0) = \rho$.*
(b) *The conditional distribution of (U^+, V^+) given $\tau < \infty$ is obtained from the Palm mark distribution of (U_0, V_0) by change of measure with likelihood ratio $U_0/\mathbb{E}^0 U_0$. That is, for each measurable function $g : \mathbb{R}_+ \times \mathbb{R}_+ \to \mathbb{R}_+$,*

$$\mathbb{E}\left[g(U^+, V^+); \tau < \infty\right] = \rho \mathbb{E}^0\Big(\frac{U_0}{\mathbb{E}^0 U_0} g(U_0, V_0)\Big). \tag{12.1.42}$$

(c) *The conditional distribution of (X^+, Y^+) given $U^+, V^+, \tau < \infty$ is that of $(U^+Z, U^+(1-Z))$, where Z is uniformly distributed on $(0,1)$ and independent of $U^+, V^+, \tau < \infty$.*

Proof (a) Putting $x = y = 0$ and $C = \mathbb{R}_+$ in (12.1.41), the assertion is immediately obtained. (b) Using integration by parts on the right-hand side of (12.1.41), formula (12.1.42) also follows easily from (12.1.41). (c) Suppose for a moment that statement (c) is already shown. Then

$$\begin{aligned}
&\mathbb{P}(X^+ > x, Y^+ > y \mid U^+ = u, V^+ \in C, \tau < \infty)\\
&= \mathbb{P}(U^+Z > x, U^+(1-Z) > y \mid U^+ = u, V^+ \in C, \tau < \infty)\\
&= \Big(1 - \frac{x+y}{u}\Big)\mathbb{I}(u > x + y).
\end{aligned}$$

Using (12.1.42), this gives

$$\begin{aligned}
&\mathbb{E}\Big[1 - \frac{x+y}{U^+}; U^+ > x + y, V^+ \in C, \tau < \infty\Big]\\
&= \lambda\beta^{-1}\mathbb{E}^0\big[U_0 - x - y; U_0 > x + y, V_0 \in C\big]\\
&= \lambda\beta^{-1}\int_{x+y}^{\infty} \mathbb{P}^0(U_0 > s, V_0 \in C)\,\mathrm{d}s.
\end{aligned}$$

In view of (12.1.41), this proves statement (c) because of the uniqueness of Radon–Nikodym derivatives. □

Remarks 1. The statements of Theorem 12.1.9 and Corollary 12.1.4 remain valid if the strict inequality in the net profit condition (12.1.38) is weakened to $\beta \geq \lambda\mu^0$. This follows from the fact that both sides of (12.1.41) are continuous functions of β in the interval $[\lambda\mu^0, \infty)$. We leave the verification to the reader.

2. A closer analysis of the proofs of Theorem 12.1.9 and Corollary 12.1.4 show that their statements can be proved in the nonergodic case as well. It suffices to assume that for almost all ergodic components of the time-stationary claim arrival process $\{(\sigma_n, X_n)\}$, the net profit condition (12.1.38) (or its slightly

weaker version mentioned in the remark above) is fulfilled and that $\{(\sigma_n, X_n)\}$ is independently marked. In particular, Theorem 12.1.9 and Corollary 12.1.4 remain valid if $\{(\sigma_n, X_n)\}$ is an independently marked mixed Poisson process for which $\mathbb{P}(\Lambda \le \beta\mu_U^{-1}) = 1$, and where Λ is the mixing random variable considered in Definition 8.5.1.

We conclude this section with a remarkable relationship between the ruin function $\psi(u)$ for the time-stationary model and the ruin function

$$\psi^0(u) = \mathbb{P}^0\Big(\sup_{t\ge 0}\Big\{\sum_{i=1}^{N(t)} U_i - \beta t\Big\} > u\Big) \qquad (12.1.43)$$

for the corresponding Palm-stationary model. Assume that there is only one type of claim (the V_n are therefore omitted) and that the claim arrival process $\{(\sigma_n, U_n)\}$ is independently marked. Put F_U for the claim size distribution and $\rho = \lambda\beta^{-1}\mu_{F_U}$.

Theorem 12.1.10 *For each $u \ge 0$,*

$$\psi(u) = \frac{\lambda}{\beta}\Big(\int_u^\infty \overline{F}_U(v)\,\mathrm{d}v + \int_0^u \psi^0(u-v)\overline{F}_U(v)\,\mathrm{d}v\Big)\,. \qquad (12.1.44)$$

The *proof* of Theorem 12.1.10 can be found, for example, in Section 9.4 of König and Schmidt (1992) where the relationship (12.1.44) is considered in the context of queueing theory and derived from a general *intensity conservation principle*. However, if we additionally assume that

$$\mathbb{E}^0\Big(\sup_{t\ge 0}\Big\{\sum_{i=1}^{N(t)} U_i - \beta t\Big\}\Big) < \infty\,, \qquad (12.1.45)$$

then (12.1.44) can be obtained by an application of the inversion formula (12.1.35) to the claim arrival process $\{(\sigma_n, U_n)\}$. As in the proof of Theorem 12.1.9, we rescale the time axis by the factor β, which is then taken equal to 1. Now use (12.1.35) with $C = \mathbb{R}_+$ and $A = \{\omega : \sup_{t\ge 0}\{\sum_{i=1}^{N(t)} U_i - t\} > u\}$, to get

$$\begin{aligned}
\psi(u) &= \lambda\mathbb{E}^0\Big(\int_0^{-\sigma_{-1}} \mathbb{I}\Big(\sup_{t\ge 0}\Big\{U_0 - x + \sum_{i=1}^{N(t)} U_i - t\Big\} > u\Big)\,\mathrm{d}x\Big)\\
&= \lambda\mathbb{E}^0\Big(\int_0^{-\sigma_{-1}} \mathbb{I}\Big(\sup_{t\ge 0}\Big\{\sum_{i=1}^{N(t)} U_i - t\Big\} + U_0 - u > x\Big)\,\mathrm{d}x\Big)\\
&= \lambda\mathbb{E}^0\Big(\Big(\sup_{t\ge 0}\Big\{\sum_{i=1}^{N(t)} U_i - t\Big\} + U_0 - u\Big)_+
\end{aligned}$$

$$
\begin{aligned}
& \qquad -\Big(\sup_{t\geq 0}\Big\{\sum_{i=1}^{N(t)} U_i - t\Big\} + U_0 + \sigma_{-1} - u\Big)_+\Big) \\
&= \lambda\Big(\mathbf{E}^0\Big(\sup_{t>0}\Big\{\sum_{i=1}^{N(t)} U_i - t\Big\} + U_0 - u\Big)_+ \\
& \qquad -\mathbf{E}^0\Big(\sup_{t\geq 0}\Big\{\sum_{i=1}^{N(t)} U_i - t\Big\} + U_0 + \sigma_{-1} - u\Big)_+\Big) \\
&= \lambda\Big(\mathbf{E}^0\Big(\sup_{t\geq 0}\Big\{\sum_{i=1}^{N(t)} U_i - t\Big\} + U_0 - u\Big)_+ \\
& \qquad -\mathbf{E}^0\Big(\sup_{t\geq 0}\Big\{\sum_{i=1}^{N(t)} U_i - t\Big\} - u\Big)_+\Big).
\end{aligned}
$$

Notice that we used (12.1.45) in the last but one equality. The last equality results from the fact that the Palm distribution $\mathbb{P}^0$ is $\boldsymbol{S}$-invariant. Thus, using the notation $M = \sup_{t\geq 0}\{\sum_{i=1}^{N(t)} U_i - t\}$, we have

$$
\begin{aligned}
\psi(u) &= \lambda\mathbf{E}^0((M + U_0 - u)_+ - (M - u)_+) \\
&= \lambda(\mathbf{E}^0 U_0 - \mathbf{E}^0 \min\{U_0, (u - M)_+\}) \\
&= \lambda\Big(\int_0^\infty \overline{F}_U(v)\,dv - \int_0^\infty \overline{F}_U(v)\mathbb{P}^0((u - M)_+ > v)\,dv\Big) \\
&= \lambda\Big(\int_0^\infty \overline{F}_U(v)\,dv - \int_0^u \overline{F}_U(v)(1 - \psi^0(u - v))\,dv\Big),
\end{aligned}
$$

where in the third equality we used that $\{(\sigma_n, U_n)\}$ is independently marked and consequently the random variables U_0 and $\sup_{t\geq 0}\{\sum_{i=1}^{N(t)} U_i - t\}$ are independent under $\mathbb{P}^0$. □

A special case of interest is the ruin function in the Sparre Andersen model. Note, however, that the ruin function $\psi(u)$ from Section 6.5 is now denoted by $\psi^0(u)$. If the underlying claim arrival process is a stationary renewal process, then we call the model a *stationary Sparre Andersen model.* We leave it to the reader to show that, for the latter, a Lundberg inequality and a Cramér–Lundberg approximation can be derived for the ruin function $\psi(u)$. They are in agreement with the results obtained in Section 6.5.

Bibliographical Notes. The introduction to point processes given in Sections 12.1.1 to 12.1.4 and in particular their representation on a canonical probability space is in the spirit of König and Schmidt (1992). Other books dealing with the general theory of point processes on the real line are, for example, Baccelli and Brémaud (1994), Daley and Vere-Jones (1988), Franken, König, Arndt and Schmidt (1982), Last and Brandt (1995) and Sigman (1995). Theorem 12.1.9 and Corollary 12.1.4 have been obtained in Asmussen and Schmidt (1995). For further related results of this type, see also Asmussen and Schmidt (1993) and Miyazawa and Schmidt (1993, 1997). From the mathematical point of view, the ruin function $\psi(u)$ in the time-stationary model is equivalent to the tail function of the stationary virtual waiting time in a G/GI/1 queue, whereas the ruin function $\psi^0(u)$ defined in (12.1.43) is equivalent to the tail function of the stationary actual waiting time in such a queue. In queueing theory relationships of the form (12.1.44) are called *Takács' formulae*; see, for example, Section 3.4.3 in Baccelli and Brémaud (1994), Section 4.5 in Franken, König, Arndt and Schmidt (1982), and Section 9.4 in König and Schmidt (1992). For the stationary Sparre Andersen model, see also Grandell (1991b), Thorin (1975) and Wikstad (1983). Mixing conditions on the point process $\{\sigma_n\}$, such that (12.1.45) is fulfilled, can be found in Daley and Rolski (1992); see also Daley, Foley and Rolski (1994).

12.2 MIXTURES AND COMPOUNDS OF POINT PROCESSES

In this section we show how the classes of mixed Poisson processes, compound Poisson processes, and renewal processes can be extended to more general point processes with a similar structure. We first introduce the notion of a nonhomogeneous Poisson process. Then we consider a general class of mixtures of nonhomogeneous Poisson processes, called Cox processes. Particular emphasis is put on two important special cases: Markov modulated Poisson processes and Björk–Grandell processes. Besides mixtures of point processes, we also discuss other methods to construct new point processes. They consist of several kinds of compounding, in particular superposition and clustering of point processes. Since in the definition of mixtures and compounds of point processes several (independent) stochastic processes occur, it is not always convenient to use the canonical point-process space as an underlying probability space.

12.2.1 Nonhomogeneous Poisson Processes

We first extend the concept of a homogeneous Poisson process to allow time-dependent arrival rates. For example, there are situations where claim occurrence epochs are likely to depend on the time of the year.

Let $\lambda(t)$ be a nonnegative, measurable and locally integrable (deterministic) function. While there are several equivalent definitions of a nonhomogeneous Poisson process, our approach via the counting measure $\{N(B)\}$ has the advantage that it can be used to introduce nonhomogeneous Poisson processes also on more general state spaces than $\mathbb{R}$. Recalling the counting measure $\{N(B)\}$, the increment $N((a,b])$ where $a < b$ is the number of points in $(a,b]$. We will say that a counting measure $\{N(B)\}$ or the corresponding point process $\{\sigma_n\}$ is a *nonhomogeneous Poisson process* with *intensity function* $\lambda(t)$ if $\{N(B)\}$ has independent increments on disjoint intervals and for all $a < b$ the random variable $N((a,b])$ is Poisson distributed with parameter $\int_a^b \lambda(x)\,\mathrm{d}x$. Then,

$$\mathbb{E}\, N((a,b]) = \int_a^b \lambda(v)\,\mathrm{d}v\,, \tag{12.2.1}$$

which means that $\lambda(t)$ plays the role of an arrival rate function. In the same vain, $\eta(t) = \int_0^t \lambda(v)\,\mathrm{d}v$ is called the *cumulative intensity function* $(t \geq 0)$ while the measure η with $\eta(B) = \int_B \lambda(v)\,\mathrm{d}v$ is called the *intensity measure* of $\{\sigma_n\}$. By $\mathbb{P}_\eta$ we denote the distribution of a nonhomogeneous Poisson process with intensity measure η.

The conditional uniformity property of homogeneous Poisson processes considered in Theorem 5.2.1 can be generalized in the following way.

Theorem 12.2.1 *A counting measure $\{N(B)\}$ is a nonhomogeneous Poisson process with intensity function $\lambda(t)$ if and only if for all $a < b$, $n = 1, 2, \ldots$ the random variable $N((a,b])$ has distribution* $\mathrm{Poi}(\int_a^b \lambda(v)\,\mathrm{d}v)$ *and, given $\{N((a,b]) = n\}$, the random vector $(\sigma_{(1)}, \ldots, \sigma_{(n)})$ of the n (ordered) locations of these points has the same distribution as the order statistics of n independent $[a,b]$-valued random variables, each with the common density function $f(v) = \lambda(v)/\int_a^b \lambda(w)\,\mathrm{d}w$.*

Proof The sufficiency part is omitted since we only have to show that $\{N(B)\}$ has independent increments. But this is fully analogous to step (b) $\Rightarrow$ (c) in the proof of Theorem 5.2.1. Assume now that $\{N(B)\}$ is Poisson with intensity function $\lambda(t)$. Then, the increment $N((t',t])$ has distribution $\mathrm{Poi}(\int_{t'}^t \lambda(v)\,\mathrm{d}v)$ for all $t' < t$. Thus, for $a = t_0 \leq t'_1 < t_1 \leq t'_2 < t_2 \leq \ldots \leq t'_n < t_n = b$,

$$\begin{aligned}
&\mathbb{P}_\eta\Big(\bigcap_{k=1}^n \{\sigma_{(k)} \in (t'_k, t_k]\} \mid N((a,b]) = n\Big) \\
&\quad = \frac{\mathbb{P}_\eta(\bigcap_{k=1}^n \{N((t'_k, t_k]) = 1\} \cap \{N((t_{k-1}, t'_k]) = 0\})}{\mathbb{P}(N((a,b]) = n)} \\
&\quad = n! \prod_{k=1}^n \Big(\int_{t'_k}^{t_k} \lambda(v)\,\mathrm{d}v \Big/ \int_a^b \lambda(v)\,\mathrm{d}v\Big)\,.
\end{aligned}$$

□

We still give another important property of nonhomogeneous Poisson processes.

Theorem 12.2.2 *Suppose that $\{N_1(B)\}$ and $\{N_2(B)\}$ are two independent nonhomogeneous Poisson processes with intensity functions $\lambda_1(t)$ and $\lambda_2(t)$, respectively. The superposition $\{N(B)\}$, where $N(B) = N_1(B) + N_2(B)$ is a nonhomogeneous Poisson process with intensity function $\lambda(t) = \lambda_1(t) + \lambda_2(t)$.*

The *proof* of Theorem 12.2.2 is left to the reader.

12.2.2 Cox Processes

If one investigates real data on the number of claims in a certain time interval, it turns out that the Poisson assumption is not always realistic. It is then often possible to fit a negative binomial distribution to the data. We have already noticed that a negative binomial distribution can be obtained by mixing the Poisson distribution with a gamma distribution, i.e. by letting the Poisson parameter be gamma distributed. As a more general variant, we can take the parameter λ of the homogeneous Poisson process to be stochastic. Such an extension has already been considered in Section 8.5, where it was called a mixed Poisson process. What is really needed, however, is more variability in the claim arrival process. In a mixed Poisson process, this variability will diminish as time progresses. In order not to lose this variability, a basic idea is to let the "expected" number of claims $\Lambda((a,b])$ in the time interval $(a,b]$ be generated by a random measure $\{\Lambda(B), B \in \mathcal{B}(\mathbb{R})\}$. Here $\Lambda(B) = \int_B \lambda(v)\,dv$ for some nonnegative stochastic process $\{\lambda(t), t \in \mathbb{R}\}$, whose sample paths are measurable and locally integrable. We call $\{\lambda(t)\}$ an *intensity process*, and $\{\Lambda(B)\}$ a *cumulative intensity measure*. Given $\{\Lambda(B)\}$, the number of claims $N((a,b])$ in the interval $(a,b]$ is assumed to be Poisson distributed with parameter $\Lambda((a,b])$. We turn to a formal description.

A counting measure $\{N(B)\}$ or the corresponding point process $\{\sigma_n\}$ is called a *Cox process* or a *doubly stochastic Poisson process* if there exists an intensity process $\{\lambda(t)\}$ such that for all $n = 1, 2, \ldots$, for $k_1, \ldots, k_n \in \mathbb{N}$, and $a_1 < b_1 \le a_2 < b_2 \le \ldots \le a_n < b_n$

$$\mathbb{P}\Big(\bigcap_{i=1}^n \{N((a_i,b_i]) = k_i\}\Big) = \mathbb{E}\Big(\prod_{i=1}^n \frac{(\int_{a_i}^{b_i} \lambda(v)\,dv)^{k_i}}{k_i!} \exp\big(-\int_{a_i}^{b_i} \lambda(v)\,dv\big)\Big). \tag{12.2.2}$$

The two-stage stochastic mechanism can be seen as follows. Consider the canonical representation $(\Omega, \mathcal{F}, \mathbb{P})$ of a point process as introduced in Section 12.1.2, where Ω is the set of all locally finite, integer-valued measures on $\mathcal{B}(\mathbb{R})$. Furthermore, let $\overline{\Omega}$ be the set of all (not necessarily integer-valued) measures $\eta : \mathcal{B}(\mathbb{R}) \to \mathbb{R}_+ \cup \{\infty\}$ such that $\eta(B) < \infty$ for all bounded $B \in \mathcal{B}(\mathbb{R})$. As in Section 12.1.2, let $\overline{\mathcal{F}}$ denote the smallest σ-algebra of subsets of $\overline{\Omega}$ containing all events of the form $\{\eta : a < \eta(B) \le b\}$, where $B \in \mathcal{B}(\mathbb{R})$ and $0 \le a < b$.

Consider the *random intensity measure* $\{\Lambda(B), B \in \mathcal{B}(\mathbb{R})\}$ given by $\Lambda(B) = \int_B \lambda(v)\,\mathrm{d}v$. As in the case of point processes, it is convenient to work on a canonical space. The *canonical representation* of the random measure $\{\Lambda(B)\}$ is then given by the triple $(\overline{\Omega}, \overline{\mathcal{F}}, Q)$, where Q is a probability measure on $\overline{\mathcal{F}}$. We can therefore identify a random measure with its distribution Q (on the canonical probability space). Using the canonical representations $(\Omega, \mathcal{F}, \mathbb{P})$ and $(\overline{\Omega}, \overline{\mathcal{F}}, Q)$, it is possible to give a definition of a Cox process as a mixture of nonhomogeneous Poisson processes. Namely, $(\Omega, \mathcal{F}, \mathbb{P})$ is said to be a *Cox process* if there is a random intensity measure with distribution Q such that

$$\mathbb{P}(A) = \int_{\overline{\Omega}} \mathbb{P}_\eta(A) Q(\mathrm{d}\eta)\,, \qquad A \in \mathcal{F}\,. \tag{12.2.3}$$

However, a formal introduction of Cox processes along these lines, requires some discussion on measurability, like for example whether the mapping $\eta \mapsto \mathbb{P}_\eta$ is measurable. In this connection, it can be useful to consider the product probability space $(\Omega \times \overline{\Omega}, \mathcal{F} \otimes \overline{\mathcal{F}}, \mathbb{P})$ with

$$\mathbb{P}(A \times \overline{A}) = \int_{\overline{A}} \mathbb{P}_\eta(A) Q(\mathrm{d}\eta), \qquad A \in \mathcal{F}\,, \overline{A} \in \overline{\mathcal{F}}\,. \tag{12.2.4}$$

Using (12.2.2), we get an alternative two-stage stochastic mechanism for Cox processes which is similar to that given in (12.2.3) and (12.2.4).

Theorem 12.2.3 *Let $\{N'(t), t \ge 0\}$ be a homogeneous Poisson process on $\mathbb{R}_+$ with intensity* 1 *and let $\{\lambda(t), t \ge 0\}$ be an intensity process. If $\{N'(t)\}$ and $\{\lambda(t)\}$ are independent, then the counting measure $\{N(B), B \in \mathcal{B}(\mathbb{R}_+)\}$ given by $N((0,t]) = N'(\int_0^t \lambda(v)\,\mathrm{d}v)$ is a Cox process with intensity process $\{\lambda(t)\}$.*

Proof We show that (12.2.2) holds. Indeed,

$$\begin{aligned}
&\mathbb{P}\Big(\bigcap_{i=1}^n \{N((a_i, b_i]) = k_i\}\Big) \\
&= \mathbb{E}\Big(\mathbb{P}\Big(\bigcap_{i=1}^n \Big\{N'\Big(\int_0^{b_i} \lambda(v)\,\mathrm{d}v\Big) - N'\Big(\int_0^{a_i} \lambda(v)\,\mathrm{d}v\Big) = k_i\Big\} \,\Big|\, \{\lambda(t)\}\Big)\Big) \\
&= \mathbb{E}\Big(\prod_{i=0}^n \mathbb{P}\Big(N'\Big(\int_0^{b_i} \lambda(v)\,\mathrm{d}v\Big) - N'\Big(\int_0^{a_i} \lambda(v)\,\mathrm{d}v\Big) = k_i \,\Big|\, \{\lambda(t)\}\Big)\Big) \\
&= \mathbb{E}\Big(\prod_{i=0}^n \mathbb{P}\Big(N'\Big(\int_{a_i}^{b_i} \lambda(v)\,\mathrm{d}v\Big) = k_i \,\Big|\, \{\lambda(t)\}\Big)\Big)\,,
\end{aligned}$$

which is equal to the right-hand side of (12.2.2). $\square$

The notion of stationarity for general (locally finite) random measures can be introduced as in the case of a time-stationary point process given in (12.1.3). We say that the random measure $\{\Lambda(B)\}$ and equivalently its canonical representation $(\overline{\Omega}, \overline{\mathcal{F}}, Q)$ is *stationary* if $Q(A) = Q(\boldsymbol{T}_x A)$ for all $A \in \overline{\mathcal{F}}$, $x \in \mathbb{R}$, where the shift operator $\boldsymbol{T}_x : \overline{\Omega} \to \overline{\Omega}$ is defined by $(\boldsymbol{T}_x \eta)(B) = \eta(B + x)$. We leave it to the reader to show that a Cox process is time-stationary, i.e. the corresponding counting process has stationary increments if and only if its random intensity measure is stationary. Furthermore, the random intensity measure $\{\Lambda(B)\}$ given by $\Lambda(B) = \int_B \lambda(v)\, \mathrm{d}v$ is stationary if and only if the intensity process $\{\lambda(t)\}$ is stationary.

Let $\{\Lambda(B)\}$ be stationary with distribution Q such that $\lambda = \int \eta((0,1])Q(\mathrm{d}\eta)$ is positive and finite. Then, for each $B \in \mathcal{B}(\mathbb{R})$ such that $0 < |B| < \infty$ we define the mapping $Q^0 : \overline{\mathcal{F}} \to [0,1]$ by

$$Q^0(A) = \frac{1}{\lambda |B|} \int_{\overline{\Omega}} \int_B \mathbb{I}(\boldsymbol{T}_x \eta \in A) \eta(\mathrm{d}x) Q(\mathrm{d}\eta), \qquad A \in \overline{\mathcal{F}}. \qquad (12.2.5)$$

In the same way as was done in Section 12.1.2 for stationary point processes, it can be shown that Q^0 is a probability measure independent of the choice of B. The probability measure Q^0 is called the *Palm distribution* corresponding to the stationary distribution Q. It can be used to describe the reduced Palm distribution $\mathbb{P}^!$ of a time-stationary Cox process. In particular, the following result shows that $\mathbb{P}^!$ again is the distribution of a Cox process.

Theorem 12.2.4 *Let* $\mathbb{P}$ *be given by* (12.2.3) *for some stationary distribution* Q *such that* $\lambda = \int \eta((0,1])Q(\mathrm{d}\eta)$ *is positive and finite. Then,*

$$\mathbb{P}^!(A) = \int_{\overline{\Omega}} \mathbb{P}_\eta(A) Q^0(\mathrm{d}\eta), \qquad A \in \mathcal{F}. \qquad (12.2.6)$$

The *proof* is omitted. It can be found, for example, in Section 5.3 of König and Schmidt (1992).

For point processes on $\mathbb{R}_+$, time stationarity and Palm distributions remain meaningful by appropriate restriction to $\mathbb{R}_+$ of corresponding objects on the whole real line, as shown in some of the examples below.

Examples 1. A special case of a Cox process is a mixed Poisson process where $\lambda(s) \equiv \Lambda$ for some nonnegative random variable Λ. From the definition (12.2.2) and (8.5.1), we immediately get that a mixed Poisson process is a time-stationary point process. Theorem 12.2.4 implies that the reduced Palm distribution of a mixed Poisson process is given by

$$\mathbb{P}^!(A) = \frac{1}{\mathbb{E}\,\Lambda} \int_0^\infty x \mathbb{P}_x(A)\, \mathrm{d}F_\Lambda(x), \qquad A \in \mathcal{F}, \qquad (12.2.7)$$

where $\mathbb{P}_x$ denotes the distribution of a homogeneous Poisson process with intensity x, and F_Λ is the distribution of Λ. In particular, (12.2.7) shows that $\mathbb{P}^!$ is again the distribution of a mixed Poisson process.

2. Let $\lambda_0 : \mathbb{R}_+ \to \mathbb{R}_+$ be a periodic (and deterministic) function with period equal to 1, say. Let $\{N'(t), t \ge 0\}$ be a homogeneous Poisson process with intensity 1. The counting process $\{N(t), t \ge 0\}$ given by $N(t) = N'(\int_0^t \lambda_0(v)\, dv)$ is sometimes called a *periodic Poisson process*; see also Section 12.4. This process does not have stationary increments. However it is possible to define a corresponding counting process with stationary increments in the class of Cox processes. Let X be uniformly distributed on $[0,1]$ and independent of $\{N'(t)\}$. Furthermore, let $\lambda(t) = \lambda_0(t + X)$ and $\Lambda(t) = \int_0^t \lambda(s)\, ds$. Then, the Cox process $\{N^*(t), t \ge 0\}$ with $N^*(t) = N'(\Lambda(t))$ has stationary increments. We leave it to the reader to prove this as an exercise.

3. Let $\{J(t)\}$ be a Markov process with state space $E = \{1, \ldots, \ell\}$ and intensity matrix $\boldsymbol{Q} = (q_{ij})_{i,j \in E}$. The process $\{J(t)\}$ models the random environment of an insurance business. If at time t the environment is $J(t) = i$, then claims are supposed to arrive according to a homogeneous Poisson process with intensity $\lambda_i \ge 0$. By a *Markov-modulated Poisson process* we mean a Cox process whose intensity process $\{\lambda(t)\}$ is given by $\lambda(t) = \lambda_{J(t)}$. We leave it to the reader to show that a Markov-modulated Poisson process has stationary increments if the environment process $\{J(t)\}$ has a stationary initial distribution. Furthermore, it follows from (12.2.5) and (12.2.6) that the reduced Palm distribution of a time-stationary Markov-modulated Poisson process is again the distribution of a Markov-modulated Poisson process. Indeed, if $\{J(t)\}$ has stationary initial distribution $\boldsymbol{\pi} = \{\pi_1, \ldots, \pi_\ell\}$, then (12.2.5) and (12.2.6) imply that

$$\mathbb{P}^! = \lambda^{-1} \sum_{i=1}^{\ell} \pi_i \lambda_i \mathbb{P}_i \,, \qquad \lambda = \sum_{i=1}^{\ell} \pi_i \lambda_i \,, \tag{12.2.8}$$

where $\mathbb{P}_i$ denotes the distribution of a Markov-modulated Poisson process governed by the same intensities $\lambda_1, \ldots, \lambda_\ell$ but by the Markov process $\{J_i(t)\}$ with intensity matrix $\boldsymbol{Q}$ and initial state $J_i(0) = i$. Ruin probabilities in risk models where the claim arrival process is a Markov-modulated Poisson process will be studied in Sections 12.2.4, 12.3 and 12.6.4.

4. We now consider the Markov-modulated process as in Example 3 above with marks added. To define the process we have to specify the number of states ℓ, the intensity matrix $\boldsymbol{Q}$, the intensities $\lambda_1, \ldots, \lambda_\ell$ and the distributions $F_1, \ldots, F_\ell$. For our purpose, the F_i are distributions on $\mathbb{R}_+$. If $J(t) = i$, then claims are arriving according to a Poisson process with intensity λ_i and the claim sizes are distributed according to F_i, independent of everything else. In this way we define a marked point process $\{(\sigma_n, X_n)\}$ called a *marked Markov-modulated Poisson process*, with $X_n = (U_n, V_n)$, where U_n is the claim related to the nth arrival σ_n and $V_n = J(\sigma_n)$. We leave it to the reader to show that $\mathbb{P}^0(U_0 \in B) = \sum_{i=1}^{\ell} \lambda^{-1} \lambda_i \pi_i F_i(B)$.

5. Let $\{(\Lambda_i, I_i), i \geq 1\}$ be a sequence of independent random vectors with $\mathbb{P}(\Lambda_i \geq 0, I_i > 0) = 1$ for all $i \geq 1$. Assume that the random vectors $(\Lambda_2, I_2), (\Lambda_3, I_3), \ldots$ are identically distributed. A *Björk–Grandell process* is a Cox process on $\mathbb{R}_+$ whose intensity process $\{\lambda(t), t \geq 0\}$ is given by $\lambda(t) = \Lambda_i$ whenever $\sum_{k=1}^{i-1} I_k \leq t < \sum_{k=1}^{i} I_k$. Thus, I_i is the duration of the intensity level Λ_i. In the special case where (Λ_1, I_1) and (Λ_2, I_2) are identically distributed we speak of an ordinary (nondelayed) Björk–Grandell process. We leave it to the reader to show that a Björk–Grandell process has stationary increments if $\mathbb{E}\, I_2 < \infty$ and, for all $B, B' \in \mathcal{B}(\mathbb{R}_+)$,

$$\mathbb{P}(\Lambda_1 \in B, I_1 \in B') = \frac{1}{\mathbb{E}\, I_2} \int_{B'} \mathbb{P}(\Lambda_2 \in B, I_2 > v)\, dv \,. \tag{12.2.9}$$

In view of (12.2.9) the stationary Björk–Grandell process is completely specified by the distribution of (Λ_2, I_2). The special case where $I_i = 1$ for $n = 2, 3, \ldots$ is called an *Ammeter process.*

12.2.3 Compounds of Point Processes

Consider the point processes $\{\sigma_{1,n}\}, \ldots, \{\sigma_{\ell,n}\}$ and the corresponding counting measures $\{N_1(B)\}, \ldots, \{N_\ell(B)\}$. By a *superposition* of these point processes we mean a point process with counting measure $\{N(B)\}$ defined by $N(B) = \sum_{i=1}^{\ell} N_i(B)$, $B \in \mathcal{B}(\mathbb{R})$; see also Theorem 12.2.2. We now state a representation formula for the Palm distribution of the superposition of ℓ independent stationary point processes $\{\sigma_{1,n}\}, \ldots, \{\sigma_{\ell,n}\}$ with positive and finite intensities $\lambda_1, \ldots \lambda_\ell$, respectively, where $\ell \in \mathbb{N}$ is fixed. By $\{N_i^0(B)\}$ we denote a *Palm version* of $\{N_i(B)\}$, i.e. the counting measure corresponding to the Palm distribution of $\{\sigma_{i,n}\}$. Assume that the sequence $\{N_1^0(B)\}, \ldots, \{N_\ell^0(B)\}$ consists of independent counting measures and is independent of $\{N_1(B)\}, \ldots, \{N_\ell(B)\}$. Consider a (product) probability space on which all these 2ℓ point processes are defined, and denote the basic (product) probability measure by $\mathbb{P}$. It is then clear that the superposition $N = \sum_{i=1}^{\ell} N_i$ is stationary and that its intensity is $\lambda = \sum_{i=1}^{\ell} \lambda_i$. In the next theorem we state a representation formula for the distribution of the Palm distribution $\mathbb{P}^0$ of N. In this connection we use the notation $N^{(i)} = N_1 + \ldots + N_{i-1} + N_i^0 + N_{i+1} + \ldots + N_\ell$ for $i = 1, \ldots, \ell$.

Theorem 12.2.5 *For each $A \in \mathcal{F}$,*

$$\mathbb{P}^0(A) = \sum_{i=1}^{\ell} \frac{\lambda_i}{\lambda} \mathbb{P}(N^{(i)} \in A) \,. \tag{12.2.10}$$

Proof By the independence assumptions, (12.2.10) easily follows from (12.1.7). We leave it to the reader to provide the details. □

In Section 12.2.4 we show how Theorem 12.2.5 can be used to derive lower and upper bounds for the time-stationary ruin function $\psi(u)$. For another application of Theorem 12.2.5, see also Section 12.6.3.

The following type of compounding leads to the notion of cluster processes. Let $\{\sigma'_n\}$ be a stationary point process with a positive and finite intensity λ'. Let $\{N_n, n \in \mathbb{Z}\}$ be a sequence of independent and identically distributed counting measures which is independent of $\{\sigma'_n\}$. Assume $0 < \mathbb{E}\, N_n(\mathbb{R}) < \infty$. The point process with counting measure $\{N(B), B \in \mathcal{B}(\mathbb{R})\}$ defined by $N(B) = \sum_{n\in\mathbb{Z}} N_n(B - \sigma'_n)$ is called a *cluster process*, where $\{\sigma'_n\}$ is called the point process of cluster centres. The counting measures $\{N_n, n \in \mathbb{Z}\}$ describe the individual clusters. It is clear that $\{N(B)\}$ is stationary and that its intensity is $\lambda = \lambda' \mathbb{E}\, N_n(\mathbb{R})$. If the point process $\{\sigma'_n\}$ of cluster centers (or parent points) is a homogeneous Poisson process, then $\{N(B)\}$ is called *Poisson cluster process.* In order to study the Palm distribution of this class of compound point processes it is convenient to introduce the notion of the generating functional of a point process.

Let I be the set of all Borel-measurable functions $f : \mathbb{R} \to \mathbb{R}$ such that $0 \le f(x) \le 1$ for all $x \in \mathbb{R}$ and $f(x) = 1$ for all $x \in \mathbb{R} \setminus B$, where $B \in \mathcal{B}(\mathbb{R})$ is some bounded set (dependent on f). Then, for any fixed point process $\{\sigma_n\}$, the mapping $\boldsymbol{G} : I \to \mathbb{R}$ defined by

$$\boldsymbol{G}(f) = \mathbb{E} \prod_n f(\sigma_n)\,, \qquad f \in I\,, \tag{12.2.11}$$

is called the *generating functional* of $\{\sigma_n\}$. The following properties of the generating functional are known.

Theorem 12.2.6 (a) *The distribution of a point process is uniquely determined by its generating functional.*
(b) *The generating functional of the superposition $\{(N_1 + N_2)(B)\}$ of two independent counting measures $\{N_1(B)\}$, $\{N_2(B)\}$ is given by the product of the generating functionals of $\{N_1(B)\}$ and $\{N_2(B)\}$.*

The *proof* of Theorem 12.2.6 goes beyond the scope of this book. We therefore omit it and refer to Daley and Vere-Jones (1988), for example.

We are now in the position to state a useful representation formula for the generating functional of the Palm version of a Poisson cluster process. It is a generalization of *Slivnyak's theorem* for Poisson processes. Let $\boldsymbol{G}$ be the generating functional of a Poisson cluster process and let $\boldsymbol{G}^0$ be the generating functional of its Palm distribution. Furthermore, let $\tilde{\boldsymbol{G}}$ denote the generating functional of the point process whose distribution $\tilde{\mathbb{P}}$ is given by

$$\tilde{\mathbb{P}}(A) = (\mathbb{E}\, N_n(\mathbb{R}))^{-1} \mathbb{E}\left(\int \mathbb{I}(\boldsymbol{T}_x N_n \in A) N_n(\mathrm{d}x) \right), \qquad A \in \mathcal{F}\,. \tag{12.2.12}$$

Theorem 12.2.7 *For each $f \in I$,*

$$\boldsymbol{G}^0(f) = \boldsymbol{G}(f)\tilde{\boldsymbol{G}}(f)\,. \tag{12.2.13}$$

The *proof* is omitted. It can be found in Section 5.5 of König and Schmidt (1992), for example.

Theorems 12.2.6 and 12.2.7 can be used to derive a lower bound for the time-stationary ruin function $\psi(u)$ if the claim arrival process is governed by an independently marked Poisson cluster process; see Section 12.2.4 below. Another application of Theorem 12.2.7 is given in Section 12.6.2, where asymptotic properties of ruin functions are studied in the case of a subexponential claim size distribution.

12.2.4 Comparison of Ruin Probabilities

The aim of this section is to develop techniques allowing comparison of the ruin function $\psi(u)$ in the time-stationary risk model to the ruin function $\psi^*(u)$ in a correspondingly averaged compound Poisson model. We begin with the stationary Markov-modulated model. So, consider the claim surplus process in the time-stationary risk model introduced in Section 12.1.5, where the claim arrival process $\{(\sigma_n, X_n)\}$ is a marked Markov-modulated Poisson process as defined in Example 3 of Section 12.2.2 and specified by ℓ, $\boldsymbol{Q}$, $\lambda_1, \ldots, \lambda_\ell$, $F_1, \ldots, F_\ell$. Moreover we assume that $\boldsymbol{Q}$ is irreducible and that $\pi_1, \ldots, \pi_\ell$ is the stationary initial distribution.

Recall that for each $i \in E$ the (conditional) claim arrival intensity is λ_i and the (conditional) claim size distribution is F_i. We assume that, given $\{J(t)\}$, the sequences $\{\sigma_n\}$ and $\{X_n\}$ are independent and that $\{\sigma_n\}$ is a Markov-modulated Poisson process governed by $\{J(t)\}$ and $\lambda_1, \ldots, \lambda_\ell$. Furthermore, we assume that, given $\{J(t)\}$ and $\{\sigma_n\}$, the claim sizes $U_1, U_2, \ldots$ are independent, where U_n has distribution F_i if $(V_n =)\ J(\sigma_n) = i$. As usual $N(t)$ is the number of claims arriving in $(0, t]$.

Here we study $\psi(u) = \mathbb{P}(\tau(u) < \infty) = \sum_{i=1}^{\ell} \pi_i \psi_i(u)$, where

$$\psi_i(u) = \mathbb{P}(\tau(u) < \infty \mid J(0) = i) = \mathbb{P}\Big(\sup_{t \ge 0}\Big\{\sum_{n=1}^{N(t)} U_n - \beta t\Big\} > u \,\Big|\, J(0) = i\Big)\,.$$

We show that under some conditions this ruin function $\psi(u)$, given by (12.1.39), is "more dangerous" than the ruin function in the following (averaged) compound Poisson model. Let $\psi^*(u)$ be the ruin function in the compound Poisson model with characteristics (λ, F) given by

$$\lambda = \sum_{i=1}^{\ell} \pi_i \lambda_i\,, \qquad F = \lambda^{-1} \sum_{i=1}^{\ell} \pi_i \lambda_i F_i\,. \tag{12.2.14}$$

Since we want to apply the results of Section 12.1.5, the net profit condition (12.1.38) will be taken for granted; see also Section 12.3.2. We observe that the relative safety loading is the same for the Markov-modulated model and for the compound Poisson model.

Theorem 12.2.8 *Let $\boldsymbol{Q}$ be stochastically monotone and let*

$$\lambda_1 \le \ldots \le \lambda_\ell , \qquad F_1 \le_{\rm st} \ldots \le_{\rm st} F_\ell . \tag{12.2.15}$$

Then $\psi(u) \ge \psi^*(u)$ for all $u \ge 0$.

The *proof* of Theorem 12.2.8 is subdivided into several steps. We also need some extra notation. Let $\{J_i(t)\}$ denote a homogeneous Markov process with intensity matrix $\boldsymbol{Q}$ and initial state $J_i(0) = i$. Furthermore, for $i = 1, \ldots, \ell$, let $\{N_i(t)\}$ be a Cox process with intensity process $\{\lambda_{J_i(t)}\}$ and let $U_1^i, U_2^i, \ldots$ be a sequence of claim sizes where the distribution of U_n^i is F_j if $J_i(\sigma_n^i) = j$. For the (conditional) ruin functions $\psi_i(u) = \mathbb{P}(\sup_{t\ge0}\{\sum_{n=1}^{N_i(t)} U_n^i - \beta t\} > u)$, $i \in E$, the following comparison holds.

Lemma 12.2.1 *Under the assumptions of Theorem* 12.2.8, *the inequality $\psi_i(u) \le \psi_j(u)$ holds for all $u \ge 0$ whenever $i \le j$.*

Proof We use a coupling argument. Let $i \le j$. Then by Theorem 8.1.8 there exists a probability space $(\Omega_{ij}, \mathcal{F}_{ij}, \mathbb{P}_{ij})$ on which $J_i(t) \le J_j(t)$ for all $t \ge 0$. By the first part of condition (12.2.15), there exists a probability space $(\Omega', \mathcal{F}', \mathbb{P}')$ on which $\{\sigma_n^i\} \subset \{\sigma_n^j\}$. We leave it to the reader to provide the details. The second part of condition (12.2.15) and a multidimensional analogue to Theorem 3.2.1 imply the existence of a probability space such that $U_n^i \le U_n^j$ for all $n = 1, 2, \ldots$. Thus, taking the product space as the basic probability space we get that on this space $\sum_{n=1}^{N_i(t)} U_n^i \le \sum_{n=1}^{N_j(t)} U_n^j$ for all $t \ge 0$. □

The following standard inequality of Chebyshev type will be useful.

Lemma 12.2.2 *If $0 \le a_1 \le \ldots \le a_\ell$, $0 \le b_1 \le \ldots \le b_\ell$ and $\alpha_i \ge 0$ for all $i \in E$, $\sum_{i=1}^\ell \alpha_i = 1$, then $\sum_{i=1}^\ell \alpha_i a_i b_i \ge \sum_{i=1}^\ell \alpha_i a_i \sum_{i=1}^\ell \alpha_i b_i$.*

Proof Let X be an E-valued random variable with probability function $\boldsymbol{\alpha} = \{\alpha_1, \ldots, \alpha_\ell\}$. If we define k by $k = \min\{i \ge 1 : b_i - \mathbb{E}\, b_X \ge 0\}$, then we have

$$\begin{aligned}\sum_{i=1}^\ell \alpha_i a_i b_i - \sum_{i=1}^\ell \alpha_i a_i \sum_{i=1}^\ell \alpha_i b_i &= \sum_{i=1}^\ell \alpha_i a_i (b_i - \mathbb{E}\, b_X)\\ &= \sum_{i=1}^{k-1} \alpha_i a_i (b_i - \mathbb{E}\, b_X) + \sum_{i=k}^{\ell} \alpha_i a_i (b_i - \mathbb{E}\, b_X)\end{aligned}$$

$$\begin{aligned}&\geq a_{k-1}\Big(\sum_{i=1}^{k-1}\alpha_i(b_i-\mathbb{E}\,b_X)+\sum_{i=k}^{\ell}\alpha_i(b_i-\mathbb{E}\,b_X)\Big)\\&= a_{k-1}\sum_{i=1}^{\ell}\alpha_i(b_i-\mathbb{E}\,b_X)=0\,.\end{aligned}$$

□

Proof of Theorem 12.2.8. By (5.3.9) we have

$$\psi^*(u)=\rho\overline{F^{\mathrm{s}}}(u)+\lambda\beta^{-1}\int_0^u\psi^*(u-v)\overline{F}(v)\,\mathrm{d}v\,. \tag{12.2.16}$$

For the Markov-modulated model we have $\mathbb{P}^0(U_0\geq v, V_0=i)=\pi_i\lambda_i\lambda^{-1}\overline{F_i}(v)$ which can be concluded from (12.2.8). Theorem 12.1.9 then implies that

$$\psi(u)=\rho\overline{F^{\mathrm{s}}}(u)+\beta^{-1}\int_0^u\sum_{i=1}^{\ell}\pi_i\lambda_i\overline{F_i}(v)\psi_i(u-v)\,\mathrm{d}v\,.$$

Since by the assumption (12.2.15) and by Lemma 12.2.1, the sequences $\{a_i\}$, $\{b_i\}$ with $a_i=\lambda_iF_i(v)$, $b_i=\psi_i(u-v)$ are increasing, Lemma 12.2.2 gives

$$\sum_{i=1}^{\ell}\pi_i\lambda_i\overline{F_i}(v)\psi_i(u-v)\geq\sum_{i=1}^{\ell}\pi_i\lambda_i\overline{F_i}(v)\sum_{i=1}^{\ell}\pi_i\psi_i(u-v)=\lambda\overline{F}(v)\psi(u-v)\,.$$

Thus,

$$\psi(u)\geq\rho\overline{F^{\mathrm{s}}}(u)+\lambda\beta^{-1}\int_0^u\psi(u-v)\overline{F}(v)\,\mathrm{d}v\,. \tag{12.2.17}$$

Comparing (12.2.16) and (12.2.17), Lemma 6.1.2 immediately implies that $\psi(u)\geq\psi^*(u)$ for all $u\geq 0$. □

The same argument as in the proof of Theorem 12.2.8 almost immediately applies when analysing the following model. Consider the ruin function in the stationary risk model where the claim arrival process is governed by a time-stationary mixed Poisson process such that, given $\Lambda=x$, the sequences $\{\sigma_n\}$ and $\{U_n\}$ are independent and the claim sizes are independent random variables with distribution F_x.

Theorem 12.2.9 *Assume that $F_x\leq_{\mathrm{st}}F_{x'}$ for $x\leq x'$ and $\mathbb{P}(\Lambda\mu_{F_\Lambda}<\beta)=1$. Then $\psi(u)\geq\psi^*(u)$ for all $u\geq 0$, where $\psi^*(u)$ is the ruin function in the compound Poisson model with arrival intensity $\mathbb{E}\,\Lambda$ and claim size distribution function $F(t)=\int_0^\infty F_x(t)\,\mathrm{d}F_\Lambda(x)$.*

Proof We use a general version of the Chebyshev-type inequality given in Lemma 12.2.2 stating that for each real-valued random variable X and for each pair of increasing functions $a(x)$ and $b(x)$,

$$\mathbb{E}\, a(X)b(X) \geq \mathbb{E}\, a(X)\mathbb{E}\, b(X)\,. \tag{12.2.18}$$

Let $\psi_x(u)$ be the ruin function in the compound Poisson model with characteristics (x, F_x). One needs to show first that $\psi_x(u) \leq \psi_y(u)$ for $x \leq y$, and this is analogous to Lemma 12.2.1. Then, using (12.2.18), the same argument as in the proof of Theorem 12.2.8 applies. The details are left to the reader. □

The ruin function $\psi(u)$ of the time-stationary Sparre Andersen model can similarly be compared with that of an appropriately chosen compound Poisson model.

Theorem 12.2.10 *Consider the time-stationary Sparre Andersen model with distribution F_T of inter-arrival times and distribution F_U of claim sizes, where $0 < \mu_T, \mu_U < \infty$. If F_T is NBUE, then $\psi(u) \leq \psi^*(u)$ for all $u \geq 0$, where $\psi^*(u)$ is the ruin function in the corresponding compound Poisson model with arrival intensity μ_T^{-1} and claim size distribution F_U. Moreover, if F_T is NWUE then $\psi(u) \geq \psi^*(u)$ for all $u \geq 0$.*

Proof Let F_T be NBUE. Then $F_T^{\mathrm{s}} \leq_{\mathrm{st}} F_T$. Therefore by Theorem 3.2.1 we can find a probability space $(\Omega, \mathcal{F}, \mathbb{P})$ and independent random variables $T_1', T_1, T_2, \ldots, U_1, U_2, \ldots$ such that $T_1 \leq T_1'$, where T_1 has distribution F_T^{s}, $T_1', T_2, T_3, \ldots$ have distribution F_T and $U_1, U_2, \ldots$ have distribution F_U. The risk reserve process in the time-stationary model is therefore always smaller than in the Palm model and hence $\psi(u) \geq \psi^0(u)$, where $\psi^0(u)$ is the ruin function in the (Palm-stationary) Sparre Andersen model. Analogously to (12.2.17), we get from (12.1.44) that for all $u \geq 0$

$$\psi(u) \leq (\mu_T\beta)^{-1}\mu_U\overline{F_U^{\mathrm{s}}}(u) + (\mu_T\beta)^{-1}\int_0^u \psi(u-v)\overline{F_U}(v)\,\mathrm{d}v\,. \tag{12.2.19}$$

As in the proof of Theorem 12.2.8, also $\psi(u) \leq \psi^*(u)$ for all $u \geq 0$, where $\psi^*(u)$ is the ruin function in the compound Poisson model with arrival intensity μ_T^{-1} and claim size distribution F_U. For F_T being NWUE, the proof is similar. □

Using the representation formula (12.2.10) for the Palm distribution, Theorem 12.2.10 can be generalized from a single renewal point process to a superposition of several renewal processes. Assume that the time-stationary claim arrival process $\{(\sigma_n, U_n)\}$ is independently marked with claim size distribution F_U and that $\{\sigma_n\}$ is the superposition of ℓ independent stationary renewal point processes with interpoint-distance distributions $F_1, \ldots, F_\ell$ and expectations $\mu_1, \ldots, \mu_\ell$, respectively. If $F_1, \ldots, F_\ell$ are NBUE, then we get from (12.1.44) and (12.2.10) that $\psi(u) \leq \psi^*(u)$ for all $u \geq 0$, where $\psi^*(u)$

is the ruin function in the compound Poisson model with arrival intensity $\sum_{i=1}^{\ell} \mu_i^{-1}$ and claim size distribution F_U. Moreover, if $F_1, \ldots, F_\ell$ are NWUE then $\psi(u) \geq \psi^*(u)$ for all $u \geq 0$.

Another natural generalization leads from a renewal point process to the class of *semi-Markov point processes* where the inter-point distances are no longer independent nor identically distributed but connected via a Markov chain with finite state space. A classical example is the so-called *alternating renewal point process*. For this class of point processes, similar conditions can be found for deriving an upper "Poisson" bound for the time-stationary ruin function $\psi(u)$. It suffices to assume that all (conditional) distributions of distances between consecutive points are NBUE. A corresponding lower bound for $\psi(u)$ is obtained if all these distributions are NWUE.

Finally, we mention that a lower bound analogous to that in Theorem 12.2.8 can be derived for the time-stationary ruin function $\psi(u)$ if the claim arrival process is governed by an independently marked Poisson cluster process. From (12.1.44) and the representation formula (12.2.13) for the generating functional of the Palm distribution of this class of stationary point processes, we get that this lower bound holds without any additional conditions.

Bibliographical Notes. Properties of Cox processes and cluster processes can be found in many books dealing with point processes on the real line; see, for example, Brémaud (1981), Daley and Vere-Jones (1988), Karr (1991), König and Schmidt (1992) and Last and Brandt (1995). The Markov-modulated risk model was first introduced by Janssen (1980) and also treated in Janssen and Reinhard (1985) and Reinhard (1984). The definition of this model using an environmental Markov chain $\{J(t)\}$ goes back to Asmussen (1989). Theorem 12.2.8 has been derived in Asmussen, Frey, Rolski and Schmidt (1995). A weak form of comparison between the ruin function in the time-stationary Markov-modulated model and the ruin function in the correspondingly averaged compound Poisson model was originally given in Rolski (1981). A survey of methods for statistical estimation of the parameters of Markov-modulated Poisson processes is given in Rydén (1994). For the queueing-theoretic analogue to the time-stationary Sparre Andersen model, the inequalities $\psi(u) \geq (\leq)\, \psi^*(u)$ for all $u \geq 0$ if F_T is NBUE (NWUE) have been proved, for example, in Franken, König, Arndt and Schmidt (1982), p. 137. The Björk–Grandell process was introduced in Björk and Grandell (1988) as a generalization to the model considered by Ammeter (1948). Another application of nonhomogeneous Poisson and Cox processes in risk theory can be found, for example, in Arjas (1989) and in Norberg (1993), where the prediction of outstanding liabilities is investigated.

12.3 THE MARKOV-MODULATED RISK MODEL VIA PDMP

We now turn to the continuous-time risk process $\{R(t)\}$. We use techniques for PDMP developed in Chapter 11, where the claim arrival process is given by a Markov-modulated Poisson process. Moreover, as in Section 12.2.4, we allow that claim size distributions are modulated by the Markov environment process $\{J(t)\}$. The stochastic process $\{(J(t), R(t))\}$ is called a *Markov-modulated risk model.* Our aim is to obtain bounds and approximations to the infinite-horizon and finite-horizon ruin functions of this model.

12.3.1 A System of Integro-Differential Equations

The risk reserve process $\{R(t)\}$ can be represented in the following way. Let $\xi_1, \xi_2, \ldots$ with $\xi_{n+1} = \inf\{t > \xi_n : J(t) \neq J(t-0)\}$ be the times where the state of the environment changes, where $\xi_0 = 0$. Consider the independent compound Poisson risk processes $\{R_1(t)\}, \ldots, \{R_\ell(t)\}$ with characteristics $(\lambda_1, F_1), \ldots, (\lambda_\ell, F_\ell)$, respectively. Let $\mu_i = \int_0^\infty v \, \mathrm{d}F_i(v)$ denote the expected claim size in state i and $\hat{m}_i(s) = \int_0^\infty \mathrm{e}^{sv} \, \mathrm{d}F_i(v)$ its moment generating function. Furthermore, the claim counting process in the i-th model is denoted by $\{N_i(t)\}$. The claim counting process $\{N(t)\}$ in the Markov-modulated risk model is then given by

$$N(t) = \sum_{i=1}^{\ell} \int_0^t \mathbb{I}(J(v) = i) \, \mathrm{d}N_i(v) , \tag{12.3.1}$$

and the corresponding risk process $\{R(t)\}$ by

$$R(t) = u + \sum_{i=1}^{\ell} \int_0^t \mathbb{I}(J(v) = i) \, \mathrm{d}R_i(v) . \tag{12.3.2}$$

This means $N(0-0) = 0$, $R(0-0) = u$ while for $\xi_n \le t < \xi_{n+1}$

$$N(t) = N(\xi_n - 0) + N_{J(t)}(t) - N_{J(t)}(\xi_n - 0) ,$$

$$R(t) = R(\xi_n - 0) + R_{J(t)}(t) - R_{J(t)}(\xi_n - 0) .$$

In particular (12.3.1) implies that, given the environment, the conditional expected number of claims in the interval $(0, t]$ is equal to $\mathbb{E}\,(N(t) \mid J(v), 0 \le v \le t) = \int_0^t \lambda_{J(v)} \, \mathrm{d}v$. Hence, $\mathbb{E}\, N(t) \le \max_{i \le \ell} \lambda_i t < \infty$.

The ruin function $\psi(u) = \mathbb{P}(\inf_{t \ge 0} R(t) < 0)$ and the conditional ruin functions $\psi_i(u) = \mathbb{P}(\inf_{t \ge 0} R(t) < 0 \mid J(0) = i)$ are expressed in terms of the risk reserve process $\{R(t)\}$ rather than using the claim surplus process $\{S(t)\}$

as was done in Section 12.2.4. Further, the conditional survival functions are denoted by $\overline{\psi}_i(u) = 1 - \psi_i(u)$.

As in Section 5.3.1, we can derive a system of integro-differential equations.

Theorem 12.3.1 *The survival functions $\overline{\psi}_i(u)$ are absolutely continuous and fulfil*

$$\beta\overline{\psi}_{i,+}^{(1)}(u) + \lambda_i\Big(\int_0^u \overline{\psi}_i(u-v)\,\mathrm{d}F_i(v) - \overline{\psi}_i(u)\Big) + \sum_{j=1}^{\ell} q_{ij}\overline{\psi}_j(u) = 0 \quad (12.3.3)$$

and

$$\beta\overline{\psi}_{i,-}^{(1)}(u) + \lambda_i\Big(\int_0^{u-0} \overline{\psi}_i(u-v)\,\mathrm{d}F_i(v) - \overline{\psi}_i(u)\Big) + \sum_{j=1}^{\ell} q_{ij}\overline{\psi}_j(u) = 0 \quad (12.3.4)$$

for $i = 1,\ldots,\ell$, where $\overline{\psi}_{i,+}(u)$ and $\overline{\psi}_{i,-}(u)$ are the right and left derivatives of $\overline{\psi}_i(u)$, respectively.

The *proof* of Theorem 12.3.1 is similar to the proof of Theorem 5.3.1 and is left to the reader.

12.3.2 Law of Large Numbers

Let $\boldsymbol{Q}$ be irreducible and write $\boldsymbol{\pi} = (\pi_1,\ldots,\pi_\ell)$ for the stationary initial distribution of $\{J(t)\}$. Recall that by Theorem 8.1.4 we have $\boldsymbol{\pi} = \boldsymbol{\pi}\exp(t\boldsymbol{Q})$ for all $t \geq 0$. In Section 8.1.2 we showed that this is equivalent to $\boldsymbol{\pi}\boldsymbol{Q} = \boldsymbol{0}$. The following *law of large numbers* for the Markov-modulated risk model extends its counterpart for the compound Poisson model as it was mentioned in the introduction to Section 5.3; see also Theorem 10.3.4. Recall that similar results for renewal processes were derived in Theorem 6.1.12.

Theorem 12.3.2 *Assume that $\boldsymbol{Q}$ is irreducible. Then*

$$\lim_{t\to\infty} \frac{1}{t}R(t) = \beta - \sum_{i=1}^{\ell} \pi_i\lambda_i\mu_i\,. \quad (12.3.5)$$

Proof Without loss of generality we can assume that $u = 0$. Let $V_i(t) = \int_0^t \mathbb{I}(J(v) = i)\,\mathrm{d}v$ denote the amount of time in $(0,t]$ that $\{J(t)\}$ spends in state i. First observe that (12.3.2) can be rewritten in the form

$$\frac{1}{t}R(t) = \sum_{i=1}^{\ell} \frac{V_i(t)}{t}\frac{1}{V_i(t)}\int_0^t \mathbb{I}(J(v) = i)\,\mathrm{d}R_i(v)\,.$$

It is left to the reader to show that $\lim_{t\to\infty} t^{-1}V_i(t) = \pi_i$. It therefore suffices to show that $(V_i(t))^{-1}\int_0^t \mathbb{I}(J(v)=i)\,\mathrm{d}R_i(v)$ tends to $\beta - \lambda_i\mu_i$ as $t\to\infty$. But $\int_0^t \mathbb{I}(J(v)=i)\,\mathrm{d}R_i(v)$ has the same distribution as $R_i(V_i(t))$ because $\{R_i(t)\}$ has independent and stationary increments. Since $\boldsymbol{Q}$ is irreducible, $V_i(t)$ tends to infinity as $t\to\infty$. The assertion now follows by the same arguments as that used in the proof of Theorem 6.3.1. □

Theorem 12.3.2 implies that $\psi(u)\equiv 1$ if $\beta \le \sum_{i=1}^{\ell}\pi_i\lambda_i\mu_i$. Indeed assume $J(0)=i$. Let $I_0=0$ and let $I_{n+1}=\inf\{t>I_n : J(t)=i, J(t-0)\ne i\}$ be the epochs, where $\{J(t)\}$ returns to state i. It is easy to see that $\{R(I_n)\}$ is a random walk. Since $n^{-1}R(I_n) = (I_n/n)(R(I_n)/I_n)$ and $I_n\to\infty$, it follows from Theorem 12.3.2 that $\{R(I_n)\}$ does not have a positive drift if $\beta\le\sum_{i=1}^{\ell}\pi_i\lambda_i\mu_i$. Thus, by Theorem 6.3.1, ruin occurs almost surely. We therefore take the *net profit condition*

$$\beta > \sum_{i=1}^{\ell}\pi_i\lambda_i\mu_i \tag{12.3.6}$$

for granted in what follows.

12.3.3 The Generator and Exponential Martingales

Thanks to the following result, the techniques for PDMP developed in Chapter 11 become available.

Theorem 12.3.3 *The process $\{(J(t),R(t),t)\}$ is a PDMP. Its generator $\boldsymbol{A}$ has the property that $g\in\mathcal{D}(\boldsymbol{A})$ and $(g,\boldsymbol{A}g)\in\boldsymbol{A}$ for each function g fulfilling the conditions of Theorem* 11.2.2, *where*

$$\begin{aligned}(\boldsymbol{A}g)(i,z,t) &= \beta\frac{\partial}{\partial z}g(i,z,t)+\frac{\partial}{\partial t}g(i,z,t)\\ &\quad+\lambda_i\Big(\int_0^\infty g(i,z-v,t)\,\mathrm{d}F_i(v)-g(i,z,t)\Big)+\sum_{j=1}^{\ell}q_{ij}g(j,z,t).\end{aligned} \tag{12.3.7}$$

Proof We leave it to the reader to show that $\{(J(t),R(t))\}$ is a PDMP. The vector field $\boldsymbol{X}$ of this PDMP is given by $(\boldsymbol{X}g)(z)=\beta(\mathrm{d}g/\mathrm{d}z)(z)$. If the environment process $\{J(t)\}$ is in state i, then jumps caused by claims occur with rate λ_i, while jumps caused by a change of the environment to $j\ne i$ have rate q_{ij}. The statement then follows from Corollary 11.2.1, where in (12.3.7) we used that $-\sum_{j\ne i}q_{ij}g(i,z,t)=q_{ii}g(i,z,t)$. □

Before we construct an exponential martingale, needed in the study of ruin probabilities in the Markov-modulated risk model, we cover some auxiliary results from matrix algebra. Let $\{\boldsymbol{B}(t),\ t \geq 0\}$ be a family of $\ell \times \ell$ matrices that satisfies the condition

$$\boldsymbol{B}(0) = \boldsymbol{I}, \qquad \boldsymbol{B}(t+t') = \boldsymbol{B}(t)\boldsymbol{B}(t') \tag{12.3.8}$$

for all $t, t' \geq 0$, and

$$\lim_{h \downarrow 0} \frac{\boldsymbol{B}(h) - \boldsymbol{I}}{h} = \boldsymbol{C}$$

for some matrix $\boldsymbol{C}$. By similar considerations as in Section 8.1.2, we can then show that

$$\boldsymbol{B}(t) = \exp(t\boldsymbol{C})\,. \tag{12.3.9}$$

We prove a lemma that is useful in its own right and that will be applied later on. Recall that for an $\ell \times \ell$ matrix $\boldsymbol{B} = (b_{ij})$ with positive entries, the trace $\mathrm{tr}\boldsymbol{B} = \sum_{j=1}^{\ell} b_{jj}$ equals the sum of the eigenvalues $\theta_1, \ldots, \theta_\ell$ of $\boldsymbol{B}$, i.e.

$$\mathrm{tr}\boldsymbol{B} = \sum_{i=1}^{\ell} \theta_i\,. \tag{12.3.10}$$

Indeed, in the characteristic polynomial $w(x) = \det(\boldsymbol{B} - x\boldsymbol{I})$, the coefficient of $x^{\ell-1}$ is $(-1)^{\ell-1} \sum_{j=1}^{\ell} b_{jj}$.

Lemma 12.3.1 *Let $\{\boldsymbol{B}'(s)\}$ be a family of $\ell \times \ell$ matrices defined for all s from a certain (possibly unbounded) interval (s_1, s_2) such that all entries $b'_{ij}(s)$ are positive. Let $b'_{ij}(s)$ be logconvex in (s_1, s_2) for all $i, j = 1, \ldots, \ell$ and let $\theta'(s)$ be the Perron–Frobenius eigenvalue of $\boldsymbol{B}'(s)$. Then $\theta'(s)$ is logconvex.*

Proof Let $s_1 < s < s_2$ and let $\theta'_1(s), \ldots, \theta'_\ell(s)$ be the eigenvalues of $\boldsymbol{B}'(s)$. The reader should verify that the class of logconvex functions is closed under addition, multiplications and raising to any positive power; moreover the limit of logconvex functions is logconvex or zero. Hence, the assertion follows from (12.3.10) because $\lim_{n\to\infty} \big(\sum_{i=1}^{\ell} (\theta'_i(s))^n\big)^{1/n}$ equals the Perron–Frobenius eigenvalue of $\boldsymbol{B}'(s)$. □

Now let $\boldsymbol{K}(s)$ be the diagonal matrix with entries $\kappa_{ii}(s) = \lambda_i(\hat{m}_i(s) - 1)$, where we put $\kappa_{ii}(s) = \infty$ if $\hat{m}_i(s) = \infty$. Furthermore let

$$\boldsymbol{C}(s) = \boldsymbol{Q} + \boldsymbol{K}(s) - \beta s\boldsymbol{I}\,. \tag{12.3.11}$$

Differentiation of $\boldsymbol{C}(s)$ tells us that $\boldsymbol{C}^{(1)}(0)$ is a diagonal matrix with entries $\lambda_i\mu_i - \beta$. Hence we get

$$\pi\boldsymbol{C}^{(1)}(0)\boldsymbol{e} = \sum_{i=1}^{\ell} \pi_i\lambda_i\mu_i - \beta\,. \tag{12.3.12}$$

Furthermore, for each $t \ge 0$, let $\boldsymbol{B}(t)$ be the $\ell \times \ell$ matrix with entries

$$b_{ij}(t) = b_{ij}(t;s) = \mathbb{E}\left(\mathrm{e}^{-s(R(t)-u)}\mathbb{I}(J(t)=j) \mid J(0)=i\right) \tag{12.3.13}$$

for $i,j = 1,\ldots,\ell$. Note that $\boldsymbol{B}(0) = \boldsymbol{I}$.

Lemma 12.3.2 *Let $s_0 > 0$ be fixed such that $\hat{m}_i(s_0) < \infty$ for all $i = 1,\ldots,\ell$. Then,*

$$\boldsymbol{B}(t) = \exp(t\boldsymbol{C}(s)) \tag{12.3.14}$$

for all $s \le s_0$ and $t \ge 0$, where $\boldsymbol{C}(s)$ is defined in (12.3.11). *In particular,* $\mathbb{P}(J(t)=j \mid J(0)=i) = (\exp(t\boldsymbol{Q}))_{ij}$.

Proof It is easily seen that the matrices $\boldsymbol{B}(t)$ given in (12.3.13) satisfy (12.3.8). Furthermore, for a small time interval $(0,h]$ we have

$$\begin{aligned}
\mathbb{P}(N(h)=0 \mid J(t)=i) &= 1-\lambda_i h + o(h)\,,\\
\mathbb{P}(N(h)=1 \mid J(t)=i) &= \lambda_i h + o(h)\,,
\end{aligned}$$

and $\mathbb{P}(J(h)=j \mid J(t)=i) = \delta_i(j) + q_{ij}h + o(h)$. Thus,

$$\begin{aligned}
b_{ij}(h) &= \delta_i(j)(1+q_{jj}h+o(h))\big((1-\lambda_j h+o(h))\mathrm{e}^{-\beta s h}\\
&\quad + (\lambda_j h + o(h))\mathrm{e}^{-\beta s h}\hat{m}_j(s)\big) + \delta_i(j)(q_{ij}h+o(h))\mathrm{e}^{-\beta s h} + o(h)\\
&= \delta_i(j)\mathrm{e}^{-\beta s h} + \delta_i(j)\kappa_{jj}(s)h + hq_{ij} + o(h)\,.
\end{aligned}$$

Rearrange the terms to obtain

$$\begin{aligned}
\frac{b_{ij}(h)-\delta_i(j)}{h} &= \delta_i(j)\frac{\mathrm{e}^{-\beta s h}-1}{h} + \delta_i(j)\kappa_{jj}(s) + q_{ij} + o(1)\\
&= (\boldsymbol{C}(s))_{ij} + o(1)\,.
\end{aligned}$$

Letting $h \to 0$, the proof is completed in view of (12.3.9). □

Let $s_0 > 0$ fulfil the conditions of Lemma 12.3.2 and let $s \le s_0$. By Lemma 12.3.2, the matrix $\exp(\boldsymbol{C}(s))$ has strictly positive entries. Let $\theta(s)$ be the logarithm of the largest absolute value of the eigenvalues of the matrix $\exp(\boldsymbol{C}(s))$. By the Perron–Frobenius theorem (see Theorem 7.2.2), $\mathrm{e}^{\theta(s)}$ is an eigenvalue of $\exp(\boldsymbol{C}(s))$. It is the unique eigenvalue with absolute value $\mathrm{e}^{\theta(s)}$ and the corresponding right eigenvector $\boldsymbol{\phi}(s) = (\phi_1(s),\ldots,\phi_\ell(s))$ has strictly positive entries. In particular $\boldsymbol{\phi}(0) = \boldsymbol{e}$, the vector with all entries equal to 1. Indeed, recall that $\boldsymbol{Q}\boldsymbol{e}^\top = \boldsymbol{0}$ and consequently $(\exp\boldsymbol{Q})\boldsymbol{e}^\top = \boldsymbol{e}^\top$. We normalize $\boldsymbol{\phi}(s)$ in such a way that $\boldsymbol{\pi}(\boldsymbol{\phi}(s))^\top = 1$.

Let $0 < s \le s_0$. By Lemma 7.1.3, $\boldsymbol{C}(s)$ can be written in the form $\boldsymbol{C}(s) = \boldsymbol{D}\boldsymbol{T}\boldsymbol{D}^{-1}$, where $\boldsymbol{T} = (t_{ij})$ is an upper triangular matrix and $\boldsymbol{D}$

is nonsingular. Hence the eigenvalues of $\boldsymbol{C}(s)$ are $\theta_i(s) = t_{ii}$ $(i = 1, \ldots, \ell)$ because

$$\det(\boldsymbol{D}\boldsymbol{T}\boldsymbol{D}^{-1} - \theta\boldsymbol{I}) = \det(\boldsymbol{T} - \theta\boldsymbol{I}) = \prod_{i=1}^{\ell}(t_{ii} - \theta) .$$

Then $\exp(\boldsymbol{C}(s)) = \boldsymbol{D}\exp(\boldsymbol{T})\boldsymbol{D}^{-1}$ and $\exp(\boldsymbol{T})$ is upper triangular too. Note that its diagonal entries $\exp(t_{ii})$ with $i = 1, \ldots, \ell$ are strictly positive. We can therefore conclude that $\theta_i(s)$ is an eigenvalue of $\boldsymbol{C}(s)$ if and only if $\exp(\theta_i(s))$ is an eigenvalue of $\exp(\boldsymbol{C}(s))$, $i = 1, \ldots, \ell$.

In Section 12.3.4, the following martingale will be used when changing the probability measure.

Theorem 12.3.4 *Assume that there exists an $s_0 > 0$ such that $\hat{m}_i(s_0) < \infty$ for all $i = 1, \ldots, \ell$. Then, the following statements are true.*
(a) *For each $s \le s_0$, the process $\{M(t), t \ge 0\}$ with*

$$M(t) = \phi_{J(t)}(s)\mathrm{e}^{-sR(t)-\theta(s)t} , \tag{12.3.15}$$

is a martingale with respect to the history of $\{(J(t), R(t))\}$.
(b) *The function $\theta(s)$ is convex on $(-\infty, s_0]$ and*

$$\theta^{(1)}(0) = -\Big(\beta - \sum_{i=1}^{\ell}\pi_i\lambda_i\mu_i\Big) < 0 . \tag{12.3.16}$$

Proof Theorem 11.1.3 tells us that (a) will follow if we find a martingale solution of the form $\{g(J(t), R(t), t), t \ge 0\}$ to the equation $\boldsymbol{A}g = 0$. We try a function g of the form $g(i, z, t) = h_i \exp(-sz - \vartheta t)$ for some $h_1, \ldots, h_\ell, \vartheta \in \mathbb{R}$. Using (12.3.7) this yields

$$-\beta r h_i - \vartheta h_i + \lambda_i(\hat{m}_i(s) - 1)h_i + \sum_{j=1}^{\ell} q_{ij}h_j = 0$$

for each $i \in \{1, \ldots, \ell\}$, i.e. $\boldsymbol{C}(s)\boldsymbol{h}^\top = \vartheta\boldsymbol{h}^\top$, where $\boldsymbol{h} = (h_1, \ldots, h_\ell)$. Thus, ϑ must be an eigenvalue of $\boldsymbol{C}(s)$ and $\boldsymbol{h}$ the corresponding eigenvector. But we already know that $\theta(s)$ is an eigenvalue of $\boldsymbol{C}(s)$ with right eigenvector $\boldsymbol{\phi}(s)$. It remains to verify that $g(i, z, t) = \phi_i(s)\exp(-sz - \theta(s)t)$ satisfies the conditions of Theorem 11.2.2. Condition (a) of Theorem 11.2.2 is obviously fulfilled since g is absolutely continuous. Condition (11.2.5) is trivial because the active boundary Γ is empty and the validity of (11.2.6) can be shown in the same way as in Section 11.3.1. This proves statement (a). To show (b), we apply Lemma 12.3.1 with $b'_{ij}(s) = b_{ij}(1; s) = \mathbb{E}\left(\mathrm{e}^{-s(R(1)-u)}\mathbb{I}(J(1) = j) \mid J(0) = i\right)$. Note that $b'_{ij}(s) = \int_{\infty}^{\infty} \mathrm{e}^{-sx}\, F'_{ij}(\mathrm{d}x)$, where $F'_{ij}(x) = \mathbb{P}(R(1) - u \le x, J(1) = j \mid J(0) = i)$. By Hölder's inequality each $b'_{ij}(s)$ is logconvex. We still have to show (12.3.16). Recall that $\theta(s)$ is an eigenvalue of $\boldsymbol{C}(s)$, i.e. $\vartheta = \theta(s)$ is a solution to the equation $\det(\boldsymbol{C}(s) - \vartheta \boldsymbol{I}) = 0$. By the implicit function theorem, see Theorems 17.1.1 and 17.4.1 in Hille (1966), $\theta(s)$ is differentiable on $(-\infty, s_0)$. Furthermore, $\boldsymbol{\phi}(s)$ is the solution to $(\boldsymbol{C}(s) - \theta(s)\boldsymbol{I})(\boldsymbol{\phi}(s))^\top = \boldsymbol{0}$, i.e. a rational function of differentiable functions. Thus $\boldsymbol{\phi}(s)$ is differentiable. Besides this, we have $\boldsymbol{\pi C}(s)(\boldsymbol{\phi}(s))^\top = \theta(s)\boldsymbol{\pi}(\boldsymbol{\phi}(s))^\top = \theta(s)$ since we normalized $\boldsymbol{\phi}(s)$ in such a way that $\boldsymbol{\pi}(\boldsymbol{\phi}(s))^\top = 1$. Hence

$$\theta^{(1)}(s) = \boldsymbol{\pi C}^{(1)}(s)(\boldsymbol{\phi}(s))^\top + \boldsymbol{\pi C}(s)(\boldsymbol{\phi}^{(1)}(s))^\top .$$

Letting $s \to 0$, (12.3.16) follows from (12.3.6) and (12.3.12) because $\boldsymbol{\pi C}(0) = \boldsymbol{\pi Q} = \boldsymbol{0}$ and $\boldsymbol{\phi}(0) = \boldsymbol{e}$. $\square$

12.3.4 Lundberg Bounds

We now use the martingale $\{M(t)\}$ given in (12.3.15), when changing the probability measure, as was done in Section 11.3.2. Let $s \in \mathbb{R}$ be such that $\hat{m}_i(s) < \infty$ for all $i = 1, \ldots, \ell$ and let $\boldsymbol{\phi}(s) = (\phi_i(s))_{i=1,\ldots,\ell}$ be the right eigenvector corresponding to the eigenvalue $\mathrm{e}^{\theta(s)}$ of $\exp(\boldsymbol{C}(s))$ introduced in Section 12.3.3. Using the martingale $\{M^{(s)}(t), t \geq 0\}$ with

$$M^{(s)}(t) = (\phi_{J(0)}(s))^{-1}\phi_{J(t)}(s)\mathrm{e}^{-s(R(t)-u)-\theta(s)t} \tag{12.3.17}$$

we define the probability measures $\{\mathbb{P}_t^{(s)}, t \geq 0\}$ by $\mathbb{P}_t^{(s)}(B) = \mathbb{E}\,[M^{(s)}(t); B]$ for $B \in \mathcal{F}_t$, where the process $\{(J(t), R(t))\}$ is assumed to be given on its canonical probability space and $\{\mathcal{F}_t\}$ denotes the history of $\{(J(t), R(t))\}$. We leave it to the reader to show that the measures $\mathbb{P}_t^{(s)}$ can be extended to a "global" measure $\mathbb{P}^{(s)}$ on $\mathcal{F} = \sigma(\bigcup_{t\geq 0} \mathcal{F}_t)$, see also Section 11.3.2. It turns out that this new measure $\mathbb{P}^{(s)}$ again describes a Markov-modulated risk model.

Lemma 12.3.3 *Under the measure $\mathbb{P}^{(s)}$, the process $\{(J(t), R(t))\}$ is a Markov-modulated risk model with intensity matrix $\boldsymbol{Q}^{(s)} = (q_{ij}^{(s)})$, where $q_{ij}^{(s)} = (\phi_i(s))^{-1}\phi_j(s)q_{ij}$ for $i \neq j$. The claim arrival intensities are $\lambda_i^{(s)} = \lambda_i \hat{m}_i(s)$ and the claim size distributions are $F_i^{(s)}(y) = \int_0^y \mathrm{e}^{sz}\,\mathrm{d}F_i(z)/\hat{m}_i(s)$. In particular, $\mathbb{P}^{(s)}(\lim_{t\to\infty} t^{-1}R(t) = -\theta^{(1)}(s)) = 1$.*

The *proof* of Lemma 12.3.3 is similar to that of Theorem 11.3.1 and consists of very long calculations. We only sketch a number of the most important constituent steps.

Step 1 Under $\mathbb{P}^{(s)}$, $\{J(t)\}$ is a Markov process with intensity matrix $\boldsymbol{Q}^{(s)}$.

Step 2 Given $J(0), J(\xi_1), \ldots$ and $\xi_1, \xi_2, \ldots$, the stochastic processes $\{R(\xi_n + t) - R(\xi_n) : 0 \le t \le \xi_{n+1} - \xi_n\}$ are (conditionally) independent and the dependence on $\mathcal{F}_{\xi_n}$ occurs via $J(\xi_n)$ only.

Step 3 Given $J(0) = i$ and $\xi_1 > v$, $N(v)$ has the conditional distribution $\mathrm{Poi}(\lambda_i^{(s)})$.

Step 4 Given $J(0) = i$, $\xi_1 > v$ and $N(v) = n$, the claim sizes $U_1, \ldots, U_n$ are independent and identically distributed with distribution $F_i^{(s)}$, independent of the claims arrival epochs.

Step 5 Given $J(0) = i$, $\xi_1 > v$ and $N(v) = n$, the first n claim arrival epochs have the same conditional distribution as under the original measure $\mathbb{P}$.

Step 6 Knowing that $\{(J(t), R(t))\}$ is a Markov-modulated risk model under $\mathbb{P}^{(s)}$, Theorems 12.3.2 and 12.3.4 give that

$$\mathbb{P}^{(s)}(\lim_{t\to\infty} t^{-1}R(t) = -\theta_s^{(1)}(0)) = 1\,,$$

where $\theta_s(h)$ corresponds to the function $\theta(h)$, but now under the measure $\mathbb{P}^{(s)}$. Thus, it remains to show that $\theta_s^{(1)}(0) = \theta^{(1)}(s)$. This is an immediate consequence of the fact that $\theta_s(h) = \theta(s+h) - \theta(s)$. An easy way to prove the latter relationship is to show that the stochastic process $\{\tilde{M}^{(h)}(t),\ t \ge 0\}$ with

$$\tilde{M}^{(h)}(t) = \frac{\phi_{J(t)}(s+h)}{\phi_{J(t)}(s)} \mathrm{e}^{-hR(t)} \mathrm{e}^{-(\theta(s+h)-\theta(s))t} \tag{12.3.18}$$

is a $\mathbb{P}^{(s)}$-martingale provided $\theta(s+h)$ exists. Indeed, in the proof of Theorem 12.3.4 we have seen that for a martingale of the form (12.3.18) it is necessary that $\tilde{\phi}_i(h) = \phi_{J(t)}(s+h)/\phi_{J(t)}(s)$ is an eigenvector and $\theta_s(h)$ is an eigenvalue. Since $\tilde{\phi}_i(h) > 0$, $\theta_s(h)$ must be the Perron–Frobenius eigenvalue. Let $t \ge v$. Then, using (10.2.26) we have

$$\begin{aligned}
&\mathbb{E}^{(s)}(M^{(h)}(t) \mid \mathcal{F}_v)\\
&= \frac{\mathbb{E}^{(0)}(M^{(h)}(t)(\mathbb{E}^{(0)}(\phi_{J(0)}(s)))^{-1}\phi_{J(t)}(s)\mathrm{e}^{-s(R(t)-u)-\theta(s)t} \mid \mathcal{F}_v)}{(\mathbb{E}^{(0)}(\phi_{J(0)}(s)))^{-1}\phi_{J(v)}(s)\mathrm{e}^{-s(R(v)-u)-\theta(s)v}}\\
&= \frac{\mathbb{E}^{(0)}((\phi_{J(t)}(s))^{-1}\phi_{J(t)}(s+h)\mathrm{e}^{-hR(t)-\theta_s(h)t}\phi_{J(t)}(s)\mathrm{e}^{-sR(t)-\theta(s)t} \mid \mathcal{F}_v)}{(\phi_{J(v)}(s))^{-1}\mathrm{e}^{-sR(v)-\theta(s)v}}\\
&= (\phi_{J(v)}(s))^{-1}\mathbb{E}^{(0)}(\phi_{J(t)}(s+h)\mathrm{e}^{-(s+h)R(t)-\theta(s+h)t} \mid \mathcal{F}_v)\mathrm{e}^{sR(v)+\theta(s)v}\\
&= (\phi_{J(v)}(s))^{-1}\phi_{J(v)}(s+h)\mathrm{e}^{-hR(v)-(\theta(s+h)-\theta(s))v} = M^{(h)}(v)\,.
\end{aligned}$$

This verifies that $\theta_s(h) = \theta(s+h) - \theta(s)$. □

In Theorem 12.3.4 it has been shown that the function $\theta(s)$ is convex while its derivative at 0 is negative. Thus, besides $s = 0$, there might be a second solution $s = \gamma > 0$ to the equation $\theta(s) = 0$. If γ exists, we again call γ the *adjustment coefficient*. The following theorem derives Lundberg bounds for $\psi(u)$ for this case. Let $x_i = \sup\{y : F_i(y) < 1\}$.

Theorem 12.3.5 *Let $\{(J(t), R(t))\}$ be a Markov-modulated risk model and assume that the adjustment coefficient γ exists. Then, for all $u \ge 0$*

$$a_- e^{-\gamma u} \le \psi(u) \le a_+ e^{\gamma u}, \tag{12.3.19}$$

where

$$\begin{aligned} a_- &= \min_{1\le i\le \ell} \inf_{0<y<x_i} \frac{e^{\gamma y}\overline{F}_i(y)\mathbb{E}^{(0)}\phi_{J(0)}(\gamma)}{\phi_i(\gamma)\int_y^\infty e^{\gamma z}\,dF_i(z)}, \\ a_+ &= \max_{1\le i\le \ell} \sup_{0<y<x_i} \frac{e^{\gamma y}\overline{F}_i(y)\mathbb{E}^{(0)}\phi_{J(0)}(\gamma)}{\phi_i(\gamma)\int_y^\infty e^{\gamma z}\,dF_i(z)}. \end{aligned}$$

Proof Let $\tau(u)$ be the time of ruin to the initial risk reserve u. By Theorem 12.3.4, $\theta(s)$ is a convex function. This yields that $\theta^{(1)}(\gamma) > 0$ and therefore $\mathbb{P}^{(\gamma)}(\tau(u) < \infty) = 1$ because $\mathbb{P}^{(\gamma)}(R(t) \to -\infty) = 1$ by Lemma 12.3.3. For the ruin probability $\psi(u)$ we get an expression under the measure $\mathbb{P}^{(\gamma)}$:

$$\psi(u) = \mathbb{E}^{(0)}\phi_{J(0)}(\gamma)\,\mathbb{E}^{(\gamma)}((\phi_{J(\tau(u))}(\gamma))^{-1}e^{\gamma R(\tau(u))})e^{-\gamma u}. \tag{12.3.20}$$

Condition on $J(\tau(u))$ and $R(\tau(u) - 0)$ to find

$$\begin{aligned} &\mathbb{E}^{(\gamma)}((\phi_{J(\tau(u))}(\gamma))^{-1}e^{\gamma R(\tau(u))} \mid J(\tau(u)) = i, R(\tau(u)-0) = y) \qquad (12.3.21) \\ &= (\phi_i(\gamma))^{-1}\mathbb{E}^{(\gamma)}(e^{\gamma(y-U^+)} \mid J(\tau(u)) = i, U^+ > y) = \frac{e^{\gamma y}\overline{F}_i(y)}{\phi_i(\gamma)\int_y^\infty e^{\gamma z}\,dF_i(z)} \end{aligned}$$

where U^+ denotes the size of the claim causing ruin. The assertion readily follows. $\square$

Note that $\min\{\phi_i(\gamma) : 1 \le i \le \ell\} > 0$ and therefore the upper bound in (12.3.19) is finite.

12.3.5 Cramér–Lundberg Approximation

We now study the question whether $\psi(u)e^{\gamma u}$ converges to a limit as $u \to \infty$, i.e. whether a Cramér–Lundberg approximation holds.

Theorem 12.3.6 *Let $\{(J(t), R(t))\}$ be a Markov-modulated risk model. Assume that the adjustment coefficient γ exists and that $\hat{m}_i^{(1)}(\gamma) < \infty$ for $i = 1, \ldots, \ell$. Then there exists a constant $c > 0$ such that*

$$\lim_{u\to\infty} \psi(u)e^{\gamma u} = c\,\mathbb{E}^{(0)}\phi_{J(0)}(\gamma).$$

Proof In view of (12.3.20) we have to show that the function

$$g(u) = \mathbb{E}^{(\gamma)}((\phi_{J(\tau(u))}(\gamma))^{-1}\mathrm{e}^{\gamma R(\tau(u))}) \tag{12.3.22}$$

converges to a constant c as $u \to \infty$. Assume first that the initial state of $\{J(t)\}$ is fixed, say $J(0) = 1$. Without loss of generality we can assume that $\lambda_1 > 0$. Let

$$\nu^- = \inf\{t > 0 : J(t) = 1 \text{ and } R(t) = \inf_{0 \le v \le t} R(v)\}$$

be the first descending ladder epoch occurring in state 1. Note that $\mathbb{P}^{(\gamma)}(\nu^- < \infty) = 1$ because $\mathbb{P}^{(\gamma)}(\lim_{t\to\infty} R(t) = -\infty) = 1$. Write $G^-(y) = \mathbb{P}^{(\gamma)}(u - R(\nu^-) \le y) = \mathbb{P}^{(\gamma)}(S(\nu^-) \le y)$ for the (modified) ladder height distribution. The function $g(u)$ fulfils the following renewal equation; see also Section 11.3.3,

$$g(u) = \int_0^u g(u-y)\,\mathrm{d}G^-(y) + \mathbb{E}^{(\gamma)}[(\phi_{J(\tau(u))}(\gamma))^{-1}\mathrm{e}^{\gamma R(\tau(u))}; R(\nu^-) < 0]\,,$$

because $\mathbb{E}^{(\gamma)}((\phi_{J(\tau(u))}(\gamma))^{-1}\mathrm{e}^{\gamma R(\tau(u))} \mid R(\nu^-) = u - y) = g(u-y)$ for $y \le u$. Define

$$z_i(u) = \mathbb{E}^{(\gamma)}[(\phi_{J(\tau(u))}(\gamma))^{-1}\mathrm{e}^{\gamma R(\tau(u))}; R(\nu^-) < 0 \mid J(0) = i]\,.$$

Note that $z_i(u)$ is bounded and $z_1(y) = (\phi_1(\gamma))^{-1}\mathrm{e}^{\gamma y}$ if $y < 0$. For $h > 0$ small

$$\begin{aligned} z_1(u) &= (1 - \lambda_1^{(\gamma)}h + q_{11}^{(\gamma)}h)z_1(u+\beta h) \\ &\quad + \lambda_1^{(\gamma)}h\int_0^\infty z_1(u-y)\,\mathrm{d}F_1^{(\gamma)}(y) + \sum_{j=2}^{\ell} q_{1j}^{(\gamma)}hz_j(u) + O(h)\,. \end{aligned}$$

Letting $h \to 0$ shows that $z_1(u)$ is right-continuous. From

$$\begin{aligned} z_1(u-\beta h) &= (1 - \lambda_1^{(\gamma)}h + q_{11}^{(\gamma)}h)z_1(u) \\ &\quad + \lambda_1^{(\gamma)}h\int_0^\infty z_1(u-y)\,\mathrm{d}F_1^{(\gamma)}(y) + \sum_{j=2}^{\ell} q_{1j}^{(\gamma)}hz_j(u) + O(h) \end{aligned}$$

it follows that $z_1(u)$ is left-continuous as well. Let $\phi_{\min}(\gamma) = \min_i \phi_i(\gamma)$. Then

$$\phi_{\min}(\gamma)z_1(u) \le \mathbb{P}^{(\gamma)}(R(\nu^-) < 0) = \mathbb{P}^{(\gamma)}(S(\nu^-) > u)\,.$$

Since the latter function is monotone, it will be directly Riemann integrable if it is integrable. But this is the case because

$$\int_0^\infty \mathbb{P}^{(\gamma)}(S(\nu^-) > u)\,\mathrm{d}u = \mathbb{E}^{(\gamma)}(S(\nu^-)) < \infty\,.$$

The expectation on the right is finite since under the measure $\mathbb{P}^{(\gamma)}$ the (conditional) expectations of the claim sizes are finite. Indeed, the latter property is implied by our assumption that $m_i^{(1)}(\gamma) < \infty$ for all $i = 1, \ldots, \ell$. The details are left to the reader. Note that, at an ordinary ladder epoch occurring in state j, there is a strictly positive probability that the next ordinary ladder epoch will occur in state 1. Thus, the function $z_1(u)$ is continuous and has a directly Riemann integrable upper bound. But this then ensures that $z_1(u)$ is directly Riemann integrable, as can be easily proved by the reader. Hence, it follows from Theorem 6.1.11 that the limit of $g(u)$ exists as $u \to \infty$. Turning to the general case, let $J(0)$ have an arbitrary distribution. Then, $g'(u) = \mathbb{E}^{(\gamma)}((\phi_{J(\tau(u))}(\gamma))^{-1}\mathrm{e}^{\gamma R(\tau(u))} \mid J(0) = 1)$ is the function considered before. Let $\tau' = \inf\{t \geq 0 : J(t) = 1\}$ and let $B'(y) = \mathbb{P}^{(\gamma)}(S(\tau') \leq y)$ be the distribution function of the claim surplus when $\{J(t)\}$ reaches state 1 for the first time. Then

$$g(u) = \int_{-\infty}^{u} g'(u-y)\,\mathrm{d}B'(y) + \mathbb{E}^{(\gamma)}[(\phi_{J(\tau(u))}(\gamma))^{-1}\mathrm{e}^{\gamma R(\tau(u))}; \tau(u) < \tau'],$$

where $g(u)$ is defined in (12.3.22). Note that the second summand on the right-hand side of this expression is bounded by $(\phi_{\min}(\gamma))^{-1}\mathbb{P}^{(\gamma)}(\tau(u) < \tau')$. We leave it to the reader to show that $\mathbb{P}^{(\gamma)}(\tau(u) < \tau')$ tends to 0 as $u \to \infty$. Since $g'(u)$ is bounded, it follows by the dominated convergence theorem that

$$\lim_{u\to\infty} g(u) = \lim_{u\to\infty} g'(u) = c,$$

which proves the theorem. □

12.3.6 Finite-Horizon Ruin Probabilities

To close Section 12.3, we extend the results of Section 10.3.2 on Lundberg bounds for finite-horizon ruin probabilities. For simplicity, we assume that $\theta(s)$ exists for all s considered below; a sufficient condition is, for example, that $s_i^+ = \sup\{s : \hat{m}_i(s) < \infty\}$ and $s_\infty = \min s_i^+$. Assume $\lim_{s\uparrow s_\infty} \hat{m}_i(s) = \infty$ for all i such that $s_\infty = s_i^+$. In this case $s_\infty > 0$ because $\hat{m}_i(0) = 1$. Moreover, we put $\theta(s) = \infty$ if $\theta(s)$ does not exist. Recall that $\theta(s) > 0$ for $s < 0$. The following inequalities hold.

Theorem 12.3.7 *Let $\gamma(y) = \sup\{s - \theta(s)y : s \in \mathbb{R}\}$ and let $s(y)$ be the argument at which the supremum is attained. Moreover, let*

$$c_s = \max_{1\leq i\leq \ell}\ \sup_{0<y<x_i} \frac{\mathrm{e}^{sy}\overline{F}_i(y)}{\phi_i(s)\int_y^\infty \mathrm{e}^{sz}\,\mathrm{d}F_i(z)}.$$

The following statements are valid:
(a) *If* $y\theta^{(1)}(\gamma) < 1$, *then* $\gamma(y) > \gamma$ *and*

$$\psi(u; yu) \le c_{s(y)} \mathbb{E}^{(0)} \phi_{J(0)}(\gamma(y))\, \mathrm{e}^{-\gamma(y)u} . \tag{12.3.23}$$

(b) *If* $y\theta^{(1)}(\gamma) > 1$, *then* $\gamma(y) > \gamma$ *and*

$$\psi(u) - \psi(u; yu) \le c_{s(y)} \mathbb{E}^{(0)} \phi_{J(0)}(\gamma(y))\, \mathrm{e}^{-\gamma(y)u} . \tag{12.3.24}$$

Proof (a) It suffices to consider $s \ge 0$ for which $\theta(s)$ exists. As usual $\tau(u)$ denotes the ruin time, starting with initial risk reserve u. Then, as in (12.3.20), we have

$$\begin{aligned}
\psi(u; yu) &= \mathbb{E}^{(0)} \phi_{J(0)}(s)\, \mathbb{E}^{(s)}[(\phi_{J(\tau(u))}(s))^{-1} \mathrm{e}^{sR(\tau(u))+\theta(s)\tau(u)}; \tau(u) \le yu] \mathrm{e}^{-su} \\
&\le \mathbb{E}^{(0)} \phi_{J(0)}(s)\, c_s\, \mathbb{E}^{(s)}[\mathrm{e}^{\theta(s)\tau(u)}; \tau(u) \le yu] \mathrm{e}^{-su} \\
&\le \mathbb{E}^{(0)} \phi_{J(0)}(s)\, c_s\, \mathrm{e}^{-\min\{s, s-y\theta(s)\}u} ,
\end{aligned}$$

where the first inequality follows as in (12.3.21). The derivative of $s - y\theta(s)$ at $s = \gamma$ is $1 - y\theta^{(1)}(\gamma) > 0$. Thus $\gamma(y) > \gamma$. Further $s - y\theta(s)$ is concave and hence the argument $s(y)$ at which the maximum is attained, is larger than γ. In particular $\theta(s(y)) > 0$. Thus $s(y) > s(y) - y\theta(s(y))$ and $\min\{s(y), s(y) - y\theta(s(y))\} = \gamma(y)$ proving the first part of the assertion.
(b) Note that for $0 \le s \le \gamma$ we have as before that

$$\begin{aligned}
&\psi(u) - \psi(u; yu) \\
&\quad = \mathbb{E}^{(0)} \phi_{J(0)}(s)\, \mathbb{E}^{(s)}[(\phi_{J(\tau(u))}(s))^{-1} \mathrm{e}^{sR(\tau(u))+\theta(s)\tau(u)}; yu < \tau(u) < \infty] \mathrm{e}^{-su} \\
&\quad \le \mathbb{E}^{(0)} \phi_{J(0)}(s)\, c_s\, \mathbb{E}^{(s)}[\mathrm{e}^{\theta(s)\tau(u)}; yu < \tau(u) < \infty] \mathrm{e}^{-su} \\
&\quad \le \mathbb{E}^{(0)} \phi_{J(0)}(s)\, c_s\, \mathrm{e}^{-(s-y\theta(s))u} ,
\end{aligned}$$

where in the last inequality we used that $\theta(s) \le 0$. Further, the derivative of the concave function $s - y\theta(s)$ at $s = \gamma$ is $1 - y\theta^{(1)}(\gamma) < 0$. Part (b) follows since $s(y) < \gamma$, and this immediately implies that $\gamma(y) > \gamma$. □

Corollary 12.3.1 *Assume there exists* $s > \gamma$ *such that* $\hat{m}_i(s) < \infty$ *for* $1 \le i \le \ell$. *Then* $\lim_{u\to\infty} u^{-1}\tau(u) = (\theta^{(1)}(\gamma))^{-1}$ *in probability on the set* $\{\tau(u) < \infty\}$.

The *proof* of Corollary 12.3.1 is omitted since it is analogous to the proof of Corollary 10.3.1. □

Bibliographical Notes. Lemma 12.3.1 is due to Kingman (1961); see also Miller (1961). Lemma 12.3.2 and Theorem 12.3.6 were first proved by Asmussen (1989). In the same paper an upper bound for the ruin probability $\psi(u)$

was derived which is larger than that given in Theorem 12.3.5. For two-sided Lundberg bounds in a Markovian environment, see also Grigelionis (1993). Matrix-algorithmic methods for the numerical computation of the ruin probability $\psi(u)$ are studied in Asmussen and Rolski (1991). A method for statistical estimation of the adjustment coefficient is given in Schmidli (1997b). In Bäuerle (1997), the expected ruin time $\mathbb{E}\,\tau(u)$ is investigated in the case of a negative safety loading. An optimal stopping problem for a Markov-modulated risk reserve process is studied in Jensen (1997). For the case that interest and cost rates are also included in the model, see Schöttl (1998). A model where the premium rate depends both on the current surplus and on the state of the Markov environment is considered in Asmussen and Kella (1996).

12.4 PERIODIC RISK MODEL

In practical situations, the claim arrival rate may vary with the time of the year or the claim size distribution may be depend on the seasons. Let $\lambda(t)$ be the intensity function of a nonhomogeneous Poisson process $\{N(t),\ t \ge 0\}$. In this section we assume that $\lambda(t)$ is periodic with period 1, say, so that $[t] = t - \lfloor t \rfloor$ is the time of season. We say that $\{N(t)\}$ is a *periodic Poisson process*. Let $\{F_t(x),\ t \ge 0\}$ be a family of distribution functions such that the mapping $t \to \int_0^\infty g(x)\,\mathrm{d}F_t(x)$ is measurable and periodic with period 1 for all integrable functions g.

Assume now that claims arrive according to the periodic Poisson process $\{N(t)\}$ with intensity function $\lambda(t)$ and that – if a claim arrives at time t – then the claim size distribution is F_t, independent of everything else. We denote the moment generating function of F_t by $\hat{m}_t(s) = \int_0^\infty \mathrm{e}^{sx}\,F_t(\mathrm{d}x)$. We also assume that the premium rate is constant and equal to β. From the construction of the risk reserve process it is apparent that $\{R(t) - u,\ t \ge 0\}$ has independent increments. We now compute the Laplace–Stieltjes transform $\hat{l}(t;s) = \mathbb{E}\,\mathrm{e}^{-s(R(t)-u)}$.

Lemma 12.4.1 *For $s, t \ge 0$,*

$$\hat{l}(t;s) = \exp\Big(-\beta st + \int_0^t \lambda(v)(\hat{m}_v(s) - 1)\,\mathrm{d}v\Big)\,. \tag{12.4.1}$$

Proof Using Theorem 12.2.1 we find

$$\begin{aligned}\mathbb{E}\,\mathrm{e}^{-s(R(t)-u)} &= \sum_{n=0}^{\infty} \mathbb{E}\,\big(\mathrm{e}^{-s(\beta t - \sum_{i=1}^{N(t)} U_i)} \mid N(t) = n\big)\mathbb{P}(N(t) = n) \\ &= \mathrm{e}^{-s\beta t} \sum_{n=0}^{\infty} \Big(\frac{\int_0^t \int_0^\infty \mathrm{e}^{sy} F_v(\mathrm{d}y)\lambda(v)\,\mathrm{d}v}{\int_0^t \lambda(v)\,\mathrm{d}v}\Big)^n \frac{(\int_0^t \lambda(v)\,\mathrm{d}v)^n}{n!} \mathrm{e}^{-\int_0^t \lambda(v)\,\mathrm{d}v}\end{aligned}$$

$$= \exp\Big(-\beta st + \int_0^t \lambda(v)(\hat{m}_v(s) - 1)\,\mathrm{d}v\Big). \qquad \square$$

Let us denote the average arrival rate by $\lambda = \int_0^1 \lambda(v)\,\mathrm{d}v$ while $F_U^0(x) = \lambda^{-1}\int_0^1 \lambda(v)F_v(x)\,\mathrm{d}v$ is the distribution function of a typical claim size. Its mean is $\mu_U^0 = \int_0^\infty x\,F_U^0(\mathrm{d}x)$ and the moment generating function is $\hat{m}_{F_U^0}(s) = \lambda^{-1}\int_0^1 \lambda(v)\hat{m}_v(s)\,\mathrm{d}v$. Let $s_0 = \sup\{s \geq 0 : \sup_{v\in[0,1)} \hat{m}_v(s) < \infty\}$. Similarly as in (11.3.2) we define $\theta^*(s) = \lambda(\hat{m}_{F_U^0}(s) - 1) - \beta s$.

Let $\gamma > 0$ be the solution to $\int_0^1 \lambda(v)(\hat{m}_v(\gamma) - 1)\,\mathrm{d}v = \beta\gamma$. Then $\gamma > 0$ fulfils $\theta^*(\gamma) = 0$. In the sequel, we assume that such a γ exists. Since $\theta^*(0) = 0$ and the derivative of $\theta^*(s)$ at zero is $\lambda\mu_U^0 - \beta < 0$, the convexity of $\theta^*(s)$ ensures that

$$\theta^{*(1)}(\gamma) = \lambda \int_0^\infty x\mathrm{e}^{\gamma x} F_U^0(\mathrm{d}x) - \beta > 0. \tag{12.4.2}$$

By $\{\mathcal{F}_t\}$ we denote the smallest (uncompleted) right-continuous filtration such that $\{R(t)\}$ is adapted. Then, a *law of large numbers* and an *exponential martingale* can be derived for the periodic Poisson risk model.

Theorem 12.4.1 *The risk reserve process $\{R(t)\}$ fulfils*

$$\lim_{t\to\infty} \frac{1}{t}R(t) = \beta - \lambda\mu_U^0 .$$

Moreover for all $s < s_0$, the process $\{M(t)\}$ given by $M(t) = \mathrm{e}^{-s(R(t)-u)}/\hat{l}(t;s)$ is a $\{\mathcal{F}_t\}$-martingale.

Proof The random variables $Y_1 = R(1) - u, Y_2 = R(2) - R(1), \ldots$ are independent and identically distributed. From the derivative of their Laplace–Stieltjes transform at $s = 0$ we obtain $\mathbb{E}\,Y = \mathbb{E}\,(R(1) - u) = \beta - \lambda\mu_U^0$. We also have

$$\begin{aligned} \frac{\lfloor t\rfloor + 1}{t}\frac{1}{\lfloor t\rfloor + 1}\sum_{i=1}^{\lfloor t\rfloor+1} Y_i - \frac{\beta(\lfloor t\rfloor + 1 - t)}{t} &\leq \frac{R(t) - u}{t} \\ &< \frac{\lfloor t\rfloor}{t}\frac{1}{\lfloor t\rfloor}\sum_{i=1}^{\lfloor t\rfloor} Y_i + \frac{\beta(t - \lfloor t\rfloor)}{t}. \end{aligned}$$

The strong law of large numbers yields the first part of the theorem. Similarly as in Example 3 of Section 10.1.3 we can prove that $\mathrm{e}^{-s(R(t)-u)}/\mathbb{E}\,\mathrm{e}^{-s(R(t)-u)}$ is an $\{\mathcal{F}_t\}$-martingale. By Lemma 12.4.1, the proof is then complete. $\square$

For all $t > 0$, we define $\mathbb{P}_t^{(s)}(A) = \mathbb{E}\,[M(t); A]$, where $A \in \mathcal{F}_t$. Since $\{M(t)\}$ is a positive martingale and $M(0) = 1$, it is easily seen that $\mathbb{P}_t^{(s)}$

is a probability measure and

$$\mathbb{P}_t^{(s)}(A) = \mathbb{E}\,[M(t'); A], \qquad A \in \mathcal{F}_t\,,$$

whenever $t \leq t'$. We leave it to the reader to show that the measures $\{\mathbb{P}_t^{(s)},\ t \geq 0\}$ can be extended to a "global" probability measure $\mathbb{P}^{(s)}$ on $\mathcal{F}_\infty = \sigma(\bigcup_{t\geq 0} \mathcal{F}_t)$. In the remaining part of this section we only consider $\mathbb{P}^{(\gamma)}$ and as usual $\mathbb{P}^{(0)} = \mathbb{P}$.

Lemma 12.4.2 *The risk reserve process $\{R(t),\ t \geq 0\}$ on $(\Omega, \mathcal{F}_\infty, \mathbb{P}^{(\gamma)})$ is again that of a periodic Poisson model specified by $(\beta, \tilde\lambda(t), \tilde F_t(x))$, where*

$$\tilde\lambda(t) = \lambda(t)\hat m_t(\gamma), \qquad \mathrm{d}\tilde F_t(x) = \frac{\mathrm{e}^{\gamma x}\,\mathrm{d}F_t(x)}{\hat m_t(\gamma)}\,.$$

Proof For $t \leq t'$ and $0 \leq s \leq s_0$ we have

$$\begin{aligned}
&\mathbb{E}_{t'}^{(\gamma)} \mathrm{e}^{-s(R(t)-u)} \\
&= \mathbb{E}\,\mathrm{e}^{-s(R(t)-u)} M(t) = \mathbb{E}\,\mathrm{e}^{-s(R(t)-u)-\gamma(R(t)-u)+\beta s t-\int_0^t \lambda(v)(\hat m_v(\gamma)-1)\,\mathrm{d}v} \\
&= \exp\Big(\int_0^t (\lambda(v)(\hat m_v(s+\gamma) - \hat m_v(\gamma)) - \beta s)\,\mathrm{d}v\Big) \\
&= \exp\Big(\int_0^t (\tilde\lambda(v)(\hat m_{\tilde F_v}(\gamma) - 1) - \beta s)\,\mathrm{d}v\Big)\,.
\end{aligned}$$

In a similar way we can prove that for $t_1 < t_2 < \ldots < t_n \leq t'$ and $0 \leq s_1, \ldots, s_n \leq s_0$

$$\mathbb{E}_{t'}^{(\gamma)} \mathrm{e}^{-(s_1(R(t_1)-u)+\ldots+s_n(R(t_n)-u))} = \prod_{i=1}^n \mathbb{E}_{t'}^{(\gamma)} \mathrm{e}^{-s_i(R(t_i)-u)}\,.$$

Thus, using Lemma 12.4.1, the assertion follows. □

Next we compute the trend of the $\{R(t)\}$ under $\mathbb{P}^{(\gamma)}$. Using (12.4.2) and Theorem 12.4.1, we obtain

$$\mathbb{E}^{(\gamma)} R(1) - u = -\theta^{*(1)}(\gamma) < 0\,.$$

Hence by Theorem 6.3.1c we have $\liminf_{t\to\infty} R(t) = -\infty$ and so ruin occurs with $\mathbb{P}^{(\gamma)}$-probability 1. Let $\tau(u)$ be the time of ruin starting with initial risk reserve u. As in (10.2.31), we can show that

$$\psi(u) = \mathbb{P}(\tau(u) < \infty) = \mathrm{e}^{-\gamma u}\mathbb{E}^{(\gamma)}\big(\hat l(\tau(u), \gamma)\mathrm{e}^{\gamma R(\tau(u))}\big)\,. \tag{12.4.3}$$

This relation brings us in a position to derive Lundberg bounds for the ruin function $\psi(u)$. Put $x_v = \sup\{y : F_v(y) < 1\}$.

Theorem 12.4.2 *For the ruin function $\psi(u)$ in the periodic Poisson model*

$$a_- \mathrm{e}^{-\gamma u} \le \psi(u) \le a_+ \mathrm{e}^{-\gamma u}, \qquad u \ge 0\,, \tag{12.4.4}$$

where

$$\begin{aligned} a_- &= \inf_{0\le v\le 1} \hat{l}(v;\gamma) \inf_{0\le x<x_v} \frac{\overline{F}_v(x)}{\int_x^\infty \mathrm{e}^{\gamma(y-x)}\, F_v(\mathrm{d}y)}\,, \\ a_+ &= \sup_{0\le v\le 1} \hat{l}(v;\gamma) \sup_{0\le x<x_v} \frac{\overline{F}_v(x)}{\int_x^\infty \mathrm{e}^{\gamma(y-x)}\, F_v(\mathrm{d}y)}\,. \end{aligned}$$

Proof Using the fact that $R(\tau) = R(\tau-0) - U^+$, where U^+ is the size of the claim causing ruin, we have

$$\begin{aligned} \psi(u) &= \mathrm{e}^{-\gamma u}\mathbb{E}^{(\gamma)}\big(\hat{l}(\tau(u);\gamma)\mathrm{e}^{\gamma(R(\tau-0)-U^+)}\big) \\ &= \mathrm{e}^{-\gamma u}\mathbb{E}^{(\gamma)}\big(\mathbb{E}^{(\gamma)}(l(\lfloor\tau\rfloor;\gamma)\mathrm{e}^{\gamma(R(\tau-0)-U^+)} \mid \lfloor\tau\rfloor, R(\tau-0))\big) \\ &\ge \mathrm{e}^{-\gamma u} \inf_{0\le v\le 1} \hat{l}(v;\gamma) \inf_{0\le x<x_v} \frac{\int_0^x \mathrm{e}^{\gamma(y-x)}\tilde{F}_v(\mathrm{d}y)}{1-\tilde{F}_v(x)} = a_- \mathrm{e}^{-\gamma u}\,. \end{aligned}$$

The upper bound in (12.4.4) can be derived in the same way. □

We remark that – under some additional conditions – it is also possible to prove a Cramér–Lundberg type approximation of the form:

$$\lim_{u\to\infty} \mathrm{e}^{\gamma u}\psi(u) = c(\hat{l}(0;\gamma))^{-1}\,. \tag{12.4.5}$$

Bibliographical Notes. The material of this section is from Asmussen and Rolski (1994). Periodic risk models were also considered by Asmussen and Rolski (1991), Beard, Pentikäinen and Pesonen (1984) and Dassios and Embrechts (1989).

12.5 THE BJÖRK–GRANDELL MODEL VIA PDMP

The Björk–Grandell model has been introduced in Example 5 of Section 12.2.2. Recall that $\{(\Lambda_i, I_i),\ i\ge 1\}$ are independent random vectors with $\mathbb{P}(\Lambda_i \ge 0, I_i > 0) = 1$ and $\{(\Lambda_i, I_i),\ i \ge 2\}$ are identically distributed. Let $\xi_n = \sum_{i=1}^n I_i$ be the time where the intensity level changes for the n-th time. For convenience we put $\xi_0 = 0$. Then $\lambda(t) = \Lambda_n$ if $\xi_{n-1} \le t < \xi_n$ and the cumulative intensity function $\eta(t)$ is given by $\eta(t) = \int_0^t \lambda(v)\,\mathrm{d}v$. Conditioned on $\{(\Lambda_i, I_i),\ i \ge 1\}$, the expected number of claims in $(0,t]$ will be $\eta(t)$. Let $\{N'(t)\}$ be a homogeneous Poisson process with intensity 1. The claim counting process $\{N(t)\}$ in the Björk–Grandell model can then be defined as $N(t) = N'(\eta(t))$; see Theorem 12.2.3. As before the risk process $\{R(t)\}$ is given by

$$R(t) = u + \beta t - \sum_{i=1}^{N(t)} U_i \,,$$

where $\{U_i\}$ is a family of independent and identically distributed random variables, independent of $\{(\Lambda_i, I_i)\}$. By (Λ, I) we denote a generic vector with the same distribution as (Λ_2, I_2). In order to exclude trivialities, we assume that $\mathbb{E}\,\Lambda > 0$ and $\mathbb{E}\,I < \infty$. Furthermore, to ensure that $\mathbb{E}\,N(t) < \infty$ for all $t > 0$, we assume $\mathbb{E}\,\Lambda < \infty$. If we want to formulate a net profit condition, we also have to assume that $\mathbb{E}\,(\Lambda I) < \infty$. If the support of Λ only consists of a finite number of points and I conditioned on Λ is exponentially distributed, then we have a Markov-modulated risk model. Note, however, that there exist Markov-modulated risk models that are not Björk–Grandell models.

12.5.1 Law of Large Numbers

We begin with the investigation of the asymptotic behaviour of $\{R(t)\}$ as $t \to \infty$.

Theorem 12.5.1 *Let $\mu = \mu_U$ denote the expected claim size. Then,*

$$\lim_{t\to\infty} t^{-1} R(t) = \beta - \mu \mathbb{E}\,(\Lambda I)/\mathbb{E}\,I \,.$$

Proof We have to show that $\lim_{t\to\infty} t^{-1} \sum_{i=1}^{N(t)} U_i = \mu \mathbb{E}\,(\Lambda I)/\mathbb{E}\,I$. First note that

$$\frac{1}{t}\sum_{i=1}^{N(t)} U_i = \frac{\eta(t)}{t}\,\frac{1}{\eta(t)} \sum_{i=1}^{N'(\eta(t))} U_i \,.$$

Because $\eta(t) \to \infty$, it follows from the law of large numbers for the sum of independent random variables and from Theorem 6.1.1a that

$$\lim_{t\to\infty} \frac{1}{\eta(t)} \sum_{i=1}^{N'(\eta(t))} U_i = \mu \,.$$

It remains to show that $\lim_{t\to\infty} t^{-1}\eta(t) = \mathbb{E}\,(\Lambda I)/\mathbb{E}\,I$. Define the counting process $N_\xi(t) = \sup\{n \ge 0 : \xi_n \le t\}$ to be the number of changes of the intensity level in the time interval $(0, t]$. Then

$$\frac{N_\xi(t)}{t}\,\frac{1}{N_\xi(t)} \int_0^{\xi_{N_\xi(t)}} \lambda(v)\,dv \le \frac{\eta(t)}{t} \le \frac{N_\xi(t)+1}{t}\,\frac{1}{N_\xi(t)+1} \int_0^{\xi_{N_\xi(t)+1}} \lambda(v)\,dv \,.$$

Since $\{N_\xi(t)\}$ is a renewal process we have $\lim_{t\to\infty} t^{-1}N_\xi(t) = (\mathbb{E}\, I)^{-1}$ by Theorem 6.1.1. Finally

$$\frac{1}{N_\xi(t)}\int_0^{\xi_{N_\xi(t)}} \lambda(v)\,\mathrm{d}v = \frac{1}{N_\xi(t)}\sum_{i=1}^{N_\xi(t)} \Lambda_i I_i$$

tends to $\mathbb{E}\,(\Lambda I)$ by the usual law of large numbers. □

The expected income in (ξ_1, ξ_2) is $\beta\mathbb{E}\, I$ and the expected aggregate claim amount during that period is $\mu\mathbb{E}\,(\Lambda I)$. Considering the random walk $\{R(\xi_n),\ n \ge 1\}$ it follows from Theorem 6.3.1 that ruin occurs almost surely if $\beta\mathbb{E}\, I \le \mu\mathbb{E}\,(\Lambda I)$. We therefore assume the *net profit condition*

$$\beta\mathbb{E}\, I > \mu\mathbb{E}\,(\Lambda I)\,, \tag{12.5.1}$$

which ensures that $R(t) \to \infty$ as $t \to \infty$.

12.5.2 The Generator and Exponential Martingales

The stochastic process $\{(R(t), t),\ t \ge 0\}$ is not a Markov process in general. In order to use the results derived in Chapter 11 for PDMP, we have to add further supplementary variables. In particular, if the conditional distribution of I given Λ is not exponential, then, as in the case of the Sparre Andersen model considered in Section 11.5, the time since the last or the time till the next change of the intensity level has to be included into the model. In Section 11.5 we showed that it is more convenient to use the time till the next change of the intensity level. Let $A(t) = \xi_n - t$ for $\xi_{n-1} \le t < \xi_n$. Then it is not difficult to see that the process $\{(R(t), \lambda(t), A(t), t)\}$ is a PDMP. The active boundary is $\Gamma = \{(y, z, 0, t) : (y, z, t) \in \mathbb{R} \times \mathbb{R}_+ \times \mathbb{R}_+\}$. For each function g fulfilling the conditions of Theorem 11.2.2 the generator $\boldsymbol{A}$ is given by

$$\begin{aligned} \boldsymbol{A}g(y,z,w,t) &= \frac{\partial}{\partial t}g(y,z,w,t) + \beta\frac{\partial}{\partial y}g(y,z,w,t) - \frac{\partial}{\partial w}g(y,z,w,t) \\ &+ z\Bigl(\int_0^\infty g(y-v,z,w,t)\,\mathrm{d}F_U(v) - g(y,z,w,t)\Bigr) \end{aligned} \tag{12.5.2}$$

and the boundary condition (11.2.5) is

$$g(y,z,0,t) = \int_0^\infty\int_0^\infty g(y,v,w,t)\,F_{\Lambda,I}(\mathrm{d}v,\mathrm{d}w)\,. \tag{12.5.3}$$

To get an idea on how to find an exponential martingale we solve the equation $\boldsymbol{A}g(y,z,w,t) = 0$. We try a function of the form $g(y,z,w,t) = a(z,w)\mathrm{e}^{-sy}\mathrm{e}^{-\theta t}$. Hence we have to solve the differential equation

$$-\theta a(z,w) - \beta s a(z,w) - a^{(01)}(z,w) + z a(z,w)(\hat{m}_U(s) - 1) = 0\,,$$

where $a^{(01)}(z,w)$ denotes the partial derivative with respect to w. The general solution to the above equation is

$$a(z,w) = a'(z)\,\mathrm{e}^{(z(\hat{m}_U(s)-1)-\beta s-\theta)w}$$

for some function a'. Since the boundary condition (12.5.3) has to be fulfilled as well, it follows that

$$\begin{aligned} a'(z) &= \int_0^\infty\int_0^\infty a'(v)\mathrm{e}^{(v(\hat{m}_U(s)-1)-\beta s-\theta)w}\,F_{\Lambda,I}(\mathrm{d}v,\mathrm{d}w) \\ &= \mathbb{E}\,(a'(\Lambda)\mathrm{e}^{(\Lambda(\hat{m}_U(s)-1)-\beta s-\theta)I})\,. \end{aligned}$$

As the right-hand side of this equation is independent of z, $a'(z)$ is constant. Without loss of generality we can assume that $a'(z)=1$. It thus follows that $\theta=\theta(s)$, where $\theta(s)$ is a solution to the equation

$$\mathbb{E}\,\exp((\Lambda(\hat{m}_U(s)-1)-\beta s-\theta(s))I)=1\,. \tag{12.5.4}$$

We leave it to the reader to show that, once it exists, this solution is unique. Without using the results of Section 11.2.3 explicitly, we get the following theorem.

Theorem 12.5.2 *Let $s_0>0$ such that $\hat{m}_U(s_0)<\infty$ and assume that for each $s\le s_0$ the solution $\theta(s)$ to (12.5.4) exists. Then, the following statements are true:*
(a) *For each $s\le s_0$, the process $\{M(t)\}$ with*

$$M(t)=\exp((\lambda(t)(\hat{m}_U(s)-1)-\beta s-\theta(s))A(t)-sR(t)-\theta(s)t) \tag{12.5.5}$$

is a martingale with respect to the history $\{\mathcal{F}_t\}$ of $\{(R(t),\lambda(t),A(t))\}$.
(b) *The function $\theta(s)$ is strictly convex on $(-\infty,s_0]$ and $\theta^{(1)}(0)=-\beta+\mu\mathbb{E}\,(\Lambda I)/\mathbb{E}\,I$.*

Proof (a) We prove the martingale property directly. There is no loss of generality to assume $u=0$. First, consider the process $\{M(t)\}$ at the epochs ξ_n only. Note that $\lambda(\xi_n)=\Lambda_{n+1}$ and $A(\xi_n)=I_{n+1}$ is independent of $\mathcal{F}_{\xi_{n-1}}$, ξ_n and $R(\xi_n)$. Then by the definition of $\theta(s)$

$$\begin{aligned} \mathbb{E}\,(M(\xi_n)\mid\mathcal{F}_{\xi_{n-1}}) &= \mathbb{E}\,(\mathrm{e}^{-sR(\xi_n)-\theta(s)\xi_n}\mid\mathcal{F}_{\xi_{n-1}}) \\ &= \mathrm{e}^{(\Lambda_n(\hat{m}_U(s)-1)-\beta s)I_n}\,\mathrm{e}^{-sR(\xi_{n-1})-\theta(s)(\xi_{n-1}+I_n)}\,, \end{aligned}$$

where the last equality follows from (5.2.7). This means $\mathbb{E}\,(M(\xi_n)\mid\mathcal{F}_{\xi_{n-1}})=M(\xi_{n-1})$, which implies

$$\mathbb{E}\,(M(\xi_n)\mid\mathcal{F}_{\xi_k})=M(\xi_k)\,,\qquad k\le n. \tag{12.5.6}$$

Now let $0 \le v < t$. Analogously as above it follows that $\mathbb{E}(M(v + A(v)) \mid \mathcal{F}_v) = M(v)$, $\mathbb{E}(M(t) \mid \mathcal{F}_{\xi_{N_\xi(t)}}) = M(\xi_{N_\xi(t)})$ and that for $t \le v + A(v)$, we have $\mathbb{E}(M(t) \mid \mathcal{F}_v) = M(v)$. We can therefore assume that $t > v + A(v)$, i.e. $v + A(v) \le \xi_{N_\xi(t)}$. The next step is to show that

$$\begin{aligned} \mathbb{E}(M(\xi_{N_\xi(t)}) \mid \mathcal{F}_{v+A(v)}) &= \mathbb{E}(\mathrm{e}^{-sR(\xi_{N_\xi(t)})-\theta(s)\xi_{N_\xi(t)}} \mid \mathcal{F}_{v+A(v)}) \\ &= M(v + A(v)) . \end{aligned} \tag{12.5.7}$$

If $s \le 0$, then the integrand in the second expression is bounded by $\mathrm{e}^{-\beta st}$ because in this case $\theta(s) \ge 0$. If $s > 0$ then the integrand is bounded by $\exp(s\sum_{i=1}^{N'(t)} U_i)$, where $N'(t) = N(\xi_{N_\xi(t)})$. To see this, it suffices to notice that $s > 0$ implies $\beta s + \theta(s) > 0$, which follows from the definition of $\theta(s)$. We want to show that the latter bound is integrable. From (5.2.7) we know that

$$\begin{aligned} \mathbb{E}\exp\Big(s\sum_{i=1}^{N'(t)} U_i\Big) &= \mathbb{E}\exp\Big(\sum_{i=1}^{N_\xi(t)} \Lambda_i I_i(\hat{m}_U(s) - 1)\Big) \\ &\le \mathrm{e}^{(\beta s+\theta(s))t}\mathbb{E}\exp\Big(\sum_{i=1}^{N_\xi(t)} (\Lambda_i(\hat{m}_U(s) - 1) - \beta s - \theta(s))I_i\Big) . \end{aligned}$$

From Fatou's lemma and the definition of $\theta(s)$ it then follows that

$$\begin{aligned} &\mathbb{E}\lim_{n\to\infty}\exp\Big(\sum_{i=1}^{N_\xi(t)\wedge n} (\Lambda_i(\hat{m}_U(s) - 1) - \beta s - \theta(s))I_i\Big) \\ &\le \liminf_{n\to\infty}\mathbb{E}\exp\Big(\sum_{i=1}^{N_\xi(t)\wedge n} (\Lambda_i(\hat{m}_U(s) - 1) - \beta s - \theta(s))I_i\Big) = 1 . \end{aligned}$$

Thus (12.5.7) follows from (12.5.6) by the dominated convergence theorem. That the process in (12.5.5) is a martingale follows from the fact that, for $t > v + A(v)$,

$$\begin{aligned} &\mathbb{E}(M(t) \mid \mathcal{F}_v) = \mathbb{E}(\mathbb{E}(M(t) \mid \mathcal{F}_{\xi_{N_\xi(t)}}) \mid \mathcal{F}_v) \\ &= \mathbb{E}(\mathbb{E}(M(\xi_{N_\xi(t)}) \mid \mathcal{F}_{v+A(v)}) \mid \mathcal{F}_v) = \mathbb{E}(M(v + A(v)) \mid \mathcal{F}_v) = M(v) . \end{aligned}$$

(b) For simplicity, we consider the nondelayed case only. So, we take $(\Lambda_1, I_1) \stackrel{\mathrm{d}}{=} (\Lambda_i, I_i)$ for $i \ge 2$. Recall from (5.2.7) that

$$\mathbb{E}(\mathrm{e}^{(\Lambda(\hat{m}_U(s)-1)-\beta s-\theta(s))I} \mid \Lambda, I) = \mathbb{E}(\mathrm{e}^{-s(R(I)-u)-\theta(s)I} \mid \Lambda, I) .$$

Thus $\theta(s)$ fulfils the equation

$$\mathbb{E}(\mathrm{e}^{-s(R(I)-u)-\theta(s)I}) = 1 . \tag{12.5.8}$$

By the implicit function theorem – see Hille (1966) – $\theta(s)$ is differentiable and

$$-\mathbb{E}\left((R(I)-u+\theta^{(1)}(s)I)\mathrm{e}^{-s(R(I)-u)-\theta(s)I}\right)=0\,,$$

from which it follows that $\theta(s)$ is infinitely often differentiable. Letting $s=0$ yields $\theta^{(1)}(0)=-\mathbb{E}\,(R(I)-u)/\mathbb{E}\,I=-\beta+\mu\mathbb{E}\,\Lambda I/\mathbb{E}\,I$. The second derivative of (12.5.8) is

$$\mathbb{E}\left(((R(I)-u+\theta^{(1)}(s)I)^2-\theta^{(2)}(s)I)\mathrm{e}^{-s(R(I)-u)-\theta(s)I}\right)=0\,,$$

from which $\theta^{(2)}(s)>0$ readily follows. □

12.5.3 Lundberg Bounds

We turn to the ruin function $\psi(u)=\mathbb{P}(\min\{t:R(t)<0\}<\infty)$. Note that the underlying Björk–Grandell claim arrival process needs neither to be time-stationary nor Palm-stationary. Nevertheless, we need the net profit condition (12.5.1).

As before, choose $s\in\mathbb{R}$ so that $\theta(s)$ is well-defined. Let $a(z,w)=\mathrm{e}^{(z(\hat{m}_U(s)-1)-\beta s-\theta)w}$. Consider the likelihood ratio process $\{L_s(t),\ t\ge 0\}$ – a nonnegative martingale with mean 1, where

$$L_s(t)=(\mathbb{E}\,a(\Lambda_1,I_1))^{-1}\mathrm{e}^{(\lambda(t)(\hat{m}_U(s)-1)-\beta s-\theta(s))A(t)}\mathrm{e}^{-s(R(t)-u)}\mathrm{e}^{-\theta(s)t}\,,$$

and define the new measures $\mathbb{P}_t^{(s)}(B)=\mathbb{E}^{(0)}[L_s(t);B]$ for $B\in\mathcal{F}_t$. We again use the smallest right-continuous filtration $\{\mathcal{F}_t\}$ such that $\{(R(t),\lambda(t),A(t))\}$ is adapted. Then the measure $\mathbb{P}_t^{(s)}$ can be extended to a "global" measure $\mathbb{P}^{(s)}$ on $\mathcal{F}_\infty$; see Remark 2 in Section 10.2.6. As one can expect, under $\mathbb{P}^{(s)}$, the process $\{(R(t),\lambda(t),A(t))\}$ is again defined by a Björk–Grandell model. But we have to slightly adapt the notation. In the new model, the process $\{\lambda(t)\}$ will no longer be the intensity process. We therefore denote the intensity process by $\{\lambda^{(s)}(t)\}$, where $\lambda^{(0)}(t)=\lambda(t)$.

Lemma 12.5.1 *Under the measure $\mathbb{P}^{(s)}$ the process $\{(R(t),\lambda(t),A(t))\}$ is defined by a Björk–Grandell model with intensity process $\{\lambda^{(s)}(t)\}$ given by $\lambda^{(s)}(t)=\hat{m}_U(s)\lambda(t)$ and with claim size distribution function $F_U^{(s)}(y)=\int_0^y \mathrm{e}^{sv}\mathrm{d}F_U(v)/\hat{m}_U(s)$. The distribution of (Λ,I) is given by*

$$F_{\Lambda,I}^{(s)}(\mathrm{d}z,\mathrm{d}w)=\mathrm{e}^{(z(\hat{m}_U(s)-1)-\beta s-\theta(s))w}\,F_{\Lambda,I}(\mathrm{d}z,\mathrm{d}w)$$

and

$$F_{\Lambda_1,I_1}^{(s)}(\mathrm{d}z,\mathrm{d}w)=(\mathbb{E}\,a(L_1,I_1))^{-1}\mathrm{e}^{(z(\hat{m}_U(s)-1)-\beta s-\theta(s))w}\,F_{\Lambda_1,I_1}(\mathrm{d}z,\mathrm{d}w)\,.$$

In particular, $\mathbb{P}^{(s)}(\lim_{t\to\infty}t^{-1}R(t)=-\theta^{(1)}(s))=1$.

The *proof* is omitted since it is similar to the proof of Lemma 12.3.3.

In Theorem 12.5.2 we have seen that $\theta(s)$ is a strictly convex function, even with a negative derivative at 0 under the net profit condition (12.5.1). There might be a second solution $s = \gamma > 0$ to $\theta(s) = 0$ besides the trivial solution $s = 0$. Such a solution $\gamma > 0$ is called the *adjustment coefficient.* Assume now that γ exists. The ruin probability $\psi(u)$ can be determined using the measure $\mathbb{P}^{(\gamma)}$. Indeed, since $\theta^{(1)}(\gamma) > 0$, we have $\mathbb{P}^{(\gamma)}(\tau(u) < \infty) = 1$. Thus, proceeding in the same way as in Section 11.3.1, we have

$$\psi(u) = \mathbb{E}^{(0)} a(\Lambda_1, I_1)\, \mathbb{E}^{(\gamma)}(\mathrm{e}^{(\beta\gamma - \lambda(\tau(u))(\hat{m}_U(\gamma)-1))A(\tau(u))}\mathrm{e}^{\gamma R(\tau(u))})\mathrm{e}^{-\gamma u}\,. \tag{12.5.9}$$

In order to use this formula we have to determine the distribution of $(\lambda(\tau(u)), A(\tau(u)))$. Rather than solving this hard problem, we obtain an upper bound for the ruin probability $\psi(u)$ but only under an additional assumption. Let $x_0 = \sup\{y : F(y) < 1\}$.

Theorem 12.5.3 *Let $\{(R(t), \lambda(t), A(t))\}$ be defined by a Björk–Grandell model such that the adjustment coefficient γ exists. The following statements are true:*

(a) *Assume there exists a constant $c > 1$ such that*

$$\inf_{\substack{x \in B \\ y \geq 0}} \mathbb{E}^{(0)}(\mathrm{e}^{(\Lambda(\hat{m}_U(\gamma)-1)-\beta\gamma)(S-y)} \mid \Lambda = x, I > y) \geq c^{-1}\,, \tag{12.5.10}$$

where $B \subset \{x : x(\hat{m}_U(\gamma) - 1) < \beta\gamma\}$ is a set such that $\mathbb{P}^{(0)}(\Lambda \in B) = \mathbb{P}^{(0)}(\Lambda(\hat{m}_U(\gamma) - 1) < \beta\gamma)$. This is particularly the case when there exists a constant $c' < \infty$ such that

$$\sup_{\substack{x \in B \\ y \geq 0}} \mathbb{E}^{(0)}(I - y \mid \Lambda = x, I > y) \leq c'$$

and $c = \mathrm{e}^{\beta\gamma c'}$. Then

$$\psi(u) \leq \mathbb{E}^{(0)} \max\{1, ca(\Lambda_1, I_1)\} \sup_{0<y<x_0} \frac{\mathrm{e}^{\gamma y}\overline{F}_U(y)}{\int_y^\infty \mathrm{e}^{\gamma v}\,\mathrm{d}F_U(v)}\mathrm{e}^{-\gamma u}\,. \tag{12.5.11}$$

(b) *Assume $\hat{m}_U^{(1)}(\gamma) < \infty$. Then*

$$\begin{aligned}\psi(u) \quad \geq \quad & \mathbb{E}^{(0)} a(\Lambda_1, I_1) \min\Big\{\inf_{y \geq 0} \frac{\mathrm{e}^{\gamma y}\overline{G}(y)}{\int_y^\infty \mathrm{e}^{\gamma v}\,\mathrm{d}G(v)}\,, \\ & \frac{1}{\mathbb{E}^{(0)}[\mathrm{e}^{-\gamma R(I_1)} \mid R(I_1) < 0]}\Big\}\,\mathrm{e}^{-\gamma u}\,, \end{aligned} \tag{12.5.12}$$

where $G(y) = \mathbb{P}^{(0)}(R(\xi_2) - R(\xi_1) \geq -y)$.

Proof (a) Assume first that a constant c' with the desired properties exists. Then by Jensen's inequality

$$\begin{aligned}
&\mathbf{E}^{(0)}(\mathrm{e}^{(\Lambda(\hat{m}_U(\gamma)-1)-\beta\gamma)(I-y)} \mid \Lambda = x, I > y) \\
&\geq \mathbf{E}^{(0)}(\mathrm{e}^{-\beta\gamma(I-y)} \mid \Lambda = x, I > y) \\
&\geq \mathrm{e}^{-\beta\gamma\mathbf{E}^{(0)}(I-y\mid\Lambda=x,I>y)} \geq \mathrm{e}^{-\beta\gamma c'} = c^{-1}.
\end{aligned}$$

Thus there exists a constant $c > 1$ such that (12.5.10) is fulfilled. Conditioning on (Λ_1, I_1) we now assume that Λ_1 and I_1 are deterministic. Later on we remove this additional assumption by integrating the obtained expression. Furthermore, conditioning on $\{\lambda(\tau(u)), A(\tau(u)), R(\tau(u) - 0)\}$ and using (12.5.9), we find

$$\begin{aligned}
\psi(u) &\leq a(\Lambda_1, I_1)\mathbf{E}^{(\gamma)}(\mathrm{e}^{(\beta\gamma-\lambda(\tau(u))(\hat{m}_U(\gamma)-1))A(\tau(u))}) \\
&\quad \times \sup_{0<y<x_0} \frac{\mathrm{e}^{\gamma y}\overline{F}_U(y)}{\int_y^\infty \mathrm{e}^{\gamma v}\,\mathrm{d}F_U(v)}\,\mathrm{e}^{-\gamma u};
\end{aligned}$$

see also (12.3.21). Taking the expectation $\mathbf{E}^{(\gamma)}$ we split the area of integration into the three subsets $\{\tau(u) < I_1\}$, $\{\tau(u) \geq I_1, \lambda(\tau(u))(\hat{m}_U(\gamma)-1) \geq \beta\gamma\}$ and $\{\tau(u) \geq I_1, \lambda(\tau(u))(\hat{m}_U(\gamma)-1) < \beta\gamma\}$. First we consider the set $\{\tau(u) < I_1\}$. Then

$$\begin{aligned}
&\mathbf{E}^{(\gamma)}(\mathrm{e}^{(\beta\gamma-\lambda(\tau(u))(\hat{m}_U(\gamma)-1))A(\tau(u))} \mid \tau(u) < I_1) \\
&\leq \max\{1, \mathrm{e}^{(\beta\gamma-\Lambda_1(\hat{m}_U(\gamma)-1))I_1}\} = \max\{1, (a(\Lambda_1, I_1))^{-1}\}.
\end{aligned}$$

On the set $\{\tau(u) \geq I_1, \lambda(\tau(u))(\hat{m}_U(\gamma)-1) \geq \beta\gamma\}$ we have

$$\mathbf{E}^{(\gamma)}(\mathrm{e}^{(\beta\gamma-\lambda(\tau(u))(\hat{m}_U(\gamma)-1))A(\tau(u))} \mid \tau(u) \geq I_1, \lambda(\tau(u))(\hat{m}_U(\gamma)-1) \geq \beta\gamma) \leq 1.$$

The most delicate case stems from the integration on the set $\{\tau(u) \geq I_1, \lambda(\tau(u))(\hat{m}_U(\gamma)-1) < \beta\gamma\}$. We first claim that

$$\mathbf{E}^{(\gamma)}(\mathrm{e}^{(\beta\gamma-\Lambda(\hat{m}_U(\gamma)-1))(I-y)} \mid \Lambda = x, I > y) \leq c$$

for any $x \in B$. It follows from (10.2.25) that

$$\begin{aligned}
&\mathbf{E}^{(\gamma)}(\mathrm{e}^{(\beta\gamma-\Lambda(\hat{m}_U(\gamma)-1))(I-y)} \mid \Lambda = x, I > y) \\
&= \frac{\mathrm{e}^{(x(\hat{m}_U(\gamma)-1)-\beta\gamma)y}}{\mathbf{E}^{(0)}(\mathrm{e}^{(\Lambda(\hat{m}_U(\gamma)-1)-\beta\gamma)I} \mid \Lambda = x, I > y)} \\
&= \frac{1}{\mathbf{E}^{(0)}(\mathrm{e}^{(\Lambda(\hat{m}_U(\gamma)-1)-\beta\gamma)(I-y)} \mid \Lambda = x, I > y)} \leq c.
\end{aligned}$$

Define the random variable τ_* as the last epoch before ruin time $\tau(u)$ where the intensity level changes. Then

$$\begin{aligned}&\mathbb{E}^{(\gamma)}(\mathrm{e}^{(\beta\gamma-\lambda(\tau(u))(\hat{m}_U(\gamma)-1))A(\tau(u))} \mid \tau(u) \geq I_1, \lambda(\tau(u))(\hat{m}_U(\gamma)-1) < \beta\gamma) \\ &= \mathbb{E}^{(\gamma)}\Big(\mathrm{e}^{(\beta\gamma-\lambda(\tau_*)(\hat{m}_U(\gamma)-1))(A(\tau_*)-(\tau(u)-\tau_*))} \,\Big|\, \\ &\qquad\qquad \tau(u) \geq I_1, \lambda(\tau_*)(\hat{m}_U(\gamma)-1) < \beta\gamma\Big).\end{aligned}$$

Conditioning on $\tau(u)$, τ_* and $\lambda(\tau_*)$ is like conditioning on $\tau(u)$, τ_*, $\lambda(\tau_*)$ and $A(\tau_*) > \tau(u) - \tau_*$. On the set $\{\lambda(\tau_*) \in B\}$ we find

$$\mathbb{E}^{(\gamma)}\Big(\mathrm{e}^{(\beta\gamma-\lambda(\tau_*)(\hat{m}_U(\gamma)-1))(A(\tau_*)-(\tau(u)-\tau_*))} \,\Big|\, \tau(u), \tau_*, \lambda(\tau_*)\Big) \leq c.$$

Thus

$$\mathbb{E}^{(\gamma)}(\mathrm{e}^{(\beta\gamma-\lambda(\tau(u))(\hat{m}_U(\gamma)-1))A(\tau(u))} \mid \tau(u) \geq I_1, \lambda(\tau(u))(\hat{m}_U(\gamma)-1) < \beta\gamma) \leq c.$$

Summarizing the above results we find that

$$\mathbb{E}^{(\gamma)}(\mathrm{e}^{(\beta\gamma-\lambda(\tau(u))(\hat{m}_U(\gamma)-1))A(\tau(u))}) \leq \max\{c, (a(\Lambda_1, I_1))^{-1}\},$$

where we used the fact that $c \geq 1$. Thus

$$\psi(u) \leq \max\{1, ca(\Lambda_1, I_1)\} \sup_{0<y<x_0} \frac{\mathrm{e}^{\gamma y}\overline{F}_U(y)}{\int_y^\infty \mathrm{e}^{\gamma v}\,\mathrm{d}F_U(v)} \mathrm{e}^{-\gamma u}.$$

If (Λ_1, I_1) is not deterministic one has to take the expectation.
(b) Define $\tau' = \inf\{\xi_n : R(\xi_n) < 0\}$ the first epoch where the intensity level changes and the risk reserve process is negative. Let $\psi'(u) = \mathbb{P}^{(0)}(\tau' < \infty)$. It is clear that $\psi(u) \geq \psi'(u)$. Since $\mathbb{P}^{(\gamma)}(\tau' < \infty) = 1$ and since $(\lambda(\tau'), A(\tau'))$ is independent of $\{R(t) : 0 \leq t \leq \tau'\}$ and has the same distribution as (Λ, I) we obtain

$$\begin{aligned}\psi'(u) &= \mathbb{E}^{(0)}a(\Lambda_1, I_1)\mathbb{E}^{(\gamma)}(\mathrm{e}^{(\beta\gamma-\lambda(\tau')(\hat{m}_U(\gamma)-1))A(\tau')})\mathbb{E}^{(\gamma)}(\mathrm{e}^{\gamma R(\tau')})\mathrm{e}^{-\gamma u} \\ &= \mathbb{E}^{(0)}a(\Lambda_1, I_1)\mathbb{E}^{(\gamma)}(\mathrm{e}^{\gamma R(\tau')})\mathrm{e}^{-\gamma u}\end{aligned}$$

where we used Lemma 12.5.1. Let τ'_* be the last epoch where the intensity level changes before τ'. We have to distinguish between the two cases: $\tau'_* = 0$ and $\tau'_* \neq 0$. Note that by (10.2.25)

$$\begin{aligned}&\mathbb{E}^{(\gamma)}(\mathrm{e}^{\gamma R(\tau')} \mid \tau'_* = 0) = \mathbb{E}^{(\gamma)}(\mathrm{e}^{\gamma R(I_1)} \mid R(I_1) < 0) \\ &= \frac{\mathbb{E}^{(0)}[\mathrm{e}^{\gamma R(I_1)}\mathrm{e}^{-\gamma(R(I_1)-u)}; R(I_1) < 0]}{\mathbb{E}^{(0)}[\mathrm{e}^{-\gamma(R(I_1)-u)}; R(I_1) < 0]} = \frac{1}{\mathbb{E}^{(0)}(\mathrm{e}^{-\gamma R(I_1)} \mid R(I_1) < 0)}.\end{aligned}$$

Conditioning on $R(\tau'_*) = y$ yields

$$\begin{aligned}&\mathbb{E}^{(\gamma)}(\mathrm{e}^{\gamma R(\tau')} \mid \tau'_* \neq 0, R(\tau'_*) = y) \\ &= \mathbb{E}^{(\gamma)}(\mathrm{e}^{\gamma(y+(R(\xi_2)-R(\xi_1)))} \mid R(\xi_2) - R(\xi_1) < -y).\end{aligned}$$

To simplify the notation let $X = R(\xi_2) - R(\xi_1)$. Then by (10.2.25)

$$\begin{aligned}\mathbb{E}^{(\gamma)}(\mathrm{e}^{\gamma(y+X)} \mid X < -y) &= \frac{\mathbb{E}^{(0)}[\mathrm{e}^{\gamma(y+X)}\mathrm{e}^{-\gamma X}; X < -y]}{\mathbb{E}^{(0)}[\mathrm{e}^{-\gamma X}; X < -y]} \\ &= \frac{\mathrm{e}^{\gamma y}\mathbb{P}^{(0)}(X < -y)}{\mathbb{E}^{(0)}[\mathrm{e}^{-\gamma X}; X < -y]},\end{aligned}$$

which proves the assertion by taking the infimum over $y \geq 0$. □

Example Consider the Ammeter risk model where $\xi_n = n$. Then $\theta(s)$ is defined by $\mathbb{E}\,(\mathrm{e}^{\Lambda(\hat{m}_U(s)-1)-\beta s-\theta(s)}) = 1$, i.e. $\theta(s) = \log \mathbb{E}\, \exp(\Lambda(\hat{m}_U(s)-1)) - \beta s$. The net profit condition (12.5.1) takes the form $\beta > \mu \mathbb{E}\,\Lambda$. The Lundberg bounds derived in Theorem 12.5.3 simplify to

$$\psi(u) \leq \mathbb{E}\,(\max\{1, \mathrm{e}^{\Lambda(\hat{m}_U(\gamma)-1)-\beta\gamma}\}) \sup_{0<y<x_0} \frac{\mathrm{e}^{\gamma y}\overline{F}_U(y)}{\int_y^\infty \mathrm{e}^{\gamma v}\,\mathrm{d}F_U(v)}\mathrm{e}^{-\gamma u} \tag{12.5.13}$$

and

$$\psi(u) \geq \inf_{0<y<x_0} \frac{\mathrm{e}^{\gamma y}\overline{G}(y)}{\int_y^\infty \mathrm{e}^{\gamma v}\,\mathrm{d}G(v)}\mathrm{e}^{-\gamma u}, \tag{12.5.14}$$

where

$$G(y) = \mathbb{P}^{(0)}\Big(\sum_{i=1}^{N(1)} U_i \leq \beta + y\Big) = \sum_{n=0}^{\infty} F_U^{*n}(\beta+y)\frac{\mathbb{E}\,(\Lambda^n \mathrm{e}^{-\Lambda})}{n!}. \tag{12.5.15}$$

In particular, if Λ has the gamma distribution $\Gamma(a,b)$ then the net profit condition is $b\beta > a\mu$ and

$$\theta(s) = a(\log b - \log(b - (\hat{m}_U(s) - 1))) - \beta s$$

provided that $\hat{m}_U(s) - 1 < b$. In this case, the distribution function $G(y)$ defined in (12.5.15) is given by

$$G(y) = \sum_{n=0}^{\infty} F_U^{*n}(\beta+y)\frac{b^a}{(b+1)^{n+a}}\frac{\Gamma(n+a)}{n!\Gamma(a)}$$

which is a shifted compound negative binomial distribution.

12.5.4 Cramér–Lundberg Approximation

We turn to the question about under what conditions a Cramér–Lundberg approximation to $\psi(u)$ would be valid. From (12.5.9) we find

$$\psi(u)\mathrm{e}^{\gamma u} = \mathbb{E}^{(0)}a(\Lambda_1, I_1)\mathbb{E}^{(\gamma)}(\mathrm{e}^{(\beta\gamma-\lambda(\tau(u)))(\hat{m}_U(\gamma)-1))A(\tau(u))}\mathrm{e}^{\gamma R(\tau(u))}).$$

The existence of a Cramér–Lundberg approximation is therefore equivalent to the existence of the limit

$$\lim_{u\to\infty} \mathbb{E}^{(\gamma)}(\mathrm{e}^{(\beta\gamma-\lambda(\tau(u))(\hat{m}_U(\gamma)-1))A(\tau(u))}\mathrm{e}^{\gamma R(\tau(u))}).$$

We leave it to the reader to verify that this limit does not depend on the distribution of (Λ_1, I_1). It is, however, not trivial to prove that the limit actually exists.

Theorem 12.5.4 *Assume that the adjustment coefficient $\gamma > 0$ exists, that $\hat{m}_U^{(1)}(\gamma) < \infty$ and that there exists a constant $c > 1$ such that*

$$\inf_{\substack{x\in B\\ y\geq 0}} \mathbb{E}^{(0)}(\mathrm{e}^{(\Lambda(\hat{m}_U(\gamma)-1)-\beta\gamma)(I-y)} \mid \Lambda = x, I > y) \geq c\,,$$

where $B \subset \{x : x(\hat{m}_U(\gamma) - 1) < \beta\gamma\}$ is a set such that $\mathbb{P}^{(0)}(\Lambda \in B) = \mathbb{P}^{(0)}(\Lambda(\hat{m}_U(\gamma) - 1) < \beta\gamma)$. Then there exists a constant c', not depending on the initial state, such that

$$\lim_{u\to\infty} \psi(u)\mathrm{e}^{\gamma u} = \mathbb{E}^{(0)} a(\Lambda_1, I_1)\, c'\,. \tag{12.5.16}$$

The *proof* of Theorem 12.5.4 is omitted. It can be found in Schmidli (1997a).

12.5.5 Finite-Horizon Ruin Probabilities

Ultimately, we turn to finite-horizon ruin probabilities. We need to be guaranteed that subsequent quantities are well-defined. Assume therefore that for every $s \in \mathbb{R}$ considered the solution $\theta(s)$ to (12.5.4) exists. We also assume that the adjustment coefficient $\gamma > 0$ exists. Furthermore, in order to get upper bounds, the technical assumption of Theorem 12.5.3a must be fulfilled for all s considered. As we showed in the proof of Theorem 12.5.3, this is most easily achieved by assuming that there exists $c' < \infty$ such that

$$\sup_{\substack{x\in B\\ y\geq 0}} \mathbb{E}^{(0)}(I - y \mid \Lambda = x, I > y) \leq c' < \infty\,,$$

where B is a set such that $\mathbb{P}^{(0)}(\Lambda \in B) = 1$. To simplify the notation let $\theta(s) = \infty$ if $\theta(s)$ is not defined.

Theorem 12.5.5 *Let $\gamma(y) = \sup\{s - y\theta(s) : s \in \mathbb{R}\}$ and denote by $s(y)$ the argument where the supremum is attained. Moreover, let*

$$c_s = \mathbb{E}^{(0)} \max\{1, \mathrm{e}^{\beta s c'} a(\Lambda_1, I_1)\} \sup_{0<y<x_0} \frac{\mathrm{e}^{sy}\overline{F}_U(y)}{\int_y^\infty \mathrm{e}^{sv}\,\mathrm{d}F_U(v)}\,.$$

The following statements are true:
(a) *If* $y\theta^{(1)}(\gamma) < 1$, *then* $\gamma(y) > \gamma$ *and*

$$\psi(u; yu) \le c_{s(y)} \mathrm{e}^{-\gamma(y)u} . \tag{12.5.17}$$

(b) *If* $y\theta^{(1)}(\gamma) > 1$, *then* $\gamma(y) > \gamma$ *and*

$$\psi(u) - \psi(u; yu) \le c_{s(y)} \mathrm{e}^{-\gamma(y)u} . \tag{12.5.18}$$

The *proof* of Theorem 12.5.5 is left to the reader since it is similar to the proof of Theorem 12.3.7.

Corollary 12.5.1 *Assume that there exists* $s > \gamma$ *such that* $\hat{m}(s) < \infty$. *Then* $\lim_{u\to\infty} u^{-1}\tau(u) = (\theta^{(1)}(\gamma))^{-1}$ *in probability on the set* $\{\tau(u) < \infty\}$.

The *proof* of Corollary 12.5.1 is analogous to the proof of Corollary 10.3.1.

Bibliographical Notes. The risk model studied in this section was introduced in Björk and Grandell (1988) as a generalization of the model considered by Ammeter (1948); see also Grandell (1995). The approach to this model via PDMP is similar to that used in Dassios (1987) for the Sparre Andersen model; see also Dassios and Embrechts (1989) and Embrechts, Grandell and Schmidli (1993). An upper bound for $\psi(u)$ which is larger than that given in (12.5.11) was obtained in Björk and Grandell (1988). Theorem 12.5.2, Theorem 12.5.5 and Corollary 12.5.1 were derived in Embrechts, Grandell and Schmidli (1993).

12.6 SUBEXPONENTIAL CLAIM SIZES

In the previous sections of this chapter we found the asymptotic behaviour of the ruin functions $\psi(u)$ and $\psi^0(u)$ given in (12.1.39) and (12.1.43), respectively, when the initial risk reserve u tends to infinity. However, our results were limited to the case of exponentially bounded claim size distributions. In the present section we assume that some of the claim size distributions are subexponential, i.e. heavy-tailed. The main idea of the approach considered below is to compare the asymptotic behaviour of $\psi(u)$ and $\psi^0(u)$ in models with a general (not necessarily renewal) claim arrival process, with that in the Sparre Andersen model, and to use Theorem 6.5.11. We first state two general theorems, one for a Palm-stationary ergodic input with independent claim sizes, the other for processes with a regenerative structure. Thereafter, we will apply the general results to several classes of claim arrival processes.

12.6.1 General Results

Consider the marked point process $\{(\sigma_n, U_n)\}$, where σ_n is the arrival epoch and U_n is the size of the n-th claim. We first assume that the point process $\{\sigma_n\}$ is Palm-stationary and ergodic and that $\{U_n\}$ consists of independent and identically distributed random variables which are independent of $\{\sigma_n\}$. We also assume that the net profit condition (12.1.38) is fulfilled in that $\lambda^{-1} = \mathbb{E}^0(\sigma_{n+1} - \sigma_n)$. For each $\varepsilon > 0$ let $Z_\varepsilon = \sup_{n\geq 1}\{n(\lambda^{-1} - \varepsilon) - \sigma_n\}$. Note that by Theorem 12.1.3 and Lemma 12.1.2 we have

$$\lim_{n\to\infty} n^{-1}\sigma_n = \lambda^{-1}\,. \tag{12.6.1}$$

Hence $Z_\varepsilon < \infty$ follows. The condition formulated in the next theorem means that, for each $\varepsilon > 0$, the tail of the distribution of the supremum Z_ε should be lighter than that of the (integrated tail) distribution F_U^{s} of claim sizes. In other words, the sequence $\{\sigma_n\}$ of claim arrival epochs should not be too bursty. Examples where this condition is fulfilled are Poisson cluster processes and superpositions of renewal processes. They will be discussed in Sections 12.6.2 and 12.6.3, respectively.

Theorem 12.6.1 *Assume that* $F_U^{\mathrm{s}} \in \mathcal{S}$ *and that* $\lim_{u\to\infty} \overline{F_{Z_\varepsilon}}(u)/\overline{F_U^{\mathrm{s}}}(u) = 0$ *for all* $\varepsilon > 0$. *Then, for the ruin function* $\psi^0(u)$ *given in* (12.1.43),

$$\lim_{u\to\infty} \frac{\psi^0(u)}{\int_u^\infty \overline{F_U}(v)\,\mathrm{d}v} = \frac{\lambda}{\beta - \lambda\mathbb{E}\,U}\,. \tag{12.6.2}$$

Proof By rescaling, we can assume without loss of generality that $\beta = 1$. Let $\varepsilon, \varepsilon' > 0$. Using (12.6.1) it is not difficult to see that there exists $c > 0$ such that

$$\mathbb{P}^0\Big(\bigcap_{n\geq 1}\{\sigma_n \leq n(\lambda^{-1} + \varepsilon) + c\}\Big) > 1 - \varepsilon'\,.$$

Hence

$$\psi(u) \geq (1 - \varepsilon')\mathbb{P}\Big(\sup_{n\geq 1}\Big\{\sum_{k=1}^n (U_k - \lambda^{-1} - \varepsilon)\Big\} > u + c\Big).$$

Thus, (2.5.7) and Theorem 6.5.11 imply

$$\begin{aligned}
\liminf_{u\to\infty} \frac{\psi(u)}{\int_u^\infty \overline{F_U}(v)\,\mathrm{d}v} &= \liminf_{u\to\infty} \frac{\psi(u)}{\int_{u+c}^\infty \overline{F_U}(v)\,\mathrm{d}v} \\
&\geq (1 - \varepsilon')\liminf_{u\to\infty} \frac{\mathbb{P}\big(\sup_{n\geq 1}\{\sum_{k=1}^n (U_k - \lambda^{-1} - \varepsilon)\} > u + c\big)}{\int_{u+c}^\infty \overline{F_U}(v)\,\mathrm{d}v} \\
&= (1 - \varepsilon')\frac{1}{\lambda^{-1} + \varepsilon - \mu}\,,
\end{aligned}$$

where $\mu = \mathbb{E}\, U$. Since $\varepsilon, \varepsilon' > 0$ are arbitrary it follows that

$$\liminf_{u\to\infty} \frac{\psi(u)}{\int_u^\infty \overline{F_U}(v)\,dv} \ge \frac{1}{\lambda^{-1} - \mu}\,.$$

Let now $0 < \varepsilon < \lambda^{-1} - \mu$ and $M_\varepsilon = \sup_{n\ge 1}\{\sum_{k=1}^n (U_k - \lambda^{-1} + \varepsilon)\}$. Then

$$\begin{aligned}
&\limsup_{u\to\infty} \frac{\psi(u)}{\int_u^\infty \overline{F_U}(v)\,dv} \\
&\quad\le \limsup_{u\to\infty} \frac{\mathbb{P}(\sup_{n\ge 1}\{\sum_{k=1}^n U_k - n(\lambda^{-1}-\varepsilon) + n(\lambda^{-1}-\varepsilon) - \sigma_n\} > u)}{\int_u^\infty \overline{F_U}(v)\,dv} \\
&\quad\le \limsup_{u\to\infty} \frac{\mathbb{P}(M_\varepsilon + Z_\varepsilon > u)}{\int_u^\infty \overline{F_U}(v)\,dv}\,.
\end{aligned}$$

From Theorem 6.5.11 we get that $\mathbb{P}(M_\varepsilon > u)/\int_u^\infty \overline{F_U}(v)\,dv$ tends to $(\lambda^{-1} - \varepsilon - \mu)^{-1}$ as $u \to \infty$. By the conditions of the theorem and by Lemma 2.5.2, this implies

$$\begin{aligned}
\limsup_{u\to\infty} \frac{\psi(u)}{\int_u^\infty \overline{F_U}(v)\,dv} &\le \limsup_{u\to\infty} \frac{\mathbb{P}(M_\varepsilon + Z_\varepsilon > u)}{\int_u^\infty \overline{F_U}(v)\,dv} \\
&= \limsup_{u\to\infty} \frac{\mathbb{P}(M_\varepsilon > u)}{\int_u^\infty \overline{F_U}(v)\,dv} = \frac{1}{\lambda^{-1} - \varepsilon - \mu}\,.
\end{aligned}$$

Since $\varepsilon > 0$ is arbitrary, the assertion follows. □

Corollary 12.6.1 *Under the assumptions of Theorem* 12.6.1, *the ruin function $\psi(u)$ in the time-stationary model given by* (12.1.39) *is asymptotically equivalent to the ruin function $\psi^0(u)$ in the Palm model. That is,* $\lim_{u\to\infty} \psi(u)/\psi^0(u) = 1$ *and, consequently,*

$$\lim_{u\to\infty} \frac{\psi(u)}{\int_u^\infty \overline{F_U}(v)\,dv} = \frac{\lambda}{\beta - \lambda \mathbb{E}\, U}\,. \tag{12.6.3}$$

Proof Using Lemma 2.5.2, the assertion is an immediate consequence of Theorems 12.1.10 and 12.6.1. □

In some applications the assumptions of Theorem 12.6.1 are too strong. For instance, in the Markov-modulated risk model, the claim sizes and the claim arrival epochs are not necessarily independent. Moreover, as can happen in other risk models as well, the claim arrival process does not always start in a time-stationary state or with a claim at the origin (Palm model). In Theorem 12.6.2 below, we therefore state another general result on the asymptotics of $\psi(u)$ as $u \to \infty$, which is applicable to these situations.

Assume that the risk reserve process $\{R(t)\}$ has a *regenerative structure*, i.e. there are stopping times $0 = \zeta_0 < \zeta_1 < \dots$ such that the processes $\{R(\zeta_k + t) - R(\zeta_k) : 0 \le t \le \zeta_{k+1} - \zeta_k\}$, $k = 0, 1, \dots$, are independent and identically distributed. One could let $\{R(t) - u : 0 \le t \le \zeta_1\}$ have a distribution different from the others, but for reasons of simplicity, we only consider the situation described above. We also assume that $\mathbf{P}(\{\xi_n, n \ge 1\} \cap \{\sigma_n, n \ge 1\} = \emptyset) = 1$, i.e. the sets of regeneration epochs and of claim arrival epochs are disjoint with probability 1.

For convenience, conditions and statements are formulated in terms of the claim surplus process $S(t) = \sum_{i=1}^{N(t)} U_i - \beta t$, rather than of the risk reserve process $\{R(t)\}$. Let $\zeta = \zeta_1$ and $S = S(\zeta)$. Assume that $-\infty < \mathbf{E}\, S < 0$ and $F^{\rm s}_{S_+} \in \mathcal{S}$, where $S_+ = \max\{0, S\}$. Considering the "embedded" ruin function $\psi'(u) = \mathbf{P}(\sup_{n \ge 1} S(\zeta_n) > u)$ and using similar arguments as in the proof of Theorem 6.5.11, we find that

$$\lim_{u \to \infty} \frac{\psi'(u)}{\int_u^\infty \overline{F_S}(v)\,{\rm d}v} = (-\mathbf{E}\, S)^{-1}\,. \tag{12.6.4}$$

In the next theorem we formulate a condition which implies that the usual "continuous-time" ruin function $\psi(u) = \mathbf{P}(\tau(u) < \infty)$ with $\tau(u) = \min\{t : S(t) > u\}$, has the same asymptotic behaviour as that of $\psi'(u)$ obtained in (12.6.4). A motivation is provided by the following observation: In many situations where ruin occurs at time $\tau(u)$, the surplus process will not be able to recover until the next regeneration epoch. In particular, if the claim size distribution is subexponential, then (2.5.7) suggests that, typically, $S(\tau(u))$ largely exceeds the level u. Thus, if the time from $\tau(u)$ to the next regeneration epoch is not long enough, the surplus process can not recover. As it turns out, a sufficient condition for this, is the asymptotic tail equivalence of the random variables S and $S' = \sum_{i=1}^{N(\zeta)} U_i$.

Theorem 12.6.2 *Assume that* $F_{S_+}, F^{\rm s}_{S_+} \in \mathcal{S}$, $\lim_{x \to \infty} \overline{F_{S'}}(x)/\overline{F_S}(x) = 1$ *and* $-\infty < \mathbf{E}\, S < 0$. *Then*

$$\lim_{u \to \infty} \frac{\psi(u)}{\int_u^\infty \overline{F_S}(v)\,{\rm d}v} = (-\mathbf{E}\, S)^{-1}\,. \tag{12.6.5}$$

In the *proof* of Theorem 12.6.2 we use the following auxiliary result for the regenerative claim surplus process $\{S(t)\}$ with negative drift and with heavy-tailed increments during the regeneration periods. Define $M = \sup_{t \ge 0} S(t)$ and $M' = \sup_{n \ge 1} S(\zeta_n)$. Further, let $\Delta_n = \sup_{\zeta_n < t \le \zeta_{n+1}} S(t) - S(\zeta_n)$, $M'_n = \sup_{1 \le k \le n} S(\zeta_k)$, and

$$\alpha(u) = \inf\{n \ge 1 : S(\zeta_n) > u\}, \quad \beta(u) = \inf\{n \ge 0 : S(\zeta_n) + \Delta_n > u\}\,.$$

Lemma 12.6.1 *Under the assumptions of Theorem* 12.6.2, *for each* $a > 0$,

$$\mathbb{P}(M' > u, S(\zeta_{\alpha(u)}) - S(\zeta_{\alpha(u)-1}) \le a) = o(\mathbb{P}(M' > u)) \tag{12.6.6}$$

and

$$\mathbb{P}(M > u, \Delta_{\beta(u)} \le a) = o(\mathbb{P}(M > u)) \tag{12.6.7}$$

as $u \to \infty$.

Proof We first show (12.6.6). Consider the random walk $\{S_n,\ n \ge 0\}$ with $S_n = S(\zeta_n)$, and the sequence of its ascending ladder epochs defined in (6.3.6). Let $N = \max\{n : \nu_n^+ < \infty\}$ be the number of finite ladder epochs. Furthermore, let $\{Y_n^+\}$ be the sequence of ascending ladder heights introduced in Section 6.3.3. Recall that N is geometrically distributed with parameter $p = \mathbb{P}(Y^+ < \infty)$. Define $\alpha'(u) = \inf\{n \ge 1 : \sum_{i=1}^n Y_i^+ > u\}$. Then,

$$\mathbb{P}(S(\zeta_{\alpha(u)}) - S(\zeta_{\alpha(u)-1}) \le a \mid M' > u) \le \mathbb{P}(Y_{\alpha'(u)}^+ \le a \mid M' > u)\,.$$

Thus, in order to prove (12.6.6), it suffices to show that

$$\lim_{u\to\infty} \mathbb{P}(Y_{\alpha'(u)}^+ \le a \mid M' > u) = 0\,. \tag{12.6.8}$$

We have

$$\begin{aligned}
&\mathbb{P}(\alpha'(u) = n, M' > u)\\
&\quad= \mathbb{P}\Big(\sum_{i=1}^{n} Y_i^+ > u\Big) - \mathbb{P}\Big(\sum_{i=1}^{n-1} Y_i^+ > u\Big)\\
&\quad= p^n\overline{G_0^{*n}}(u) - p^{n-1}\overline{G_0^{*(n-1)}}(u)\,,
\end{aligned} \tag{12.6.9}$$

where $G_0(x) = \mathbb{P}(Y^+ \le x \mid Y^+ < \infty)$. From Lemma 6.5.2 we know that $G_0 \in \mathcal{S}$. Later on, we need an upper bound for $\mathbb{P}(\alpha'(u) = n \mid M' > u)$, which is obtained in the following way. Choose $\varepsilon > 0$ such that $\mathbb{E}\,(1+\varepsilon)^N < \infty$. From Lemma 2.5.3 we know that there exists a constant $c < \infty$ such that $\overline{G_0^{*n}}(u) \le c(1+\varepsilon)^n\overline{G_0}(u)$ for all $n \ge 1$, $u \ge 0$. Thus, using (12.6.9) we have $\mathbb{P}(\alpha'(u) = n, M' > u) \le p^n c(1+\varepsilon)^n\overline{G_0}(u)$. On the other hand, from Theorem 6.3.3 and from the inequality $\overline{G_0^{*k}}(u) > (\overline{G_0}(u))^k$ we obtain

$$\mathbb{P}(M' > u) = (1-p)\sum_{k=1}^{\infty} p^k\overline{G_0^{*k}}(u) \ge \frac{(1-p)p\overline{G_0}(u)}{1-p\overline{G_0}(u)}\,.$$

This yields the bound

$$\mathbb{P}(\alpha'(u) = n \mid M' > u) \le \frac{p^{n-1}}{1-p}c(1+\varepsilon)^n \tag{12.6.10}$$

for all $u \geq 0$. Furthermore,

$$\begin{aligned}
&\mathbb{P}(Y_n^+ \leq a, \alpha'(u) = n, M' > u) \\
&= \mathbb{P}\Big(Y_n^+ \leq a, \sum_{i=1}^{n-1} Y_i^+ \leq u, \sum_{i=1}^{n} Y_i^+ > u\Big) \\
&\leq \mathbb{P}\Big(Y_n^+ \leq a, \sum_{i=1}^{n-1} Y_i^+ > u - a\Big) - \mathbb{P}\Big(Y_n^+ \leq a, \sum_{i=1}^{n-1} Y_i^+ > u\Big) \\
&= p^n G_0(a)(\overline{G_0^{*(n-1)}}(u-a) - \overline{G_0^{*(n-1)}}(u)) .
\end{aligned}$$

Using (2.5.7) and (2.5.11), this and (12.6.9) imply that

$$\lim_{u\to\infty} \mathbb{P}(Y_{\alpha'(u)}^+ \leq a \mid M' > u, \alpha'(u) = n) = 0 .$$

Now, (12.6.8) follows from (12.6.10) and the bounded convergence theorem, since

$$\begin{aligned}
&\mathbb{P}(Y_{\alpha'(u)}^+ \leq a \mid M' > u) \\
&= \sum_{n=1}^{\infty} \mathbb{P}(\alpha'(u) = n \mid M' > u)\, \mathbb{P}(Y_{\alpha'(u)}^+ \leq a \mid M' > u, \alpha'(u) = n) .
\end{aligned}$$

To prove (12.6.7) note that

$$\mathbb{P}(M > u, \Delta_{\beta(u)} \leq a) \leq \mathbb{P}(M' > u - a, S(\zeta_{\alpha(u-a)}) \in (u-a, u)) .$$

Since $\mathbb{P}(M' > u) \leq \mathbb{P}(M > u)$, and $\mathbb{P}(M' > u - a) \sim \mathbb{P}(M > u)$ as $u \to \infty$, it thus suffices to show that

$$\lim_{u\to\infty} \mathbb{P}(S(\zeta_{\alpha(u-a)}) \in (u-a, u) \mid M' > u - a) = 0 . \tag{12.6.11}$$

Using that

$$\begin{aligned}
&\mathbb{P}(M' > u - a, \alpha'(u-a) = n, S(\zeta_{\alpha(u-a)}) \in (u-a, u)) \\
&\leq \mathbb{P}\Big(\sum_{i=1}^{n} Y_i^+ > u - a\Big) - \mathbb{P}\Big(\sum_{i=1}^{n} Y_i^+ > u\Big) ,
\end{aligned}$$

the proof of (12.6.11) is analogous to the proof of (12.6.6). □

Proof of Theorem 12.6.2 Recall that $\psi(u) = \mathbb{P}(M > u)$ and $\psi'(u) = \mathbb{P}(M' > u)$. Using (12.6.4), it therefore suffices to show that

$$\lim_{u\to\infty} \mathbb{P}(M' > u)/\mathbb{P}(M > u) = 1 . \tag{12.6.12}$$

The assumed asymptotic tail equivalence of S and S' implies that, for each $\varepsilon > 0$, one can find $a > 0$ such that $\mathbb{P}(S > x) \geq (1-\varepsilon)\mathbb{P}(\Delta_0 > x)$ for all $x \geq a$. Using (12.6.6) and (12.6.7), this gives

$$\begin{aligned}
\mathbb{P}(M' > u) &\sim \mathbb{P}(M' > u, S(\zeta_{\alpha(u)}) - S(\zeta_{\alpha(u)-1}) > a) \\
&= \sum_{n=1}^{\infty} \mathbb{P}(M'_n \leq u, S(\zeta_{n+1}) - S(\zeta_n) > a \vee (u - S(\zeta_n))) \\
&\geq (1-\varepsilon) \sum_{n=1}^{\infty} \mathbb{P}(M'_n \leq u, \Delta_n > a \vee (u - S(\zeta_n))) \\
&\geq (1-\varepsilon) \sum_{n=1}^{\infty} \mathbb{P}(\max_{0<t\leq\zeta_n} S(t) \leq u, \Delta_n > a \vee (u - S(\zeta_n))) \\
&= (1-\varepsilon)\mathbb{P}(M > u, \Delta_{\beta(u)} > a) \\
&\sim (1-\varepsilon)\mathbb{P}(M > u).
\end{aligned}$$

Thus, letting $u \to \infty$ and next $\varepsilon \downarrow 0$ yields

$$\liminf_{u\to\infty} \frac{\mathbb{P}(M' > u)}{\mathbb{P}(M > u)} \geq 1.$$

On the other hand, it is clear that

$$\limsup_{u\to\infty} \frac{\mathbb{P}(M' > u)}{\mathbb{P}(M > u)} \leq 1$$

since $M' \leq M$. Hence, (12.6.12) follows. □

12.6.2 Poisson Cluster Arrival Processes

Assume that the claim arrival process $\{\sigma_n\}$ is a time-stationary Poisson cluster process introduced in Section 12.2.3. Furthermore, assume that the sequence $\{U_n\}$ of claim sizes consists of independent and identically distributed random variables, independent of $\{\sigma_n\}$. Using Theorem 12.6.1 and Corollary 12.6.1, we derive conditions under which the asymptotic behaviour of the ruin functions $\psi(u)$ and $\psi^0(u)$ is given by (12.6.3) and (12.6.2), respectively.

We leave it to the reader to show that a time-stationary Poisson cluster process is ergodic. By Theorem 12.1.6, this is equivalent to ergodicity of the corresponding Palm distribution. We also remark that the distribution of the homogeneous Poisson process $\{\sigma'_n\}$ of cluster centres remains invariant under independent and identically distributed shifting of points. Indeed, if $\{Y_n\}$ is a sequence of independent and identically distributed random variables, independent of $\{\sigma'_n\}$, then $\{\sigma'_n - Y_n\}$ has the same distribution as $\{\sigma'_n\}$. The proof of this fact is left to the reader. Thus, without loss of generality we can and will assume that for each $n \in \mathbb{Z}$ the cluster centre σ'_n is the leftmost point of the n-th cluster, i.e. $N_n((-\infty, 0)) = 0$.

Since, under the conditions of Theorem 12.6.1, the ruin functions $\psi(u)$ and $\psi^0(u)$ are asymptotically equivalent (see Corollary 12.6.1), we can pass to a Palm-stationary model for the claim arrival process. That is, we assume that the point process of claim arrival epochs is distributed according to the Palm distribution of $\{\sigma_n\}$. Since no confusion is possible, we will further use the same notation $\{\sigma_n\}$ for this Palm-stationary cluster process.

In Theorem 12.2.7 we saw that the Palm distribution of a time-stationary Poisson cluster process can be obtained by the superposition of this Poisson cluster process with an independent single cluster $\tilde{N}$ in such a way, that a randomly chosen point of $\tilde{N}$ is at the origin; here the distribution of $\tilde{N}$ is given by (12.2.12). Thus, for each $n \geq 1$, the claim arrival epoch σ_n in $(0, \infty)$ can come from three different types of clusters: (i) from a cluster with $\sigma'_n \in (0, \infty)$, (ii) from a cluster with $\sigma'_n \in (-\infty, 0)$, (iii) from the independent cluster $\tilde{N}$. For each $n \geq 1$, let $k(n) = \sup\{i : \sigma'_i \leq \sigma_n\}$ denote the number of the last cluster centre before σ_n. Then obviously $n \leq \sum_{i=1}^{k(n)} N_i(\mathbb{R}) + \sum_{i=-\infty}^{0} N_i((-\sigma'_i, \infty)) + \tilde{N}(\mathbb{R})$ and $\sigma_n \geq \sigma'_{k(n)}$. Thus, for the random variable Z_ε appearing in Theorem 12.6.1 we have

$$Z_\varepsilon \leq Z^{(1)} + Z^{(2)} + Z^{(3)}, \tag{12.6.13}$$

where

$$Z^{(1)} = \sup_{n \geq 1} \sum_{i=1}^{n} N_i(\mathbb{R})(\lambda^{-1} - \varepsilon) - \sigma'_n,$$

$$Z^{(2)} = \sum_{i=-\infty}^{0} N_i(-\sigma'_i, \infty)(\lambda^{-1} - \varepsilon),$$

and $Z^{(3)} = \tilde{N}(\mathbb{R})(\lambda^{-1} - \varepsilon)$. Let $V_n = \sup\{x : N_n(\{x\}) > 0\}$ and notice that the distribution of V_n does not depend on n since the clusters $\{N_n\}$ are identically distributed. We therefore arrive at the following result.

Theorem 12.6.3 *Let the net profit condition* (12.1.38) *be fulfilled. Assume that* $F_U^{\mathrm{s}} \in \mathcal{S}$, $\mathbb{E}\,\mathrm{e}^{\delta N_n(\mathbb{R})} < \infty$ *and* $\mathbb{E}\,\mathrm{e}^{\delta V_n} < \infty$ *for some* $\delta > 0$. *Then,* $\lim_{u\to\infty} \psi(u)/\psi^0(u) = 1$ *and*

$$\lim_{u\to\infty} \frac{\psi(u)}{\int_u^\infty \overline{F_U}(v)\,\mathrm{d}v} = \frac{\lambda}{\beta - \lambda \mathbb{E}\,U}. \tag{12.6.14}$$

Proof Choosing $\delta_3 = \delta/(\lambda^{-1} - \varepsilon)$, it follows that $\mathbb{E}\,\mathrm{e}^{\delta_3 Z^{(3)}} < \infty$. Note that $\lambda = \lambda' \mathbb{E}\,N_i(\mathbb{R})$, where λ' is the intensity of the cluster centres. Then $\mathbb{E}\,N_i(\mathbb{R})(\lambda^{-1} - \varepsilon) = (\lambda')^{-1} - \varepsilon \mathbb{E}\,N_i(\mathbb{R})$. Thus $\{\sum_{i=1}^{n} N_i(\mathbb{R})(\lambda^{-1} - \varepsilon) - \sigma'_n\}$

is a compound Poisson risk model where the adjustment coefficient exists when $\varepsilon < \lambda^{-1}$. Thus by (5.4.4) we have $\mathbb{E}\, e^{\delta_1 Z^{(1)}} < \infty$ for some $\delta_1 > 0$. Now let $m = \sup\{i \ge 0 : V_{-i} + \sigma'_{-i} \ge 0\}$. Then, using $N_i((-\infty, 0)) = 0$, we have $Z^{(2)} \le \sum_{i=0}^m N_{-i}(\mathbb{R})(\lambda^{-1} - \varepsilon)$. We first want to show that $\mathbb{P}(m > n) \le a e^{-\delta' n}$ for some $a, \delta' > 0$ and for all $n \ge 1$. By (5.4.4) there exists γ' such that

$$\mathbb{P}(-\sigma'_{-k} \le \lambda' k/4) \le \mathbb{P}(\sup_n\{\lambda' n/2 + \sigma'_{-n}\} \ge \lambda' k/4) \le e^{-\gamma'\lambda' k/4}.$$

By our assumptions there exists a constant $a' > 0$ such that $\mathbb{P}(V_{-k} > \lambda' k/4) \le a' e^{-k\delta}$ for all $k \ge 1$, where $\delta' = \min\{\gamma'\lambda'/4, \delta/2\}$. Then

$$\mathbb{P}(m > n) \le \sum_{k=n}^{\infty}\Big(\mathbb{P}(V_{-k} > \lambda'\frac{k}{4}) + \mathbb{P}(-\sigma'_{-k} \le \lambda'\frac{k}{4})\Big) \le (a'+1)e^{-\delta' n}/(1 - e^{-\delta'}).$$

Now choose $\delta_2 > 0$ such that $b = \mathbb{E}\, e^{2\delta_2(\lambda^{-1}-\varepsilon)N_i(\mathbb{R})} < e^{\delta'}$. Then

$$\begin{aligned}\mathbb{E}\, e^{\delta_2 Z^{(2)}} &\le \sum_{n=0}^{\infty} \mathbb{E}\left[e^{\delta_2(\lambda^{-1}-\varepsilon)(N_0(\mathbb{R}) + \ldots + N_{-n}(\mathbb{R}))}; m = n\right] \\ &\le \sum_{n=0}^{\infty} b^{(n+1)/2}\sqrt{\mathbb{P}(m = n)} \le c\sum_{n=0}^{\infty} b^{(n+1)/2} e^{-\delta'(n-1)/2} < \infty,\end{aligned}$$

where $c = \sqrt{(a'+1)/(1-e^{-\delta'})}$. Thus for $\delta_0 = \min\{\delta_1, \delta_2, \delta_3\}$ we have $\mathbb{E}\, e^{\delta_0 Z_\varepsilon} < \infty$. Hence, using (12.6.13) and the assumption that $F_U^{\mathrm{s}} \in \mathcal{S}$, Theorem 2.5.2 and (2.5.2) give that $\lim_{u\to\infty} \overline{F_{Z_\varepsilon}}(u)/\overline{F_U^{\mathrm{s}}}(u) = 0$ for all $\varepsilon > 0$. The assertion now follows from Theorem 12.6.1. $\square$

12.6.3 Superposition of Renewal Processes

Assume now that claims arrive according to the superposition $N = N_1 + \ldots + N_\ell$ of ℓ independent time-stationary renewal processes $N_1, \ldots, N_\ell$ as considered in Section 12.2.3. Using Corollary 12.1.3, it is not difficult to show that the point process N is ergodic. By Theorem 12.1.6, its Palm distribution is also ergodic. Furthermore, we assume that the sequence $\{U_n\}$ of claim sizes consists of independent and identically distributed random variables, independent of N. If their integrated tail distribution is subexponential, we obtain the following result without further assumptions.

Theorem 12.6.4 *Let the net profit condition* (12.1.38) *be fulfilled. Assume* $F_U^{\mathrm{s}} \in \mathcal{S}$. *Then*

$$\lim_{u\to\infty} \frac{\psi(u)}{\int_u^\infty \overline{F_U}(v)\, dv} = \frac{\lambda}{\beta - \lambda \mathbb{E}\, U} \tag{12.6.15}$$

where $\lambda = \lambda_1 + \ldots + \lambda_\ell$.

Proof As in Section 12.6.2, we pass to the Palm-stationary claim arrival process to show that the conditions of Theorem 12.6.1 are fulfilled. In particular, we show that the tail of the distribution of Z_ε is exponentially bounded. Note that

$$\mathbf{P}^0(Z_\varepsilon > u) \le \mathbf{P}^0\Big(\bigcup_{n\ge 1}\{N(n(\lambda^{-1}-\varepsilon)-u)\ge n\}\Big).$$

Thus from Theorem 12.2.5 we can conclude that it suffices to show that for some $c, \delta > 0$

$$\mathbf{P}\Big(\bigcup_{n\ge 1}\{N_j(n(\lambda^{-1}-\varepsilon)-u)\ge n\lambda_j/\lambda\}\Big) \le c\mathrm{e}^{-\delta u},$$

$$\mathbf{P}\Big(\bigcup_{n\ge 1}\{N_j^0(n(\lambda^{-1}-\varepsilon)-u)\ge n\lambda_j/\lambda\}\Big) \le c\mathrm{e}^{-\delta u}$$

for all $j = 1, \ldots, \ell$ and $u > 0$. We will only show the first inequality as the second follows analogously. Indeed, the tail of the distribution of Z_ε will then be exponentially bounded and the conditions of Theorem 12.6.1 will be fulfilled. Recall that the points of the j-th renewal process N_j we denoted by $\{\sigma_{j,n}\}$. Furthermore, let $\varepsilon_j = \lambda_j^{-1} - \varepsilon\lambda/\lambda_j$. Then, using the notation $\varepsilon_j = \lambda_j^{-1} - \varepsilon\lambda/\lambda_j$, we have

$$\begin{aligned}
&\mathbf{P}\Big(\bigcup_{n\ge 1}\{N_j(n(\lambda^{-1}-\varepsilon)-u)\ge n\lambda_j/\lambda\}\Big)\\
&\quad\le \mathbf{P}\Big(\bigcup_{n\ge 1}\{\sigma_{j,\lfloor n\lambda_j/\lambda\rfloor}\le n(\lambda^{-1}-\varepsilon)-u\}\Big)\\
&\quad\le \mathbf{P}\Big(\bigcup_{m\ge 1}\{\sigma_{j,m} < (m+1)(\lambda^{-1}-\varepsilon)\lambda/\lambda_j-u\}\Big)\\
&\quad\le \mathbf{P}\Big(\bigcup_{m\ge 1}\{\sigma_{j,m}-\sigma_{j,1} < (m-1)\varepsilon_j-(u-2\varepsilon_j)\}\Big)\\
&\quad= \mathbf{P}\big(\sup_{m\ge 2}\{(m-1)\varepsilon_j-(\sigma_{j,m}-\sigma_{j,1}) > u-2\varepsilon_j\}\big)\\
&\quad\le a_j\mathrm{e}^{-\gamma_j u},
\end{aligned}$$

for some $a_j, \gamma_j > 0$, where the last inequality follows from Theorem 6.5.11. This completes the proof. □

12.6.4 The Markov-Modulated Risk Model

In the last two sections, we considered ruin probabilities in time-stationary and Palm-stationary models. In what follows this does not need to be the case. Let $\psi(u)$ be the ruin function in the Markov-modulated risk model with an arbitrary initial distribution of the Markov environment process $\{J(t)\}$. Assume that the intensity matrix $\boldsymbol{Q}$ of $\{J(t)\}$ is irreducible. Let F be a distribution on $\mathbb{R}_+$ such that for each $i \in \{1,\ldots,\ell\}$ we have $\lim_{x\to\infty} \overline{F}_i(x)/\overline{F}(x) = c_i$ for some constant $c_i \in [0,\infty)$, where F_i denotes the conditional claim size distribution in state i. As before, the conditional claim arrival intensity in state i is denoted by λ_i, and $\{\pi_i\}$ is the stationary initial distribution of $\{J(t)\}$. Then we arrive at the following result for the asymptotic behaviour of $\psi(u)$ as $u \to \infty$.

Theorem 12.6.5 *Let the net profit condition* (12.3.6) *be fulfilled. Assume that $F, F^{\mathrm{s}} \in \mathcal{S}$ and $c = \sum_{i=1}^{\ell} \pi_i\lambda_i c_i > 0$. Then, for each distribution of the initial state $J(0)$,*

$$\lim_{u\to\infty} \frac{\psi(u)}{\int_u^\infty \overline{F}(v)\,\mathrm{d}v} = \frac{c}{\beta - \sum_{i=1}^{\ell} \pi_i\lambda_i\mu_i}\,. \tag{12.6.16}$$

Proof Suppose $J(0) = i$ and let $\zeta = \inf\{t > 0 : J(t) = i, J(t-0) \neq i\}$ be the regeneration epoch in the model of Theorem 12.6.2. We leave it to the reader to show that the distribution of ζ is phase-type and therefore the tail of $N(\zeta)$ is exponentially bounded. Let $\mathcal{F}_\infty^J$ denote the σ-algebra generated by $\{J(t) : t \geq 0\}$. Let $\{U_n'\}$ be a sequence of independent and identically distributed random variables with distribution $F'(x) = \inf_{1\leq j\leq \ell} F_j(x)$. Then $U_n \leq_{\mathrm{st}} U_n'$. Furthermore, the quotient $\overline{F'}(x)/\overline{F}(x)$ tends to $c' = \max\{c_j : j = 1,\ldots,\ell\} > 0$ as $x \to \infty$. In particular $F' \in \mathcal{S}$ by Lemma 2.5.4. Let $\delta > 0$ such that $\mathbb{E}\,(1+\delta)^{N(\zeta)} < \infty$. By Lemma 2.5.3, there exists a constant c'' such that

$$\mathbb{P}(S > x \mid \mathcal{F}^J) \leq \mathbb{P}\Big(\sum_{n=1}^{N(\zeta)} U_n' > x \,\Big|\, \mathcal{F}_\infty^J\Big) \leq c''\overline{F}(x)(1+\delta)^{N(\zeta)}\,. \tag{12.6.17}$$

Because the U_n are conditionally independent given $\mathcal{F}_\infty^J$, Lemma 2.5.2 implies that

$$\lim_{x\to\infty} \frac{\mathbb{P}(S' > x \mid \mathcal{F}_\infty^J)}{\overline{F}(x)} = \sum_{j=1}^{\ell} c_j N_j\,,$$

where $N_j = \sum_{n=1}^{N(\zeta)} \mathbb{I}(J(\sigma_n) = j)$. In particular, by Lemma 2.5.4 the conditional distribution of S' given $\mathcal{F}_\infty^J \cap \{\sum_{j=1}^{\ell} c_j N_j > 0\}$ is subexponential. Thus by (2.5.7)

$$\lim_{x\to\infty} \frac{\mathbb{P}(S > x \mid \mathcal{F}_\infty^J)}{\overline{F}(x)} = \lim_{x\to\infty} \frac{\mathbb{P}(S' > x + \beta\zeta \mid \mathcal{F}_\infty^J)}{\overline{F}(x)} = \sum_{j=1}^{\ell} c_j N_j\,.$$

Now, in view of the integrable bound given in (12.6.17), the dominated convergence theorem yields

$$\lim_{x\to\infty} \frac{\overline{F_S}(x)}{\overline{F}(x)} = \lim_{x\to\infty} \mathbb{E}\left(\frac{\mathbb{P}(S > x \mid \mathcal{F}^J_\infty)}{\overline{F}(x)}\right) = \mathbb{E}\left(\sum_{j=1}^{\ell} c_j N_j\right) = c\,.$$

The last equation follows from the law of large numbers and Theorem 12.3.2 by considering a Markov-modulated risk model with claim sizes $U_n = \sum_{j=1}^{\ell} c_j \mathbb{I}(J(\sigma_n) = j)$. Letting $\beta = 0$ it immediately follows that $\overline{F_S}(x)/\overline{F}(x)$ tends to c as well. Thus, $F_{S_+} \in \mathcal{S}$ and $\lim_{x\to\infty} \overline{F_{S'}}(x)/\overline{F_S}(x) = 1$. By L'Hospital's rule also $F^{\mathrm{s}}_{S_+} \in \mathcal{S}$. Therefore, for the case $J(0) = i$, the assertion follows from Theorems 12.3.2 and 12.6.2. Because the limit in (12.6.16) does not depend on i the result holds for any initial distribution. □

At first sight, it might be surprising that the constant in (12.6.16) does not depend on the initial state. Intuitively, starting with a large initial reserve u, ruin will be caused by a large claim. However, it takes some time before such a claim occurs. Thus, the process $\{J(t)\}$ has enough time to reach the steady state. Alternatively, the independence of the initial state conforms with the fact that the constant in (12.6.14) and (12.6.15) only depends on the arrival intensity λ, not on the specific structure of the underlying point process of claim occurrence epochs.

12.6.5 The Björk–Grandell Risk Model

Consider the nondelayed Björk–Grandell risk model of Section 12.5, i.e. assume that $(\Lambda_1, I_1) \stackrel{\mathrm{d}}{=} (\Lambda, I)$. Furthermore, assume that the claim size distribution is subexponential. To state the result on the asymptotic behaviour of the ruin function $\psi(u)$, we also have to make an assumption on the distribution of (Λ, I). The distribution of the number of claims in the interval (ξ_i, ξ_{i+1}) must have an exponentially bounded tail if we want to avoid the aggregate claim amount in (ξ_i, ξ_{i+1}) becoming large due to the number of claims.

Theorem 12.6.6 *Let* $(\Lambda_1, I_1) \stackrel{\mathrm{d}}{=} (\Lambda, I)$ *and let the net profit condition* (12.5.1) *be fulfilled. Assume that* $F_U, F^{\mathrm{s}}_U \in \mathcal{S}$ *and that there exists* $\delta > 0$ *such that* $\mathbb{E}\,\mathrm{e}^{\delta\Lambda I} < \infty$. *Then,*

$$\lim_{u\to\infty} \frac{\psi(u)}{\int_u^\infty \overline{F_U}(v)\,\mathrm{d}v} = \frac{\mathbb{E}\,(\Lambda I)}{\beta\mathbb{E}\,I - \mathbb{E}\,(\Lambda I)\mathbb{E}\,U_i}\,. \tag{12.6.18}$$

Proof We choose $\zeta_n = \xi_n$. From

$$\mathbb{E}\,(1+\delta)^{N(\zeta)} = \mathbb{E}\,(\mathbb{E}\,((1+\delta)^{N(\zeta)} \mid \Lambda, I)) = \mathbb{E}\,\mathrm{e}^{\delta\Lambda I} < \infty\,,$$

we conclude that $N(\zeta)$ is light-tailed. The rest of the proof is analogous to the proof of Theorem 12.6.5. □

Note that by the result of Theorem 2.5.6, the assumption that $F_U, F_U^{\mathrm{s}} \in \mathcal{S}$ made in Theorem 12.6.6 can be replaced by $F_U \in \mathcal{S}^*$. An analogously modified assumption can be made in Theorem 12.6.5.

Bibliographical Notes. The material of this section is from Asmussen, Schmidli and Schmidt (1999), where also the case of heavy-tailed Λ and I in the Björk–Grandell model is considered. For the special case of Pareto type distributions, see also Grandell (1997). The proof of Lemma 12.6.1 follows an idea developed in Asmussen and Klüppelberg (1996); see also Klüppelberg and Mikosch (1997). A version of Theorem 12.6.5 was already obtained by Asmussen, Fløe Henriksen and Klüppelberg (1994). Some related results can be found in Asmussen, Klüppelberg and Sigman (1998), Rolski, Schlegel and Schmidt (1999) and Schmidli (1998).

CHAPTER 13

Diffusion Models

In this chapter we begin with a short introduction to stochastic differential equations. As an application we consider models with stochastic interest rates which are described by diffusion processes. An insurance risk reserve process with such a diffusion component is called a *perturbed risk process.* It turns out that, for perturbed risk processes, methods can be used which are similar to those applied to unperturbed risk processes in earlier chapters. Other examples from actuarial and financial mathematics follow. We discuss the popular Black–Scholes model and consider a life-insurance model under stochastic interest rates. As before, all processes are assumed to have càdlàg sample paths. We start from a given filtration $\{\mathcal{F}_t\}$ and assume that all stochastic processes considered below are adapted to this filtration. Unless stated otherwise, we assume in this chapter that the underlying probability space $(\Omega, \mathcal{F}, \mathbb{P})$ and the filtration $\{\mathcal{F}_t\}$ are complete.

13.1 STOCHASTIC DIFFERENTIAL EQUATIONS

13.1.1 Stochastic Integrals and Itô's Formula

For a stochastic process $\{X(t)\}$ we define the variation over $[0, t]$ by

$$V(t) = \sup\Big\{\sum_{i=1}^{n} |X(t_i) - X(t_{i-1})|\Big\},$$

where the supremum is taken over all sequences $0 = t_0 < t_1 < \ldots < t_n = t$ and $n \in \mathbb{N}$. Let $\{X(t)\}$ be a process with *bounded variation*, i.e. $V(t) < \infty$ for each $t > 0$. Writing

$$\overline{X}(t) = \sup\Big\{\sum_{i=1}^{n} (X(t_i) - X(t_{i-1}))\mathbb{I}(X(t_i) - X(t_{i-1}) > 0)\Big\}$$

and $\underline{X}(t) = \overline{X}(t) - (X(t) - X(0))$, we have

$$X(t) = X(0) + \overline{X}(t) - \underline{X}(t)\,. \tag{13.1.1}$$

Here $\{\overline{X}(t)\}$ and $\{\underline{X}(t)\}$ are increasing processes and (13.1.1) is called the *Jordan decomposition* of $\{X(t)\}$. For example the risk reserve process in the Sparre Andersen model has bounded variation. As we will see later in this section, this is no longer true for Brownian motion. If the process $\{X(t)\}$ has bounded variation, the stochastic integral $\int_0^t Y(v)\,\mathrm{d}X(v)$ for an arbitrary stochastic process $\{Y(t)\}$ is defined pathwise as a Lebesgue–Stieltjes integral, provided that $\{Y(t)\}$ is integrable with respect to $\{X(t)\}$. It however turns out that the standard Brownian motion $\{W(t)\}$ has unbounded variation and therefore the integral $\int_0^t Y(v)\,\mathrm{d}W(v)$ cannot be defined in the preceding sense. Indeed, one can show that

$$\sum_{i=1}^{2^n}\Big|W\Big(\frac{it}{2^n}\Big)-W\Big(\frac{(i-1)t}{2^n}\Big)\Big|\longrightarrow\infty \tag{13.1.2}$$

with probability 1 as $n\to\infty$. To prove (13.1.2), define

$$Z_n=\sum_{i=1}^{2^n}\Big(W\Big(\frac{it}{2^n}\Big)-W\Big(\frac{(i-1)t}{2^n}\Big)\Big)^2-t\,.$$

Then $\mathbb{E}\,Z_n=0$. Also $\mathbb{E}\,(Z_n^2)=t^2 2^{-n+1}$, as can be easily shown by the reader, using the fact that, for each constant $c>0$, the scaled process $\{\sqrt{c}W(t/c),\ t\ge 0\}$ is again a standard Brownian motion. For each $\varepsilon>0$, Chebyshev's inequality implies

$$\mathbb{P}(|Z_n|\ge\varepsilon)\le\varepsilon^{-2}\mathbb{E}\,(Z_n^2)=(t/\varepsilon)^2 2^{-n+1}\,.$$

Hence $\sum_{n=1}^{\infty}\mathbb{P}(|Z_n|\ge\varepsilon)<\infty$. By the Borel–Cantelli lemma we can then conclude that $\lim_{n\to\infty}Z_n=0$ with probability 1. From this we have

$$t\le\liminf_{n\to\infty}\max_{1\le k\le 2^n}\Big|W\Big(\frac{kt}{2^n}\Big)-W\Big(\frac{(k-1)t}{2^n}\Big)\Big|\sum_{i=1}^{2^n}\Big|W\Big(\frac{it}{2^n}\Big)-W\Big(\frac{(i-1)t}{2^n}\Big)\Big|.$$

The sample-path continuity of $\{W(t)\}$ entails

$$\max_{1\le k\le 2^n}\Big|W\Big(\frac{kt}{2^n}\Big)-W\Big(\frac{(k-1)t}{2^n}\Big)\Big|\longrightarrow 0$$

with probability 1 as $n\to\infty$, and so (13.1.2) follows.

We are facing the problem of defining an integral with respect to a process with unbounded variation. For example, if we want to define $\int_0^t W(v)\,\mathrm{d}W(v)$, we already have the two different Riemann sums

$$W_{1,n}(t)=\sum_{i=1}^{2^n}W\Big(\frac{(i-1)t}{2^n}\Big)\Big(W\Big(\frac{it}{2^n}\Big)-W\Big(\frac{(i-1)t}{2^n}\Big)\Big) \tag{13.1.3}$$

and

$$W_{2,n}(t) = \sum_{i=1}^{2^n} W\Big(\frac{it}{2^n}\Big)\Big(W\Big(\frac{it}{2^n}\Big) - W\Big(\frac{(i-1)t}{2^n}\Big)\Big) \tag{13.1.4}$$

that could be used. Note that

$$W_{1,n}(t) + W_{2,n}(t) = \sum_{i=1}^{2^n}\Big(W^2\Big(\frac{it}{2^n}\Big) - W^2\Big(\frac{(i-1)t}{2^n}\Big)\Big) = W^2(t)\,.$$

Taking the difference of (13.1.3) and (13.1.4), we also obtain

$$W_{2,n}(t) - W_{1,n}(t) = \sum_{i=1}^{2^n}\Big(W\Big(\frac{it}{2^n}\Big) - W\Big(\frac{(i-1)t}{2^n}\Big)\Big)^2$$

and this converges to t with probability 1, as was shown above. Thus,

$$\lim_{n\to\infty} W_{2,n}(t) = 2^{-1}(W^2(t) + t)$$

and

$$\lim_{n\to\infty} W_{1,n}(t) = 2^{-1}(W^2(t) - t)\,.$$

We can choose either one of the expressions above as the value of the stochastic integral $\int_0^t W(v)\,\mathrm{d}W(v)$. In Theorem 10.1.2 we have seen that both $\{W(t)\}$ and $\{W^2(t) - t\}$ are martingales. Since the martingale property is desirable, we will choose $\int_0^t W(v)\,\mathrm{d}W(v) = 2^{-1}(W^2(t) - t)$ as our stochastic integral. This choice is inspired by a general theory for defining integrals with respect to Brownian motion. Due to its complexity, we will only sketch this theory. Readers interested in more details should consult the appropriate references to the literature.

Let $\{Y(t)\}$ be an arbitrary stochastic process. We call $\{Y(t)\}$ a *piecewise constant process* if it can be represented in the form

$$Y(t) = \sum_{i=0}^{\infty} Y(\tau_i)\mathbb{I}(\tau_i \le t < \tau_{i+1})$$

for some increasing sequence $\{\tau_n\}$ of stopping times for which $0 = \tau_0 < \tau_1 < \ldots$, $\lim_{n\to\infty}\tau_n = \infty$ and where for each n the random variable $Y(\tau_n)$ is $\mathcal{F}_{\tau_n}$-measurable. Note that we do not assume that the sequence $\{Y(\tau_n)\}$ is bounded, nor that it consists of square integrable random variables. For a piecewise constant process $\{Y(t)\}$ it is natural to define

$$\int_0^t Y(v)\,\mathrm{d}W(v) = \sum_{i=0}^{N(t)-1} Y(\tau_i)\,(W(\tau_{i+1}) - W(\tau_i)) + Y(t)\,\big(W(t) - W(\tau_{N(t)})\big)\,, \tag{13.1.5}$$

where $N(t) = \max\{i : \tau_i \le t\}$. The stochastic process $\{\int_0^t Y(v)\,\mathrm{d}W(v),\ t \ge 0\}$ so defined is continuous and has the following important property. Let $\tau'_n = \inf\{t \ge 0 : |\int_0^t Y(v)\,\mathrm{d}W(v)| \ge n\}$; then $\{\int_0^{t\wedge\tau'_n} Y(v)\,\mathrm{d}W(v),\ t \ge 0\}$ is a uniformly integrable martingale. This fact, to be proved by the reader, induces the concept of a local martingale. A stochastic process $\{X(t)\}$ adapted to the filtration $\{\mathcal{F}_t\}$ is called a *local martingale* if there exists a sequence of $\{\mathcal{F}_t\}$-stopping times $\{\tau_n\}$ such that $\lim_{n\to\infty}\tau_n = \infty$ and if for each $n = 1, 2, \ldots$ the process $\{X_n(t),\ t \ge 0\}$ with $X_n(t) = X(t \wedge \tau_n)$ is a martingale. The sequence $\{\tau_n\}$ is called a *localization sequence.* In many cases it is possible to choose $\{\tau_n\}$ in such a way that each process $\{X_n(t)\}$ is a uniformly integrable martingale. Moreover, the following is true.

Lemma 13.1.1 *Let $\{X(t)\}$ and $\{Y(t)\}$ be two piecewise constant processes, and $\{W(t)\}$ a Brownian motion. Then the processes $\{M(t)\}$ and $\{M'(t)\}$ with $M(t) = \int_0^t Y(v)\,\mathrm{d}W(v)$ and*

$$M'(t) = \int_0^t Y(v)\,\mathrm{d}W(v) \int_0^t X(v)\,\mathrm{d}W(v) - \int_0^t Y(v)X(v)\,\mathrm{d}v \tag{13.1.6}$$

are local martingales. If $\{X(t)\}$ and $\{Y(t)\}$ are bounded, then $\{M(t)\}$ and $\{M'(t)\}$ are martingales.

Proof First consider the case where $\{X(t)\}$ and $\{Y(t)\}$ are bounded. In this case it is clear that $\{M(t)\}$ is a martingale, as can readily be seen from (13.1.5). Furthermore, for $0 \le v < t$,

$$\mathbb{E}\left(\int_v^t Y(w)\,\mathrm{d}W(w)\,\Big|\,\mathcal{F}_v\right) = 0\,,$$

where we put $\int_v^t Y(w)\,\mathrm{d}W(w) = \int_0^t Y(w)\,\mathrm{d}W(w) - \int_0^v Y(w)\,\mathrm{d}W(w)$. Thus,

$$\begin{aligned}\mathbb{E}(M'(t) - M'(v) \mid \mathcal{F}_v) &= \mathbb{E}\left(\int_v^t Y(w)\,\mathrm{d}W(w)\int_v^t X(w)\,\mathrm{d}W(w)\right.\\ &\qquad \left.-\int_v^t Y(w)X(w)\,\mathrm{d}w\,\Big|\,\mathcal{F}_v\right).\end{aligned} \tag{13.1.7}$$

Denote by $\{\tau_n\}$ the increasing sequence of jump epochs both of $\{X(t)\}$ and $\{Y(t)\}$. Since for $k > n$ and $\tau_{n+1} \ge v$,

$$\begin{aligned}&\mathbb{E}\,(X(\tau_k)(W(\tau_{k+1}) - W(\tau_k))Y(\tau_n)(W(\tau_{n+1}) - W(\tau_n \vee v)) \mid \mathcal{F}_v)\\ &= \mathbb{E}\,(\mathbb{E}\,(X(\tau_k)(W(\tau_{k+1}) - W(\tau_k))\\ &\qquad \times Y(\tau_n)(W(\tau_{n+1}) - W(\tau_n \vee v)) \mid \mathcal{F}_{\tau_k}) \mid \mathcal{F}_v)\\ &= \mathbb{E}\,(X(\tau_k)\mathbb{E}\,(W(\tau_{k+1}) - W(\tau_k) \mid \mathcal{F}_{\tau_k})\\ &\qquad \times Y(\tau_n)(W(\tau_{n+1}) - W(\tau_n \vee v)) \mid \mathcal{F}_v) = 0\end{aligned}$$

we obtain

$$
\begin{aligned}
& \mathbb{E}\Big(\int_v^t Y(w)\,\mathrm{d}W(w)\int_v^t X(w)\,\mathrm{d}W(w)\;\Big|\;\mathcal{F}_v\Big) \\
&= \mathbb{E}\Big(X(v)Y(v)(W(\tau_{N(v)+1})-W(v))^2 \\
&\qquad + \sum_{i=N(v)+1}^{N(t)-1} Y(\tau_i)X(\tau_i)\left(W(\tau_{i+1})-W(\tau_i)\right)^2 \\
&\qquad + X(t)Y(t)\left(W(t)-W(\tau_{N(t)})\right)^2\;\Big|\;\mathcal{F}_v\Big) \\
&= \mathbb{E}\Big(X(v)Y(v)(\tau_{N(v)+1}-v) + \sum_{i=N(v)+1}^{N(t)-1} Y(\tau_i)X(\tau_i)\left(\tau_{i+1}-\tau_i\right) \\
&\qquad + X(t)Y(t)\left(t-\tau_{N(t)}\right)\;\Big|\;\mathcal{F}_v\Big) \\
&= \mathbb{E}\Big(\int_v^t Y(w)X(w)\,\mathrm{d}w\;\Big|\;\mathcal{F}_v\Big).
\end{aligned}
$$

Using (13.1.7), it follows that $\{M'(t)\}$ is a martingale. For the general case we consider the stopping times $\tau_n' = \inf\{t \ge 0 : \max\{|X(t)|, |Y(t)|\} \ge n\}$. In this case the processes $\{X_n(t)\}$ and $\{Y_n(t)\}$ with

$$X_n(t) = X(t)\mathbb{I}(t < \tau_n') \qquad \text{and} \qquad Y_n(t) = Y(t)\mathbb{I}(t < \tau_n')$$

are bounded. By the first part of the proof, $\{M(t)\}$ and $\{M'(t)\}$ are local martingales. □

From the result of Lemma 13.1.1 we can conclude that, in particular, $\{(\int_0^t Y(v)\,\mathrm{d}W(v))^2 - \int_0^t Y^2(v)\,\mathrm{d}v,\ t \ge 0\}$ is a local martingale if $\{Y(t)\}$ is piecewise constant. This indicates that the right class of processes, for which we can define a stochastic integral with respect to Brownian motion, is the class L^2_{loc} of all càdlàg processes. The notation L^2_{loc} is motivated by the fact that for each càdlàg process $\{X(t)\}$, $\int_0^t X^2(v)\,\mathrm{d}v < \infty$ for all $t \ge 0$. We also define the smaller class $L^2 \subset L^2_{\mathrm{loc}}$ of càdlàg processes $\{X(t)\}$ for which $\int_0^t \mathbb{E}\,X^2(v)\,\mathrm{d}v < \infty$ for all $t \ge 0$. Note that, if $\{Y(t)\} \in L^2_{\mathrm{loc}}$ and $\tau_n = \inf\{t \ge 0 : \max\{|Y(t)|, |Y(t-0)|\} \ge n\}$, then $\{Y_n(t)\} \in L^2$, where

$$Y_n(t) = Y(t)\mathbb{I}(t < \tau_n). \tag{13.1.8}$$

It is shown in Lemma 5.2.2 of Ethier and Kurtz (1986) that each process $\{Y(t)\}$ in L^2 can be approximated by bounded piecewise constant processes $\{Y_n(t)\}$ such that, for each $t \ge 0$,

$$\lim_{n\to\infty} \mathbb{E}\Big(\int_0^t (Y_n(v) - Y(v))^2\,\mathrm{d}v\Big) = 0. \tag{13.1.9}$$

The following theorem extends the notion of the stochastic integral with respect to $\{W(t)\}$ to arbitrary processes in L^2 and L^2_{loc}, respectively.

Theorem 13.1.1 (a) *Let $\{Y(t)\} \in L^2$. Then there exists a (up to indistinguishability) unique continuous martingale denoted by $\{\int_0^t Y(v)\,\mathrm{d}W(v),\ t \geq 0\}$ such that, for all $x \geq 0$,*

$$\lim_{n\to\infty} \sup_{0\leq t\leq x} \left| \int_0^t Y_n(v)\,\mathrm{d}W(v) - \int_0^t Y(v)\,\mathrm{d}W(v) \right| = 0$$

and

$$\lim_{n\to\infty} \mathbb{E}\left(\sup_{0\leq t\leq x} \left| \int_0^t Y_n(v)\,\mathrm{d}W(v) - \int_0^t Y(v)\,\mathrm{d}W(v) \right| \right)^2 = 0$$

whenever the approximating sequence of bounded piecewise constant processes $\{Y_n(t)\}$ satisfies

$$\sum_{n=1}^{\infty} \left(\mathbb{E}\left(\int_0^t (Y_n(v) - Y(v))^2\,\mathrm{d}v \right) \right)^{1/2} < \infty, \qquad t \geq 0. \tag{13.1.10}$$

Moreover, if $\{X(t)\} \in L^2$ then $\mathbb{E}\int_0^t |Y(v)X(v)|\,\mathrm{d}v < \infty$ and $\{M'(t)\}$ defined in (13.1.6) *is a martingale.*
(b) *Let $\{Y(t)\} \in L^2_{\mathrm{loc}}$. Then there exists a (up to indistinguishability) unique continuous local martingale denoted by $\{\int_0^t Y(v)\,\mathrm{d}W(v),\ t \geq 0\}$ such that, for each stopping time τ with $\{Y(t)\mathbb{I}(\tau > t),\ t \geq 0\} \in L^2$,*

$$\int_0^{\tau\wedge t} Y(v)\,\mathrm{d}W(v) = \int_0^t Y(v)\mathbb{I}(\tau > v)\,\mathrm{d}W(v).$$

Moreover, if $\{X(t)\} \in L^2_{\mathrm{loc}}$ then $\int_0^t |Y(v)X(v)|\,\mathrm{d}v < \infty$ and $\{M'(t)\}$ defined in (13.1.6) *is a local martingale.*

The *proof* of Theorem 13.1.1 goes beyond the scope of this book and can be found, for example, in Ethier and Kurtz (1986), Theorems 5.2.3 and 5.2.6.

Remarks 1. In textbooks it is usually required that $\{X(t)\}$ and $\{Y(t)\}$ are progressively measurable. However, the reader should note that a right-continuous process is progressively measurable. Since we exclusively work with càdlàg processes, the (weaker) condition of progressive measurability is automatically fulfilled.

2. Note that by (13.1.8) we can always find localization times $\{\tau_n\}$ such that $\{Y(t)\mathbb{I}(t < \tau_n)\} \in L^2$, provided that $\{Y(t)\} \in L^2_{\mathrm{loc}}$. Furthermore, for $\{Y(t)\} \in L^2$, by (13.1.9) we can always choose a subsequence of approximating processes for which (13.1.10) holds. Thus, it is quite natural that the process

$\{\int_0^t Y(v)\,\mathrm{d}W(v)\}$ appearing in Theorem 13.1.1 is called the *stochastic integral* of $\{Y(t)\}$ with respect to the Brownian motion $\{W(t)\}$.

3. It is often the case that $\{Y(t)\}$ is continuous and that (13.1.10) is fulfilled for the approximating processes $\{Y_n(t)\}$ with $Y_n(t) = \sum_{j=0}^{\infty} Y(jm_n)\mathbb{I}(jm_n \leq t < (j+1)m_n)$, where $\{m_n\}$ is some sequence converging to 0 as $n \to \infty$. In this case, the stochastic integral $\int_0^t Y(v)\,\mathrm{d}W(v)$ can be approximated by the Riemann sums $\sum_{i<tm_n^{-1}} Y(im_n)(W((i+1)m_n) - W(im_n))$. This suggests a possibility to simulate the process $\{\int_0^t Y(v)\,\mathrm{d}W(v)\}$ using the Riemann sum for a "large" n.

The next theorem is a special case of a result known as *Itô's formula.*

Theorem 13.1.2 *Let $n \in \mathbb{N}$ be fixed. For $1 \leq i \leq n$, let $\{Y_i(t)\} \in L^2_{\mathrm{loc}}$, $\{V_i(t)\}$ a process with bounded variation, and $X_i(t) = X_i(0) + \int_0^t Y_i(v)\,\mathrm{d}W_i(v) + V_i(t)$, where the processes $\{W_i(t)\}$ are Brownian motions such that either $W_i = W_j$, or W_i and W_j are independent. Consider the process $\{X(t)\}$ with*

$$X(t) = g(t, X_1(t), X_2(t), \ldots, X_n(t)),$$

where $g(t, x_1, \ldots, x_n)$ is a function continuously differentiable with respect to t and twice continuously differentiable with respect to $(x_1, \ldots, x_n)$. Then

$$\{g_{x_i}(t, X_1(t), X_2(t), \ldots, X_n(t))Y_i(t)\} \in L^2_{\mathrm{loc}}, \tag{13.1.11}$$

all the integrals below exist, and

$$\begin{aligned} X(t) - X(0) &= \int_0^t g_t(v, X_1(v), X_2(v), \ldots, X_n(v))\,\mathrm{d}v \\ &+ \sum_{i=1}^n \int_0^t g_{x_i}(v, X_1(v), X_2(v), \ldots, X_n(v))\,\mathrm{d}V_i(v) \\ &+ \sum_{i=1}^n \int_0^t g_{x_i}(v, X_1(v), X_2(v), \ldots, X_n(v))Y_i(v)\,\mathrm{d}W_i(v) \\ &+ \tfrac{1}{2}\sum_{i=1}^n \sum_{j=1}^n \eta_{ij} \int_0^t g_{x_i x_j}(v, X_1(v), X_2(v), \ldots, X_n(v))Y_i(v)Y_j(v)\,\mathrm{d}v \end{aligned} \tag{13.1.12}$$

for all $t \geq 0$, where $\eta_{ij} = 1$ if $W_i = W_j$ and $\eta_{ij} = 0$ otherwise, and where g_{x_i} (g_t, respectively) is the partial derivative with respect to x_i (t, respectively).

The *proof* of Theorem 13.1.2 can be found in Ethier and Kurtz (1986), Theorem 5.2.9.

If $X(t) = X(0) + \int_0^t Y(v)\,\mathrm{d}W(v) + \int_0^t Z(v)\,\mathrm{d}v$ for some process $\{Z(t)\}$ then we use the short notation

$$\mathrm{d}X(t) = Y(t)\,\mathrm{d}W(t) + Z(t)\,\mathrm{d}t\,, \tag{13.1.13}$$

and (13.1.13) is called a *stochastic differential.* For example, in the case that $n = 1$, $Y_1(t) = 1$ and $V_1(t) = 0$, we write Itô's formula (13.1.12) in the following way:

$$\mathrm{d}g(t, W(t)) = g_t(t, W(t))\,\mathrm{d}t + g_x(t, W(t))\,\mathrm{d}W(t) + \tfrac{1}{2} g_{xx}(t, W(t))\,\mathrm{d}t\,. \tag{13.1.14}$$

If $g(t, x) = x^2/2$, the differential form (13.1.14) of Itô's formula reads

$$\mathrm{d}(W^2(t)/2) = W(t)\,\mathrm{d}W(t) + \tfrac{1}{2}\,\mathrm{d}t\,. \tag{13.1.15}$$

This is different from the bounded variation case where $\mathrm{d}(x^2(t)/2) = x(t)\,\mathrm{d}x(t)$. The term $\frac{1}{2}\,\mathrm{d}t$ in (13.1.15) is due to the unbounded variation of $\{W(t)\}$ and is called the *quadratic variation part.* Note that from (13.1.15) we have

$$\frac{W^2(t)}{2} = \int_0^t W(v)\,\mathrm{d}W(v) + \frac{t}{2}$$

and therefore

$$\int_0^t W(v)\,\mathrm{d}W(v) = \frac{W^2(t) - t}{2}\,.$$

From Itô's formula (13.1.12) it also follows that the well-known formula for integration by parts of processes with bounded variation is no longer valid in the case of unbounded variation. If for instance $\mathrm{d}X_i(t) = Y_i(t)\,\mathrm{d}W(t) + V_i(t)\,\mathrm{d}t$, $i = 1, 2$, then (13.1.12) gives that

$$\mathrm{d}(X_1(t)X_2(t)) = X_1(t)\,\mathrm{d}X_2(t) + X_2(t)\,\mathrm{d}X_1(t) + Y_1(t)Y_2(t)\,\mathrm{d}t\,. \tag{13.1.16}$$

Remark The stochastic integral introduced in Theorem 13.1.1 is the so-called Itô stochastic integral. If $\mathcal{F}_t = \mathcal{F}_{t+}$ for all $t \geq 0$ one could consider (13.1.4), instead of (13.1.3), as another possible candidate for the Riemann sums to define a stochastic integral. Alternatively, we could define the Stratonovich stochastic integral $\int_0^t Y(v) \circ \mathrm{d}W(v)$, which is obtained via the Riemann sums

$$\sum_{i=1}^n Y\Big(\frac{(i - \frac{1}{2})t}{n}\Big)\Big(W\Big(\frac{it}{n}\Big) - W\Big(\frac{(i-1)t}{n}\Big)\Big)\,.$$

For the Stratonovich integral, the integration rules for processes with bounded variation remain valid in the case of unbounded variation, but generally $\{\int_0^t Y(v) \circ \mathrm{d}W(v),\ t \geq 0\}$ is no longer a martingale. For instance $\int_0^t W(v) \circ \mathrm{d}W(v) = W^2(t)/2$.

13.1.2 Diffusion Processes

By the *stochastic differential equation*

$$\mathrm{d}X(t) = a(t, X(t))\,\mathrm{d}t + \sigma(t, X(t))\,\mathrm{d}W(t)\,, \tag{13.1.17}$$

we mean the equation

$$X(t) = X(0) + \int_0^t a(v, X(v))\,\mathrm{d}v + \int_0^t \sigma(v, X(v))\,\mathrm{d}W(v)\,, \quad t \geq 0\,, \tag{13.1.18}$$

and a stochastic process $\{X(t)\}$ fulfilling (13.1.18) is its solution. If the solution is unique, then the process $\{X(t)\}$ is called a *diffusion process* with *infinitesimal drift function* $a(t,x)$ and *infinitesimal variance* $\sigma^2(t,x)$ at (t,x), provided that $\sigma^2(t,x) > 0$ for all $t \geq 0$ and $x \in E$, where $E \subset \mathbb{R}$ is the state space of $\{X(t)\}$. We also say that $\{X(t)\}$ is an $(a(t,x), \sigma^2(t,x))$-diffusion.

The following conditions imply the existence of a unique solution to (13.1.17).

Theorem 13.1.3 *Assume that* $\mathbb{E}\,X^2(0) < \infty$ *and that for any* $x > 0$ *there exists a constant* $c_x \in (0,\infty)$ *such that for all* $y, z \in \mathbb{R}$,

$$|\sigma(t,z) - \sigma(t,y)| + |a(t,z) - a(t,y)| \leq c_x|z - y| \tag{13.1.19}$$

and

$$\sigma^2(t,y) + a^2(t,y) \leq c_x(1 + y^2) \tag{13.1.20}$$

whenever $0 \leq t \leq x$. *Then there exists a unique solution* $\{X(t)\}$ *to* (13.1.17). *The solution* $\{X(t)\}$ *is continuous and fulfils*

$$\mathbb{E}\,X^2(t) \leq k_x \mathrm{e}^{k_x t}(1 + \mathbb{E}\,X^2(0)) \tag{13.1.21}$$

for some constant k_x *and all* $0 \leq t \leq x$. *Moreover, the solution* $\{X(t)\}$ *to* (13.1.17) *is an* $\{\mathcal{F}_t\}$*-Markov process, and a strong Markov process with respect to the filtration* $\{\mathcal{F}_{t+}\}$.

Again, the *proof* of Theorem 13.1.3 goes beyond the scope of this book. We therefore omit it and refer to the books of Ethier and Kurtz (1986) and Karatzas and Shreve (1991).

Examples 1. If $\sigma(t,x) = \sigma x$ and $a(t,x) = \mu x$ for some constants $\sigma > 0$ and $\mu \in \mathbb{R}$, then the solution $\{X(t)\}$ to (13.1.17) can be guessed from Itô's formula (13.1.12):

$$X(t) = X(0)\exp\big((\mu - \sigma^2/2)t + \sigma W(t)\big)\,. \tag{13.1.22}$$

The process $\{X(t)\}$ given in (13.1.22) is called a *geometric Brownian motion.* Note that the solution to the (deterministic) differential equation $x^{(1)}(t) =$

$a(t, x(t))$ is $x(t) = x(0)\exp(\mu t) = \mathbb{E}(X(t) \mid X(0) = x(0))$, which is the value of a capital $x(0)$ after time t if the force of interest is constant and equal to μ. Moreover, $X(t) > 0$ provided that $X(0) > 0$. Therefore the geometric Brownian motion is often used to model prices of financial securities, as for example stocks and bonds. We will study this process later in Section 13.3.

2. In the next example we study a diffusion process which is a Gaussian process. A process $\{X(t),\ t \in \mathcal{T}\}$, where $\mathcal{T}$ is an arbitrary set of parameters is called a *Gaussian process* if for all $t_1, \ldots, t_n \in \mathcal{T}$ the random vector $(X(t_1), \ldots, X(t_n))$ has a multivariate normal distribution. Notice that a random vector $(Z_1, \ldots, Z_n)$ is said to have a *multivariate normal distribution* if $\sum_{j=1}^n s_j Z_j$ has a (univariate) normal distribution for all $s_1, \ldots, s_n \in \mathbb{R}$. In particular, if the covariance matrix $\boldsymbol{C}$ of $(Z_1, \ldots, Z_n)$ is nonsingular, then the density $f(x_1, \ldots, x_n)$ of $(Z_1, \ldots, Z_n)$ has the form

$$f(x_1, \ldots, x_n) = \frac{\sqrt{\det \boldsymbol{C}'}}{(2\pi)^{n/2}} \exp\Big(-\frac{1}{2}\sum_{i,j=1}^n c'_{ij}(x_i - \mu_i)(x_j - \mu_j)\Big),$$

where $(\mu_1, \ldots, \mu_n) \in \mathbb{R}$ is the expectation vector and $\boldsymbol{C}' = \boldsymbol{C}^{-1}$. We leave it to the reader to show that for a deterministic function $h(t)$ from the class L^2, the random variable $\int_0^t h(v)\,\mathrm{d}W(v)$ is $\mathrm{N}(0, \int_0^t h^2(v)\,\mathrm{d}v)$-distributed. If $\sigma(t, x) = \sigma$ and $a(t, x) = -\alpha(x - \bar{\delta})$ for some $\alpha, \sigma > 0$ and $\bar{\delta} \in \mathbb{R}$, then the process $\{X(t)\}$ with

$$X(t) = \mathrm{e}^{-\alpha t}X(0) + \bar{\delta}(1 - \mathrm{e}^{-\alpha t}) + \sigma \mathrm{e}^{-\alpha t}\int_0^t \mathrm{e}^{\alpha v}\,\mathrm{d}W(v) \qquad (13.1.23)$$

is the unique solution to (13.1.17). It is called an *Ornstein–Uhlenbeck process*. If $X(0)$ is deterministic or if $(X(0), \{W(t)\})$ is jointly Gaussian, then the process $\{X(t)\}$ given by (13.1.23) is a Gaussian process. We leave it to the reader to show that the expectation and covariance functions of this process have the form

$$\mathbb{E}\,X(t) = \bar{\delta} + \mathrm{e}^{-\alpha t}(\mathbb{E}\,X(0) - \bar{\delta}), \qquad (13.1.24)$$

$$\mathrm{Cov}\,(X(t), X(t+h)) = \frac{\sigma^2}{2\alpha}\big(\mathrm{e}^{-\alpha h} - \mathrm{e}^{-\alpha(2t+h)}\big) \qquad (13.1.25)$$

for all $t, h \geq 0$. Note that a Gaussian process is uniquely defined if the mean and covariance functions are known. To show that the stochastic differential equation

$$\mathrm{d}X(v) = -\alpha(X(v) - \bar{\delta})\,\mathrm{d}v + \sigma\,\mathrm{d}W(v)$$

has indeed the solution given by (13.1.23), we multiply the above equation by $\mathrm{e}^{\alpha v}$ and use formula (13.1.16) to obtain

$$\mathrm{d}(\mathrm{e}^{\alpha v}X(v)) = \bar{\delta}\alpha \mathrm{e}^{\alpha v}\,\mathrm{d}v + \sigma \mathrm{e}^{\alpha v}\,\mathrm{d}W(v)$$

from which we get (13.1.23) after integration from 0 to t. The Ornstein–Uhlenbeck process has the property that it is mean-reverting, i.e. it always tries to come back to its asymptotic mean value $\bar{\delta}$. It is therefore sometimes used as a model for the force of interest. In connection with this, the integral $\int_0^t X(v)\,\mathrm{d}v$ is considered; see Section 13.4. It follows from (13.1.23) that the Riemann sums for the integral $\int_0^t X(v)\,\mathrm{d}v$ follow a normal distribution. Taking into account that the limit of normally distributed random variables is normal, it suffices to compute the expectation and variance of the integral $\int_0^t X(v)\,\mathrm{d}v$ to determine its distribution. Using (13.1.24) and (13.1.25) we have

$$\begin{aligned} \mathbb{E}\int_0^t X(v)\,\mathrm{d}v &= \int_0^t \mathbb{E}\,X(v)\,\mathrm{d}v \\ &= \bar{\delta}t + (X(0)-\bar{\delta})\Big(\frac{1-\mathrm{e}^{-\alpha t}}{\alpha}\Big) \end{aligned} \tag{13.1.26}$$

and

$$\begin{aligned} &\operatorname{Var}\int_0^t X(v)\,\mathrm{d}v \\ &= \int_0^t\int_0^t \mathbb{E}\,(X(v)X(w))\,\mathrm{d}w\,\mathrm{d}v - \Big(\int_0^t \mathbb{E}\,X(w)\,\mathrm{d}w\Big)^2 \\ &= \int_0^t\int_0^t \operatorname{Cov}(X(v),X(w))\,\mathrm{d}w\,\mathrm{d}v \\ &= \frac{\sigma^2}{\alpha}\int_0^t\Big(\int_0^w \mathrm{e}^{-\alpha(w-v)} - \mathrm{e}^{-\alpha(v+w)}\,\mathrm{d}v\Big)\,\mathrm{d}w \\ &= \frac{\sigma^2}{\alpha^2}t - \frac{\sigma^2}{\alpha^3}\big(1-\mathrm{e}^{-\alpha t}\big) - \frac{\sigma^2}{2\alpha^3}\big(1-\mathrm{e}^{-\alpha t}\big)^2 . \end{aligned} \tag{13.1.27}$$

3. Let $\sigma(t,x)=\sigma\sqrt{x}$ and $a(t,x)=-\alpha(x-\bar{\delta})$ for $\sigma,\alpha,\bar{\delta}>0$. Then, (13.1.17) becomes

$$\mathrm{d}X(t) = -\alpha(X(t)-\bar{\delta})\,\mathrm{d}t + \sigma\sqrt{X(t)}\,\mathrm{d}W(t)\,. \tag{13.1.28}$$

It can be shown – see for instance Cox, Ingersoll and Ross (1985) or Feller (1951) – that there exists a unique solution $\{X(t)\}$ to (13.1.28) such that $X(t)\ge 0$ for all $t>0$, where $\{\int_0^t \sqrt{X(v)}\,\mathrm{d}W(v)\}$ and $\{\int_0^t (X(v))^{3/2}\,\mathrm{d}W(v)\}$ are martingales and $\{X(t)\}$ is a strong Markov process with respect to the filtration $\{\mathcal{F}_{t+}\}$. Notice, however, that we cannot apply Theorem 13.1.3 because the square root function is only defined on $\mathbb{R}_+$ and because it is not Lipschitz continuous, i.e. condition (13.1.19) is not fulfilled. The process $\{X(t)\}$ is called a *Cox–Ingersoll–Ross model.* From $\mathbb{P}(X(t)\ge 0)=1$ it follows that $X(t)$ cannot be normally distributed. Since the process $\{X(t)\}$ cannot be negative, it is popular to model interest rates as a Cox–Ingersoll–Ross model; see also Section 13.4. For the expectation and the variance of $X(t)$ we can derive the following formulae. As above we assume that $X(0)$ is deterministic. From (13.1.28) we obtain

$$X(t) = X(0) - \alpha \int_0^t (X(v) - \bar{\delta})\,\mathrm{d}v + \sigma \int_0^t \sqrt{X(v)}\,\mathrm{d}W(v)$$

and therefore

$$\mathbb{E}\,X(t) = X(0) - \alpha \int_0^t (\mathbb{E}\,X(v) - \bar{\delta})\,\mathrm{d}v$$

because $\{\int_0^t \sqrt{X(v)}\,\mathrm{d}W(v)\}$ is a martingale. Solving this integral equation we find that

$$\mathbb{E}\,X(t) = \bar{\delta} + \mathrm{e}^{-\alpha t}(X(0) - \bar{\delta})\,. \tag{13.1.29}$$

For the second moment of $X(t)$, Itô's formula (13.1.12) yields

$$\begin{aligned} X^2(t) &= X^2(0) - \int_0^t 2\alpha X(v)(X(v) - \bar{\delta})\,\mathrm{d}v + \int_0^t \sigma^2 X(v)\,\mathrm{d}v \\ &\quad + \int_0^t 2\alpha\sigma(X(v))^{3/2}\,\mathrm{d}W(v)\,. \end{aligned}$$

Since $\{\int_0^t (X(v))^{3/2}\,\mathrm{d}W(v)\}$ is a martingale, this gives

$$\begin{aligned} \mathbb{E}\,X^2(t) &= X^2(0) - 2\alpha \int_0^t \mathbb{E}\,(X(v)(X(v) - \bar{\delta}))\,\mathrm{d}v + \sigma^2 \int_0^t \mathbb{E}\,X(v)\,\mathrm{d}v \\ &= X^2(0) - 2\alpha \int_0^t \mathbb{E}\,X^2(v)\,\mathrm{d}v + (\sigma^2 + 2\alpha\bar{\delta}) \int_0^t \mathbb{E}\,X(v)\,\mathrm{d}v\,. \end{aligned}$$

Hence, using (13.1.29), the solution of this integral equation is

$$\mathbb{E}\,X^2(t) = X^2(0)\mathrm{e}^{-2\alpha t} + (\sigma^2 + 2\alpha\bar{\delta})\Big(\frac{\bar{\delta}}{2\alpha}(1 - \mathrm{e}^{-2\alpha t}) + \frac{X(0) - \bar{\delta}}{\alpha}(\mathrm{e}^{-\alpha t} - \mathrm{e}^{-2\alpha t})\Big)\,.$$

Furthermore,

$$\mathrm{Var}\,X(t) = \frac{\sigma^2}{2\alpha}\left(\bar{\delta}(1 - \mathrm{e}^{-2\alpha t}) + 2X(0)\mathrm{e}^{-\alpha t}(1 - \mathrm{e}^{-\alpha t})\right)\,. \tag{13.1.30}$$

In the remaining part of the present section we give a few technical remarks concerning the generator of the Markov process $\{X(t)\}$ which is the solution to (13.1.17).

Assume $\sigma(t,x)$ and $a(t,x)$ are chosen such that (13.1.17) admits a unique solution. Then, if $g(t,x)$ fulfils the requirements of Theorem 13.1.2, it follows from Itô's formula (13.1.12) that

$$\begin{aligned}\mathrm{d}g(t,X(t)) &= g_t(t,X(t))\,\mathrm{d}t + g_x(t,X(t))a(t,X(t))\,\mathrm{d}t \\ &\quad + g_x(t,X(t))\sigma(t,X(t))\,\mathrm{d}W(t) + \tfrac{1}{2}g_{xx}(t,X(t))\sigma^2(t,X(t))\,\mathrm{d}t\end{aligned}$$

and hence

$$\begin{aligned}g(t,X(t)) - g(0,X(0)) &- \int_0^t \boldsymbol{A}g(v,X(v))\,\mathrm{d}v \\ &= \int_0^t g_x(v,X(v))\sigma(v,X(v))\,\mathrm{d}W(v)\end{aligned} \tag{13.1.31}$$

where

$$\boldsymbol{A}g(t,x) = \frac{\partial}{\partial t}g(t,x) + \frac{\sigma^2(t,x)}{2}\frac{\partial^2}{\partial x^2}g(t,x) + a(t,x)\frac{\partial}{\partial x}g(t,x)\,. \tag{13.1.32}$$

Suppose now that $g(t,x) = g(x)$, where $g : \mathbb{R} \to \mathbb{R}$ is twice continuously differentiable. Furthermore, if we would know that g is such that the right-hand side of (13.1.31) is a martingale, then we would know that $(g, \boldsymbol{A}g)$ belongs to the full generator of the Markov process $\{X(t)\}$. However, it may be difficult to give conditions ensuring that a function g is in the domain $\mathcal{D}(\boldsymbol{A})$ of the full generator. We therefore generalize the definition of the full generator introduced in Section 11.1.4. The *extended generator* is the multivalued operator $\boldsymbol{A}$ consisting of all pairs $(g,\tilde{g})$ for which

$$\left\{g(X(t)) - g(X(0)) - \int_0^t \tilde{g}(X(v))\,\mathrm{d}v,\ t \geq 0\right\} \tag{13.1.33}$$

is a local martingale. The set of all functions g such that there exists a $\tilde{g}$ with $(g,\tilde{g}) \in \boldsymbol{A}$ is called the *domain* of the extended generator. As before in Section 11.1.4, we denote the domain of the extended generator by $\mathcal{D}(\boldsymbol{A})$. Note that, obviously, the full generator is contained in the extended generator. This is the reason why we use the same symbol for the extended and for the full generator. For the rest of this chapter we simply say generator if we mean extended generator. We will write $\boldsymbol{A}g$ for a version of the functions $\tilde{g}$ such that $(g,\tilde{g}) \in \boldsymbol{A}$. Note that by the dominated convergence theorem, the process given in (13.1.33) is a martingale if both g and $\boldsymbol{A}g$ are bounded.

13.1.3 Lévy's Characterization Theorem

The following auxiliary result will be useful in the proof of Theorem 13.1.4 below. Note, however, that Lemma 13.1.2 is frequently used in other characterization theorems as well.

Lemma 13.1.2 *Let $\{M(t)\}$ be a continuous local martingale with bounded variation. Then $\{M(t)\}$ is constant.*

Proof Without loss of generality we assume $M(0) = 0$. Denote by $\{V(t)\}$ the variation process, which can be written as

$$V(t) = \lim_{n\to\infty} \sum_{i=1}^{k_n} |M(t_i^{(n)}) - M(t_{i-1}^{(n)})|,$$

where $k_n \in \mathbb{N}$ and $0 = t_0^{(n)} < t_1^{(n)} < \ldots < t_{k_n}^{(n)} = t$ is a series of partitions such that

$$\lim_{n\to\infty} \sum_{i=1}^{k_n} |M(t_i^{(n)}) - M(t_{i-1}^{(n)})| = \sup \sum_{i=1}^{m} |M(v_i) - M(v_{i-1})|.$$

Here the supremum is taken over all partitions $0 = v_0 < v_1 < \ldots < v_m = t$ and $m \in \mathbb{N}$. Without loss of generality we can assume that $\sup\{t_i^{(n)} - t_{i-1}^{(n)}\} \to 0$ as $n \to \infty$. This is possible because by the triangle inequality we have $|M(t_i^{(n)}) - M(t_{i-1}^{(n)})| \le |M(t') - M(t_{i-1}^{(n)})| + |M(t_i^{(n)}) - M(t')|$ for any $t' \in (t_{i-1}^{(n)}, t_i^{(n)})$ and therefore one can consider a sequence of nested partitions. Since $\{M(t)\}$ is continuous, it is uniformly continuous on bounded intervals, i.e. $|M(t_i^{(n)}) - M(t_{i-1}^{(n)})| < \varepsilon$ whenever $|t_i^{(n)} - t_{i-1}^{(n)}| < \delta = \delta(\omega)$. Thus, for $\sup |t_i^{(n)} - t_{i-1}^{(n)}| < \delta$ we have

$$\sum_{i=1}^{k_n} |M(t_i^{(n)}) - M(t_{i-1}^{(n)})|^2 \le \varepsilon \sum_{i=1}^{k_n} |M(t_i^{(n)}) - M(t_{i-1}^{(n)})|$$

and therefore

$$\lim_{n\to\infty} \sum_{i=1}^{k_n} |M(t_i^{(n)}) - M(t_{i-1}^{(n)})|^2 = 0.$$

Assume first that $\{V(t)\}$ and $\{M(t)\}$ are bounded. Then $\{M(t)\}$ is a martingale which can be easily shown by the reader. From

$$\begin{aligned}
&\mathbb{E}\,(M(t_i^{(n)}) - M(t_{i-1}^{(n)}))^2 \\
&= \mathbb{E}\,M^2(t_i^{(n)}) + \mathbb{E}\,M^2(t_{i-1}^{(n)}) - 2\mathbb{E}\left(\mathbb{E}\,(M(t_i^{(n)}) \mid \mathcal{F}_{t_{i-1}^{(n)}})M(t_{i-1}^{(n)})\right) \\
&= \mathbb{E}\,(M^2(t_i^{(n)}) - M^2(t_{i-1}^{(n)})),
\end{aligned}$$

it follows that

$$\mathbb{E}\,M^2(t) = \mathbb{E}\sum_{i=1}^{n}(M^2(t_i^{(n)}) - M^2(t_{i-1}^{(n)})) = \mathbb{E}\sum_{i=1}^{n}(M(t_i^{(n)}) - M(t_{i-1}^{(n)}))^2.$$

The random variable on the right-hand side is bounded by $2\sup\{M(v) : 0 \le v \le t\}V(t)$, and therefore the dominated convergence theorem yields $\mathbb{E}\, M^2(t) = 0$, that is $M(t) = 0$. In the general case, let $T = \inf\{t > 0 : |M(t)| \ge 1 \text{ or } V(t) \ge 1\}$. This is a stopping time because $\{M(t)\}$ and $\{V(t)\}$ are continuous processes. The stopped process $\{M(T \wedge t)\}$ is bounded and has a bounded variation process. Thus $M(T \wedge t) = 0$ and, consequently, $V(T \wedge t) = 0$. By the definition of T, this is only possible if $T > t$. Hence $T = \infty$ and $M(t) = 0$ follows. □

The next result is known as *Lévy's theorem*. The theorem and the idea of its proof are used later in Section 13.4.

Theorem 13.1.4 *Let $M(t) = \int_0^t \sigma_1(v)\,\mathrm{d}W_1(v) + \int_0^t \sigma_2(v)\,\mathrm{d}W_2(v)$, where the processes $\{W_1(t)\}, \{W_2(t)\}$ are independent standard Brownian motions, while $\{\sigma_1(t)\}, \{\sigma_2(t)\}$ are stochastic processes in L^2_{loc}. Assume that $\{M^2(t) - t\}$ is a local martingale. Then $\{M(t)\}$ is a standard Brownian motion.*

Proof Note that, by Theorem 13.1.1, $\{M(t)\}$ is a local martingale. From Itô's formula (13.1.12) we obtain

$$\begin{aligned} M^2(t) - t &= \sum_{j=1}^{2}\Big(2\int_0^t M(v)\sigma_j(v)\,\mathrm{d}W_j(v) + \int_0^t \sigma_j^2(v)\,\mathrm{d}v\Big) - t \\ &= \sum_{j=1}^{2} 2\int_0^t M(v)\sigma_j(v)\,\mathrm{d}W_j(v) + \int_0^t (\sigma_1^2(v) + \sigma_2^2(v))\,\mathrm{d}v - t\,. \end{aligned}$$

The process $\{\int_0^t (\sigma_1^2(v) + \sigma_2^2(v))\,\mathrm{d}v - t\}$ is a local martingale. For, by Theorem 13.1.1b and (13.1.11), the processes $\{\int_0^t M(v)\sigma_j(v)\,\mathrm{d}W_j(v)\}$ for $j = 1, 2$ are local martingales, while the sum and the difference of two local martingales is a local martingale. Moreover, the process is continuous and of bounded variation. By Lemma 13.1.2, it needs to be constant. This means that $\sigma_1^2(t) + \sigma_2^2(t) = 1$. Now, let $s \in \mathbb{R}$ and consider the process $\{X(t)\}$ with

$$X(t) = \exp(\mathrm{i}sM(t) + s^2 t/2)\,, \tag{13.1.34}$$

where i denotes the imaginary unit. If for a complex-valued process we consider the real and the imaginary part separately, we see that Itô's formula (13.1.12) is also valid for complex-valued processes. Thus, for the process defined in (13.1.34) we have

$$\begin{aligned} \mathrm{d}X(t) &= \frac{s^2}{2}X(t)\,\mathrm{d}t + \mathrm{i}sX(t)\,\mathrm{d}M(t) - \frac{s^2}{2}X(t)(\sigma_1^2(t) + \sigma_2^2(t))\,\mathrm{d}t \\ &= \mathrm{i}sX(t)\sigma_1(t)\,\mathrm{d}W_1(t) + \mathrm{i}sX(t)\sigma_2(t)\,\mathrm{d}W_2(t)\,. \end{aligned}$$

But then $\{X(t)\}$ can be represented as the sum of stochastic integrals that fulfil the conditions of Theorem 13.1.1b by (13.1.11). The process $\{X(t)\}$

therefore is a ($\mathbb{C}$-valued) local martingale as a sum of two local martingales. Moreover, for any $x > 0$ we have $\sup_{t \le x} |X(t)| \le \exp(s^2 x/2)$. Thus $\{X(t)\}$ is bounded on bounded intervals and therefore must be a martingale. Let $0 \le v < t$. Then

$$\mathbb{E}\,(\mathrm{e}^{\mathrm{i}s(M(t)-M(v))} \mid \mathcal{F}_v) = \mathrm{e}^{-s^2(t-v)/2}(X(v))^{-1}\mathbb{E}\,(X(t) \mid \mathcal{F}_v) = \mathrm{e}^{-s^2(t-v)/2}\,.$$

Thus $M(t) - M(v)$ is independent of $\mathcal{F}_v$ and has distribution $\mathrm{N}(0, t-v)$. Since $M(0) = 0$ and $\{M(t)\}$ is continuous, $\{M(t)\}$ must be a standard Brownian motion. □

Bibliographical Notes. There are many books on stochastic calculus. The material presented in Section 13.1 is taken from Ethier and Kurtz (1986), Karatzas and Shreve (1991) and Ikeda and Watanabe (1989). Techniques for numerics and simulation of stochastic differential equations are described in Kloeden and Platen (1992); see also Rogers and Talay (1997) in the context of applications to financial mathematics.

13.2 PERTURBED RISK PROCESSES

In previous chapters we considered the risk reserve process introduced in Section 5.1.4, where we assumed that premiums were collected at a constant rate $\beta > 0$. The cumulative income was a linear function of time. So far, the only model with a more general income function has been the compound Poisson model in an economic environment which was considered in Section 11.4. In the latter case, the income was typically a nonlinear, but still a deterministic, function of time. In reality, the income of an insurer is not deterministic. There are fluctuations in the number of customers, the claim arrival intensity may depend on time, the insurer invests the surplus, and claims as well as premiums increase with inflation. Moreover, the difference of interest and inflation rates is not always constant in time and yields still another source of uncertainty. To model these additional uncertainties, we consider a *perturbed risk process* $\{X(t)\}$ defined by $X(t) = R(t) + Z(t)$, where $\{R(t)\}$ is the risk reserve process introduced in Section 5.1.4 and $\{Z(t)\}$ is some stochastic perturbation process.

13.2.1 Lundberg Bounds

By way of example, in this section we only consider the case that

$$X(t) = R(t) + \varepsilon W(t)\,, \tag{13.2.1}$$

where $\{R(t)\}$ is the (unperturbed) risk reserve process in a compound Poisson model as studied in Section 5.3, $\{W(t)\}$ is a standard Brownian motion independent of $\{R(t)\}$, and $\varepsilon \in \mathbb{R}$ a constant. The reader should convince himself that $\{X(t)\}$ is a process with stationary and independent increments, hence $\{X(t)\}$ is a homogeneous Markov process with respect to its history $\{\mathcal{F}_t^X\}$. As in our investigation in Chapter 11 of infinite-horizon ruin probabilities, we again do not assume that the filtration $\{\mathcal{F}_t^X\}$ is complete. Notice, however, that $\{X(t)\}$ is not a PDMP since the paths of the Brownian motion $\{W(t)\}$ are nowhere differentiable. The martingale methods for PDMP, developed in Chapter 11, will show us a way how to study the ruin function $\psi(u) = \mathbb{P}(\inf_{t\ge 0} X(t) < 0)$ of the perturbed risk process $\{X(t)\}$ in the case of light-tailed claim sizes. The only difference will be that, while we do not necessarily obtain martingales, we still get local martingales. In most cases, however, it will be easy to prove that the local martingales actually are martingales.

We first derive a representation for the extended generator of the Markov process $\{X(t)\}$. We use the same notation as in Section 5.3.

Lemma 13.2.1 *Let $g : \mathbb{R} \to \mathbb{R}$ be a twice continuously differentiable function such that*

$$\mathbb{E} \sum_{i=1}^{N(t)\wedge n} |g(X(\sigma_i)) - g(X(\sigma_i - 0))| < \infty \tag{13.2.2}$$

for all $t \ge 0$, $n \in \mathbb{N}$. Then g is in the domain $\mathcal{D}(\boldsymbol{A})$ of the extended generator $\boldsymbol{A}$ of $\{X(t)\}$, where

$$\boldsymbol{A}g(y) = \frac{\varepsilon^2}{2} g^{(2)}(y) + \beta g^{(1)}(y) + \lambda\Bigl(\int_0^\infty g(y-v)\,\mathrm{d}F_U(v) - g(y)\Bigr)\,. \tag{13.2.3}$$

Proof Consider the perturbed risk process $\{X(t)\}$ up to time σ_n. Then

$$X(\sigma_n \wedge t) = u + \sum_{i=1}^{N(t)\wedge n}\Bigl(\int_{\sigma_{i-1}}^{\sigma_i} (\beta\,\mathrm{d}v + \varepsilon\,\mathrm{d}W(v)) - U_i\Bigr) + \int_{\sigma_{N(t)\wedge n}}^{\sigma_n \wedge t} (\beta\,\mathrm{d}v + \varepsilon\,\mathrm{d}W(v))\,.$$

For every $n \in \mathbb{N}$, let the process $\{M_n(t)\}$ be defined by

$$M_n(t) = g(X(\sigma_n \wedge t)) - g(u) - \int_0^{\sigma_n \wedge t} \boldsymbol{A}g(X(v))\,\mathrm{d}v\,,$$

where $\boldsymbol{A}g$ is given by (13.2.3). We apply Itô's formula (13.1.12) to the process $\{g(X(\sigma_n \wedge t)) - g(u),\ t \ge 0\}$ to obtain the formula

$$\begin{aligned} M_n(t) &= \int_0^{\sigma_n\wedge t} \varepsilon g^{(1)}(X(v))\,\mathrm{d}W(v) + \sum_{i=1}^{N(t)\wedge n} g(X(\sigma_i)) - g(X(\sigma_i - 0)) \\ &\quad - \lambda \int_0^{\sigma_n\wedge t}\int_0^\infty g(X(v) - y) - g(X(v))\,\mathrm{d}F_U(y)\,\mathrm{d}v\,. \end{aligned}$$

By Theorem 13.1.1b, the process $\{\int_0^t \varepsilon g^{(1)}(X(v))\,\mathrm{d}W(v),\ t \ge 0\}$ is a local martingale. We can thus find a sequence of stopping times $\{\sigma'_n\}$ with $\sigma'_n \to \infty$ such that $\{\int_0^{t\wedge\sigma'_n} \varepsilon g^{(1)}(X(v))\,\mathrm{d}W(v),\ t \ge 0\}$ is a martingale. Furthermore, the expression

$$\sum_{i=1}^{N(t)\wedge n} g(X(\sigma_i)) - g(X(\sigma_i - 0)) - \lambda \int_0^{\sigma_n \wedge t} \int_0^\infty (g(X(v)-y) - g(X(v)))\,\mathrm{d}F_U(y)\,\mathrm{d}v\,,$$

seen as a function of t, gives a martingale. This follows as in the proof of Theorem 11.2.2, where condition (13.2.2) is exploited. Thus, by Theorem 10.2.4, $\{M_n(t\wedge\sigma'_n)\}$ is a martingale. Since $\{\sigma_n \wedge \sigma'_n\}$ is a sequence of stopping times with $\lim_{n\to\infty} \sigma_n \wedge \sigma'_n = \infty$, the assertion follows. □

Our next goal is to show how to obtain an upper Lundberg-type bound for the ruin function $\psi(u) = \mathbb{P}(\inf_{t\ge 0} X(t) < 0)$. We proceed as in Section 11.3.1. For this purpose we assume in the rest of this section that the tail function of claim sizes is "light", i.e. there exists $s > 0$ such that $\hat{m}_U(s) < \infty$. To get an idea how to find a (local) martingale of the form $\{\mathrm{e}^{-sX(t)}\}$, we apply Lemma 13.2.1 and solve the equation $\boldsymbol{A}g(y) = 0$ for $g(y) = \mathrm{e}^{-sy}$. After dividing by $g(y)$, (13.2.3) yields that we have to solve the equation $\theta(s) = 0$, where

$$\theta(s) = \frac{\varepsilon^2}{2}s^2 - \beta s + \lambda(\hat{m}_U(s) - 1)\,. \tag{13.2.4}$$

Notice that this generalizes the function obtained in (11.3.2) for the unperturbed case, where $\varepsilon = 0$. Let now $s = \gamma > 0$ such that $\theta(\gamma) = 0$, provided that this solution exists. Then, condition (13.2.2) is fulfilled because

$$\begin{aligned}
\mathbb{E}\Big(\sum_{i=1}^{N(t)\wedge n} \mathrm{e}^{-\gamma X(\sigma_i)} - \mathrm{e}^{-\gamma(X(\sigma_i)+U_i)}\Big) &\le \sum_{i=1}^{n} \mathbb{E}\,\mathrm{e}^{-\gamma X(\sigma_i)} \\
&= \sum_{i=1}^{n} \mathbb{E}\,(\mathbb{E}\,(\mathrm{e}^{-\gamma X(\sigma_i)} \mid \sigma_i)) \\
&= \sum_{i=1}^{n} \int_0^\infty \mathrm{e}^{-\gamma(u+\beta v)} \hat{m}_U(\gamma)^i \mathbb{E}\,\mathrm{e}^{-\gamma\varepsilon W(v)} \frac{\lambda^i v^{i-1}}{(i-1)!}\mathrm{e}^{-\lambda v}\,\mathrm{d}v \\
&= \mathrm{e}^{-\gamma u} \sum_{i=1}^{n} \hat{m}_U(\gamma)^i \lambda^i \int_0^\infty \mathrm{e}^{(\varepsilon^2\gamma^2/2 - \beta\gamma - \lambda)v} \frac{v^{i-1}}{(i-1)!}\,\mathrm{d}v < \infty
\end{aligned}$$

where $\varepsilon^2\gamma^2/2 - \beta\gamma - \lambda = -\lambda\hat{m}_U(\gamma) < 0$. Thus, by Lemma 13.2.1, $\{\mathrm{e}^{-\gamma X(t)}\}$ is a local martingale. It turns out that $\{\mathrm{e}^{-\gamma X(t)}\}$ is even a martingale.

Lemma 13.2.2 *Assume that the equation $\theta(s) = 0$ has a solution $s = \gamma > 0$, where $\theta(s)$ is given in (13.2.4). Then, $\{\mathrm{e}^{-\gamma X(t)},\ t \ge 0\}$ is a martingale with respect to the (uncompleted) right-continuous filtration $\{\mathcal{F}^X_{t+}\}$.*

Proof For all $t, h \geq 0$ we have

$$\mathbb{E}\,(\mathrm{e}^{-\gamma X(t+h)} \mid \mathcal{F}^X_{t+}) = \mathrm{e}^{-\gamma X(t)}\mathbb{E}\,(\mathrm{e}^{-\gamma(X(t+h)-X(t))} \mid \mathcal{F}^X_{t+})\,.$$

Since $\{X(t)\}$ has independent and stationary increments, it is easily seen that $\mathbb{E}\,(\mathrm{e}^{-\gamma(X(t+h)-X(t))} \mid \mathcal{F}^X_{t+}) = \mathbb{E}\,\mathrm{e}^{-\gamma X(h)}$. Furthermore,

$$\begin{aligned}\mathbb{E}\,\mathrm{e}^{-\gamma X(h)} &= \mathbb{E}\,\exp\Big(-\gamma\Big(\beta h - \sum_{i=1}^{N(h)} U_i + \varepsilon W(h)\Big)\Big)\\ &= \mathrm{e}^{-\gamma\beta h}\mathbb{E}\,\exp\Big(\gamma\sum_{i=1}^{N(h)} U_i\Big) + \mathbb{E}\,\mathrm{e}^{-\gamma\varepsilon W(h)}\\ &= \mathrm{e}^{(-\gamma\beta+\lambda(\hat{m}_U(\gamma)-1)+\gamma^2\varepsilon^2/2)h} = \mathrm{e}^{\theta(\gamma)h} = 1\,,\end{aligned}$$

which proves the lemma. □

Theorem 13.2.1 *Under the condition of Lemma* 13.2.2,

$$\psi(u) \leq \mathrm{e}^{-\gamma u}\,, \qquad u \geq 0\,. \tag{13.2.5}$$

Proof We use the martingale $\{\mathrm{e}^{-\gamma X(t)}\}$ obtained in Lemma 13.2.2 to change the probability measure $\mathbb{P}$. For each $t \geq 0$, we define

$$\mathbb{P}^{(\gamma)}_t(A) = \mathbb{E}\,[\mathrm{e}^{-\gamma(X(t)-u)}; A] \tag{13.2.6}$$

for any $A \in \mathcal{F}^X_{t+}$. Recall that $\tau(u) = \inf\{t : X(t) < 0\}$ denotes the time of ruin. Since the filtration $\{\mathcal{F}^X_{t+}\}$ is right-continuous, $\tau(u)$ is a stopping time with respect to this filtration, see Theorem 10.1.1. As in Section 11.3.1 it follows that

$$\psi(u) = \mathbb{E}^{(\gamma)}(\mathrm{e}^{\gamma X(\tau(u))})\mathrm{e}^{-\gamma u}\,, \tag{13.2.7}$$

where $\mathbb{E}^{(\gamma)}$ denotes the expectation with respect to the "global" measure $\mathbb{P}^{(\gamma)}$ on $\mathcal{F}^X_\infty = \sigma(\bigcup_{t\geq 0}\mathcal{F}^X_{t+}) = \sigma(\bigcup_{t\geq 0}\mathcal{F}^X_t)$ corresponding to the family $\{\mathbb{P}^{(\gamma)}_t,\ t \geq 0\}$ of probability measures defined in (13.2.6); see also Section 10.2.6. The upper bound in (13.2.5) immediately follows because $X(\tau(u)) \leq 0$. □

Remark In contrast with the unperturbed case, the case $\varepsilon \neq 0$ does not lead to an upper bound of the form

$$\psi(u) \leq c\mathrm{e}^{-\gamma u} \tag{13.2.8}$$

for all $u \geq 0$ and where $c < 1$. Indeed

$$\begin{aligned}\psi(u) &\geq \mathbb{P}(\tau(u) \leq 1) \geq \mathbb{P}\Big(\inf_{0<t\leq 1} u + \varepsilon W(t) + \beta t < 0\Big)\\ &= \Phi\Big(\frac{-u-\beta}{\varepsilon}\Big) + \exp\Big(\frac{-2\beta}{\varepsilon^2}u\Big)\Phi\Big(\frac{\beta-u}{\varepsilon}\Big)\,,\end{aligned}$$

where the last equality follows from Theorem 10.3.2. But the latter expression tends to 1 as $u \downarrow 0$. Thus, $\psi(0) = \lim_{u\downarrow 0} \psi(u) = 1$. In particular, there is no $c < 1$ such that (13.2.8) holds for all $u \geq 0$.

Turning to a lower Lundberg-type bound, we investigate the distribution of $\{X(t)\}$ under the new probability measure $\mathbb{P}^{(\gamma)}$ corresponding to the "local" measures $\mathbb{P}_t^{(\gamma)}$, as defined in (13.2.6). The results for the unperturbed case obtained in previous chapters suggest the conjecture that $\{X(t)\}$ remains a perturbed compound Poisson risk process. The next lemma shows that this conjecture is true. However, in the present case also the premium rate changes and this is in contrast with the unperturbed compound Poisson model considered in Theorem 11.3.1.

Lemma 13.2.3 *Under the measure $\mathbb{P}^{(\gamma)}$, the stochastic process $\{X(t)\}$ is a perturbed compound Poisson risk process with claim arrival intensity $\lambda^{(\gamma)} = \lambda\hat{m}_U(\gamma)$, claim size distribution $F_U^{(\gamma)}(x) = \int_0^x \mathrm{e}^{\gamma y}\,\mathrm{d}F_U(y)/\hat{m}_U(\gamma)$ and premium rate $\beta^{(\gamma)} = \beta - \varepsilon^2\gamma$, and with an independent Brownian perturbation process $\{\varepsilon W'(t)\}$. Moreover,*

$$\mathbb{E}^{(\gamma)}X(1) - u = -(\lambda\hat{m}_U^{(1)}(\gamma) - \beta + \varepsilon^2\gamma) = -\theta^{(1)}(\gamma) < 0$$

and therefore $\mathbb{P}^{(\gamma)}(\tau(u) < \infty) = 1$ for each $u \geq 0$.

Proof Let $R'(t) = R(t) - \gamma\varepsilon^2 t$, $W'(t) = W(t) + \gamma\varepsilon t$ and note that $X(t) = R'(t) + \varepsilon W'(t)$. Fix $t > 0$ and let $A_1 \in \mathcal{F}_t^R = \mathcal{F}_t^{R'}$ and $A_2 \in \mathcal{F}_t^W = \mathcal{F}_t^{W'}$ be two events. Then, by (13.2.6) we have

$$\begin{aligned}\mathbb{P}^{(\gamma)}(A_1 \cap A_2) &= \mathbb{E}^{(0)}[\mathrm{e}^{-\gamma(R'(t)-u)}\mathrm{e}^{-\gamma\varepsilon W'(t)}; A_1 \cap A_2] \\ &= \mathbb{E}^{(0)}[\mathrm{e}^{-\gamma(R'(t)-u)-\gamma^2\varepsilon^2 t/2}; A_1]\mathbb{E}^{(0)}[\mathrm{e}^{-\gamma\varepsilon W'(t)+\gamma^2\varepsilon^2 t/2}; A_2],\end{aligned}$$

since $\{R(t)\}$ and $\{W(t)\}$ are independent under $\mathbb{P}^{(0)}$. The same argument implies that $\{R'(t)\}$ and $\{W'(t)\}$ are independent under $\mathbb{P}^{(\gamma)}$. Moreover, Theorem 11.3.1 shows that under $\mathbb{P}^{(\gamma)}$ the process $\{R'(t)\}$ is a compound Poisson risk process with the desired parameters. Note that $W'(0) = 0$ and that the sample paths of $\{W'(t)\}$ are from $C(\mathbb{R}_+)$. By Theorem 10.1.2, $\{\exp(-\gamma\varepsilon W'(t) + \gamma^2\varepsilon^2 t/2)\} = \{\exp(-\gamma\varepsilon W(t) - \gamma^2\varepsilon^2 t/2)\}$ is a positive martingale under $\mathbb{P}^{(0)}$. Using (10.2.26) we find for $0 \leq v < t$ and $x \in \mathbb{R}$

$$\begin{aligned}&\mathbb{P}^{(\gamma)}(W'(t) - W'(v) \leq x \mid \mathcal{F}_v) \\ &= \frac{\mathbb{E}^{(0)}(\mathrm{e}^{-\gamma\varepsilon W'(t)+\gamma^2\varepsilon^2 t/2}\mathbb{I}(W'(t) - W'(v) \leq x) \mid \mathcal{F}_v)}{\mathrm{e}^{-\gamma\varepsilon W'(v)+\gamma^2\varepsilon^2 v/2}} \\ &= \mathbb{E}^{(0)}(\mathrm{e}^{-\gamma\varepsilon(W(t)-W(v))-\gamma^2\varepsilon^2(t-v)/2}\mathbb{I}(W(t) - W(v) \leq x - \gamma\varepsilon(t-v) \mid \mathcal{F}_v) \\ &= \frac{1}{\sqrt{2\pi(t-v)}}\int_{-\infty}^{x-\gamma\varepsilon(t-v)} \mathrm{e}^{-\gamma\varepsilon y-\gamma^2\varepsilon^2(t-v)/2-y^2/(2(t-v))}\,\mathrm{d}y = \Phi\Big(\frac{x}{\sqrt{t-v}}\Big),\end{aligned}$$

where $\Phi(y)$ denotes the distribution function of the standard normal distribution. Thus, under the measure $\mathbb{P}^{(\gamma)}$, the process $\{W'(t)\}$ fulfils the conditions in the definition of the standard Brownian motion in Section 10.1.4. □

In a previous remark, we noted another major difference between the unperturbed and the perturbed models, in that $\psi(0) = 1$ in the perturbed case. Moreover, one can show that for $u = 0$, i.e. $X(0) = 0$, the perturbed risk process $\{X(t)\}$ crosses the level 0 infinitely often in any time interval $(0, h)$. This follows from the following well-known property of the Brownian motion $\{W(t)\}$. Given $W(t) = x$, almost every trajectory of $\{W(t)\}$ crosses the level x infinitely often in any interval $(t, t + h)$. Thus ruin can occur in two ways:

- by an arriving claim causing a negative surplus (in which case we have $X(\tau(u)) < 0$),
- by the Brownian motion (in which case we have $X(\tau(u)) = 0$).

The value of $\mathbb{E}^{(\gamma)}(\mathrm{e}^{\gamma X(\tau(u))})$ can therefore be split into two components. Using the abbreviation $\tau = \tau(u)$ we can write

$$\mathbb{E}^{(\gamma)}(\mathrm{e}^{\gamma X(\tau)}) = \mathbb{E}^{(\gamma)}[\mathrm{e}^{\gamma X(\tau)}; X(\tau) < 0] + \mathbb{E}^{(\gamma)}[\mathrm{e}^{\gamma X(\tau)}; X(\tau) = 0], \quad (13.2.9)$$

where $\mathbb{E}^{(\gamma)}[\mathrm{e}^{\gamma X(\tau)}; X(\tau) = 0] = \mathbb{P}^{(\gamma)}(X(\tau) = 0)$. Unfortunately, we do not have an explicit expression for the latter probability. But, using Lemma 13.2.3, we obtain the following lower bound for $\psi(u)$.

Theorem 13.2.2 *Under the conditions of Lemma* 13.2.2, *for all* $u \geq 0$

$$\psi(u) \geq \inf_{0<y<x_0} \Big\{ \frac{\mathrm{e}^{\gamma y}\overline{F}_U(y)}{\int_y^\infty \mathrm{e}^{\gamma v}\, \mathrm{d}F_U(v)} \Big\} \mathrm{e}^{-\gamma u}, \quad (13.2.10)$$

where $x_0 = \sup\{y : F_U(y) < 1\}$.

Proof The expectation $\mathbb{E}^{(\gamma)}[\mathrm{e}^{\gamma X(\tau)}; X(\tau) < 0]$ can be estimated by a conditioning on $X(\tau - 0)$; see also (12.3.21). By the result of Lemma 13.2.3, this gives

$$\begin{aligned} &\inf_{0\leq y<x_0} \Big\{ \frac{\mathrm{e}^{\gamma y}\overline{F}_U(y)}{\int_y^\infty \mathrm{e}^{\gamma v}\, \mathrm{d}F_U(v)} \Big\} \mathbb{P}^{(\gamma)}(X(\tau) < 0) \\ \leq\ & \mathbb{E}^{(\gamma)}[\mathrm{e}^{\gamma X(\tau)}; X(\tau) < 0] \leq \sup_{0\leq y<x_0} \Big\{ \frac{\mathrm{e}^{\gamma y}\overline{F}_U(y)}{\int_y^\infty \mathrm{e}^{\gamma v}\, \mathrm{d}F_U(v)} \Big\} \mathbb{P}^{(\gamma)}(X(\tau) < 0). \end{aligned}$$

Now, (13.2.10) immediately follows from (13.2.7) and (13.2.9). □

From (13.2.5) and (13.2.10), we see that γ fulfils (11.4.5) and (11.4.6). Thus, γ is the *adjustment coefficient* for the ruin function $\psi(u)$. We also remark that if we could estimate $\mathbb{P}^{(\gamma)}(X(\tau) = 0)$ then the last inequality in the proof of Theorem 13.2.2 would lead to a refined upper bound for $\psi(u)$. However, the problem is that $\mathbb{P}^{(\gamma)}(X(\tau) = 0)$ tends to 1 as $u \downarrow 0$.

13.2.2 Modified Ladder Heights

We turn to a more general perturbed risk model than in Section 13.2.1. We still assume that $\{X(t)\}$ has the form given in (13.2.1). But we admit now that the claim arrival process $\{(\sigma_n, U_n)\}$ which generates the (unperturbed) risk process $\{R(t)\}$, is an arbitrary ergodic time-stationary marked point process for which the net profit condition (12.1.38) is fulfilled. The initial risk reserve u is taken to be 0. We extend Theorem 12.1.9 on the joint distribution of the surplus prior to ruin and the severity of ruin to the case of risk processes perturbed by the Brownian motion $\{\varepsilon W(t)\}$. Since $\inf\{t > 0 : X(t) < 0\} = 0$ we cannot use the same definition of ladder epochs as in the unperturbed case. We therefore consider the random variable

$$\tau^+ = \inf\{\sigma_i : X(\sigma_i) < \inf\{X(t) : 0 \le t < \sigma_i\}\} \tag{13.2.11}$$

which is the first time when a jump leads to a new minimum of $\{X(t)\}$. Furthermore, let

$$Y_\mathrm{c}^+ = -\inf\{X(t) : 0 < t < \tau^+\}, \qquad Y_\mathrm{d}^+ = -X(\tau^+) - Y_\mathrm{c}^+ \tag{13.2.12}$$

and $X^+ = X(\tau^+ - 0) + Y_\mathrm{c}^+$. Then the size of the first claim leading to a new minimum is $U^+ = Y_\mathrm{d}^+ + X^+$. Note that Y_c^+ is also well-defined if $\tau^+ = \infty$. For an illustration of these quantities, see Figure 13.2.1. Notice that in the special case $\varepsilon = 0$, i.e. if there is no perturbation, then $Y_\mathrm{c}^+ = 0$, $\tau^+ = \tau(0)$ is the usual ruin time, and (X^+, Y_d^+) coincides with the random vector (X^+, Y^+) considered in Section 12.1.5.

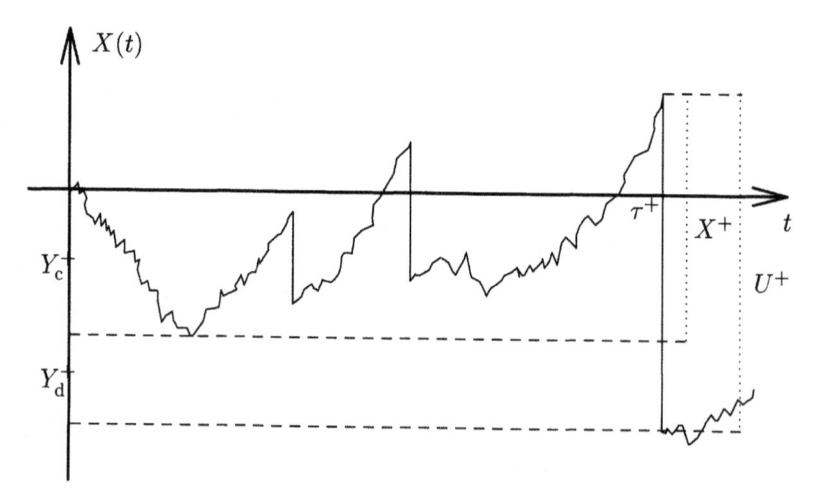

Figure 13.2.1 Modified ladder heights

Theorem 13.2.3 *Let $\varepsilon \neq 0$. Then, for $x, y_c, y_d \geq 0$*

$$\mathbb{P}(X^+ \geq x, Y_c^+ \geq y_c, Y_d^+ \geq y_d, \tau^+ < \infty) = e^{-2\beta y_c/\varepsilon^2} \frac{\lambda}{\beta} \int_{x+y_d}^{\infty} \overline{F_U^0}(v)\, dv \tag{13.2.13}$$

and

$$\mathbb{P}(Y_c^+ \geq y_c, \tau^+ = \infty) = e^{-2\beta y_c/\varepsilon^2} \frac{\beta - \lambda\mu^0}{\beta}, \tag{13.2.14}$$

where F_U^0 denotes the Palm mark distribution of claim sizes, and μ^0 its expectation.

Proof Let $N = \{(\sigma_n, U_n),\ n \in \mathbb{Z}\}$ denote the arrival process of past and future claims. Consider the perturbed risk process $\{X(t),\ t \in \mathbb{R}\}$ on the whole real line, where $\{X(t),\ t \geq 0\}$ is defined as before. For $t < 0$, let

$$X(t) = \beta t + \sum_{i=0}^{\infty} U_{-i} \mathbb{I}(\sigma_{-i} > t) + \varepsilon W(-t),$$

where $\{-W(-t),\ t \geq 0\}$ is an independent standard Brownian motion. Furthermore, let

$$g(N, v) = \mathbb{P}(X^+ \geq x, Y_c^+ \geq y_c, Y_d^+ \geq y_d, \tau^+ = v \mid N)$$

and

$$\begin{aligned} p(t) \quad = \quad & \mathbb{P}^0\Big(\sup_{-t<w<0} (X(-t) - X(w)) \geq y_c, \inf_{-t<w<0} X(w) \geq y_d, \\ & \sup_{-t<w<0} (X(0-0) - X(w)) \geq x, \\ & \bigcap_{-t<w<0} \Big\{ X(w) \geq \inf_{-t<v<w} X(v) \Big\} \Big). \end{aligned}$$

Here, $\mathbb{P}^0$ denotes the product measure built by the Palm distribution of N and by the distribution of $\{\varepsilon W(t)\}$. Then, by Campbell's formula (12.1.34), we have

$$\begin{aligned} \mathbb{P}(X^+ \geq x, Y_c^+ \geq y_c, Y_d^+ \geq y_d, \tau^+ < \infty) &= \mathbb{E}[g(N, \tau^+); \tau^+ < \infty] \\ &= \lambda \int_0^{\infty} p(t)\, dt. \end{aligned}$$

Consider the process $\{X(t)\}$ backwards, i.e. let $X'(t) = -X(-t-0)$ and let $M'(t) = \sup_{0<w<t} X'(w)$. Then $X'(0) = -U_0$ under the measure $\mathbb{P}^0$. Note that the function $p(t)$ can be written as

$$\begin{aligned} p(t) \quad = \quad & \mathbb{P}^0\Big(U_0 + M'(t) \geq x, M'(t) - X'(t-0) \geq y_c, -M'(t) \geq y_d, \\ & \bigcap_{0<w<t} \Big\{ X'(w-0) \leq \sup_{w<v<t} X'(v) \Big\} \Big). \end{aligned}$$

The event $\bigcap_{0<w<t}\{X'(w-0) \le \sup_{w<v<t} X'(v)\}$ certainly occurs for all t in the interval $(0, -\sigma_{-1})$. Thereafter it does not occur until the first epoch $t \ge -\sigma_{-1}$ such that $X'(t) = X'(-\sigma_{-1}-0)$. If we cut out all intervals on which this event does not occur and put all the pieces together we get a Brownian motion with drift β starting in U_0 because $\{W(t)\}$ is a strong Markov process. Define $\overline{M}(t) = \sup_{0\le w\le t}\{\varepsilon W(w) + \beta w\}$ and put

$$\overline{p}(t) = \mathbf{P}^0(\overline{M}(t) - \varepsilon W(t) - \beta t \ge y_{\rm c}, -\overline{M}(t) \ge y_{\rm d} - U_0, \overline{M}(t) \ge x)\,.$$

Then

$$\int_0^\infty p(t)\,{\rm d}t = \int_0^\infty \overline{p}(t)\,{\rm d}t\,.$$

The condition $\overline{M}(t) \ge x$ is not fulfilled until $\varepsilon W(t) + \beta t = x$ for the first time and is fulfilled thereafter. Cutting out the piece where $\overline{M}(t) < x$ yields

$$\int_0^\infty \overline{p}(t)\,{\rm d}t = \int_0^\infty p'(t)\,{\rm d}t\,, \tag{13.2.15}$$

where

$$p'(t) = \mathbf{P}^0(\overline{M}(t) - \varepsilon W(t) - \beta t \ge y_{\rm c}, -\overline{M}(t) \ge y_{\rm d} + x - U_0)\,.$$

Obviously $U_0 > x + y_{\rm d}$ has to be fulfilled for $p'(t) \ne 0$. Equation (13.2.13) now readily follows using the well-known formula

$$\begin{aligned}\mathbf{P}(\overline{M}(t) \le x, \varepsilon W(t) + \beta t \le y) &= \frac{1}{\sqrt{2\pi\varepsilon^2 t}}\int_{-\infty}^{y} \exp\{-(v-\beta t)^2/(2\varepsilon^2 t)\} \\ &\quad \times \exp\{2x(x-v)/(\varepsilon^2 t)\}\,{\rm d}v\,; \end{aligned}\tag{13.2.16}$$

see, for example, Theorem 4.2 in Anderson (1960). Further details are left to the reader. We now sketch the proof of (13.2.14). Using (13.2.13), it suffices to show that

$$\mathbf{P}(Y_{\rm c}^+ \ge y_{\rm c}) = {\rm e}^{-2\beta y_{\rm c}/\varepsilon^2} \tag{13.2.17}$$

for all $y_{\rm c} > 0$. Put $\tau_h^1 = \inf\{t \ge 0 : X(t) \le -h\}$, $\tau_h^2 = \inf\{t \ge 0 : \varepsilon W(t) + \beta t \le -h\}$ for $h > 0$, and for $i = 1, 2$

$$X_h^i(t) = \begin{cases} X(t+\tau_h^i) - X(\tau_h^i) & \text{if } \tau_h^i < \infty, \\ 0 & \text{otherwise.}\end{cases}$$

Further, consider the shifted process $\{X_h^i(t),\ t \ge 0\}$ instead of $\{X(t),\ t \ge 0\}$ and let $Y_{{\rm c},h}^i$ denote the ladder height defined in the same way as $Y_{\rm c}^+$, but now with respect to $\{X_h^i(t),\ t \ge 0\}$. Then we obtain the following representation of the tail function of $Y_{\rm c}^+$. For each $x > 0$ we have

$$\begin{aligned}\mathbf{P}(Y_{\rm c}^+ \ge x + h) &= \mathbf{P}(\tau_h^1 < \infty, X(\tau_h^1) = X(\tau_h^1 - 0), Y_{{\rm c},h}^1 \ge x) \\ &= \mathbf{P}(\tau_h^2 < \infty, Y_{{\rm c},h}^2 \ge x) + o(h) \\ &= \mathbf{P}(\tau_h^2 < \infty)\mathbf{P}(Y_{\rm c}^+ \ge x) + o(h)\end{aligned}$$

as $h \to 0$. Here the last equality is obtained from the law of total probability using the facts that $\{W(t)\}$ and $\{(\sigma_n, U_n)\}$ are independent, that $\{W(t)\}$ has stationary and independent increments and that $\{(\sigma_n, U_n)\}$ is a stationary point process. Using (10.3.7) this gives

$$\mathbb{P}(Y_{\rm c}^+ \geq x+h) = {\rm e}^{-2\beta h/\varepsilon^2}\mathbb{P}(Y_{\rm c}^+ \geq x) + o(h) \tag{13.2.18}$$

as $h \to 0$. Hence, the tail function of $Y_{\rm c}^+$ is continuous on $(0,\infty)$. Moreover, using (13.2.18) we get for the right-hand derivative

$$\frac{{\rm d}^+}{{\rm d}x}\mathbb{P}(Y_{\rm c}^+ \geq x) = \frac{-2\beta}{\varepsilon^2}\mathbb{P}(Y_{\rm c}^+ \geq x)\,,$$

which yields (13.2.17) since $\mathbb{P}(Y_{\rm c}^+ > 0) = 1$ by the definition of $Y_{\rm c}^+$. □

Theorem 13.2.3 shows that $Y_{\rm c}^+$ has the same distribution as the minimum of a (β, ε^2)-Brownian motion $\{\varepsilon W(t) + \beta t\}$; see (10.3.7). The joint distribution of the overshoot $Y_{\rm d}^+$ and the size U^+ of the first claim leading to a new minimum is the same as in the unperturbed case; see Theorem 12.1.9. In particular, the probability $\mathbb{P}(\tau^+ < \infty)$ is the same as in the unperturbed case. For the perturbed compound Poisson risk model, this immediately leads to the following formula of the *Pollaczek–Khinchin type.*

Corollary 13.2.1 *Let $\{(\sigma_n, U_n)\}$ be an independently marked Poisson process and assume that $\beta > \lambda\mu_U$. Then, for all $u \geq 0$*

$$\psi(u) = \Big(1 - \frac{\lambda\mu_U}{\beta}\Big)\sum_{n=0}^{\infty}\Big(\frac{\lambda\mu_U}{\beta}\Big)^n\,\overline{F_U^{{\rm s}*n} * F_{\rm c}^{*(n+1)}}(u)\,, \tag{13.2.19}$$

where $F_{\rm c}(x) = 1 - \exp(-2\beta x/\varepsilon^2)$.

Proof Let ν be the number of (modified) ladder epochs τ^+ such that $\tau^+ < \infty$. Let $Y_{\rm c}^k$, $Y_{\rm d}^k$ for $k \leq \nu$ and $Y_{\rm c}^{\nu+1}$ be the corresponding (modified) ladder heights. Note that

$$-\inf_{t\geq 0}\{X(t) - u\} = \Big(\sum_{k=1}^{\nu} Y_{\rm c}^k + Y_{\rm d}^k\Big) + Y_{\rm c}^{\nu+1}\,.$$

Furthermore, note that by the stationary and independent increments of the compound Poisson process (see Theorem 5.2.1) and the Brownian motion (see Section 10.1.4), the random variables $Y_{\rm c}^k$, $Y_{\rm d}^k$ for $k \leq \nu$ and $Y_{\rm c}^{\nu+1}$ are independent. Formula (13.2.19) is then a simple consequence of Theorem 13.2.3. □

13.2.3 Cramér–Lundberg Approximation

Assume that $\{X(t)\}$ is given by (13.2.1), where $\{(\sigma_n, U_n)\}$ is an independently marked Poisson process. Let $\hat{m}_U(s) < \infty$ for some $s > 0$ and let $\theta(s) = 0$ have a solution $s = \gamma > 0$, where $\theta(s)$ is given in (13.2.4). We now turn to the Cramér–Lundberg approximation to the ruin function $\psi(u) = \mathbb{P}(\inf_{t\geq 0} X(t) < 0)$. For this purpose, we split the event of ruin into the two events: *ruin caused by a claim* and *ruin caused by the Brownian motion.* For the ruin probability $\psi(u)$, this leads to the representation $\psi(u) = \psi_1(u)+\psi_2(u)$ with $\psi_1(u) = \mathbb{P}(\tau < \infty, X(\tau) < 0)$ and $\psi_2(u) = \mathbb{P}(\tau < \infty, X(\tau) = 0)$, where $\tau = \inf\{t : X(t) < 0\}$.

Theorem 13.2.4 *Let $\hat{m}_U^{(1)}(\gamma) < \infty$. Then,*

$$\lim_{u\to\infty} \psi_1(u)\mathrm{e}^{\gamma u} = \frac{\beta - \lambda\mu_U - \gamma\varepsilon^2/2}{\lambda\hat{m}_U^{(1)}(\gamma) - \beta + \varepsilon^2\gamma} \tag{13.2.20}$$

and

$$\lim_{u\to\infty} \psi_2(u)\mathrm{e}^{\gamma u} = \frac{\gamma\varepsilon^2/2}{\lambda\hat{m}_U^{(1)}(\gamma) - \beta + \varepsilon^2\gamma}\,. \tag{13.2.21}$$

Proof From (13.2.7) and (13.2.9) we get

$$\psi_1(u) = \mathbb{E}^{(\gamma)}[\mathrm{e}^{\gamma X(\tau)}; X(\tau) < 0]\mathrm{e}^{-\gamma u}$$

and

$$\psi_2(u) = \mathbb{E}^{(\gamma)}[\mathrm{e}^{\gamma X(\tau)}; X(\tau) = 0]\mathrm{e}^{-\gamma u}\,.$$

We therefore have to show that the expectations $\mathbb{E}^{(\gamma)}[\mathrm{e}^{\gamma X(\tau)}; X(\tau) < 0]$ and $\mathbb{E}^{(\gamma)}[\mathrm{e}^{\gamma X(\tau)}; X(\tau) = 0]$ converge to the limits stated in (13.2.20) and (13.2.21) respectively, as $u \to \infty$. Put $g(u) = \mathbb{E}^{(\gamma)}[\mathrm{e}^{\gamma X(\tau)}; X(\tau) < 0]$. In Lemma 13.2.3 we proved that under $\mathbb{P}^{(\gamma)}$ the process $\{X(t)\}$ has a negative drift. Hence, $\mathbb{P}^{(\gamma)}(\tau^+ < \infty) = 1$, where τ^+ is defined in (13.2.11). Let G_c and G_d denote the distributions of Y_c^+ and Y_d^+ under the measure $\mathbb{P}^{(\gamma)}$, where Y_c^+ and Y_d^+ are defined in (13.2.12). By Lemma 13.2.3 and Theorem 13.2.3

$$\begin{aligned}\mathbb{P}^{(\gamma)}(Y_\mathrm{c}^+ \in B_\mathrm{c}, Y_\mathrm{d}^+ \in B_\mathrm{d}) &= \mathbb{E}^{(0)}[\mathrm{e}^{\gamma(Y_\mathrm{c}^+ + Y_\mathrm{d}^+)}; Y_\mathrm{c}^+ \in B_\mathrm{c}, Y_\mathrm{d}^+ \in B_\mathrm{d}, \tau_+ < \infty]\\ &= \mathbb{E}^{(0)}[\mathrm{e}^{\gamma Y_\mathrm{c}^+}; Y_\mathrm{c}^+ \in B_\mathrm{c}]\,\mathbb{E}^{(0)}[\mathrm{e}^{\gamma Y_\mathrm{d}^+}; Y_\mathrm{d}^+ \in B_\mathrm{d}, \tau_+ < \infty]\\ &= \mathbb{P}^{(\gamma)}(Y_\mathrm{c}^+ \in B_\mathrm{c})\,\mathbb{P}^{(\gamma)}(Y_\mathrm{d}^+ \in B_\mathrm{d})\end{aligned}$$

for any Borel sets $B_\mathrm{d}, B_\mathrm{c} \in \mathcal{B}(\mathbb{R})$. Therefore Y_c^+ and Y_d^+ are independent under $\mathbb{P}^{(\gamma)}$. Thus, $G = G_\mathrm{c} * G_\mathrm{d}$ is the distribution of $Y_\mathrm{c}^+ + Y_\mathrm{d}^+$. Hence, $g(u)$ fulfils the renewal equation

$$g(u) = \int_0^u g(u-y)\,\mathrm{d}G(y) + \int_0^u \int_{u-x}^\infty \mathrm{e}^{\gamma(u-x-y)}\,\mathrm{d}G_\mathrm{d}(y)\,\mathrm{d}G_\mathrm{c}(x)\,. \tag{13.2.22}$$

Integrating the last term over $u \in (0, \infty)$, we find

$$\begin{aligned}
&\int_0^\infty \int_0^u \int_{u-x}^\infty \mathrm{e}^{\gamma(u-x-y)}\,\mathrm{d}G_{\mathrm{d}}(y)\,\mathrm{d}G_{\mathrm{c}}(x)\,\mathrm{d}u \\
&\quad = \int_0^\infty \int_x^\infty \int_{u-x}^\infty \mathrm{e}^{\gamma(u-x-y)}\,\mathrm{d}G_{\mathrm{d}}(y)\,\mathrm{d}u\,\mathrm{d}G_{\mathrm{c}}(x) \\
&\quad = \int_0^\infty \int_0^\infty \int_u^\infty \mathrm{e}^{\gamma(u-y)}\,\mathrm{d}G_{\mathrm{d}}(y)\,\mathrm{d}u\,\mathrm{d}G_{\mathrm{c}}(x) \\
&\quad = \int_0^\infty \int_0^y \mathrm{e}^{\gamma(u-y)}\,\mathrm{d}u\,\mathrm{d}G_{\mathrm{d}}(y) = \gamma^{-1}(1 - \mathbb{E}^{(\gamma)}(\mathrm{e}^{-\gamma Y_{\mathrm{d}}^+}))\,,
\end{aligned}$$

where $\mathbb{E}^{(\gamma)}(\mathrm{e}^{-\gamma Y_{\mathrm{d}}^+}) = \mathbb{E}\,[\mathrm{e}^{\gamma Y_{\mathrm{c}}^+}; \tau^+ < \infty]$. Thus, by Theorem 13.2.3 we have

$$\begin{aligned}
\int_0^\infty \int_0^u \int_{u-x}^\infty \mathrm{e}^{\gamma(u-x-y)}\,\mathrm{d}G_{\mathrm{d}}(y)\,\mathrm{d}G_{\mathrm{c}}(x)\,\mathrm{d}u &= \gamma^{-1}(1 - \mathbb{E}\,[\mathrm{e}^{\gamma Y_{\mathrm{c}}^+}; \tau^+ < \infty]) \\
&= \frac{\beta - \lambda\mu_U - \gamma\varepsilon^2/2}{\gamma(\beta - \gamma\varepsilon^2/2)} < \infty\,.
\end{aligned}$$

In particular, we see that the conditions of Lemma 5.4.2 are fulfilled with $z_1(u) = \mathrm{e}^{\gamma u}$. Thus, we can conclude that

$$\lim_{u\to\infty} g(u) = \frac{\int_0^\infty \int_0^u \int_{u-x}^\infty \mathrm{e}^{\gamma(u-x-y)}\,\mathrm{d}G_{\mathrm{d}}(y)\,\mathrm{d}G_{\mathrm{c}}(x)\,\mathrm{d}u}{\mathbb{E}^{(\gamma)}(Y_{\mathrm{d}}^+ + Y_{\mathrm{c}}^+)}\,.$$

The denominator simplifies since

$$\mathbb{E}^{(\gamma)}(Y_{\mathrm{d}}^+ + Y_{\mathrm{c}}^+) = \mathbb{E}\,((Y_{\mathrm{d}}^+ + Y_{\mathrm{c}}^+)\mathrm{e}^{\gamma(Y_{\mathrm{d}}^+ + Y_{\mathrm{c}}^+)}) = \frac{\lambda\hat{m}_U^{(1)}(\gamma) - \beta + \varepsilon^2\gamma}{\gamma(\beta - \gamma\varepsilon^2/2)}\,,$$

where we used $\lambda(\hat{m}_U(\gamma) - 1) = \beta\gamma - \gamma^2\varepsilon^2/2$. This proves (13.2.20). To show (13.2.21), consider the function $g'(u) = \mathbb{E}^{(\gamma)}[\mathrm{e}^{\gamma X(\tau)}; X(\tau) = 0] = \mathbb{P}^{(\gamma)}(X(\tau) = 0)$. Then

$$g'(u) = \int_0^u g'(u-y)\,\mathrm{d}G(y) + \overline{G}_{\mathrm{c}}(u)\,.$$

By Lemma 5.4.2, we then have

$$\lim_{u\to\infty} g'(u) = \frac{\mathbb{E}^{(\gamma)} Y_{\mathrm{c}}^+}{\mathbb{E}^{(\gamma)}(Y_{\mathrm{d}}^+ + Y_{\mathrm{c}}^+)}$$

and (13.2.21) follows in the same way as (13.2.20). □

13.2.4 Subexponential Claim Sizes

We return to the more general perturbed risk process $\{X(t)\}$ considered in Section 13.2.2. Let $\delta_0 > 0$ be fixed and sufficiently small. The claim arrival process $\{(\sigma_n, U_n)\}$ is taken to be an arbitrary marked point process. The (unperturbed) ruin function is $\psi_\delta(u) = \mathbb{P}(\sup_{t\geq 0}\{\sum_{i=1}^{N(t)} U_i - (\beta - \delta)t\} > u)$ with $N(t) = \max\{n : \sigma_n \leq t\}$. We assume that for all $\delta \in (-\delta_0, \delta_0)$,

$$\lim_{u\to\infty} \frac{\psi_\delta(u)}{\overline{F}(u)} = c(\delta) \tag{13.2.23}$$

for some distribution $F \in \mathcal{S}$ which does not depend on δ and where $c(\delta) \in (0, \infty)$. Furthermore, $c(\delta)$ is taken to be continuous at $\delta = 0$.

Recall that various examples of marked point processes fulfil the above conditions, as was discussed in Section 12.6. It turns out that, in this case, the ruin function

$$\psi(u) = \mathbb{P}\Big(\sup_{t\geq 0}\Big\{\sum_{i=1}^{N(t)} U_i - \beta t + \varepsilon W(t)\Big\} > u\Big)$$

of the corresponding perturbed risk process has the same asymptotic behaviour as $\psi_0(u)$.

Theorem 13.2.5 *Under the above conditions,*

$$\lim_{u\to\infty} \frac{\psi(u)}{\overline{F}(u)} = c(0)\,. \tag{13.2.24}$$

Proof Let $\delta \in (0, \beta)$ be sufficiently small. Then,

$$\psi(u) \leq \mathbb{P}\Big(\sup_{t\geq 0}\Big\{\sum_{i=1}^{N(t)} U_i - (\beta - \delta)t\Big\} + \sup_{t\geq 0}\Big\{\varepsilon W(t) - \delta t\Big\} > u\Big)\,.$$

By Lemma 2.5.2 and Theorem 10.3.1, this and (13.2.23) yield

$$\limsup_{u\to\infty} \frac{\psi(u)}{\overline{F}(u)} \leq c(\delta)\,. \tag{13.2.25}$$

To get a lower bound, observe that

$$\psi(u) \geq \mathbb{P}\Big(\sup_{t\geq 0}\Big\{\sum_{i=1}^{N(t)} U_i - (\beta + \delta)t\Big\} - \sup_{t\geq 0}\Big\{-\varepsilon W(t) - \delta t\Big\} > u\Big)\,.$$

Thus, in the same way as above we obtain

$$\liminf_{u\to\infty} \frac{\psi(u)}{\overline{F}(u)} \geq c(-\delta)\,. \tag{13.2.26}$$

Using the assumption that $c(\delta)$ is continuous at $\delta = 0$, the assertion follows from (13.2.25) and (13.2.26). □

Notice that the conditions of Theorem 13.2.5 are especially fulfilled if $\{(\sigma_n, U_n)\}$ is an independently marked Poisson process. This immediately follows from Theorem 5.4.3. Besides this, an alternative way to investigate the asymptotic behaviour of the ruin function $\psi(u)$ in this specific perturbed risk model is given by the Pollaczek–Khinchin formula (13.2.19).

Corollary 13.2.2 *Let $\psi(u)$ be the ruin function in the perturbed compound Poisson risk model with positive safety loading. Assume that $F_U^{\mathrm{s}} \in \mathcal{S}$. Then*

$$\lim_{u\to\infty} \frac{\psi(u)}{\overline{F_U^{\mathrm{s}}}(u)} = \frac{\lambda\mu_U}{\beta - \lambda\mu_U}\,. \tag{13.2.27}$$

Proof Recall that $F_{\mathrm{c}}(x) = 1 - \exp\{-2\beta x/\varepsilon^2\}$. By Lemma 2.5.2 and Theorem 2.5.2 we have $\lim_{x\to\infty} \overline{F_U^{\mathrm{s}} * F_{\mathrm{c}}}(x)/\overline{F_U^{\mathrm{s}}}(x) \to 1$. By Theorem 2.5.4 we find

$$\lim_{u\to\infty} \frac{(1-(\lambda\mu_U/\beta))\sum_{n=0}^{\infty}(\lambda\mu_U/\beta)^n\overline{F_U^{\mathrm{s}*n} * F_{\mathrm{c}}^{*n}}(u)}{\overline{F_U^{\mathrm{s}}}(u)} = \frac{\lambda\mu_U}{\beta-\lambda\mu_U}\,.$$

Thus, using (13.2.19) and applying Lemma 2.5.2 again, assertion (13.2.27) follows. □

Bibliographical Notes. The perturbed compound Poisson risk model has been introduced in Gerber (1970), where the perturbation process is a Brownian motion. For this model, the case of light-tailed claim sizes has been comprehensively analysed in Dufresne and Gerber (1991). Moreover, from Dufresne and Gerber (1991) the idea emerged to distinguish between the ladder heights Y_{c}^+ due to the continuous part of the perturbed risk process and Y_{d}^+ due to jumps; see also Gerber and Landry (1998). The present approach follows Furrer and Schmidli (1994) and Schmidli (1995), where also other models like the perturbed Markov modulated risk model and the perturbed Björk–Grandell model are treated. The lower Lundberg bound (13.2.10) seems to be new. For the perturbed compound Poisson risk model, the ladder height formulae (13.2.13) and (13.2.14) have been derived in Dufresne and Gerber (1991). For the more general model considered in Theorem 13.2.3, they correspond to a result derived in Asmussen and Schmidt (1995) for unperturbed risk processes with general stationary claim arrival processes; see also Theorem 12.1.9. Formula (13.2.19) is of the *Pollaczek–Khinchin type*; see also Section 5.3. Using Laplace transforms, a related formula has been discussed in Bingham (1975), Harrison (1977b) and Zolotarev (1964) for the supremum of negatively drifted *Lévy processes* with no negative jumps. The case of heavy-tailed claim sizes in the perturbed compound Poisson risk model was first studied in Veraverbeke (1993). For more general perturbed risk models this has been done in Schlegel (1998), where Theorem 13.2.5 and various extensions of it have been derived. Notice that in Schlegel (1998) it is neither assumed that the claim arrival process is Poisson nor that the perturbation process is a Brownian motion. The compound Poisson risk model perturbed by an *α-stable Lévy motion* has been studied in Furrer (1998). A barrier strategy for dividend payments in a perturbed compound Poisson risk model with state-dependent stochastic premiums is investigated in Paulsen and Gjessing (1997a). They model the perturbation by two independent Brownian motions that can be interpreted as stochastic models for the fluctuations of both the rate of inflation and the return on investments; see also Gjessing and Paulsen (1997) and Paulsen and Gjessing (1997b). For the case where the risk reserve process and the perturbation process are independent Lévy processes, see Paulsen (1993,1998).

13.3 OTHER APPLICATIONS TO INSURANCE AND FINANCE

Apart from the perturbed risk processes considered in Section 13.2, there are many other models with diffusion components that have found applications in insurance and finance. In the present section, we treat some typical examples. Further applications to model stochastic interest rates will be considered in Section 13.4. The reader should be aware that a comprehensive overview of diffusion models in actuarial and financial mathematics would fill more than one book.

If not stated otherwise, $\{\mathcal{F}_t\}$ denotes the smallest complete and right-continuous filtration such that the standard Brownian motion $\{W(t)\}$ is adapted with respect to $\{\mathcal{F}_t\}$.

13.3.1 The Black–Scholes Model

The Black–Scholes model is one of the most popular models in financial mathematics. It describes a market with two financial goods: a risky security (such as stocks or other risky assets) and a riskless bond. Let us first consider the *risky security* and denote its price at time t by $X(t)$. Assume that $\{X(t)\}$ is an $\{\mathcal{F}_t\}$-adapted stochastic process which fulfils the stochastic differential equation

$$\mathrm{d}X(t) = \mu X(t)\,\mathrm{d}t + \sigma X(t)\,\mathrm{d}W(t)\,, \qquad X(0) > 0. \tag{13.3.1}$$

Since $\mathbb{P}(A) = 0$ or $\mathbb{P}(A) = 1$ for all $A \in \mathcal{F}_0$, we have $\mathbb{P}(X(0) = c) = 1$ for some constant $c > 0$. In Example 1 of Section 13.1.2 we showed, see (13.1.22), that the solution to (13.3.1) is the geometric Brownian motion given by

$$X(t) = X(0)\exp\big((\mu - \sigma^2/2)t + \sigma W(t)\big)\,. \tag{13.3.2}$$

It follows from (13.3.2) that the filtration $\{\mathcal{F}_t\}$ defined at the beginning of Section 13.3 is also the natural filtration of the price process $\{X(t)\}$. Furthermore, we find $\mathbb{E}\,X(t) = X(0)\mathrm{e}^{\mu t}$, since $\mathbb{E}\,\mathrm{e}^{\sigma W(t)} = \exp(\sigma^2 t/2)$. Thus μ is similar to the force of interest considered in Section 11.4. We call μ the expected *rate of return.* Moreover, the *return* is the stochastic process $\{X'(t)\}$ defined via the stochastic differential equation

$$\mathrm{d}X'(t) = \frac{\mathrm{d}X(t)}{X(t)} = \mu\,\mathrm{d}t + \sigma\,\mathrm{d}W(t)\,. \tag{13.3.3}$$

The infinitesimal variance $\sigma > 0$ of the return is called the *volatility.*

We also consider a *riskless bond,* i.e. a possibility to invest without any risk. The deterministic price of the bond at time t is denoted by $I(t)$ and we assume that the function $I : \mathbb{R}_+ \to \mathbb{R}_+$ fulfils the differential equation

$$\mathrm{d}I(t) = \delta I(t)\,\mathrm{d}t\,, \qquad I(0) = 1\,, \tag{13.3.4}$$

for some constant $\delta \ge 0$ which is called the *force of interest.* The ordinary differential equation (13.3.4) has the solution $I(t) = \mathrm{e}^{\delta t}$. The discounting factor for a time interval of length t is $1/I(t) = \mathrm{e}^{-\delta t}$. Notice that the assumption $\delta \ge 0$ is quite natural. Otherwise it would be possible to borrow money (by selling bonds) at a negative interest rate, i.e. to pay back less than one has borrowed. Furthermore, because the price process $\{X(t)\}$ is more risky than that of the riskless bond, a risk premium has to be added. We model this by assuming that $\mu > \delta$. We also consider the discounted price process $\{X^*(t)\}$ with $X^*(t) = \mathrm{e}^{-\delta t}X(t)$. By (13.1.16) we have

$$\mathrm{d}X^*(t) = (\mu - \delta)X^*(t)\,\mathrm{d}t + \sigma X^*(t)\,\mathrm{d}W(t)\,, \qquad X^*(0) = X(0)\,. \tag{13.3.5}$$

Imagine an agent who is buying and selling risky assets and riskless bonds. Assume that the agent can do this continuously in time, that any fraction of the asset and the bond can be traded, that he can take a short position (holding a negative amount of the security) in both risky asset and riskless bond, that there is always an agent to deal with and that there are no transaction costs. The agent may use a strategy. A *trading strategy* is a pair $\{(\alpha(t), \gamma(t))\}$ of adapted stochastic processes. The pair has the meaning that at time t the agent holds $\alpha(t)$ units of the asset and $\gamma(t)$ units of the bond in his portfolio. In order to introduce the notion of a self-financing trading strategy, we consider the stochastic process $\{M(t)\}$ with

$$M(t) = \exp\big(-((\mu - \delta)/\sigma)W(t) - \frac{1}{2}((\mu - \delta)/\sigma)^2 t\big)\,. \tag{13.3.6}$$

Notice that $\{M(t)\}$ is a martingale by the result of Theorem 10.1.2c. A *self-financing trading strategy* is a trading strategy such that for all $t \ge 0$

$$\begin{aligned}\alpha(t)X(t) &+ \gamma(t)I(t) \\ &= \alpha(0)X(0) + \gamma(0)I(0) + \int_0^t \alpha(v)\,\mathrm{d}X(v) + \int_0^t \gamma(v)\,\mathrm{d}I(v) \quad (13.3.7)\end{aligned}$$

and

$$\mathbb{E}\, M(t) \int_0^t \alpha^2(v) X^2(v)\,\mathrm{d}v < \infty\,. \qquad (13.3.8)$$

Remarks 1. Condition (13.3.7) states that, for a self-financing strategy, the value of the portfolio at time t is the value invested at time 0 plus the gains from the asset and the bond, where no money can be put into or taken out of the portfolio after time 0.

2. Condition (13.3.8) is technical. It is related to the notion of arbitrage. Let $V(t) = \alpha(t)X(t) + \gamma(t)I(t)$ denote the value of the portfolio at time t. A trading strategy is called an *arbitrage* if $V(0) \le 0$, $V(t) \ge 0$ and $\mathbb{P}(V(t) - V(0) > 0) > 0$ for some $t > 0$. A basic requirement for an economy is that there is no arbitrage, i.e. there is no strategy that allows to achieve a positive gain without any risk. Note, however, that arbitrage is generally possible. However, the expected amount of money needed to play such a strategy is infinite; see Chapter 6 of Duffie (1996). We will see in Theorem 13.3.1 below that one way to exclude such strategies is the technical condition (13.3.8) in the definition of a self-financing strategy.

For the rest of this section and in Section 13.3.2 we consider a finite horizon $t_0 > 0$ and we restrict the stochastic process to the interval $[0, t_0]$. Proceeding in the same way as in Section 10.2.6, define the new measure $\tilde{\mathbb{P}}$ on $\mathcal{F}_{t_0}$ by

$$\tilde{\mathbb{P}}(A) = \mathbb{E}\,[M(t); A]\,, \qquad A \in \mathcal{F}_t\,, \qquad (13.3.9)$$

for each $0 \le t \le t_0$, where $\{M(t)\}$ is the positive martingale given in (13.3.6). Then $\{W^*(t)\}$ with $W^*(t) = W(t) + ((\mu - \delta)/\sigma)\,t$ is a standard Brownian motion under the new measure $\tilde{\mathbb{P}}$. This can be verified as in the proof of Lemma 13.2.3. Moreover, by (13.3.5) we have

$$\mathrm{d}X^*(t) = \sigma X^*(t)\,\mathrm{d}W^*(t) \qquad (13.3.10)$$

which means that $X^*(t) = X(0)\exp(\sigma W^*(t) - \sigma^2 t/2)$. By Theorem 10.1.2c, $\{X^*(t)\}$ is a martingale under $\tilde{\mathbb{P}}$.

Lemma 13.3.1 *Let $\{(\alpha(t), \gamma(t))\}$ be a self-financing strategy and define $V^*(t) = V(t)/I(t)$ as the discounted value of the portfolio at time t. Then the process $\{V^*(t)\}$ is a martingale under $\tilde{\mathbb{P}}$.*

Proof By (13.1.12) and (13.3.7), the process $\{V^*(t)\}$ fulfils the stochastic differential equation

$$\begin{aligned} dV^*(t) &= dV(t)/I(t) - V(t)/I^2(t)\,dI(t) \\ &= \alpha(t)dX(t)/I(t) + (\gamma(t) - \alpha(t)X^*(t) - \gamma(t))\,dI(t)/I(t) \\ &= \alpha(t)X^*(t)(\sigma\,dW(t) + (\mu - \delta)\,dt)\,. \end{aligned}$$

Condition (13.3.7) translates into

$$\alpha(t)X^*(t) + \gamma(t) = \alpha(0)X^*(0) + \gamma(0) + \int_0^t \alpha(v)X^*(v)\sigma\,dW^*(v)\,. \quad (13.3.11)$$

Hence, the discounted process $\{V^*(t)\}$, given by $V^*(t) = \alpha(t)X^*(t) + \gamma(t)$, is a $\tilde{\mathbb{P}}$-local martingale. Since condition (13.3.8) ensures that $\tilde{\mathbb{E}} \int_0^t \alpha^2(v)X^2(v)\,dv = \mathbb{E}\,M(t) \int_0^t \alpha^2(v)X^2(v)\,dv < \infty$, $\{V^*(t)\}$ is a martingale under $\tilde{\mathbb{P}}$; see Theorem 13.1.1. □

As was already mentioned, a feasible requirement for an economy is the impossibility to produce money out of nothing. In other words, if one wants to achieve a profit there will also be a risk of losing money. The next theorem shows that the Black–Scholes model is arbitrage-free, i.e. a profit without risk is not possible in this model.

Theorem 13.3.1 *There exists no self-financing trading strategy such that*

$$V(0) \le 0\,, \qquad V(t) \ge 0\,, \qquad \mathbb{P}(V(t) - V(0) > 0) > 0 \quad (13.3.12)$$

for some $t > 0$.

Proof Suppose that there is a self-financing strategy such that (13.3.12) holds for some $t > 0$. Choose $t_0 \ge t$. Note that it is no loss of generality to assume $V(0) = 0$, otherwise we replace $V(t)$ by $V(t) - V(0)$. Since $\mathbb{P}(V(t) \ge 0) = 1$ and since the measures $\mathbb{P}$ and $\tilde{\mathbb{P}}$ are equivalent on $\mathcal{F}_{t_0}$, we also have $\tilde{\mathbb{P}}(V(t) \ge 0) = 1$ and $\tilde{\mathbb{P}}(V^*(t) \ge 0) = 1$. Since $\{V^*(t)\}$ is a martingale under $\tilde{\mathbb{P}}$, see Lemma 13.3.1, we have $\tilde{\mathbb{E}}\,V^*(t) = \tilde{\mathbb{E}}\,V^*(0) = \tilde{\mathbb{E}}\,V(0) = 0$ and therefore $\tilde{\mathbb{P}}(V^*(t) = 0) = 1$. Thus $\tilde{\mathbb{P}}(V(t) = 0) = 1$. This gives $\mathbb{P}(V(t) = 0) = 1$ because the measures $\mathbb{P}$ and $\tilde{\mathbb{P}}$ are equivalent. □

In addition to the two primary securities considered above, we now introduce a *derivative security* into our market, i.e. a security depending on the price process $\{X(t),\ 0 \le t \le t_0\}$. Let us consider the following special case. Imagine that at time $t \le t_0$ an agent sells an option on the risky asset, where the buyer has the right (but not the obligation) to buy an asset for a fixed price K at time t_0. Such a security is called a *European call option.* The holder of the option will exercise the option if and only if $X(t_0) > K$, i.e. there will only be a cashflow $g(X(t_0)) = (X(t_0) - K)_+$ at time t_0. The random variable $g(X(t_0)) = (X(t_0)-K)_+$ is said to be a *Black–Scholes claim.* In Theorem 13.3.2 below, we will see that there is a self-financing strategy $\{(\alpha(t), \gamma(t))\}$, called a *duplication strategy,* that yields the value

$$\alpha(t_0)X(t_0) + \gamma(t_0)I(t_0) = g(X(t_0)) \tag{13.3.13}$$

of the European call option at time t_0. A duplication strategy is also called a *hedging stratcgy.* Such a strategy is important for the seller of a derivative security, who therefore has the possibility to hedge away the financial risk involved.

We first want to find out how a fair (arbitrage-free) price of a European call option can be defined. Suppose that $\{(\alpha(t), \gamma(t))\}$ is a duplication strategy of the European call option. Then, according to the no-arbitrage principle, $V(t)$ is the value of this option at time $t \le t_0$. Indeed, if the price would be larger than $V(t)$, then one could sell the option and the duplication strategy would lead to a riskless profit. If the price would be lower than $V(t)$, then one could buy the option and use the strategy $\{(-\alpha(t), -\gamma(t))\}$ to make a riskless profit. Notice that a similar argument makes it plausible that there cannot be two different duplication strategies. If so, the value of the two portfolios would be different at some time point with positive probability. Then a change from the more expensive portfolio to the cheaper one would yield an arbitrage.

Since by the result of Lemma 13.3.1 the discounted process $\{V^*(t)\}$ is a martingale under $\tilde{\mathbb{P}}$, we can write

$$\begin{aligned} V(t) &= e^{\delta t}\tilde{\mathbb{E}}\left(e^{-\delta t_0}V(t_0) \mid \mathcal{F}_t\right) \\ &= \tilde{\mathbb{E}}\left(e^{-\delta(t_0-t)}(X(t_0) - K)_+ \mid \mathcal{F}_t\right) \end{aligned} \tag{13.3.14}$$

for $t \le t_0$, where we used (13.3.13) in the last equality. The conditional expectation in (13.3.14) can be determined more explicitly, leading to the *Black–Scholes formula* (13.3.15) for the option price $V(t)$. Moreover, the duplication strategy of the European call option can be given by similar formulae. It turns out that $V(t)$ only depends on the price $X(t)$ of the underlying primary security at time t, but not on earlier prices of this security. This is a consequence of the fact that the geometric Brownian motion $\{X(t)\}$ is a Markov process. Moreover, formula (13.3.15) below shows that $V(t)$ does not depend on the rate of return μ, which is important because this parameter is difficult to estimate.

Theorem 13.3.2 *The value $V(t)$ at time $t < t_0$ of the European call option is given by*

$$\begin{aligned} V(t) &= X(t)\Phi\Big(\frac{\log(X(t)/K) + (\delta + \sigma^2/2)(t_0 - t)}{\sigma\sqrt{t_0 - t}}\Big) \\ &\quad - e^{-\delta(t_0-t)}K\Phi\Big(\frac{\log(X(t)/K) + (\delta - \sigma^2/2)(t_0 - t)}{\sigma\sqrt{t_0 - t}}\Big). \end{aligned} \tag{13.3.15}$$

Moreover, the duplication strategy $\{(\alpha(t),\gamma(t))\}$ *has the form*

$$\alpha(t)=\begin{cases}\Phi\Big(\frac{\log(X(t)/K)+(\delta+\sigma^2/2)(t_0-t)}{\sigma\sqrt{t_0-t}}\Big) & \text{if } 0\le t<t_0\,,\\ \mathbb{I}(X(t_0)>K) & \text{if } t=t_0\end{cases} \tag{13.3.16}$$

and

$$\gamma(t)=\begin{cases}-\mathrm{e}^{-\delta t_0}K\Phi\Big(\frac{\log(X(t)/K)+(\delta-\sigma^2/2)(t_0-t)}{\sigma\sqrt{t_0-t}}\Big) & \text{if } 0\le t<t_0\,,\\ -\mathrm{e}^{-\delta t_0}K\mathbb{I}(X(t_0)>K) & \text{if } t=t_0\,,\end{cases} \tag{13.3.17}$$

where $\Phi(x)$ *denotes the distribution function of the* $\mathrm{N}(0,1)$*-distribution.*

Proof We first show that the right-hand sides of (13.3.14) and (13.3.15) coincide for $t=0$. Indeed, by (13.3.10) we have

$$\begin{aligned}\tilde{\mathbb{E}}\left(\mathrm{e}^{-\delta t_0}(X(t_0)-K)_+\mid\mathcal{F}_0\right)&=\mathrm{e}^{-\delta t_0}\tilde{\mathbb{E}}\,(X(t_0)-K)_+\\ &=\mathrm{e}^{-\delta t_0}\tilde{\mathbb{E}}\left(X(0)\exp(\sigma W^*(t_0)+\delta t_0-\sigma^2 t_0/2)-K\right)_+\\ &=\mathrm{e}^{-\delta t_0}\tilde{\mathbb{E}}\,(c\mathrm{e}^Z-K)_+\,,\end{aligned}$$

where Z is $\mathrm{N}((\delta-\sigma^2/2)t_0,\sigma^2t_0)$-distributed under $\tilde{\mathbb{P}}$ and $c=X(0)$. Using the formula for the density of normal distributions, an easy substitution yields

$$\tilde{\mathbb{E}}\,(c\mathrm{e}^Z-K)_+=c\mathrm{e}^{a+b^2/2}\Phi\Big(\frac{\log(c/K)+a+b^2}{b}\Big)-K\Phi\Big(\frac{\log(c/K)+a}{b}\Big)\,,$$

where $a=(\delta-\sigma^2/2)t_0$ and $b=\sigma\sqrt{t_0}$. This proves (13.3.15) for $t=0$. If $0<t<t_0$, we can write

$$\begin{aligned}&\tilde{\mathbb{E}}\left(\mathrm{e}^{-\delta(t_0-t)}(X(t_0)-K)_+\mid\mathcal{F}_t\right)\\ &=\mathrm{e}^{-\delta(t_0-t)}\tilde{\mathbb{E}}\left((X(t)\mathrm{e}^{\sigma(W^*(t_0)-W^*(t))+\delta(t_0-t)-\sigma^2(t_0-t)/2}-K)_+\mid\mathcal{F}_t\right).\end{aligned}$$

Since the random variable $W^*(t_0)-W^*(t)$ is independent of $\mathcal{F}_t$, formula (13.3.15) follows in the same way as before if 0 and t_0 are replaced by t and t_0-t, respectively. We still have to show that the trading strategy $\{(\alpha(t),\gamma(t))\}$ given by (13.3.16) and (13.3.17) is a duplication strategy for the European call option. Since the validity of $\alpha(t_0)X(t_0)+\gamma(t_0)I(t_0)=(X(t_0)-K)_+$ is obvious, it remains to prove that $\{(\alpha(t),\gamma(t))\}$ is self-financing. Consider the function $f:(0,\infty)\times[0,t_0]\to\mathbb{R}$ defined by $f(x,t)=x\alpha(x,t)+\gamma(x,t)\mathrm{e}^{\delta t}$, where

$$\alpha(x,t)=\begin{cases}\Phi\Big(\frac{\log(x/K)+(\delta+\sigma^2/2)(t_0-t)}{\sigma\sqrt{t_0-t}}\Big) & \text{if } 0\le t<t_0,\\ \mathbb{I}(x>K) & \text{if } t=t_0\end{cases}$$

and

$$\gamma(x,t)=\begin{cases}-\mathrm{e}^{-\delta t_0}K\Phi\Big(\frac{\log(x/K)+(\delta-\sigma^2/2)(t_0-t)}{\sigma\sqrt{t_0-t}}\Big) & \text{if } 0\le t<t_0,\\ -\mathrm{e}^{-\delta t_0}K\mathbb{I}(x>K) & \text{if } t=t_0.\end{cases}$$

Then, $V(t) = f(X(t), t)$ and Itô's formula (13.1.12) give

$$\mathrm{d}V(t) = f_x(X(t), t)\,\mathrm{d}X(t) + \big(f_t(X(t), t) + \frac{1}{2} f_{xx}(X(t), t)\sigma^2 X^2(t)\big)\,\mathrm{d}t\,.$$

On the other hand, since $\gamma(X(t), t)\,\mathrm{d}\mathrm{e}^{\delta t} = \delta \mathrm{e}^{\delta t}\gamma(X(t), t)\,\mathrm{d}t$, condition (13.3.7) is fulfilled if

$$\mathrm{d}V(t) = \alpha(X(t), t)\,\mathrm{d}X(t) + \delta \mathrm{e}^{\delta t}\gamma(X(t), t)\,\mathrm{d}t\,.$$

Thus, we have to show that $f_x(x, t) = \alpha(x, t)$ and

$$f_t(x, t) + \frac{1}{2} f_{xx}(x, t)\sigma^2 x^2 = \delta \mathrm{e}^{\delta t}\gamma(x, t)\,,$$

which is left to the reader. Using (13.3.2) and (13.3.6), the validity of (13.3.8) follows from the facts that $0 \le \alpha(x, t) \le 1$ and $\mathbb{E}\,\mathrm{e}^{sW(t)} = \exp(s^2 t/2)$. □

We need to remark that the existence of a duplication strategy can be proved for a much broader class of Black–Scholes claims. Indeed, one can show that any $\mathcal{F}_{t_0}$-measurable claim Y with $\tilde{\mathbb{E}}\,Y^2 < \infty$ can be duplicated, i.e. there is a self-financing trading strategy $\{(\alpha(t), \gamma(t))\}$ such that

$$Y = \tilde{\mathbb{E}}\,Y/I(t_0) + \int_0^{t_0} \alpha(v)\,\mathrm{d}X(v) + \int_0^{t_0} \gamma(v)\,\mathrm{d}I(v)\,; \tag{13.3.18}$$

see Duffie (1996). In this case, we say that $\{(\alpha(t), \gamma(t))\}$ *hedges* the claim Y. A market which allows duplication strategies for all $\mathcal{F}_{t_0}$-measurable claims with finite variance under $\tilde{\mathbb{P}}$ is said to be *complete*. In extension of (13.3.14), the correct price at time t of the claim Y is then given by the random variable $I(t)\tilde{\mathbb{E}}\,(Y/I(t_0) \mid \mathcal{F}_t)$. Thus, option prices in the Black–Scholes model can be calculated under the changed measure $\tilde{\mathbb{P}}$ defined in (13.3.9), instead of using the original probability measure $\mathbb{P}$.

13.3.2 Equity Linked Life Insurance

For the life insurance models treated in this section, net premiums will be calculated using the ideas developed in Section 13.3.1 on the Black–Scholes model. However, the reader has to keep in mind that the real premium will also include administration costs and a security loading.

In classical life insurance, interest is paid at a fixed (technical) rate. In reality, however, insurance companies invest parts of the reserve in financial markets. The expected return is then higher than by investing into riskless bonds, but there is a financial risk involved. The idea of an equity linked life insurance contract is to transfer some of this risk to the policy-holder. At the same time, one can offer an interest rate that is higher than in a classical contract, but stochastic. This might be more attractive to some customers.

Assume now that the payoff of a life insurance contract is linked to an equity like a portfolio, a security price or a financial index. We will only consider the case where the payoff is the maximum of the value of the equity and of a guaranteed amount b. This will lower the risk for the policy-holder. Furthermore, we assume that the value of the equity follows a geometric Brownian motion $\{X(t)\}$ as in the Black–Scholes model studied in Section 13.3.1. Suppose that the payoff is at time T, where in general T is a random variable. The time T can be related to the death of the policy-holder or to another event specified in the contract. The payoff at time T is then $X(T) \vee b$. There are two types of contracts and in both of them a finite (deterministic) time horizon t_0 is considered:

- *term insurance*, where the policy-holder gets the value $X(T) \vee b$ upon death at time $T \in [0, t_0]$,
- *pure endowment insurance*, where the policy-holder gets the value $X(t_0) \vee b$ provided he is still alive at time t_0.

Usually these two types of life insurances are combined (maybe with different bs) in one contract. However, we can consider the two types separately. In order to treat a combined insurance contract, one simply has to add the premiums and/or the trading strategies.

For simplicity, we assume that T is the lifetime of the insured (after policy issue) and that the price process $\{X(t)\}$ and the lifetime T are independent. In particular, let $T = T_a$ denote the remaining lifetime of an a-year-old policy-holder. The distribution of T_a will depend on different external factors, such as sex, health, country of residence, etc. Let $1 - F_a(t) = \mathbb{P}(T_a > t)$ denote the probability that the policy-holder survives the next t years. We assume that the distribution of T_a is absolutely continuous with density $f_a(t)$ and we denote the hazard rate function of F_a by $\{m_a(t)\}$, where $m_a(t) = f_a(t)/(1 - F_a(t))$. Note that $1 - F_a(t) = \exp(-\int_0^t m_a(v)\,\mathrm{d}v)$; see also Section 2.4.2.

There are two types of risks involved in an equity linked life insurance contract:

- the financial risk due to the development of the equity,
- the mortality risk due to the death (survival) of the policy-holder.

This is reflected by the fact that the payoff depends on two sources of randomness, the stochastic price process $\{X(t)\}$ and the remaining lifetime T_a. Note that the form of the payoff $X(T_a) \vee b = b + (X(T_a) - b)_+$ is similar to the value of a European call option studied in Section 13.3.1, but this payoff is now paid at a random time. We have seen in Section 13.3.1 that in some cases the financial risk can be removed. Indeed, we showed that there is a duplication strategy which hedges the payout at a deterministic point in time. However, there is no possibility to hedge mortality as well because we assumed independence of mortality and financial assets. Thus, our life insurance market is not complete.

We want to get rid of the financial risk in order to be left with the mortality risk only. Let us first consider the case of a *term insurance*. Suppose we know that the insured dies in the interval $[v, v + \mathrm{d}v)$, where $v < t_0$. Then, in accordance with the procedure developed in Section 13.3.1, we would use a strategy $\{(\alpha(t,v), \gamma(t,v))\}$ which duplicates the claim $X(v) \vee b = b + (X(v) - b)_+$. The insurer should therefore hold a portfolio $(\alpha(t,v), \gamma(t,v))$ for each possible time of death v, weighted with the probability of death in the interval $[v, v + \mathrm{d}v)$. This leads to the strategy $\{(\alpha^{\mathrm{t}}(t), \gamma^{\mathrm{t}}(t))\}$, where the portfolio $(\alpha^{\mathrm{t}}(t), \gamma^{\mathrm{t}}(t))$ to hold at time $t < t_0$ is

$$
\begin{aligned}
\alpha^{\mathrm{t}}(t) &= \mathbb{I}(T_a > t) \int_t^{t_0} \alpha(t,v) m_a(v)(1 - F_{a+t}(v - t))\, \mathrm{d}v\,, \\
\gamma^{\mathrm{t}}(t) &= \mathbb{I}(T_a > t) \int_t^{t_0} \gamma(t,v) m_a(v)\,(1 - F_{a+t}(v - t))\, \mathrm{d}v\,,
\end{aligned}
$$

since $\mathbf{P}(v < T_a < v + \mathrm{d}v \mid T_a > t) = m_a(v)(1 - F_{a+t}(v - t))\,\mathrm{d}v$. Note that the insurer only has to hold the portfolio as long as the insured is alive. We therefore have to condition on the event $\{T_a > t\}$ that the insured is still alive at time t.

For the *pure endowment insurance* the situation is easier. There is only one possible payout time and therefore the strategy $\{(\alpha^{\mathrm{e}}(t), \gamma^{\mathrm{e}}(t))\}$ for this contract is given by

$$
\begin{aligned}
\alpha^{\mathrm{e}}(t) &= \mathbb{I}(T_a > t)\alpha(t, t_0)(1 - F_{a+t}(t_0 - t))\,, \\
\gamma^{\mathrm{e}}(t) &= \mathbb{I}(T_a > t)\gamma(t, t_0)(1 - F_{a+t}(t_0 - t))\,.
\end{aligned}
$$

Note that, due to the mortality risk, the strategies $\{(\alpha^{\mathrm{t}}(t), \gamma^{\mathrm{t}}(t))\}$ and $\{(\alpha^{\mathrm{e}}(t), \gamma^{\mathrm{e}}(t))\}$ are not self-financing. Hence, the portfolios $(\alpha^{\mathrm{t}}(t), \gamma^{\mathrm{t}}(t))$ and $(\alpha^{\mathrm{e}}(t), \gamma^{\mathrm{e}}(t))$ have to be adjusted continuously in time if the policy-holder does not die.

Let us now compute the *net premium* of equity linked life insurance contracts. We first consider the case of a term insurance. Recall from (13.3.14) that, given $T_a = v$, the arbitrage-free value $V(0)$ of the contract at time $t = 0$ (seen from the policy issue at time a) is

$$
V(0) = \alpha(0,v)X(0) + \gamma(0,v) = \tilde{\mathbb{E}}\,((b \vee X(v))/I(v))
$$

and therefore the single net premium Π^{t} of a term insurance contract is given by

$$
\Pi^{\mathrm{t}} = \int_0^{t_0} \tilde{\mathbb{E}}\,((b \vee X(v))/I(v)) m_a(v)(1 - F_a(v))\, \mathrm{d}v\,, \tag{13.3.19}
$$

while for the single net premium Π^{e} of a pure endowment contract, we have

$$\Pi^{\mathrm{e}} = \tilde{\mathbb{E}}\,((b \vee X(t_0))/I(t_0))(1 - F_a(t_0)) \,. \tag{13.3.20}$$

Using arguments as in the proof of (13.3.15), a simple calculation yields the formula

$$\begin{aligned}\tilde{\mathbb{E}}\,((b \vee X(v))/I(v)) &= X(0)\Phi\Big(\frac{\log(X(0)/b) + (\delta + \sigma^2/2)v}{\sigma\sqrt{v}}\Big) \\ &\quad + be^{-\delta v}\Phi\Big(\frac{\log(b/X(0)) - (\delta - \sigma^2/2)v}{\sigma\sqrt{v}}\Big)\,.\end{aligned}$$

Thus, if the mortality rate function $\{m_a(t)\}$ is known, then the net premiums Π^{t} and Π^{e} can be calculated explicitly. However, in the case of a term insurance, this does not lead to a closed formula for Π^{t}, even in the case where $m_a(t)$ is constant.

Assume now that the policy-holder wants to pay his premiums Π^{t} and Π^{e} at a *constant rate* β, where payments will have to be made as long as the policy-holder is alive, but not longer than t_0. The cumulative discounted payment up to time $t \le t_0$ is then

$$\int_0^{T_a \wedge t} I(v)^{-1}\beta\,\mathrm{d}v = \beta\int_0^t I(v)^{-1}\mathbb{I}(T_a > v)\,\mathrm{d}v\,.$$

The hedging strategy for a constant amount, say 1, to be paid out at time v is simply $\gamma(t) = 1/I(v)$. The value at time t is $\gamma(t)I(t) = I(t)\tilde{\mathbb{E}}\,(1/I(v) \mid \mathcal{F}_t)$. Thus, if the premium payments are invested in the riskless bond, the strategy of the insurer is given by

$$\begin{aligned}\alpha^{\mathrm{t}}(t) &= \mathbb{I}(T_a > t)\int_t^{t_0} \alpha(t,v)m_a(v)(1 - F_{a+t}(v - t))\,\mathrm{d}v\,, \\ \gamma^{\mathrm{t}}(t) &= \mathbb{I}(T_a > t)\Big(-\int_t^{t_0} \beta I(w)^{-1}\,\mathrm{d}w(1 - F_{a+t}(t_0 - t)) \\ &\quad + \int_t^{t_0}\Big(\gamma(t,v) - \int_t^v \beta I(w)^{-1}\,\mathrm{d}w\Big)m_a(v)(1 - F_{a+t}(v - t))\,\mathrm{d}v\Big)\end{aligned}$$

in the term insurance case, and by

$$\begin{aligned}\alpha^{\mathrm{e}}(t) &= \mathbb{I}(T_a > t)\alpha(t,t_0)(1 - F_{a+t}(t_0 - t))\,, \\ \gamma^{\mathrm{e}}(t) &= \mathbb{I}(T_a > t)\Big(\Big(\gamma(t,t_0) - \int_t^{t_0} \beta I(w)^{-1}\,\mathrm{d}w\Big)(1 - F_{a+t}(t_0 - t)) \\ &\quad - \int_t^{t_0}\int_t^v \beta I(w)^{-1}\,\mathrm{d}w\, m_a(v)(1 - F_{a+t}(v - t))\,\mathrm{d}v\Big)\end{aligned}$$

in the pure endowment insurance case. Furthermore, having in mind that the value $\beta\tilde{\mathbb{E}}\int_0^{T_a\wedge t_0} I(v)^{-1}\,dv = \beta\mathbb{E}\int_0^{T_a\wedge t_0} I(v)^{-1}\,dv$ should be equal to the net premiums Π^t and Π^e determined in (13.3.19) and (13.3.20), respectively, the *net premium rate* β is given by

$$\beta = \frac{\int_0^{t_0} \tilde{\mathbb{E}}\,((b \vee X(v))/I(v)) m_a(v)(1 - F_a(v))\,dv}{\mathbb{E}\int_0^{T_a\wedge t_0} I(v)^{-1}\,dv} \tag{13.3.21}$$

in the term insurance case, and by

$$\beta = \frac{\tilde{\mathbb{E}}\,((b \vee X(t_0))/I(t_0))(1 - F_a(t_0))}{\mathbb{E}\int_0^{T_a\wedge t_0} I(v)^{-1}\,dv} \tag{13.3.22}$$

in the pure endowment insurance case.

Remarks 1. As we have already mentioned, the disadvantage of the hedging strategies $\{(\alpha^t(t), \gamma^t(t))\}$ and $\{(\alpha^e(t), \gamma^e(t))\}$ is that the portfolio has to be adjusted continuously in time.

2. An alternative hedging possibility can be given because the claim $X(v) \vee b = b + (X(v) - b)^+$ is similar to the value of a European call option. Let us discuss this for the term insurance case. The reserves for the constant value b have to be built up classically, i.e. at time t one holds $\mathbb{I}(T_a > t) b \int_t^{t_0} m_a(v)(1 - F_{a+t}(v - t))\,dv$ units of the bond. Furthermore, for the event $T_a \in [v, v + dv)$ one holds $m_a(v)(1 - F_{a+t}(v - t))\,dv$ units of a European call option with exercise date v and strike price b. Of course, in practice one divides the interval $(t, t_0]$ into small subintervals. Then, for $T_a \in (v, v + \Delta v]$ one would hold $\int_v^{v+\Delta v} m_a(w)(1 - F_{a+t}(w - t))\,dw$ units of a European call option with exercise date $v + \Delta v$.

3. From the theoretical point of view, it would be possible to get completely rid of the risk. In the case of a term insurance, the insurer could buy an American type option where one can exercise the option at any time point in the interval $(0, t_0]$. The price of such an option at time 0 would be $\sup_\tau \tilde{\mathbb{E}}\,(X(\tau) \vee b)/I(\tau)$, where the supremum is taken over all stopping times $\tau \le t_0$; see Chapter 8 of Duffie (1996). Then there will be a duplication strategy ensuring that $V(t) \ge \max\{X(t), b\}$ for all $t \in [0, t_0]$. In the case of pure endowment insurance, a European type option would do the job. However, it would be cheaper for the insured to buy such a product directly on the financial market. In this case he would avoid the administration costs of the insurer. However, the direct investment in the market would then not only cover death or survival, but also other risks which could be associated with the family of all possible stopping times $\tau \le t_0$. Thus, this product should be more expensive than an equity linked life insurance.

13.3.3 Stochastic Interest Rates in Life Insurance

Consider the general life insurance model of Section 8.4.3 and assume that the Markov process $\{X(t)\}$ and the cumulative loss $B'(t)$ defined by $B'(t) = B(t) - \Pi(t)$ fulfil the conditions formulated there. In the present section, however, we consider the case that the (riskless) force of interest $\delta(t)$ is a random variable which is independent of the state of the insured, that is independent of $\{X(t)\}$. Furthermore, we assume that the stochastic process $\{\delta(t)\}$ is given by a stochastic differential equation of the form

$$\mathrm{d}\delta(t) = a(t, \delta(t))\,\mathrm{d}t + \sigma(t, \delta(t))\,\mathrm{d}W(t)\,, \tag{13.3.23}$$

where the functions $a(t,x)$ and $\sigma(t,x)$ are such that the solution $\{\delta(t)\}$ to (13.3.23) exists, is unique and Markovian. This is, for instance, the case under the conditions of Theorem 13.1.3. Examples of stochastic processes satisfying (13.3.23) have been studied in Section 13.1.2; see also Section 13.4. Notice that to have $\{\delta(t)\}$ and $\{X(t)\}$ independent, we must assume that $\delta(0)$ and $\{W(t)\}$ are independent of $\{X(t)\}$. Let $\{\mathcal{F}_t\}$ be the smallest complete and right-continuous filtration such that $\delta(0)$ is $\mathcal{F}_0$-measurable and the processes $\{X(t)\}$ and $\{W(t)\}$ are adapted with respect to $\{\mathcal{F}_t\}$. Instead of the net prospective premium reserve $\mu_i(t)$ given in (8.4.29), we now consider the reserve

$$\mu_i(t,z) = \mathbb{E}\Big(\int_t^\infty v(t,x)\,\mathrm{d}B'(x) \mid X(t) = i, \delta(t) = z\Big)\,, \tag{13.3.24}$$

where $v(t,x) = \exp(-\int_t^x \delta(w)\,\mathrm{d}w)$. Let $b_i'(t) = b_i(t) - \beta_i(t)$. Recall that by our assumptions in Section 8.4.3, the functions $b_i'(x)$, $b_{ij}(x)$ and $q_{ij}(x)$ are bounded. Thus, $\int_0^\infty v(0,x)\,\mathrm{d}B'(x)$ is a well-defined and integrable random variable if

$$\mathbb{E}\int_0^\infty v(0,x)\,\mathrm{d}x < \infty\,. \tag{13.3.25}$$

Notice that the latter holds in all cases of practical interest. Indeed, one expects an exponential decay of $\mathbb{E}\,v(0,x)$ as $x \to \infty$. Otherwise it would not at all be favourable to invest money; see also Section 13.4.

Assume that condition (13.3.25) is fulfilled. Furthermore, suppose that the functions $\mu_i(t,z)$, $i = 1,\ldots,\ell$, fulfil the smoothness assumptions of Theorem 13.1.2. Whether or not this will be the case depends on the functions $b_i'(x)$, $b_{ij}(x)$, $q_{ij}(x)$, $\sigma(t,z)$ and $a(t,z)$. For explicit conditions, see for instance Appendix E of Duffie (1996) or Karatzas and Shreve (1991).

Theorem 13.3.3 *Under the above conditions, the net prospective premium reserve $\mu_i(t,z)$ fulfils the system of partial differential equations*

$$b_i'(t) + \sum_{j\neq i}(b_{ij}(t) + \mu_j(t,z) - \mu_i(t,z))q_{ij}(t) - z\mu_i(t,z)$$

$$+ \quad \frac{\partial \mu_i}{\partial t}(t,z) + a(t,z)\frac{\partial \mu_i}{\partial z}(t,z) + \frac{\sigma^2(t,z)}{2}\frac{\partial^2 \mu_i}{\partial z^2}(t,z) = 0\,. \tag{13.3.26}$$

Proof Take the martingale $\{M(t)\}$, where $M(t) = \mathbb{E}\,(\int_0^\infty v(0,x)\,\mathrm{d}B'(x) \mid \mathcal{F}_t)$. Notice that

$$\begin{aligned} M(t) &= \int_0^t v(0,x)\,\mathrm{d}B'(x) + \mathbb{E}\Big(\int_t^\infty v(0,x)\,\mathrm{d}B'(x) \mid \mathcal{F}_t\Big) \\ &= \int_0^t v(0,x)\,\mathrm{d}B'(x) \\ &\quad + \sum_{j\neq i}\int_0^t v(0,x)\big(\mu_j(x,\delta(x)) - \mu_i(x,\delta(x))\big)\mathrm{d}N_{ij}(x) \\ &\quad + v(0,t)\,\mu_{X(t)}(t,\delta(t)) \\ &\quad - \sum_{j\neq i}\int_0^t v(0,x)\big(\mu_j(x,\delta(x)) - \mu_i(x,\delta(x))\big)\mathrm{d}N_{ij}(x)\,. \end{aligned}$$

The first two terms on the right-hand side of this equation can be written in the form

$$\begin{aligned} &\int_0^t v(0,x)\Big(b'_{X(x)}(x)\,\mathrm{d}x \\ &\qquad + \sum_{j\neq i}\big(b_{ij}(x) + \mu_j(x,\delta(x)) - \mu_i(x,\delta(x))\big)\,\mathrm{d}N_{ij}(x)\Big) \\ &= M'(t) + \int_0^t v(0,x)\Big(b'_{X(x)}(x) \\ &\qquad + \sum_{j\neq X(x)}\big(b_{X(x),j}(x) + \mu_j(x,\delta(x)) - \mu_{X(x)}(x,\delta(x))\big)\,q_{X(x)j}(x)\Big)\,\mathrm{d}x\,, \end{aligned}$$

where

$$\begin{aligned} M'(t) &= \int_0^t v(0,x)\sum_{j\neq i}\big(b_{ij}(x) \\ &\qquad + \mu_j(x,\delta(x)) - \mu_i(x,\delta(x))\big)\big(\mathrm{d}N_{ij}(x) - \mathbb{I}(X(x)=i)q_{ij}(x)\,\mathrm{d}x\big)\,. \end{aligned}$$

Notice that $\{M'(t)\}$ is a local martingale. A proof of this fact can be found for instance in Brémaud (1981), p. 27. Furthermore, by Itô's formula (13.1.12), we have

$$\begin{aligned} &v(0,t)\,\mu_{X(t)}(t,\delta(t)) \\ &\qquad - \int_0^t \sum_{j\neq i} v(0,x)\big(\mu_j(x,\delta(x)) - \mu_i(x,\delta(x))\big)\mathrm{d}N_{ij}(x) \end{aligned}$$

$$
\begin{aligned}
&= \mu_{X(0)}(0,\delta(0)) + M''(t) + \int_0^t v(0,x)\Big(-\delta(x)\mu_{X(x)}(x,\delta(x)) \\
&+ \frac{\partial \mu_{X(x)}}{\partial t}(x,\delta(x)) + a(x,\delta(x))\frac{\partial \mu_{X(x)}}{\partial z}(x,\delta(x)) \\
&+ \frac{\sigma^2(x,\delta(x))}{2}\,\frac{\partial^2 \mu_{X(x)}}{\partial z^2}(x,\delta(x))\Big)\,\mathrm{d}x\,,
\end{aligned}
$$

where

$$
M''(t) = \int_0^t v(0,x)\sigma(x,\delta(x))\frac{\partial \mu_{X(x)}}{\partial z}(x,\delta(x))\,\mathrm{d}W(x)\,.
$$

By Theorem 13.1.1, the stochastic process $\{M''(t)\}$ is a local martingale. Thus, the process $\{M^*(t)\}$ with $M^*(t) = M(t) - M'(t) - M''(t)$ is also a local martingale. Since

$$
\begin{aligned}
M^*(t) &= \mu_{X(0)}(0,\delta(0)) \\
&+ \int_0^t v(0,x)\Big(b'_{X(x)}(x) + \sum_{j\neq X(x)} \big(b_{X(x),j}(x) + \mu_j(x,\delta(x)) \\
&- \mu_{X(x)}(x,\delta(x))\big)q_{X(x),j}(x) - \delta(x)\mu_{X(x)}(x,\delta(x)) + \frac{\partial \mu_{X(x)}}{\partial t}(x,\delta(x)) \\
&+ a(x,\delta(x))\frac{\partial \mu_{X(x)}}{\partial z}(x,\delta(x)) + \frac{\sigma^2(x,\delta(x))}{2}\,\frac{\partial^2 \mu_{X(x)}}{\partial z^2}(x,\delta(x))\Big)\,\mathrm{d}x\,,
\end{aligned}
$$

the local martingale $\{M^*(t)\}$ is continuous and has bounded variation. By Lemma 13.1.2, $\{M^*(t)\}$ must be constant. Using the fact that $v(0,x) > 0$, this gives (13.3.26). □

Remarks 1. The partial differential equation (13.3.26) is a generalization of Thiele's differential equation (8.4.32). Note also that equation (13.3.26) is similar to (11.2.13).

2. The terms in (13.3.26) can be interpreted as follows: $\frac{1}{2}\sigma^2(t,z)\,\partial^2\mu_i/\partial z^2(t,z)$ is the diffusion term, $\partial\mu_i/\partial t(t,z) + a(t,\delta)\partial\mu_i/\partial z(t,z)$ is the drift term, $q_{ij}(t)$ is the intensity of a jump from i to j, $b_{ij}(t) + \mu_j(t,z)$ is the value of $\mu_{X(t)}$ just after a jump from i to j, and $\mu_i(t,z)$ is the value of $\mu_{X(t)}$ immediately before the jump. Finally, the factor z in (13.3.26) corresponds to the function κ in Theorem 11.2.3, while $b'_i(t)$ corresponds to the function γ there.

3. Equation (13.3.26) is difficult to solve analytically. If we are only interested in $\mu_i(0,\delta(0))$, then the problem can be simplified. Since the processes $\{\delta(t)\}$ and $\{X(t)\}$ are independent, we have

$$
\mu_i(0,\delta(0)) = \mathbb{E}\left(\int_0^\infty \mathbb{E}\,v(0,x)\,\mathrm{d}B'(x) \mid X(0) = i\right).
$$

Let $0 \le t \le x$. If we put $v'(t,x) = \mathbb{E}\, v(t,x)$ and

$$\delta'(t) = -\frac{\partial}{\partial t} \log \mathbb{E} \exp\Big(-\int_0^t \delta(v)\,\mathrm{d}v\Big)\,, \tag{13.3.27}$$

then

$$v'(t,x) = \exp\Big(-\int_t^x \delta'(v)\,\mathrm{d}v\Big)\,.$$

Note that by the assumptions made in this section, the random variable $\int_0^\infty v'(0,t)\,\mathrm{d}B'(t)$ is well-defined and integrable. We can now consider the functions

$$\mu_i'(t) = \mathbb{E}\Big(\int_t^\infty v'(t,x)\,\mathrm{d}B'(x) \mid X(t) = i\Big).$$

Note that $\mu_{X(0)}(0,\delta(0)) = \mu'_{X(0)}(0)$. Moreover, Theorem 13.3.3 implies that the functions $\mu_i'(t)$ fulfil the ordinary system of differential equations

$$b_i'(t) + \sum_{j\ne i}(b_{ij}(t) + \mu_j'(t) - \mu_i'(t))q_{ij}(t) - \delta'(t)\mu_i'(t) + (\mu_i')^{(1)}(t) = 0\,. \tag{13.3.28}$$

The advantage of (13.3.28) is that this system of differential equations is easier to solve than (13.3.26). But if one is interested in reserves at another point in time than at $t = 0$, one has to solve the equations (13.3.28) again with a different function $\delta'(t)$; for then, the initial condition $\delta(0)$ has changed, whereas (13.3.26) gives a solution for each initial value $\delta(0)$.

Example Let $\delta(0)$ be deterministic and assume that $\{\delta(t)\}$ is an Ornstein–Uhlenbeck process as considered in Example 2 of Section 13.1.2. We showed there that then the integral $\int_0^t \delta(v)\,\mathrm{d}v$ has a normal distribution with expectation $\mu = \mathbb{E}\int_0^t \delta(v)\,\mathrm{d}v$ and variance $\sigma^2 = \mathrm{Var}\int_0^t \delta(v)\,\mathrm{d}v$ given in (13.1.26) and (13.1.27), respectively. Thus, the random variable $Z = \exp\Big(-\int_0^t \delta(v)\,\mathrm{d}v\Big)$ is lognormally distributed with $\mathbb{E}\, Z = \exp(-\mu + \sigma^2/2)$. Using (13.1.26) and (13.1.27) it follows that

$$\begin{aligned} &-\log \mathbb{E} \exp\Big(-\int_0^t \delta(v)\,\mathrm{d}v\Big) \\ &= \Big(\bar\delta - \frac{\sigma^2}{2\alpha^2}\Big)t + \frac{\delta(0) - \big(\bar\delta - \sigma^2/(2\alpha^2)\big)}{\alpha}(1 - e^{-\alpha t}) + \frac{\sigma^2}{4\alpha^3}(1 - e^{-\alpha t})^2 \end{aligned}$$

and therefore

$$\delta'(t) = \bar\delta - \frac{\sigma^2}{2\alpha^2} + (\delta(0) - \bar\delta)\mathrm{e}^{-\alpha t} + \frac{\sigma^2}{2\alpha^2}(2\mathrm{e}^{-\alpha t} - \mathrm{e}^{-2\alpha t})\,. \tag{13.3.29}$$

Notice that condition (13.3.25) for $\{\delta(t)\}$ is fulfilled if $\sigma^2 \le 2\alpha^2\bar\delta$. We will see in Section 13.4.2 that this assumption is quite natural. For t large, the force of interest $\delta'(t)$ used is smaller than the long-run expectation $\bar{\delta}$ of the real force of interest $\delta(t)$. This is due to the stochastic fluctuations of $\{\delta(t)\}$. Solving the differential equations (13.3.28) for the function $\delta'(t)$ given in (13.3.29) will yield the reserve $\mu_i(0, \delta(0))$ for $i = 1, \ldots, \ell$.

Bibliographical Notes. The correct option price formula (13.3.15) was first obtained by Black and Scholes (1973). The connection to equivalent martingale measures leading to (13.3.14) and (13.3.18) was treated in Harrison and Kreps (1979) and Harrison and Pliska (1981). Examples of introductory textbooks on option pricing in the Black–Scholes model include Bingham and Kiesel (1998), Lamberton and Lapeyre (1997), Irle (1998) and Mikosch (1998). For a broader introduction to diffusion models in financial mathematics, see for instance Duffie (1996) and Musiela and Rutkowski (1997). Recall that in the Black–Scholes model, the price process $\{X(t)\}$ given in (13.3.2) has continuous sample paths. However, in some cases, the price of a risky asset can change discontinuously from time to time where the jump epochs and the jump sizes are random. In the intervals between the jumps the price behaves like the sample path of a diffusion process. This situation can be modelled by a *jump-diffusion process*. Notice that the structure of jump-diffusion processes is similar to that of perturbed risk processes considered in Section 13.2. Their usage typically leads to incomplete financial markets. More details on this class of stochastic processes and on pricing and hedging of derivatives in incomplete markets can be found, for instance, in Aase (1988), Bakshi, Cao and Chen (1997), Bardhan and Chao (1993), Bladt and Rydberg (1998), Colwell and Elliott (1993), El Karoui and Quenez (1995), Föllmer and Schweizer (1991), Schweizer (1992,1994), Lamberton and Lapeyre (1997) and Mercurio and Runggaldier (1993). Furthermore, recall that, in the Black–Scholes model, the return $\{X'(t)\}$ defined in (13.3.3) is simply a Brownian motion with drift. But the normal distribution does not always fit empirical return data sufficiently well. In Barndorff–Nielsen (1997,1998) and Eberlein and Keller (1995), alternative distributions are proposed which are based on a background driving *Lévy process*. Examples of papers where option price formulae have been derived when the underlying security price follows a *pure jump process* (without a diffusion component) include Elliot and Kopp (1990), Korn, Kreer and Lenssen (1998) and Page and Sanders (1986). Results on equity linked insurance contracts can be found in papers like Aase and Persson (1994), Delbaen (1990), Møller (1998), Nielsen and Sandmann (1995) and Russ (1998). The generalization (13.3.26) of Thiele's differential equation goes back to Norberg and Møller (1996), and equation (13.3.28) is due to Schmidli (1996); see also Levikson and Mizrahi (1994).

13.4 SIMPLE INTEREST RATE MODELS

We continue the discussion of two simple classes of diffusion processes which have been mentioned before in Examples 2 and 3 of Section 13.1.2. They can be used as stochastic models of the force of interest. In particular, we determine the no-arbitrage price of a zero-coupon bond in these models.

Since the only source of randomness considered in the present section is the standard Brownian motion $\{W(t)\}$, we assume that $\{\mathcal{F}_t\}$ is the smallest complete and right-continuous filtration such that $\{W(t)\}$ is adapted.

13.4.1 Zero-Coupon Bonds

In Section 8.4.3 and also in Section 13.3.3 of the present chapter, we have modelled interest rates via the (time-dependent) force of interest $\{\delta(t)\}$, i.e. the value at time $t \geq 0$ of a unit invested at time 0 is $\exp(\int_0^t \delta(v)\,\mathrm{d}v)$. In practice, the force of interest is piecewise constant. However, since it usually changes in small steps, we assume in this section that $\{\delta(t)\}$ is a stochastic process with continuous sample paths. More specifically, let $\{\delta(t)\}$ be a diffusion process given by the stochastic differential equation (13.3.23), i.e.

$$\mathrm{d}\delta(t) = a(t,\delta))\,\mathrm{d}t + \sigma(t,\delta(t))\,\mathrm{d}W(t)\,. \tag{13.4.1}$$

A *zero-coupon bond* is a security that pays a dividend of one unit at time $t_0 > 0$. We denote the value at time $t \in [0,t_0]$ of a zero-coupon bond by $D(t,t_0)$. It is clear that $D(t_0,t_0) = 1$. If $\{\delta(t)\}$ is deterministic, we would have $D(t,t_0) = \exp(-\int_t^{t_0} \delta(v)\,\mathrm{d}v)$. Indeed, by borrowing the amount $D(t,t_0)$, at time t_0 one would have to pay back the amount 1. To get an idea of how to determine the (arbitrage-free) value $D(t,t_0)$ in the case of a stochastic force of interest $\{\delta(t)\}$, we recall the option price formula (13.3.15) in the Black–Scholes model. This formula shows that for pricing one can use a changed measure $\tilde{\mathbb{P}}$ defined on $\mathcal{F}_{t_0}$ which is equivalent to the restriction of the original measure $\mathbb{P}$ to $\mathcal{F}_{t_0}$, such that $\{D^*(t,t_0),\ t \in [0,t_0]\}$ becomes a martingale under $\tilde{\mathbb{P}}$. Here

$$D^*(t,t_0) = \exp\Big(-\int_0^t \delta(v)\,\mathrm{d}v\Big) D(t,t_0)$$

is the discounted value of the zero-coupon bond price $D(t,t_0)$. Then,

$$\begin{aligned}
D(t,t_0) &= \exp\Big(\int_0^t \delta(v)\,\mathrm{d}v\Big) D^*(t,t_0) = \exp\Big(\int_0^t \delta(v)\,\mathrm{d}v\Big)\tilde{\mathbb{E}}\,(D^*(t_0,t_0)\mid \mathcal{F}_t) \\
&= \exp\Big(\int_0^t \delta(v)\,\mathrm{d}v\Big)\tilde{\mathbb{E}}\Big(\exp\Big(-\int_0^{t_0} \delta(v)\,\mathrm{d}v\Big) D(t_0,t_0)\;\Big|\; \mathcal{F}_t\Big) \\
&= \tilde{\mathbb{E}}\Big(\exp\Big(-\int_t^{t_0} \delta(v)\,\mathrm{d}v\Big)\;\Big|\; \mathcal{F}_t\Big)\,.
\end{aligned} \tag{13.4.2}$$

Equation (13.4.2) means that the zero-coupon bond price is the conditional expectation under a changed measure of the same discount factor as in the deterministic case. At first glance, one could think that the set of possible changed measures $\tilde{\mathbb{P}}$ is large and that one therefore would be able to choose the price of the zero-coupon bond from a continuum of possible prices. Usually there are however other securities in the market where the discounted values of these securities should also be martingales under $\tilde{\mathbb{P}}$. This means that the set of measures $\tilde{\mathbb{P}}$ to choose from may be quite small. The actual measure $\tilde{\mathbb{P}}$ chosen for pricing, if there are several, will be selected by the preferences of the agents present in the market. In the rest of this section we assume that there is at least one measure $\tilde{\mathbb{P}}$ which is equivalent to $\mathbb{P}$ on $\mathcal{F}_{t_0}$ and such that $\{D^*(t, t_0),\ t \in [0, t_0]\}$ is a martingale under $\tilde{\mathbb{P}}$.

The following lemma is useful in many situations. Theorem 13.4.1 below illustrates how the zero-coupon price $D(t, t_0)$, given in (13.4.2), can be expressed by the original probability measure $\mathbb{P}$.

Lemma 13.4.1 *Let $\{M(t)\}$ be an $\{\mathcal{F}_t\}$-local martingale. Then there exists a unique adapted process $\{Y(t)\} \in L^2_{\mathrm{loc}}$ such that $M(t) = M(0) + \int_0^t Y(v)\,\mathrm{d}W(v)$ for all $t \ge 0$. If $\mathbb{E}\, M^2(t) < \infty$ for all $t \ge 0$ then $\{Y(t)\} \in L^2$.*

The *proof* of Lemma 13.4.1 goes beyond the scope of this book. It can be found, for instance, in Karatzas and Shreve (1991).

We first use the result of Lemma 13.4.1 in the following context. Recall that we consider the finite time horizon t_0. Let $L(t_0)$ denote the Radon–Nikodym derivative $\mathrm{d}\tilde{\mathbb{P}}/\mathrm{d}\mathbb{P}$ on $\mathcal{F}_{t_0}$, i.e. for all $B \in \mathcal{F}_{t_0}$ we have $\tilde{\mathbb{P}}(B) = \mathbb{E}\,[L(t_0); B]$. Let $\{L(t),\ t \in [0, t_0]\}$ be the martingale defined by $L(t) = \mathbb{E}\,(L(t_0) \mid \mathcal{F}_t)$. By Lemma 10.2.2, the random variable $L(t)$ is the Radon–Nikodym derivative of the measures $\mathbb{P}$ and $\tilde{\mathbb{P}}$ restricted to $\mathcal{F}_t$. Then the martingale $\{L(t)\}$ has the following representation.

Lemma 13.4.2 *There exists a process $\{Z(t)\} \in L^2_{\mathrm{loc}}$ such that for each $t \in [0, t_0]$*

$$L(t) = \exp\Big(-\int_0^t Z(v)\,\mathrm{d}W(v) - \tfrac{1}{2}\int_0^t Z^2(v)\,\mathrm{d}v\Big). \tag{13.4.3}$$

Proof By Lemma 13.4.1, there is a process $\{Y(t)\}$ such that $L(t) = 1 + \int_0^t Y(v)\,\mathrm{d}W(v)$. Itô's formula (13.1.12) implies that

$$\log L(t) = \int_0^t \frac{1}{L(v)} Y(v)\,\mathrm{d}W(v) - \tfrac{1}{2}\int_0^t \frac{1}{L^2(v)} Y^2(v)\,\mathrm{d}v\,.$$

The choice $Z(t) = -Y(t)/L(t)$ gives (13.4.3). □

Theorem 13.4.1 *The zero-coupon price $D(t,t_0)$ given in* (13.4.2) *can be expressed as*

$$D(t,t_0) = \mathbb{E}\left(\exp\Big(-\int_t^{t_0} \delta(v)\,\mathrm{d}v - \int_t^{t_0} Z(v)\,\mathrm{d}W(v) - \tfrac{1}{2}\int_t^{t_0} Z^2(t)\,\mathrm{d}v\Big) \;\Big|\; \mathcal{F}_t\right) \tag{13.4.4}$$

where $\{Z(t)\}$ is the process appearing in Lemma 13.4.2.

Proof Using (10.2.26), for any nonnegative $\mathcal{F}_{t_0}$-measurable random variable X we have

$$\tilde{\mathbb{E}}\,(X \mid \mathcal{F}_t) = \frac{\mathbb{E}\,(XL(t_0) \mid \mathcal{F}_t)}{L(t)} \tag{13.4.5}$$

for all $t \in [0,t_0]$. Now, putting $X = \exp\big(-\int_t^{t_0} \delta(v)\,\mathrm{d}v\big)$, formula (13.4.4) is obtained from (13.4.5) if we use (13.4.2) and (13.4.3). □

Define the process $\{W^*(t)\}$ by $W^*(t) = W(t) + \int_0^t Z(v)\,\mathrm{d}v$, where $\{Z(t)\}$ is from Lemma 13.4.2. Using similar arguments as in the proof of Theorem 13.1.4, one can show that $\{W^*(t)\}$ is a standard Brownian motion under $\tilde{\mathbb{P}}$. The details are left to the reader. Equation (13.4.1) can be written as

$$\mathrm{d}\delta(t) = \big(a(t,\delta(t)) - Z(t)\sigma(t,\delta(t))\big)\,\mathrm{d}t + \sigma(t,\delta(t))\,\mathrm{d}W^*(t)\,, \tag{13.4.6}$$

for $t \in [0,t_0]$. Using the notation $a^*(t,x) = a(t,x) - Z(t)\sigma(t,x)$, we see that under $\tilde{\mathbb{P}}$ the process $\{\delta(t)\}$ fulfils equation (13.4.1) with respect to the Brownian motion $\{W^*(t)\}$ and now with the new drift $a^*(t,x)$. Since the term $a(t,\delta(t))$ in (13.4.6) can be interpreted as the return of the bond at time t (under $\mathbb{P}$), the random variable $Z(t)$ can be seen as a *risk premium* per unit of volatility.

13.4.2 The Vasicek Model

Next we consider a special case of a stochastic force of interest $\{\delta(t)\}$ which is usually called the *Vasicek model.* Let $\delta(0)$ be deterministic and assume that $\{\delta(t)\}$ is an Ornstein–Uhlenbeck process. More precisely, we assume that the process $\{\delta(t)\}$ satisfies the stochastic differential equation

$$\mathrm{d}\delta(t) = -\alpha\Big(\delta(t) - \big(\bar{\delta} + \frac{\beta\sigma}{\alpha}\big)\Big)\,\mathrm{d}t + \sigma\,\mathrm{d}W(t)$$

for some constants $\alpha,\beta,\bar{\delta},\sigma > 0$. We also assume that the risk premium $Z(t)$ is constant and equal to β. Using (13.4.6) we see that $\{\delta(t)\}$ remains an Ornstein–Uhlenbeck process when passing from $\mathbb{P}$ to the changed measure $\tilde{\mathbb{P}}$. Indeed, under $\tilde{\mathbb{P}}$ we have

$$\mathrm{d}\delta(t) = -\alpha(\delta(t) - \bar{\delta})\,\mathrm{d}t + \sigma\,\mathrm{d}W^*(t)\,, \tag{13.4.7}$$

where the standard Brownian motion $\{W^*(t)\}$ is given by $W^*(t) = W(t) + \beta t$. We have seen in Example 2 of Section 13.1.2 that then $\{\delta(t)\}$ is a Gaussian Markov process under $\tilde{\mathbb{P}}$. This makes it possible to specify the general formulae (13.4.2) and (13.4.4) for the zero-coupon bond price $D(t, t_0)$.

Theorem 13.4.2 *For* $0 \le t \le t_0$,

$$
\begin{aligned}
D(t,t_0) &= \exp\Big(-\Big(\bar{\delta} - \frac{\sigma^2}{2\alpha^2}\Big)(t_0 - t) - \frac{\delta(t) - \big(\bar{\delta} - \frac{\sigma^2}{2\alpha^2}\big)}{\alpha}(1 - e^{-\alpha(t_0-t)}) \\
&\quad - \frac{\sigma^2}{4\alpha^3}(1 - e^{-\alpha(t_0-t)})^2\Big). \qquad (13.4.8)
\end{aligned}
$$

Proof Using the fact that $\{\delta(t)\}$ is Markov under $\tilde{\mathbb{P}}$, formula (13.4.2) yields

$$
D(t,t_0) = \tilde{\mathbb{E}}\left(\exp\Big(-\int_t^{t_0} \delta(v)\,\mathrm{d}v\Big) \,\Big|\, \delta(t)\right).
$$

Furthermore, using (13.1.26) and (13.1.27) we have

$$
\tilde{\mathbb{E}}\left(\int_t^{t_0} \delta(v)\,\mathrm{d}v \,\Big|\, \delta(t)\right) = \bar{\delta}(t_0 - t) + (\delta(t) - \bar{\delta})\left(\frac{1 - \mathrm{e}^{-\alpha(t_0-t)}}{\alpha}\right)
$$

and

$$
\begin{aligned}
&\widetilde{\mathrm{Var}}\left(\int_t^{t_0} \delta(v)\,dv \,\Big|\, \delta(t)\right) \\
&= \frac{\sigma^2}{\alpha^2}(t_0 - t) - \frac{\sigma^2}{\alpha^3}(1 - \mathrm{e}^{-\alpha(t_0-t)}) - \frac{\sigma^2}{2\alpha^3}(1 - \mathrm{e}^{-\alpha(t_0-t)})^2.
\end{aligned}
$$

From the considerations in Example 2 of Section 13.1.2 we have that, given $\delta(t)$, the random variable $D(t, t_0)$ is lognormally distributed. Thus, proceeding in a similar way to that in the example of Section 13.3.3 we get

$$
\begin{aligned}
D(t,t_0) &= \tilde{\mathbb{E}}\left(\exp\Big(-\int_t^{t_0} \delta(v)\,\mathrm{d}v\Big) \,\Big|\, \delta(t)\right) \\
&= \exp\Big(-\tilde{\mathbb{E}}\Big(\int_t^{t_0} \delta(v)\,\mathrm{d}v \,\Big|\, \delta(t)\Big) + \tfrac{1}{2}\widetilde{\mathrm{Var}}\Big(\int_t^{t_0} \delta(v)\,\mathrm{d}v \,\Big|\, \delta(t)\Big)\Big) \\
&= \exp\Big(-\Big(\bar{\delta} - \frac{\sigma^2}{2\alpha^2}\Big)(t_0 - t) - \frac{\delta(t) - \big(\bar{\delta} - \frac{\sigma^2}{2\alpha^2}\big)}{\alpha}(1 - e^{-\alpha(t_0-t)}) \\
&\quad - \frac{\sigma^2}{4\alpha^3}(1 - e^{-\alpha(t_0-t)})^2\Big).
\end{aligned}
$$

This proves the theorem. □

Remarks 1. If $t_0 - t$ is large, then $D(t, t_0)$ is approximately $\exp(-(\bar{\delta} - \sigma^2/(2\alpha^2)))(t_0 - t)$. This implies that $\sigma^2 \leq 2\alpha^2\bar{\delta}$ should hold. If not, then the price of a zero-coupon bond would ultimately be larger than 1. That would mean that one has to pay more than one monetary unit in order to get a monetary unit at time t_0. Therefore, one would keep a monetary unit at home until time t_0, rather than investing it.

2. The Vasicek model has the following disadvantage. Since $\{\delta(t)\}$ is a Gaussian process we have $\mathbb{P}(\delta(t) < 0) > 0$. Now if $\delta(t) < 0$ for some time t then $D(t, t') > 1$ for t' small enough; $t' > t$. We saw above that this should not happen. Nevertheless the model is often used because a Gaussian process is easy to handle. One then argues that the probability of a negative $\delta(t)$ for $t \in [0, t_0]$ is small, provided that $\delta(0)$ is not too small.

3. From the considerations stated in Section 13.3.1 we know that the price $D(t, t_0)$ given in (13.4.8) should be arbitrage-free. But since $D(t, t_0) > 1$ with positive probability we can find the following arbitrage. If $D(t, t_0) > 1$, sell a zero-coupon bond and keep the money. Then at time t_0 pay the dividend. This gives a riskless profit of $D(t, t_0) - 1$ monetary units. The reason for this arbitrage is that we here introduced a new security called "money". Together with this new security the market allows arbitrage. Indeed, the discounted value $\exp(-\int_0^t \delta(v)\,\mathrm{d}v)$ of "money" is not a martingale under the pricing measure $\tilde{\mathbb{P}}$. If the agents are only allowed to buy zero-coupon bonds and keep money on a bank account, then there is no arbitrage as indicated in Section 13.3.1.

13.4.3 The Cox–Ingersoll–Ross Model

The difficulty with $\delta(t) < 0$ mentioned in Section 13.4.2 can be overcome by choosing the model of Example 3 of Section 13.1.2. Assume that $\{\delta(t)\}$ satisfies

$$\mathrm{d}\delta(t) = -(\alpha - \sigma)\Big(\delta(t) - \frac{\alpha\bar{\delta}}{\alpha - \sigma}\Big)\,\mathrm{d}t + \sigma\sqrt{\delta(t)}\,\mathrm{d}W(t) \tag{13.4.9}$$

for some constants $\alpha, \bar{\delta}, \sigma > 0$ such that $\alpha > \sigma$. Furthermore, let the risk premium process $\{Z(t)\}$ be given by $Z(t) = \sqrt{\delta(t)}$. Then, the stochastic differential equation (13.4.6) takes the form

$$\mathrm{d}\delta(t) = -\alpha(\delta(t) - \bar{\delta})\,\mathrm{d}t + \sigma\sqrt{\delta(t)}\,\mathrm{d}W^*(t)\,. \tag{13.4.10}$$

Under the changed measure $\tilde{\mathbb{P}}$, the diffusion process $\{\delta(t)\}$ satisfying (13.4.10) belongs to the same class of processes as considered in Example 3 of Section 13.1.2. Thus, under $\tilde{\mathbb{P}}$, the solution $\{\delta(t)\}$ to (13.4.10) is a Markov process, such that $\delta(t) \geq 0$ and both the processes $\{\int_0^t \sqrt{\delta(v)}\,\mathrm{d}W^*(v)\}$ and

$\{\int_0^t (\delta(v))^{3/2}\,\mathrm{d}W^*(v)\}$ are martingales. However, the zero-coupon bond price is much harder to obtain than in the case of the Vasicek model considered in Section 13.4.2.

We first answer the question under which condition the force of interest never becomes zero, that is $\delta(t) > 0$ for all $t > 0$. It turns out that this event has probability 1 (both under $\mathbb{P}$ and $\tilde{\mathbb{P}}$) if and only if the inequality $\sigma^2 \le 2\alpha\bar{\delta}$ holds. Let $\tau_\varepsilon = \inf\{t \ge 0 : \delta(t) = \varepsilon\}$ and $\tau_{\varepsilon,\varepsilon'} = \tau_\varepsilon \wedge \tau_{\varepsilon'}$.

Lemma 13.4.3 *Assume that* $\delta(0) > 0$.
(a) *If* $\sigma^2 \le 2\alpha\bar{\delta}$, *then* $\tilde{\mathbb{P}}(\tau_0 < \infty) = 0$.
(b) *If* $\sigma^2 > 2\alpha\bar{\delta}$, *then* $\tilde{\mathbb{P}}(\tau_0 < \infty) = 1$.

Proof Consider the extended generator $\boldsymbol{A}$ of the Markov process $\{\delta(t)\}$ under $\tilde{\mathbb{P}}$. We want to solve $\boldsymbol{A}g(y) = 0$ for some twice continuously differentiable function $g : (0,\infty) \to \mathbb{R}$, where $\boldsymbol{A}$ is given by (13.1.32). Thus, we search for a solution to

$$\frac{\sigma^2}{2} y g^{(2)}(y) - \alpha(y - \bar{\delta}) g^{(1)}(y) = 0\,.$$

Such a solution on $(0,\infty)$ is given by

$$g(y) = \int_1^y \mathrm{e}^{2\alpha v/\sigma^2} v^{-2\alpha\bar{\delta}/\sigma^2}\,\mathrm{d}v\,. \tag{13.4.11}$$

Let $0 < \varepsilon < \delta(0) < \varepsilon' < \infty$. Using (13.1.31) we find

$$g(\delta(\tau_{\varepsilon,\varepsilon'} \wedge t)) = g(\delta(0)) + \sigma \int_0^{\tau_{\varepsilon,\varepsilon'}\wedge t} g^{(1)}(\delta(v))\sqrt{\delta(v)}\,\mathrm{d}W^*(v) \tag{13.4.12}$$

for the function g given in (13.4.11). Since the integrand in (13.4.12) is bounded, $\{g(\delta(\tau_{\varepsilon,\varepsilon'} \wedge t))\}$ is a martingale by the result of Theorem 13.1.1a. Moreover, consider the process $\{M'(t)\}$ defined in (13.1.6) with $X(v) = Y(v) = g^{(1)}(\delta(v))\sqrt{\delta(v)}$. Then, using (13.4.12) and Theorem 13.1.1a, we get

$$\begin{aligned}
\infty &> \tilde{\mathbb{E}}\left(g(\delta(\tau_{\varepsilon,\varepsilon'} \wedge t)) - g(\delta(0))\right)^2 \\
&= \sigma^2 \tilde{\mathbb{E}}\left(\int_0^{\tau_{\varepsilon,\varepsilon'}\wedge t} g^{(1)}(\delta(v))\sqrt{\delta(v)}\,\mathrm{d}W^*(v)\right)^2 \\
&= \sigma^2 \tilde{\mathbb{E}}\left(\int_0^{\tau_{\varepsilon,\varepsilon'}\wedge t} \left(g^{(1)}(\delta(v))\right)^2 \delta(v)\,\mathrm{d}v\right) \ge c\sigma^2 \tilde{\mathbb{E}}\left(\tau_{\varepsilon,\varepsilon'} \wedge t\right),
\end{aligned}$$

where $c = \inf\{v(g^{(1)}(v))^2 : \varepsilon \le v \le \varepsilon'\} > 0$. The left-hand side of the above equations is bounded, uniformly with respect to t. Thus, it follows by the monotone convergence theorem, that $\tilde{\mathbb{E}}\,\tau_{\varepsilon,\varepsilon'} < \infty$ and hence $\tau_{\varepsilon,\varepsilon'} < \infty$. Since ε can be chosen arbitrarily close to $\delta(0)$, it follows that $\tau_{\varepsilon'} < \infty$. Moreover, since the martingale $\{g(\delta(\tau_{\varepsilon,\varepsilon'} \wedge t))\}$ is bounded, we have

$$g(\delta(0)) = \tilde{\mathbb{E}}\,g(\delta(\tau_{\varepsilon,\varepsilon'})) = g(\varepsilon)\tilde{\mathbb{P}}(\tau_\varepsilon < \tau_{\varepsilon'}) + g(\varepsilon')(1 - \tilde{\mathbb{P}}(\tau_\varepsilon < \tau_{\varepsilon'}))$$

and, equivalently,

$$\tilde{\mathbf{P}}(\tau_\varepsilon < \tau_{\varepsilon'}) = \frac{g(\varepsilon') - g(\delta(0))}{g(\varepsilon') - g(\varepsilon)}. \tag{13.4.13}$$

Assume now that $\sigma^2 \le 2\alpha\bar{\delta}$. Then $\lim_{\varepsilon\to 0} g(\varepsilon) = -\infty$. Thus, letting $\varepsilon \to 0$ in (13.4.13) yields $\tilde{\mathbf{P}}(\tau_0 < \tau_{\varepsilon'}) = 0$. Now letting $\varepsilon' \to \infty$ shows (a). If $\sigma^2 > 2\alpha\bar{\delta}$, we have $\lim_{\varepsilon\to 0} g(\varepsilon) > -\infty$. Letting first $\varepsilon \to 0$ and then $\varepsilon' \to \infty$ in (13.4.13) gives (b) because $g(\infty) = \infty$. □

We now show that the class of Cox–Ingersoll–Ross models with a fixed pair of parameters $\alpha, \sigma > 0$ is closed under convolution. In other words, the sum of two independent diffusion processes of the Cox–Ingersoll–Ross type with identical parameters $\alpha, \sigma > 0$ is a diffusion process of the same type. Lemma 13.4.4 is used in the proof of Theorem 13.4.3 below. We therefore formulate this closure property in terms of the changed probability measure $\tilde{\mathbf{P}}$ starting from the stochastic differential equation (13.4.10). But it is obvious that the same result is true under the original measure $\mathbf{P}$.

Let $\{\delta(t)\}$ be given by (13.4.10). Let $\{W'(t)\}$ be a standard Brownian motion independent of $\{W^*(t)\}$ and consider the Cox–Ingersoll–Ross model $\{\delta'(t)\}$ given by

$$\mathrm{d}\delta'(t) = -\alpha(\delta'(t) - \bar{\delta}')\,\mathrm{d}t + \sigma\sqrt{\delta'(t)}\,\mathrm{d}W'(t) \tag{13.4.14}$$

for some $\bar{\delta}' > 0$.

Lemma 13.4.4 *The process $\{\delta''(t)\}$ with $\delta''(t) = \delta(t) + \delta'(t)$ is a Cox–Ingersoll–Ross model fulfilling*

$$\mathrm{d}\delta''(t) = -\alpha(\delta''(t) - (\bar{\delta} + \bar{\delta}'))\,\mathrm{d}t + \sigma\sqrt{\delta''(t)}\,\mathrm{d}W''(t), \tag{13.4.15}$$

for some standard Brownian motion $\{W''(t)\}$.

Proof Adding (13.4.10) and (13.4.15) sidewise yields

$$\mathrm{d}\delta''(t) = -\alpha(\delta''(t) - (\bar{\delta} + \bar{\delta}'))\,\mathrm{d}t + \sigma\sqrt{\delta''(t)}\frac{\sqrt{\delta(t)}\,\mathrm{d}W^*(t) + \sqrt{\delta'(t)}\,\mathrm{d}W'(t)}{\sqrt{\delta''(t)}}.$$

Let $\mathrm{d}W''(t) = (\sqrt{\delta(t)}\,\mathrm{d}W^*(t) + \sqrt{\delta'(t)}\,\mathrm{d}W'(t))/\sqrt{\delta''(t)}$. It remains to show that $\{W''(t)\}$ is a standard Brownian motion. Using the representation formula

$$\begin{aligned}
(W''(t))^2 - t &= \Big(\int_0^t \sqrt{\frac{\delta(v)}{\delta''(v)}}\,\mathrm{d}W^*(v)\Big)^2 - \int_0^t \frac{\delta(v)}{\delta''(v)}\,\mathrm{d}v \\
&\quad + \Big(\int_0^t \sqrt{\frac{\delta'(v)}{\delta''(v)}}\,\mathrm{d}W'(v)\Big)^2 - \int_0^t \frac{\delta'(v)}{\delta''(v)}\,\mathrm{d}v \\
&\quad + 2\int_0^t \sqrt{\frac{\delta(v)}{\delta''(v)}}\,\mathrm{d}W^*(v) \int_0^t \sqrt{\frac{\delta'(v)}{\delta''(v)}}\,\mathrm{d}W'(v)
\end{aligned}$$

and Theorem 13.1.1b, one can prove that $\{(W''(t))^2 - t\}$ is a local martingale. Hence, the assertion follows from Theorem 13.1.4. □

We are now ready to determine the price $D(t, t_0)$ of a zero-coupon bond in the Cox–Ingersoll–Ross model.

Theorem 13.4.3 *For* $0 \le t \le t_0$,

$$D(t, t_0) = \left(\frac{2c e^{(c+\alpha)(t_0-t)/2}}{c - \alpha + e^{c(t_0-t)}(c+\alpha)}\right)^{2\alpha\bar{\delta}/\sigma^2} \exp\left(-\delta(t)\frac{2(e^{c(t_0-t)} - 1)}{c - \alpha + e^{c(t_0-t)}(c+\alpha)}\right), \tag{13.4.16}$$

where $c = \sqrt{\alpha^2 + 2\sigma^2}$.

Proof Since $\{\delta(t)\}$ is a Markov process, it is enough to consider the case $t = 0$. Write the price $D(0, t_0)$ of a zero-coupon bond as $D(0, t_0; \delta(0), \bar{\delta})$. Suppose that

$$D(0, t_0; \delta(0), \bar{\delta}) = \exp(-\bar{\delta} g_1(t_0) - \delta(0) g_2(t_0)) \tag{13.4.17}$$

for some functions g_1 and g_2 (this guess is justified by (13.4.2) and Lemma 13.4.4). Let $\zeta(x, y) = \exp(-\bar{\delta} g_1(x) - y g_2(x))$. It will be useful to find $g_1(x)$ and $g_2(x)$ such that $\zeta(x, y)$ fulfils the partial differential equation

$$\frac{\partial \zeta}{\partial x}(x, y) = \frac{\sigma^2}{2} y \frac{\partial^2 \zeta}{\partial y^2}(x, y) - \alpha(y - \bar{\delta})\frac{\partial \zeta}{\partial y}(x, y) - y\zeta(x, y). \tag{13.4.18}$$

Hence, we consider the equation

$$-\bar{\delta} g_1^{(1)}(x) - y g_2^{(1)}(x) = \frac{\sigma^2}{2} y (g_2(x))^2 + \alpha(y - \bar{\delta}) g_2(x) - y$$

and since the terms in y have to match we get the two equations

$$-\, g_2^{(1)}(x) = \frac{\sigma^2}{2}(g_2(x))^2 + \alpha g_2(x) - 1, \qquad g_1^{(1)}(x) = \alpha g_2(x). \tag{13.4.19}$$

We choose as boundary condition $\zeta(0, y) = 1$, i.e. $g_1(0) = g_2(0) = 0$. Then the solution to (13.4.19) with $g_1(0) = g_2(0) = 0$ is

$$g_1(x) = -\frac{2\alpha}{\sigma^2}\log\left(\frac{2c e^{(c+\alpha)x/2}}{c - \alpha + e^{cx}(c+\alpha)}\right), \quad g_2(x) = \frac{2(e^{cx} - 1)}{c - \alpha + e^{cx}(c+\alpha)}.$$

For this solution (g_1, g_2) to (13.4.19), consider the process $\{M(t),\ t \in [0, t_0]\}$ with

$$M(t) = \exp\left(-\int_0^t \delta(v)\, dv\right)\zeta(t_0 - t, \delta(t)).$$

Recall that

$$\delta(t) = \delta(0) - \int_0^t \alpha(\delta(v) - \bar{\delta})\,\mathrm{d}v + \int_0^t \sigma\sqrt{\delta(v)}\,\mathrm{d}W^*(v)\,.$$

Hence, from Itô's formula (13.1.12) we find

$$\begin{aligned}
M(t) &= \zeta(t_0,\delta(0)) + \int_0^t \exp\Big(-\int_0^w \delta(v)\,\mathrm{d}v\Big)\Big(-\delta(w)\zeta(t_0-w,\delta(w)) \\
&\quad - \frac{\partial\zeta}{\partial x}(t_0-w,\delta(w)) - \alpha(\delta(w)-\bar{\delta})\frac{\partial\zeta}{\partial y}(t_0-w,\delta(w)) \\
&\quad + \frac{\sigma^2}{2}\delta(w)\frac{\partial^2\zeta}{\partial y^2}(t_0-w,\delta(w))\Big)\,\mathrm{d}w \\
&\quad + \int_0^t \exp\Big(-\int_0^w \delta(v)\,\mathrm{d}v\Big)\sigma\sqrt{\delta(w)}\frac{\partial\zeta}{\partial y}(t_0-w,\delta(w))\,\mathrm{d}W^*(w) \\
&= M(0) + \sigma\int_0^t \exp\Big(-\int_0^w \delta(v)\,\mathrm{d}v\Big)\sqrt{\delta(w)}\frac{\partial\zeta}{\partial y}(t_0-w,\delta(w))\,\mathrm{d}W^*(w),
\end{aligned}$$

where we used (13.4.18) in the last equation. Thus $\{M(t)\}$ is a local martingale by Theorem 13.1.1. Note that $g_2(w) \ge 0$ for $w \ge 0$ and $\tilde{\mathbb{E}}\,\delta(w)$ is bounded for $0 \le w \le t_0$. Thus

$$\tilde{\mathbb{E}}\left(\exp\Big(-2\int_0^w \delta(v)\,\mathrm{d}v\Big)\delta(w)\Big(\frac{\partial\zeta}{\partial y}(t_0-w,\delta(w))\Big)^2\right)$$

is bounded for $0 \le w \le t_0$. Now, using Theorem 13.1.1 again, we see that $\{M(t)\}$ is a martingale. Hence,

$$\begin{aligned}
\zeta(t_0,\delta(0)) &= M(0) = \tilde{\mathbb{E}}\,M(t_0) = \tilde{\mathbb{E}}\left(\exp\Big(-\int_0^{t_0} \delta(v)\,\mathrm{d}v\Big)\zeta(0,\delta(0))\right) \\
&= D(0,t_0)
\end{aligned}$$

because $\zeta(0,\delta(0)) = 1$. This proves the theorem. □

Remark The method used in the proof of Theorem 13.4.3 can be applied to obtain the price $D(t,t_0)$ not only for a zero-coupon bond, but for any sufficiently integrable claim $g(\delta(t_0))$ such that $g:(0,\infty)\to\mathbb{R}_+$ is some appropriately smooth function. Indeed, an extension of (13.4.2) and (13.4.16) gives

$$\begin{aligned}
D(t,t_0) &= \tilde{\mathbb{E}}\left(\exp\Big(-\int_t^{t_0} \delta(v)\,\mathrm{d}v\Big)g(\delta(t_0))\;\Big|\;\delta(t)\right) \\
&= \zeta(t_0-t,\delta(t))\,,
\end{aligned}$$

where the function $\zeta(w,y)$ is the solution to the *Feynman–Kac partial differential equation* (13.4.18) with the boundary condition $\zeta(0,y) = g(y)$ for all $y > 0$.

Bibliographical Notes. The approach to the pricing of zero-coupon bonds in models with stochastic interest rates presented in Section 13.4 follows Artzner and Delbaen (1989); see also Lamberton and Lapeyre (1997). A broader introduction to stochastic interest rate models can be found in Chen (1996), Duffie (1996), Musiela and Rutkowski (1997) and Rebonato (1997), where also *multifactor models* are discussed such as models with *stochastic volatility*, and the Heath–Jarrow–Morton model of *forward rates*. For numerics and simulation of interest rate models, see Pliska and Rogers (1995). In the financial context, the Vasicek model has been introduced in Vasicek (1977). Parker (1994) studied the present value of a portfolio when the force of interest is modelled by an Ornstein–Uhlenbeck process. The Cox–Ingersoll–Ross model can be found, for instance, in Cox, Ingersoll and Ross (1985); see also Feller (1951). Recently, this model has been used in Paulsen (1997) to study the present value of an insurance portfolio.

APPENDIX

Distribution Tables

The tables on the following pages review the distributions used frequently in this book. For each of the distributions the mean, the variance and the coefficient of variation (c.v.) defined as $\sqrt{\mathrm{Var}\, X}/\mathbb{E}\, X$ are given. The probability generating function (p.g.f.) defined as $\mathbb{E}\, s^X$ for discrete distributions as well as the moment generating function (m.g.f.) defined as $\mathbb{E}\, \exp(sX)$ for absolutely continuous distributions can be found in the table, provided a closed expression for the m.g.f. can be obtained. For the lognormal, the Pareto and the Weibull distribution no closed form of the m.g.f. is available. This is indicated by the symbol $*$. For a more detailed discussion of these distributions, see Chapter 2.

Distribution	Parameters	Probability function $\{p_k\}$
$\mathrm{B}(n,p)$	$n=1,2,\ldots$ $0\le p\le 1, q=1-p$	$\binom{n}{k}p^k q^{n-k}$ $k=0,1,\ldots,n$
$\mathrm{Ber}(p)$	$0\le p\le 1, q=1-p$	$p^k q^{1-k}; k=0,1$
$\mathrm{Poi}(\lambda)$	$\lambda>0$	$\frac{\lambda^k}{k!}\mathrm{e}^{-\lambda}; k=0,1,\ldots$
$\mathrm{NB}(r,p)$	$r>0, 0<p<1$ $q=1-p$	$\frac{\Gamma(r+k)}{\Gamma(r)k!}q^r p^k$ $k=0,1,\ldots$
$\mathrm{Geo}(p)$	$0<p\le 1, q=1-p$	$qp^k; k=0,1,\ldots$
$\mathrm{Log}(p)$	$0<p<1$ $r=\frac{-1}{\log(1-p)}$	$r\frac{p^k}{k}; k=1,2,\ldots$
$\mathrm{UD}(n)$	$n=1,2,\ldots$	$\frac{1}{n}; k=1,\ldots,n$

Table 1 Discrete distributions

Mean	Variance	c.v.	p.g.f.
np	npq	$\left(\frac{q}{np}\right)^{\frac{1}{2}}$	$(ps+q)^n$
p	pq	$\left(\frac{q}{p}\right)^{\frac{1}{2}}$	$ps+q$
λ	λ	$\left(\frac{1}{\lambda}\right)^{\frac{1}{2}}$	$\exp[\lambda(s-1)]$
$\frac{rp}{q}$	$\frac{rp}{q^2}$	$(rp)^{-\frac{1}{2}}$	$q^r(1-ps)^{-r}$
$\frac{p}{q}$	$\frac{p}{q^2}$	$p^{-\frac{1}{2}}$	$q(1-ps)^{-1}$
$\frac{rp}{1-p}$	$\frac{rp(1-rp)}{(1-p)^2}$	$\left(\frac{1-rp}{rp}\right)^{\frac{1}{2}}$	$\frac{\log(1-ps)}{\log(1-p)}$
$\frac{n+1}{2}$	$\frac{n^2-1}{12}$	$\left(\frac{n-1}{3(n+1)}\right)^{\frac{1}{2}}$	$\frac{1}{n}\sum_{k=1}^{n} s^k$

Table 1 Discrete distributions (cont.)

Distribution	Parameters	Density $f(x)$
$\mathrm{U}(a,b)$	$-\infty < a < b < \infty$	$\frac{1}{b-a}; x \in (a,b)$
$\Gamma(a,\lambda)$	$a, \lambda > 0$	$\frac{\lambda^a x^{a-1} \mathrm{e}^{-\lambda x}}{\Gamma(a)}; x \geq 0$
$\mathrm{Erl}(n,\lambda)$	$n = 1, 2 \ldots, \lambda > 0$	$\frac{\lambda^n x^{n-1} \mathrm{e}^{-\lambda x}}{(n-1)!}; x \geq 0$
$\mathrm{Exp}(\lambda)$	$\lambda > 0$	$\lambda \mathrm{e}^{-\lambda x}; x \geq 0$
$\mathrm{N}(\mu, \sigma^2)$	$-\infty < \mu < \infty$, $\sigma > 0$	$\frac{\exp\left(\frac{(x-\mu)^2}{2\sigma^2}\right)}{\sqrt{2\pi}\sigma}; x \in \mathbb{R}$
$\chi^2(n)$	$n = 1, 2, \ldots$	$\left(2^{n/2}\Gamma(n/2)\right)^{-1} x^{\frac{n}{2}-1} \mathrm{e}^{-\frac{x}{2}}; x \geq 0$
$\mathrm{LN}(a,b)$	$-\infty < a < \infty, b > 0$ $\omega = \exp(b^2)$	$\frac{\exp\left(-\frac{(\log x - a)^2)}{2b^2}\right)}{xb\sqrt{2\pi}}; x \geq 0$

Table 2 Absolutely continuous distributions

Mean	Variance	c.v.	m.g.f.
$\frac{a+b}{2}$	$\frac{(b-a)^2}{12}$	$\frac{b-a}{\sqrt{3}(b+a)}$	$\frac{e^{bs}-e^{as}}{s(b-a)}$
$\frac{a}{\lambda}$	$\frac{a}{\lambda^2}$	$a^{-\frac{1}{2}}$	$\left(\frac{\lambda}{\lambda-s}\right)^a; s<\lambda$
$\frac{n}{\lambda}$	$\frac{n}{\lambda^2}$	$n^{-\frac{1}{2}}$	$\left(\frac{\lambda}{\lambda-s}\right)^n; s<\lambda$
$\frac{1}{\lambda}$	$\frac{1}{\lambda^2}$	1	$\frac{\lambda}{\lambda-s}; s<\lambda$
μ	σ^2	$\frac{\sigma}{\mu}$	$e^{\mu s+\frac{1}{2}\sigma^2 s^2}$
n	$2n$	$\sqrt{\frac{2}{n}}$	$(1-2s)^{-n/2}; s<1/2$
$\exp\left(a+\frac{b^2}{2}\right)$	$e^{2a}\omega(\omega-1)$	$(\omega-1)^{\frac{1}{2}}$	*

Table 2 Absolutely continuous distributions (cont.)

Distribution	Parameters	Density $f(x)$
$\mathrm{Par}(\alpha, c)$	$\alpha, c > 0$	$\frac{\alpha}{c}\left(\frac{c}{x}\right)^{\alpha+1}; x > c$
$\mathrm{W}(r, c)$	$r, c > 0$ $\omega_n = \Gamma\left(\frac{r+n}{r}\right)$	$rcx^{r-1}\mathrm{e}^{-cx^r}; x \geq 0$
$\mathrm{IG}(\mu, \lambda)$	$\mu > 0, \lambda > 0$	$\left[\frac{\lambda}{2\pi x^3}\right]^{\frac{1}{2}} \exp\left(\frac{-\lambda(x-\mu)^2}{2\mu^2 x}\right); x > 0$
$\mathrm{PME}(\alpha)$	$\alpha > 1$	$\int_{\frac{\alpha-1}{\alpha}}^{\infty} \alpha\left(\frac{\alpha-1}{\alpha}\right)^{\alpha} y^{-(\alpha+2)}\mathrm{e}^{-x/y}\mathrm{d}y$ $x > 0$

Table 2 Absolutely continuous distributions (cont.)

Mean	Variance	c.v.	m.g.f.
$\frac{\alpha c}{\alpha - 1}$ $\alpha > 1$	$\frac{\alpha c^2}{(\alpha-1)^2(\alpha-2)}$ $\alpha > 2$	$[\alpha(\alpha-2)]^{-\frac{1}{2}}$ $\alpha > 2$	*
$\omega_1 c^{-1/r}$	$(\omega_1 - \omega_2^2)c^{-2/r}$	$\left[\frac{\omega_2}{\omega_1^2} - 1\right]^{\frac{1}{2}}$	*
μ	$\frac{\mu^3}{\lambda}$	$\left(\frac{\mu}{\lambda}\right)^{\frac{1}{2}}$	$\frac{\exp\left(\frac{\lambda}{\mu}\right)}{\exp\left(\sqrt{\frac{\lambda^2}{\mu^2} - 2\lambda s}\right)}$
1	$1 + \frac{2}{\alpha(\alpha-2)}$	$\sqrt{1 + \frac{2}{\alpha(\alpha-2)}}$	$\frac{\int_0^{\frac{\alpha}{\alpha-1}} \frac{x^\alpha}{x-s} \mathrm{d}x}{\frac{\alpha^{\alpha-1}}{(\alpha-1)^\alpha}}$ $s < 0$

Table 2 Absolutely continuous distributions (cont.)

References

Aase K.K. (1988). Contingent claim valuation when the security price is a combination of an Itô process and a random point process. *Stochastic Processes and their Applications* **28**, 185–220.

Aase K.K. and Persson S.A. (1994). Pricing of unit-linked life insurance policies. *Scandinavian Actuarial Journal*, 26–52.

Abate J., Choudhury G.L. and Whitt W. (1994). Waiting-time probabilities in queues with long-tail service-time distributions. *Queueing Systems* **16**, 311–338.

Abate J. and Whitt W. (1992). The Fourier-series method for inverting transforms of probability distributions. *Queueing Systems* **10**, 5–88.

Abate J. and Whitt W. (1995). Numerical inversion of Laplace transforms of probability distributions. *ORSA Journal on Computing* **7**, 36–43.

Abramowitz M. and Stegun I.A. (1965). *Handbook of Mathematical Functions with Formulas, Graphs and Mathematical Tables*. Dover Publications, New York.

Adelson R. (1966). Compound Poisson distributions. *Operations Research Quarterly* **17**, 73–75.

Adler R., Feldman R. and Taqqu M.S. (1997). *A Practical Guide to Heavy Tails*. Birkhäuser, Boston.

Albrecht P. (1984). Laplace transforms, Mellin transforms and mixed Poisson processes. *Scandinavian Actuarial Journal*, 58–64.

Albrecht P. (1992). Premium calculation without arbitrage? A note on a contribution by G. Venter. *ASTIN Bulletin* **22**, 254–283.

Ammeter H. (1948). A generalization of the collective theory of risk in regard to fluctuating basic probabilities. *Skandinavisk Aktuarietidskrift* **31**, 171–198.

Anderson T.W. (1960). A modification of the sequential probability ratio test to reduce the sample size. *Annals of Mathematical Statistics* **31**, 165–197.

Andersson H. (1971). An analysis of the development of the fire losses in the northern countries after the second world war. *ASTIN Bulletin* **6**, 25–30.

Arfwedson G. (1950). Some problems in the collective theory of risk. *Skandinavisk Aktuarietidskrift* **33**, 1–38.

Arfwedson G. (1955). Research in collective risk theory. Part 2. *Skandinavisk Aktuarietidskrift* **38**, 53–100.

Arjas E. (1989). The claims reserving problem in non-life insurance — some structural ideas. *ASTIN Bulletin* **19**, 139–152.

Artzner P. and Delbaen F. (1989). Term structure of interest rates: the martingale approach. *Advances in Applied Mathematics* **10**, 95–129.

Asmussen S. (1982). Conditioned limit theorems relating a random walk to its associate, with applications to risk reserve process and the GI/G/1 queue. *Advances in Applied Probability* **14**, 143–170.

Asmussen S. (1984). Approximations for the probability of ruin within finite time. *Scandinavian Actuarial Journal*, 31–57.
Asmussen S. (1987). *Applied Probability and Queues*. J. Wiley & Sons, New York.
Asmussen S. (1989). Risk theory in a Markovian environment. *Scandinavian Actuarial Journal*, 66–100.
Asmussen S. and Binswanger K. (1997). Simulation of ruin probabilities for subexponential claims. *ASTIN Bulletin* **27**, 297–318.
Asmussen S. and Bladt M. (1996). Phase-type distributions and risk processes with state-dependent premiums. *Scandinavian Actuarial Journal*, 19–36.
Asmussen S., Frey A., Rolski T. and Schmidt V. (1995). Does Markov-modulation increase the risk? *ASTIN Bulletin* **25**, 49–66.
Asmussen S., Henriksen L.F. and Klüppelberg C. (1994). Large claims approximations for risk processes in a Markovian environment. *Stochastic Processes and their Applications* **54**, 29–43.
Asmussen S. and Kella O. (1996). Rate modulation in dams and ruin problems. *Journal of Applied Probability* **33**, 523–535.
Asmussen S. and Klüppelberg C. (1996). Large deviation results for subexponential tails, with applications to insurance risk. *Stochastic Processes and their Applications* **64**, 103–125.
Asmussen S., Klüppelberg C. and Sigman K. (1998). Sampling at subexponential times, with queueing applications. *Preprint*, University of Lund.
Asmussen S. and Nielsen H.M. (1995). Ruin probabilites via local adjustment coefficients. *Journal of Applied Probability* **32**, 736–755.
Asmussen S. and Petersen S.S. (1988). Ruin probabilities expressed in terms of storage processes. *Advances in Applied Probability* **20**, 913–916.
Asmussen S. and Rolski T. (1991). Computational methods in risk theory: a matrix-algorithmic approach. *Insurance: Mathematics and Economics* **10**, 259–274.
Asmussen S. and Rolski T. (1994). Risk theory in a periodic environment: the Cramér–Lundberg approximation and Lundberg inequality. *Mathematics of Operations Research* **19**, 410–433.
Asmussen S. and Rubinstein R.Y. (1995). Steady-state rare events simulation in queueing models and its complexity properties. In: Dshalalov J.H. (ed.), *Advances in Queueing*, CRC Press, Boca Raton, 429–461.
Asmussen S., Schmidli H. and Schmidt V. (1999). Tail probabilities for non-standard risk and queueing processes with subexponential jumps. *Advances in Applied Probability* **31**, to appear.
Asmussen S. and Schmidt V. (1993). The ascending ladder height distribution for a certain class of dependent random walks. *Statistica Neerlandica* **47**, 269–277.
Asmussen S. and Schmidt V. (1995). Ladder height distributions with marks. *Stochastic Processes and their Applications* **58**, 105–119.
Athreya K.B. and Ney P.E. (1972). *Branching Processes*. Springer-Verlag, Berlin.
Aven T. and Jensen U. (1997). Information based hazard rates for ruin times of risk processes. *Preprint*, Universität Ulm.
Aven T. and Jensen U. (1998). *Stochastic Models for Reliability*. Springer-Verlag, New York.
Baccelli F. and Brémaud P. (1994). *Elements of Queueing Theory, Palm-Martingale Calculus and Stochastic Recurrences*. Springer-Verlag, Berlin.
Bachelier L. (1900). Théorie de la spéculation. *Annales Scientifiques de l'École Normale Supérieure* **17**, 21–86.
Bakshi G., Cao C. and Chen Z. (1997). Empirical performance of alternative option pricing models. *Journal of Finance* **52**, 2003–2049.

Barbour A.D., Chen L.H.Y. and Loh W.L. (1992). Compound Poisson approximation for nonnegative random variables via Stein's method. *Annals of Probability* **20**, 1843–1866.

Barbour A.D. and Hall P. (1984). On the rate of Poisson convergence. *Mathematical Proceedings of the Cambridge Philosophical Society* **95**, 473–480.

Barbour A.D., Holst L. and Janson S. (1992). *Poisson Approximation.* Oxford University Press, New York.

Bardhan I. and Chao X. (1993). Pricing options on securities with discontinuous returns. *Stochastic Processes and their Applications* **48**, 123–137.

Barlow R.E. and Proschan F. (1965). *Mathematical Theory of Reliability.* J. Wiley & Sons, New York.

Barlow R.E. and Proschan F. (1975). *Statistical Theory of Reliability and Life Testing: Probability Models.* Holt, Rinehart & Winston, New York.

Barndorff-Nielson O. (1997). Normal inverse Gaussian distributions and stochastic volatility modelling. *Scandinavian Journal of Statistics* **24**, 1–13.

Barndorff-Nielson O. (1998). Processes of normal inverse Gaussian type. *Finance & Stochastics* **2**, 41–68.

Bäuerle N. (1997). Some results about the expected ruin time in Markov-modulated risk models. *Insurance: Mathematics and Economics* **18**, 119–127.

Beard R.E., Pentikäinen T. and Pesonen E. (1984). *Risk Theory.* 3rd ed., Chapman & Hall, London.

Beekman J. (1969). A ruin function approximation. *Transactions of the Society of Actuaries* **21**, 41–48 and 275–279.

Beirlant J. and Teugels J.L. (1992). Modeling large claims in non-life insurance. *Insurance: Mathematics and Economics* **11**, 17–30.

Beirlant J. and Teugels J.L. (1996). A simple approach to classical extreme value theory. In: *Exploring Stochastic Laws, Festschrift for Academician V.S. Korolyuk*, Zeist, 457–468.

Benckert L.G. and Jung L. (1974). Statistical models of claim distribution in fire insurance. *ASTIN Bulletin* **8**, 9–23.

Benckert L.G. and Sternberg I. (1958). An attempt to find an expression for the distribution of fire damage account. In: *Transactions 15-th International Congress of Actuaries*, vol. II.

Benjamin B. and Pollard J.H. (1993). *The Analysis of Mortality and Other Actuarial Statistics.* Institute of Actuaries, The Chameleon Press Ltd., London.

Benktander G. (1963). A note on the most "dangerous" and the skewest class of distributions. *ASTIN Bulletin* **2**, 387–390.

Benktander G. and Segerdahl C.O. (1960). On the analytical representation of claim distributions with special reference to excess of loss reinsurance. In: *Transactions 16-th International Congress of Actuaries*, 626–646.

Berger M.A. (1993). *An Introduction to Probability and Stochastic Processes.* Springer-Verlag, New York.

Bergmann R. and Stoyan D. (1976). On exponential bounds for the waiting time distribution function in GI/G/1. *Journal of Applied Probability* **13**, 411–417.

Bharucha-Reid A.T. (1960). *Elements of the Theory of Markov Processes and Their Applications.* McGraw Hill, New York.

Billingsley P. (1995). *Probability and Measure.* 3rd ed., J. Wiley & Sons, New York.

Bingham N. (1975). Fluctuation theory in continuous time. *Advances in Applied Probability* **7**, 705–766.

Bingham N.H., Goldie C.M. and Teugels J.L. (1987). *Regular Variation.* Cambridge University Press, Cambridge.

Bingham N.H. and Kiesel R. (1998). *Risk-Neutral Valuation: Pricing and Hedging of Financial Derivatives*. Springer-Verlag, Berlin.

Björk T. and Grandell J. (1988). Exponential inequalities for ruin probabilities in the Cox case. *Scandinavian Actuarial Journal*, 77–111.

Black F. and Scholes M. (1973). The pricing of options and corporate liabilities. *Journal of Political Economy* **81**, 637–654.

Bladt M. and Rydberg T. (1998). An actuarial approach to option pricing under the physical measure and without market assumptions. *Insurance: Mathematics and Economics* **22**, 65–73.

Bohman H. (1975). Numerical inversions of characteristics functions. *Scandinavian Actuarial Journal*, 121–124.

Bohman H. and Esscher F. (1963). Studies in risk theory with numerical illustrations concerning distribution function and stop loss premiums. *Skandinavisk Aktuarietidskrift* **46**, 173–225.

Bonsdorff H. (1992). On the convergence rate of bonus-malus systems. *ASTIN Bulletin* **22**, 217–223.

Boogaert P. and Crijns V. (1987). Upper bounds on ruin probabilities in case of negative loadings and positive interest rates. *Insurance: Mathematics and Economics* **6**, 221–232.

Boogaert P., Delbaen F. and Haezendonck J. (1988). Macro-economic influences on the crossing of dividend barriers. *Scandinavian Actuarial Journal*, 231–245.

Boogaert P., Haezendonck J. and Delbaen F. (1988). Limit theorems for the present value of the surplus of an insurance portfolio. *Insurance: Mathematics and Economics* **7**, 131–138.

Boos A. (1991). *Effizienz von Bonus-Malus-Systemen*. Gabler-Verlag, Wiesbaden.

Borch K. (1960). An attempt to determine the optimal amount of stop loss insurance. In: *Transactions of the 16-th International Congress of Actuaries*, vol. I, 597–610.

Borgan O., Hoem J.M. and Norberg R. (1981). A nonasymptotic criterion for the evaluation of automobile bonus systems. *Scandinavian Actuarial Journal* , 165–178.

Borovkov A.A. (1976). *Stochastic Processes in Queueing Theory*. Springer-Verlag, New York.

Bowers N., Gerber H., Hickman J., Jones D. and Nesbitt C. (1986). *Actuarial Mathematics*. Society of Actuaries, Schaumburg.

Boyce W.E. and Di Prima A.B. (1969). *Elementary Differential Equations*. J. Wiley & Sons, New York.

Breiman L. (1992). *Probability*. SIAM, Philadelphia.

Brémaud P. (1981). *Point Processes and Queues*. Springer-Verlag, Berlin.

Brémaud P. (1988). *An Introduction to Probabilistic Modeling*. Springer-Verlag, New York.

Buchwalder M., Chevallier E. and Klüppelberg C. (1993). Approximation methods for the total claimsize distribution — An algorithmic and graphical presentation. *Mitteilungen der Schweizerischen Vereinigung der Versicherungsmathematiker*, 187–227.

Bühlmann H. (1970). *Mathematical Methods in Risk Theory*. Springer-Verlag, New York.

Bühlmann H. (1980). An economic premium principle. *ASTIN Bulletin* **11**, 52–60.

Bühlmann H. (1984). Numerical evaluation of the compound. *Scandinavian Actuarial Journal*, 116–126.

Centeno M.L. (1997). Excess of loss reinsurance and the probability of ruin in finite horizon. *ASTIN Bulletin* **27**, 59–70.

Chatelin F. (1993). *Eigenvalues of Matrices*. J. Wiley & Sons, Chichester.
Chen L. (1996). *Interest Rate Dynamics, Derivative Pricing and Risk Management*. Springer-Verlag, Berlin.
Chistyakov V.P. (1964). A theorem on sums of independent positive random variables and its applications to branching random processes. *Theory of Probability and its Applications* **9**, 640–648.
Choudhury G.L., Lucantoni D.M. and Whitt W. (1994). Multidimensional transform inversion with applications to the transient M/G/1 queue. *Annals of Applied Probability* **4**, 719–740.
Chover J., Ney P.E. and Wainger S. (1973). Functions of probability measures. *Journal d'Analyse Mathématique* **26**, 255–302.
Christ R. and Steinebach J. (1995). Estimating the adjustment coefficient in an ARMA(p,q) risk model. *Scandinavian Actuarial Journal* , 149–161.
Chung K.L. (1967). *Markov Chains with Stationary Probabilities*. Springer-Verlag, New York.
Chung K.L. (1974). *A Course in Probability Theory*. 2nd ed., Academic Press, New York.
Chung K.L. (1982). *Lectures from Markov Processes to Brownian Motion*. Springer-Verlag, New York.
Çinlar E. (1975). *Introduction to Stochastic Processes*. Prentice-Hall, Englewood Cliffs, N.J.
Cline D.B.H. (1987). Convolutions of distributions with exponential and subexponential tails. *Journal of the Australian Mathematical Society. Series A* **43**, 347–365.
Cline D.B.H. and Resnick S. (1988). Distributions that are both subexponential and in the domain of attraction of an extreme-value distribution. *Advances in Applied Probability* **20**, 706–718.
Cohen J.W. (1973). Some results on regular variation in queueing and fluctuation theory. *Journal of Applied Probability* **10**, 343–353.
Colwell D. and Elliott R. (1993). Discontinuous asset prices and non-attainable contingent claims. *Mathematical Finance* **3**, 295–308.
Conti B. (1992). *Familien von Verteilungsfunktionen zur Beschreibung von Großschäden: Anforderungen eines Praktikers*. Winterthur-Versicherungen, Winterthur.
Cox J.C., Ingersoll J.E. and Ross S.A. (1985). A theory of the term structure of interest rates. *Econometrica* **51**, 385–407.
Cramér H. (1930). *On the Mathematical Theory of Risk*. Skandia Jubilee Volume, Stockholm.
Cramér H. (1955). *Collective Risk Theory*. Skandia Jubilee Volume, Stockholm.
Crane M.A. and Lemoine A.J. (1977). *An Introduction to the Regenerative Method for Simulation Analysis*. Springer-Verlag, New York.
Csörgő M. and Steinebach J. (1991). On the estimation of the adjustment coefficient in risk theory via intermediate order statistics. *Insurance: Mathematics and Economics* **10**, 37–50.
Csörgő S. and Teugels J.L. (1990). Empirical Laplace transform and approximation of compound distributions. *Journal of Applied Probability* **27**, 88–101.
Daduna H. and Szekli R. (1996). A queueing theoretical proof of an increasing property of Pólya frequency functions. *Statistics & Probability Letters* **26**, 233–242.
d'Agostino and Stephens M. (1986). *Goodness-of-fit Techniques*. Marcel Dekker, New York.

Daley D.J., Foley R.D. and Rolski T. (1994). Conditions for finite moments of waiting times in G/G/1 queues. *Queueing Systems* **17**, 89–106.

Daley D.J. and Rolski T. (1984). Some comparability results for certain tandem and multiple-server queues. *Journal of Applied Probability* **21**, 887–900.

Daley D.J. and Rolski T. (1992). Finiteness of waiting-time moments in general stationary single-server queues. *Annals of Applied Probability* **2**, 987–1008.

Daley D.J. and Vere-Jones D. (1988). *An Introduction to the Theory of Point Processes*. Springer-Verlag, New York.

Dassios A. (1987). *Insurance, Storage and Point Processes: An Approach via Piecewise Deterministic Markov Processes*. Ph.D. thesis, Imperial College, London.

Dassios A. and Embrechts P. (1989). Martingales and insurance risk. *Communications in Statistics – Stochastic Models* **5**, 181–217.

Davidson A. (1969). On the ruin problem in the collective theory of risk under the assumption of variable safety loading. *Scandinavian Actuarial Journal, Supplement* , 70–83.

Davis M.H.A. (1984). Piecewise-deterministic Markov processes: a general class of non-diffusion stochastic models. *Journal of the Royal Statistical Society. Series B.* **46**, 353–388.

Davis M.H.A. (1993). *Markov Models and Optimization*. Chapman & Hall, London.

Davis M.H.A. and Vellekoop M.H. (1995). Permanent health insurance: a case study in piecewise-deterministic Markov modelling. *Mitteilungen der Schweizerischen Vereinigung der Versicherungsmathematiker*, 177–212.

Daykin C.D., Pentikäinen T. and Pesonen M. (1994). *Practical Risk Theory for Actuaries*. Chapman & Hall, London.

De Haan L. (1970). On regular variation and its application to the weak convergence of sample extremes. *Mathematical Centre Tracts* **32**, Amsterdam.

De Pril N. (1986). On the exact computation of the aggregate claims distribution in the individual life model. *ASTIN Bulletin* **16**, 109–112.

De Pril N. (1988). Improved approximations for the aggregate claims distribution of a life insurance portfolio. *Scandinavian Actuarial Journal* , 61–68.

De Pril N. (1989). The aggregate claims distribution in the individual life model with arbitrary positive claims. *ASTIN Bulletin* **19**, 9–24.

De Pril N. and Dhaene J. (1992). Error bounds for compound Poisson approximations of the individual risk model. *ASTIN Bulletin* **22**, 135–148.

De Vylder F. (1977). Martingales and ruin in a dynamical risk process. *Scandinavian Actuarial Journal*, 217–225.

De Vylder F. (1978). A practical solution to the problem of ultimate ruin probability. *Scandinavian Actuarial Journal*, 114–119.

De Vylder F. and Goovaerts M.J. (1988). Recursive calculation of finite-time ruin probabilities. *Insurance: Mathematics and Economics* **7**, 1–7.

Deheuvels P. and Steinebach J. (1990). On some alternative estimates of the adjustment coefficient. *Scandinavian Actuarial Journal* , 135–159.

Delaporte P.J. (1960). Un problème de tarification de l'assurance accidents d'automobiles examiné par la statistique mathématique. In: *Transactions of the 16-th International Congress of Actuaries*, vol. II, 121–135.

Delaporte P.J. (1965). Tarification du risque individuel d'accidents d'automobiles par la prime modelé sur le risque. *ASTIN Bulletin* **3**, 251–271.

Delbaen F. (1990). Equity linked policies. *Bulletin Association Royal Actuaires Belges*, 33–52.

Delbaen F. and Haezendonck J. (1985). Inversed martingales in risk theory. *Insur-*

ance: Mathematics and Economics **4**, 201–206.
Delbaen F. and Haezendonck J. (1986). Martingales in Markov processes applied to risk theory. *Insurance: Mathematics and Economics* **5**, 201–215.
Delbaen F. and Haezendonck J. (1987). Classical risk theory in an economic environment. *Insurance: Mathematics and Economics* **6**, 85–116.
Dellacherie C. (1972). *Capacités et Processus Stochastiques*. Springer-Verlag, Berlin.
Dellacherie C. and Meyer P.A. (1982). *Probabilities and Potential*. North-Holland, Amsterdam.
Dellaert N.P., Frenk J.B.G. and van Rijsoort L.P. (1993). Optimal claim behaviour for vehicle damage insurance. *Insurance: Mathematics and Economics* **12**, 225–244.
den Iseger P.W., Smith M.A.J. and Dekker R. (1997). Computing compound distributions faster! *Insurance: Mathematics and Economics* **20**, 23–34.
Denneberg D. (1990). Premium calculation: why standard deviation should be represented by absolute deviation. *ASTIN Bulletin* **20**, 181–190.
Dhaene J. and De Pril N. (1994). On the class of approximative computation methods in the individual risk model. *Insurance: Mathematics and Economics* **14**, 181–196.
Dhaene J. and Goovaerts M.J. (1996). Dependency of risks and stop-loss order. *ASTIN Bulletin* **26**, 201–212.
Dhaene J. and Goovaerts M.J. (1997). On the dependency of risks in the individual life model. *Insurance: Mathematics and Economics* **19**, 243–253.
Dhaene J. and Sundt B. (1997). On error bounds for approximations to aggregate claims distribution. *ASTIN Bulletin* **27**, 243–262.
Dhaene J. and Vandebroek M. (1995). Recursions for the individual model. *Insurance: Mathematics and Economics* **16**, 31–38.
Dhaene J., Willmot G. and Sundt B. (1996). Recursions for distribution functions and stop-loss transforms. In: *Proceedings of the 27-th ASTIN Colloquium, Copenhagen*, vol. 2, 497–515.
di Lorenzo E. and Sibillo M. (1994). Some results on a model in risk theory with constant dividend barrier. *Mitteilungen der Schweizerischen Vereinigung der Versicherungsmathematiker*, 209–218.
Dickson D.C.M. (1991). The probability of ultimate ruin with a variable premium loading — special case. *Scandinavian Actuarial Journal*, 75–86.
Dickson D.C.M. (1992). On the distribution of the surplus prior to ruin. *Insurance: Mathematics and Economics* **11**, 191–207.
Dickson D.C.M. (1995). A review of Panjer's recursion formula and its applications. *British Actuarial Journal* **1**, 107–124.
Dickson D.C.M. and dos Reis A.D.E. (1996). On the distribution of the duration of negative surplus. *Scandinavian Actuarial Journal*, 148–164.
Dickson D.C.M. and dos Reis A.D.E. (1997). The effect of interest on negative surplus. *Insurance: Mathematics and Economics* **12**, 23–38.
Dickson D.C.M., dos Reis A.D.E. and Waters H.R. (1995). Some stable algorithms in ruin theory and their applications. *ASTIN Bulletin* **25**, 153–175.
Dickson D.C.M. and Gray J.R. (1984). Exact solutions for the ruin probability in the presence of an absorbing upper barrier. *Scandinavian Actuarial Journal* , 174–186.
Dickson D.C.M. and Waters H.R. (1991). Recursive calculation of survival probabilities. *ASTIN Bulletin* **21**, 199–221.
Dickson D.C.M. and Waters H.R. (1992). The probability and severity of ruin in finite and infinite time. *ASTIN Bulletin* **22**, 177–190.
Dickson D.C.M. and Waters H.R. (1993). Gamma processes and finite time survival

probabilities. *ASTIN Bulletin* **23**, 259–272.
Dickson D.C.M. and Waters H.R. (1997). Relative reinsurance retention levels. *ASTIN Bulletin* **27**, 207–227.
Dobrushin R.L. (1953). Generalization of Kolmogorov's equations for Markov processes with a finite number of possible states. *Matematicheskiĭ Sbornik. Novaya Seriya.* **33 (75)**, 567–596 (in Russian).
Doetsch G. (1950). *Handbuch der Laplace-Transformation I*. Birkhäuser, Basel.
Doob J.L. (1953). *Stochastic Processes*. J. Wiley & Sons, New York.
dos Reis A.D.E. (1993). How long is the surplus below zero? *Insurance: Mathematics and Economics* **12**, 23–38.
Duffie D. (1996). *Dynamic Asset Pricing Theory*. 2nd ed., Princeton University Press, Princeton.
Dufresne F. (1984). Distributions stationaires d'un système bonus-malus probabilité de ruine. *ASTIN Bulletin* **18**, 31–46.
Dufresne F. (1996). An extension of Kornya's method with application to pension funds. *Mitteilungen der Schweizerischen Vereinigung der Versicherungsmathematiker*, 171–181.
Dufresne F. and Gerber H.U. (1988). The surpluses immediately before and at ruin, and the amount of the claim causing ruin. *Insurance: Mathematics and Economics* **7**, 193–199.
Dufresne F. and Gerber H.U. (1991). Risk theory for the compound Poisson process that is perturbed by diffusion. *Insurance: Mathematics and Economics* **10**, 51–59.
Dufresne F. and Gerber H.U. (1993). The probability of ruin for the inverse Gaussian and related processes. *Insurance: Mathematics and Economics* **12**, 9–22.
Dufresne F., Gerber H.U. and Shiu E. (1991). Risk theory with the Gamma process. *ASTIN Bulletin* **21**, 177–192.
Eberlein E. and Keller M. (1995). Hyperbolic distributions in finance. *Bernoulli* **1**, 281–299.
Efron B. (1965). Increasing properties of Pólya frequency functions. *Annals of Mathematical Statistics* **36**, 272–279.
Einstein A. (1905). Die von der molekularkinetischen Theorie der Wärme geforderte Bewegung von in ruhenden Flüssigkeiten suspendierten Teilchen. *Annalen der Physik* **17**, 549–560.
El Karoui N. and Quenez M.C. (1995). Dynamic programming and pricing of contingent claims in an incomplete market. *SIAM Journal of Control and Optimization* **33**, 29–66.
Elliot R. and Kopp P. (1990). Option pricing and hedge portfolios for Poisson processes. *Stochastic Analysis and Applications* **8**, 157–167.
Elliot R.J. (1982). *Stochastic Calculus and Applications*. Springer-Verlag, New York.
Embrechts P., Goldie C.M. and Veraverbeke N. (1979). Subexponentiality and infinite divisibility. *Zeitschrift für Wahrscheinlichkeitstheorie und Verwandte Gebiete* **49**, 335–347.
Embrechts P., Grandell J. and Schmidli H. (1993). Finite-time Lundberg inequalities in the Cox case. *Scandinavian Actuarial Journal* , 17–41.
Embrechts P., Grübel R. and Pitts S. (1994). Some applications of the fast Fourier transform algorithm in insurance mathematics. *Statistica Neerlandica* **47**, 59–75.
Embrechts P. and Klüppelberg C. (1994). Some aspects of insurance mathematics. *Theory of Probability and its Applications* **38**, 262–295.
Embrechts P., Klüppelberg C. and Mikosch T. (1997). *Modelling Extremal Events for Insurance and Finance*. Springer-Verlag, Heidelberg.
Embrechts P., Maejima M. and Omey E. (1984). A renewal theorem of Blackwell

type. *Journal of Applied Probability* **12**, 561–570.
Embrechts P., Maejima M. and Teugels J. (1985). Asymptotic behaviour of compound distributions. *ASTIN Bulletin* **15**, 45–48.
Embrechts P. and Mikosch T. (1991). A bootstrap procedure for estimating the adjustment coefficient. *Insurance: Mathematics and Economics* **10**, 181–190.
Embrechts P. and Schmidli H. (1994). Ruin estimation for a general insurance risk model. *Advances in Applied Probability* **26**, 404–422.
Embrechts P. and Veraverbeke N. (1982). Estimates for the probability of ruin with special emphasis on the possibility of large claims. *Insurance: Mathematics and Economics* **1**, 55–72.
Embrechts P. and Villaseñor J. (1988). Ruin estimates for large claims. *Insurance: Mathematics and Economics* **7**, 269–274.
Esscher F. (1932). On the probability function in the collective theory of risk. *Skandinavisk Aktuarietidskrift* **15**, 175–195.
Ethier S.N. and Kurtz T.G. (1986). *Markov Processes*. J. Wiley & Sons, New York.
Feller W. (1951). Two singular diffusion problems. *Annals of Mathematics* **54**, 173–182.
Feller W. (1968). *An Introduction to Probability Theory and its Applications*, vol. I. 3rd ed., J. Wiley & Sons, New York.
Feller W. (1971). *An Introduction to Probability Theory and its Applications*, vol. II. 2nd ed., J. Wiley & Sons, New York.
Fishman G.S. (1996). *Monte-Carlo*. Springer-Verlag, New York.
Föllmer H. and Schweizer M. (1991). Hedging of contingent claims under incomplete information. In: Davies M. and Elliot R. (eds.), *Applied Stochastic Analysis*, Gordon and Breach, New York, 389–414.
Forsyth A.R. (1906). *Theory of Differential Equations*, vol. V. Cambridge University Press, Cambridge.
Franken P., König D., Arndt U. and Schmidt V. (1982). *Queues and Point Processes*. J. Wiley & Sons, Chichester.
Frey A. and Schmidt V. (1996). Taylor-series expansion for multivariate characteristics of classical risk processes. *Insurance: Mathematics and Economics* **18**, 1–12.
Furrer H.J. (1998). Risk processes perturbed by α-stable Lévy motion. *Scandinavian Actuarial Journal*, 59–74.
Furrer H.J. and Schmidli H. (1994). Exponential inequalities for ruin probabilities of risk processes perturbed by diffusion. *Insurance: Mathematics and Economics* **15**, 23–26.
Gänssler P. and Stute W. (1977). *Wahrscheinlichkeitstheorie*. Springer-Verlag, Berlin.
Gerber H.U. (1970). An extension of the renewal equation and its application in the collective theory of risk. *Skandinavisk Aktuarietidskrift* **53**, 205–210.
Gerber H.U. (1971). Der Einfluss von Zins auf die Ruinwahrscheinlichkeit. *Mitteilungen der Schweizerischen Vereinigung der Versicherungsmathematiker*, 63–70.
Gerber H.U. (1973). Martingales in risk theory. *Mitteilungen der Schweizerischen Vereinigung der Versicherungsmathematiker*, 205–216.
Gerber H.U. (1975). The surplus process as a fair game — utilitywise. *ASTIN Bulletin* **8**, 307–322.
Gerber H.U. (1979). *An Introduction to Mathematical Risk Theory*. S.S. Huebner Foundation, Wharton School, Philadelphia.
Gerber H.U. (1981). On the probability of ruin in the presence of a linear dividend barrier. *Scandinavian Actuarial Journal*, 105–115.

Gerber H.U. (1982). On the numerical evaluation of the distribution of aggregate claims and its stop-loss premiums. *Insurance: Mathematics and Economics* **1**, 13–18.

Gerber H.U. (1984). Error bounds for the compound Poisson approximation. *Insurance: Mathematics and Economics* **3**, 191–194.

Gerber H.U. (1988). Mathematical fun with ruin theory. *Insurance: Mathematics and Economics* **7**, 15–23.

Gerber H.U. (1994). Martingales and tail probabilities. *ASTIN Bulletin* **24**, 145–146.

Gerber H.U. (1995). *Life Insurance Mathematics*. Springer-Verlag, Zürich.

Gerber H.U., Goovaerts M.J. and Kaas R. (1987). On the probability and severity of ruin. *ASTIN Bulletin* **17**, 152–163.

Gerber H.U. and Shiu E.S.W. (1996). Actuarial bridges to dynamic hedging and option pricing. *Insurance: Mathematics and Economics* **17**, 152–163.

Gerber H.U. and Shiu E.S.W. (1997). The joint distribution of the time of ruin, the surplus immediately before ruin, and the deficit at ruin. *Insurance: Mathematics and Economics* **21**, 129–137.

Gill R.D. and Johansen S. (1990). A survey of product-integration with a view toward application in survival analysis. *Annals of Statistics* **18**, 1501–1555.

Gjessing H. and Paulsen J. (1997). Present value distributions with applications to ruin theory and stochastic equations. *Stochastic Processes and their Applications* **71**, 123–144.

Glynn P. and Iglehart D.I. (1989). Importance sampling for stochastic simulations. *Management Sciences* **35**, 1367–1392.

Gnedenko B.V. and Kovalenko I.N. (1989). *Introduction to Queueing Theory*. 2nd ed., Birkhäuser, Boston.

Goovaerts M.J., De Vylder F. and Haezendonck J. (1982). Ordering of risks: a review. *Insurance: Mathematics and Economics* **1**, 131–163.

Goovaerts M.J., De Vylder F. and Haezendonck J. (1984). *Insurance Premiums*. North-Holland, Amsterdam.

Goovaerts M.J., Kaas R., van Heerwarden A.E. and Bauwelinckx T. (1990). *Effective Actuarial Methods*. North-Holland, Amsterdam.

Graham A. (1981). *Kronecker Products and Matrix Calculus with Applications*. Ellis Horwood, Chichester.

Grandell J. (1991a). Finite time ruin probabilities and martingales. *Informatica* **2**, 3–32.

Grandell J. (1991b). *Aspects of Risk Theory*. Springer-Verlag, New York.

Grandell J. (1995). Some results on the Ammeter risk process. *Mitteilungen der Schweizerischen Vereinigung der Versicherungsmathematiker*, 43–72.

Grandell J. (1997). *Mixed Poisson Processes*. Chapman & Hall, London.

Grigelionis B. (1993). Two-sided Lundberg inequalities in a Markovian environment. *Lithuanian Mathematical Journal* **33**, 23–32.

Grübel R. (1983). Über unbegrenzt teilbare Verteilungen. *Archiv der Mathematik* **41**, 80–88.

Grübel R. (1984). *Asymptotic Analysis in Probability Theory Using Banach-Algebra Techniques*. Habilitationsschrift, Universität Essen.

Hald A. (1987). On the early history of life insurance mathematics. *Scandinavian Actuarial Journal*, 4–18.

Hardy G.H., Littlewood J.E. and Pólya G. (1929). Some simple inequalities satisfied by convex functions. *Messenger of Mathematics* **58**, 145–152.

Harris T.E. (1963). *The Theory of Branching Processes*. Springer-Verlag, Berlin.

Harrison J.M. (1977a). Ruin problems with compounding assets. *Stochastic Process-*

es and their Applications **5**, 67–79.
Harrison J.M. (1977b). The supremum distribution of a Lévy process with no negative jumps. *Advances in Applied Probability* **9**, 417–422.
Harrison J.M. and Kreps D. (1979). Martingales and arbitrage in multiperiod security markets. *Journal of Economic Theory* **20**, 381–408.
Harrison J.M. and Pliska S. (1981). Martingales and stochastic integrals in the theory of continuous trading. *Stochastic Processes and their Applications* **11**, 215–260.
Harrison J.M. and Resnick S.I. (1978). The recurrence classification of risk and storage processes. *Mathematics of Operations Research* **3**, 57–66.
Heilmann W.R. (1988). *Fundamentals of Risk Theory*. Verlag Versicherungswirtschaft e.V., Karlsruhe.
Heilmann W.R. and Schröter K.J. (1991). Ordering of risks and their actuarial applications. In: *Stochastic Orders and Decision under Risk*, vol. 19, IMS Lecture Notes Monograph, 157–173.
Helbig M. and Milbrodt H. (1998). *Personenversicherungsmathematik*. Book manuscript, Universität Köln.
Herkenrath U. (1986). On the estimation of the adjustment coefficient in risk theory by means of stochastic approximation procedures. *Insurance: Mathematics and Economics* **5**, 305–313.
Hesselager O. (1993). Extensions of Ohlin's lemma with applications to optimal reinsurance structures. *Insurance: Mathematics and Economics* **13**, 83–97.
Hesselager O. (1994). A recursive procedure for calculation of some compound distributions. *ASTIN Bulletin* **24**, 19–32.
Hille E. (1966). *Analysis II*. Blaisdell Publishing Company, Waltham, Massachusetts.
Hipp C. (1985). Approximation of aggregate claims distributions by compound Poisson distribution. *Insurance: Mathematics and Economics* **4**, 227–232.
Hipp C. (1986). Improved approximations for the aggregate claims distribution in the individual model. *ASTIN Bulletin* **16**, 89–100.
Hipp C. and Michel R. (1990). *Risikotheorie: Stochastische Modelle und Methoden*. Verlag Versicherungswirtschaft e.V., Karlsruhe.
Hoem J.M. (1969). Markov chain models in life insurance. *Blätter der Deutschen Gesellschaft für Versicherungsmathematik* **9**, 91–107.
Hoem J.M. (1983). The reticent trio: Some little-known early discoveries in life insurance mathematics by L.H.F. Oppermann, T.N. Thiele and J.P. Gram. *International Statistical Review* **51**, 213–221.
Hogg R.U. and Klugman S.A. (1984). *Loss Distributions*. J. Wiley & Sons, New York.
Hürlimann W. (1994a). A note on experience rating, reinsurance and premium principles. *Insurance: Mathematics and Economics* **14**, 197–204.
Hürlimann W. (1994b). Splitting risk and premium calculation. *Mitteilungen der Schweizerischen Vereinigung der Versicherungsmathematiker*, 167–197.
Hürlimann W. (1995). Transforming, ordering and rating risks. *Mitteilungen der Schweizerischen Vereinigung der Versicherungsmathematiker*, 213–236.
Ikeda N. and Watanabe S. (1989). *Stochastic Differential Equations and Diffusion Processes*. 2nd ed., North-Holland, Amsterdam.
Iosifescu M. (1980). *Finite Markov Processes and Their Applications*. J. Wiley & Sons, Chichester.
Iosifescu M. and Tautu P. (1973). *Stochastic Processes and Applications in Biology and Medicine: Theory*, vol. I. Springer-Verlag, Berlin.
Irle A. (1998). *Finanzmathematik*. Teubner-Verlag, Stuttgart.

Islam M.N. and Consul P.C. (1992). A probabilistic model for automobile claims. *Mitteilungen der Schweizerischen Vereinigung der Versicherungsmathematiker*, 85–93.

Ito K. and McKean H.P. (1974). *Diffusion Processes and Their Sample Paths*. Springer-Verlag, Berlin.

Jagerman D. (1978). An inversion technique for the Laplace transform with applications. *Bell System Technical Journal* **57**, 669–710.

Jagerman D. (1982). An inversion technique for the Laplace transform. *Bell System Technical Journal* **61**, 1995–2002.

Janssen J. (1980). Some transient results on the M/SM/1 special semi-Markov model in risk and queueing theories. *ASTIN Bulletin* **11**, 41–51.

Janssen J. and Reinhard J.M. (1985). Probabilités de ruine pour une classe de modèles de risque semi-Markoviens. *ASTIN Bulletin* **15**, 123–133.

Jean-Marie A. and Liu Z. (1992). Stochastic comparisons for queueing models via random sums and intervals. *Advances in Applied Probability* **24**, 960–985.

Jensen J.L. (1995). *Saddlepoint Approximations*. Clarendon Press, Oxford.

Jensen U. (1997). An optimal stopping problem in risk theory. *Scandinavian Actuarial Journal*, 149–159.

Johnson N.L. and Kotz S. (1972). *Distributions in Statistics: Continuous Multivariate Distributions*. J. Wiley & Sons, New York.

Johnson N.L., Kotz S. and Balakrishnan N. (1994). *Continuous Univariate Distributions*, vol. I. 2nd ed., J. Wiley & Sons, New York.

Johnson N.L., Kotz S. and Balakrishnan N. (1995). *Continuous Univariate Distributions*, vol. II. 2nd ed., J. Wiley & Sons, New York.

Johnson N.L., Kotz S. and Kemp A.W. (1992). *Univariate Discrete Distributions*. 2nd ed., J. Wiley & Sons, Chichester.

Johnson N.L., Kotz S. and Kemp A.W. (1996). *Discrete Multivariate Distributions*. J. Wiley & Sons, New York.

Johnsonbaugh R. (1979). Summing an alternating series. *American Mathematical Monthly* **86**, 637–648.

Kaas R. (1987). *Bounds and approximations for some risk theoretical quantities*. Ph.D. thesis, University of Amsterdam.

Kaas R., van Heerwaarden A.E. and Goovaerts M.J. (1994). *Ordering of Actuarial Risks*. Caire Education Series 1, Brussels.

Kalashnikov V. (1997). *Geometric Sums: Bounds for Rare Events with Applications*. Kluwer Academic Publishers, Dordrecht.

Karamata J. (1932). Sur une inégalité rélative aux fonctions convexes. *Institut Mathématique. Publications. (Beograd)* **1**, 145–148.

Karatzas I. and Shreve S.E. (1991). *Brownian Motion and Stochastic Calculus*. Springer-Verlag, New York.

Karlin S. and Novikoff A. (1963). Generalized convex inequalities. *Pacific Journal of Mathematics* **13**, 1251–1279.

Karlin S. and Taylor H.M. (1981). *A Second Course in Stochastic Processes*. Academic Press, New York.

Karr A.F. (1991). *Point Processes and Their Statistical Inference*. Marcel Dekker, New York.

Karr A.F. (1993). *Probability*. Springer-Verlag, New York.

Keller B. and Klüppelberg C. (1991). Statistical estimation of large claim distributions. *Mitteilungen der Schweizerischen Vereinigung der Versicherungsmathematiker*, 203–216.

Kemeny J.G. and Snell J.L. (1990). *Finite Markov Chains*. Springer-Verlag, New

York.
Kendall D.G. (1948). On the generalized birth-and-death process. *Annals of Mathematical Statistics* **19**, 1–15.
Kennedy J. (1994). Understanding the Wiener-Hopf factorization for the simple random walk. *Journal of Applied Probability* **31**, 561–563.
Kingman J.F.C. (1961). A convexity property of positive matrices. *The Quarterly Journal of Mathematics, Oxford, Second Series* **12**, 283–284.
Kingman J.F.C. (1964). A martingale inequality in the theory of queues. *Mathematical Proceedings of the Cambridge Philosophical Society* **59**, 359–361.
Kingman J.F.C. (1970). Inequalities in the theory of queues. *Journal of the Royal Statistical Society. Series B.* **32**, 102–110.
Kingman J.F.C. (1972). *Regenerative Phenomena*. J. Wiley & Sons, London.
Kloeden P.E. and Platen E. (1992). *The Numerical Solution of Stochastic Differential Equations*. Springer-Verlag, Berlin.
Klüppelberg C. (1988). Subexponential distributions and integrated tails. *Journal of Applied Probability* **25**, 132–141.
Klüppelberg C. (1989). Estimation of ruin probabilites by means of hazard rates. *Insurance: Mathematics and Economics* **8**, 279–285.
Klüppelberg C. (1993). Asymptotic ordering of risks and ruin probabilities. *Insurance: Mathematics and Economics* **12**, 259–264.
Klüppelberg C. and Mikosch T. (1997). Large deviations of heavy-tailed random sums with applications to insurance and finance. *Journal of Applied Probability* **34**, 293–308.
Klüppelberg C. and Stadtmüller U. (1998). Ruin probabilities in the presence of heavy-tails and interest rates. *Scandinavian Actuarial Journal* , 49–58.
Kolmogorov A.N. (1931). Über die analytischen Methoden in der Wahrscheinlichkeitsrechnung. *Mathematische Annalen* **104**, 415–458.
König D. and Schmidt V. (1992). *Zufällige Punktprozesse*. Teubner-Verlag, Stuttgart.
Korn G.A. and Korn T.M. (1968). *Mathematical Handbook for Scientists and Engineers*. McGraw-Hill, New York.
Korn R., Kreer M. and Lenssen M. (1998). Pricing of European options when the underlying stock price follows a linear birth-death process. *Communications in Statistics - Stochastic Models* **14**, 647–662.
Kornya P.S. (1983). Distribution of aggregate claims in the individual risk theory model. *Transactions of the Society of Actuaries* **35**, 823–858.
Krengel U. (1985). *Ergodic Theorems*. De Gruyter, Berlin.
Krengel U. (1991). *Einführung in die Wahrscheinlichkeitstheorie und Statistik*. Vieweg, Braunschweig.
Kulkarni V.G. (1995). *Modeling and Analysis of Stochastic Systems*. Chapman & Hall, London.
Kuon S., Radtke M. and Reich A. (1991). The right way to switch from the individual risk model to the collective one. *Proceedings of the XXIII ASTIN Colloquium, Stockholm*.
Kuon S., Reich A. and Reimers L. (1987). Panjer vs. Kornya vs. De Pril: a comparison from a practical point of view. *ASTIN Bulletin* **17**, 183–191.
Kupper J. (1963). Some aspects of cumulative risk. *ASTIN Bulletin* **3**, 85–103.
Kupper J. (1971). Methoden zur Berechnung der Verteilungsfunktion des Totalschadens. *Mitteilungen der Schweizerischen Vereinigung der Versicherungsmathematiker*, 279–315.
Lamberton D. and Lapeyre B. (1997). *Introduction to Stochastic Calculus Applied*

to Finance. Chapman & Hall, London.

Last G. and Brandt A. (1995). *Marked Point Processes on the Real Line*. Springer-Verlag, New York.

Latouche G. and Ramaswami V. (1998). *Homogeneous Quasi-Birth-and-Death Processes*. Book Manuscript, University of Brussels/Bell Laboratories.

Lehtonen T. and Nyrhinen H. (1992a). Simulating level-crossing probabilities by importance sampling. *Advances in Applied Probability* **24**, 858–874.

Lehtonen T. and Nyrhinen H. (1992b). On asymptotically efficient simulation of ruin probabilities in a Markovian environment. *Scandinavian Actuarial Journal* , 60–75.

Lemaire J. (1985). *Automobile Insurance*. Kluwer-Nijhoff Publishing, Boston.

Lemaire J. (1995). *Bonus-Malus Systems in Automobile Insurance*. Kluwer Academic Publishers, Boston.

Levikson B. and Mizrahi G. (1994). Pricing long term care contracts. *Insurance: Mathematics and Economics* **14**, 1–18.

Lindvall T. (1992). *Lectures on the Coupling Method*. J. Wiley & Sons, New York.

Liptser R.S. and Shiryayev A.N. (1977). *Statistics of Random Processes. Vol.I: General Theory*. Springer-Verlag, New York.

Loimaranta K. (1972). Some asymptotic properties of bonus systems. *ASTIN Bulletin* **6**, 233–245.

Lundberg F. (1903). *I. Approximerad Framställning av Sannolikhetsfunktionen. II. Återförsäkering av Kollektivrisker*. Almqvist & Wiksell, Uppsala.

Lundberg F. (1926). *Försäkringsteknisk Riskutjämning*. F. Englunds boktryckeri A. B., Stockholm.

Lundberg F. (1930). Über die Wahrscheinlichkeitsfunktion einer Risikenmasse. *Skandinavisk Aktuarietidskrift* **13**, 1–83.

Lundberg F. (1932). Some supplementary researches on the collective risk theory. *Skandinavisk Aktuarietidskrift* **15**, 137–158.

Lundberg F. (1934). *On the Numerical Application of the Collective Risk Theory*. De Förenade Jubilee Volume, Stockholm.

Lundberg O. (1964). *On Random Processes and Their Application to Sickness and Accident Statistics*. Almquist & Wicksell, Uppsala.

Mack T. (1997). *Schadenversicherungsmathematik*. Verlag Versicherungswirtschaft e.V., Karlsruhe.

Malinovskii V.K. (1994). Corrected normal approximation for the probability of ruin within finite time. *Scandinavian Actuarial Journal* , 161–174.

Mammitzsch V. (1986). A note on the adjustment coefficient in ruin theory. *Insurance: Mathematics and Economics* **5**, 147–149.

Mandelbrot B. (1964). Random walks, fire damage and other Paretian risk phenomena. *Operations Research* **12**, 582–585.

Marshall A.W. and Olkin I. (1979). *Inequalities: Theory of Majorization and Its Applications*. Academic Press, New York.

Massey W.A. (1987). Stochastic orderings for Markov processes on partially ordered spaces. *Mathematics of Operations Research* **12**, 350–367.

McFadden J. (1965). The mixed Poisson process. *Sankhyā. Series A* **27**, 83–92.

Mercurio F. and Runggaldier W. (1993). Option pricing for jump diffusions: approximation and their interpretation. *Mathematical Finance* **3**, 191–200.

Meyer P.A. (1966). *Probability and Potentials*. Blaisdell Publishing Company, Waltham (Mass.).

Michaud F. (1994). The p-th power variance principle. *Mitteilungen der Schweizerischen Vereinigung der Versicherungsmathematiker*, 203–207.

Mikosch T. (1997). Heavy-tailed modelling in insurance. *Communications in Statistics – Stochastic Models* **13**, 799–815.
Mikosch T. (1998). *Elementary Stochastic Calculus With Finance in View*. World Scientific Publishers, Singapore.
Milbrodt H. and Stracke A. (1997). Markov models and Thiele's integral equations for the prospective reserve. *Insurance: Mathematics and Economics* **19**, 187–235.
Miller H.D. (1961). A convexity property in the theory of random variables defined on a finite Markov chain. *Annals of Mathematical Statistics* **32**, 1260–1270.
Miyazawa M., Schassberger R. and Schmidt V. (1995). On the structure of an insensitive generalized semi-Markov process with reallocation and point-process input. *Advances in Applied Probability* **27**, 203–225.
Miyazawa M. and Schmidt V. (1993). On the ladder height distributions of general risk processes. *Annals of Applied Probability* **3**, 763–776.
Miyazawa M. and Schmidt V. (1997). Level crossings of stochastic processes with stationary bounded variations and continuous decreasing components. *Probability and Mathematical Statistics* **17**, 79–93.
Moler C. and Van Loan C. (1978). Nineteen dubious ways to compute the exponential of a matrix. *SIAM Review* **20**, 801–836.
Møller C.M. (1992). Numerical evaluation of Markov transition probabilities based on the discretized product integral. *Scandinavian Actuarial Journal*, 76–87.
Møller C.M. (1993). A stochastic version of Thiele's differential equation. *Scandinavian Actuarial Journal*, 1–16.
Møller C.M. (1996). Aspects of prospective mean values in risk theory. *Insurance: Mathematics and Economics* **18**, 173–181.
Möller H.G. and Zwiesler H.J. (1996). Mehrzustandsmodelle in der Berufsunfähigkeitsversicherung. *Blätter der Deutschen Gesellschaft für Versicherungsmathematik* **22**, 479–499.
Møller T. (1998). Risk minimizing hedging strategies for unit linked life insurance contracts. *ASTIN Bulletin* **28**, 17–47.
Mosler K. and Scarsini M. (1993). *Stochastic Orders and Applications, a Classified Bibliography*. Springer-Verlag, Berlin.
Müller A. (1997). Stop-loss order for portfolios of dependent risks. *Insurance: Mathematics and Economics* **21**, 219–223.
Musiela M. and Rutkowski M. (1997). *Martingale Methods in Financial Modelling*. Springer-Verlag, Berlin.
Neuts M. (1981). *Matrix-Geometric Solutions in Stochastic Models*. Johns Hopkins University Press, Baltimore.
Neuts M. (1989). *Structured Stochastic Matrices of the M/G/1 Type and their Applications*. Marcel Dekker, New York.
Nielsen J. and Sandmann K. (1995). Equity-linked life insurance: a model with stochastic interest rates. *Insurance: Mathematics and Economics* **16**, 225–253.
Nilsen T. and Paulsen J. (1996). On the distribution of a randomly discounted coumpound Poisson process. *Stochastic Processes and their Applications* **61**, 305–310.
Norberg R. (1976). A credibility theory for automobile bonus systems. *Scandinavian Actuarial Journal*, 92–107.
Norberg R. (1991). Reserves in life and pension insurance. *Scandinavian Actuarial Journal*, 3–24.
Norberg R. (1992). Hattendorff's theorem and Thiele's differential equation generalized. *Scandinavian Actuarial Journal*, 2–14.
Norberg R. (1993). Prediction of outstanding liabilities in non-life insurance. *ASTIN*

Bulletin **23**, 95–115.
Norberg R. (1995). Differential equations for higher-order moments of present values in life insurance. *Insurance: Mathematics and Economics* **17**, 171–180.
Norberg R. and Møller C.M. (1996). Thiele's differential equation with stochastic interest of diffusion type. *Scandinavian Actuarial Journal* , 37–49.
O'Cinneide C. (1997). Euler summation for Fourier series and Laplace transform inversion. *Communications in Statistics – Stochastic Models* **13**, 315–337.
Ohlin J. (1969). On a class of measures for dispersion with application to optimal insurance. *ASTIN Bulletin* **5**, 22–26.
Omey E. and Willekens E. (1987). Second order behavior of distributions subordinate to a distribution with finite mean. *Communications in Statistics – Stochastic Models* **3**, 311–342.
Page F. and Sanders A. (1986). General derivation of the jump process option pricing formula. *Journal of Financial & Quantitative Analysis* **21**, 437–446.
Pakes A.G. (1975). On the tails of waiting-time distributions. *Journal of Applied Probability* **12**, 555–564.
Panjer H.H. (1980). The aggregate claims distribution and stop-loss reinsurance. *Transactions of the Society of Actuaries* **32**, 523–545.
Panjer H.H. (1981). Recursive evaluation of a family of compound distributions. *ASTIN Bulletin* **12**, 22–26.
Panjer H.H. and Lutek B.W. (1983). Practical aspects of stop-loss calculations. *Insurance: Mathematics and Economics* **2**, 159–177.
Panjer H.H. and Wang S. (1993). On the stability of recursive formulas. *ASTIN Bulletin* **23**, 227–258.
Panjer H.H. and Willmot G.E. (1981). Finite sum evaluation of the negative binomial-exponential model. *ASTIN Bulletin* **12**, 133–137.
Panjer H.H. and Willmot G.E. (1986). Computational aspects of recursive evaluation of compound distribution. *Insurance: Mathematics and Economics* **5**, 113–116.
Panjer H.H. and Willmot G.E. (1992). *Insurance Risk Models*. American Society of Actuaries, Schaumburg (Illinois).
Parker G. (1994). Moments of the present value of a portfolio of policies. *Scandinavian Actuarial Journal*, 53–67.
Paulsen J. (1993). Risk theory in a stochastic economic environment. *Stochastic Processes and their Applications* **46**, 327–361.
Paulsen J. (1997). Present value of some insurance portfolios. *Scandinavian Actuarial Journal*, 11–37.
Paulsen J. (1998). Sharp conditions for certain ruin in a risk process with stochastic return on investments. *Stochastic Processes and their Applications* **75**, 135–148.
Paulsen J. and Gjessing H.K. (1997a). Ruin theory with stochastic return on investments. *Advances in Applied Probability* **29**, 965–985.
Paulsen J. and Gjessing H.K. (1997b). Optimal choice of dividend barriers for a risk process with stochastic return on investments. *Insurance: Mathematics and Economics* **20**, 215–223.
Pellerey F. (1995). On the preservation of some orderings of risks under convolution. *Insurance: Mathematics and Economics* **16**, 23–30.
Petersen S.S. (1990). Calculation of ruin probabilities when the premium depends on the current reserve. *Scandinavian Actuarial Journal*, 147–159.
Petrov V.V. (1975). *Sums of Independent Random Variables*. Springer-Verlag, New York.
Philipson C. (1960). The theory of confluent hypergeometric functions and its application to compound Poisson processes. *Skandinavisk Aktuarietidskrift* **43**,

136–162.
Picard P. (1994). On some measures of the severity of ruin in the classical Poisson model. *Insurance: Mathematics and Economics* **14**, 107–116.
Picard P. and Lefevre C. (1997). The probability of ruin in finite time with discrete claim size distribution. *Scandinavian Actuarial Journal*, 58–69.
Pitman E.J.G. (1980). Subexponential distribution functions. *Australian Mathematical Society. Journal. Series A* **29**, 337–347.
Pitman J. (1993). *Probability*. Springer-Verlag, New York.
Pitts S., Grübel R. and Embrechts P. (1996). Confidence bounds for the adjustment coefficient. *Advances in Applied Probability* **28**, 802–827.
Pliska S.R. and Rogers C. (1995). *Mathematical Theory and Practical Implementation of Interest Rate Models*. RISK, London.
Prabhu N.U. (1965). *Queues and Inventories*. J. Wiley & Sons, New York.
Prabhu N.U. (1980). *Stochastic Storage Processes*. Springer-Verlag, New York.
Press W.H., Flannery B.P., Teukolsky S.A. and Vetterling W.T. (1988). *Numerical Recipes. The Art of Scientific Computing*. Cambridge University Press, Cambridge.
Ramlau-Hansen H. (1990). Thiele's differential equation with stochastic interest of diffusion type. *Scandinavian Actuarial Journal*, 97–104.
Ramsay C.M. (1994). Loading gross premiums for risk without using utility theory, with discussions. *Transactions of the Society of Actuaries* **45**, 305–349.
Ramsay C.M. and Usabel M.A. (1997). Calculating ruin probabilities via product integration. *ASTIN Bulletin* **27**, 263–271.
Rebonato R. (1997). *Interest-Rate Option Models*. J. Wiley & Sons, Chichester.
Reich A. (1986). Properties of premium calculation principles. *Insurance: Mathematics and Economics* **5**, 97–101.
Reinhard J.M. (1984). On a class of semi-Markov risk models obtained as classical risk models in a Markovian environment. *ASTIN Bulletin* **14**, 23–43.
Reiss R.D. and Thomas M. (1997). *Statistical Analysis of Extreme Values*. Birkhäuser, Basel.
Resnick S.I. (1987). *Extreme Values, Regular Variation and Point Processes*. Springer-Verlag, New York.
Resnick S.I. (1992). *Adventures in Stochastic Processes*. Birkhäuser, Boston.
Resnick S.I. (1997). Discussion of the Danish data on large fire insurance losses. *ASTIN Bulletin* **27**, 139–151.
Richter W.D., Steinebach J. and Taube S. (1993). On a class of estimators for an exponential tail coefficient with applications in risk theory. *Statistics & Decision, Supplement Issue* **3**, 145–173.
Rogers L.C.G. and Talay D. (1997). *Numerical Methods in Finance*. Cambridge University Press, Cambridge.
Rogers L.C.G. and Williams D. (1994). *Diffusions, Markov Processes, and Martingales. Vol. I, Foundations*. 2nd ed., J. Wiley & Sons, Chichester.
Rolski T. (1976). Order relations in the set of probability distribution functions and their applications in queueing theory. *Dissertationes Mathematicae* **132**.
Rolski T. (1981). Queues with non-stationary input stream: Ross's conjecture. *Advances in Applied Probability* **13**, 603–618.
Rolski T., Schlegel S. and Schmidt V. (1999). Asymptotics of Palm-stationary buffer content distributions in fluid flow queues. *Advances in Applied Probability* **31**, to appear.
Ross S.M. (1974). Bounds on the delay distribution in GI/G/1 queues. *Journal of Applied Probability* **11**, 417–421.

Ross S.M. (1997a). *An Introduction to Probability Models*. Academic Press, San Diego.

Ross S.M. (1997b). *Simulation*. Academic Press, San Diego.

Rudin W. (1986). *Real and Complex Analysis*. McGraw-Hill, Singapore.

Runnenburg J.T. and Goovaerts M.J. (1985). Bounds on compound distribution and stop loss premiums. *Insurance: Mathematics and Economics* **4**, 287–293.

Ruohonen M. (1988). On a model for the claim number process. *ASTIN Bulletin* **18**, 57–68.

Russ J. (1998). *Die aktienindexgebundene Lebensversicherung mit garantierter Mindestverzinsung in Deutschland*. Ph.D. thesis, Universität Ulm.

Rydén T. (1994). Parameter estimation for Markov modulated Poisson processes. *Communications in Statistics – Stochastic Models* **10**, 795–829.

Schassberger R. (1973). *Warteschlangen*. Springer-Verlag, Wien.

Schassberger R. (1978). Insensitivity of steady-state distributions of generalized semi-Markov processes with speeds. *Advances in Applied Probability* **10**, 836–851.

Scheike T.H. (1992). A general risk process and its properties. *Journal of Applied Probability* **29**, 73–81.

Schlegel S. (1998). Ruin probabilities in perturbed risk models. *Insurance: Mathematics and Economics* **22**, 93–104.

Schmidli H. (1992). *A General Insurance Risk Model*. Ph.D. thesis, ETH Zürich.

Schmidli H. (1994). Risk theory in an economic environment and Markov processes. *Mitteilungen der Schweizerischen Vereinigung der Versicherungsmathematiker*, 51–69.

Schmidli H. (1995). Cramér–Lundberg approximations for ruin probabilities of risk processes perturbed by diffusion. *Insurance: Mathematics and Economics* **16**, 135–149.

Schmidli H. (1996). Martingales and insurance risk. In: Obretenov A. (ed.), *Lecture Notes of the 8th International Summer School on Probability Theory and Mathematical Statistics*, Science Culture Technology Publishing, Singapore, 155–188.

Schmidli H. (1997a). An extension to the renewal theorem and an application to risk theory. *Annals of Applied Probability* **7**, 121–133.

Schmidli H. (1997b). Estimation of the Lundberg coefficient for a Markov modulated risk model. *Scandinavian Actuarial Journal*, 48–57.

Schmidli H. (1998). Compound sums and subexponentiality. *Bernoulli* , to appear.

Schmidt K.D. (1996). *Lectures on Risk Theory*. Teubner-Verlag, Stuttgart.

Schöttl A. (1998). Optimal stopping of a risk reserve process with interest and cost rates. *Journal of Applied Probability* **35**, 115–123.

Schröter K.J. (1990). On a family of counting distributions and recursions for related compound distributions. *Scandinavian Actuarial Journal*, 161–175.

Schweizer M. (1992). Mean-variance hedging for general claims. *Annals of Applied Probability* **2**, 171–179.

Schweizer M. (1994). On the minimal martingale measure and the Föllmer–Schweizer decomposition. *Stochastic Analysis and Applications* **13**, 573–599.

Seal H.L. (1974). The numerical calculation of $U(w,t)$, the probability of non-ruin in an interval $(0,t)$. *Scandinavian Actuarial Journal*, 121–139.

Seal H.L. (1980). Survival probabilities based on Pareto claim distributions. *ASTIN Bulletin* **11**, 61–71.

Segerdahl C.O. (1942). Über einige risikotheoretische Fragestellungen. *Skandinavisk Aktuarietidskrift* **25**, 43–83.

Segerdahl C.O. (1954). A survey of results in the collective theory of risk. In:

Grenander U. (ed.), *Probability and Statistics. The Harald Cramér volume*, J. Wiley & Sons, Stockholm, 276–299.

Segerdahl C.O. (1955). When does ruin occur in the collective theory of risk? *Skandinavisk Aktuarietidskrift* **38**, 22–36.

Seneta E. (1981). *Non-Negative Matrices and Markov Chains*. Springer-Verlag, New York.

Serfling R.J. (1980). *Approximation Theorems of Mathematical Statistics*. J. Wiley & Sons, New York.

Sevastyanov A.B. (1973). *Verzweigungsprozesse*. Akademie-Verlag, Berlin.

Shaked M. and Shanthikumar J.G. (1993). *Stochastic Orders and Their Applications*. Academic Press, New York.

Shanthikumar G. (1987). On stochastic comparison of random vectors. *Journal of Applied Probability* **24**, 123–136.

Shiryayev A.N. (1984). *Probability*. Springer-Verlag, New York.

Shiu E.S.W. (1989). On Gerber's fun. *Scandinavian Actuarial Journal* , 65–68.

Shorack G.R. and Wellner J.A. (1986). *Empirical Processes with Applications to Statistics*. J. Wiley & Sons, New York.

Shpilberg D.C. (1977). The probability distribution of fire loss amount. *The Journal of Risk and Insurance* **44**, 103–115.

Sichel H.S. (1971). On a family of discrete distributions particular suited to represent long tailed frequency data. In: Laubscher N.F. (ed.), *Proceedings of the Third Symposium of Mathematical Statistics*, CSIR, Pretoria, 51–97.

Sichel H.S. (1974). On a distribution representing sentence-length in written prose. *Journal of the Royal Statistical Society. Series A.* **137**, 25–34.

Sichel H.S. (1975). On a distribution law for word frequencies. *Journal of the American Statistical Association* **70**, 542–547.

Siegmund D. (1975). The time until ruin in collective risk theory. *Mitteilungen der Schweizerischen Vereinigung der Versicherungsmathematiker*, 157–166.

Siegmund D. (1976). Importance sampling in the Monte Carlo study of sequential tests. *Annals of Statistics* **4**, 673–684.

Sigman K. (1995). *Stationary Marked Point Processes: an Intuitive Approach*. Chapman & Hall, New York.

Simmons G.F. (1991). *Differential Equations with Applications and Historical Notes*. 2nd ed., McGraw-Hill, New York.

Stanford D.A. and Stroinski K.J. (1994). Recursive methods for computing finite-time ruin probabilities for phase-distributed claim sizes. *ASTIN Bulletin* **24**, 235–254.

Steinebach J. (1997). Personal communication.

Stoyan D. (1983). *Comparison Methods for Queues and Other Stochastic Models*. J. Wiley & Sons, Chichester.

Straub E. (1988). *Non-Life Insurance Mathematics*. Springer-Verlag, Berlin.

Sundt B. (1992). On some extensions of Panjer's class of counting distributions. *ASTIN Bulletin* **22**, 61–80.

Sundt B. (1993). *An Introduction to Non-Life Insurance Mathematics*. 3rd ed., Verlag Versicherungswirtschaft e.V., Karlsruhe.

Sundt B. and Jewell W. (1981). Further results on recursive evaluation of compound distributions. *ASTIN Bulletin* **12**, 27–39.

Sundt B. and Teugels J.L. (1995). Ruin estimates and interest force. *Insurance: Mathematics and Economics* **16**, 7–22.

Sundt B. and Teugels J.L. (1997). The adjustment function in ruin estimates under interest force. *Insurance: Mathematics and Economics* **19**, 85–94.

Szekli R. (1995). *Stochastic Ordering and Dependence in Applied Probability.* Springer-Verlag, New York.

Szynal D. and Teugels J.L. (1993). A stop loss experience rating scheme for fleets of cars, Part II. *Insurance: Mathematics and Economics* **13**, 255–262.

Takács L. (1962). *Introduction to the Theory of Queues.* Oxford University Press, New York.

Taylor G.C. (1976). Use of differential and integral inequalities to bound ruin and queueing probabilities. *Scandinavian Actuarial Journal*, 197–208.

Taylor G.C. (1980). Probability of ruin with variable premium rate. *Scandinavian Actuarial Journal*, 57–76.

Tempelman A.A. (1992). *Ergodic Theorems for Group Actions.* Kluwer Academic Publishers, London.

Teugels J.L. (1975). The class of subexponential distributions. *Annals of Probability* **3**, 1000–1011.

Teugels J.L. (1982). Estimation of ruin probabilities. *Insurance: Mathematics and Economics* **1**, 163–175.

Teugels J.L. and Veraverbeke N. (1973). Cramér-type estimates for the probability of ruin,. *CORE Discussion Paper 7316* .

Teugels J.L. and Willmot G.E. (1987). Approximations for stop-loss premiums. *Insurance: Mathematics and Economics* **6**, 195–202.

Thépaut A. (1950). Une nouvelle forme de réassurance. Le traité d'excédent du coût moyen relatif (ECOMOR). *Bulletin Trimestriel Institute Actuaire France* **49**, 273–343.

Thiele T.N. (1880). Om Anvendelse af mindste Kvadraters Methode i nogle Tilfælde, hvor Komplikation af visse Slags uensartede tilfældige Fejlkilder giver Fejlene en 'systematisk' Karakter. *Videnskabernes Selskab Skrifter, 5. række* **12**, 381–408.

Thorin O. (1975). Stationarity aspects for the Sparre Andersen risk process and the corresponding ruin probabilities. *Scandinavian Actuarial Journal* , 87–98.

Thorin O. and Wikstad N. (1977). Calculation of ruin probabilities when the claim distribution is lognormal. *ASTIN Bulletin* **9**, 231–246.

Thyrion P. (1959). Sur une propriété des processus de Poisson généralisés. *Bulletin Association Royal Actuaires Belges* **59**, 35–46.

Tijms H.C. (1994). *Stochastic Models: An Algorithmic Approach.* J. Wiley & Sons, Chichester.

Vasicek O. (1977). An equilibrium characterization of the term structure. *Journal of Financial Economics* **5**, 177–188.

Venter G.G. (1991). Premium calculation implications of reinsurance without arbitrage. *ASTIN Bulletin* **21**, 223–230.

Vepsäläinen S. (1972). Applications to a theory of bonus systems. *ASTIN Bulletin* **6**, 212–221.

Veraverbeke N. (1977). Asymptotic behaviour of Wiener–Hopf factors of a random walk. *Stochastic Processes and their Applications* **5**, 27–37.

Veraverbeke N. (1993). Asymptotic estimates for the probability of ruin in a Poisson model with diffusion. *Insurance: Mathematics and Economics* **13**, 57–62.

Ville J. (1939). *Etude Critique de la Notion du Collectif.* Gauthier-Villars, Paris.

Vittal P.R. and Vasudevan R. (1987). Some first passage time problems with restricted reserve and two components of income. *Scandinavian Actuarial Journal*, 198–210.

von Bahr B. (1974). Ruin probabilities expressed in terms of ladder height distributions. *Scandinavian Actuarial Journal*, 190–204.

von Bahr B. (1975). Asymptotic ruin probabilities when exponential moments do

not exist. *Scandinavian Actuarial Journal*, 6–10.
Waldmann K.H. (1994). On the exact calculation of the aggregate claims distribution in the individual life model. *ASTIN Bulletin* **24**, 89–96.
Waldmann K.H. (1995). Exact calculation of the aggregate claims distribution in the individual life model by use of an n-layer model. *Blätter der Deutschen Gesellschaft für Versicherungsmathematik* **22**, 279–287.
Wang S. (1995). Insurance pricing and increased limits ratemaking by proportional hazards transforms. *Insurance: Mathematics and Economics* **17**, 109–114.
Wang S. (1996). Premium calculation by transforming the layer premium density. *Preprint*, University of Waterloo, Ontario.
Wang S. and Panjer H.H. (1994). Proportional convergence and tail-cutting techniques in evaluating aggregate claim distributions. *Insurance: Mathematics and Economics* **14**, 129–138.
Waters H.R. (1983). Some mathematical aspects of reinsurance. *Insurance: Mathematics and Economics* **2**, 17–26.
Widder D.V. (1971). *An Introduction to Transform Theory*. Academic Press, New York.
Wiener N. (1923). Differential space. *Journal of Mathematical Physics* **2**.
Wiener N. (1924). Un problème de probabilitès dènombrables. *Bulletin de la Société Mathématique de France* **52**, 569–578.
Wikstad N. (1983). A numerical illustration of differences between ruin probabilities originated in the ordinary and in the stationary cases. *Scandinavian Actuarial Journal*, 47–48.
Wilkinson J.H. (1965). *The Algebraic Eigenvalue Problem*. Clarendon Press, Oxford.
Willekens E. and Teugels J.L. (1992). Asymptotic expansions for waiting time probabilities in an M/G/1 queue with long-tailed service time. *Queueing Systems* **10**, 295–312.
Williams D. (1991). *Probability with Martingales*. Cambridge University Press, Cambridge.
Willmot G.E. (1986). Mixed compound Poisson distributions. *ASTIN Bulletin* **16**, 59–79.
Willmot G.E. (1987). The Poisson-inverse Gaussian distribution as an alternative to the negative binomial. *Scandinavian Actuarial Journal* , 113–127.
Willmot G.E. (1988). Sundt and Jewell's faily of discrete distributions. *ASTIN Bulletin* **18**, 17–29.
Willmot G.E. (1994). Refinements and distributional generalizations of Lundberg's inequality. *Insurance: Mathematics and Economics* **15**, 49–63.
Willmot G.E. (1997a). On the relationship between bounds on the tails of compound distributions. *Insurance: Mathematics and Economics* **19**, 95–103.
Willmot G.E. (1997b). Bounds for compound distributions based on mean residual lifetimes and equlibrium distributions. *Insurance: Mathematics and Economics* **21**, 25–42.
Willmot G.E. and Lin X.D. (1994). Lundberg bounds on the tails of compound distributions. *Journal of Applied Probability* **31**, 743–756.
Willmot G.E. and Lin X.D. (1997a). Simplified bounds on the tails of compound distributions. *Journal of Applied Probability* **34**, 127–133.
Willmot G.E. and Lin X.D. (1997b). Upper bounds for the tail of the compound negative binomial distribution. *Scandinavian Actuarial Journal*, 138–148.
Willmot G.E. and Sundt B. (1989). On evaluation of the Delaporte distribution and related distributions. *Scandinavian Actuarial Journal*, 101–113.
Winkler G. (1995). *Image Analysis, Random Fields and Dynamic Monte Carlo*

Methods. Springer-Verlag, Berlin.
Wolthuis H. (1987). Hattendorff's theorem for a continuous-time Markov model. *Scandinavian Actuarial Journal*, 157–175.
Wolthuis H. (1994). *Life Insurance Mathematics (The Markovian Model)*. Caire Education Series 2, Brussels.
Zolotarev V. (1964). The first passage time of a level and the behavior at infinity for a class of processes with independent increments. *Theory of Probability and its Applications* **9**, 653–662.

Index